W9-BHE-633

RECEPTOR ACTIVATION BY ANTIGENS, CYTOKINES, HORMONES, AND GROWTH FACTORS

ANNALS OF THE NEW YORK ACADEMY OF SCIENCES

Volume 766

RECEPTOR ACTIVATION BY ANTIGENS, CYTOKINES, HORMONES, AND GROWTH FACTORS

Edited by David Naor, Pierre De Meyts, Marc Feldmann, and Joseph Schlessinger

The New York Academy of Sciences
New York, New York
1995

Copyright © 1995 by the New York Academy of Sciences. All rights reserved. Under the provisions of the United States Copyright Act of 1976, individual readers of the Annals *are permitted to make fair use of the material in them for teaching and research. Permission is granted to quote from the* Annals *provided that the customary acknowledgment is made of the source. Material in the* Annals *may be republished only by permission of the Academy. Address inquiries to the Executive Editor at the New York Academy of Sciences.*

Copying fees: *For each copy of an article made beyond the free copying permitted under Section 107 or 108 of the 1976 Copyright Act, a fee should be paid through the Copyright Clearance Center, Inc., 222 Rosewood Drive, Danvers, MA 01923. For articles of more than 3 pages, the copying fee is $1.75.*

∞The paper used in this publication meets the minimum requirements of American National Standard for Information Sciences—Permanence of Paper for Printed Library Materials, ANSI Z39.48-1984.

Cover designer: Avi Eisenstein

Library of Congress Cataloging-in-Publication Data

Receptor activation by antigens, cytokines, hormones, and growth factors / edited by David Naor . . . [et al.].
p. cm. — (Annals of the New York Academy of Sciences, ISSN 0077-8923 ; v. 766)
Includes bibliographical references and index.
ISBN 0-89766-951-7 (cloth : alk. paper). — ISBN 0-89766-952-5 (paper : alk. paper)
1. Cellular signal transduction—Congresses. 2. Cell receptors—Congresses. 3. Hormone receptors—Congresses. 4. Growth factors—Receptors—Congresses. 5. Antigen recognition—Congresses. I. Naor, David. II. Series.
[DNLM: 1. Signal Transduction—physiology. 2. Receptors, Immunologic. 3. Receptors, Growth Factor. W1 AN626YL v. 766 1995 / QH 601 R294 1995]
Q11.N5 vol. 766
[QP517.C45]
500 s—dc20
[612′.01583]
DNLM/DLC
for Library of Congress 95-31753
CIP

PCP
Printed in the United States of America
ISBN 0-89766-951-7 (cloth)
ISBN 0-89766-952-5 (paper)
ISSN 0077-8923

ANNALS OF THE NEW YORK ACADEMY OF SCIENCES

Volume 766
September 7, 1995

RECEPTOR ACTIVATION BY ANTIGENS, CYTOKINES, HORMONES, AND GROWTH FACTORS[a]

Editors and Conference Organizers
DAVID NAOR, PIERRE DE MEYTS, MARC FELDMANN, AND JOSEPH SCHLESSINGER

CONTENTS

Preface. *By* DAVID NAOR and PIERRE DE MEYTS xiii

Part I. Receptor Structures and Signaling Mechanisms Common to Multiple Biological Systems

Definition of Signals for Neuronal Differentiation. *By* AXEL OBERMEIER, RALPH A. BRADSHAW, KLAUS SEEDORF, AXEL CHOIDAS, JOSEPH SCHLESSINGER, and AXEL ULLRICH 1

Structure and Catalytic Properties of Protein Tyrosine Phosphatases. *By* JACK E. DIXON 18

The G Protein–coupled Receptor Family and One of Its Members, the TSH Receptor. *By* G. VASSART, F. DESARNAUD, L. DUPREZ, D. EGGERICKX, O. LABBÉ, F. LIBERT, C. MOLLEREAU, J. PARMA, R. PASCHKE, M. TONACCHERA, P. VANDERHAEGHEN, J. VAN SANDE, J. DUMONT, and M. PARMENTIER 23

Inositol Trisphosphate and Calcium Signaling. *By* MICHAEL J. BERRIDGE 31

Interaction of Signaling and Trafficking Proteins with the Carboxyterminus of the Epidermal Growth Factor Receptor. *By* GRAHAM CARPENTER, CONCEPCIÓ SOLER, JOSEP BAULIDA, LAURA BEGUINOT, and ALEXANDER SORKIN 44

[a] This volume is the result of a conference entitled **Receptor Activation by Antigens, Cytokines, Hormones, and Growth Factors,** held in Orlando, Florida on October 21–25, 1994, by the New York Academy of Sciences.

Part II. Cell Receptors of the Immune System: Development, Structure, Function, and Diversity

Control of T-Cell Development by the Pre-T and αβ T-Cell Receptor. *By* HARALD VON BOEHMER 52

T-Cell Recognition of Antigen: A Process Controlled by Transient Intermolecular Interactions. *By* J. JAY BONIFACE and MARK M. DAVIS 62

Receptors That Regulate T-Cell Susceptibility to Apoptotic Cell Death. *By* LAWRENCE H. BOISE, ANDY J. MINN, and CRAIG B. THOMPSON 70

Signaling Difference between Class IgM and IgD Antigen Receptors. *By* KWANG-MYONG KIM and MICHAEL RETH 81

Interaction of $p56^{lck}$ with CD4 in the Yeast Two-hybrid System. *By* KERRY S. CAMPBELL, ANNIE BUDER, and ULRICH DEUSCHLE .. 89

Induction of Terminal Differentiation of Promyelocytic HL-60 Leukemic Cells Implanted with Lymphocyte Receptors and Stimulated with Various Lymphocyte Stimulators. *By* IRIT ALTBOUM and ISRAEL ZAN-BAR 93

Independent Signaling for Growth Arrest and Apoptosis by Igα and Igβ Subunits of the B-Cell Antigen Receptor Complex. *By* XIAO-RUI YAO and DAVID W. SCOTT 96

Part III. Signal Transduction and Second Messengers in Activated Cells of the Immune System

The Kinase-dependent Function of Lck in T-Cell Activation Requires an Intact Site for Tyrosine Autophosphorylation. *By* HUA XU and DAN R. LITTMAN 99

Coreceptors and Adapter Proteins in T-Cell Signaling. *By* K. S. RAVICHANDRAN, J. C. PRATT, S. SAWASDIKOSOL, H. Y. IRIE, and S. J. BURAKOFF 117

Symmetry of the Activation of Cyclin-dependent Kinases in Mitogen and Growth Factor–stimulated T Lymphocytes. *By* JAIME F. MODIANO, JOANNE DOMENICO, AGOTA SZEPESI, NAOHIRO TERADA, JOSEPH J. LUCAS, and ERWIN W. GELFAND 134

Molecular and Genetic Insights into T-Cell Antigen Receptor Signaling. *By* ARTHUR WEISS, THERESA KADLECEK, MAKIO IWASHIMA, ANDREW CHAN, and NICOLAI VAN OERS 149

Signal Transduction Mediated by the T-Cell Antigen Receptor. *By* LAWRENCE E. SAMELSON, JERALD A. DONOVAN, NOAH ISAKOV, YASUO OTA, and RONALD L. WANGE 157

Genetic Dissection of the Transducing Subunits of the T-Cell Antigen Receptor. *By* BERNARD MALISSEN, GRACE KU, MIRJAM HERMANS, ERIC VIVIER, and MARIE MALISSEN........ 173

The Cyclosporin-sensitive Transcription Factor NFATp Is Expressed in Several Classes of Cells in the Immune System. *By* DON Z. WANG, PATRICIA G. MCCAFFREY, and ANJANA RAO 182

Signal Transduction by the B-Cell Antigen Receptor. *By* ANTHONY L. DEFRANCO, JAMES D. RICHARDS, JONATHAN H. BLUM, TRACY L. STEVENS, DEBBIE A. LAW, VIVIEN W.-F. CHAN, SANDIP K. DATTA, SHAUN P. FOY, SHARON L. HOURIHANE, MICHAEL R. GOLD, and LINDA MATSUUCHI 195

Interaction of Shc with Grb2 Regulates the Grb2 Association with mSOS. *By* K. S. RAVICHANDRAN, U. LORENZ, S. E. SHOELSON, and S. J. BURAKOFF 202

A 115 kDa Tyrosine Phosphorylated Protein Associates with Grb-2 in Activated Jurkat Cells. *By* HERMAN MEISNER and MICHAEL P. CZECH 204

Csk Associates with the TCR ζ and ϵ Chains through Its SH2 Domain. *By* THORUNN RAFNAR, JONATHAN P. SCHNECK, MARY E. BRUMMET, DAVID G. MARSH, and BRANIMIR ČATIPOVIĆ 206

Fc Receptor Stimulation of PI 3-Kinase in NK Cells Is Associated with Protein Kinase C–independent Granule Release and Cell-mediated Cytotoxicity. *By* JOY D. BONNEMA, LARRY M. KARNITZ, RENEE A. SCHOON, ROBERT T. ABRAHAM, and PAUL J. LEIBSON 209

Potentiation of B-Cell Antigen Receptor-mediated Signal Transduction by the Heterologous *src* Family Protein Tyrosine Kinase, *src*. *By* JIEJIAN LIN, ANNE L. BURKHARDT, JOSEPH B. BOLEN, and LOUIS B. JUSTEMENT 214

Regulation of Human Natural Killer-Cell Lytic Activity by Serine/Threonine Phosphatases and Kinases. *By* ANIL BAJPAI and ZACHARIE BRAHMI 216

Mechanisms of Enhanced Nuclear Translocation of the Transcription Factors c-Rel and NF-κB by CD28 Costimulation in Human T Lymphocytes. *By* JENN-HAUNG LAI, GYÖRGYI HORVATH, YONGQIN LI, and TSE-HUA TAN..... 220

Part IV. Cytokine Receptors

Cytokine Signal Transduction through a Homo- or Heterodimer of gp130. *By* TADAMITSU KISHIMOTO, TAKASHI TANAKA, KANJI YOSHIDA, SHIZUO AKIRA, and TETSUYA TAGA............... 224

IL-2 Signaling Involves Recruitment and Activation of Multiple Protein Tyrosine Kinases by the IL-2 Receptor. *By* TADATSUGU TANIGUCHI, TADAAKI MIYAZAKI, YASUHIRO MINAMI, ATSUO KAWAHARA, HODAKA FUJII, YOKO NAKAGAWA, MASANORI HATAKEYAMA, and ZHAO-JUN LIU 235

Costimulation Requirement for AP-1 and NF-κB Transcription Factor Activation in T Cells. *By* STEFFEN JUNG, AVRAHAM YARON, IRIT ALKALAY, ADA HATZUBAI, AYELET AVRAHAM, and YINON BEN-NERIAH 245

Structural Aspects of Cytokine/Receptor Interactions. *By* NICOS A. NICOLA 253

Cell Growth Signal Transduction Is Quantal. *By* KENDALL A. SMITH 263

TNFα Is an Effective Therapeutic Target for Rheumatoid Arthritis. *By* MARC FELDMANN, FIONULA M. BRENNAN, MICHAEL J. ELLIOTT, RICHARD O. WILLIAMS, and RAVINDER N. MAINI 272

Investigation of Ligand Binding to Members of the Cytokine Receptor Family within a Microbial System. *By* K. H. YOUNG and B. A. OZENBERGER 279

The Proline-rich Motif Is Necessary but Not Sufficient for Prolactin Receptor Signal Transduction. *By* KEVIN D. O'NEAL, LI-YUAN YU-LEE, and WILLIAM T. SHEARER 282

Activation of Multiple Protein Kinases by Interleukin-1. *By* J. E. DUNFORD, N. R. CORBETT, C. L. VARLEY, C. H. DAWSON, B. L. BROWN, and P. R. M. DOBSON 285

IL-8 Signal Transduction in Human Neutrophils. *By* CINDY KNALL, G. SCOTT WORTHEN, ANNE METTE BUHL, and GARY L. JOHNSON 288

IL-8 Induces Calcium Mobilization in Interleukin-2-activated Natural Killer Cells Independently of Inositol 1,4,5 Trisphosphate. *By* ALA AL-AOUKATY, ADEL GIAID, and AZZAM A. MAGHAZACHI 292

Interleukin-11 Induces Tyrosine Phosphorylation, and c-jun and c-fos mRNA Expression in Human K562 and U937 Cells. *By* SAMUEL E. ADUNYAH, GLENDORA C. SPENCER, ROLAND S. COOPER, JUAN A. RIVERO, and KARAMBA CEESAY 296

Analysis of the Interaction between Two TGF-β-binding Proteins and Three TGF-β Isoforms Using Surface Plasmon Resonance. *By* M. D. O'CONNOR-MCCOURT, P. SEGARINI, S. GROTHE, M. L.-S. TSANG, and J. A. WEATHERBEE 300

Part V. Growth Factor and Hormone Receptors: Structure and Function

The Stress-activated Protein Kinases: A Novel ERK Subfamily Responsive to Cellular Stress and Inflammatory Cytokines. *By* JOHN M. KYRIAKIS, JAMES R. WOODGETT, and JOSEPH AVRUCH 303

Protein Serine/Threonine Kinases of the MAPK Cascade. *By* J. D. GRAVES, J. S. CAMPBELL, and E. G. KREBS 320

Structural Aspects of Receptor Dimerization: c-Kit as an Example. *By* JANNA M. BLECHMAN and YOSEF YARDEN 344

Signal Transduction Interception as a Novel Approach to Disease Management. *By* ALEXANDER LEVITZKI 363

Common and Distinct Elements in Insulin and PDGF Signaling. *By* MARTIN G. MYERS JR., BENTLEY CHEATHAM, TRACEY L. FISHER, BOZENA R. JACHNA, C. RONALD KAHN, JONATHAN M. BACKER, and MORRIS F. WHITE 369

Mechanism of Insulin and IGF-I Receptor Activation and Signal Transduction Specificity: Receptor Dimer Cross-linking, Bell-shaped Curves, and Sustained versus Transient Signaling. *By* PIERRE DE MEYTS, BIRGITTE URSØ, CLAUS T. CHRISTOFFERSEN, and RONALD M. SHYMKO 388

The Role of the Insulin-like Growth Factor-I Receptor in Cancer. *By* D. LEROITH, H. WERNER, S. NEUENSCHWANDER, T. KALEBIC, and L. J. HELMAN 402

Mitogenic Potential of Insulin on Lymphoma Cells Lacking IGF-1 Receptor. *By* DVORAH ISH-SHALOM, GURI TZIVION, CLAUS T. CHRISTOFFERSEN, BIRGITTE URSØ, PIERRE DE MEYTS, and DAVID NAOR 409

Platelet-derived Growth Factor: Distinct Signal Transduction Pathways Associated with Migration versus Proliferation. *By* KARIN E. BORNFELDT, ELAINE W. RAINES, LEE M. GRAVES, MICHAEL P. SKINNER, EDWIN G. KREBS, and RUSSELL ROSS... 416

Thrombin and Its Receptor in Growth Control. *By* ELLEN VAN OBBERGHEN-SCHILLING, VALÉRIE VOURET-CRAVIARI, YAO-HUI CHEN, DOMINIQUE GRALL, JEAN-CLAUDE CHAMBARD, and JACQUES POUYSSÉGUR 431

Structural and Functional Properties of the TRK Family of Neurotrophin Receptors. *By* MARIANO BARBACID 442

Protein Kinase C Mediates Short- and Long-term Effects on Receptor Tyrosine Kinases: Regulation of Tyrosine Phosphorylation and Degradation. *By* KLAUS SEEDORF, MARK SHERMAN, and AXEL ULLRICH 459

Flag-Insulin Receptor Mutants. *By* H. JOSEPH GOREN 463

Localization of Specific Amino Acids Contributing to Insulin Specificity of the Insulin Receptor. *By* ASSER S. ANDERSEN, FINN C. WIBERG, and THOMAS KJELDSEN 466

The Acute Insulin-like Effects of Growth Hormone in Primary Adipocyte-signaling Mechanisms. *By* HANS TORNQVIST, MARTIN RIDDERSTRÅLE, HANS ERIKSSON, and EVA DEGERMAN.. 469

Regulatory Interaction between Calmodulin and the Epidermal Growth Factor Receptor. *By* ALBERTO BENGURÍA, JOSÉ MARTÍN-NIETO, GUSTAVO BENAIM, and ANTONIO VILLALOBO . 472

Phosphorylation of Connexin-32 by the Epidermal Growth Factor Receptor Tyrosine Kinase. *By* JUAN ANTONIO DÍEZ, MARIBEL ELVIRA, and ANTONIO VILLALOBO.......................... 477

Molecular Dissection of the Growth Hormone Receptor: Identification of Distinct Cytoplasmic Domains Corresponding to Different Signaling Pathways. *By* J. H. NIELSEN, N. BILLESTRUP, G. ALLEVATO, A. MØLDRUP, E. D. PETERSEN, J. AMSTRUP, J. A. HANSEN, and C. SVENSSON 481

Endothelin Stimulates MAP Kinase Activity and Protein Synthesis in Isolated Adult Feline Cardiac Myocytes. *By* LINDA G. JONES, KATRINA C. GAUSE, and KATHRYN E. MEIER.......... 484

Index of Contributors 487

Financial assistance was received from:

Supporters

- AMGEN INC.
- HAGEDORN RESEARCH INSTITUTE
- PFIZER CENTRAL RESEARCH

Contributors

- ABBOTT, PHARMACEUTICAL PRODUCTS DIVISION
- AMERICAN CYANAMID COMPANY
- AMERSHAM LIFE SCIENCE INC.
- BACHEM BIOSCIENCE INC.
- BEHRINGWERKE AKTIENGESELLSCHAFT
- BOEHRINGER INGELHEIM PHARMACEUTICALS, INC.
- BRISTOL-MYERS SQUIBB PHARMACEUTICAL RESEARCH INSTITUTE
- BURROUGHS WELLCOME COMPANY
- DNAX RESEARCH INSTITUTE OF MOLECULAR AND CELLULAR BIOLOGY, INC.
- THE DUPONT MERCK PHARMACEUTICAL COMPANY
- GENENTECH, INC.
- GLAXO RESEARCH INSTITUTE
- HOFFMANN-LA ROCHE INC.
- IMMUNEX RESEARCH AND DEVELOPMENT CORPORATION
- R. W. JOHNSON PHARMACEUTICAL RESEARCH INSTITUTE
- MERCK RESEARCH LABORATORIES
- MILES INC.
- NATIONAL INSTITUTE OF ALLERGY AND INFECTIOUS DISEASES–NATIONAL INSTITUTES OF HEALTH
- ONCOGENE SCIENCE, INC.
- PARKE-DAVIS PHARMACEUTICAL RESEARCH, WARNER-LAMBERT COMPANY
- RHÔNE-POULENC ROVER CENTRAL RESEARCH
- SANDOZ PHARMACEUTICALS CORPORATION
- SCHERING AG PHARMACEUTICAL DIVISION
- THE UPJOHN COMPANY

The New York Academy of Sciences believes it has a responsibility to provide an open forum for discussion of scientific questions. The positions taken by the participants in the reported conferences are their own and not necessarily those of the Academy. The Academy has no intent to influence legislation by providing such forums.

Preface

Upon ligand binding, cell-surface receptors transmit activity signals into the cell, activating multiple metabolic and mitogenic pathways. Although the nature of the stimulatory molecules is infinitely varied, the principles involved in propagating the signals are ubiquitous. Thus, such different stimulators as antigens, cytokines, hormones, and growth factors use similar transduction principles, including ligand-induced receptor internalization of receptor-ligand complexes, activation of G proteins, oligomerization, increases in Ca^{2+} uptake, activation of tyrosine kinases or other kinases and phosphatases, protein phosphorylation or dephosphorylation, and the triggering of oncogenes and transcription factors. The signaling molecules interact with each other through universal "zippers," such as SH2 and SH3 domains.

Scientists from many different fields of research have, in recent years, made significant contributions toward resolving the mechanism(s) involved in the signal-transduction pathways. The unknown in this integrated discipline, however, remains far greater than the known. The immense progress accomplished in molecular cloning and identification of many signaling molecules has defined many details of the molecular anatomy of signaling, but quantitative analysis remains scarce, and we do not always understand the functional relationships or the molecular physiology.

We do not fully understand how receptors are coupled to downstream signal-transduction pathways to promote cell growth, cell differentiation, cell metabolism, or programmed cell death. We do not appreciate the downstream pathways shared by different receptors and those confined to specific stimulating signals. Do cytokine receptors and growth-factor receptors share a common signal transduction pathway leading to cell growth? Do distinctive cellular functions, such as proliferation and differentiation, use the same or different signaling pathways? If they use the same pathway, how does the cell "know" which function to follow? Does the intensity or duration of the stimulus, the number of available receptors, or perhaps the availability of end products, such as transcription factors, influence the cell's decision to proliferate or differentiate? The regulatory mechanism of receptor activation (*e.g.*, dephosphorylation) has been unraveled, at least partly, for certain (but not all) receptors. Does the regulation of receptor activity use similar principles in different biological systems? We do not know the basic rules that regulate the interconnections of the downstream receptor signal network. How are the postreceptor pathways organized? Do distinct multiple signals delivered by different receptors merge into a common pathway, or do they split when arriving at a crossroads? Does a signal released by a certain receptor have the option of using different pathways, or is there only one? If multiple pathways are available to a single stimulatory signal, which one will be used, and why? Is the decision irrevocable or flexible? Is memory or learning inculcated through repeatedly used pathways? Some of these questions were addressed directly or indirectly in this symposium, and a few answers may be found in articles collected in this volume.

The aim of the symposium was to bring together scientists from such diverse fields as immunology, molecular genetics, endocrinology, neurobiology, pharmacology, and cell biology, but with the common interest in the signaling machinery mediated by cell-surface receptors. The objective was to share the progress made in each one of

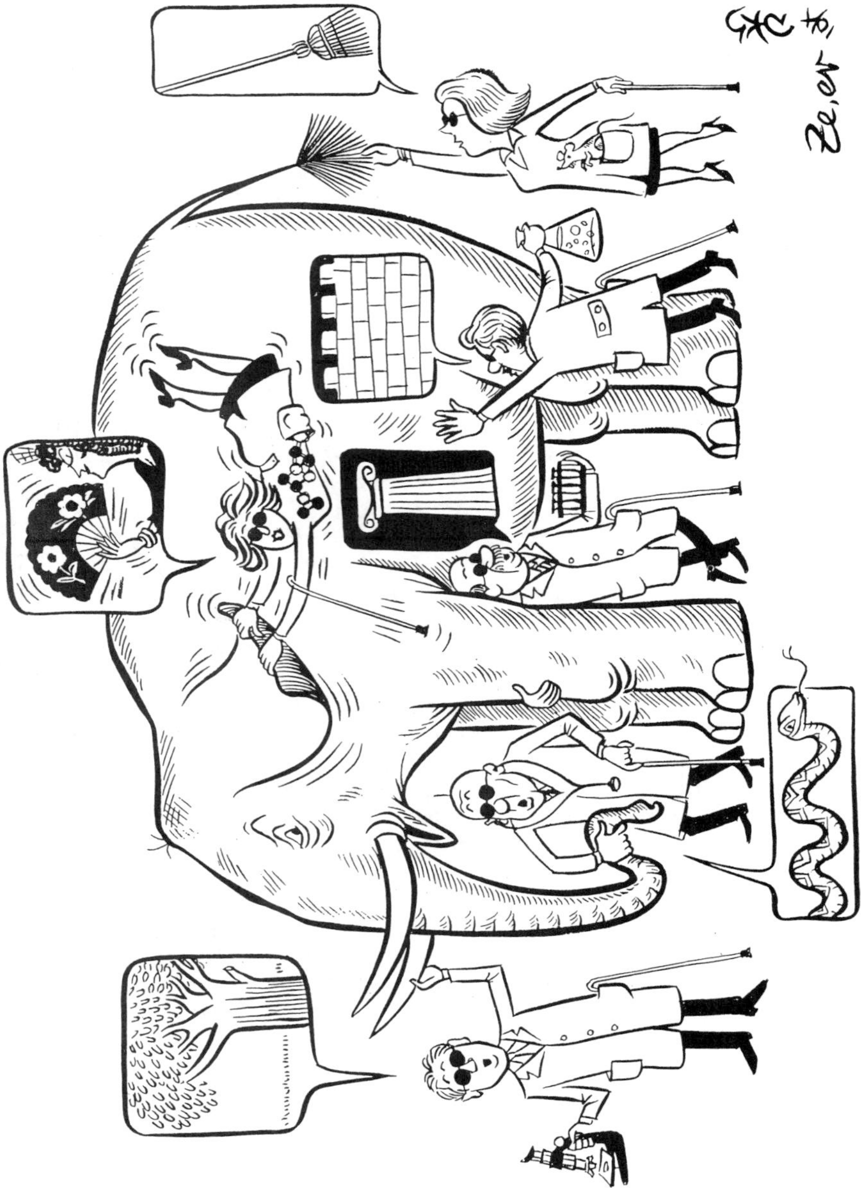

these disciplines with scientists from other fields. As a result, the presentation of different concepts and technical strategies led to fruitful discussions of mutual benefit and the creation of acceptable models. Common concepts emerged, such as the newly recognized importance of signal timing, kinetics, and duration in determining pathway selectivity.

Above all, we hope that this volume will help in the fight against the "elephant syndrome" (FIG. 1), a disease afflicting many scientists who miss the significance of their findings because they do not grasp the entire situation.

DAVID NAOR and PIERRE DE MEYTS

FIGURE 1. The "elephant syndrome." This is an old Indian tale, which describes six blind beggars who stumble upon an elephant. One of them touches the trunk and says, "This must be a snake." The second feels a tusk and decides that it is a branch of a tree. The third palpates the ear and claims it is a fan. The fourth fingers the end of the tail and says, "No doubt, this is a broom." The fifth beggar touches the creature's side and declares, "This is the wall of a fortress." The last one embraces the animal's leg and decides that he is leaning on a pillar. The cartoon was done by Ze'ev, a leading Israeli cartoonist, who draws a daily political cartoon for *Ha'aretz*, the Tel Aviv-based newspaper.

Definition of Signals for Neuronal Differentiation

AXEL OBERMEIER,[a] RALPH A. BRADSHAW,[b]
KLAUS SEEDORF,[a] AXEL CHOIDAS,[a]
JOSEPH SCHLESSINGER,[c] AND AXEL ULLRICH[a]

[a]*Department of Molecular Biology*
Max-Planck-Institut für Biochemie
Am Klopferspitz 18A
82152 Martinsried, Germany

[b]*Department of Biological Chemistry*
College of Medicine
University of California
Irvine, California 92717

[c]*Department of Pharmacology*
New York University Medical Center
550 First Avenue
New York, New York 10016

Tyrosine phosphorylation-mediated cellular signals are generated by extracellular interaction of receptor tyrosine kinase (RTK) binding domains with specific ligands, followed by receptor dimerization and autophosphorylation of tyrosine residues within the intracellular signaling domain. This triggers the recruitment of src homology 2 (SH2) domain-containing proteins to specific phosphotyrosines (pY) and leads to the assembly of these primary signal transfer factors at the inner face of the plasma membrane in proximity to molecules involved in subsequent steps of signal transduction.[1–3] This initial interaction between the RTK cytoplasmic domain and cellular substrates involves amino acid sequences flanking the receptor pY docking sites and the variant residues within the SH2 consensus motif of the substrate, which together define the binding affinity and thereby the specificity of the signal.[2]

For nerve growth factor receptor (NGF-R) Trk, phospholipase Cγ (PLCγ) and the noncatalytic p85 subunit of phosphatidylinositol 3′ kinase (PI3′-K) have been described to be direct tyrosine kinase substrates.[4,5] Moreover, SHC, an oncogenic SH2 domain-containing molecule,[6] was recently implicated in linking Trk to the $p21^{ras}$ signaling pathway,[7] which had been shown previously to be essential for NGF-induced neurite formation by PC12 cells.[8,9] To investigate in more detail the Trk substrate interaction properties, we sought to identify the sites on Trk that are responsible for association with these three signal-transducing proteins. We showed that tyrosine residue Y-490, located in the juxtamembrane region, Y-751 in the kinase core domain, and Y-785 in the short carboxy-terminus of Trk are the interaction sites for SHC, p85/PI3′-K, and PLCγ, respectively. Moreover, we demonstrated that p85 and PLCγ can bind simultaneously to the Trk cytoplasmic domain *in vitro,* in spite of very close proximity of their specific binding sites.[10]

Several substrate interaction sites on RTKs, such as platelet-derived growth factor receptor (PDGF-R), fibroblast growth factor receptor (FGF-R), colony-stimulating factor-1 receptor (CSF1-R), epidermal growth factor receptor (EGF-R), hepatocyte growth factor receptor (HGF-R)/Met, and others, have been identified, with the βPDGF-R being the most closely examined. Tyrosine autophosphorylation sites Y-1021 and -1009, located in its carboxy-terminus, are binding sites of PLCγ and the SH2 domain-containing phosphotyrosine phosphatase (PTP) 1D/Syp, respectively.[11,12] Y-771 in the kinase insertion domain binds the Ras-GTPase-activating protein (GAP), whereas Y-740 and Y-751 represent interaction sites for p85/PI3′-K,[13–15] the latter being also a specific site of association with an adapter protein named Nck.[16] Src, Fyn, and Yes, three closely related cytoplasmic tyrosine kinases, bind to Y-579 and Y-581 in the βPDGF-R juxtamembrane domain.[17]

Although such detailed information about RTK-substrate interactions is rapidly increasing, only little is known about the biological significance of these interactions. Using the neuron-like rat pheochromocytoma cell line PC12,[18] we investigated the roles of the NGF-R/Trk-associated primary signal transducers SHC, p85/PI3′-K, and PLCγ in neuronal differentiation.

TRK Y-785 IS A HIGH-AFFINITY BINDING SITE FOR PLCγ

To examine the signaling capacity of Trk, we constructed a chimeric receptor, ET-R, which consists of the EGF-R extracellular domain fused to the transmembrane and cytoplasmic sequences of the NGF receptor, Trk, previously shown to be fully functional when activated by EGF.[5] We employed ET-R in order to circumvent the need for NGF to activate the Trk tyrosine kinase, because it is not clear whether the low-affinity NGF receptor, $p75^{LNGFR}$, is necessary in addition to Trk for high-affinity binding of NGF. In addition, a carboxy-terminal truncation mutant, ET-ΔCT, and a point mutant, ET-YF, containing a phenylalanine in place of the carboxy-terminal domain tyrosine residue at position 785, were generated.

Upon transfection of respective expression plasmids in 293 cells, EGF-R and ET-R displayed different association patterns, reflecting distinct substrate interaction and, hence, different signal transduction capacities of EGF-R and Trk (FIG. 1A). The most striking signal was a strong band migrating at 145 kDa, which was exclusively associated with stimulated ET-R and identified as PLCγ by an αPLC immunoblot of the anti-receptor precipitates (FIG. 1B, upper panel). This high amount of coprecipitated PLCγ, which is derived from endogenous levels only, indicates an unusually high affinity of Trk cytoplasmic sequence for this protein. A comparatively small amount of PLCγ (~1%) was found to coprecipitate with EGF-R, clearly visible after longer exposures of the autoradiograph (A) and the αPLCγ immunoblot (B, upper panel) (not shown). This dramatic difference in affinity could be corroborated by an analogous *in vitro* association experiment, in which either normal 293 cell lysate or lysate from transfected, PLCγ-overexpressing 293 cells, was added to lysates containing high amounts of autophosphorylated EGF-R or autophosphorylated or non-phosphorylated ET-R. PLCγ coprecipitation with the Trk cytoplasmic domain (FIG. 2, lane 2) was almost as efficient as immunoprecipitation with an excess of αPLCγ antiserum (lane 4), whereas PLCγ coprecipitated by EGF-R was almost undetectable

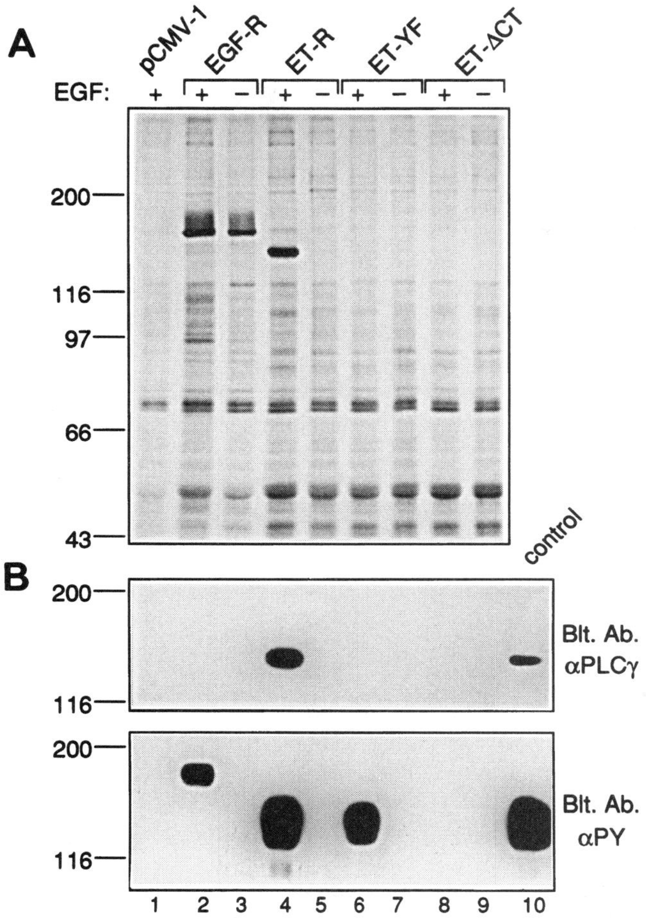

FIGURE 1. *In vitro* association with PC12 cell proteins. Receptor-expressing 293 cells were treated with EGF (as well as with Na-vanadate), where indicated (+), and lysed, and the precleared lysates were mixed with equal aliquots of precleared lysate from [^{35}S]methionine-labeled PC12 cells. Receptors were immunoprecipitated with anti-human EGF-R extracellular domain mAb 108.1;[54] the samples were halved and separately electrophoresed on 7.5% SDS gels. One gel was processed for autoradiography (**A**) to visualize total coimmunoprecipitated proteins from labeled PC12 cell lysate, whereas proteins of the other gel were electrophoretically transferred to nitrocellulose and probed with polyclonal αPLCγ antiserum and with mAb 5E2 (αPY)[55] (**B**). The control lane differs from the ET-R (+) lane only in that 293 cell lysate containing autophosphorylated ET-R was left without radiolabeled PC12 cell lysate. The difference in the signal intensity of PLCγ bands between control and ET-R (+) lanes demonstrates that associated PLCγ is mainly derived from PC12 cells. (Blt. Ab., blotting antibodies). (Obermeier *et al.*[5] With permission from the *EMBO Journal*.)

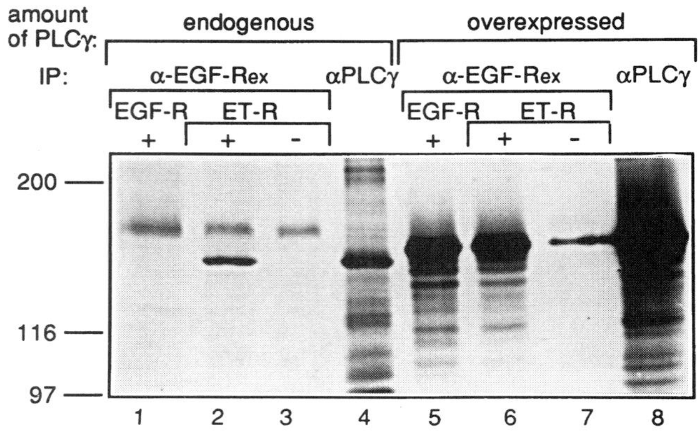

FIGURE 2. Trk and EGF-R display different affinities for PLCγ. 293 cell lysates containing either autophosphorylated EGF-R (lanes 1, 5), autophosphorylated ET-R (lanes 2, 6), or nonphosphorylated ET-R (lanes 3, 7) were mixed with equal aliquots of lysate from untreated (lanes 1-3) or transfected PLCγ-overexpressing 293 cells (lanes 5-7), and receptors precipitated with anti-EGF-R extracellular domain mAb 108 (αEGF-Rex). After separation on a 7.5% SDS gel, precipitates were blotted with an αPLCγ antiserum. Equal lysate mixtures as used for lanes 2 and 6 were subjected in parallel to αPLCγ immunoprecipitation (lanes 4 and 8, respectively), in order to monitor the amounts of PLCγ present in both cases.

(lane 1). Under PLCγ overexpression conditions (lanes 5-8), however, this difference was no longer evident due to saturation of receptor binding sites.

PLCγ association with ET-R was abrogated by a single Y/F point mutation at position 785 of Trk. This, together with the results from receptor/PLCγ coexpression experiments,[5] showing that Trk Y-785 is necessary for phosphorylation of PLCγ by the Trk tyrosine kinase, strongly suggested that Y-785 of Trk is indispensable for high-affinity binding to PLCγ. This was subsequently confirmed by *in vitro* association experiments with ET-R and [^{35}S]methionine-labeled PC12 cell lysates, involving a pentadecapeptide identical in sequence to the Trk carboxy-terminus in which the tyrosine residue corresponding to Trk-Y-785 was phosphorylated.

At a concentration of 10 nM, the phosphopeptide significantly reduced the amount of PLCγ that coprecipitated with ET-R (Fig. 3, lane 4). Substrate-receptor association was further decreased at increasing phosphopeptide concentrations and was completely abolished to undetectable amounts at 500 nM (Fig. 3, lane 10). By contrast, under the same conditions the nonphosphorylated peptide had no capacity to inhibit PLCγ binding. Comparable effects, however, were obtained at concentrations four orders of magnitude higher than that employed for the phosphopeptide (data not shown). These results demonstrate that the carboxy-terminal sequence of Trk, containing phosphorylated Y-785, is the binding site for PLCγ.

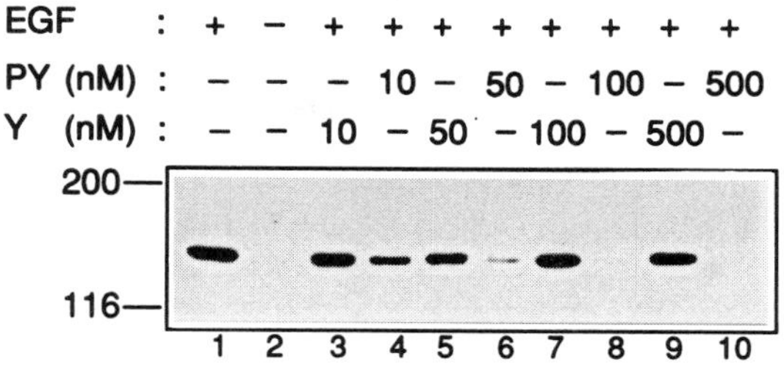

FIGURE 3. Phosphotyrosine peptide inhibition of PLCγ binding to ET-R. ET-R-expressing 293 cells were treated with EGF (as well as with Na-vanadate), where indicated (+), and lysed, and the precleared lysates were mixed with precleared lysate from PC12 cells. Then different amounts of phosphopeptide (PY) QALAQAPPVY(PO_3H_2)LDVLG or nonphosphorylated peptide (Y) of the same sequence was added. ET-R was immunoprecipitated with mAb 108.1; the precipitates were separated by 7.5% SDS-PAGE and electrophoretically transferred to nitrocellulose. Amounts of coprecipitated PLCγ were detected with αPLCγ-polyclonal antiserum (Obermeier *et al.*[5] With permission from the *EMBO Journal.*)

IDENTIFICATION OF TRK BINDING SITES FOR SHC AND p85/PI3′-K AND FORMATION OF A MULTIMERIC SIGNALING COMPLEX

In addition to PLCγ, p85/PI3′-K has been described to be a direct substrate of the Trk tyrosine kinase,[4,5] and a novel SH2 domain-containing molecule, termed SHC,[6] has been shown to induce neurite outgrowth when overexpressed in PC12 cells.[7] To determine the binding sites of p85 and SHC, a series of peptides was synthesized, which corresponded to all but one of the eleven tyrosine residues of the Trk receptor cytoplasmic domain flanked by 3-6 amino acids on each side.[10] The peptide containing Y-634 was omitted because of the hydrophobic nature of surrounding amino acids. Trk peptides phosphorylated on tyrosine and nonphosphorylated controls were then examined for their ability to prevent the *in vitro* association between the autophosphorylated Trk cytoplasmic domain and p85 or SHC substrates.

The association potential of p85 and SHC to Trk cytoplasmic sequences was examined by *in vitro* associations, combining ET-R-containing 293 cell lysates with either p85 or SHC lysates from 293 cells overexpressing either of the two substrates, followed by immunoprecipitation with anti-receptor antibody 108 (FIG. 4). Both p85 and SHC specifically associated with the autophosphorylated (FIG. 4, lane 1), but not with nonphosphorylated (lane 2) receptors. Coprecipitation of p85 with ET-R was inhibited by the peptide encompassing Trk-pY-751 (FIG. 4A, lane 13), whereas all other phosphopeptides had no effect. Coprecipitation of SHC with ET-R was solely inhibited by the phosphopeptide pY-490 (FIG. 1B, lane 3). Moreover, addition of all ineffective phosphopeptides to one immunoprecipitation at the same time did not influence the association (data not shown). These data strongly suggest that p85 and SHC specifically bind to Trk by way of pY-751 and -490, respectively. Y-751 is the most C-terminal tyrosine residue of the kinase domain, and Y-490 is the only tyrosine located in the cytoplasmic juxtamembrane domain of Trk.

For confirmation, ET-R mutants containing phenylalanine in place of Y-490 and Y-751 (ET-Y490F and ET-Y751F) were tested together with ET-R and the EGF-R

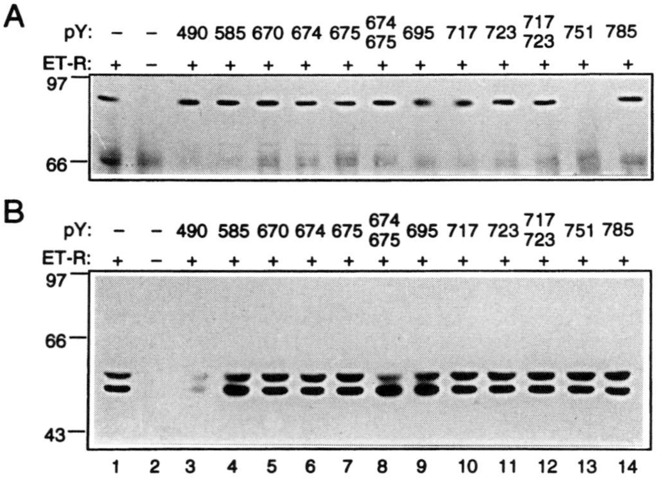

FIGURE 4. Association of p85 (**A**) and SHC (**B**) with ET-R and specific inhibition by phosphopeptides. Lysates of either EGF-stimulated (ET-R+) or -unstimulated (ET-R–) ET-R chimera-expressing 293 cells (3×10^5 cells) were mixed with lysates of 293 cells overexpressing p85 (5×10^5 cells) (**A**) or SHC (8×10^4 cells) (**B**). Where indicated, phosphopeptides were added to a final concentration of 50 μM before immunoprecipitation with anti-receptor antibody (108.1). Precipitates were subjected to SDS-PAGE and immunoblotted with αp85- (**A**) or αSHC (**B**) antiserum. (Obermeier *et al.*[10] With permission from the *Journal of Biological Chemistry.*)

for their binding capacities towards p85 and SHC (Fig. 5). In contrast to ET-R, ET-Y490F failed to bind SHC, and the association of p85 with ET-Y751F was strongly impaired, corroborating the data obtained with the phosphopeptides. A schematic representation of Trk-substrate interaction topography is shown in Figure 6. The EGF-R, which was included for comparison, appears to have a higher affinity for p85 than has Trk (represented by ET-R), whereas both receptors bind similar amounts of SHC under these experimental conditions. Moreover, as compared with ET-R, ET-Y490F retained full binding capacity towards p85, and ET-Y751F coprecipitated SHC, comparable to ET-R. Similarly, a receptor mutant lacking the PLCγ binding site, ET-Y785F, had lost its binding affinity for this signal-transducing protein (Fig. 1). However, equal amounts of SHC and p85 coprecipitated with ET-Y785F as compared with ET-R (Fig. 4), suggesting simultaneous association of PLCγ and p85 or SHC with Trk. This is remarkable, inasmuch as the binding sites of p85 and PLCγ are only 34 amino acids apart.

To determine whether PLCγ and p85 can simultaneously bind to the same receptor molecule, lysates from 293 cells overexpressing p85, PLCγ, or ET-R were mixed, followed by immunoprecipitation with αPLCγ antibody. Different portions of the precipitates, as indicated in the legend to Figure 7, were subjected to SDS-PAGE, blotted on to nitrocellulose, and probed with either αPLCγ, αTrk, or αp85 antibody. As expected, autophosphorylated but not nonphosphorylated receptor was coprecipitated

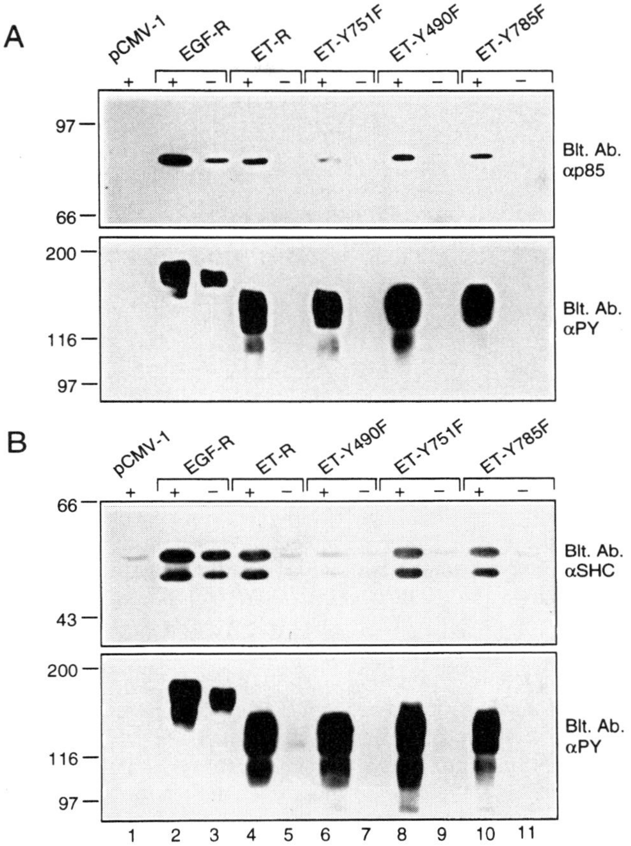

FIGURE 5. Association of p85 (**A**) and SHC (**B**) with EGF-R, ET-R, and ET-YF mutants. Lysates containing equal amounts of autophosphorylated (+) or nonphosphorylated (−) receptors were mixed with p85 (**A**) or SHC lysates (**B**), and receptors immunoprecipitated with anti-receptor antibody (108.1). Precipitates were subjected to SDS-PAGE and immunoblotted with αp85 (**A**) or αSHC (**B**) antiserum (upper panels) and reprobed with antiphosphotyrosine antibody (αPY; lower panels). pCMV-1 denotes a control, where lysates of cells, transfected with only the expression vector and further treated like cells producing autophosphorylated receptors, were mixed with substrate containing lysates. (Obermeier *et al.*[10] With permission from the *Journal of Biological Chemistry.*)

with PLCγ. p85 was only detected with precipitates containing autophosphorylated and therefore coprecipitated receptor, clearly demonstrating that the receptor served as a link between PLCγ and p85, binding both substrates at the same time.

INHIBITION OF PC12 CELL NEURITE OUTGROWTH BY ELIMINATION OF TRK BINDING SITES FOR SHC AND PLCγ

The use of receptor Y/F mutants allows dissection of signal-transduction pathways, yielding important clues regarding the relative contributions of different substrates

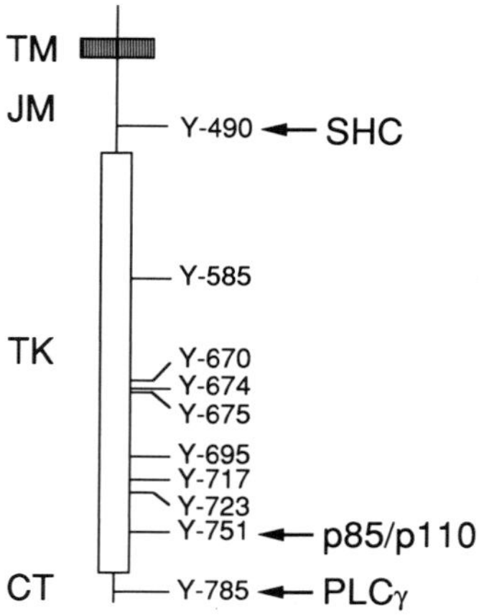

FIGURE 6. Substrate binding sites within the Trk cytoplasmic domain. Receptor subdomains are indicated. TM: transmembrane domain. JM: juxtamembrane domain. TK: tyrosine kinase domain. CT: carboxy terminus. (Obermeier *et al.*[10] With permission from the *Journal of Biological Chemistry.*)

for exerting various biological effects. The PC12 rat pheochromocytoma cell line is ideally suited for the investigation of molecular mechanisms underlying NGF-R/Trk-mediated neuronal differentiation signals resulting in the induction of neurite outgrowth, in contrast to the EGF-R, which is known to mediate cell proliferation in this cell line.

Taking advantage of the absence of PDGF receptors in PC12 cells,[19] we employed the RTK chimera approach[20] to investigate the Trk signal under PDGF control within its normal cell environment rather than in PC12-derived, genetically altered mutant cell lines, which may be altered in their signal conversion properties due to random mutagenesis.[21] By introducing tyrosine to phenylalanine (YF) mutations either singly at positions 785, 751, and 490 or in all possible combinations into PT-R (FIG. 8), a chimeric receptor consisting of the βPDGF-R extracellular domain and the transmembrane and intracellular domains of Trk, we analyzed the significance of signaling pathways connected to SHC, PLCγ, and p85/PI3′-K phosphotyrosine target sites for

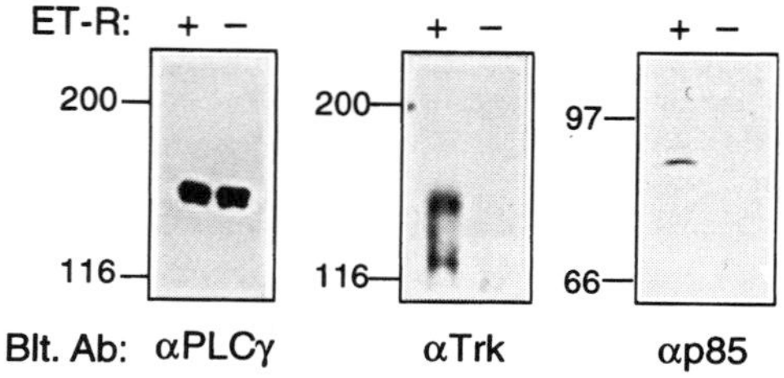

FIGURE 7. Simultaneous binding of PLCγ and p85 to ET-R. Lysates containing either EGF-stimulated (+) or -unstimulated (−) ET-R (5×10^5 cells), p85 (1×10^6 cells), and PLCγ (2×10^5 cells) were mixed and precipitated with α-PLCγ-antiserum. 5% (lanes 1, 2), 25 % (lanes 3, 4), and 70% (lanes 5, 6) of the precipitates were subjected in parallel to SDS-PAGE, blotted onto nitrocellulose, and probed with α-PLCγ, αTrk, or αp85 antiserum, respectively. (Obermeier *et al.*[10] With permission from the *Journal of Biological Chemistry.*)

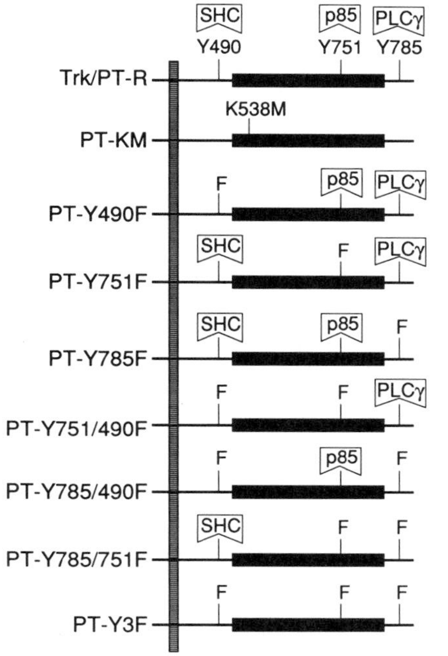

FIGURE 8. Schematic representation of substrate-binding capacities of receptor constructs. The cytoplasmic portions of the PT-R chimera mutants are shown to the right of the schematically indicated plasma membrane. Black boxes denote the tyrosine kinase domain. PT-Y3F is an abbreviation for PT-Y490/751/785F. (Obermeier *et al.*[23] With permission from the *EMBO Journal.*)

NGF-R/Trk-specific signal transduction, which ultimately leads to neuronal differentiation and is essential for the maintenance of the neuronal network. As gene transfer vehicles, we employed the Moloney murine leukemia virus-based retrovirus vector system, pLEN,[22] which uses the neomycin resistance gene as a selection marker. PC12 cells were engineered by infection with respective recombinant retroviruses to stably express the wild-type chimera, PT-R, tyrosine binding site-deficient PT-YF receptor mutants, and, as a control, the kinase-negative PT-KM receptor, carrying a point mutation generated by replacing lysine 538 in the consensus ATP-binding sequence with a methionine.[23]

Upon infection it was necessary to reduce fetal calf serum (FCS) from 5% to 2.5% of the medium and to use horse serum (HS) from platelet-poor plasma to prevent constitutive differentiation of the cells due to the presence of PDGF in the sera. Two months after infection and G418 selection, resulting polyclonal cell lines were tested for expression of chimeric receptors, or overexpression of Trk. Prior to lysis, cells were incubated for 10 min with 30 ng/mL PDGF-BB or 100 ng/mL NGF in order to determine the autophosphorylation capacities of the receptors. Equal amounts of total cellular proteins were subjected to immunoprecipitation with monoclonal antibody B2 directed against the βPDGF-R extracellular domain,[24] or with the polyclonal antibody Ab-1 raised against a C-terminal Trk peptide. After SDS-PAGE, receptors were immunoblotted with Ab-1 (FIG. 9, upper panel) and reprobed with the monoclonal antiphosphotyrosine antibody, 5E2 (FIG. 9, lower panel). Taking into consideration the somewhat different expression levels, PT-R and all YF mutants,

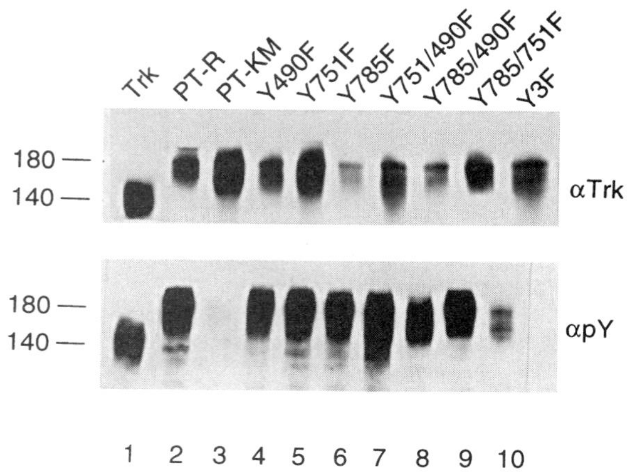

FIGURE 9. Expression of PT receptors and Trk in PC12 cells. PC12 cells were infected with retroviruses containing cDNAs for Trk, PT-R, PT-KM, and PT receptors bearing either one, two, or all three YF mutations at amino acid positions 490, 751, and 785. After G418 selection and propagation, about 10^7 cells of each line were stimulated with 30 ng/mL PDGF-BB (lanes 2–10) or 100 ng/mL NGF (lane 1) for 10 min and collected in lysis buffer. Receptors were immunoprecipitated with either monoclonal antibody B2[24] against the human βPDGF-R extracellular domain (lanes 2–10) or with an affinity-purified polyclonal antibody against the C-term of Trk (Ab-1; Oncogene Science) (lane 1). Precipitated proteins were separated by 7.5% SDS-PAGE, immunoblotted with Ab-1 (upper panel) and reprobed with monoclonal αphosphotyrosine antibody 5E2[55] (lower panel). (Obermeier *et al.*[23] With permission from the *EMBO Journal.*)

except the triple mutant, displayed comparable autophosphorylation activities. The dramatically reduced tyrosine phosphorylation of PT-Y3F was not due only to the lack of three autophosphorylation sites, inasmuch as the effect was not additive, as is apparent from single and double YF mutants. It rather appears that at least any one of the three mutated tyrosine residues is necessary for full tyrosine kinase activity. As expected, PT-KM did not display any detectable kinase activity and was therefore used as a negative control in subsequent differentiation assays.

PC12 cells expressing PT-R readily differentiated upon stimulation with PDGF (Fig. 10). This result provides another argument in favor of Trk being sufficient for mediating neuronal differentiation of PC12 cells, inasmuch as $p75^{LNGFR}$ supposedly does not participate in PDGF binding. Not surprisingly, in contrast to PT-R, PT-KM was not capable of mediating any differentiation in PC12 cells (Fig. 10). The effects of Trk cytoplasmic domain Y/F mutations on PC12 cells were evaluated by visual examination of the frequency and length of ligand-induced neurite outgrowth in direct comparison with PC12/PT-R and PC12/Trk controls. PC12 cells expressing PT-YF mutants display different neuronal differentiation responses upon stimulation with PDGF (Fig. 10; summarized in Table 1). PC12 cells expressing PT-Y785F or PT-Y751F fully differentiated upon addition of PDGF. Thus, elimination of the PLCγ or PI3′-K binding site does not impair Trk-mediated neurite outgrowth. However,

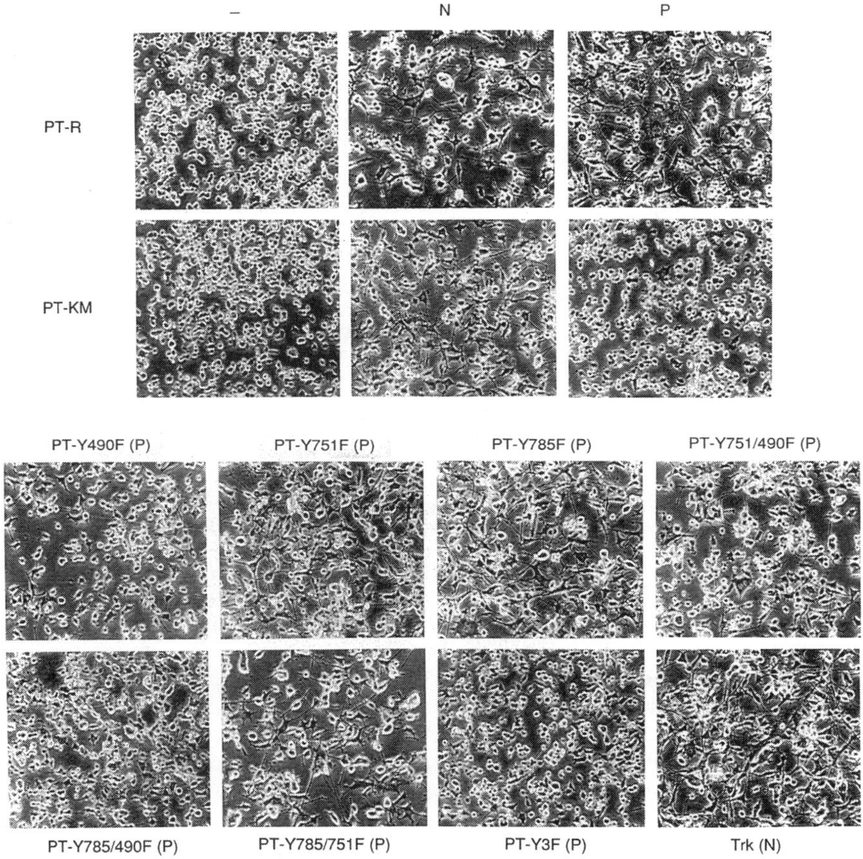

FIGURE 10. Differentation of PC12 transfectants. PC12 cells stably overexpressing PT receptors (or overexpressing Trk) were seeded into six well dishes, and 1 d later either no ligand (–), NGF (N), or PDGF-BB (P) were added to final concentrations of 100 and 30 ng/mL, respectively. Photographs of the cells were taken at 0, 1 d, and 3 d after addition of ligand. Medium and ligand were replaced every two days. Depicted photographs show cells 3 d after addition of ligand, are representative of 2-4 independently generated polyclonal cell lines per receptor construct, and represent those cell lines for which receptor expression levels are presented in FIGURE 9. Experiments were performed twice, and each time receptor expression levels were monitored. (Obermeier *et al.*[23] With permission from the *EMBO Journal.*)

neurite outgrowth in PC12 cells expressing SHC binding site-defective PT-Y490F was severely diminished. Only a few PC12/PT-Y490F cells developed neurites after three or more days of PDGF treatment, and the average length of neurites was significantly reduced. Mutation of both binding sites for SHC and PI3′-K (PT-Y751/490F) did not further reduce the PDGF-induced differentiation response, but rather enhanced it somewhat. Receptors bearing only the PI3′-K binding site and lacking

TABLE 1. Differentiation of PC12 Cell Transfectants[a]

Receptor Overexpressed	Binding Capacity			Differentiation response to:	
	PLCγ	PI3′-K	SHC	PDGF	NGF
PT-R	+	+	+	++++	++++
PT-KM	−	−	−	−	++++
PT-Y490F	+	+	−	+	++++
PT-Y751F	+	−	+	++++	++++
PT-Y785F	−	+	+	++++	++++
PT-Y751/490F	+	−	−	++	++++
PT-Y785/490F	−	+	−	±	++++
PT-Y785/751F	−	−	+	+++	++++
PT-Y3F	−	−	−	±	++++
Trk	+	+	+	−	++++

[a] The degree of differentiation, as judged by the frequency of cells bearing neurites in combination with the number and length of the neurites, is given on an arbitrary scale, ranging from − (no detectable differentiation) to ++++ (maximal observed differentiation).[23]

sites of interaction with both SHC and PLCγ (PT-Y785/490F) were not capable of triggering any significant differentiation of PC12. Only very few PC12/PT-Y785/490F cells displayed short neurite-like spikes. The same was true for PC12/PT-Y3F cells expressing chimeric receptors lacking all three binding sites. A slight but significant decrease in neurite outgrowth was evident in cells expressing receptors that bound SHC but lacked sites of association with both PLCγ and PI3′-K (PT-Y785/751F). Thus, compared with the single mutants PT-Y785F and PT-Y751F, SHC-initiated signaling pathways appeared to be slightly less potent in transducing a Trk-specific signal if PLCγ and PI3′-K pathways were abolished.

The observed effects were not due to different receptor expression levels, because PC12/PT-Y785F cells, which display the lowest expression level, fully differentiated upon addition of PDGF. Moreover, several independently derived cell lines expressing different levels of the same receptors reproducibly presented essentially the same biological response (not shown). In addition, while transphosphorylation of endogenous Trk by overexpressed chimeric receptors did not occur (not shown), normal differentiation of all cell lines was inducible by NGF through activation of endogenous receptor, demonstrating that all cell lines principally retained their full differentiation capacity. Furthermore, in no case was there a significant differentiation in the absence of ligand after three or more days, clearly indicating that the observed effects were strictly ligand-dependent and mediated by the chimeric receptors. Cells treated with NGF or no ligand are shown exemplarily for PT-R and PT-KM only (FIG. 10). As expected, cells overexpressing Trk did not differentiate upon PDGF addition (not shown) but, compared to PC12 cells with normal Trk levels, displayed a more rapid differentiation upon NGF addition, consistent with previously reported findings.[25] Accordingly, differentiation in all PC12/PT lines was almost completed after one to two days of PDGF treatment (data not shown).

Taken together, we conclude that the SHC binding site plays a dominant role in Trk-specific signaling. PLCγ, however, also appears to be involved in Trk-specific signal transduction, as revealed by the double-mutant receptor PT-Y785/490F, which was unable to mediate any significant neurite outgrowth. In addition, a receptor able to bind PLCγ but not SHC or PI3′-K (PT-Y751/490F) still mediated a weak differentiation response. By contrast, the p85/PI3′-K binding site did not provide Trk with any obvious significant signal-transducing capacity in neuronal differentiation. If at all, it slightly modified pathways triggered by SHC and PLCγ and/or possibly other proteins that might bind to tyrosine residues 490 and 785.

DISCUSSION

Growth, differentiation, movement, and the metabolic homeostasis of cells is regulated by ligand-induced pleiotropic signals that are generated by autophosphorylated receptor tyrosine kinases through the interaction with cellular SH2 domain-containing signal transducers.[3] Although this basic concept is well established, the question of how an RTK-characteristic signal is defined remains poorly understood. Recently, a new RTK subfamily has been described that consists of the closely related Trk, TrkB, and TrkC receptors (reviewed in ref. 26). These structurally similar tyrosine kinases are functional receptors for the neurotrophic factors, NGF, brain-derived neurotrophic factor (BDNF), and neurotrophin-3 (NT-3), respectively (reviewed in ref. 27). Each of these receptor-ligand pairs are implicated in neuronal survival and differentiation. In PC12 cells, the NGF-R/Trk but not the EGF-R is able to promote the extension of neurites from cell bodies and thus induce a differentiated phenotype, which *in vivo* is essential for establishment and maintenance of nerve functions.

We and others have previously shown that PLCγ and p85 associate with activated Trk and become phosphorylated on tyrosine by the Trk kinase.[4,5,28,29] We show here that SHC specifically binds autophosphorylated Trk cytoplasmic sequences (FIG. 4). Moreover, SHC has recently been demonstrated to be phosphorylated by ligand-activated Trk (our unpublished observations; ref. 30).

In order to investigate the neuronal signaling mechanisms employed by Trk, our first aim was to determine the sites of interaction of this RTK with the signal transducers PLCγ, p85/PI3′-K, and SHC. We identified tyrosines Y-490, -751, and -785 as docking sites for SHC, p85/PI3′-K, and PLCγ, respectively. PLCγ has been generally found to associate within the carboxy-termini of the RTKs PDGF-R,[11,12] EGF-R,[31] and FGF-R,[32] consistent with its binding site on Trk. In contrast to many other RTKs, Trk has only two tyrosine residues outside the catalytic domain: Y-490 in the juxtamembrane domain and Y-785 in the short C-terminus. Because these two tyrosines mediate binding to SHC and PLCγ, respectively, additional Trk-associating substrates must either compete with SHC and PLCγ for their binding sites or interact with autophosphorylation sites within the kinase domain. The latter alternative applies to p85: Trk, lacking a typical kinase insertion sequence, where p85 binding sites have generally been mapped (as in the case of PDGF-R,[13–15] colony-stimulating factor-1 receptor,[33] and Kit[34]), interacts with p85 through Y-751, located in the C-terminal-most portion within the kinase core domain.

We have demonstrated that p85 and PLCγ bind simultaneously to the same receptor molecule.[10] If the active form of Trk is a homodimer, as suggested by Jing *et al.*,[35] it could be that the simultaneous binding of p85 and PLCγ occurs only on one monomer each. In this case, however, one would expect a significant increase in the amount of p85 associated with ET-Y785F as compared with ET-R, which cannot be detected (FIG. 4A). We cannot formally exclude, however, the possibilities that p85 association is facilitated by PLCγ binding or that structural differences between ET-Y785F and ET-R might counteract such an increase. If p85 and PLCγ can simultaneously and independently associate with the same monomer, as is suggested by our results, this has some implications for the three-dimensional structure of the carboxy-terminal portion of the receptor. Because the respective binding sites Y-751 and Y-785 are only 34 amino acids apart, they very likely reside on opposite sides of the molecule's surface, and the stretch of amino acids between the two tyrosines must be relatively elongated.

We next investigated the biological significance of the Trk-associated primary signal transducers, SHC, p85/PI3′-K, and PLCγ in neuronal differentiation. Our experimental observations with PC12 cells expressing a variety of Trk SH2 protein binding site mutants demonstrate distinct roles for the PLCγ, p85/PI3′-K, and SHC binding sites in this signaling process. The severe neurite outgrowth signaling defect displayed by the SHC association-incompetent PT-mutant receptor chimera in PC12/PT-Y490F cells suggests an essential role for this protein and the respective downstream signaling pathway in the regulation of neuronal differentiation. SHC has been implicated in linking receptor and cytoplasmic tyrosine kinases to the Ras signaling pathway by virtue of the ability of its tyrosine-phosphorylated form to bind to GRB-2, the connector protein for the Ras guanine nucleoside exchange factor SOS.[7,36] In conjunction with previous reports of neurite outgrowth stimulation by SHC overexpression,[7] constitutive Ras activation,[8] and Raf overexpression[37] in PC12 cells, our results provide strong evidence that SHC binding to Trk/Y490 initiates an intracellular cascade of protein-protein interactions, likely involving GRB-2/Sem5, SOS, and Ras, which directly connects with the downstream Ser/Thr kinases Raf, microtubule-associated protein (MAP) kinase (MAPKK), and MAP kinase (MAPK),[38–45] and that this signaling cascade is essential for full neuronal differentiation of PC12 cells. Moreover, our Trk signaling domain mutants provide support for a complementary or cooperative role of PLCγ, which exhibits a remarkably high affinity for Trk.[5] The significance of the PLCγ binding site, Tyr 785, for Trk-specific signal transduction is revealed, on one hand, by the double-mutant receptor PT-Y785/490F, which is unable to mediate any appreciable neurite outgrowth, and the PT-R mutant PT-Y751/490F, which is able to bind PLCγ but not SHC or PI3′-K, yet still mediates a moderate differentiation response. Activation of this signal transducer by NGF/Trk may, among possibly other effects related to cytoskeletal rearrangements,[46,47] recruit the protein kinase C (PKC) signaling system through an increase in membrane-proximal diacylglyercol concentrations and eventually lead to modulation of MAPK function, as suggested by a previous report.[48] This downstream effect might be mediated by PKC-induced activation of the Raf kinase,[4,49] which then could be considered a converging point of SHC and PLCγ pathways originating at Trk tyrosines 490 and 785, respectively. Thus, in PC12 cells, phosphatidylinositol hydrolysis and PKC activation may cooperate with the Ras signaling pathway and may be involved in the implementation of differentiation-related changes in the cytoskeletal organization and physiology of the cell.

A significant role for the PLCγ signaling pathway in PDGF-induced growth regulation is suggested by recent experiments in which reintroduction of its tyrosine binding site into a PI3′-K, GAP, PTP1D, and PLCγ binding-deficient receptor mutant restored a significant portion of the PDGF-R-mediated signal,[50] although, as previously shown, mutation of the PLCγ binding site had no effect on PDGF-induced DNA synthesis.[11] Analogously, the role of PLCγ in Trk-specific signaling was not obvious from experiments with the PLCγ binding site mutant Y785F, but only from those with an "add-back" mutation of the PLCγ binding site to a receptor lacking association sites for PI3′-K, SHC, and PLCγ.

In view of previous reports that suggested Src as an essential element in neuronal differentiation of PC12 cells[51] and the recently unraveled chain of events of the mitogenic signal,[7,36,38–45] our findings raise the possibility of a link between the Src signaling system and the pathway triggered by SHC/Trk and possibly PLCγ/Trk interaction. Such a cooperative signal amplification scenario may in fact be critical for the generation of a more sustained activation of PLCγ and downstream signal transducers like Ras and MAP kinase[52,53] and could be the feature that distinguishes the Trk signal from that of the EGF-R in PC12 cells. Alternatively, differential signal duration of individual receptors, as defined by negative regulation systems involving phosphotyrosine-specific phosphatases, may be critical for distinct cell responses such as differentiation and growth.

REFERENCES

1. Cantley, L. C., K. R. Auger, C. Carpenter, B. Duckworth, A. Graziani, R. Kapeller & S. Soltoff. 1991. Cell **64:** 281-302.
2. Koch, A. C., D. Anderson, M. F. Moran, C. Ellis & T. Pawson. 1991. Science **252:** 668-674.
3. Schlessinger, J. & A. Ullrich. 1992. Neuron **9:** 1-20.
4. Soltoff, S. P., S. Rabin, L. C. Cantley & D. R. Kaplan. 1992. J. Biol. Chem. **267:** 17472-17477.
5. Obermeier, A., H. Halfter, K.-H. Wiesmüller, G. Jung, J. Schlessinger & A. Ullrich. 1993a. EMBO J. **12:** 933-941.
6. Pelicci, G., L. Lanfrancone, F. Grignani, J. McGlade, F. Cavallo, G. Forni, I. Nicoletti, F. Grignani, T. Pawson & P. G. Pelicci. 1992. Cell **70:** 93-104.
7. Rozakis-Adcock, M., J. McGlade, G. Mbamalu, G. Pelicci, R. Daly, W. Li, A. Batzer, S. Thomas, J. Brugge, P. G. Pelicci, J. Schlessinger & T. Pawson. 1992. Nature **360:** 689-692.
8. Bar-Sagi, D. & J. R. Feramisco. 1985. Cell **42:** 841-848.
9. Hagag, N., S. Halegoua & M. Viola. 1986. Nature **319:** 680-682.
10. Obermeier, A., R. Lammers, K-H. Wiesmüller, G. Jung, J. Schlessinger & A. Ullrich. 1993b. J. Biol. Chem. **268:** 22963-22966.
11. Rönnstrand, L., S. Mori, A-K. Arridsson, A. Eriksson, C. Wernstedt, U. Hellman, L. Claesson-Welsh & C-H. Heldin. 1992. EMBO J. **11:** 3911-3919.
12. Valius, M., C. Bazenet & A. Kazlauskas. 1993. Mol. Cell. Biol. **13:** 133-143.
13. Escobedo, J. A., D. R. Kaplan, M. W. Kavanaugh, C. W. Turck & L. T. Williams. 1991. Mol. Cell. Biol. **11:** 1125-1132.
14. Fantl, W. J., J. A. Escobedo, G. A. Martin, C. W. Turck, M. del Rosario, F. McCormick & L. T. Williams. 1992. Cell **69:** 413-423.

15. Kashishian, A., A. Kazlauskas & J. A. Cooper. 1992. EMBO J. **11:** 1373-1382.
16. Nishimura, R., W. Li, A. Kashishian, A. Mondino, M. Zhou, J. Cooper & J. Schlessinger. 1993. Mol. Cell. Biol. **13:** 6889-6896.
17. Mori, S., L. Rönnstrand, K. Yokote, A. Engström, S. A. Courtneidge, L. Claesson-Welsh & C-H. Heldin. 1993. EMBO J. **12:** 2257-2264.
18. Tischler, A. S. & L. A. Greene. 1975. Nature **258:** 341-342.
19. Heasley, L. E. & G. L. Johnson. 1992. Mol. Biol. Cell **3:** 545-553.
20. Riedel, H., T. J. Dull, J. Schlessinger & A. Ullrich. 1986. Nature **324:** 68-70.
21. Green, S. H., R. E. Rydel, J. L. Connolly & L. A. Greene. 1986. J. Cell Biol. **102:** 830-843.
22. Adam, M. A., N. Ramesh, D. A. Miller & W. R. A. Osborne. 1991. J. Virol. **65:** 4985-4990.
23. Obermeier, A., R. A. Bradshaw, K. Seedorf, A. Choidas, J. Schlessinger & A. Ullrich. 1994. EMBO J. **13:** 1585-1590.
24. Rönnstrand, L., L. Terracio, L. Claesson-Welsh, C-H. Heldin & K. Rubin. 1988. J. Biol. Chem. **263:** 10429-10435.
25. Hempstead, B. L., S. J. Rabin, L. Kaplan, S. Reid, L. F. Parada & D. Kaplan. 1992. Neuron **9:** 883-896.
26. Barbacid, M., F. Lamballe, D. Pulido & R. Klein. 1991. Biochim. Biophys. Acta **1072:** 115-127.
27. Meakin, S. O. & E. M. Shooter. 1992. Trends Neurosci. **15:** 323-331.
28. Raffioni, S. & R. A. Bradshaw. 1992. Proc. Natl. Acad. Sci. USA **89:** 9121-9125.
29. Vetter, M. L., D. Martin-Zanca, L. F. Parada, M. J. Bishop & D. R. Kaplan. 1991. Proc. Natl. Acad. Sci. USA **88:** 5650-5654.
30. Stephens, R. M., D. M. Loeb, T. D. Copeland, T. Pawson, L. A. Greene & D. R. Kaplan. 1994. Neuron **12:** 691-705.
31. Rotin, D., B. Margolis, M. Mohammadi, R. J. Daly, G. Daum, N. Li, E. H. Fischer, W. H. Burgess, A. Ullrich & J. Schlessinger. 1992. EMBO J. **11:** 559-567.
32. Mohammadi, M., A. M. Honegger, D. Rotin, R. Fishcer, F. Bellot, W. Li, C. A. Dionne, M. Jaye, M. Rubinstein & J. Schlessinger. 1991. Mol. Cell. Biol. **11:** 5068-5078.
33. Reedijk, M., X. Liu, P. van der Geer, K. Letwin, M. D. Waterfield, T. Hunter & T. Pawson. 1992. EMBO J. **11:** 1365-1372.
34. Serve, H., Y.-C. Hu & P. Besmer. 1994. J. Biol. Chem. **269:** 6026-6030.
35. Jing, S., P. Tapley & M. Barbacid. 1992. Neuron **9:** 1067-1079.
36. Egan, S. E., B. W. Giddings, M. W. Brooks, L. Buday, A. M. Sizeland & R. A. Weinberg. 1993. Nature **363:** 45-51.
37. Wood, K. W., Q. Haiqing, G. D'Arcangelo, R. C. Armstrong, T. M. Roberts & S. Halegoua. 1993. Proc. Natl. Acad. Sci. USA **90:** 5016-5020.
38. Moodie, S. A., B. M. Willumsen, M. J. Weber & A. Wolfman. 1993. Science **260:** 1658-1661.
39. Vojtek, A. B., S. M. Hollenberg & J. Cooper. 1993. Cell **74:** 205-214.
40. Zhang, X.-F., J. Settleman, J. M. Kyriakis, E. Takeuchi-Suzuki, S. J. Elledge, M. S. Marshall, J. T. Bruder, U. R. Rapp & J. Avruch. 1993. Nature **364:** 308-313.
41. Warne, P., P. R. Viciana & J. Downward. 1993. Nature **364:** 352-355.
42. vanAelst, L., M. Barr, S. Marcus, A. Polverino & M. Wigler. 1993. Proc. Natl. Acad. Sci. USA **90:** 6213-6217.
43. Dent, P., W. Haser, T. Haystead, L. A. Vincent, T. M. Roberts & T. W. Sturgill. 1992. Science **257:** 1404-1407.
44. Kyriakis, J. M., H. App, X. F. Zhang, P. Banerjee, D. L. Brautigan, U. R. Rapp & J. Avruch. 1992. Nature **358:** 417-421.
45. Howe, L. R., S. Leevers, N. Gomez, S. Nakielny, P. Cohen & C. J. Marshall. 1992. Cell **71:** 335-342.

46. GOLDSCHMIDT-CLERMONT, P. J., J. W. KIM, L. M. MACHESKY, S. G. RHEE & T. D. POLLARD. 1990. Science **251:** 1231-1233.
47. SHARIFF, A. & E. J. LUNA. 1992. Science **256:** 245-247.
48. GOTOH, Y., E. NISHIDA, T. YAMASHITA, M. HOSHI, M. KAWAKAMI & H. SAKAI. 1990. Eur. J. Biochem. **193:** 661-669.
49. SOZERI, O., K. VOLLMER, M. LIYANAGE, D. FRITH, G. KOUR, G. E. I. MARK & S. STABLE. 1992. Oncogene **7:** 2259-2262.
50. VALIUS, M. & A. KAZLAUSKAS. 1993. Cell **73:** 321-334.
51. KREMER, N. E., G. D'ARCANGELO, S. M. THOMAS, M. DEMARCO, J. S. BRUGGE & S. HALEGOUA. 1991. J. Cell Biol. **115:** 809-819.
52. QIU, M-S. & S. H. GREEN. 1992. Neuron **9:** 705-717.
53. TRAVERSE, S., N. GOMEZ, H. PATERSON, C. MARSHALL & P. COHEN. 1992. Biochem. J. **288:** 351-355.
54. HONEGGER, A. M., R. M. KRIS, A. ULLRICH & J. SCHLESSINGER. 1989. Proc. Natl. Acad. Sci. USA **86:**925-929.
55. FENDLY, B. M., M. WINGET, R. HUDZIAK, M. T. LIPARI, M. A. NAPIER & A. ULLRICH. 1990. Cancer Res. **50:** 1550-1558.

Structure and Catalytic Properties of Protein Tyrosine Phosphatases[a]

JACK E. DIXON

Department of Biological Chemistry
The University of Michigan Medical School
Ann Arbor, Michigan 48109-0606

The level of protein tyrosine phosphorylation is controlled within the cell by both tyrosine kinase and phosphatase activities. Our laboratory has focused its energy on the protein tyrosine phosphatases (PTPases). To understand the latter, we and others have cloned a number of PTPases.[1,2] Now, more than 40 cDNAs are known that encode a wide spectrum of PTPases. The PTPases can be subdivided into two major families. One family of the PTPases resembles receptors: the PTPases have an intracellular tyrosine phosphatase domain(s), a transmembrane domain, and an extracellular domain that may function in cell-cell or ligand-cell interactions.[1,2] This family has been further classified into five subtypes based largely upon differences in their extracellular domains (FIG. 1). We have a limited understanding of the potential ligands for the receptor-like PTPases. However, the sequence similarity of type II PTPases to cell adhesion molecules (CAM) suggests that these PTPases may exhibit homophilic binding. Expression of PTPμ in SF9 insect cells results in cell aggregation, suggesting that the extracellular domains of PTPμ mediate this self-aggregation.[3,4] The expression of cytoplasmic domain-deleted constructs indicates that PTPase catalytic activity is not required for adhesion; only the extracellular domain is necessary for cell-cell interactions. The purified extracellular domain conjugated to beads can also mediate bead-bead adhesion. Work from Schlessinger's laboratory also indicates that the closely related molecule PTPκ also displays homophilic adhesive properties.[5] When cells expressing PTPμ are incubated with cells expressing PTPκ, the cells segregate and adhere in a homophilic fashion. This shows that, although PTPμ and PTPκ are structurally similar, they display a high degree of specificity in their cell-cell adhesion.

The second family of PTPases is found exclusively within a cell.[1,2,6] The distinguishing feature of the intracellular PTPase is the diversity of sequences flanking a single catalytic domain (FIG. 2). One of the functions of these flanking sequences appears to be the targeting of the enzyme to specific intracellular locations. These include (1) membrane-association domains, (2) nuclear-localization domains, (3) Src homology 2 (SH2) domains, and (4) cytoskeletal-association domains. Locating specific PTPases to define subcellular locations will undoubtedly restrict and define their substrate specificity.[6]

[a] This work was supported by the National Institutes of Health (NIDDKD 18024) and the Walther Cancer Institute.

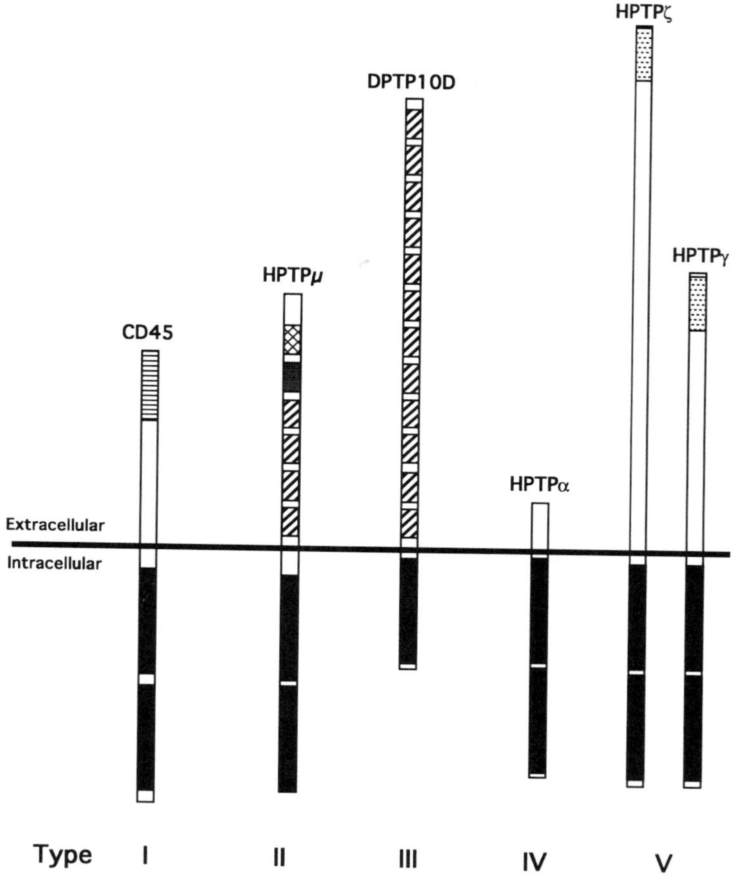

FIGURE 1. Transmembrane receptor-like PTPases. The members of this family of PTPases possess a single transmembrane domain and one or two intracellular PTPase catalytic domains (black bar) (see ref. 16 for details). They can be subdivided into five types on the basis of their extracellular domain structures: I, CD45; II, HPTPμ; III, DPTP10D; IV, HPTPα; and V, HPTPζ and HPTPγ. The extracellular domain structures are shown: amino terminus isoforms (horizontal lines) resulting from differential splicing; immunoglobulin-like (vertical lines); fibronectin type III-like (shaded bar); Meprin amino acid motif (MAM) adhesive protein homology-like (diagonal lines); and carbonic anhydrase-like (stippled). (Mauro & Dixon.[6] With permission from Elsevier Trends Journals.)

We discovered that certain pathogenic bacteria also have a protein that has protein tyrosine phosphatase activity.[7] This is unusual because bacteria are not known to contain proteins that are phosphorylated on tyrosine. The bacteria that have the tyrosine phosphatase activity are from the genus *Yersinia.* This bacterium was responsible for the plague or the Black Death that reduced the population of Europe on several

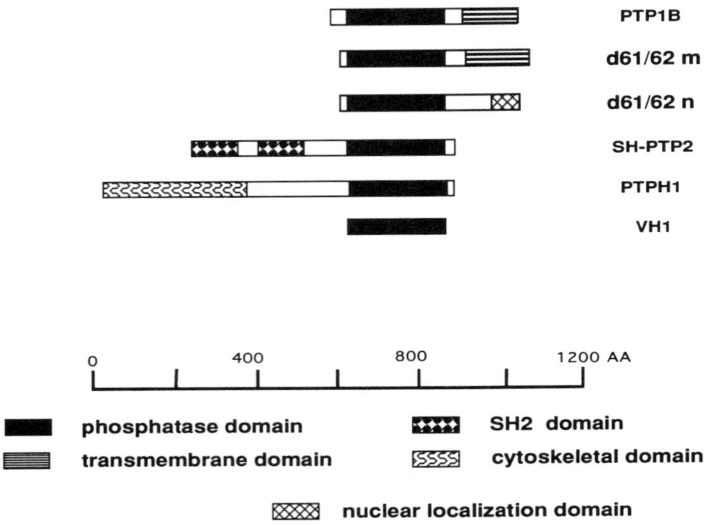

FIGURE 2. Structural features of representative intracellular PTPase. Each PTPase possesses one conserved catalytic domain. The PTPase pictured with their corresponding noncatalytic sequences include (a) membrane-association domains in PTP1B; (b) an Src homology 2 (SH2) domain in SH-PTP2; and (c) cytoskeleton-association domains in PTPH1. (Mauro & Dixon.[6] With permission from Elsevier Trends Journals.)

occasions by more than 25 million people. As you are well aware, the plague has recently broke out again in India. This simply points out that such diseases, although treatable by antibodies, have not been eliminated as human pathogens. We have shown that the *Yersinia* PTPase is a virulent gene in bacterium and that it functions to dephosphorylate tyrosine residues necessary for regulating mechanisms within the mammalian host.[8]

We also observed that the *vaccinia* virus and other members of the pox virus family (including small pox) have encoded within their genomes a phosphatase. The activity of the *vaccinia* phosphatase is not restricted to the dephosphorylation of tyrosine; it will also dephosphorylate serine- and threonine-containing substrates.[9] The *vaccinia* phosphatase has an amino acid sequence identity to the yeast cell cycle gene product, cdc 25, and a group of related mammalian dual-specific phosphatases. TABLE 1 lists representatives of the various dual-specific phosphatases that have been described to date.[10]

All PTPases possess at least one 230 amino acid catalytic domain containing a highly conserved active-site region with the consensus motif [I/V]HCXAGXXR[S/T]G (where X is any amino acid). This region is the signature sequence of most PTPases.[11,12] Site-directed mutagenesis and trapping experiments have shown that the cysteinyl residue within this motif is essential for phosphatase activity. The enzyme appears to form a thiol-phosphate intermediate during catalytic turnover.[13] Interestingly, this catalytic domain of the PTPases bears no resemblance to that of the serine-threonine phosphatases or of the alkaline or acid phosphatases. Recently,

TABLE 1. The Dual-specific PTPases

Dual-specific PTPases	Substrate Motif	Induction	Function
HVH1, CL100, SCH134, MKP-1	-(pT)E(pY)-	Serum, growth factors, phorbol esters	Deactivation of ERK[a], MAP[b] kinase
PAC1	-(pT)E(pY)-	T-cell activation	Deactivation of ERK-MAP kinase
Cdc25	-(pT)(pY)-	Cell cycle-dependent	Activation of Cdc2; entry into mitosis
KAP, Cdi1	-(pT)(pY)-	Cell cycle-dependent	Cell-cycle regulation; associates with CDK2, cdc2
YVH1	—	Nitrogen starvation	Growth control
MSG5	-(pT)E(pY)-	Mating pheromone	Inactivation of FUSS: adaptation to pheromone response in yeast

[a] Extracellular signal-regulated kinase.
[b] Mitogen-activated protein.

the structures of two PTPases have been solved.[14,15] Their structures are quite similar and support the idea that the PTPase most likely employs a common catalytic mechanism. Details of the structure and the catalytic mechanism will be discussed.

REFERENCES

1. WALTON, K. M. & J. E. DIXON. 1993. Protein tyrosine phosphatases. Annu. Rev. Biochem. **62:** 101-20.
2. FISCHER, E. H., H. CHARBONNEAU & N. K. TONKS. 1991. Protein tyrosine phosphatases: a diverse family of intracellular and transmembrane enzymes. Science **253**(5018): 401-6.
3. BRADY-KALNAY, S. M., A. J. FLINT & N. K. TONKS. 1993. Homophilic binding of PTPμ, a receptor-type protein tyrosine phosphatase, can mediate cell-cell aggregation. J. Cell Biol. **122:** 961-972.
4. GEBBINK, M. F. B. G., G. C. M. ZONDAG, R. W. WUBBOLTS, R. L. BEIJERSBERGEN, I. VAN ETTEN & W. H. MOOLENAAR. 1993. Cell-cell adhesion mediated by a receptor-like protein tyrosine phosphatase. J. Biol. Chem. **268:** 16101-16104.
5. SAP, J., Y. P. JIANG, D. FRIEDLANDER, M. GRUMET & J. SCHLESSINGER. 1994. Receptor tyrosine phosphatase R-PTP-kappa mediates homophilic binding. Mol. Cell. Biol. **14**(1): 1-9.
6. MAURO, L. J. & J. E. DIXON. 1994. "Zip codes" direct intracellular protein tyrosine phosphatases to the correct cellular "address." TIBS **19**(4): 151-155.
7. GUAN, K. & J. E. DIXON. 1990. Protein tyrosine phosphatase activity of an essential virulence determinant in *Yersinia.* Science **249:** 553-556.
8. BLISKA, J. B., K. GUAN, J. E. DIXON & S. FALKOW. 1991. Tyrosine phosphate hydrolysis of host proteins by an essential *Yersinia* virulence determinant. Proc. Natl. Acad. Sci. USA **88:** 1187-1191.

9. Guan, K., S. S. Broyles & J. E. Dixon. 1991. A Tyr/Ser protein phosphatase encoded by *Vaccinia* virus. Nature **350:** 359-362.
10. Tonks, N. K., A. J. Flint, M. F. B. G. Gebbink, H. Sun & Q. Yang. 1993. Signal transduction and protein tyrosine dephosphorylation. Adv Second Messenger Phosphoprotein Res. **28:** 203-210.
11. Zhang, Z.-Y. & J. E. Dixon. 1994. Protein tyrosine phosphatases: Mechanism of catalysis and substrate specificity. Adv. Enzymol. **68:** 1-36.
12. Zhang, Z.-Y. & J. E. Dixon. 1993. Active site labeling of the *Yersinia* protein tyrosine phosphatase: The determination of the pKa of the active site cysteine and the function of the conserved histidine 402. Biochemistry **32**(36): 9340-9345.
13. Guan, K.-L. and J. E. Dixon. 1991. Evidence for protein tyrosine phosphatase catalysis proceeding via a cysteine-phosphate intermediate. J. Biol. Chem. **266**(26): 17026-17030.
14. Barford, D., A. J. Flint & N. K. Tonks. 1994. Crystal structure of human protein tyrosine phosphatase 1B. Science **263**(5152): 1397-1404.
15. Stuckey, J., H. Schubert, E. Fauman, Z.-Y. Zhang, J. E. Dixon & M. Saper. 1994. Crystal structure of *Yersinia* protein tyrosine phosphatase at 2.5 Å and the complex with tungstate. Nature **370:** 571-575.
16. Mourey, R. J. & J. E. Dixon. 1994. Protein tyrosine phosphatases: characterization of extracellular and intracellular domains. Curr. Opinion Genet. Dev. **4:** 31-39.

The G Protein–coupled Receptor Family and One of Its Members, the TSH Receptor[a]

G. VASSART, F. DESARNAUD, L. DUPREZ,
D. EGGERICKX, O. LABBÉ, F. LIBERT,
C. MOLLEREAU, J. PARMA, R. PASCHKE,
M. TONACCHERA, P. VANDERHAEGHEN,
J. VAN SANDE, J. DUMONT, AND M. PARMENTIER

Institut de Recherche Interdisciplinaire
and
Department of Medical Genetics
Faculty of Medicine
University of Brussels
Campus Erasme
808 route de Lennik
B-1070 Brussels, Belgium

INTRODUCTION

G protein-coupled receptors constitute the largest family of receptors identified to date. From their primary structures, they can be subdivided into three main groups: the first has sequence similarity with the adrenergic-rhodopsin gene family; the second has similarity with the secretin-VIP receptor genes; and the third has similarity with the glutamate metabotropic receptors. All three subfamilies display an analogous structure containing seven transmembrane segments, with the amino-terminus in the extracellular space, three extracellular loops, three intracytoplasmic loops, and a cytoplasmic carboxylterminus. They are implicated in the transduction of signals as diverse as light, odorants, bioactive amines, neuropeptides, and glycoprotein hormones (for a review, see ref. 1). They share a common mode of action: upon activation by the binding of their ligand, they promote exchange of GDP for GTP on the alpha subunit of G proteins. The result is activation of effectors, either by α_{GTP}, or by the released beta/gamma of the G protein (see ref. 2 for a review). A given receptor may be coupled to a single or to several different G proteins, resulting in the possibility

[a] This study was supported by the Belgian Programme of University Poles of Attraction initiated by the Belgian State, Prime Minister's office, Science programming. The scientific responsibility is assumed by the authors. This study was also supported by Grants from the Fonds de la Recherche Scientifique Médicale, Télévie and Biomed (EC) program.

to regulate a variety of regulatory cascades. For example, the human thyrotropin (TSH) receptor couples mainly to G_s, but, when activated by high concentrations of TSH, it couples also to G_q, thereby activating the inositolphosphate diacylglycerol cascade.[3]

Once adrenergic and muscarinic receptors had been cloned, low stringency PCR was established as an extraordinary powerful means of cloning additional members of this gene subfamily. The method relies on the strong conservation of segments of the receptors, in particular in the transmembrane regions, to amplify by means of degenerate primers, cDNAs or gene fragments of new receptors.[4,5] The result is the availability of a host of subtypes for most receptors and the cloning of numerous receptors whose ligands are completely unknown (orphan receptors).

SUBTYPES OF KNOWN RECEPTORS AND ORPHAN RECEPTORS

In the frame of our search for the TSH receptor, we have cloned a series of such orphan receptors. Among them, we have previously identified adenosine receptors A1 and A2,[6,7] serotonin 5HT1d receptor,[8] the human cannabinoid receptor,[9] an interleukin 8 receptor,[10] bradykinin B2 receptor,[11] complement C5a receptor,[12] and members of the olfactory receptor gene family expressed in male germ cells of the dog.[13] More recently we have identified two subtypes of MSH receptors, MC3 and MC5.[14,15]

Expression of olfactory receptors[16] in the male canine germ cells has been further investigated. The presence of mRNA encoding a subset of olfactory receptors in late spermatids has been confirmed by an RNAase protection assay. "Testicular" receptors are expressed at very low levels in the olfactory mucosa, whereas a selection of truly "olfactory" receptors are not expressed in the testis. However, the mere presence of transcripts in the testis is no guarantee that the protein would be synthesized. An antibody has been prepared against a bacterially expressed recombinant receptor that allows the identification of the receptor protein by immunohistochemistry. Its use on epididymal and mature sperm demonstrates the presence of immunoreactive material in the cytoplasmic drop and the proximal part of the sperm flagella, respectively.[17] The presence of members of the olfactory receptor proteins in sperm does not exclude that the observation would correspond to "transcription leakage" during gametogenesis, a phenomenon with no physiological relevance. However, the possibility that they could be implicated in chemotaxis of spermatozoa within the female genital tract deserves further study.

A current survey of sequence data bases identifies more than 25 orphan G protein-coupled receptors. Considering the reluctance researchers have to publishing such sequences (but this may be changing, see below), this is certainly a gross underestimate. Orphan receptors may be classified into two qualitatively different subgroups: those that are truly orphans, meaning that all that is known about them is their belonging to the family of G protein-coupled receptors; and those that belong to a cluster of receptors sharing sequence similarity with known receptors. An interesting example of this last category is provided by a clone displaying significant sequence similarity with the opioid receptors: ORL1.[18] The search for the ligands of true orphans is a daunting task because most receptors corresponding to known ligands have now been cloned. In comparison, the search of the ligands for partial orphans is orientated by the nature of the ligands of their mates in the similarity dendrogram. Applying this reasoning to ORL1, we could demonstrate that CHO cells transfected with ORL1 cDNA did not bind or respond to classical endogenous opioids.[18] However, its belonging to the opioid subfamily was confirmed when we could demonstrate that ethorphin, a nonspecific opioid agonist, was capable of decreasing cAMP production in the tranfected cells.[18] This lead, together with the information about its localization within the brain, constitute a useful starting point for the search of the true endogenous ligand.

The accumulation of information about expressed sequences[19] will lead to the identification of additional orphan receptors at a fast pace. This will constitute an impetus for the publication of orphan receptor sequences that were kept dormant until now.

STRUCTURE-FUNCTION RELATIONSHIPS OF THE TSH RECEPTOR

The structure-function relationships of G protein-coupled receptors have been the subject of extensive studies, the major approach being that of site-directed mutagenesis (see ref. 20 for a review). In the emerging picture for the mechanism of receptor activation, the ligand would interact with specific residues of some of the transmembrane helices. In doing so, it would release a built-in constraint that maintains the unliganded receptor in an inactive state. The activated receptor would then achieve a conformation allowing it to activate its cognate G protein by way of interaction with the intracellular loops. Most of this picture comes from the data accumulated on the activation of rhodopsin and adrenergic receptors. A subset of spontaneous mutations in transmembrane segments of rhodopsin in humans was shown to activate it constitutively and cause retinitis pigmentosa or night blindness.[21–23] A seminal observation by the group of R. Lefkowitz[24,25] demonstrated that substitution of a single residue (Ala293) of the third intracellular loop of the alpha1b adrenergic receptor by any of the other 19 amino acids resulted in its constitutive activation.[24] It is this observation that led to the notion that unliganded receptors are normally under negative constraint. The gain of function of the mutants would correspond to the release of this constraint.

The model for receptor activation as outlined above has been elalorated for receptors with small, usually nonpeptidic, ligands. There are arguments that peptidic ligands may interact with the extracellular loops of their receptor as part of the activation mechanism.[26,27] The problem of how the bulky glycoprotein hormones and their receptors would fit within this model is of particular interest. Experiments with the TSH, LH, and FSH receptors have clearly demonstrated that the hormones bind to the large extracellular amino terminal extension that is characteristic of this receptor subfamily.[28,29] The nature of the peptide segment from the hormones (if any) that would insert in the receptor transmembrane pocket to activate it is completely unknown.

Despite these considerations, the sequence similarity between all members of the G protein-coupled receptor family makes it extremely likely that they will share a common activating mechanism. It was therefore tempting to assume that some mutations within the TSH receptor would have the effect of activating it constitutively. The fact that TSH controls positively the function, expression of differentiation, and growth of thyrocytes by cyclic AMP-dependent mechanisms suggests that somatic mutations of that kind would cause a readily identifiable phenotype in the form of a monoclonal hyperfunctional benign tumor.[30,31] This type of "toxic adenoma" constitutes, in fact, a relatively frequent cause of hyperthyroidism in humans. Consistent with this hypothesis, mutations of Gs interfering with its GTPase activity have been found in a minority of such tumors.[32]

In a total of eleven hyperfunctioning adenomas that were investigated, nine were found to harbor a mutation at various positions of exon ten of the TSH receptor gene (ref. 33 and our unpublished results). The adjacent nontumoral tissue was free of mutations. The mutations were distributed within the segment encoding the carboxyl portion of the receptor with homology to other G protein-coupled receptors. One of the mutated residues, Ala623 in the third cytoplasmic loop, was the exact homologue of Ala293 of the alpha1b adrenergic receptor.[33] The others involved another amino acid of the third loop (Asp619), residues of the sixth transmembrane helix and, more surprising, of the first extracellular loop. When the mutated receptors are transfected in COS cells, they display a higher constitutive activity than the wild-type receptor, in terms of adenylyl cyclase stimulation.[33] Interestingly, however, the wild-type receptor itself stimulates cAMP accumulation in transfected COS cells in the absence of TSH.[33,34] Some of the mutants with a strong effect on cAMP accumulation stimulated also inositol phosphate accumulation.

Germline mutations with a similar activity would be expected to cause hereditary hyperthyroidism with thyroid hyperplasia, or sporadic congenital hyperthyroidism, in case of a neomutation. When two large pedigrees from France[35] and an isolated case of congenital hyperthyroidism were analyzed for the presence of TSH receptor gene mutations, three different mutations were identified. The two hereditary diseases segregated with mutations in transmembrane segments three (Val509) and seven (Cys672), respectively.[36] The sporadic case harbored a mutation in transmembrane segment six (Phe631), leading to an identical amino acid substitution as the one observed in three cases of toxic adenomas.[37] When tested by transfection in COS cells, the mutant receptors exhibited an increase in constitutive activity for adenylyl cyclase stimulation.[36,37]

Altogether, ten different residues have been found mutated in toxic adenomas or toxic thyroid hyperplasia[33,36–38] (FIG. 1). Considering that they all cause gain-of-function, this number came as a surprise. Whereas the observation is fully compatible with the model of receptor activation by release of a negative constraint, it is still unexpected, considering what is known from other receptors. In the LH receptor a more limited number of mutations have been found to cause precocious puberty of the male as a result of constitutive activation of adenylyl cyclase.[39,40] In the MSH receptor of the mice, three mutations have been found associated with variant colors of the coat.[41] An interesting hypothesis to account for this peculiarity of the TSH receptor would be to relate it to the significant constitutive activity of the wild-type receptor. Not all Gs-coupled receptors exhibit constitutive activity; under the conditions for which the TSH receptor gene clearly increases cAMP when transfected in COS cells, the MSH receptor (MC3) and the LH receptor do not (our unpublished

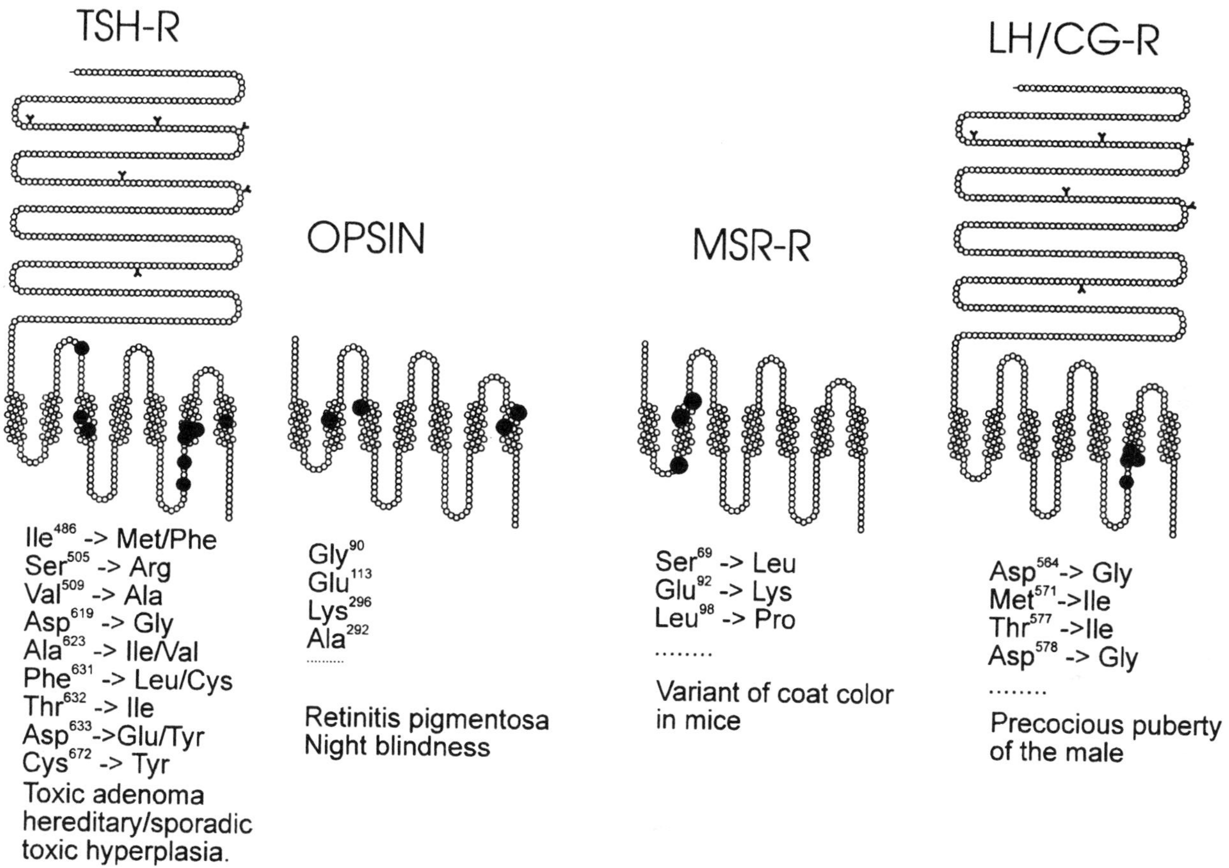

FIGURE 1. Spontaneous mutations activating G protein-coupled receptors. The structure of the opsin, TSH, MSH, and LH/chorionic gonadotrophin (CG) receptors are only schematically represented, with the approximate positions of individual mutations. The data are compiled from references 22, 23, 33, 36, 37, and 39-43.

results). It is tempting to speculate that being already "noisy" in its wild-type state, the TSH receptor would be more readily activated by mutations than others, whose silence would be enforced by more stringent structural constraints. Future studies will explore the possibility that the putative lability of the TSH receptor could play a role in its frequent activation by autoantibodies as observed in Graves' disease.

ACKNOWLEDGMENTS

We thank Stephane Swillens for frequent helpful discussions.

REFERENCES

1. Lefkowitz, R. J., S. Cotecchia, M. A. Kjelsberg, J. Pitcher, W. J. Koch, J. Inglese & M. G. Caron. 1993. Adrenergic receptors: recent insights into their mechanism of activation and desensitization. Adv. Second Messenger Phosphoprotein Res. **28:** 1-9.
2. Bourne, H. R. & R. Nicoll. 1993. Molecular machines integrate coincident synaptic signals. Cell **72** Suppl: 65-75.
3. Laurent, E., J. Mockel, J. Van Sande, I. Graff & J. E. Dumont. 1987. Dual activation by thyrotropin of the phospholipase C and cyclic AMP cascades in human thyroid. Mol. Cell Endocrinol. **52:** 273-278.
4. Parmentier, M., F. Libert, C. Maenhaut, A. Lefort, C. Gerard, J. Perret, J. Van Sande, J. E. Dumont & G. Vassart. 1989. Molecular cloning of the thyrotropin receptor. Science **246:** 1620-1622.
5. Libert, F., M. Parmentier, A. Lefort, C. Dinsart, J. Van Sande, C. Maenhaut, M. J. Simons, J. E. Dumont & G. Vassart. 1989. Selective amplification and cloning of four new members of the G protein-coupled receptor family. Science **244:** 569-572.
6. Maenhaut, C., J. Van Sande, F. Libert, M. Abramowicz, M. Parmentier, J. J. Vanderhaegen, J. E. Dumont, G. Vassart & S. Schiffmann. 1990. RDC8 codes for an adenosine A2 receptor with physiological constitutive activity. Biochem. Biophys. Res. Commun. **173:** 1169-1178.
7. Libert, F., S. N. Schiffmann, A. Lefort, M. Parmentier, C. Gerard, J. E. Dumont, J. J. Vanderhaeghen & G. Vassart. 1991. The orphan receptor cDNA RDC7 encodes an A1 adenosine receptor. EMBO J. **10:** 1677-1682.
8. Maenhaut, C., J. Van Sande, C. Massart, C. Dinsart, F. Libert, E. Monferini, E. Giraldo, H. Ladinsky, G. Vassart & J. E. Dumont. 1991. The orphan receptor cDNA RDC4 encodes a 5-HT1D serotonin receptor. Biochem. Biophys. Res. Commun. **180:** 1460-1468.
9. Gerard, C. M., C. Mollereau, G. Vassart & M. Parmentier. 1991. Molecular cloning of a human cannabinoid receptor which is also expressed in testis. Biochem. J. **279:** 129-134.
10. Mollereau, C., E. Passage, M. G. Mattei, G. Vassart & M. Parmentier. 1992. The high affinity interleukin 8 receptor gene maps to the 2q333-2q36 region of the human genome. Cloning of a pseudogene for the low affinity receptor. Genomics **16:** 248-251.
11. Eggerickx, D., E. Raspe, D. Bertrand, G. Vassart & M. Parmentier. 1992. Molecular cloning, functional expression and pharmacological characterization of a human bradykinin B2 receptor gene. Biochem. Biophys. Res. Commun. **187:** 1306-1313.
12. Perret, J., E. Raspe, G. Vassart & M. Parmentier. 1992. Cloning and functional expression of the canine anaphylatoxin C5a receptor: evidence for high interspecies variability. Biochem. J. **288:** 911-917.

13. PARMENTIER, M., F. LIBERT, S. SCHURMANS, S. SCHIFFMANN, A. LEFORT, D. EGGERICKX, C. LEDENT, C. MOLLEREAU, C. GERARD, J. PERRET, A. GROOTEGOED & G. VASSART. 1992. Expression of members of the putative olfactory receptor gene family in mammalian germ cells. Nature **355:** 453-455.
14. DESARNAUD, F., O. LABBE, D. EGGERICKX, G. VASSART & M. PARMENTIER. 1994. Molecular cloning, functional expression and pharmacological characterization of a mouse melanocortin receptor gene. Biochem. J. **299:** 367-373.
15. LABBE, O., F. DESARNAUD, D. EGGERICKX, G. VASSART & M. PARMENTIER. 1994. Molecular cloning of a mouse melanocortin 5 receptor gene widely expressed in peripheral tissues. Biochemistry **33:** 4543-4549.
16. BUCK, L. & R. AXEL. 1991. A novel multigene family may encode odorant receptors: a molecular basis for odor recognition. Cell **65:** 175-187.
17. VANDERHAEGHEN, P., S. SCHURMANS, G. VASSART & M. PARMENTIER. 1993. Olfactory receptors are displayed on dog mature sperm cells. J. Cell Biol. **123:** 1441-1452.
18. MOLLEREAU, C., M. PARMENTIER, P. MAILLEUX, J. L. BUTOUR, C. MOISAND, P. CHALON, D. CAPUT, G. VASSART & J. C. MEUNIER. 1994. ORL1, a novel member of the opioid receptor family. Cloning, functional expression and localization. FEBS Lett. **341:** 33-38.
19. ADAMS, M. D., A. R. KERLAVAGE, J. M. KELLEY, J. D. GOCAYNE, C. FIELDS, C. M. FRASER & J. C. VENTER. 1994. A model for high-throughput automated DNA sequencing and analysis core facilities. Nature **368:** 474-475.
20. SAVARESE, T. M. & C. M. FRASER. 1992. *In vitro* mutagenesis and the search for structure-function relationships among G protein-coupled receptors. Biochem. J. **283:** 1-19.
21. COHEN, G. B., T. YANG, P. R. ROBINSON & D. D. OPRIAN. 1993. Constitutive activation of opsin: influence of charge at position 134 and size at position 296. Biochemistry **32:** 6111-6115.
22. ROBINSON, P. R., G. B. COHEN, E. A. ZHUKOVSKY & D. D. OPRIAN. 1992. Constitutively active mutants of rhodopsin. Neuron **9:** 719-725.
23. RAO, V. R., G. B. COHEN & D. D. OPRIAN. 1994. Rhodopsin mutation G90D and a molecular mechanism for congenial night blindness. Nature **367:** 639-642.
24. KJELSBERG, M. A., S. COTECCHIA, J. OSTROWSKI, M. G. CARON & R. J. LEFKOWITZ. 1992. Constitutive activation of the alpha 1B-adrenergic receptor by all amino acid substitutions at a single site. Evidence for a region which constrains receptor activation. J. Biol. Chem. **267:** 1430-1433.
25. ALLEN, L. F., F. J. LEFKOWITZ, M. G. CARON & S. COTECCHIA. 1991. G-protein-coupled receptor genes as protooncogenes: constitutively activating mutation of the alpha 1B-adrenergic receptor enhances mitogenesis and tumorigenicity. Proc. Natl. Acad. Sci. USA **88:** 11354-11358.
26. YOKOTA, Y., C. AKAZAWA, H. OHKUBO & S. NAKANISHI. 1992. Delineation of structural domains involved in the subtype specificity of tachykinin receptors through chimeric formation of substance P/substance K receptors. EMBO J. **11:** 3585-3591.
27. FONG, T. M., H. YU, R. R. HUANG & C. D. STRADER. 1992. The extracellular domain of the neurokinin-1 receptor is required for high-affinity binding of peptides. Biochemistry **31:** 11806-11811.
28. NAGAYAMA, Y., H. L. WADSWORTH, G. D. CHAZENBALK, D. RUSSO, P. SETO & B. RAPOPORT. 1991. Thyrotropin-luteinizing hormone/chorionic gonadotropin receptor extracellular domain chimeras as probes for thyrotropin receptor function. Proc. Natl. Acad. Sci. USA **88:** 902-905.
29. SEGALOFF, D. L. & M. ASCOLI. 1993. The lutropin/choriogonadotropin receptor . . . 4 years later. Endocr. Rev. **14:** 324-347.
30. DUMONT, J. E., F. LAMY, P. ROGER & C. MAENHAUT. 1992. Physiological and pathological regulation of thyroid cell proliferation and differentiation by thyrotropin and other factors. Physiol. Rev. **72:** 667-697.

31. Vassart, G. & J. E. Dumont. 1992. The thyrotropin receptor and the regulation of thyrocyte function and growth. Endocr. Rev. **13:** 596-611.
32. Lyons, J., C. A. Landis, G. Harsh, L. Vallar, K. Grunewald, H. Feichtinger, Q. Y. Duh, O. H. Clark, E. Kawasaki, & H. R. Bourne *et al.* 1990. Two G protein oncogenes in human endocrine tumors. Science **249:** 655-659.
33. Parma, J., L. Duprez, J. Van Sande, P. Cochaux, C. Gervy, J. Mockel, J. E. Dumont & G. Vassart. 1993. Somatic mutations in the thyrotropin receptor gene cause hyperfunctioning thyroid adenomas. Nature **365:** 649-651.
34. Kosugi, S., F. Okajima, T. Ban, A. Hidaka, A. Shenker & L. D. Kohn. 1994. Substitutions of different regions of the third cytoplasmic loop of the TSH receptor have selective effects on constitutive, TSH-, and TSH receptor autoantibody-stimulated phosphoinositide and cAMP signal generation. Mol. Endocrinol. **7:** 1009-1020.
35. Thomas, J. S., J. Leclere, P. Hartemann, J. Duheille, J. Orgiazzi, M. Petersen, C. Janot & J. C. Guedenet. 1982. Familial hyperthyroidism without evidence of autoimmunity. Acta Endocrinol. (Copenhagen) **100:** 512-518.
36. Duprez, L., J. Parma, J. Van Sande, A. Allegeier, J. Leclere, C. Schvartz, M. J. Delisle, M. Decoulx, J. Orgiazzi, J. E. Dumont, & G. Vassart. 1994. Germline mutations in the thyrotropin receptor gene cause non autoimmune autosomal dominant hyperthyroidism. Nat. Genet. **7:** 396-407.
37. Kopp, P., J. Van Sande, J. Parma, L. Duprez, K. Zuppinger, J. L. Jameson, & G. Vassart. 1995. Congenital non-autoimmune hyperthyroidism caused by a neomutation in the thyrotropin receptor gene. New Engl. J. Med. **19:** 150-154.
38. Porcellini, A., I. Ciullo, L. Laviola, A. Amabile, G. Fenzi & V. Avvedimento. 1994. Novel mutations of thyrotropin receptor gene in thyroid hyperfunctioning adenomas. J. Clin. Endocrinol. Metab. **79:** 657-661.
39. Shenker, A., L. Laue, S. Kosugi, J. J. Merendino, T. Minegishi & G. B. Cutler. 1993. A constitutively activating mutation of the luteinizing hormone receptor in familial male precocious puberty. Nature **365:** 652-654.
40. Kremer, H., E. Mariman, B. J. Otten, G. W. J. Moll, G. B. Stoelinga, J. M. Wit, M. Jansen, S. L. Drop, B. Faas, & H. H. Ropers *et al.* 1993. Cosegregation of missense mutations of the luteinizing hormone receptor gene with familial male-limited precocious puberty. Hum. Mol. Genet. **2:** 1779-1783.
41. Robbins, L. S., J. H. Nadeau, K. R. Johnson, M. A. Kelly, L. Roselli-Rehfuss, E. Baack, G. K. Mountjoy & R. D. Cone. 1993. Pigmentation phenotypes of variant extension locus alleles results from point mutations that alter MSH receptor function. Cell **72:** 827-834.
42. Cohen, G. B., D. D. Oprian & P. R. Robinson. 1992. Mechanism of activation and inactivation of opsin: role of Glu113 and Lys296. Biochemistry **31:** 12592-12601.
43. Paschke, R., M. Tonacchera, J. Van Sande, J. Parma & G. Vassart. 1994. Identification and functional characterization of two new somatic mutations causing constitutive activation of the TSH receptor in hyperfunctioning autonomous adenomas of the thyroid. J. Clin. Endocrinol. Metab. **79:** 1785-1789.

Inositol Trisphosphate and Calcium Signaling

MICHAEL J. BERRIDGE

The Babraham Institute
Laboratory of Molecular Signaling
Department of Zoology
Downing Street
Cambridge CB2 3EJ, United Kingdom

INTRODUCTION

Calcium plays a central role in the control of many cellular processes including cell proliferation. Perhaps the most dramatic example of this role in proliferation occurs at the time of fertilization, when a large surge of calcium is responsible for triggering the developmental program of the embryo. An important aspect of this program is the rapid growth of cells early in development. Once cells have differentiated to form specific functions, they often retain the ability to return to the cell cycle. The stimulus to begin growing again is usually provided by specific growth factors. Once again, calcium had been implicated as one of the signals responsible for mitogenesis, with lymphocyte activation being a particularly good example. For the two examples cited so far, fertilization and lymphocyte activation, the calcium signal often appears in the form of repetitive calcium spikes. Such spiking behavior has now been described in many different cell types in response to a wide range of stimuli and seems to represent the way in which the calcium signaling pathway is organized.[1–4] In order to understand how calcium might contribute to mitogenesis, therefore, it is necessary to describe how such calcium signals are generated and how they might contribute to the onset of DNA synthesis.

SPATIOTEMPORAL ASPECTS OF SIGNALING

The calcium signal observed in single cells responding to external signals is not uniform but appears as discrete spikes that occur repetitively. Such spiking has been described in mammalian eggs following fertilization,[5] in activated lymphocytes,[6,7] and in many cultured cells stimulated with growth factors. At least two separate mechanisms exist for generating these spikes, depending on whether the cells are using either internal or external sources of calcium. In some cells, such as mammalian eggs or cultured cells responding to growth factors, the spikes are derived from the periodic release of internal calcium. In the case of such a cytosolic oscillator, the individual spikes often reveal a distinct spatial organization, in that they initiate in one region of the cell and then sweep through the cytoplasm in the form of a regenerative wave. In other cells, typified by the lymphocyte, there is a periodic influx of calcium across the plasma membrane. The initiation of either the cytosolic or membrane oscillator is critically dependent upon the action of the calcium-mobilizing second messenger, inositol 1,4,5-trisphosphate ($InsP_3$).

INOSITOL TRISPHOSPHATE AND CALCIUM RELEASE

Both the mobilization of calcium from internal stores as well as the entry of external calcium are critically dependent upon the formation of $InsP_3$. There are two major pathways responsible for hydrolyzing phosphatidylinositol 4,5-bisphosphate to release $InsP_3$.[1] First, G protein-linked receptors activate phospholipase C β1 (PLC β1). Second, tyrosine kinase-linked receptors act through PLC γ1, and this represents one of the signaling pathways employed by many growth factors. It is important to stress that the tyrosine kinase-linked receptors can activate a number of other signaling pathways such as the ras/raf1/MAPK cascade and the stimulation of PI 3-kinase. It seems likely that these different signaling pathways regulate separate downstream events that must be activated to varying degrees or in varying combinations before the cell will commit itself to proliferate. The primary purpose of this article is to examine the calcium signaling pathway with particular emphasis on the way in which $InsP_3$ acts to generate calcium spikes and calcium waves.

Once $InsP_3$ has been released from the membrane, it diffuses into the cytosol where it binds to specific receptors located on the intracellular calcium stores. The $InsP_3$ receptor is a very large protein that is embedded in the membrane of the endoplasmic reticulum through distinctive membrane spanning regions located in the C-terminal domain.[8] The $InsP_3$-binding site is located at the N-terminal end that projects out into the cytoplasm in the form of a large bulbous head. Four of these subunits come together to form a functional $InsP_3$-regulated calcium channel. The original concept was that the external signal acted on a cell-surface receptor to stimulate the formation of $InsP_3$ that then acted on its receptor to release calcium from the internal stores. Although this sequence of events seems to be correct, it is clear that the action of $InsP_3$ may be far more subtle than originally envisaged. Much of this subtlety seems to depend upon the fact that the receptor for $InsP_3$ is very sensitive to calcium itself.

Perhaps the single most important property of the $InsP_3$ receptor is that it is under the dual control of both $InsP_3$ and calcium. The phenomenon of calcium-induced calcium release (CICR) was first described for the ryanodine receptor of cardiac muscle, where it plays a central role in excitation-contraction coupling. Voltage-operated calcium channels in the plasma membrane gate a small pulse of trigger calcium that then acts on the underlying ryanodine receptors to release the large amounts of calcium necessary for contraction. The process of CICR is not restricted to the ryanodine receptors of the heart but is also a general property of $InsP_3$ receptors. The effect of calcium on the $InsP_3$ receptor is complex in that low levels are stimulatory, whereas at concentrations above about 300 nM calcium becomes inhibitory.[9–12] The ability of calcium to promote its own release is regulated by $InsP_3$, such that in the absence of this second messenger, calcium has little effect. As the concentration of $InsP_3$ rises, it increases the sensitivity of the receptor to the stimulatory effect of calcium. What $InsP_3$ does, therefore, is to convert the cytoplasm into an "excitable medium" in which calcium spikes and waves can occur.

INOSITOL TRISPHOSPHATE AND CALCIUM ENTRY

One of the mechanisms for setting up oscillations depends upon a periodic opening and closing of a capacitative calcium entry pathway located in the plasma membrane. The basic phenomenon of capacitative calcium entry is that there are channels in the plasma membrane that are switched on once the internal stores are emptied of their calcium.[13] A calcium release-activated current (I_{CRAC}) has been characterized.[14,15] Any agent or treatment that empties the internal stores (*e.g.*, $InsP_3$, incubation in calcium-free medium, inhibition of the endoplasmic reticulum calcium pump by thapsigargin) will activate I_{CRAC}. This inward calcium current is inactivated by calcium, and this negative feedback effect might contribute to the onset of oscillations in membrane conductance (see below).

Even though the I_{CRAC} channel is independent of voltage, the flux of calcium through this channel is strongly influenced by membrane potential that alters the driving force for entry. The influx rate is enhanced by hyperpolarization and depressed by depolarization. This effect of membrane potential on calcium entry may account for the observation that modulation of potassium channels can exert a profound effect on cell proliferation. In HeLa cells there are changes in a calcium-dependent potassium channel during the course of the cell cycle; it is maximal early in G1, reaches a minimum during S, and then rises again towards the next mitosis.[16] A large-conductance potassium channel undergoes a similar sequence of changes during the course of the first cell cycle in mouse embryos.[17] Potassium channel modulators (TEA, 4-aminopyridine, and diazoxide) and high potassium inhibited the growth of two human brain tumors.

A study of potassium channels in lymphocytes has led to the idea that an increase in channel activity may maintain membrane hyperpolarization, thus enhancing calcium signaling.[18] Charybdotoxin can block both voltage- and calcium-dependent K^+ channels in T cells, and this might account for its antiproliferative effects.[19] A similar mechanism may function in human melanoma cells. Inhibiting these channels with TEA or addition of cyclic AMP reduces cell growth. They propose that the reduction in driving force for calcium may prevent the stimulation of p34 at G_1/S.[20] The enhanced expression of K channels may help provide the calcium signal necessary for the G_1/S transition.[21]

Although there is general agreement that emptying of the internal stores is responsible for promoting the influx of external calcium, there is less consensus concerning the coupling mechanism whereby an empty store sends a signal to the channels in the plasma membrane. A number of mechanisms have been proposed. The empty stores may generate a calcium influx factor (CIF) that then diffuses from the stores to the plasma membrane.[22] A role for G proteins has also been proposed because capacitative calcium entry is sensitive to GTPγS.[23–25] The G protein could function either to produce CIF, to be the diffusible messenger, or it could act at the level of the channel itself. Yet another mechanism for capacitative calcium entry is that the $InsP_3$ receptor itself might function to transmit information between the two membranes through a process of conformational coupling (FIG. 1).[26,27] The idea is that the $InsP_3$ receptor in the endoplasmic reticulum (ER) would sense the state of filling of the pool and would then transmit the information to the I_{CRAC} channel by means of its large bulbous head structure.[26,27] Whatever the mechanism turns out to be, the

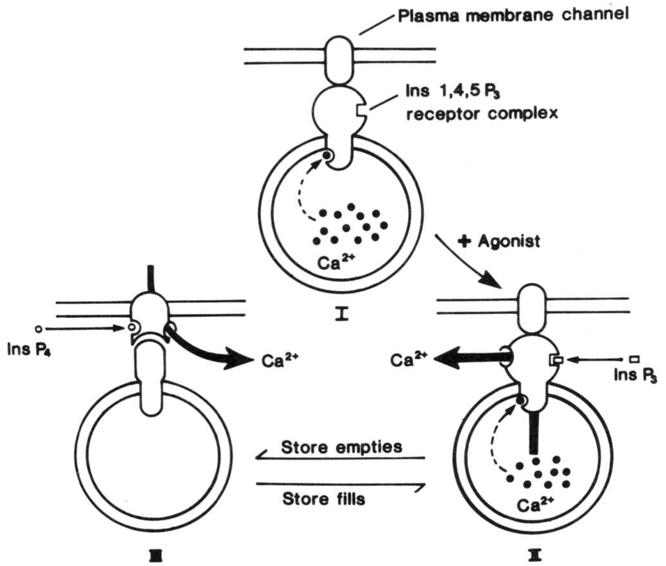

FIGURE 1. Conformational-coupling model of capacitative calcium entry. (Berridge.[26] With permission from the *Journal of Biological Chemistry*.)

important point is that $InsP_3$ is able to activate calcium entry indirectly through its ability to empty intracellular calcium stores. This $InsP_3$-mediated control of calcium entry has two important consequences with regard to the generation of calcium oscillations: it provides a steady input of calcium to charge up the stores of those cells that have a cytosolic oscillator, or it might be the basis of the membrane oscillator, as has been described in lymphocytes.[28,29]

THE CYTOSOLIC OSCILLATOR

The main component of the cytosolic oscillator is the $InsP_3$ receptor located on the internal stores that are responsible for the pulsatile release of calcium. A large number of models have been proposed to explain these cytosolic calcium oscillations.[2–4,30–35] All these models stress the importance of a positive feedback process whereby calcium enhances its own release. In one of these models, the feedback is represented by calcium stimulating the formation of $InsP_3$,[30] whereas all the other models are based on the process of CICR whereby calcium promotes its own release at the level of the $InsP_3$ receptor. The release cycle has four main components: charging up the stores, initiation of release, regenerative release as the wave sweeps through the cell by CICR, and finally the recovery phase as calcium is pumped out of the cell or back into the stores. This sequence of events is summarized in FIGURE 2.

Charging up the Stores

Although the cytosolic oscillator depends upon the release of internal calcium, the stores have to be recharged following each cycle. Cells seems to vary concerning their reliance on external calcium. Although some can continue to oscillate for some time in the absence of external calcium, the majority are very dependent upon continuous calcium entry, usually mediated through I_{CRAC}. This entry component is relatively small and thus has little effect on the resting level of calcium because it can be quickly removed by uptake into the internal stores. This buffering of the influx component can continue until the stores become saturated, at which point the resting level will begin to rise and probably account for the pacemaker elevation of calcium that precedes the onset of the calcium spike in many cells. An important aspect of this hypothesis, therefore, is that the interspike interval is determined, at least in part, by the rate of store loading. If the influx rate is slow, it will take a long time to load up the stores, resulting in a low frequency of spiking. Therefore, any maneuver that enhances the rate of loading will accelerate the oscillator. Consistent with this interpretation is the observation that varying the concentration of external calcium can exert a profound effect on oscillator frequency in a number of different cell types such as endothelial cells,[36] smooth muscle cells,[37] and the insect salivary gland.[38] Likewise, enhancing the driving force for calcium entry by hyperpolarizing the membrane can accelerate oscillations in *Xenopus* oocytes.[39]

Initiation

Perhaps the most difficult aspect of the cycle to explain is the process of spike initiation. As noted earlier, spikes often initiate from discrete regions where the stores appear to be particularly sensitive. Evidence for such a variation in $InsP_3$ sensitivity may account for the phenomenon of quantal calcium release, which can be demonstrated in single HeLa cells responding to different concentrations of histamine.[40] As the level of stimulation increases, more and more of the stores contribute to the calcium signal. This differential sensitivity may thus account for the process of initiation, in that the most sensitive stores function as trigger zones by releasing calcium into a localized area that then sets off the secondary regenerative phase by activating neighboring stores that are somewhat less sensitive. Returning to the concept of an excitable medium, one can imagine that certain areas are more excitable than others, and it is the former that acts as the initiation sites.

Further insight into the nature of initiation has come from the study of calcium sparks in cardiac muscle cells[41] and calcium puffs in *Xenopus* oocytes.[42] These small localized bursts of calcium have been observed by using a confocal microscope to observe a narrow two-dimensional slice of cytoplasm. It has been calculated that the cardiac calcium sparks represent the opening of a single ryanodine receptor.[41] In the case of the *Xenopus* oocyte, the puffs seem to result from the opening of a small collection of $InsP_3$ receptors that begin to release puffs of calcium at regular intervals

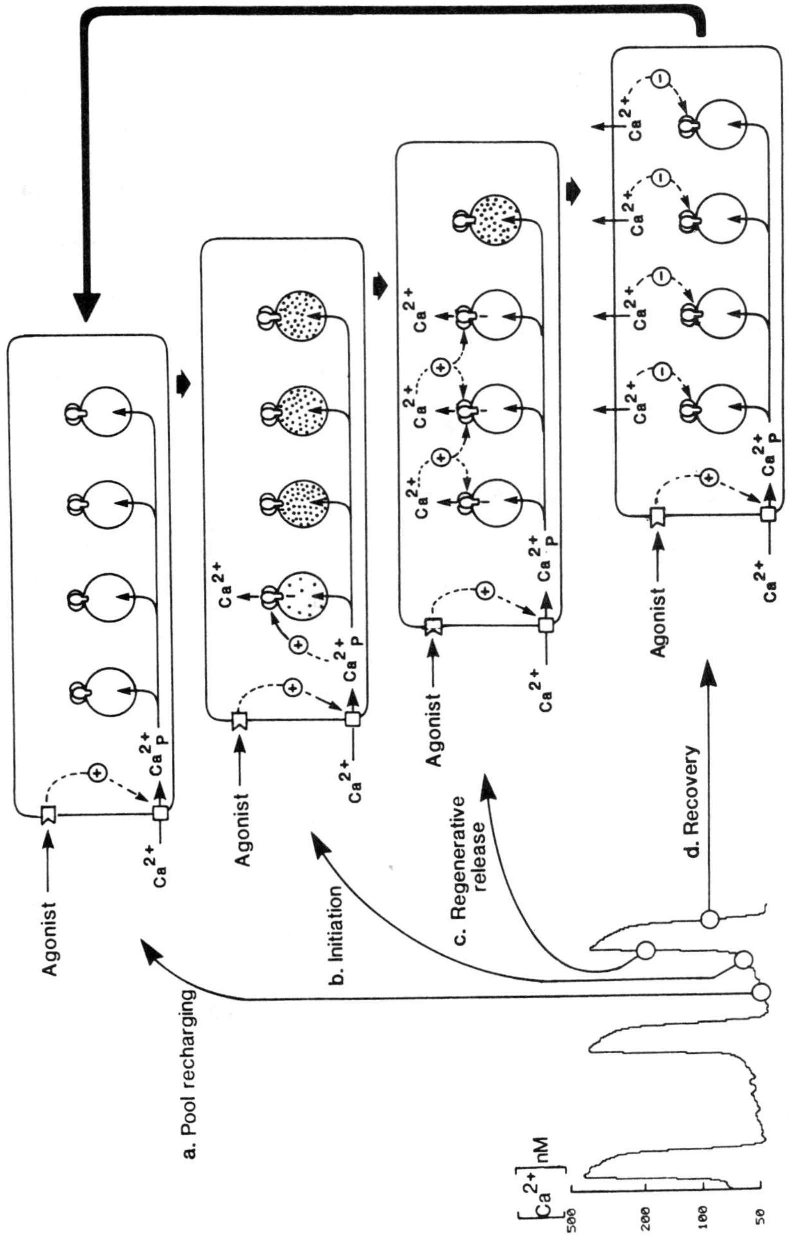

FIGURE 2. Model of a cytosolic calcium oscillator. See text for further details. (Berridge.[1] With permission from *Nature*.)

when the cell is injected with a low dose of $InsP_3$.[42] At this low dose, these pulsing puff sites remain localized because the neighboring $InsP_3$ receptors are not sufficiently sensitive to set off a regenerative calcium wave, that is, the medium is not sufficiently excitable. However, further elevation of the level of $InsP_3$ does enhance the excitability of all the receptors, such that they will begin to respond to the calcium coming from the puff sites and will thus begin to transmit a calcium wave.

Regenerative Release

Waves spread away from the initiation site through a process of CICR. The calcium that is released at the initiation site functions as a diffusible messenger to activate neighboring receptors, thereby setting up a regenerative calcium wave that spreads at rates between 10-100 μm/s. Perhaps the most important factor determining the rate of wave propagation is the excitability of the medium, which, in turn, is determined by the level of stimulation and hence the level of $InsP_3$.

Recovery

Once the wave has spread through the cytoplasm, release ceases, and the calcium is rapidly removed from the cytoplasm by extrusion from the cell or by resequestration back into the stores. Just why the receptor shuts off is not fully understood, but it can probably be explained by the bell-shaped calcium activation curve described earlier. Indeed, there are a number of models suggesting that the complex response of the $InsP_3$ receptor to calcium is solely responsible for setting up calcium spikes and waves.[33,34,43] During the onset of the spike, calcium promotes its own release, but once the concentration reaches a certain level (approximately 300 nM), the inhibitory phase takes over and release is terminated. As soon as the channels close, the calcium spike recovers rapidly as the mechanisms responsible for removing calcium take over.

THE MEMBRANE OSCILLATOR

A good example of a cell that uses a membrane oscillator is the T lymphocyte. Activation of the T-cell receptor (TCR) induces the appearance of repetitive calcium spikes with a periodicity of about 100 seconds that can persist for hours. These calcium spikes are critically dependent upon capacitative calcium entry, in that they can be duplicated by treating T cells with thapsigargin that empties the stores by inhibiting the pump on the ER.[29] Oscillations are only observed at intermediate thapsigargin concentrations, which cause a partial inhibition of the pumps. Because complete inhibition results in an elevated plateau level of calcium devoid of ER pump activity, it has been proposed that the periodic influx of external calcium through I_{CRAC} channels in the plasma membrane is entrained to a cycle of store filling and emptying.[29] As for the cytosolic oscillator, the membrane oscillatory cycle can be separated into four events (FIG. 3).

Step 1. Interspike Interval

The driving force for this membrane oscillator is the emptying of the store by $InsP_3$ generated by the activation of the TCR. The store is thought to be empty during the interspike interval because, in the absence of calcium entry, the activity of the plasma membrane calcium pumps has served to drain the store.

Step 2. Onset of the Spike

The empty store begins to pull calcium into the cell across the plasma membrane using the I_{CRAC} channels, resulting in the upstroke of the spike.

Step 3. Spike Termination

As the spike develops, the increase in calcium concentration will enable the pumps on the ER to counteract the $InsP_3$-induced calcium leak, resulting in a gradual recovery in the calcium content of the store up to the point when it begins to switch off capacitative calcium entry, thereby terminating the spike.

Step 4. Recovery

As soon as entry ceases, calcium will be pumped down and $InsP_3$ will once again begin to drain the stores, thus setting up the conditions for the cycle to begin again. In order for this oscillatory cycle to occur, it is necessary to assume that there is a time-dependent delay between the emptying of the store (step 4) and the subsequent activation of I_{CRAC} (step 2).[29] This delay may involve the processes required to send a signal from the empty store to the plasma membrane channel. Inasmuch as the I_{CRAC} channel is known to be inhibited by calcium,[15,29] the delay may also be explained by the time taken to recover from such inhibition.

CONCLUSION

There are two major mechanisms for setting up calcium oscillations in cells. One mechanism depends upon the pulsatile release of calcium stored up in internal reservoirs. An important component of this internal release is the process of CICR, which is responsible for the generation of calcium waves that are particularly evident during the fertilization of mammalian eggs. The other oscillatory mechanism depends upon the periodic influx of calcium across the plasma membrane as has been described in lymphocytes. These two oscillatory mechanisms may provide important mitogenic signals at critical points during the cell cycle (FIG. 4). The oscillatory release of

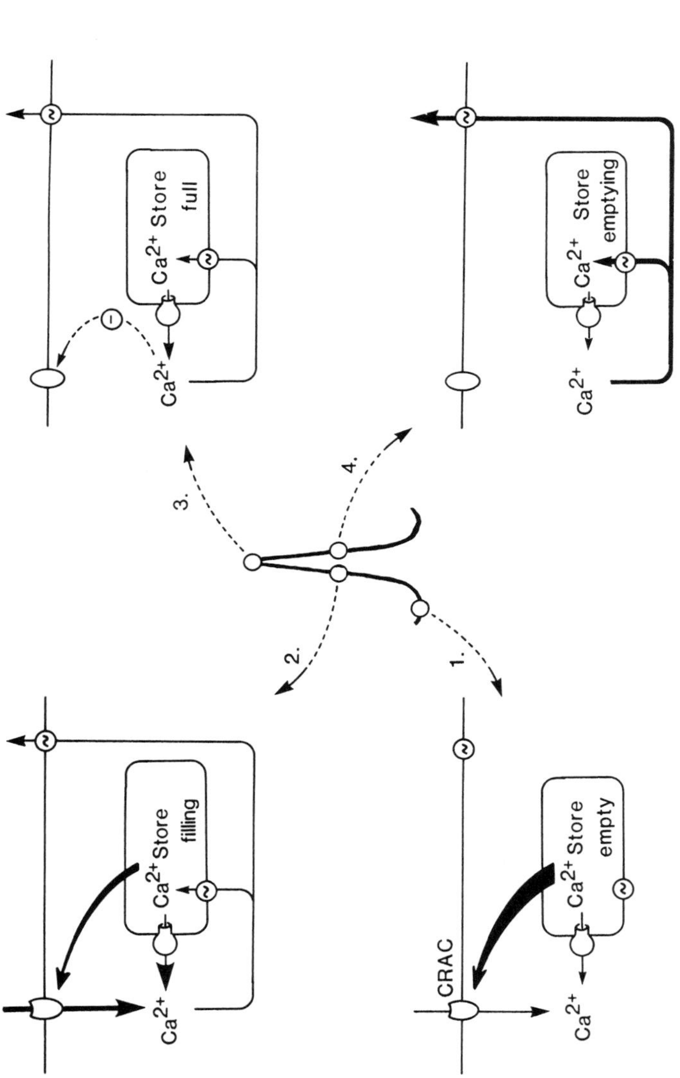

FIGURE 3. Model of a membrane oscillator based on the periodic opening of a calcium release-activated channel (CRAC).

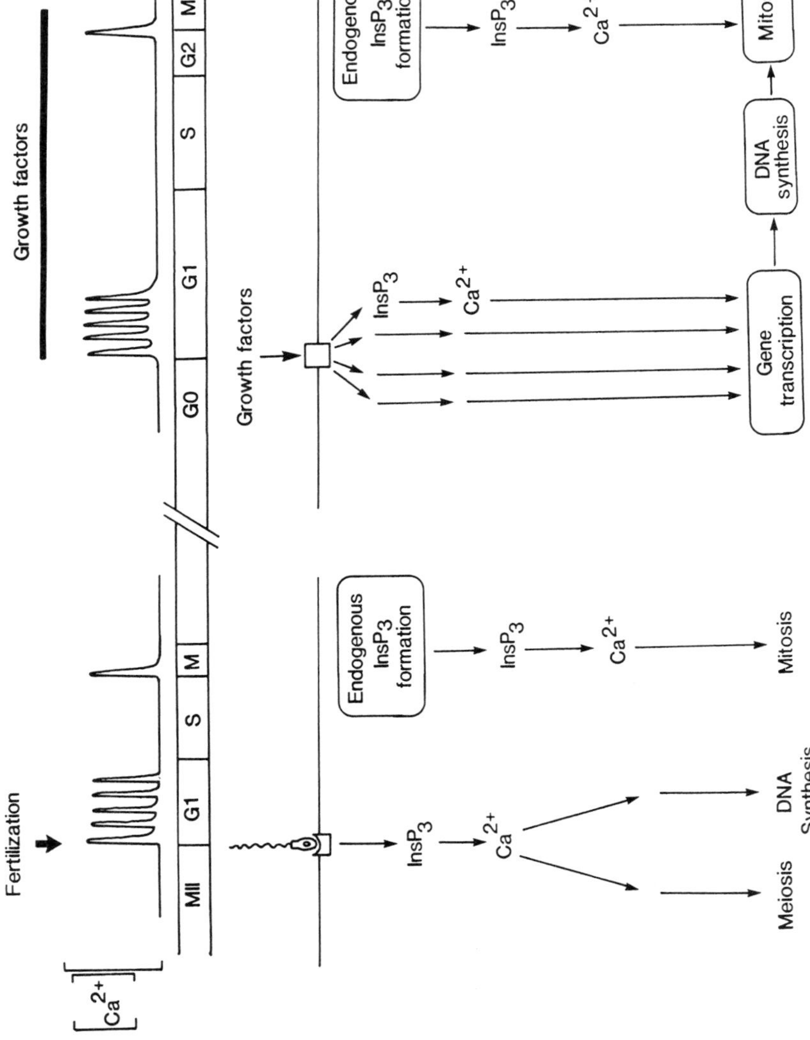

FIGURE 4. Proposed role of calcium oscillations in the control of key events during the cell cycle.

calcium at the time of fertilization of mammalian eggs seems to be responsible for the completion of meiosis and for the subsequent initiation of DNA synthesis. Repetitive calcium spikes are also an important component of the mitogenic signals produced by quiescent cells as they respond to growth factors. In lymphocytes, this calcium signaling is responsible for activating the production of interleukin-2 and its receptor. Such early gene activation then culminates in the onset of DNA synthesis. On a more speculative note, endogenous pulses of $InsP_3$ originating at the G2/M interface may be responsible for initiating the process of mitosis.

REFERENCES

1. BERRIDGE, M. J. 1993. Inositol trisphosphate and calcium signalling. Nature **361:** 315-325.
2. FEWTRELL, C. 1993. Ca^{2+} oscillations in non-excitable cells. Annu. Rev. Physiol. **55:** 427-454.
3. MEYER, T. & L. STRYER. 1991. Calcium spiking. Annu. Rev. Biophys. Biophys. Chem. **20:** 153-174.
4. BERRIDGE, M. J. & G. DUPONT. 1994. Spatial and temporal signalling by calcium. Curr. Opinion Cell. Biol. **6:** 267-274.
5. MIYAZAKI, S., H. SHIRAKAWA, K. NAKADA & Y. HONDA. 1993. Essential role of the inositol 1,4,5-trisphosphate receptor/Ca^{2+} release channel in Ca^{2+} waves and Ca^{2+} oscillations at fertilization of mammalian eggs. Dev. Biol. **158:** 62-78.
6. LEWIS, R. S. & M. D. CAHALAN. 1989. Mitogen-induced oscillations of cytosolic Ca^{2+} and transmembrane Ca^{2+} current in human leukemic T cells. Cell Regul. **1:** 199-212.
7. CHOQUET, D., G. KU, S. CASSARD, B. MALISSEN, H. KORN, W. H. FRIDMAN & C. BONNEROT. 1994. Different patterns of calcium signaling triggered through two components of the B lymphocyte antigen receptor. J. Biol. Chem. **269:** 6491-6397.
8. MIKOSHIBA, K. 1993. Inositol 1,4,5-trisphosphate receptor. Trends Pharmacol. Sci. **14:** 86-89.
9. BEZPROZVANNY, I., J. WATRAS & B. E. EHRLICH. 1991. Bell-shaped calcium-response curves of Ins(1,4,5)P_3- and calcium-gated channels from endoplasmic reticulum of cerebellum. Nature **351:** 751-754.
10. PARKER, I. & I. IVORRA. 1990. Inhibition by Ca^{2+} of inositol trisphosphate-mediated Ca^{2+} liberation: a possible mechanism for oscillatory release of Ca^{2+}. Proc. Natl. Acad. Sci. USA **87:** 260-264.
11. IINO, M. 1990. Biphasic Ca^{2+} dependence of inositol 1,4,5-trisphosphate-induced Ca release in smooth muscle cells of the guinea pig *Taenia caeci.* J. Gen. Physiol. **95:** 1103-1122.
12. FINCH, E. A., T. J. TURNER & S. M. GOLDIN. 1991. Calcium as a coagonist of inositol 1,4,5-trisphosphate-mediated Ca^{2+} release. Science **252:** 443-446.
13. PUTNEY, JR., J. W. 1990. Capacitative calcium entry revisited. Cell Calcium **11:** 611-624.
14. HOTH, M. & R. PENNER. 1992. Depletion of intracellular calcium stores activates a calcium current in mast cells. Nature **355:** 353-356.
15. HOTH, M. & R. PENNER. 1993. Calcium release-activated calcium current in mast cells. J. Physiol. **465:** 359-386.
16. TAKAHASHI, A., H. YAMAGUCHI & H. MIYAMOTO. 1993. Change in K^+ current of HeLa cells with progression of the cell cycle studied by patch-clamp technique. Am. J. Physiol. **265:** C328-C336.
17. DAY, M. L., S. J. PICKERING, M. H. JOHNSON & D. J. COOK. 1993. Cell cycle control of a large-conductance K^+ channel in mouse early embryos. Nature **365:** 560-562.
18. LEWIS, R. S. & M. D. CAHALAN. 1988. The plasticity of ion channels: parallels between the nervous and immune systems. Trends Neurosci. **11:** 214-218.

19. LEONARD, R. J., M. L. GARCIA, R. S. SLAUGHTER & J. P. REUBEN. 1992. Selective blockers of voltage-gated K^+ channels depolarize human T lymphocytes: Mechanism of the antiproliferative effects of charybdotoxin. Proc. Natl. Acad. Sci. USA **89:** 10094-10098.
20. NILIUS, B. & W. WOHLRAB. 1992. Potassium channels and regulation of proliferation of human melanoma cells. J. Physiol. **445:** 537-548.
21. NILIUS, B., G. SCHWARZ & G. DROOGMANS. 1993. Control of intracellular calcium by membrane potential in human melanoma cells. Am. J. Physiol. **265:** C1501-C1510.
22. RANDRIAMAMPITA, C. & R. Y. TSIEN. 1993. Emptying of intracellular Ca^{2+} stores releases a novel small messenger that stimulates Ca^{2+} influx. Nature **364:** 809-814.
23. BIRD, G. S. J. & J. W. PUTNEY, JR. 1993. Inhibition of thapsigargin-induced calcium entry by microinjected guanine nucleotide analogues. Evidence for the involvement of a small G-protein in capacitative calcium entry. J. Biol. Chem. **268:** 21486-21488.
24. FASOLATO, C., M. HOTH & R. PENNER. 1993. A GTP-dependent step in the activation mechanism of capacitative calcium influx. J. Biol. Chem. **268:** 20737-20740.
25. JACONI, M. E. E., D. P. LEW, A. MONOD & K.-H. KRAUSE. 1993. The regulation of store-dependent Ca^{2+} influx in HL-60 granulocytes involves GTP-sensitive elements. J. Biol. Chem. **268:** 26075-26078.
26. BERRIDGE, M. J. 1990. Calcium oscillations. J. Biol. Chem. **265:** 9583-9586.
27. IRVINE, R. F. 1990. 'Quantal' Ca^{2+} release and the control of Ca^{2+} entry by inositol phosphates—a possible mechanism. FEBS Lett. **263:** 5-9.
28. ZWEIFACH, A. & R. S. LEWIS. 1993. Mitogen-regulated Ca^{2+} current of T lymphocytes is activated by depletion of intracellular Ca^{2+} stores. Proc. Natl. Acad. Sci. USA **90:** 6295-6299.
29. DOLMETSCH, R. E. & R. S. LEWIS. 1994. Signaling between intracellular Ca^{2+} stores and depletion-activated Ca^{2+} channels generates $[Ca^{2+}]_i$ oscillations in T lymphocytes. J. Gen. Physiol. **103:** 365-388.
30. MEYER, T. & L. STRYER. 1988. Molecular model for receptor-stimulated calcium spiking. Proc. Natl. Acad. Sci. USA **85:** 5051-5055.
31. GOLDBETER, A., G. DUPONT & M. J. BERRIDGE. 1990. Minimal model for signal-induced calcium oscillations and for their frequency encoding through protein phosphorylation. Proc. Natl. Acad. Sci. USA **87:** 1461-1465.
32. GIRARD, S., A. LÜCKHOFF, J. LECHLEITER, J. SNEYD & D. CLAPHAM. 1992. Two-dimensional model of calcium waves reproduces the patterns observed in *Xenopus* oocytes. Biophys. J. **61:** 509-517.
33. DEYOUNG, G. W. & J. KEIZER. 1992. A single pool IP_3-receptor-based model for agonist stimulated Ca^{2+} oscillations. Proc. Natl. Acad. Sci. USA **89:** 9895-9899.
34. ATRI, A., J. AMUNDSON, D. CLAPHAM & J. SNEYD. 1993. A single-pool model for intracellular calcium oscillations and waves in the *Xenopus laevis* oocyte. Biophys. J. **65:** 1727-1739.
35. BEZPROZVANNY, I. 1994. Theoretical analysis of calcium wave propagation based on inositol (1,4,5)-trisphophsate ($InsP_3$) receptor functional properties. Cell Calcium **16:** 151-166.
36. JACOB, R., J. E. MERRITT, T. J. HALLAM & T. J. RINK. 1988. Repetitive spikes in cytoplasmic calcium evoked by histamine in human endothelial cells. Nature **335:** 40-45.
37. MAHONEY, M. G., C. J. RANDALL, J. J. LINDERMAN, D. J. GROSS & L. L. SLAKEY 1992. Independent pathways regulate the cytosolic $[Ca^{2+}]$ initial transient and subsequent oscillations of individual cultured arterial smooth muscle cells responding to extracellular ATP. Mol. Biol. Cell **3:** 493-505.
38. BERRIDGE, M. J. 1994. Relationship between latency and period for 5-hydroxytryptamine-induced membrane responses in the *Calliphora* salivary gland. Biochem. J. **302:** 545-550.
39. GIRARD, S. & D. CLAPHAM. 1993. Acceleration of intracellular calcium waves in *Xenopus* oocytes by calcium influx. Science **260:** 229-232.

40. Bootman, M. 1994. Intracellular calcium: Questions about quantal Ca^{2+} release. Curr. Biol. **4:** 169-172.
41. Cheng, H., W. J. Lederer & M. B. Cannell. 1993. Calcium sparks: Elementary events underlying excitation-contraction coupling in heart muscle. Science **262:** 740-744.
42. Parker, I. & Y. Yao. 1991. Regenerative release of calcium from functionally discrete stores by inositol trisphosphate. Proc. R. Soc. Lond. B. **246:** 269-274.
43. Lechleiter, J. D. & D. Clapham. 1992. Molecular mechanisms of intracellular calcium excitability in *X. laevis* oocytes. Cell **69:** 283-294.

Interaction of Signaling and Trafficking Proteins with the Carboxyterminus of the Epidermal Growth Factor Receptor[a]

GRAHAM CARPENTER,[b] CONCEPCIÓ SOLER,[b,d] JOSEP BAULIDA,[b] LAURA BEGUINOT,[c] AND ALEXANDER SORKIN[b,e]

[b]*Department of Biochemistry*
Vanderbilt University School of Medicine
Nashville, Tennessee 37232-0146

[c]*Laboratorio di Oncologia Molecolare*
Dibit and Instituto di Neuroscience e Bioímmagini del CNR
H.S. Raffaele
Milano, Italy

The cytoplasmic domain of the epidermal growth factor (EGF) receptor has several significant features or functional domains.[1] The juxtamembrane region contains several serine/threonine residues that serve as phosphorylation sites for other protein kinases, such as protein kinase C[2] and MAP kinase.[3,4] Immediately distal to the juxtamembrane region is the tyrosine kinase domain. The activity of this domain is regulated by ligand binding to the extracellular domain by a mechanism(s) that is not understood. Distal to the kinase domain is a region of approximately 200 residues, known as the carboxyterminal domain or cytoplasmic tail. The focus of this article is the features of this carboxyterminal domain and the cellular proteins that associate with it.

SIGNALING MOLECULES

Distinctive of the carboxyterminal domain are five tyrosyl residues (Tyr 992, 1068, 1086, 1148, and 1173) that are known to be sites of autophosphorylation.[1,5]

[a]Support from NIH Grants CA24071 and CA43720 (G. Carpenter) and DK46817 (A. Sorkin) is gratefully acknowledged, as well as Grant support from the Associazione Italiana per la recerca sul cancro (AIRC) (L. B. Beguinot) and fellowships (J. Baulida, C. Soler) from the Ministerio de Educación y Ciencia, Spain.

[d]Present address: Unitat de Fisiologia, Departament de Bioquímica i Fisiologia, Universitat de Barcelona, Diagonal 645, 0828 Barcelona, Spain.

[e]Present address: Department of Pharmacology, University of Colorado Health Science Center, Denver, CO 80262.

However, stoichiometric data relevant to interpreting the relative significance of these sites are lacking. Hence, the autophosphorylation sites are probably not quantitatively equivalent, as some may be phosphorylated to a much higher extent than others. Also, the number of tyrosyl residues that are simultaneously phosphorylated on one receptor molecule is unknown. Autophosphorylated receptors, therefore, are likely to be a very heterogeneous group of molecules. These quantitative issues aside, it is clear that autophosphorylation sites have a significant function in terms of forming association complexes with molecules that contain SH2 or *src* homology domains.[6] At least seven SH2-containing molecules have been detected to associate with the activated EGF receptor.[7] These SH2-containing molecules include the following: phospholipase C-γl (PLC-γ1), a phosphotyrosine phosphatase known as syp, the GTPase activating protein for ras (rasGAP), the p85 regulatory subunit of phosphatidylinositol 3-kinase, and adaptor molecules known simply as nck, SHC, and GRB-2. The mechanistic basis for the association of these proteins with receptor autophosphorylation sites is the affinity of SH2 sequence motifs for proteins containing phosphotyrosine. Specificity seems to be provided by the recognition of particular phosphotyrosine-containing sequences by specific SH2 motifs. However, in contrast to data obtained with autophosphorylation site mutants of several other growth factor receptors, particularly the platelet-derived growth factor (PDGF) receptor, single autophosphorylation site mutants (Y $\rightarrow$ F) of the EGF receptor do not stringently define individual phosphotyrosyl residues as critical for the association of several SH2-containing protein with the activated receptor.[8] The sequences surrounding each of the five EGF receptor autophosphorylations are quite distinct. Rather, the autophosphorylation sites, which are clustered in 180 residues of the receptor carboxyterminal domain, seem flexible or compensatory with regard to the recognition of individual SH2-containing molecules. A second exception to the notion that specific autophosphorylation site sequences define interaction sites for different SH2 molecules is the hepatocyte growth factor receptor. This receptor contains two autophosphorylation sites with highly related sequences, yet the sites are capable of associating with several different SH2-containing proteins.[9] Results with the EGF and hepatocyte growth factor receptors, therefore, are distinct for those obtained earlier with other receptors and suggest that additional factors beyond autophosphorylation site sequences govern association of SH2 proteins with some receptors.

Nearly all the SH2 proteins that associate with activated growth factor receptors are also phosphorylated by these receptors. This has suggested that the SH2 domain interaction with autophosphorylation sites represents a highly efficient "docking" mechanism that in most cases is necessary for efficient tyrosine phosphorylation of the docked protein. The "adapter" SH2 protein GRB-2,[10] which mediates formation of a ternary complex composed of the activated EGF receptor, GRB-2, and the ras GDP/GTP exchanger,[11–15] is, however, not a tyrosine kinase substrate. The mechanism of GRB-2 action has suggested a second function for SH2 association with receptor autophosphorylation sites, that is, this association may produce a biologically significant translocation of a cytoplasmic SH2 protein, such as GRB-2, to a position close to the cytoplasmic leaflet of the plasma membrane. In the case of GRB-2, this may be significant because this molecule is involved in the activation of ras, a membrane-localized GTP binding protein with an important role in mitogenic signal transduction.

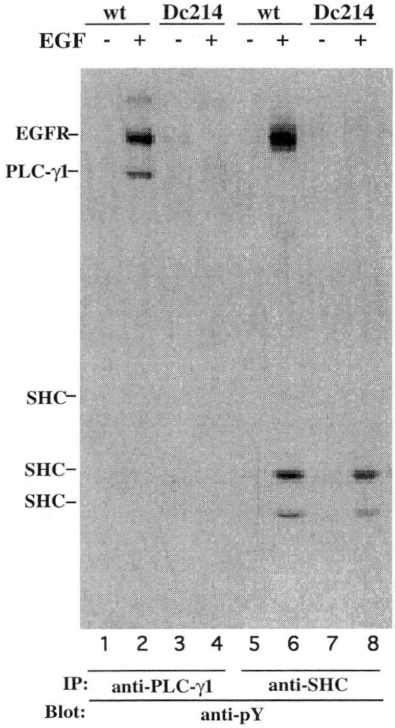

FIGURE 1. Tyrosine phosphorylation of PLC-γ1 and SHC by EGF-treated NR6 cells expressing wild-type (wt) or truncated (Dc214) EGF receptors. Cells having equivalent numbers of wild-type or Dc214 mutant EGF receptor were incubated for 1 h at 5°C in the absence or presence of EGF (100 ng/mL). The cells were then lysed, and PLC-γ1 and SHC were immunoprecipitated from separate aliquots. The immunoprecipitated samples were subjected to SDS gel electrophoresis, and proteins were transferred to nitrocellulose. Western blots were then performed with phosphotyrosine antibody, and the amount of bound antiphosphotyrosine was detected.

A test of the docking and phosphorylation model for SH2 protein complexes with an activated receptor tyrosine kinase is shown in the experiment presented in FIGURE 1 and elsewhere.[16] Cells expressing the wild-type (wt) EGF receptor or an autophosphorylation site-negative deletion mutant of the EGF receptor (Dc214) were used. The mutant is truncated at residue 972; hence all known autophosphorylation sites are deleted and control studies with antiphosphotyrosine blots show that no major new autophosphorylation sites are present in this altered receptor. After the cells were treated with EGF, two SH2-containing proteins, PLC-γ1 and the src homology collagen-like (SHC) adaptor protein, were immunoprecipitated. Following electrophoresis and transfer to nitrocellulose, each sample was then Western blotted with antibody to phosphotyrosine. The results, presented in lanes 1-4, demonstrate that when cells are treated with EGF, the wild-type EGF receptor, but not the Dc214 truncation mutant, is able to increase the phosphotyrosine level on PLC-γ1. Therefore, the data indicate that the receptor carboxyterminal domain is necessary for PLC-γ1 tyrosine phosphorylation by the EGF receptor. As previous data have shown that Dc214 encodes an EGF-activatable tyrosine kinase,[16,17] the lack of PLC-γ1 phosphorylation by this receptor mutant is not due to a general deficit in receptor kinase activity. The results obtained with PLC-γ1 would support the docking and phosphorylation model. However, in contrast to the inability of Dc214 to phosphorylate PLCγ1, the results shown in lanes 5–8 of FIGURE 1 demonstrate that this receptor mutant is able to effectively phosphorylate SHC. Therefore, efficient SHC phosphorylation by the EGF does not require the receptor carboxyterminus.

These studies with PLC-γ1 and SHC have led to the conclusion that among the different phosphorylation substrates of the EGF receptor, there is not always an obligatory requirement of receptor : substrate association, mediated by autophosphorylation-site interaction with SH2 domains, for efficient substrate tyrosine phosphorylation. Inasmuch as protein kinases, in general, do not have recognized docking mechanisms for efficient substrate phosphorylation, it is actually somewhat curious that certain tyrosine kinase substrates, such as PLC-γ1, do seem to require association with the receptor for efficient phosphorylation. The behavior of SHC is more like the general mechanism of protein kinase action, that is, independent of an association mechanism. In addition, another SH2 substrate, rasGAP, is also efficiently phosphorylated by the Dc214 EGF receptor mutant.[17] Also, it is important to note that the Dc214 mutant receptor is competent to produce biological responses, that is, transformation and mutagenesis, in cells exposed to EGF.[17–20] Therefore, the functional significance of EGF receptor autophosphorylation sites is complex and may not follow generalizations made from other growth factor receptors.

TRAFFICKING MOLECULES

Although signal transduction occurs rapidly after EGF interaction with its receptor, in the context of an intact cell, the activated EGF receptor also rapidly undergoes sorting decisions at the plasma membrane followed by internalization of ligand : receptor complexes.[21] The intracellular complexes are subsequently delivered to lysosomes and both EGF and its receptor are degraded. This endocytotic pathway is generally thought to be a mechanism for the attenuation or desensitization of receptor signaling complexes. However, it may also provide a mechanism by which signals are generated at various sites within the cell. These putative intracellular signals could be qualitatively or quantitatively distinct from signals initially generated by EGF : receptor complexes at the cell surface.

The fact that the EGF receptor is subject to endocytosis has been known for nearly 20 years.[22] A mechanistic understanding of molecular interactions in this process, however, has only recently begun to emerge. At an early step in this pathway, EGF : receptor complexes, moving in the plane of the plasma membrane, are selectively retained in specialized regions of the plasma membrane, termed coated pits, which represents an entry step to the endocytosis pathway. A recent study[23] has focused on EGF receptor interaction with a known component of coated pits, the tetrameric clathrin adaptor protein complex AP-2. The AP-2 complex, which binds clathrin, a major structural constituent of the coated pit, may serve as an organizing component of the coated pit. Data now show that activated EGF receptors form an association complex with AP-2. In intact cells this occurs rapidly following the addition of EGF and requires that the cells are incubated at 37°C. At low temperature (*i.e.*, 5°C), activated receptors do not associate with AP-2. By contrast, receptor association with SH2-containing proteins, such as PLC-γ1 and SHC, takes place efficiently when cells are incubated with EGF at 5° or 37°C.

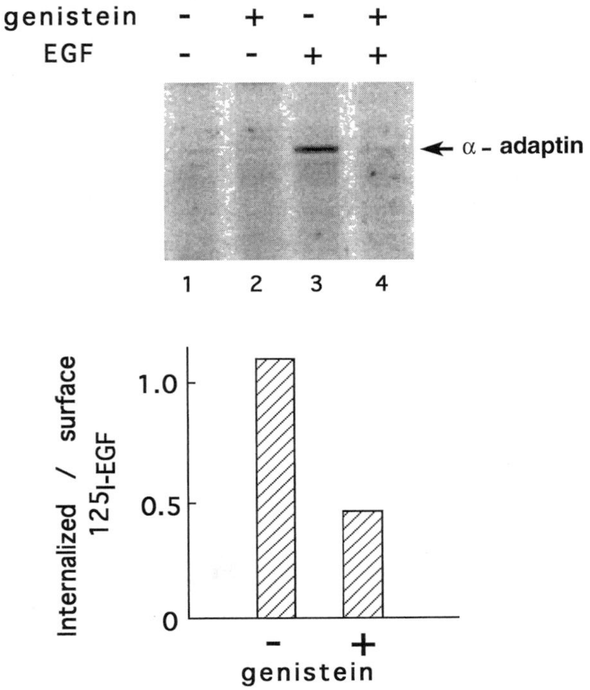

FIGURE 2. Influence of the tyrosine kinase inhibitor genistein on AP-2 association with the EGF receptor and [^{125}I]EGF internalization. Upper panel: A-431 cells were incubated without or with genistein (150 μM) for a total of 38 min at 37°C. The indicated cultures were also incubated with EGF (300 ng/mL) for the last 14 min. The EGF receptor was then immunoprecipitated from cell lysates and, following SDS electrophoresis and transfer to nitrocellulose, each sample was blotted with antisera to AP-2. This was then used to measure the amount of AP-2 that coprecipitated with EGF receptors. Lower panel: Cells were similarly pretreated without or with genistein at 37°C, and then [^{125}I]EGF was added for 1-6 min at 37°C. An acid wash technique was then used to measure the amount of [^{125}I]EGF present on cell surface and internalized to intracellular compartments.

Several questions arise as to what activities or structural features of the EGF receptor are necessary for its association with AP-2. The data presented in FIGURE 2 (upper panel) show that the tyrosine kinase inhibitor genistein prevents AP-2 association with the EGF receptor in EGF-treated cells. The results shown in the lower panel of FIGURE 2 demonstrate that this kinase inhibitor also significantly reduces the rate of EGF internalization by the EGF receptor. A conclusion from these data is that the tyrosine kinase activity of the receptor must be active for AP-2 association and rapid endocytosis to take place. This result is consistent with prior data demonstrating that simultaneous mutagenesis of EGF receptor autophosphorylation sites to phenylalanine reduced the rate of receptor-mediated internalization of [^{125}I]EGF.[24,25]

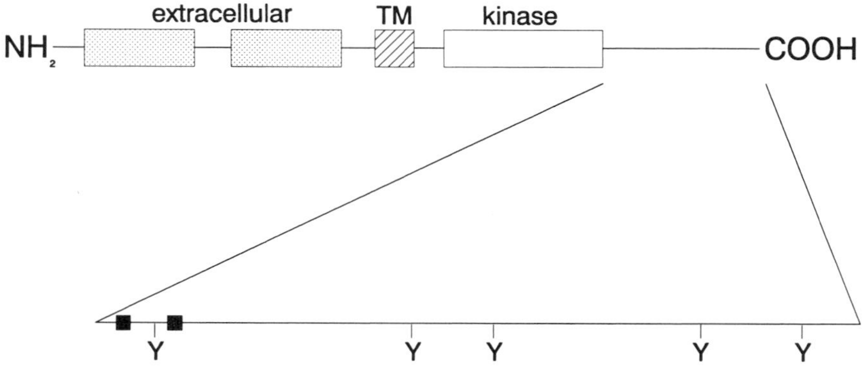

FIGURE 3. Schematic presentation of structural features within the EGF receptor carboxyterminus. The full-length receptor contains an extracellular ligand-binding domain characterized by two cysteine-rich regions (dotted boxes). A single transmembrane (TM) domain (hatched box) separates the extracellular and cytoplasmic domains. Within the cytoplasmic domain are the tyrosine kinase domain (open box) and carboxyterminal region. Expansion of the carboxyterminal regions is presented to depict tyrosine (Y) residues known to be autophosphorylated and sequences (25) that constitute internalization motifs (closed boxes).

If receptor tyrosine kinase activity is required for AP-2 association, this may indicate that receptor autophosphorylation sites are necessary or that a kinase-dependent conformational change in the receptor is required. It may, however, implicate a combination of both factors. The Dc214 truncation mutant of the EGF receptor, presented previously in FIGURE 1, fails to form an association complex with AP-2 and is not rapidly internalized.[26] Hence, the receptor carboxyterminus is required for receptor association with AP-2 as well as SH2-containing signaling proteins. Because none of the subunits of AP-2 contain an SH2 sequence motif, AP-2 association with the activated receptor likely occurs by a mechanism distinct from that employed by SH2-containing molecules. In addition to autophosphorylation sites, at least two sequences (FIG 3) that define internalization codes for the EGF receptor have been proposed to be present within the receptor carboxyterminus.[27] Internalization sequences are present, in various forms, in several different receptors that are subject to endocytosis and thought to mediate interactions between the receptors and proteins involved in endocytosis. A plausible notion for the EGF receptor is that kinase activation and perhaps autophosphorylation of tyrosyl residues in the receptor carboxyterminus alters the conformation of specific regions within the carboxyterminal domain, such that an internalization motif becomes more exposed and able to interact with AP-2. Inasmuch as other experiments have shown by antiphosphotyrosine blotting and ^{32}P-labeling that EGF does not increase the phosphorylation of AP-2 subunits (A. Sorkin, J. Baulida, and G. Carpenter, unpublished data), it does not seem likely that the association of AP-2 and the EGF receptor could be regulated by AP-2 phosphorylation in response to growth factors.

Previous experiments[28] have demonstrated that the erbB-2 receptor-like transmembrane tyrosine kinase is internalized slowly compared to the EGF receptor. This result was unexpected, as the EGF receptor and the erbB-2 molecule are very similar—approximately 50% sequence identity overall. The difference in internalization efficiency of these two molecules was attributed to differences in carboxyterminal sequences of the two molecules. We have used a chimeric EGF/erbB-2 molecule to test whether the cytoplasmic domain of erbB-2 is able to form an association complex with AP-2. The results indicate that the previously noted inefficient internalization of EGF/erbB-2 chimeric receptors is paralleled by a lack of efficient AP-2 association (J. Baulida and G. Carpenter, unpublished data).

CONCLUSION

Key features of the EGF receptor carboxyterminal domain are schematically presented in Figure 3. This domain is relatively small, approximately 25 kDa, compared to the many proteins with which it interacts. The results described above indicate that the carboxyterminus of the receptor interacts with at least two classes of intracellular proteins, each having different functions: signal transduction and protein trafficking. A major question in the future will be to determine how these interactions influence each other. For example, does AP-2 compete with SH2-containing molecules for association with the receptor carboxyterminus or can both classes of molecules simultaneously associate with the EGF receptor? This would lead to questions as to whether trafficking molecules regulate the signaling capacity of the EGF receptor, or vice versa.

ACKNOWLEDGMENT

The authors appreciate the efforts of Sue Carpenter in preparation of the manuscript.

REFERENCES

1. Carpenter, G. & S. Cohen. 1990. J. Biol. Chem. **265**(14): 7709-7712.
2. Hunter, T., N. Ling & J. A. Cooper. 1985. Nature **311**(5985): 480-483.
3. Takishima, K., I. Griswold-Prenner, T. Ingebritsen & M. R. Rosner. 1991. Proc. Natl. Acad. Sci. USA **88**(6): 2520-2524.
4. Northwood, I. C., F. A. Gonzalez, M. Wartmann, D. L. Raden & R. J. Davis. 1991. J. Biol. Chem. **266**(23): 15266-15276.
5. Walton, G. M., W. S. Chen, M. G. Rosenfeld & G. N. Gill. 1990. J. Biol. Chem. **265**(3): 1750-1754.
6. Koch, C. A., D. Anderson, M. F. Moran, C. Ellis & T. Pawson. 1991. Science **252**(5006): 668-674.
7. Carpenter, G. 1993. FORUM: Trends Exp. Clin. Med. **3**(6): 616-634.
8. Soler, C., L. Beguinot & G. Carpenter. 1994. J. Biol. Chem. **269**(16): 12320-12324.
9. Ponzetto, C., A. Bardelli, Z. Zhen, F. Marina, P. dalla Zonca, S. Giordano, A. Graziani, G. Panayotou & P. M. Comoglio. 1994. Cell **77**(2): 261-271.
10. Lowenstein, E. J., R. J. Daly, A. G. Batzer, W. Li, B. Margolis, R. Lammers, A. Ullrich, E. Y. Skolnik, D. Bar-Sagi & J. Schlessinger. Cell **70**(3): 431-443.
11. Egan, S. E., B. W. Giddings, M. W. Brooks, L. Buday, A. W. Sizeland & R. A. Weinberg. 1993. Nature **363**(6424): 45-51.

12. ROZAKIS-ADCOCK, M., R. FERNLEY, J. WADE, W. PAWSON & D. BOWTELL. 1993. Nature **363**(6424): 83-85.
13. GALE, N. W., S. KAPLAN, E. J. LOWENSTEIN, J. SCHLESSINGER & D. BAR-SAGI. 1993. Nature **363**(6424): 88-92.
14. LI, N., A. BATZER, R. DALY, V. YAJNIK, E. SKOLNIK, P. CHARDIN, D. BAR-SAGI, B. MARGOLIS & J. SCHLESSINGER. 1993. Nature **363**(6424): 85-87.
15. BUDAY, L. & J. DOWNWARD. 1993. Cell **73**(3): 611-620.
16. SOLER, C., C. V. ALVAREZ, L. BEGUINOT & G. CARPENTER. 1994. Oncogene **9**(8): 2207-2215.
17. SOLER, C., L. BEGUINOT, A. SORKIN & G. CARPENTER. 1993. J. Biol. Chem. **268**(29): 22010-22019.
18. CHEN, S. W., C. S. LAZAR, K. A. LUND, J. B. WELSH, C.-P. CHANG, G. M. WALTON, C. J. DER, H. S. WILEY, G. N. GILL & M. G. ROSENFELD. 1989. Cell **59**(1): 33-43.
19. WELLS, A., J. B. WELSH, C. S. LAZAR, H. S. WILEY, G. N. GILL & M. G. ROSENFELD. 1990. Science **247:** 962-964.
20. MASUI, H., A. WELLS, C. S. LAZAR, M. G. ROSENFELD & G. N. GILL. 1991. Cancer Res. **51:** 6170-6175.
21. SORKIN, A. & C. M. WATERS. 1993. BioEssays **15**(6): 375-382.
22. CARPENTER, G. & S. COHEN. 1976. J. Cell Biol. **71**(1): 159-171.
23. SORKIN, A. & G. CARPENTER. 1993. Science **261**(5121): 612-615.
24. SORKIN, A., C. WATERS, K. A. OVERHOLSER & G. CARPENTER. 1991. J. Biol. Chem. **266**(13): 8355-8362.
25. SORKIN, A., K. HELIN, C. M. WATERS, G. CARPENTER & L. BEGUINOT. 1992. J. Biol. Chem. **267**(12): 8672-8678.
26. SORKIN, A., T. MCKINSEY, W. SHIH, T. KINCHHAUSEN & G. CARPENTER. 1995. J. Biol. Chem. **270**(2): 619-625.
27. CHANG, C.-P., C. S. LAZAR, B. J. WALSH, M. KOMURO, J. F. COLLAWN, L. A. KUHN, J. A. TAINER, I. S. TROWBRIDGE, M. G. FARGUHAR, M. G. ROSENFELD, H. S. WILEY & G. N. GILL. 1993. J. Biol. Chem. **268**(26): 19312-19320.
28. SORKIN, A., P. P. DIFIORE & G. CARPENTER. 1993. Oncogene **8**(11): 3021-3028.

Control of T-Cell Development by the Pre-T and $\alpha\beta$ T-Cell Receptor

HARALD VON BOEHMER

Basel Institute for Immunology
Grenzacherstrasse 487
Postfach, CH-4005 Basel, Switzerland

INTRODUCTION

For many years, it had been implicated that lymphocyte development represents a crucial phase in which the immune system adapts to its environment. In particular, it was postulated that tolerance to self was established in immature lymphocytes,[1,2] and it was also hypothesized that by some mechanism the immune system adaptively differentiated such as to be able to effectively interact with antigen-presenting MHC molecules in its environment.[3–5] It was thought that tolerance might result from negative selection of certain immune cells,[1,2] involving either suppression or cell death, whereas adaptive differentiation might involve positive selection either in the form of cellular expansion of certain T-cell clones or perhaps both proliferation and somatic mutation.[3,5] Revising a discussion of our earlier paper,[3] I had proposed that positive selection actually did not only represent the bending of an emerging repertoire by expansion or suppression of some, but not other, T-cell clones[6,7] but represented an essential step in lymphocyte development that involved rescue from programmed cell death, that is, all developing lymphocytes were programmed to die unless rescued by binding of the T-cell receptor to MHC molecules in the thymus.[8]

The dilemma of the various hypotheses was that they were difficult to address experimentally, and in fact the concepts of both negative and positive selection had both supporters, as well as opponents, simply reflecting the shortcomings of the various experimental approaches. It was clear that distinction of the various proposals required a study of negative and positive selection of lymphocytes with a single receptor of known specificity. In this way, one should be able to distinguish whether or not negative or positive selection depended on other regulating cells, expressing, for instance, receptors that were antiidiotypic. One should also be able to find out about molecular mechanisms of selection.

At the time these various theories were ventilated, immunologists were jealous of scientists working with the fruitfly *Drosophila*, who because of the large number of mutations in their flies could reach unambiguous conclusions from their experiments. Immunologists, however, could hardly explain their concepts to anybody but immunologists, in part also because they had developed immunological jargon that was not understood by others. This unfavorable situation was radically changed by the possibility of working with mutant mice, and more importantly, creating mutant mice either by providing them with transgenes[9] or by changing certain genes by homologous recombination.[10] In fact, at present, gene targeting works in mammals but not in *Drosophila,* and for that reason mammalian genetics caught up.

In our initial approach to the controversial questions of negative and positive selection of immature T cells, we made use of both natural mutants as well as transgenic mice. We figured that it should be possible to produce a mouse with a single T-cell receptor (TCR) for antigen by introducing productively rearranged α and β genes into mutants that were defective in TCR gene rearrangement. In this way, we should be able to address conclusively the question of whether or not ligation of the TCR by distinct ligands was both necessary and sufficient to mediate either negative or positive selection. Although this was the original aim of our experiments, we were lucky to be confronted with some unexpected results that provided new insights into developmentally regulated processes that serve the purpose to effectively generate an immune system of cells that are induced to different effector function depending on the antigen specificity of their receptors. Our studies were complemented by a series of other experiments in mouse mutants that confirmed and extended original results and even made us change some of our initial, simple-minded interpretations. Such experiments are still ongoing to get down to the molecular details of negative and positive selection, and some of them are hotly debated as to their meaning with regard to lymphocyte selection under more physiological conditions.

In the following, I will briefly discuss the various experimental results of our own and other laboratories and, whenever appropriate, will bring out the controversial interpretation of some of the findings. This is perhaps done best by discussing lymphocyte selection in ontogenic order from the stage onwards when they begin to rearrange their TCR gene segments in the thymus.

POSITIVE SELECTION BY THE PRE–T-CELL RECEPTOR

One of the surprises that we found when we produced rearrangement-deficient mice harboring productively rearranged TCR transgenes was that TCR β genes that normally rearrange before TCR α genes already produced very significant changes in the thymus of rearrangement-deficient mice:[11–13] the few $CD4^-8^-$ $CD25^+$ cells of rearrangement-deficient mice expanded, and many cells expressed CD4 or CD8 coreceptors. These original experiments were repeated later by using, instead of the natural scid mutant, artificial RAG gene-deficient mice, with basically the same results.[14,15] Our initial observation showed that a TCR β protein alone, expressed in rearrangement-deficient mice, was sufficient to promote T-cell development, and the question was in what form could a putative pre-T-cell receptor mediate these events. Analysis of thymocytes from TCR β transgenic, rearrangement-deficient mice showed that the TCR β protein was present on the surface of thymocytes as a gp1-linked monomer[16,17] as well as a disulfide-linked complex of about 70 kDa. After initial difficulties, because of poor labeling with iodine,[12] a TCR β-associated protein of 33 kDa was identified.[18] The gene encoding this pre-T-cell receptor α chain (pTα) was recently cloned and shown to encode a type I transmembrane protein, with one extracellular Ig-like domain that shows little homology with any particular antigen-receptor chain; a cysteine just above the transmembrane part that serves to form a disulfide bridge with the TCR β chain; a transmembrane part that contains two polar

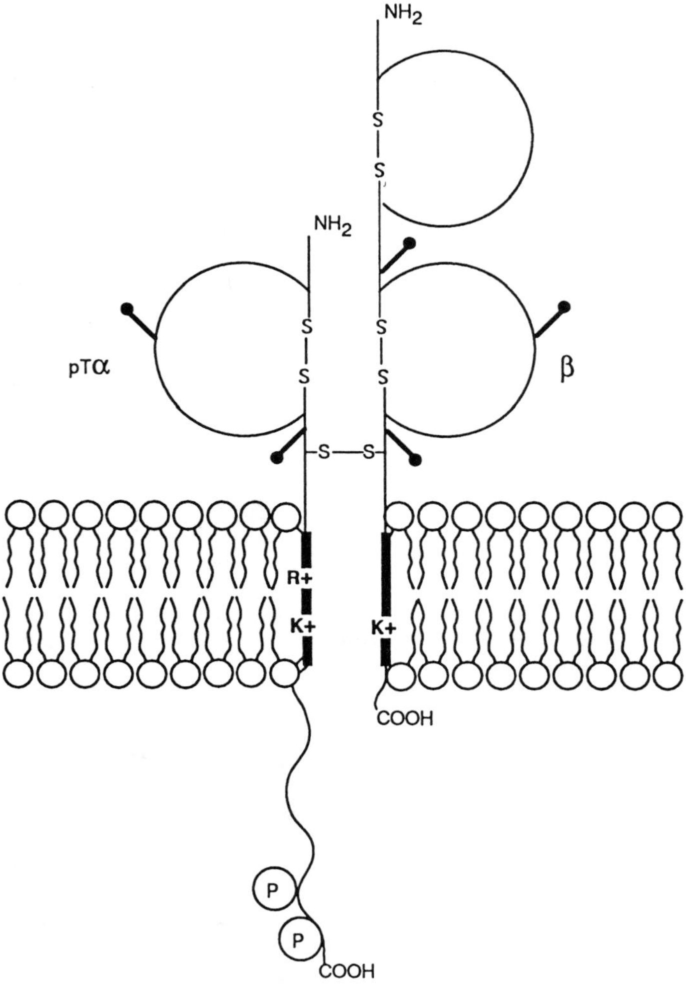

FIGURE 1. Current view of the pre-TCR. pTα = pre-T-cell receptor α chain; ʔ, glycosylation sites; P = protein kinase sites. The receptor may have a noncovalently attached V pre-T protein.

residues, just like the TCR α or TCR δ chain and that believed to be essential for proper assembly of the TCR αβ or TCR γδ complexes with CD3 proteins in mature T cells; and finally a 32 aa (amino acid)-long cytoplasmic tail with potential phosphorylation sites and motifs similar to those believed to be involved in CD2-mediated T-cell activation as well as a proline-rich sequence that could bind SH_3 domains[19] (FIG. 1). The pre-TCR complex was associated with CD3 proteins, and cross-linking resulted in Ca^{2+} mobilization.[13] In the meantime, other studies were reported that likewise supported the idea that a signal transducing pre-TCR is sufficient to promote thymocyte development by positively selecting cells that have productive TCR β rearrangements: it was shown that CD ε antibodies could induce T-cell development much like a TCR β chain in rearrangement-deficient mice,[20–22] that a dominant negative $p56^{lck}$ mutation arrested T-cell development,[23] whereas overexpression of an active $p56^{lck}$ gene promoted T-cell development in rearrangement-deficient mice.[24] These experiments suggested that CD3 molecules were important in signal transduction by the pre-TCR (and could be expressed on the cell surface in the absence of the pre-TCR) and that $p56^{lck}$ played an essential role further downstream in the signaling pathway. In fact, the cytoplasmic tail of the pTα chain may be directly involved in activation of $p56^{lck}$.[19] Experiments in TCR β-deficient mice indicated that a pre-TCR was not essential for expression of CD4 or CD8 molecules in developing thymocytes,[15] whereas experiments that analyzed the frequency of productive rearrangements in thymocyte subsets concluded that cells with productive TCR β rearrangements had a greater expansion potential than TCR β chain-negative cells.[25,26] Thus the pre-TCR may regulate survival and expansion rather than regulate directly CD4 and CD8 coreceptor expression and thereby promote expansion of cells that have productive TCR β rearrangements, mediate allelic exclusion of the TCR β locus,[27] and enhance TCR α rearrangement.[12] In this way, immature thymocytes lacking TCR β chains are discarded, probably by programmed cell death, whereas TCR β-positive cells are positively selected for further development (FIG. 2). The pre-TCR may bind to ligands on stromal cells that still need to be identified.

POSITIVE SELECTION BY THE αβ T-CELL RECEPTOR

Reconstitution of rearrangement-deficient mice with productive TCR β and α genes showed for the first time that the binding of the TCRαβ to intrathymic ligands was necessary for the development of mature T cells from immature precursors.[28] In order for this developmental transition to occur, the TCRαβ that is specific for a certain peptide presented by certain MHC molecules has to bind to these MHC molecules in the absence of the specific peptide.[29,30] In addition it was shown that as a rule positive selection yielded mature $CD4^+8^-$ or $CD4^-8^+$ cells, depending on whether the TCRαβ was class II or class I MHC-restricted, respectively.[28–33] Both the ligand that induced positive selection as well as the mechanisms that generated $CD4^-8^+$ and $CD4^+8^-$ cells have been subject of further speculation and experimentation: with respect to the ligand, it was shown that positive selection of cells with certain transgenic receptors was influenced by mutations in the floor of the MHC-binding groove and hence that peptides may be able to influence positive selection.[34,35] Further experiments in which defined peptides were supplied together with β_2 microglobulin (β_2m) to embryonic thymuses from β_2m-deficient mice or peptide-transporter (TAP)-deficient mice concluded that peptides could positively contribute to positive selection in that either peptide antagonists[36] or small amounts of the specific peptide[37] appeared to enhance positive selection. Inasmuch as these experiments were conducted under grossly unphysiological conditions (MHC densities that were two orders of magnitude below normal[36] and in some experiments the functional potential of the positively selected cells was not analyzed[37]), it remains to be seen whether a peptide is required at all (except for stabilizing MHC molecules) under physiological condi-

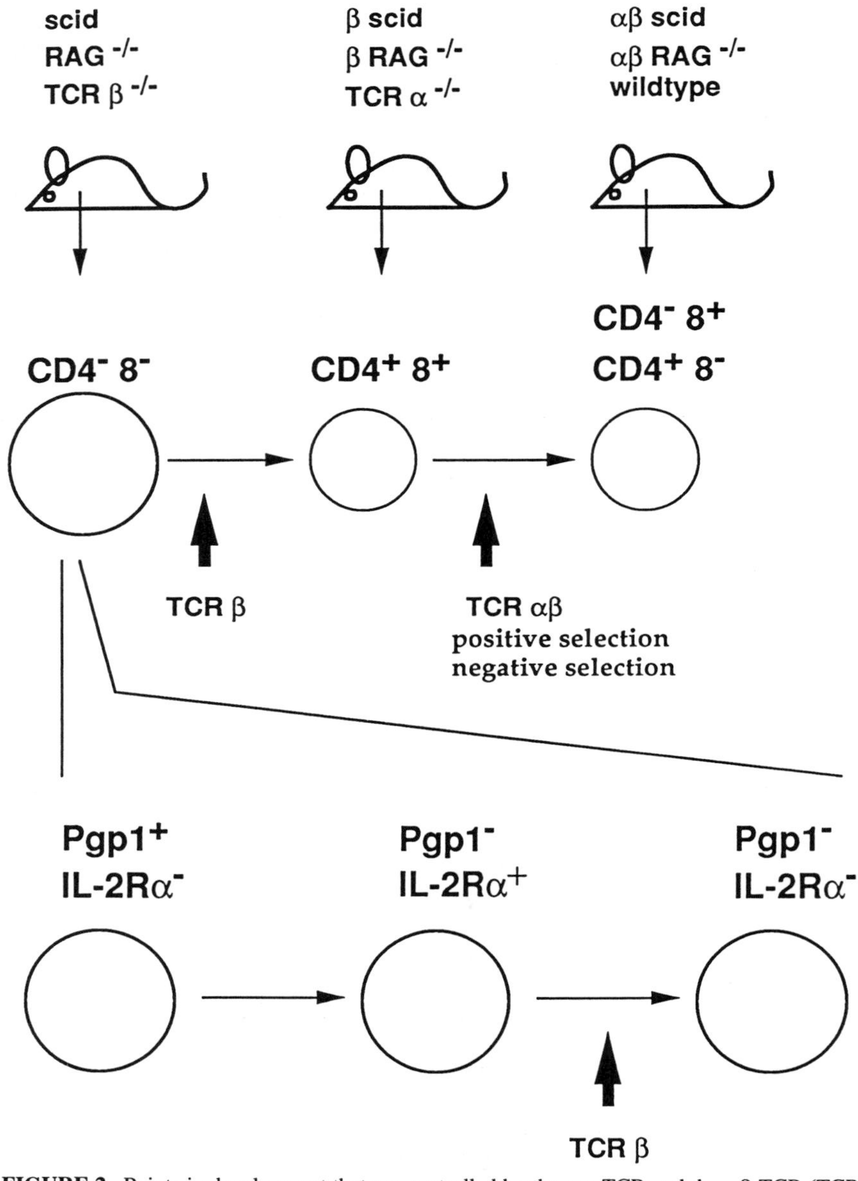

FIGURE 2. Points in development that are controlled by the pre-TCR and the αβ TCR (TCR αβ). (β SCID, β RAG = TCR β transgenic, rearrangement-deficient mice. αβ = TCR αβ transgenic mice.)

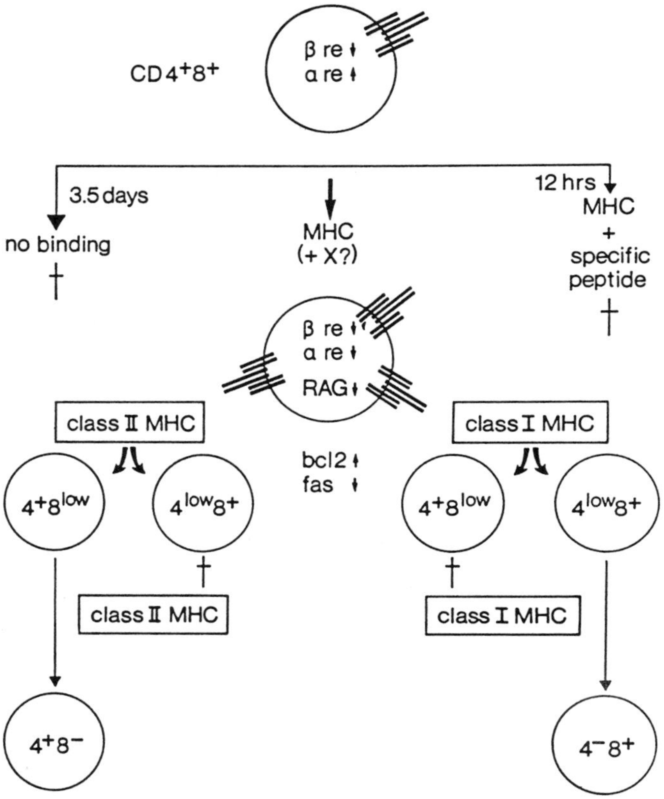

FIGURE 3. Developmental control by αβ TCR; positive and negative selection in the thymus. Binding of the αβ TCR decides whether cells die a programmed death after 3-5 days (no binding), are positively selected by 8 MHC molecules in the absence of the specific peptide, or die within 12 hours in the presence of the specific peptide. re = rearrangement.

tions and whether indeed small amounts of the specific peptide yield mature functional T cells that respond to higher doses of the same peptide. Perhaps the most significant result of these experiments is the fact that the specific peptide is not needed for positive selection. Because of the other questionable conclusions, FIG. 3 contains a question mark with regard to positively selecting ligands.

The other yet unresolved question is whether lineage commitment to either $CD4^+8^-$ or $CD4^-8^+$ cells occurs stochastically or is determined by the type of TCR-ligand interaction (FIG. 3). With regard to this question, it was found that some cells with a class II MHC-restricted receptor assume a $CD4^{low}8^+$ and cells with a class I MHC-restricted receptor assume a $CD4^+8^{low}$ phenotype.[38] It is not clear, however, whether $CD4^{low}8^+$ and $CD4^+8^{low}$ cells go on and become exclusively $CD4^-8^+$ and $CD4^+8^-$ cells, respectively. Nevertheless, CD4 and CD8 transgenes can rescue some class II

MHC-restricted CD4$^-$8$^+$ and class I MHC-restricted CD4$^+$8$^-$ cells, indicating that there is some stochastic component of CD4/CD8 lineage commitment.[39–41] It is, however, not clear whether commitment is only stochastic or whether the stochastic component reflects the failure of an instructive mechanism. If the former were true, it is astonishing that the unorthodox subsets are relatively small in the various CD4 or CD8 transgenic mice.[32,34] A more interesting variant of the stochastic-versus-instructive issue is represented by positive selection of class II MHC I-E-restricted cells. In rearrangement-deficient mice that solely expressed an I-E-restricted TCR, the majority of cells were CD4$^+$8$^-$, but there was a definite subset that was CD4$^-$8$^+$, and the latter could be induced to develop potent cytotoxic activity when recognizing I-E-presented peptides. Positive selection of the CD4$^+$8$^-$ as well as CD4$^-$8$^+$ T cells required expression of thymic class II I-E molecules, but full maturation of the latter subset also required class I MHC molecules, apparently of any sort.[42] This indicates that positive selection can proceed in two steps: an early one that requires only class II MHC molecules, and a latter one that requires either class II and class I or class I MHC molecules alone. It is interesting to note that the I-E-restricted CD4$^-$8$^+$ develop into potent cytotoxic T cells, whereas the CD4$^+$8$^-$ cells with the same specificity do not. This indicates that the functional potential of T cells correlates more closely with the CD4/8 phenotype rather than with the restriction specificity. In this context, it is interesting to note that I-E molecules have been implicated in inducing immune suppression.[43] It is likely that this suppression is mediated by I-E-restricted CD4$^-$8$^+$ T cells that either kill peptide-presenting APC and/or B cells or produce certain lymphokines that suppress immune responses. At present, one can only speculate why I-E CD4$^-$8$^+$ T cells have been detected in various transgenic mice.[32,42] Perhaps I-E-restricted TCRs show some cross-reaction on class I molecules, and/or CD8 molecules can somehow bind to I-E molecules. The former possibility has been extensively analyzed by Dr. Kajalainen (personal communication) in our experimental system, and he showed that a soluble form of the I-E-restricted TCR bound to class II but not to class I MHC molecules (unpublished results).

NEGATIVE SELECTION BY THE αβ TCR

The combined use of TCR transgenic and rearrangement-deficient mice has also established that immature thymocytes will undergo rapid apoptosis when their receptor binds to MHC plus specific peptide.[44,45] Actually, one can demonstrate *in vitro* that the same dose of antigen will induce apoptosis in immature thymocytes and cell proliferation in mature T cells.[46] In addition, it was shown that CD4$^+$8$^+$ outer cortical blasts as well as their small CD4$^+$8$^+$ descendants were susceptible to antigen-induced deletion, indicating that there was not any stage in T-cell development in which a T cell was either susceptible to deletion or positive selection only, but that simply the ligand of the TCR decides whether a cell will die a rapid apoptotic cell death or will be rescued from programmed cell death.[46,47]

CONCLUDING REMARKS

The combined use of TCR transgenic mice and various gene-deficient mice has led to the discovery of a novel pre-TCR that promotes expansion and differentiation of thymocytes that have productive TCR β but not TCRα rearrangement; it has likewise led to the discovery that positive selection by the αβ TCR represents an essential step in T-cell development, involving rescue from programmed cell death through binding of the αβ TCR to thymic MHC molecules in the absence of the specific peptide. Furthermore, genetically manipulated mice have shown that generally the receptor specificity for either class I or class II MHC molecules determines the CD4/CD8 phenotype of mature T cells, with the notable exception of I-E-restricted TCRs that can promote selection of both $CD4^+8^-$ helper and $CD4^-8^+$ killer cells with class II MHC-restricted TCRs. Finally, studies in transgenic mice have shown that binding of the TCR to MHC plus specific peptide results in deletion of large and small immature $CD4^+8^+$ thymocytes and that a given cell at a given stage in development may undergo either positive or negative selection solely depending on whether the TCR binds to MHC in the absence or presence of the specific peptide.

REFERENCES

1. BURNET, F. M. 1959. The clonal selection theory of acquired immunity. Vanderbilt University Press. Nashville, TN.
2. LEDERBERG, J. 1959. Genes and antibodies. Science **129:** 1649-1653.
3. VON BOEHMER, H., W. HAAS & N. K. JERNE. 1978. Major histocompatibility complex linked immune-responsiveness is acquired by lymphocytes of low-responder mice differentiating in thymus of high-responder mice. Proc. Natl. Acad. Sci. USA **75:** 2439-2442.
4. ZINKERNAGEL, R. M., G. N. CALLAHAN, A. ALTHAGE *et al.* 1978. On the thymus in the differentiation of "H-2 self-recognition" by T cells: Evidence for dual recognition? J. Exp. Med. **147:** 882-96.
5. BEVAN, M. J. & R. T. HÜNIG. 1978. T cells respond preferentially to antigens that are similar to self H-2. Proc. Natl. Acad. Sci. USA **78:** 1843.
6. MATZINGER, P. 1981. A one receptor view of T-cell behavior. Nature **292:** 497-501.
7. SMITH, F. I. & J. F. A. P. MILLER. 1980. Suppression of T cells specific for the nonthymic parental H-2 haplotype in thymus-grafted chimeras. J. Exp. Med. **151:** 246-251.
8. VON BOEHMER, H. 1986. The selection of the α,β heterodimeric T cell receptor for antigen. Immunol. Today **7:** 333-336.
9. PALMITER, R. D. & R. L. BRINSTER. 1986. Germ-line transformation of mice. Annu. Rev. Genet. **20:** 465-499.
10. CAPECCHI, M. R. 1989. Altering the genome by homologous recombination. Science **244:** 1288-1292.
11. VON BOEHMER, H. 1990. Developmental biology of T-cells in T cell receptor transgenic mice. Annu. Rev. Immunol. **8:** 531.
12. KISHI, H., P. BORGULYA, B. SCOTT *et al.* 1991. Surface expression of the β T cell receptor (TCR) chain in the absence of other TCR or CD3 proteins on immature T cells. EMBO Journal **11:** 93.
13. GROETTRUP, M., A. BARON, G. GRIFFITHS *et al.* 1992. T cell receptor b chain homodimers on the surface of immature but not mature a,g,d chain deficient T cell lines. EMBO J. **11:** 2735-2746.
14. SHINKAI, Y., S. KOYASU, K. NAKAYAMA *et al.* 1993. Restoration of T cell development in RAG-2-deficient mice by functional TCR transgenes. Science **259:** 822-5.
15. MOMBAERTS, P., A. R. CLARKE, M. A. RUDNICKI *et al.* 1992. Mutations in T-cell antigen receptor genes alpha and beta block thymocyte development at different stages. Nature **360:** 225-31.
16. GROETTRUP, M. & H. VON BOEHMER. 1993. T cell receptor beta chain dimers on immature thymocytes from normal mice. Eur. J. Immunol. **23:** 1393-6.

17. von Boehmer, H., M. Bonneville, I. Ishida *et al.* 1988. Early expression of a T-cell receptor β-chain transgene suppresses rearrangement of the Vγ4 gene segment. Proc. Natl. Acad. Sci. USA **85:** 9729-9732.
18. Groettrup, M., K. Ungewiss, O. Azogui *et al.* 1993. A novel disulfide-linked heterodimer on pre-T cells consists of the T cell receptor beta chain and a 33 kd glycoprotein. Cell **75:** 283-94.
19. Saint-Ruf, C., K. Ungewiss, M. Groettrup *et al.* 1994. Science **266:** 1208-1212.
20. Levelt, C. N., A. Ehrfeld & K. Eichmann. 1993. Regulation of thymocyte development through CD3. I. Timepoint of ligation of CD3 epsilon determines clonal deletion or induction of developmental program. J. Exp. Med. **177:** 707-16.
21. Levelt, C. N., P. Mombaerts, A. Iglesias *et al.* 1993. Restoration of early thymocyte differentiation in T cell receptor-β chain-deficient mutation mice by transmembrane signaling through CD3ε. Proc. Natl. Acad. Sci. USA **90:** 11401-05.
22. Jacobs, H., D. Vandeputte, L. Tolkamp *et al.* 1994. CD3 components at the surface of pro-T cells can mediate pre-T cell development *in vivo.* Eur. J. Immunol. **24:** 934-939.
23. Levin, S. D., S. J. Anderson, K. A. Forbush *et al.* 1993. A dominant-negative transgene defines a role for p56lck in thymopoiesis. EMBO J. **12:** 1671-80.
24. Mombaerts, P., S. J. Anderson, R. M. Perlmutter *et al.* 1994. An activated Ick transgene promotes thymocyte development in RAG-1 mutant mice. Immunity **1:** 261-267.
25. Mallick, C. A., E. C. Dudley, J. I. Viney *et al.* 1993. Rearrangement and diversity of T cell receptor beta chain genes in thymocytes: a critical role for the beta chain in development. Cell **73:** 513-9.
26. Dudley, E. C., H. T. Petrie, L. M. Shah *et al.* 1994. T cell receptor β chain gene rearrangement and selection during thymocyte development in adult mice. Immunity **1:** 83.
27. Uematsu, Y., S. Ryser, Z. Dembic *et al.* 1988. In transgenic mice the introduced functional T cell receptor β gene prevents expression of endogenous β genes. Cell **52:** 831.
28. Scott, B., H. Blüthmann, H. S. Teh *et al.* 1989. The generation of mature T cells requires an interaction of the aβ T cell receptor with major histocompatibility antigens. Nature **338:** 591-593.
29. Teh, H. S., P. Kisielow, B. Scott *et al.* 1988. Thymic major histocompatibility complex antigens and the alpha beta T-cell receptor determine the CD4/CD8 phenotype of T cells. Nature **335:** 229-33.
30. Kisielow, P., H. S. Teh, H. Blüthmann *et al.* 1988. Positive selection of antigen-specific T cells in thymus by restricting MHC molecules. Nature **335:** 730-3.
31. Sha, W. C., C. A. Nelson, R. D. Newberry *et al.* 1988. Selective expression of an antigen receptor on CD8-bearing T lymphocytes in transgenic mice. Nature **335:** 271-4.
32. Berg, L. J., A. M. Pullen, B. de St. Groth Fazekas *et al.* 1989. Antigen/MHC-specific T cells are preferentially exported from the thymus in the presence of their MHC ligand. Cell **58:** 1035-46.
33. Kaye, J., M. L. Hsu, M. E. Sauron *et al.* 1989. Selective development of $CD4^+$ T cells in transgenic mice expressing a class II MHC-restricted antigen receptor. Nature **341:** 746-9.
34. Jacobs, H., H. von Boehmer, C. J. Melief *et al.* 1990. Mutations in the major histocompatibility complex class I antigen-presenting groove affect both negative and positive selection of T cells. Eur. J. Immunol. **20:** 2333-7.
35. Sha, W. C., C. A. Nelson, R. D. Newberry *et al.* 1990. Positive selection of transgenic receptor-bearing thymocytes by Kb antigen is altered by Kb mutations that involve peptide binding. Proc. Natl. Acad. Sci. USA **87:** 6186-6190.
36. Hogquist, K. A., M. A. Gavin & M. J. Bevan. 1993. Positive selection of $CD8^+$ T cells induced by major histocompatibility complex binding peptides in fetal thymic organ culture. J. Exp. Med. **177:** 1469-73.

37. ASHTON-RICKARDT, P. G., A. BANDEIRA, J. R. DELANEY *et al.* 1994. Evidence for a differential avidity model of T cell selection in the thymus. Cell **76:** 651-663.
38. CHAN, S. H., D. COSGROVE, C. WALTZINGER *et al.* 1993. A new view of the selective model of thymocyte selection. Cell **73:** 225-236.
39. DAVIS, C. B., N. KILLEEN, M. E. CROOKS *et al.* 1993. Evidence for a stochastic mechanism in the differentiation of mature subsets of T lymphocytes [see comments]. Cell **73:** 237-47.
40. ITANO, A., D. KIOUSSIS & E. ROBEY. 1994. Stochastic component to development of class I major histocompatibility complex-specific T cells. Proc. Natl. Acad. Sci. USA **91:** 220-224.
41. BARON, A., K. HAFEN & H. VON BOEHMER. 1994. A human CD4 transgene rescues $CD4^-8^+$ cell in β2-microblobulin deficient mice. Eur. J. Immunol. **24:** 1933.
42. KIRBERG, J., A. BARON, S. JAKOB *et al.* 1994. Thymic selection of $CD8^+$ single positive cells with a class II MHC-restricted receptor. J. Exp. Med. **180:** 25-34.
43. WASSOM, E. L., C. J. KRCO & C. S. DAVID. 1978. I-E expression and susceptibility to parasite infection. Immunol. Today **8:** 39-43.
44. KISIELOW, P., H. BLÜTHMANN, U. D. STAERZ *et al.* 1988. Tolerance in T-cell-receptor transgenic mice involves deletion of nonmature $CD4^+8^+$ thymocytes. Nature **333:** 742-6.
45. SWAT, W., L. IGNATOWICZ, H. VON BOEHMER *et al.* 1991. Clonal deletion of immature $CD4^+8^+$ thymocytes in suspension culture by extrathymic antigen-presenting cells. Nature **351:** 150-3.
46. SWAT, W., H. VON BOEHMEER & P. KISIELOW. 1994. Central tolerance: clonal deletion or clonal arrest? Eur. J. Immunol. **24:** 485-487.
47. SWAT, W., H. VON BOEHMER & P. KISIELOW. 1994. Small $CD4^+8^+TCR^{low}$ thymocytes contain precursors of mature T cells. Eur. J. Immunol. **24:** 1010-1012.

T-Cell Recognition of Antigen

A Process Controlled by Transient Intermolecular Interactions[a]

J. JAY BONIFACE[b] AND MARK M. DAVIS[b,c]

[b]*Department of Microbiology and Immunology and*
[c]*Howard Hughes Medical Institute*
Beckman Center, Room B221
Stanford University School of Medicine
Stanford, California 94305-5428

Recognition of antigen by T cells is clearly crucial to a full understanding of the function of the immune system, yet information in this area lags considerably behind what is known about immunoglobulins. For example, only recently have affinity and kinetic measurements been possible, and complete crystal structures are not yet available for T-cell receptors. Nevertheless, the recently determined physicochemical parameters that describe the interactions occurring during T cell-mediated antigen recognition suggest distinct differences when compared to immunoglobulins. These data will be reviewed and incorporated into a model for T-cell immunity that illustrates the importance of interactions of a transient nature, as is the case in other systems of cell-cell association.

BACKGROUND

The discrimination of "self" and "nonself" allows vertebrate immune systems to recognize and defend against foreign substances such as bacteria, viruses, and transplanted tissues or grafts from nonidentical individuals. The specificity of the immune system is provided by B and T lymphocytes, which mature from common progenitor cells and produce unique but structurally similar antigen receptors. Immunoglobulins, which occur as either cell-surface or secreted molecules are produced by B cells, whereas T cells possess a T-cell antigen receptor (TCR) on their cell surface that is not secreted. T cells have both effector and regulatory functions, and the two arms of the defense are tightly coupled by the fact that most B cells will not produce immunoglobulin without T-cell help in the form of secreted lymphokines. TCRs and immunoglobulins are derived from evolutionarily related gene products and generate the diversity required to bind all potential antigens by conserved mechanisms. Despite these similarities, they differ in that immunoglobulins act primarily in the aqueous phase and are capable of recognizing intact antigenic surfaces, whereas the

[a]This work was supported by NIH Grant 5 R01 AI 22511 and the Howard Hughes Medical Institute.

TCRs bind to small fragments of antigen that are associated with major histocompatibility (MHC) molecules on the surface of antigen-presenting cells (APC) (reviewed in ref 1). The fact that T-cell recognition is dependent on cell-cell contact has interesting implications with regard to the affinity and kinetics of the antigen-specific and nonspecific events. Additionally, because specific binding involves a self-protein, the affinity of the interaction must be tightly regulated to prevent autoimmune events.

MHC molecules exist in two types. Class I MHC molecules exist on all nucleated cells and present antigenic peptides to CD8-positive cytotoxic T cells, whereas class II MHC molecules are present on specialized cells of immune function and interact with CD4-positive helper T cells (reviewed in ref. 2). X-ray crystallographic studies have now clearly demonstrated that peptides from processed antigen bind to both classes of MHC molecules in a membrane distal domain containing a tight groove formed by two adjacent antiparallel α helices that rest on a floor of β pleated sheet.[3,4] Polymorphic residues line the inside of this peptide-binding groove and form pockets that are critical for the binding of particular peptides, whereas some antigenic peptide side chains "point up" and "out" of the groove for recognition by TCRs.

In addition to the recognition of antigen/MHC complexes by the TCR during cell-cell association, other signals are produced by coreceptors and accessory molecules. CD4[5,6] and CD8[7,8] molecules on the surfaces of T cells function as coreceptors by binding in an antigen-independent fashion to class II and class I proteins, respectively, and affect signaling by recruiting the $p56^{lck}$ tyrosine kinase to the site of TCR engagement.[9,10] Thus, CD4 and CD8 may serve both to elevate the overall affinity (or avidity) of the interaction and to provide additional signals. Nevertheless, in addition to the above, maximal T-cell activation requires a second antigen-independent signal delivered by the APC. The second signal is delivered, at least in part, by the interaction of CD28 on the surface of T cells with the B71 or B72 molecules on the APC.[11]

Finally, cell-adhesion molecules (CAMS) are also involved in the association of T cells with APCs (reviewed in ref 12). Some of these CAMs are T-cell specific with expression levels that vary depending on the activation or developmental state of the cell, whereas others appear to be expressed more ubiquitously (reviewed in ref 13). Although CAMs clearly are important for T cell-mediated immune recognition by increasing the overall avidity of cell-cell contact, their role in signaling T cells is not yet understood.

Antigen recognition by T cells is thus a complicated interaction of two cells that involves the interplay of antigen-specific and nonspecific events. Although many of these interactions are qualitatively understood, our understanding of their affinity and kinetics and how these parameters relate to T-cell activation is still in an embryonic state. This paper will discuss our current understanding of the physicochemical properties of TCR/peptide-MHC interaction and how these data are in line with a model for antigen recognition that occurs by cell-cell association. A cartoon illustrating T cell-mediated recognition of antigen is shown in FIGURE 1.

EXPRESSION OF SOLUBLE PROTEINS

Obtaining affinity and kinetic measurements for TCR/MHC interactions has been hampered because of the complexity of the trimolecular complex, the fact that both

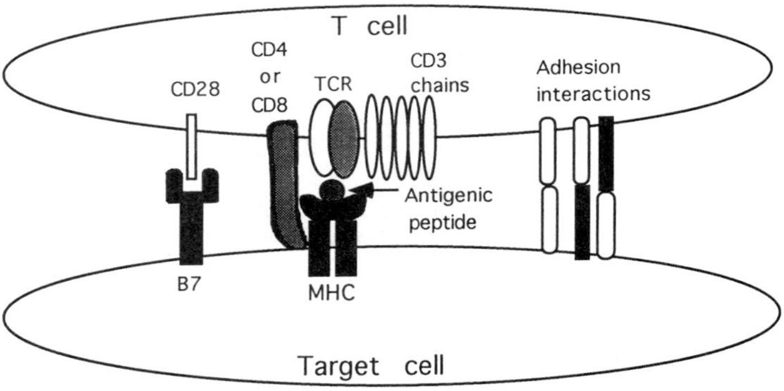

FIGURE 1. T cell-mediated recognition of antigen. Shown is a cartoon illustrating some of the interactions important for T-cell specific recognition of antigen. The antigen-specific event is the binding of the TCR to the antigenic peptide/MHC complex. Coreceptors include the CD4 or CD8 molecule, whereas the CD3 chains are required to transduce signals. Costimulation can be provided by the interaction of B7 and CD28, and adhesion interactions of many forms help to stabilize the conjugate and provide additional signals. Here a "target cell" refers to any cell that can be specifically engaged by a T cell.

proteins occur naturally as integral membrane proteins and interact with a very low affinity. In our system IE^k, a murine class II MHC molecule,[14] and a TCR[15] that binds IE^k when associated with the C-terminal fragment of moth cytochrome c peptide (MCC) are expressed as glycosylphosphoinositide-anchored molecules that are recovered from CHO transfectants by digestion with phospholipase c. IE^k has also been expressed in *E. coli* and refolded from inclusion bodies.[16] Similar results have been obtained with both GPI-IE^k and *E. coli* IE^k, but this discussion will be limited only to GPI-IE^k (J. Altman, unpublished observation). In all cases, soluble proteins are purified on immunoaffinty columns. IE^k loaded with specific antigenic peptides is prepared by coincubation at pH 5.0 for 3 days and subsequent FPLC gel filtration.[17]

EARLY AFFINITY ESTIMATES FOR MHC/TCR INTERACTIONS

The first estimates of the affinity for TCR binding to an antigen/MHC complex derived from a competitive binding strategy. An affinity of 40–60 μM for the interaction of MCC/IE^k and the 2B4 TCR was obtained by determining the concentration of soluble MCC/IE^k required to inhibit the binding of a labeled anti-TCR Fab to T cells.[17] In related studies with a flu peptide, I-E^d-restricted T cell, Weber *et el.*[18] obtained an affinity estimate of 10 μM by inhibiting a T-cell response with a soluble TCR-immunoglobulin chimera. Working with a class I-restricted alloreactive T cell, 2C, and using an antibody-competition scheme similar to Matsui *et al.*,[17] Sykulev *et al.*[19] found affinities ranging from 0.1 to 100 μM, depending on which peptide analogue was loaded into the MHC. From these data, TCR affinities for peptide/MHC complexes seemed generally low when compared to that of most immunoglobulins, and these observations made kinetic analysis an interesting pursuit.

KINETICS OF MHC/TCR INTERACTIONS

Although knowledge of the affinity of an interaction can be helpful in understanding its role, a complete description requires the determination of the kinetic rate constants. Additionally, the rates at which a reaction occurs often shed more light on the biological relevance of a reaction because the relationship, in time, between the event and previous or subsequent events is then established. The kinetic rate constants for the interaction of the 2B4 TCR and MCC/IE^k have been determined by surface plasmon resonance, using the BIAcore™ instrument by Pharmacia Biosensor. The 2B4 TCR was immobilized onto the dextran matrix of the biosensor by standard amine-directed chemistry, and binding was monitored in real time to soluble MCC/IE^k complexes.[20] By this approach, Matsui *et al.*[20] confirmed the affinity estimate of 40–60 μM for the 2B4 TCR for its MCC/IE^k ligand in a completely cell-free system. Additionally, reaction rates were estimated at 1,000 M^{-1} s^{-1} and 0.05 s^{-1}, for the association and dissociation rate constants, respectively. Therefore, the low affinity of this interaction seems characterized by a very slow on-rate and a rapid off-rate ($t_{1/2}$ of 12 seconds). By comparison, good antibodies will often bind antigen with on-rates of 50,000 to 250,000 $M^{-1}s^{-1}$ and with off-rates that are 10- to 100-fold slower than 0.05 s^{-1}.

The fast off-rate for this MHC/TCR pair indicates that any single interaction will have a lifetime of less than 1 minute. As discussed in more detail below, this may reflect the fact that the recognition event involves cell-cell contact that must remain reversible, inasmuch as T cells are not consumed during an immune response but rather engage many cells. The very slow on-rate of this interaction could indicate that the reaction is somehow intrinsically limited. The on-rate is so much slower than the diffusion-controlled reaction rate that it cannot be explained simply by specific orientation requirements. One possible explanation is that a conformational change in one or both proteins limits the association event. In support of this possibility, the binding of the 2B4 TCR to MCC/IE^k has been measured by near-UV circular dichroism (D. Lyons, unpublished observation). Preliminary data indicate that this assay may be monitoring a conformational change occurring as result of binding. Inasmuch as a conformation change in the TCR could affect its interaction with a coreceptor (CD4) or a signaling element, this is currently a focus of investigation.

Further kinetic analysis has been performed for the binding of the 2B4 TCR to MCC peptide variants complexed with IE^k. Pigeon cytochrome c peptide (PCC), which has an alanine insertion at the penultimate residue, and MCC(102S), containing a Thr to Ser substitution, have similar (50–100 μM) affinities for IE^k but are approximately 4- and 500-fold poorer, respectively, at stimulating 2B4 T cells when incubated with APCs.[20] Interestingly, all three complexes (MCC/IE^k, PCC/IE^k, and MCC(102S)/IE^k) were found to have similar affinities for the 2B4 TCR by the competitive binding strategy.[20] Therefore, neither the affinity of the peptides for IE^k nor of the peptide/IE^k complexes for the TCR can explain their range of activities in the T-cell activation assay. However, kinetic analysis with these complexes revealed that their off-rate correlated with their differences in activity in the bioassay. Although similar in affinity, MCC/IE^k, PCC/IE^k, and MCC(102S)/IE^k dissociate from the TCR with apparent off-rates of approximately 0.05, 0.09, and 0.1–0.3 s^{-1}, respectively.[20] This corresponds to half-lives that range from about 12 seconds to as short as 2 seconds.

The marked differences in T-cell activation (2 to 3 orders of magnitude) across a narrow range of off-rates may indicate that the stability of the complex is a key determinant of T-cell responsiveness and that an important event in the generation of a signal from the TCR may occur on a time scale of a few seconds postengagement. Finally, these kinetic data may provide an explanation for T-cell antagonism, which results in a diminished (or different) T-cell response when some antigenic peptides with altered T-cell contacts are coincubated with wild-type peptide in T-cell activation assays.[21,22] Altered peptide/MHC molecules could signal T cells poorly due to off-rates that have dropped below some critical thresholds, but compete effectively with wild-type peptide/MHC complexes for TCR because of similar affinities. In some cases antagonistic peptides are required at large excess over wild-type peptides. These peptide/MHC complexes could have substantially lower affinities for TCR, a product of both a slower on-rate and the rapid off-rate that diminishes T-cell signaling.

KINETIC MEASUREMENTS OF OTHER PEPTIDE/MHC/TCR COMBINATIONS

A number of other laboratories have estimated the kinetic rate constants for the interactions of peptide/MHC complexes with TCRs. By an indirect approach, Sykulev *et al.*[19] have determined an on-rate of 11,000 M^{-1} s^{-1} and an off rate of 0.005 s^{-1} for the interactions of a TCR with a complex of the class I L^d molecule and a peptide (termed p2Ca). The on-rate was found to increase 5-fold when a different peptide was used that is more efficient at stimulating T cells.[23] Corr *et al.*,[24] using the BIAcore technology, reported a faster on-rate (210,000 $M^{-1}s^{-1}$) and a faster off-rate (0.026 s^{-1}) for the same interaction discussed above (p2Ca/L^d). A number of other affinities and rate constants are summarized in TABLE 1.

ADHESION INTERACTIONS

The interactions discussed above are generally of low-affinity and characterized by rapid dissociation rates. This kinetic behavior is not limited to TCR-antigen/MHC interactions but is also common among many adhesion interactions. These similarities suggest that transient interactions are a characteristic of cell-cell association, particularly when the cells are not fixed in a tissue for their lifetime, and even when the interaction is one of immune recognition and requires a high degree of specificity. A notable example of a transient adhesion interaction is the binding of CD2 to CD48 and CD58, which occurs with an extremely rapid dissociation rate of 6 s^{-1}. van der Merwe *et al.*[25] have pointed out that a parallel exists between the affinity of a number of CAM interactions and the cell type. Those cells requiring more transient contact contain CAMS of lower affinity. Of course the affinity and kinetics of these interactions will not be the only factors that govern the strength and duration of cell-cell contact, inasmuch as it is known that the overall avidity can be regulated by the cell for at least some CAM interactions.[13] Other mechanisms, such as controlled proteolytic shedding (reviewed in ref. 26), exist to dissociate cells.

TABLE 1. Affinity and Kinetics of Protein-Protein Interactions Occurring during T cell–Mediated Immune Recognition

Interacting Proteins: T-cell "receptor"	Interacting Proteins: Ligand	Parameters: K_D (μM)	Parameters: k_{on} ($M^{-1}s^{-1}$)	Parameters: k_{off} (s^{-1})	T°C	Ref.
2B4 TCR	MCC/IE^k	40–60	1,000	0.05	25	17,20
2B4 TCR	PCC/IE^k	50	1,700	0.09	25	20
2B4 TCR	102S/IE^k	~100		0.1–0.3	25	20,27
2B4 TCR	IE^k only	~430			25	27
14.3.dTCR	HM/IE^d	~10			37	18
2C TCR	p2CL/L^d	0.1	210,000	0.026	25	24
2C TCR	p2CL/L^d	0.5	11,000	0.0055	25	19
2C TCR	p2CL^a/L^d	0.5–70			25	19
2C TCR	p2CL/K^b	330			25	19
2C TCR	QL9/L^d	0.06–0.07	53,000	0.003	25	23
		0.17	90,000	0.015	37	23
4G3 TCR	pOV/K^b	0.6–0.7	22,000	0.02	25	23
HA1.7 TCR	HA/DR1	>25				34
HA1.7 TCR	SEB^b	0.82	13,000	0.001		34
SEB^b	HA/DR1	1.7				34
CD4	MHC	>100			25	28
CD4	MHC	3.2				29
CD2	LFA-3	0.4			4	30
CD2	LFA-3	100–1000				31
CD2	LFA-3	9–22	≥400,000	≥4.0	37	32
CD2	CD48	60–90	≥100,000	≥6.0	37	26
CD28	B7	0.2			23	11
LFA-1	ICAM-1	≤90^c				33

[a] The authors report an affinity range for different p2CL peptide variants.
[b] SEB is a staphylococcal enterotoxin.
[c] Estimated to be around 90 μM on resting T cells and of higher affinity on activated T cells.

SUMMARY

As recently as ten years ago, the nature of the T-cell receptor for antigen was a mystery, as was the precise role of histocompatibility molecules in antigen-presentation to T cells. Although T-cell receptors have now been cloned and crystal structures of MHC/peptide molecules exist, our understanding of the parameters that characterize this interaction and other interactions relevant to T-cell immunity are still unclear. The engineering of soluble forms of proteins that mediate T-cell recognition of antigen has allowed the first measurements of these parameters. Interestingly, many of these interactions are of a transient nature, with very rapid off-rates. These data suggest a model whereby highly reversible intermolecular interactions mediate the cell-cell association. The association of adhesion molecules is probably the first step in the stabilization of a conjugate, because they are more numerous than any antigen-specific interaction, followed later by TCR-MHC engagements.[17,20] Diffusion within each lipid bilayer should allow the congregation of MHC/TCR interactions at the cell-cell interface, with peptide-specific TCR interactions outcompeting irrelevant interactions. Rapid off-rates for both the antigen-specific and nonspecific interactions may be necessary to maintain reversibility, yet allow a rapid approach to equilibrium and consequent signaling when a specific antigen is present or disengagement when it is not.

ACKNOWLEDGMENTS

We thank Dan Lyons and John Altman for contributing unpublished data and Brenda Robertson for expert clerical assistance. J. J. Boniface was a fellow of the Irvington Institute of Medical Research.

REFERENCES

1. Unanue, E. R. & P. M. Allen. 1987. Science **236:** 551-556.
2. Yewdell, J. W. & J. R. Bennink. 1990. Cell **62:** 203-209.
3. Madden, D. R., J. C. Gorga, J. L. Strominger & D. C. Wiley. 1991. Nature **353:** 321-325.
4. Brown, J. H., T. S. Jardetzky, J. C. Gorga, L. J. Stern, R. G. Urban, J. L. Strominger & D. C. Wiley. 1993. Nature **364:** 33-39.
5. Konig, R., L.-Y. Huang & R. N. Germain. 1992. Nature **356:** 796-798.
6. Cammarota, G., A. Scheirle, B. Takacs, D. M. Doran, R. Knorr, W. Bannwarth, J. Guardiola & F. Sinigaglia. 1992. Nature **356:** 799-802.
7. Salter, R. D., R. J. Benjamin, P. K. Wesley, S. E. Buxton, T. P. Garrett, C. Clayberger, A. M. Krensky, A. M. Norment, D. R. Littman & P. Parham. 1990. Nature **345:** 41-46.
8. Potter, T. A., T. V. Rajan, R. F. Dick & J. A. Bluestone. 1989. Nature **337:** 73-76.
9. Rudd, C. E., J. M. Trevillyan, J. D. Dasgupta, L. L. Wong & S. F. Schlossman. 1988. Proc. Natl. Acad. Sci. USA **85:** 5190-5194.
10. Veillette, A., M. A. Bookman, E. M. Horak & J. B. Bolen. 1988. Nature **338:** 257-260.
11. Linsley, P. S., W. Brady, L. Grosmaire, A. Aruffo, N. K. Damle & J. A. Ledbetter. 1991. J. Exp. Med. **173:** 721-730.
12. Singer, S. J. 1992. Science **255:** 1671-1677.
13. Diamond, M. S. & T. A. Springer. 1994. Curr. Biology **4:** 506-516.
14. Wettstein, D. A., J. J. Boniface, P. A. Reay, H.-J. Schild & M. M. Davis. 1991. J. Exp. Med. **174:** 219-228.
15. Lin, A. Y., B. Devaux, A. Green, C. G. Sagerstrom, J. F. Elliot & M. M. Davis. 1990. Science **249:** 677-681.
16. Altman, J. A., P. A. Reay & M. M. Davis. 1993. Proc. Natl. Acad. Sci. USA **90:** 10330-10334.
17. Matsui, K., J. J. Boniface, P. A. Reay, H. Schild, B. Fazekas de St. Groth & M. M. Davis. 1991. Science **254:** 1788-1791.
18. Weber, S., A. Traunecker, F. Oliveri, W. Gerhard & K. Karjalainen. 1992. Nature **356:** 793-796.
19. Sykulev, Y., A. Brunmark, M. Jackson, R. J. Cohen, P. A. Peterson & H. N. Eisen. 1994. Immunity **1:** 15-22.

20. Matsui, K., J. J. Boniface, P. Steffner, P. A. Rey & M. M. Davis. 1994. Proc. Natl. Acad. Sci. USA **91:** 12862-12866.
21. Evavold, B. D., J. Sloan-Lancaster & P. M. Allen. 1993. Immunol. Today **14:** 602-609.
22. De Magistris, M. T., J. Alexander, M. Coggeshall, A. Altman, F. C. A. Gaeta, H. M. Grey & A. Sette. 1992. Cell **68:** 625-629.
23. Sykulev, Y., A. Brunmark, T. J. Tsomides, S. Kageyama, M. Jackson, P. A. Peterson & H. N. Eisen. 1994. Proc. Natl. Acad. Sci. USA **91:** 11487-11491.
24. Corr, M., A. E. Slanetz, L. F. Boyd, M. T. Jelonek, S. Khilko, B. K. Al-Ramadi, Y. S. Kim, S. E. Maher, A. L. M. Bothwell & D. H. Margulies. 1994. Science **265:** 946-949.
25. Van der Merwe, P. A. & A. N. Barclay. 1994. Trends Biol. Sci. **19:** 354-358.
26. Van der Merwe, P. A., M. H. Brown, S. J. Davis & A. N. Barclay. 1993. EMBO J. **12:** 4945-4954.
27. D. Lyons. In preparation.
28. Weber, S. & K. Karjalainen. 1993. Int. Immunol. **5:** 695-698.
29. Cammarota, G., A. Scheirle, B. Takacs, D. M. Doran, R. Knorr, W. Bannwarth, J. Guardiola & F. Sinigaglia. 1992. Nature **356:** 799-801.
30. Sayre, P. H., R. E. Hussey, H.-C. Chang, T. L. Ciardelli & E. L. Reinherz. 1989. J. Exp. Med. **169:** 995-1009.
31. Dustin, M. L., D. Olive & T. A. Springer. 1989. J. Exp. Med. **169:** 503-509.
32. Van der Merwe, P. A., A. N. Barclay, D. W. Mason, E. A. Davies, B. P. Morgan, M. Tone, A. K. Krishnam, C. Ianelli & S. J. Davis. 1994. Biochemistry **33:** 10149-10153.
33. Lollo, B. A., K. W. H. Chan, E. M. Hanson, V. T. Moy & A. A. Brian. 1993. J. Biol. Chem. **268:** 21693-21700.
34. Seth, A., L. J. Stern, T. H. M. Ottenhoff, I. Engel, M. J. Owen, J. R. Lamb, R. D. Klausner & D. C. Wiley. 1994. Nature **369:** 324-327.

Receptors That Regulate T-Cell Susceptibility to Apoptotic Cell Death

LAWRENCE H. BOISE,[a,b] ANDY J. MINN,[a,c] AND CRAIG B. THOMPSON[a–f]

[a]*The Gwen Knapp Center for Lupus and Immunology Research*

[b]*Department of Medicine*

[c]*The Committee on Immunology*

[d]*Department of Molecular Genetics and Cell Biology*

[e]*Howard Hughes Medical Institute*
The University of Chicago
Chicago, Illinois 60637-5420

THE MAJORITY OF PERIPHERAL BLOOD LYMPHOCYTES ARE IN A RESTING OR QUIESCENT STATE

Resting T cells circulate between the various lymphoid compartments and are only induced to leave their quiescent state by signal transduction through the T-cell receptor/CD3 complex. T-cell receptors can be activated by binding a peptide antigen presented in the context of a self-encoded MHC molecule on the surface of an antigen-presenting cell (FIG. 1). T-cell receptor signal transduction induces the cell to enter the G1 stage of the cell cycle where it becomes competent to respond to the cell-cycle progression effects of lymphokines.[1] Perhaps the best characterized of these lymphokines is IL-2.[2] IL-2 stimulation of the IL-2Rαβγ on an antigen-activated T cell is sufficient to drive the T cell through all the subsequent phases of the cell cycle leading to clonal expansion of the cell and the acquisition of effector functions.

Although a great deal has been learned about the molecular events associated with T-cell activation and cell-cycle progression, increasing evidence suggests that there is an equally complex set of events that regulate the survival of T cells at various stages within their life cycle. Regulation of cell survival is critical to the maintenance of the homeostasis of the immune system. Over time, the total number of lymphocytes in an individual is held relatively constant despite continuous expansion and contraction of individual T-cell subsets as they respond to antigenic challenge.

Recently, it has become clear that cell survival is regulated by factors that control a central death effector system present in all complex eukaryotic cells. The genes involved in this death-effector system leave a cell capable of controlling its own death

[f]Corresponding author: Craig B. Thompson, M.D., The University of Chicago, Gwen Knapp Center, 924 E. 57th Street, Rm. R413A, Chicago, IL 60637-5420.

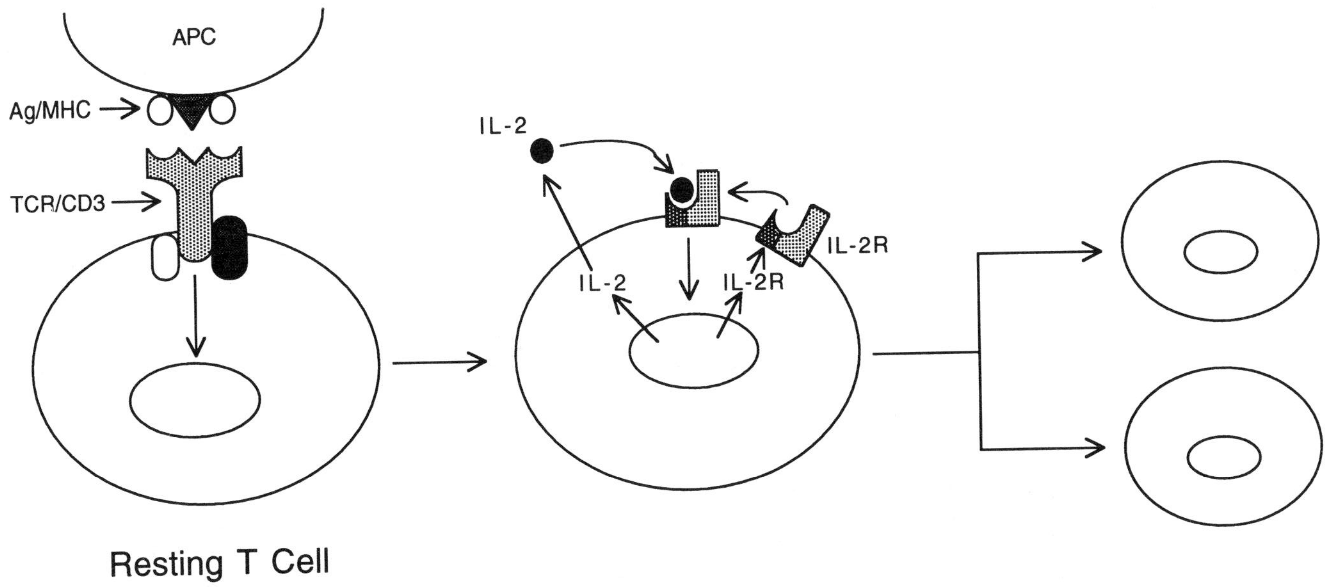

FIGURE 1. Activation of T cells to a proliferative state requires two steps. Resting T cells are activated by the engagement of their T-cell receptor/CD3 complex (TCR/CD3) by peptide antigen-bearing MHC molecules (Ag/MHC) present on antigen-presenting cells (APC). This signal will activate many molecular changes within the cell, including expression of high-affinity IL-2 receptor complexes on the surface of the cell, as well as production of the growth progression factor IL-2. Engagement of IL-2 receptors on antigen-activated T cells is sufficient to drive the cell through the rest of the cell cycle.

through a form of cellular suicide, termed apoptosis.[3] Cells can undergo apoptosis in response to a wide variety of both external and internal stimuli. The ability of cells to be eliminated by the process of apoptosis is particularly important in the immune system.[3–6] Apoptosis involves a set of classical morphological changes, including blebbing of cytoplasmic membrane, destruction of the cytoskeleton, proteolytic degradation of cytoplasmic proteins, and endonucleolytic cleavage of nuclear DNA. These events result in the autodigestion of the cell. Despite these extensive internal changes, the plasma membrane remains intact, and molecules are expressed on the surface of the plasma membrane to signal neighboring phagocytic cells to rapidly eliminate the dying cell. This highly regulated process results in the elimination of unwanted or unneeded cells without the induction of an inflammatory response. By contrast, necrotic cell death, which results in rapid cellular swelling and leakage of cytoplasmic contents, is a proinflammatory signal. Recent evidence suggests that the survival of resting T cells is regulated at no fewer than three distinct phases of their life cycle. Resting T cells have the ability to survive at multiple locations within the body for prolonged periods of time.[7] Recently, the ability of lymphocytes to survive in a quiescent state has been shown to be dependent on the expression of the *bcl-2* gene.[8,9] Bcl-2 has been shown to be a critical regulator of cellular apoptosis.[10,11] The level of expression of *bcl-2* seems to either directly or indirectly determine the apoptotic threshold of a cell. High-level expression of *bcl-2* seems to render cells relatively resistant to the requirement for continuous stimulation by survival signals provided by growth factors, cell-cell contact, or contact with the extracellular matrix.[12–14] For T cells to acquire the effector function in regulating cell-mediated immunity, they must be able to migrate to and survive at sites of inflammation. This led us to compare the susceptibility of resting T cell and T cells activated through cross-linking of the T-cell receptor/CD3 complex for their susceptibility to undergoing apoptosis in response to extracellular signals.

ACTIVATED T CELLS DISPLAY ENHANCED RESISTANCE TO RADIATION-INDUCED APOPTOSIS

As an initial experiment, the susceptibility of resting peripheral blood T cells and T cells stimulated by cross-linking the T-cell receptor/CD3 complex with immobilized anti-CD3 monoclonal antibody to undergo apoptosis in response to irradiation was examined. Highly purified human peripheral blood T cells were subjected to 1500 rads of γ-irradiation. This is a dose of irradiation sufficient to cause significant DNA damage and preclude further cell proliferation. Therefore, we can measure the percent viability in the cultures over time as a direct indication of the intrinsic susceptibility of cells to undergo apoptosis in response to irradiation. The results of these experiments demonstrate that there is a consistent and significantly enhanced rate of survival of activated T cells compared to resting peripheral blood T cells following treatment with γ-irradiation (FIG. 2). For example, four days after treatment with γ-irradiation, 30% of anti-CD3-stimulated cells survived, whereas less than 1% of resting peripheral blood T cells were still alive at this same point. Similar data have been obtained with other apoptotic stimuli (data not shown). Together these data suggest that in preparation for mounting an immune response, activated T cells can upregulate their resistance to programmed cell death.

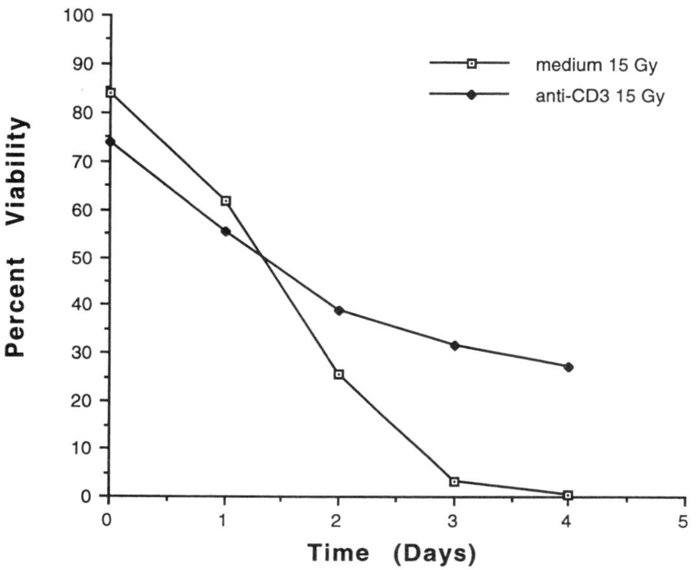

FIGURE 2. Anti-CD3-induced activation can enhance T-cell survival. Purified T cells were treated with medium alone (medium) or plate-adhered anti-CD3 (anti-CD3) for 12 h prior to treatment with 1500 rad γ irradiation from a Cs source irradiator. Viability was assessed by propidium iodide exclusion at the indicated time points following irradiation. The data points are the means of two independent determinations and are representative of five independent experiments.

ACTIVATION OF PERIPHERAL BLOOD T CELLS RESULTS IN THE INDUCTION OF BCL-X_L GENE EXPRESSION

Recently, we cloned a gene related to *bcl-2, bcl-x.*[15] Although *bcl-x* has been shown to be capable of encoding several proteins with distinct biological function,[15,16] Bcl-x_L is encoded by the largest open-reading frame in the *bcl-x* gene, encompassing 233 amino acids. Bcl-x_L is at least as potent as Bcl-2 in increasing a cell's resistance to undergoing apoptotic cell death.[15–18] By contrast, an alternatively spliced form of *bcl-x* created by usage of an alternative 5′ splice donor site within the first coding exon is a protein product that can oppose the function of either Bcl-2 or Bcl-x_L in cells (ref. 15 and A. Minn, unpublished observations). In view of the marked difference in the ability between resting and activated T cells to undergo irradiation-induced programmed cell death, we investigated the expression of Bcl-2 and Bcl-x in resting T cells and in T cells activated for various periods by cross-linking of the T-cell receptor/CD3 complex.

Consistent with the data that *bcl-2* is a required gene for the prolonged survival of resting T cells in various cellular compartments,[8,9] Bcl-2 protein was found to be expressed at high levels even in quiescent T cells (FIG. 3). In fact, Bcl-2 was found

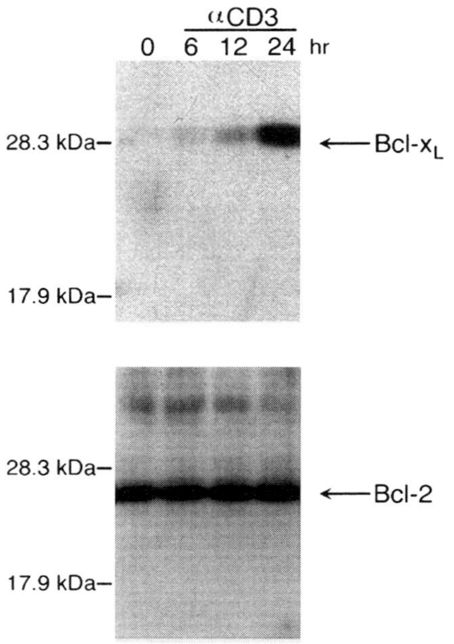

FIGURE 3. Bcl-x$_L$ and Bcl-2 expression following activation of T cells with anti-CD3. T cells were activated for the indicated times (in hours) with immobilized anti-CD3 (αCD3). Cells were harvested at the indicated time points and the resulting lysates were split into two equal portions. Each portion was then immunoprecipitated and immunoblotted with antibodies against Bcl-x (upper panel) or Bcl-2 (lower panel). Only the 29 kDa Bcl-x$_L$ form of the *bcl-x* gene products was detected in these experiments, despite the fact that the antibodies used can recognize other forms of the protein (data not shown). The data are representative of two independent experiments.

to be constitutively expressed in peripheral blood T cells despite the fact that *bcl-2* mRNA has been shown to undergo significant changes in its level of expression during T-cell activation response.[19] By contrast, Bcl-x$_L$ proteins were found to be absent from resting peripheral blood T cells; however, there was a dramatic induction of Bcl-x$_L$ protein within the first 24 hours after T-cell activation by cross-linking of the T-cell receptor/CD3 complex. These data suggest that the enhanced ability of activated T cells to resist programmed cell death is at least in part due to the enhanced expression of Bcl-x$_L$, which results in a higher apoptotic threshold for an activated cell. These data also provide one important reason for the existence for two gene products with relatively similar biological function in peripheral blood T cells. Bcl-2 appears to be a constitutively expressed protein that sets a stable level of resistance to physiologic cell death. By contrast, Bcl-x expression can be dramatically modulated in response to cellular activation, and this provides dynamic control of the apoptotic threshold in response to extracellular events.

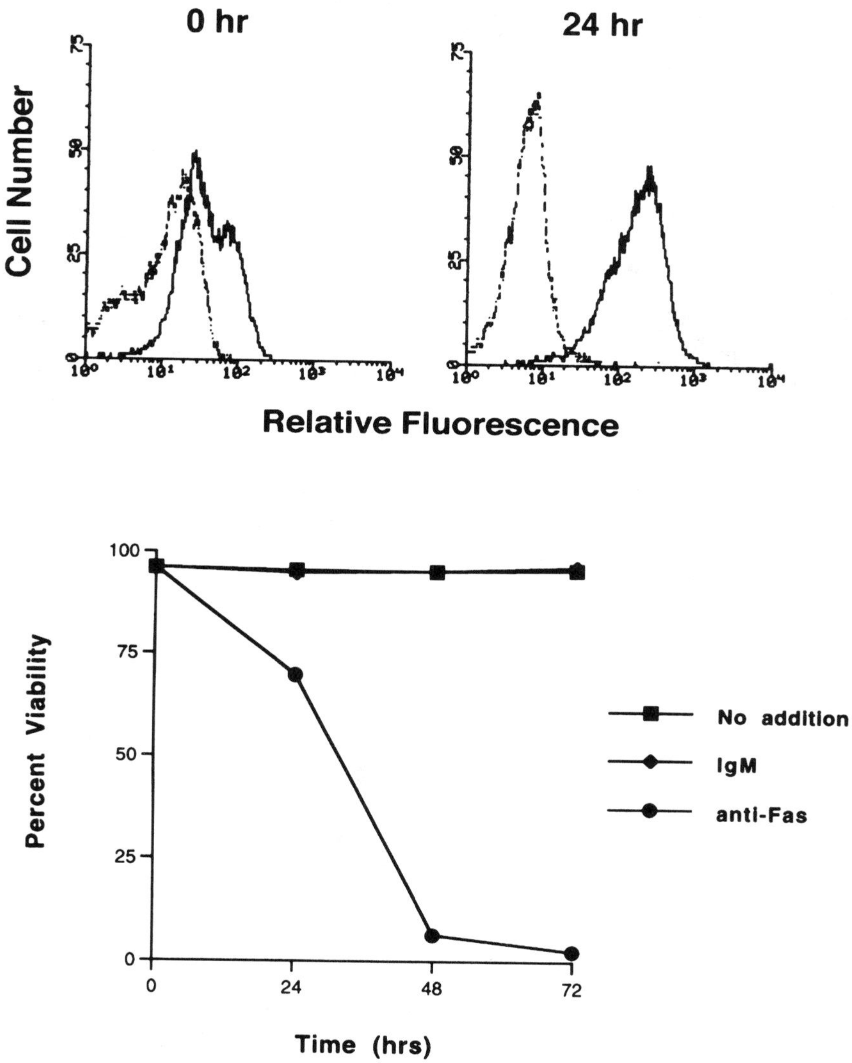

FIGURE 4. Fas expression is up-regulated by anti-CD3 activation T cells and can result in elimination of previously activated T cells. Upper panels: Purified T cells were treated for 0 (left panel) or 24 hours (right panel) with immobilized anti-CD3, and surface expression of Fas was determined by flow cytometry. Following stimulation, cells were incubated with an antibody to Fas (solid line) or an isotype-matched control antibody (broken line). Surface staining was analyzed with Lysis II software. Lower panel: The Jurkat T-cell line, which constitutively expresses Fas on its surface, was treated with an antibody to Fas or isotype control at 10 ng/mL for the time points indicated in the FIGURE. At each point, cells were removed from the culture and viability assessed by propidium iodide exclusion.

LYMPHOKINES REGULATE T-CELL SURVIVAL WITHOUT AFFECTING THE EXPRESSION OF BCL-2 AND BCL-X

T-cell survival at sites of inflammatory lesions is dependent not only on the intrinsic cell survival genes, *bcl-2* and *bcl-x,* but also on the expression of progression growth factors, such as IL-2. Lymphokines have also been shown to be important in maintaining T-cell survival by maintaining their continuous proliferation following antigen activation.[20–22] However, an important distinction can be made concerning the effects of growth factors, such as IL-2, on activated T-cell survival when compared to cell survival genes, such as *bcl-2* or *bcl-*x_L. For growth factors, cell survival goes hand in hand with the proliferative expansion of cells. This suggests that at least some growth factors may maintain cell survival only by being able to maintain the orderly cell-cycle progression of an activated cell. IL-2, for example, maintains T-cell survival during a proliferative response without apparently altering the expression of *bcl-2* or *bcl-x.*[22] In this way, the supply of extracellular growth factors controls the actual extent of clonal amplification that occurs in response to antigenic challenge. By contrast, the intrinsic cell-survival genes, *bcl-2* and *bcl-*x_L, have no observable ability to maintain a proliferative response of an antigen-initiated cell. Instead, these gene products appear only to enhance the resistance of cells to the induction of programmed cell death in response to either intrinsic or extrinsic signals.

SPECIFIC RECEPTORS INDUCE APOPTOSIS IN ACTIVATED T CELLS

The preceding data suggested that the cell-survival genes, *bcl-2* and *bcl-*x_L, are critical in maintaining the intrinsic survival of resting and recently activated T cells, whereas the actual extent of clonal expansion in response to antigen activation is dependent on extrinsic growth factors that control cell-cycle progression. These factors working in concert following antigen-activation are sufficient to lead to the large clonal expansions of T cells necessary to initiate an effective cell-mediated immune response against a foreign pathogen. However, the regulation of these genes does not provide an explanation for how activated T cells are cleared at the end of an immune response. Recently, it has been shown that activated lymphocytes express a cell-surface receptor, termed Fas or Apo1, which when cross-linked can directly induce programmed cell death.[23,24] Peripheral deletion of previously activated T cells is dependent on Fas.[4,25,26] In mice, mutations in the *fas* gene occur in the *lpr* strain of mice.[4] Fas deficiency results in large accumulation of T cells with activated morphology in both peripheral and central lymphoid tissue. The physiologic ligand for Fas is a member of the TNF family, which is rapidly expressed on the surface of activated helper T cells following antigen receptor cross-linking.[27] This has led to the hypothesis that naive T cells undergoing their first activation event express Fas ligand on their surface, and in doing so can induce cell death in neighboring previously activated T cells that express Fas. Consistent with this hypothesis, resting T cells express little to no Fas antigen, and Fas is up-regulated on the cell surface

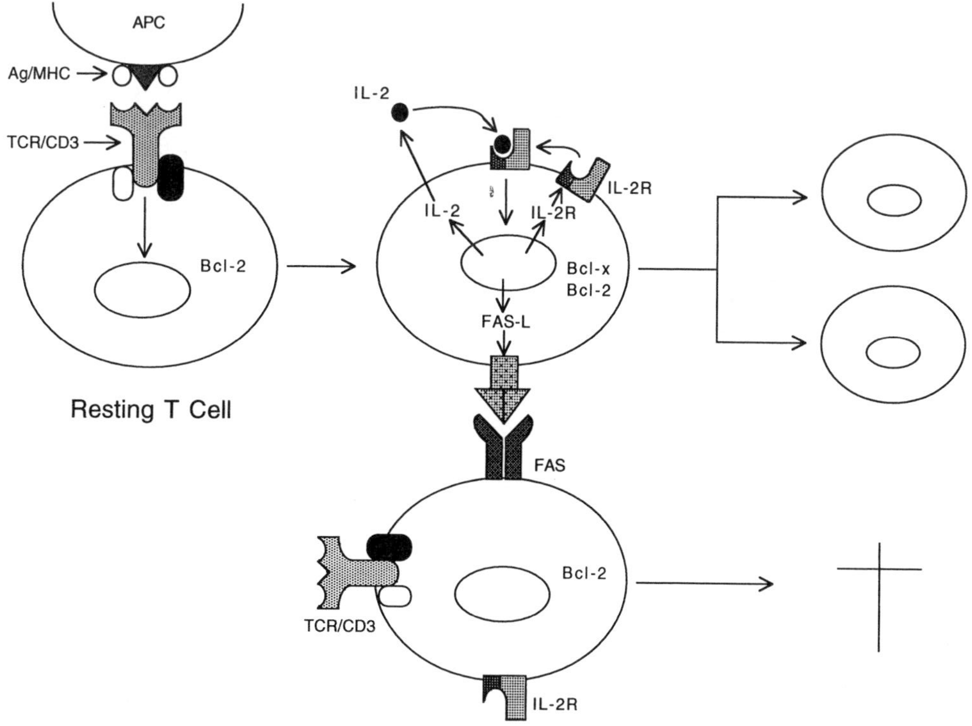

FIGURE 5. T-cell survival is regulated in antigen-activated T cells by multiple survival and death factors. Resting T cells are maintained in the periphery by expressing the intrinsic cell survival factor Bcl-2. Upon activation of the T cell through the TCR/CD3 complex by way of antigen/MHC (Ag/MHC) complexes on the surface of antigen-presenting cells (APC), T cells acquire the ability to proliferate by the expression of high-affinity IL-2 receptors and IL-2. IL-2 functions as an extrinsic survival factor by promoting proliferation in these cells and is required for survival of the proliferating cells. Activated T cells also enhance their survival capacity through the induction of Bcl-x. Antigen activation can also up-regulate the death-promoting molecule Fas ligand, which facilitates the removal of previously activated cells that express the Fas surface receptor. The ability of Fas ligation to induce cell death does not occur immediately following antigen activation and correlates with the down-regulation of Bcl-x expression.

within the first 24 hours following T-cell receptor cross-linking (FIG. 4). T-cell lines chronically expressing Fas, such as Jurkat, will specifically undergo programmed cell death in response to Fas-receptor cross-linking (FIG. 4). Based on this work and similar observations by a number of laboratories, signal transduction through the Fas receptor plays a critical role in determining the long-term survival of recently activated T cells.[28–32]

CONCLUSIONS

Recent evidence suggests that T-cell survival is just as critically regulated as T-cell activation in determining the outcome of peripheral immune responses. At least three classes of gene products have been shown in addition to the T-cell receptor to regulate cell survival (FIG. 5). These include intrinsic survival genes that are members of the *bcl-2* family and extrinsic survival factors, such as IL-2, that can initiate signal transduction events that maintain the proliferative status of cells and thus may secondarily prevent a cell from initiating apoptosis. Lymphocytes also appear to have specific receptors that can directly induce programmed cell death upon ligand binding and may play a critical role in removing excess immune cells once an infection is effectively irradicated. The biochemical mechanisms by which these three types of genes operate and the way in which the systems become integrated to lead ultimately to lymphoid homeostasis is unknown at the present time. However, further studies of these important gene products will undoubtedly play an important role in furthering our understanding of the molecular pathways associated with the control of apoptosis.

SUMMARY

The immune system provides a useful model system in which to study the signal transduction events involved in the regulation of programmed cell death. Mature lymphocytes have the capacity to survive in the body for prolonged periods of time. During an immune response, cells of the appropriate antigenic specificity must undergo clonal amplification to mount a protective response, and cells participating in inflammatory immune responses need to have the capacity to survive at sites of inflammation. However, upon completion of a successful inflammatory response, the majority of cells produced must die off in order to maintain the homeostasis of the organism. Over the last several years we have learned a great deal about how mature lymphocytes regulated their susceptibility to undergo programmed cell death. Three types of information appear to be used by the lymphocyte to control its susceptibility to undergo programmed cell death. The intrinsic susceptibility of a cell to undergo programmed cell death is determined by members of the Bcl-2 gene family. In addition, extrinsic survival factors, such as IL-2, can initiate signal transduction events that can prevent a cell from initiating apoptosis. Finally, lymphocytes appear to have specific receptors, such as Fas, that can directly induce programmed cell death upon ligand binding. The integration of these three systems is discussed.

ACKNOWLEDGMENTS

The authors would like to thank Therese Conway for her assistance in preparation and Brian Chang for critical review of the manuscript. L. H. Boise is a fellow of the Leukemia Society of America.

REFERENCES

1. Schwartz, R. H. 1990. A cell culture model for T lymphocyte clonal anergy. Science **248:** 1349-1356.
2. Taniguchi, T. & Y. Minami. 1993. The IL-2/IL-2 receptor system: A current overview. Cell **73:** 5-8.
3. Cohen, J. J., R. C. Duke, V. A. Fadok & K. S. Sellins. 1992. Apoptosis and programmed cell death in immunity. Annu. Rev. Immunol. **10:** 267-293.
4. Watanabe-Fukunaga, R., C. I. Brannan, N. G. Copeland, N. A. Jenkins & S. Nagata. 1992. Lymphoproliferation disorder in mice explained by defects in Fas antigen that mediates apoptosis. Nature **356:** 314-317.
5. Strasser, A., S. Whittingham, D. L. Vaux, M. L. Bath, J. M. Adams, S. Cory & A. W. Harris. 1991. Enforced *Bcl-2* expression in B-lymphoid cells prolongs antibody responses and elicits autoimmune disease. Proc. Natl. Acad. Sci. USA **88:** 8661-8665.
6. Webb, S., C. Morris & J. Sprent. 1990. Extrathymic tolerance of mature T cells: Clonal elimination as a consequence of immunity. Cell **63:** 1249-1256.
7. Sprent, J., M. Schaefer, M. Hurd, C. D. Surh & Y. Ron. 1991. Mature murine B and T cells transferred to SCID mice can survive indefinitely and many maintain a virgin phenotype. J. Exp. Med. **174:** 717-728.
8. Nakayama, K.-I., K. Nakayama, I. Negishi, K. Kuida, Y. Shinkai, M. C. Louie, L. E. Fields, P. J. Lucas, V. Stewart, F. W. Alt & D. Y. Loh. 1993. Disappearance of the lymphoid system in Bcl-2 homozygous mutant chimeric mice. Science **261:** 1584-1588.
9. Veis, D. J., C. M. Sorenson, J. R. Shutter & S. J. Korsmeyer. 1993. *Bcl-2*-deficient mice demonstrate fulminant lymphoid apoptosis, polycystic kidneys, and hypopigmented hair. Cell **75:** 229-240.
10. Vaux, D. L., S. Cory & J. M. Adams. 1988. *Bcl-2* gene promotes haemopoietic cell survival and cooperates with c-*myc* to immortalize pre-B cells. Nature **335:** 440-442.
11. Hockenbery, D., G. Nuñez, C. Milliman, R. D. Schreiber & S. J. Korsmeyer. 1990. *Bcl-2* is an inner mitochondrial membrane protein that blocks programmed cell death. Nature **348:** 334-336.
12. Nuñez, G., L. London, D. Hockenbery, M. Alexander, J. P. Mckearn & S. J. Korsmeyer. 1990. Deregulated *bcl-2* gene expression selectively prolongs survival of growth factor-deprived hemopoietic cell lines. J. Immunol. **144:** 3602-3610.
13. Sentman, C. L., J. R. Shutter, D. Hockenbery, O. Kanagawa & S. J. Korsmeyer. 1991. *bcl-2* inhibits multiple forms of apoptosis but not negative selection in thymocytes. Cell **67:** 879-888.
14. Strasser, A., A. W. Harris & S. Cory. 1991. *bcl-2* transgene inhibits T cell death and perturbs thymic self-censorship. Cell **67:** 889-899.
15. Boise, L. H., M. González-García, C. E. Postema, L. Ding, T. Lindsten, L. A. Turka, X. Mao, G. Nuñez & C. B. Thompson. 1993. *bcl-x,* a *bcl-2*-related gene that functions as a dominant regulator of apoptotic cell death. Cell **74:** 597-608.
16. González-García, M., R. Pérez-Ballestero, L. Ding, L. Duan, L. H. Boise, C. B. Thompson & G. Nuñez. 1994. *bcl-*x_L is the major *bcl-x* mRNA form expressed during murine development and its product localizes to mitochondria. Development **120:** 3033-3042.

17. GOTTSCHALK, A. R., L. H. BOISE, C. B. THOMPSON & J. QUINTÁNS. 1994. Identification of immunosuppressant-induced apoptosis in a murine B-cell line and its prevention by *bcl-x* but not *bcl-2*. Proc. Natl. Acad. Sci. USA **91:** 7350-7354.
18. YANG, E., J. ZHA, J. JOCKEL, L. H. BOISE, C. B. THOMPSON & S. J. KORSMEYER. 1995. Bad, a heterodimeric partner for Bcl-x_L and Bcl-2, displaces Bax and promotes cell death. Cell **80:** 285-291.
19. REED, J. C., Y. TSUJIMOTO, J. D. ALPERS, C. M. CROCE & P. C. NOWELL. 1987. Regulation of *bcl-2* proto-oncogene expression during normal human lymphocyte proliferation. Science **236:** 1295-1299.
20. GILLIS, S. & K. A. SMITH. 1977. Long term culture of tumour-specific cytotoxic T cells. Nature **268:** 154-156.
21. GROUX, H., D. MONTE, B. PLOUVIER, A. CAPRON & J.-C. AMEISEN. 1993. CD3-mediated apoptosis of human medullary thymocytes and activated peripheral T cells: respective roles of interleukin-1, interleukin-2, interferon-γ and accessory cells. Eur. J. Immunol. **23:** 1623-1629.
22. BOISE, L. H., A. J. MINN, P. J. NOEL, C. H. JUNE, M. A. ACCAVITTI, T. LINDSTEN & C. B. THOMPSON. 1995. CD28 costimulation can promote T cell survival by enhancing the expression of Bcl-$_{XL}$. Immunity. In press.
23. YONEHARA, S., A. ISHII & M. YONEHARA. 1989. A cell-killing monoclonal antibody (ANTI-Fas) to a cell surface antigen co-downregulated with the receptor of tumor necrosis factor. J. Exp. Med. **169:** 1747-1756.
24. TRAUTH, B. C., C. KLAS, A. M. J. PETERS, S. MATZKU, P. MÖLLER, W. FALK, K.-M. DEBATIN & P. H. KRAMMER. 1989. Monoclonal antibody-mediated tumor regression by induction of apoptosis. Science **245:** 301-305.
25. SINGER, G. G. & A. K. ABBAS. 1994. The Fas antigen is involved in peripheral but not thymic deletion of T lymphocytes in T cell receptor transgenic mice. Immunity **1:** 365-371.
26. CRISPE, I. N. 1994. Fatal interactions: Fas-induced apoptosis of mature T cells. Immunity **1:** 347-349.
27. SUDA, T., T. TAKAHASHI, P. GOLSTEIN & S. NAGATA. 1993. Molecular cloning and expression of the Fas ligand, a novel member of the tumor necrosis factor family. Cell **75:** 1169-1178.
28. RUSSELL, J. H., B. RUSH, C. WEAVER & R. WANG. 1993. Mature T cells of autoimmune *lpr/lpr* mice have a defect in antigen-stimulated suicide. Proc. Natl. Acad. Sci. USA **90:** 4409-4413.
29. ALDERSON, M. R., T. W. TOUGH, T. DAVIS-SMITH, S. BRADDY, B. FALK, K. A. SCHOOLEY, R. G. GOODWIN, C. A. SMITH, F. RAMSDELL & D. H. LYNCH. 1995. Fas ligand mediates activation-induced cell death in human T lymphocytes. J. Exp. Med. **181:** 71-77.
30. DHEIN, J., H. WALCZAK, C. BÄUMLER, K.-M. DEBATIN & P. H. KRAMMER. 1995. Autocrine T-cell suicide mediated by APO-1 (Fas/CD95). Nature **373:** 438-441.
31. BRUNNER, T., R. J. MOGIL, D. LAFACE, N. J. YOO, A. MAHBOUBI, F. ECHEVERRI, S. J. MARTIN, W. R. FORCE, D. H. LYNCH, C. F. WARE & D. R. GREEN. 1995. Cell-autonomous Fas (CD95)/Fas-ligand interaction mediates activation-induced apoptosis in T-cell hybridomas. Nature **373:** 441-444.
32. JU, S.-T., D. J. PANKA, H. CUI, R. ETTINGER, M. EL-KHATIB, D. H. SHERR, B. Z. STANGER & A. MARSHAK-ROTHSTEIN. 1995. Fas (CD95)/FasL interactions required for programmed cell death after T-cell activation. Nature **373:** 444-448.

Signaling Difference between Class IgM and IgD Antigen Receptors[a]

KWANG-MYONG KIM[b] AND MICHAEL RETH[c]

Max-Planck Institut für Immunbiologie
Stübeweg 51, D79108 Freiburg, Germany

INTRODUCTION

The B-cell antigen receptor (BCR) is a multimeric protein complex consisting of the membrane-bound immunoglobulin (mIg) molecule and the Ig-α/β heterodimer.[1] Ig-α (CD79a) and Ig-β (CD79b) are glycosylated transmembrane proteins encoded by the B cell-specific genes mb-1 and B29, respectively.[2,3] Whereas the immunoglobulin molecule serves as the ligand-binding component of the receptor, the noncovalently associated Ig-α/β heterodimer has been shown to be the signal transduction unit of the BCR.[4,5]

Signal transduction from the cross-linked BCR involves the rapid activation of two types of protein tyrosine kinases (PTK): the src-related PTKs Lyn, Fyn, Lck, and Blk[6–8] (for a review, see ref. 9) as well as the cytoplasmic PTK Syk/PTK72.[10–13] These enzymes phosphorylate several substrate proteins in B cells, including the BCR components Ig-α and Ig-β.[14] The Ig-α/Ig-β heterodimer plays an important role in the activation of these PTKs[4,5,15–19] and, after, its tyrosine phosphorylation becomes a target for SH2-carrying proteins.[20–22] Ig-α and Ig-β contain a signaling motif[23] within their cytoplasmic sequence, which consists of six amino acids occurring with a precise spacing (D/ExxxxxxD/ExxYxxLxxxxxxxYxxL/I). It comprises two negatively charged amino acids and two tyrosines followed at position Y + 3 by either a leucine or an isoleucine residue. This motif is found in the signaling subunits of antigen and Fc receptors as well as in some viral transmembrane proteins.[24] A sequence alignment of the relevant sequence in different receptors is depicted in FIGURE 1. This motif is known under several names (TAM, ARAM, ARH1).[15,25–27] At a recent signaling meeting (Eighth Symposium on Signals and Signal Processing in the Immune System; Kecskemet, Hungary), this motif was called *i*mmunoglobulin (superfamily) *t*yrosine based *a*ctivation *m*otif (ITAM). The signaling function of ITAM has been established by different groups.[15,28–30] Using chimeric receptors carrying the cytoplasmic portion of the T-cell receptor (TCR) or BCR signaling subunits, it was shown that even conserved mutations of the tyrosine or leucine residues of ITAM abrogate signal tranduction from these receptors. If an Ig-α protein with a double mutation of the two tyrosines of ITAM is introduced into

[a] This work was supported by a Grant from the Deutsche Forschungsgemeinschaft Re 571-31. K. M. Kim was a recipient of a fellowship from the Humboldt Foundation.

[b] Present address: Department of Pediatrics, Kyoto University, Sakyo-ku, Kyoto 606-01, Japan.

[c] To whom correspondence should be addressed.

1.	BCR/mIg-α	EDENL YEGL NLDDCSM YEDI
2.	BCR/hIg-α	EDENL YEGL NLDDCSM YEDI
3.	BCR/mIg-β	EEDHT YEGL NIDQTAT YEDI
4.	BCR/hIg-β	EEDHT YEGL DIDQTAT YEDI
5.	CD3-γ	QNDQL YQPL KDREYDQ YSHL
6.	CD3-δ	KNEQV YQPL RDREDTQ YSSL
7.	CD3-ε	VPNPD YEPI RKGQRDL YSGL
8.	TCR-ζ 1	GQNQL YNEL NLGRREE YDVL
9.	TCR-ζ 2	PQEGV YNAL QKDKMAEA YSEI
10.	TCR-ζ 3	GHDGL YQGL STATKDT YDAL
11.	FcεRI-g	KADAV YTGL VTRSQET YETL
12.	FcεRI-b	PDDRL YEEL NVYSPI. YSAL
13.	BLV-gp30-1	KPDSD YQAL LPSAPEI YSHL
14.	BLV-gp30-2	SAPEI YSHL SPVKPD. YINL
15.	EBV-LMP2A	DRHSD YQPL GTQDQSL YLGL
	Consensus:	xxDxx YxxL xxxxxxx YxxL
		E I

FIGURE 1. Alignment of the ITAM sequences of the signaling components of the B-cell receptor, T-cell receptor, and the high-affinity IgE Fc-receptor. This sequence motif is also found in transmembrane proteins of two viruses that infect B lymphocytes, namely in the gp30 envelope protein of bovine leukemia virus (BLV) and in the LMP2A protein of Epstein Barr virus (EBV). mIg, murine immunoglobulin; hIg, human immunoglobulin.

an Ig-α-deficient cell line, a BCR is expressed, but its signal transducing ability is drastically reduced. Furthermore if the same mutation is introduced into the murine Ig-α gene locus, no B cells are generated by these mutant mice, confirming that Ig-α provides an important signal for the proliferation and/or maturation of B cells in the mouse. These results also show that the ITAM of Ig-β cannot function independently of Ig-α.

RESULTS AND DISCUSSION

Most mature B lymphocytes coexpress two classes of antigen receptor, IgM and IgD. The differences in the signal transduction from the two receptors are still a matter of controversy. It has been shown that the early events in signal transduction like the PTK activation[31–33] and the release of calcium ions[34] are similar in the two receptors. However, cellular responses of activated B cells, like tolerance induction, apoptosis, or interleukin secretion, can be different, depending on whether the cells are

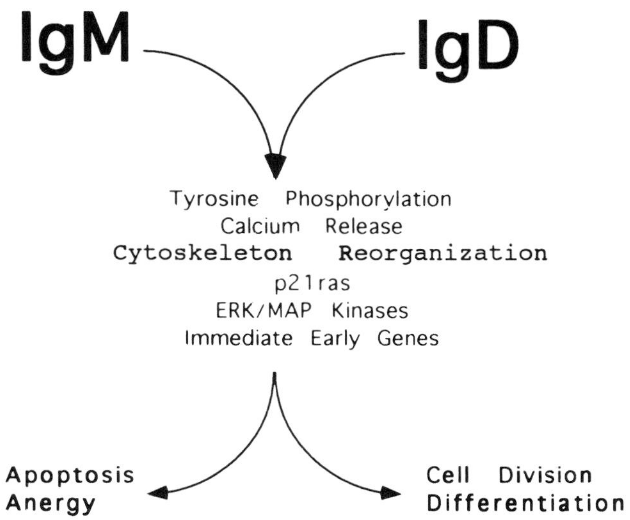

FIGURE 2. IgM- and IgD-BCR use the same signaling route, although their activation can have a different biological outcome.

stimulated through the IgM-BCR (FIG. 2) or the IgD-BCR. Specifically, engagement of IgM on immature B cells results in their deletion, whereas engagement of the IgD-BCR on these cells does not have this effect.[35] Similarly, data from δm transfectants of the WEHI231[36] or CH33[37] cell lines show that cross-linking of the IgM-BCR, but not of the IgD-BCR, can activate an apoptosis program in these cells.

We have analyzed Ig transfectants of the myeloma line J558L expressing either IgM or IgD antigen receptors with the same antigen specificity (subclones 7-1 and 7-14 of J558Lμm3/δm7). Cross-linking of these receptors with either antigen or class-specific antibodies results in the activation of protein tyrosine kinases and the phosphorylation of the same substrate proteins. The kinetics and intensity of phosphorylation, however, were different between the two receptors when these were cross-linked by antigen. In mIgM-expressing cells, the substrate phosphorylation reached a maximum after one minute and diminished after 60 minutes, whereas in the mIgD-expressing cells, the substrate phosphorylation increased further over time, reaching its maximum at 60 minutes and persisting longer than 240 minutes after exposure to antigen. As a result, the intensity of protein tyrosine phosphorylation induced by cross-linking of membrane IgD was stronger than that induced by mIgM.

Studies of chimeric receptors demonstrate that only the membrane-proximal C domain and/or the transmembrane part of mIgD molecule is required for the long-lasting substrate phosphorylation. Together, these data suggest that the signal emission from the two receptors is controlled differently and that the IgM-BCR, but not the IgD-BCR, is under a negative feedback control. Such a control could either operate by way of the modulation of the receptor on the cell surface or through the silencing of the receptor by modification (*i.e.,* serine/threonine phosphorylation) of the part of

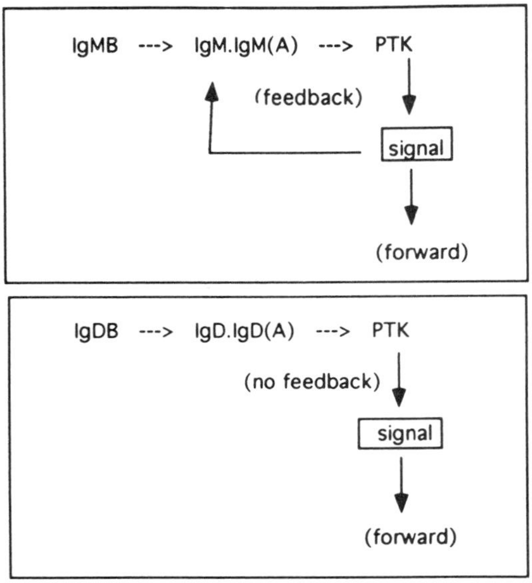

FIGURE 3. One model to explain the different duration of the activation signal generated through IgM or IgD. The cross-linked (top) IgM-BCR may take a conformation that results in PTK activation and susceptibility for a negative feedback signal, whereas the cross-linked (bottom) IgD-BCR is not allowing a negative feedback and therefore can signal for a prolonged time.

the molecule responsible for signal emission. When cultured in the presence of antigen, the IgM- and IgD-expressing J558L cells down-modulate their respective antigen receptor to the same extent (5- to 7-fold). The difference between the two classes of antigen receptors may thus be due rather to a specific silencing than to a down-modulation of the BCR. Apparently the IgM, but not the IgD antigen, receptor is under negative feedback control (FIG. 3).

The different behavior of the IgM and IgD antigen receptors on B-cell lines is reminiscent of what is observed with tyrosine-kinase receptors (TKR) on the neuronal cell line PC12. These cells express two types of TKR, namely the epidermal growth factor receptor (EGFR) and the neuronal growth factor receptor (NGF). Signal transduction from the two receptors involves the same signaling pathway; yet the outcome of these signals is different, resulting either in growth or in neural differentiation. How the two receptors could achieve these differential responses was a puzzle so far. Two recent publications[38,39] provide an answer to this riddle. The activation of the NGFR has been shown to result in a stronger and more prolonged MAP2 kinase activation than the activation of the EGFR. Furthermore, overexpression of the EGFR or mutation of serines in the cytoplasmic part of the EGFR involved in feedback regulation of the signal from this receptor no longer resulted in growth but in the neuronal differentiation of PC12, thus giving the same result as signal transduction by way of the NGFR. These experiments clearly demonstrate that an increase and prolongation of signals can dictate the outcome of the signal. The results obtained by up-regulating or mutating the EGFR may also offer an explanation for the mild phenotype seen in knock-out mice lacking an IgD-BCR. In these mice the IgM-BCR possibly mimics IgD-BCR signaling by either up-modulation of IgM-BCR or changes in the feed-back control mechanism.

The functional difference between the IgM-BCR and IgD-BCR is more obvious when the two receptors are cross-linked by antigen or antiidiotypic antibody than by class-specific antibodies. This may be the reason why these differences have remained undetected so far because most previous studies used the latter reagents. Our results are in line with studies from other receptors that show that most antireceptor antibodies can only imperfectly mimic the ligand-receptor interaction.[40] Apparently, for efficient signal induction to occur, the BCR has to be cross-linked with a precise orientation. For the antigen or for idiotype-specific antibodies that bind the receptor from the top, this criterium may be easier to fulfil than for isotype-specific antibodies that bind the receptor from the side.

The different signaling behavior of IgM and IgD should be reflected by structural differences of the two receptors. Apart from a differential glycosylation of the extracellular Ig domain of IgM-α and IgD-α, however, the same Ig-α/Ig-β heterodimer is coupled to the two mIg classes.[41–43] Furthermore, the mIgM and mIgD molecules have the same cytoplasmic tail consisting of only three amino acids (Lys, Val, Lys). The cytoplasmic parts of the mIgM/Ig-α/Ig-β and mIgD/Ig-α/Ig-β antigen receptor complexes should thus be identical to each other. However, it is still possible that the topological arrangement between mIg and the Ig-α/Ig-β heterodimer is different in the two classes of receptors and that this influences their function. Alternatively, the two antigen receptor complexes may contain a new and yet undefined class-specific component.

We have searched for such new mIg-associated molecules in mIg transfectants of the myeloma line J558L with a protein labeling protocol developed for this purpose. This protocol allows the identification of components of protein complexes independently of their surface expression or rate of turnover. The mIg complexes were first bound to a sorbent and thereafter biotinylated on the sorbent (BOS). Using this protocol we have found two proteins (BAP37 and BAP32) that are specifically associated with mIgM[44] and two proteins (BAP31 and BAP29) that are preferentially associated with mIgD and to some extent with membrane IgM.[45] These *B* cell receptor-*a*ssociated *p*roteins (BAP) are nonglycosylated transmembrane proteins with a hydrophobic N-terminus and a charged, α-helical C-terminus. The four BAPs are encoded by separate genes that are all ubiquitously expressed and that are lying on four different chromosomes. Interestingly, BAP31 is lying on the X chromosome in both mouse and human (Adachi *et al.*, unpublished). BAP31 and BAP29 are related proteins (47% identity) as are BAP37 and BAP32 (57% identity), but only little homology exists between the mIgM- and mIgD-associated BAPs. The C-terminal parts of BAP31 and BAP29 are predicted from α-helices and contain leucine zipper that may be mediating the dimerization of these proteins. According to this model, the BAP31/BAP29 pair would form a coiled-coil structure.

Little is known at present about the cellular function of these proteins. Their evolutionary conservation and ubiquitous expression suggest that they play an important role in all cell types. For example, these molecules could be involved in the specific export or import of receptor molecules. In line with this hypothesis is the finding that fluorescence-labeled anti-BAP antibodies show a strong staining of intracellular vesicles (Kim *et al.*, unpublished). This finding does not, however, exclude that some BAPs also come on the cell surface, for example, together with the mIg molecule. The C-terminal sequence of BAP31 and BAP29 has a weak homology with spectrin, a cytoskeleton protein forming a coiled-coil structure. The BAPs may therefore function as linkers between receptors and cytoskeleton elements. In an exon-swapping experiment we have shown that the specific binding of BAPs to mIg occurs by way of the transmembrane parts (Tm) of the mIg molecule.[45] The Tm amino acids of mIg molecules cross the lipid bilayer as α-helix. Interestingly one side of this α-helix is nearly identical between mIgM and mIgD and presumably in contact with the Ig-α/Ig-β heterodimer, whereas the other side differs between the two receptor classes and is therefore likely to be the BAP binding side. This assumption is supported by a mutational analysis that shows that a point mutation on the class-specific side of the Tm helix drastically reduces BAP binding (Adachi *et al.*, unpublished).

SUMMARY

Most mature B lymphocytes coexpress two classes of antigen receptor, IgM and IgD. The differences in the signal transduction from the two receptors are still a matter of controversy. We have analyzed B-cell lines expressing IgM or IgD antigen receptors with the same antigen specificity. Cross-linking of these receptors with either antigen or class-specific antibodies results in the activation of protein tyrosine kinases and the phosphorylation of the same substrate proteins. The kinetic and intensity of phosphorylation, however, was quite different between the two receptors when they were cross-linked by antigen. In membrane IgM-expressing cells, the substrate phosphorylation reached a maximum after one minute and diminished after 60 minutes, whereas in the membrane IgD-expressing cells, the substrate phosphorylation increased further over time, reaching its maximum at 60 minutes and persisting longer than 240 minutes after exposure to antigen. Recently prolonged signaling has been found to be responsible for signaling differences between tyrosine kinase receptors using otherwise similar signaling routes. Thus, the duration of a signal may be an important biological feature of signal-transducing cascades.

REFERENCES

1. Hombach, J., T. Tsubata, L. Leclercq, H. Stappert & M. Reth. 1990. Nature **343:** 760-2.
2. Hermanson, G. G., D. Eisenberg, P. W. Kincade & R. Wall. 1988. Proc. Natl. Acad. Sci. USA **85:** 6890-4.
3. Sakaguchi, N., S. Kashiwamura, M. Kimoto, P. Thalmann & F. Melchers. 1988. EMBO J. **7:** 3457-64.
4. Kim, K. M., G. Alber, P. Weiser & M. Reth. 1993. Eur. J. Immunol. **23:** 911-6.
5. Sanchez, M., Z. Misulovin, A. L. Burkhardt, S. Mahajan, T. Costa, J. B. Bolen & M. Nussenzweig. 1993. J. Exp. Med. **178:** 1049-1055.
6. Yamanashi, Y., T. Kakiuchi, J. Mizuguchi, T. Yamamoto & K. Toyoshima. 1991. Science **251:** 192-4.

7. Clark, M. R., K. S. Campbell, A. Kazlauskas, S. A. Johnson, M. Hertz, T. A. Potter, C. Pleiman & J. C. Cambier. 1992. Science **258:** 123-6.
8. Dymecki, S. M., P. Zwollo, K. Zeller, F. P. Kuhajda & S. V. Desiderio. 1992. J. Biol. Chem. **267:** 4815-23.
9. Rudd, C. E., O. Janssen, K. V. Prasad, M. Raab, A. Da Silva, J. C. Telfer & M. Yamamoto. 1993. Biochem. Biophys. Acta **1155:** 239-66.
10. Hutchcroft, J. E., M. L. Harrison & R. L. Geahlen. 1992. J. Biol. Chem. **267:** 8613-9.
11. Taniguchi, T., T. Kobayashi, J. Kondo, K. Takahashi, H. Nakamura, J. Suzuki, K. Nagai, T. Yamada, S. Nakamura & H. Yamamura. 1991. J. Biol. Chem. **266:** 15790-6.
12. Kurosaki, T., M. Takata, Y. Yamanashi, T. Inazu, T. Taniguchi, T. Yamamoto & H. Yamamura. 1994. J. Exp. Med. **179:** 1725-9.
13. Takata, M., H. Sabe, A. Hata, T. Inazu, Y. Homma, T. Nukada, H. Yamamura & T. Kurosaki. 1994. EMBO J. **13:** 1341-9.
14. Gold, M. R., L. Matsuuchi, R. B. Kelly & A. L. DeFranco. 1991. Proc. Natl. Acad. Sci. USA **88:** 3436-40.
15. Flaswinkel, H. & M. Reth. 1994. EMBO J. **13:** 83-9.
16. Clark, M. R., S. A. Johnson & J. C. Cambier. 1994. EMBO J. **13:** 1911-9.
17. Law, D. A., M. R. Gold & A. L. DeFranco. 1992. Mol. Immunol. **29:** 917-26.
18. Matsuo, T., J. Nomura, K. Kuwahara, H. Igarashi, S. Inui, M. Hamaguchi, M. Kimoto & N. Sakaguchi. 1993. J Immunol. **150:** 3766-75.
19. Weiser, P., C. Riesterer & M. Reth. 1994. Eur. J. Immunol. **24:** 665-671.
20. Songyang, Z., S. E. Shoelson, J. McGlade, P. Olivier, T. Pawson, X. R. Bustelo, M. Barbacid, H. Sabe, H. Hanafusa, T. Yi *et al.* 1994. Mol. Cell. Biol. **14:** 2777-85.
21. Baumann, G., D. Maier, F. Freuler, C. Tschopp, K. Baudisch & J. Wienands. 1994. Eur. J. Immunol. **24:** 1799-1807.
22. Pleiman, C. M., C. Abrams, L. T. Gauen, W. Bedzyk, J. Jongstra, A. S. Shaw & J. C. Cambier. 1994. Proc. Natl. Acad. Sci. USA. **91:** 4268-72.
23. Reth, M. 1989. Nature **338:** 383.
24. Alber, G., K.-M. Kim, P. Weiser, C. Riesterer, R. Carsetti & M. Reth. 1993. Curr. Biol. **3:** 333-39.
25. Samelson, L. E. & R. D. Klausner. 1992. J. Biol. Chem. **267:** 24913-6.
26. Weiss, A. 1993. Cell **73:** 209-212.
27. Cambier, J. C. & K. S. Campbell. 1992. FASEB J. **6:** 3207-17.
28. Irving, B. A. & A. Weiss. 1991. Cell **64:** 891-901.
29. Letourneur, F. & R. D. Klausner. 1991. Proc. Natl. Acad. Sci. USA **88:** 8905-9.
30. Romeo, C. & B. Seed. 1991. Cell **64:** 1037-46.
31. Gold, M. R., D. A. Law & A. L. DeFranco. 1990. Nature **345:** 810-3.
32. Burkhardt, A. L., M. Brunswick, J. B. Bolen & J. J. Mond. 1991. Proc. Natl. Acad. Sci. USA **88:** 7410-4.
33. Ales-Martinez, J. E., D. W. Scott, R. P. Phipps, J. E. Casnellie, G. Kroemer, C. Martinez & L. Pezzi. 1992. Eur. J. Immunol. **22:** 845-50.
34. Justement, L. B., J. Wienands, J. Hombach, M. Reth & J. C. Cambier. 1990. J. Immunol. **144:** 3272-80.
35. Carsetti, R., G. Kohler & M. C. Lamers. 1993. Eur. J. Immunol. **23:** 168-78.
36. Tisch, R., C. M. Roifman & N. Hozumi. 1988. Proc. Natl. Acad. Sci. USA **85:** 6914-8.
37. Ales-Martinez, J. E., G. L. Warner & D. W. Scott. 1988. Proc. Natl. Acad. Sci. USA **85:** 6919-23.
38. Dikic, I., J. Schlessinger & I. Lax. 1994. Curr. Biol. **4:** 702-708.
39. Traverse, S., K. Seedorf, H. Paterson, C. Marshall, P. Cohen & A. Ullrich. 1994. Curr. Biol. **4:** 694-701.

40. Ortega, E., S. R. Schweitzer & I. Pecht. 1988. EMBO J. **7:** 4101-9.
41. Venkitaraman, A. R., G. T. Williams, P. Dariavach & M. S. Neuberger. 1991. Nature **352:** 777-81.
42. Campbell, K. S., E. J. Hager & J. C. Cambier. 1991. J Immunol. **147:** 1575-80.
43. Wienands, J. & M. Reth. 1991. Eur. J. Immunol. **21:** 2373-8.
44. Terashima, M., K.-M. Kim, T. Adachi, P. J. Nielsen, M. Reth, G. Köhler & M. C. Lamers. 1994. EMBO J. **13:** 3782-3792.
45. Kim, K.-M., T. Adachi, P. J. Nielsen, M. Terashima, M. C. Lamers, G. Köhler & M. Reth. 1994. EMBO J. **13:** 3793-3800.

Interaction of p56^{lck} with CD4 in the Yeast Two-hybrid System[a]

KERRY S. CAMPBELL, ANNIE BUDER, AND ULRICH DEUSCHLE

Basel Institute for Immunology
Grenzacherstrasse 487
CH-4005 Basel, Switzerland

Association of the amino terminal domain of p56^{lck} with the cytoplasmic domain of CD4 in T cells is one of the best characterized intracellular protein interactions involving immune cell receptors. p56^{lck} is an *src* family protein tyrosine kinase (PTK) and is activated upon CD4 ligation. Coimmunoprecipitation studies have shown that the noncovalent interaction is mediated by way of cysteine pairs in each protein.[1,2] Interestingly, this interaction has not been successfully reproduced with fusion proteins *in vitro*, presumably due to the inability to mimic exact intracellular conditions or due to involvement of additional cofactors (*e.g.*, a third coupling protein). To determine if the protein interaction was indeed binary, we made use of the yeast two-hybrid system.

Plasmids and the *Saccharomyces cerevisiae* strain were from Dr. R. Brent and colleagues at Harvard.[3,4] Constructs of hybrid proteins were created by polymerase chain reaction (FIG. 1A). The cytoplasmic domain of murine CD4 was fused to the LexA DNA binding protein. Full-length (Lck) and the amino-terminal domain (aLck; amino acids #1–67) of murine p56^{lck} were fused to the B42 acid patch domain, which exhibits moderate transcriptional activity. Transcription of the B42 hybrid protein is induced only in a galactose-based medium. Interacting cysteine pairs were eliminated by truncation of CD4 (tCD4; amino acids #420–442) and mutation of aLck (maLck; cysteines #20 and 23 changed to alanine). Constructs were introduced into yeast-bearing chromosomal LEU2 and episomal *lacZ* reporter genes controlled by upstream LexA operator sequences.

Transfected yeast were tested for growth on leucine-deficient medium (FIG. 1B) and for β-galactosidase activity (TABLE 1). Yeast containing all combinations of plasmids grew well on glucose-based agarose plates containing leucine, whereas none grew on glucose-based medium lacking leucine. Full-length p56^{lck} significantly inhibited growth when induced on galactose-based medium containing leucine. This is presumably due to nonspecific inhibitory effects of protein tyrosine phosphorylation by p56^{lck}. Only yeast transfected with both aLck or Lck in combination with CD4 grew on galactose-based leucine-deficient medium, indicating that these proteins interact in a binary manner. Similarly, increased β-galactosidase activity above control

[a] The Basel Institute for Immunology was founded and is supported by F. Hoffmann-La Roche, Ltd., Basel, Switzerland.

A.

Hybrid Proteins

```
CD4    LexA-RCRHQQRQAARMSQIKRLLSEKKTCQCPHRMQKSHNLI
tCD4   LexA-RCRHQQRQAARMSQIKRLLSEKK
Lck    B42-FULL LENGTH Lck PROTEIN
aLck   B42-MGCVCSSNPEDDWMENIDVCENCHYPIVPLDSKISLPIR
        NGSEVRDPLVTYEGSLPPASPLQDNLVI
maLck  B42-MGCVCSSNPEDDWMENIDVAENAHYPIVPLDSKISLPIR
        NGSEVRDPLVTYEGSLPPASPLQDNLVI
```

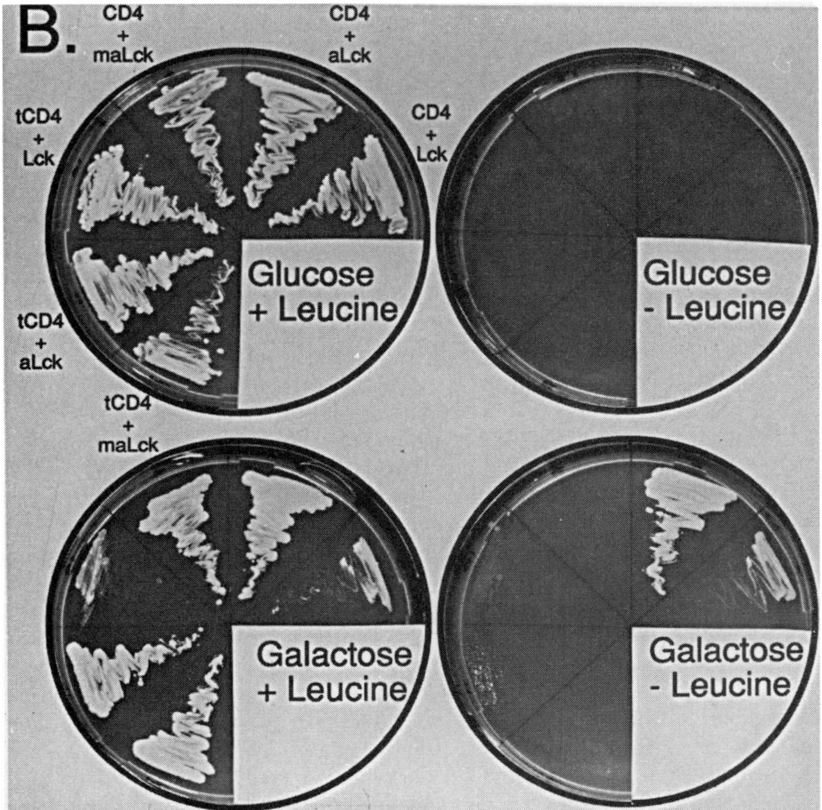

FIGURE 1. Growth of yeast transfectants producing hybrid proteins. **A.** Sequences of the hybrid proteins expressed in yeast. The proteins were all fused onto the COOH-terminal end of the DNA-binding protein LexA (plasmid pEG202) or the B42 acid patch transcriptional activation domain (plasmid pJG4-5). Interacting cysteine pairs on each protein (or their mutations) are boxed. **B.** Growth of yeast transfectants on glucose- or galactose-based agarose plates in the presence or absence of leucine. Transfectants were spread on each plate as labeled on the upper left plate. The LexA hybrid proteins are constitutively produced, whereas production of the B42 transcriptional domain hybrid proteins requires galactose. The EGY48 yeast strain cannot survive in the absence of leucine, and, therefore, growth on leucine-deficient plates can only occur if the hybridized proteins interact, thereby driving the LEU2 reporter.

TABLE 1. β-Galactosidase Assay of Yeast Transfectants[a]

Plasmids		Units β-Galactosidase		
LexA Hybrid	TA Hybrid	Experiment #1	Experiment #2	Experiment #3
+ control		4050	915	
− control		15	17	
CD4	Lck	44	140	27
CD4	aLck	978	511	674
tCD4	Lck	13	14	4.4
tCD4	aLck	34	18	0.7
CD4	maLck			0.9
tCD4	maLck			1.1

[a] Yeast transfectants also containing a *lacZ* reporter plasmid (pSH18-34) were grown in galactose-based liquid medium. β-galactosidase activity was measured from log-phase liquid cultures using *o*-nitrophenyl-β-D-galactopyranoside as a substrate, and units of galactose activity were calculated as normalized to cell concentration (as described in ref. 5). The positive (+) control is LexA hybridized with the Gal4 transcriptional activation domain (pSH17-4), and the negative (−) control is LexA hybridized with bicoid protein, which has no transcriptional activity (pRFHM1). TA = transcriptional activation domain (B42 acid patch).

was observed with the combinations of CD4 with either Lck or aLck. The growth and enzyme activity were, however, significantly reduced with full-length $p56^{lck}$ as compared to aLck. Elimination of the interacting cysteine pairs on tCD4 or maLck eliminated growth and β-galactosidase activity, thereby confirming previous biochemical data indicating their intimate involvement in the interaction. There was slight growth and occasional increased enzyme activity, however, with the combination of tCD4 and aLck, suggesting a weak cysteine-independent interaction of these regions.

Our results strongly suggest that the interaction of CD4 with $p56^{lck}$ is binary and does not require additional T cell-specific cofactors. Additionally, full-length $p56^{lck}$ inhibits the growth of yeast, indicating that full-length *src* PTKs may be difficult to detect from cDNA libraries in the two-hybrid system.

ACKNOWLEDGMENTS

We thank F. Letourneur, E. Palmer, and C. Edfjäll for critical reading of the manuscript.

REFERENCES

1. SHAW, A. S., J. CHALUPNY, J. A. WHITNEY, C. HAMMOND, K. E. AMREIN, P. KAVATHAS, B. A. SEFTON & J. K. ROSE. 1990. Short related sequences in the cytoplasmic domains of CD4 and CD8 mediate binding to the amino-terminal domain of the $p56^{lck}$ tyrosine protein kinase. Mol. Cell. Biol. **10:** 1853-1862.
2. TURNER, J. M., M. H. BRODSKY, B. A. IRVING, S. D. LEVIN, R. M. PERLMUTTER & D. R. LITTMAN. 1990. Interaction of the unique N-terminal region of tyrosine kinase $p56^{lck}$ with cytoplasmic domains of CD4 and CD8 is mediated by cysteine motifs. Cell **60:** 755-765.
3. GOLEMIS, E. A., J. GYURIS & R. BRENT. 1994. Interaction trap/two-hybrid system to identify interacting proteins. *In* Current Protein and Molecular Biology. F. M. Ausubel *et al.*, Eds.: **Suppl. 27:** 13.14.1-17. Current Protocols, New York.
4. GYURIS, J., E. A. GOLEMIS, H. CHERTKOV & R. BRENT. 1993. Cdi1, a human B1 and S phase protein phosphatase that associates with Cdk2. Cell **75:** 791-803.
5. REYNOLDS, A. & V. LUNDBLAD. 1989. Yeast vectors and assays for expression of cloned genes. *In* Current Protein and Molecular Biology. F. M. Ausubel *et al.*, Eds.: **Suppl. 6:** 13.6.1-4. Current Protocols, New York.

Induction of Terminal Differentiation of Promyelocytic HL-60 Leukemic Cells Implanted with Lymphocyte Receptors and Stimulated with Various Lymphocyte Stimulators

IRIT ALTBOUM AND ISRAEL ZAN-BAR

Department of Human Microbiology
Sackler Faculty of Medicine
Tel-Aviv University
Ramat Aviv, 69978, Israel

Stimulation of human promyelocyte (HL-60) leukemic cells with retinoic acid and DMSO, or with vitamin D and TPA, induces differentiation to monocytes or granulocytes, respectively.[1] These reagents induce concomitantly new enzyme activities and blockage of the cells in the G0/G1 phase.[2,3] We intended to examine whether stimulation of the HL-60 cells by way of foreign receptors, operating through the same or different signal transduction delivery systems, would lead to the differentiation of these cells. Thus we inserted lymphocyte plasma membrane (PM) onto HL-60 cells and examined the ability of various lymphocyte-specific differentiating "pushers" to induce long-lasting differentiation effects on these cells.

The insertion of the foreign PM was carried out by fusion of normal functional murine lymphocyte PM vesicles to HL-60 cells by the intact noninfectious Sendai virus or by the Sendai virus envelop glycoproteins.[4,5] This temporal implantation of crude plasma membrane operates successfully in induction of T- and B-cell proliferation, in secretion of immunoglobulins or IL-2, both by normal and functionally impaired lymphocytes.[5] Using this technique we induced serotonin secretion by mast cells, Ig secretion by lymphocytes originating from immunodeficient patients,[6,7] and differentiation and proliferation arrest in acute myelocyte and acute lymphoblastoid leukemic cells (in preparation).

PM vesicles were prepared by the disruption of murine splenic T and B cells or HeLa cells, solubilization with Triton X-100, purification by ultracentrifugation on 22% sucrose cushion, and removal of the detergent with Bio-beads. The HL-60 cells were fused with the foreign PM in a two-step procedure whereby Sendai virus glycoproteins (SV) are fused first to the plasma membrane vesicles, and the PM + SV vesicles are then fused to the cells.[6] The HL-60 cells implanted with lymphocyte PM, or with HeLa PM as controls, were stimulated with anti-Ig, anti-T-cell receptor (TCR) antibodies, Con-A, LPS, or with retinoic acid.

The differentiation of the stimulated implanted HL-60 cells was determined by cell-cycle analysis, measurements of cell-proliferation rate, cell-adhesion rate, uptake

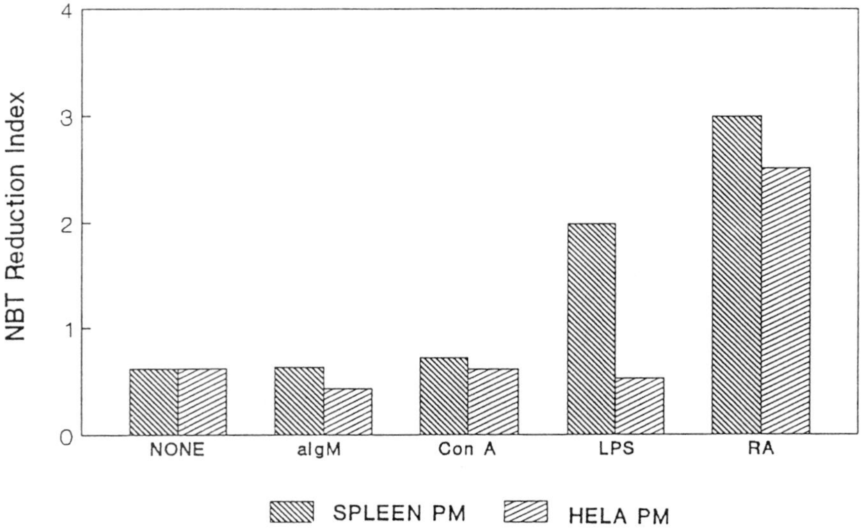

FIGURE 1. Induction of peroxide radicals in PM-implanted HL-60 cells. 5×10^4 HL-60 cells implanted with lymphocyte PM or with HeLa PM were cultured with 1 μg/mL anti-IgM antibody, 10 μg/mL LPS, 2 μg/mL Con-A, or with 1000 nM retinoic acid. Five days later the cells were harvested, and the reduction of NBT was determined. The results shown represent the mean of reduction index of four experiments carried out in triplicate.

of latex beads, production of peroxide radicals, and activation of various differentiating enzymes.

Cell-cycle analysis and uptake of FITC-latex particles were performed by flow cytometry, rate of proliferation was measured by uptake [^{3}H]thymidine, and induction of peroxide radicals was determined by measuring the reduction rate of nitroblue tetrazolium (NBT).[8]

Modified HL-60 cells differentiate in response to lipopolysaccharide (LPS). There was no differentiation response to anti-Ig antibodies or to Con-A, although these did induce elevation in the concentration of free cytosolic Ca^{2+}. Upon stimulation of the modified cells by way of the foreign implanted receptors, peroxide radicals, and other enzyme activities (FIG. 1), uptake of latex particles (FIG. 2) and cell adhesion were induced. However, no effect on cell proliferation could be detected.

Thus HL-60 cells can differentiate by way of foreign stimulators, which, operating through the implanted PM, induce the proper signal transduction delivery system.

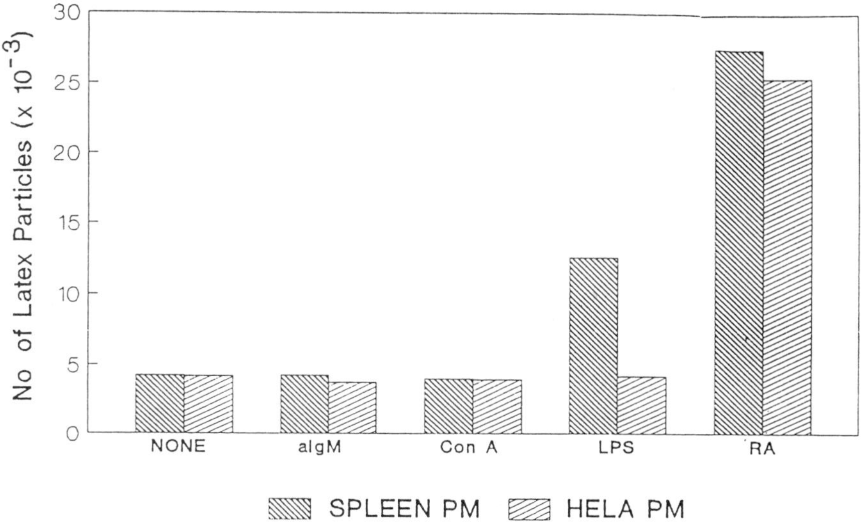

FIGURE 2. Induction of uptake of FITC-latex particles by PM-implanted HL-60 cells. 5×10^4 HL-60 cells implanted with lymphocyte PM or with HeLa PM were cultured with 1 μg/mL anti-IgM antibody, 10 μg/mL LPS, 2 μg/mL Con-A, or with 1000 nM retinoic acid. Five days later the cells were harvested and incubated with FITC-labeled latex beads, and the number of beads taken up by the cells was then determined. The results shown represent the mean of the number of beads taken up by 5×10^3 cells in four experiments carried out in triplicate.

REFERENCES

1. Abe, E. C. Miyaura, H. Sakagami *et al.* 1981. Proc. Natl. Acad. Sci. USA **78:** 4990-4994.
2. Tanaka, H., E. Abe, C. Miyaura *et al.* 1983. Biophys. Res. Commun. **117:** 86-92.
3. Brelvi, Z. S., S. Christakos & G. P. Studzinski. 1986. Lab. Invest. **55:** 269-274.
4. Prujansky-Jakobovits, A., D. J. Volsky, A. Loyter *et al.* 1980. Proc. Natl. Acad. Sci. USA **77:** 7247-7251.
5. Jakobovits, A., N. Sharon & I. Zan-Bar. 1982. J. Exp. Med. **156:** 1274-1279.
6. Abramovits, B., I. Altboum, M. Lapidot *et al.* 1991. Exp. Cell Res. **194:** 228-231.
7. Levy, D., Y. Porat, Z. T. Hendzel *et al.* 1993. Immunodeficiency **4:** 43-46.
8. Collin, S., F. W. Ruscetti, R. E. Gallagher & R. C. Gallo. 1979. J. Exp. Med. **149:** 969-974.

Independent Signaling for Growth Arrest and Apoptosis by Igα and Igβ Subunits of the B-Cell Antigen Receptor Complex[a]

XIAO-RUI YAO AND DAVID W. SCOTT

Department of Immunology
Holland Laboratory
American Red Cross
Rockville, Maryland 20855

B-cell antigen receptors are expressed in association with a Igα/Igβ heterodimer. The cytoplasmic tails of Igα and Igβ both contain a conserved sequence of ~26 amino acids, termed the antigen-receptor homology 1(ARH1) motif. This motif is shared by a number of signal transducing molecules and is involved in the specific interaction with SH2 domain-containing protein tyrosine kinases. These structural and functional features suggest that Igα/Igβ dimer may jointly or individually transduce receptor-mediated signals. Recently, a number of studies have shown that the chimeric receptor with the entire cytoplasmic tail or ARH1 motif of Igα or Igβ can mediate signal transduction as evaluated by tyrosine phosphorylation of intracellular proteins and Ca^{2+} influx. These data suggest that the ARH1 motif contains sufficient structural information to mediate signal transduction.

The functional outcomes of the ARH1 motif-mediated signal transduction have been evaluated in a variety of model systems. We have previously established a model system to study the inhibitory signal pathways in B-lymphoma cells, in which cross-linking of the B-cell receptor by anti-μ induces growth arrest and subsequent cell death by apoptosis. In the present study, we used this model to examine the ability and independence of Igα or Igβ to mediate the inhibitory signaling. We generated cell lines that express a chimeric receptor composed of extracellular and transmembrane domains of murine CD8 with the cytoplasmic domain of the murine B-cell receptor subunit Igα or Igβ, respectively; the cytoplasmic tail of murine IgG2a (γ_{2a}) was used as a control. The chimeras were introduced into CH31, an immature B-cell lymphoma line, by retroviral infection. The stable G418-resistant cells were selected and analyzed for expression of chimeric molecules on the cell surface. We found that the chimera was expressed as a homodimer on the cell surface; there was no interaction between the chimeric receptor and endogenous Igα or Igβ. To determine whether ligation of the chimera would result in intracellular signaling events, we stimulated the transfectants with rat anti-CD8 monoclonal antibody (mAb). The enhanced tyrosine-phosphorylation of intracellular substrates, including the chimeric

[a]This study was supported by NIH Grants CA55644 and AI29691.

TABLE 1. Growth Inhibition Mediated by Chimeric Receptors[a]

Stimuli	CH31	CD8:γ2a	CD8:Igα	CD8:Igβ
Bet-2	89.6 ± 4.7[b]	91.6 ± 5.2	87.5 ± 6.3	90.3 ± 5.2
53-6.72	3.1 ± 4.5	5.8 ± 2.9	39.8 ± 9.8	47.1 ± 7.6
Bet-2/GαRat IgG	90.2 ± 5.2	93.2 ± 4.5	88.3 ± 4.7	94.1 ± 3.2
53-6.72/GαRat IgG	11.3 ± 3.6	9.6 ± 8.9	87.6 ± 8.5	89.6 ± 4.0

[a] The CH31 parental cell line or its transfectants were incubated with 1.0 μg/mL of Bet-2 rat anti-mouse IgM or 53-6.72 rat anti-mouse CD8α mAb in the absence or presence of 5 μg/mL goat anti-rat IgG at 37°C for 24 h. Cell growth was measured by incorporation of tritiated thymidine added during the last 8 h of culture. Data are expressed as a percentage of growth inhibition as compared to control growth (no anti-μ or anti-CD8). Results are mean values (± SEM) of five to seven separate experiments.

[b] Percent inhibition ± SEM, n = 5–7.

receptor per se, was observed in both CD8:Igα, and CD8:Igβ lines but not in CH31 parental cells and CD8:γ_{2a} cells (data not shown), a result suggesting that the cytoplasmic region of Igα or Igβ in the chimeras is required for this signaling event. To obtain functional information about the signal transduction, we investigated whether chimeras could mediate signal transduction leading to growth arrest and apoptosis. The growth inhibition was assayed by using a [^{3}H]thymidine uptake assay. Results summarized in TABLE 1 show that soluble anti-CD8 mAb can induce partial growth arrest through CD8:Igα or CD8:Igβ receptors; when cross-linked by a secondary anti-Ig, anti-CD8 can induce growth inhibition to ~90% in those two cell lines, a level comparable with that obtained by cross-linking of BCR with anti-μ. We have shown that the growth-arrested cells undergo apoptosis by a DNA fragmentation gel assay (data not shown here). This process of apoptosis was also quantitated by

TABLE 2. Thymidine Release Mediated by Chimeric Receptors[a]

Stimuli	CH31	CD8:γ2a	CD8:Igα	CD8:Igβ
Bet-2	26.5 ± 5.7[b]	24.0 ± 4.9	22.7 ± 8.2	19.0 ± 4.8
53-6.72	1.0 ± 4.0	0.3 ± 2.5	2.6 ± 1.7	2.6 ± 1.1
Bet-2/GαRat IgG	19.5 ± 1.3	23.0 ± 4.5	25.6 ± 2.3	18.8 ± 3.9
53-6.72/GαRat IgG	2.6 ± 1.6	1.3 ± 2.6	31.0 ± 9.0	27.0 ± 8.7

[a] The CH31 parental cell line or its transfectants were prelabeled with 10 μci/mL [^{3}H]thymidine at 37°C for 2 h. The washed cells were then plated into a microculture plate and incubated at 37°C with 1.0 μg/mL of Bet-2 rat anti-mouse IgM or 53-6.72 rat anti-mouse CD8α mAb in the absence or presence of 5 μg/mL goat anti-rat IgG. After incubation for 24 h, cells were harvested onto glass fiber filters and counted for remaining radioactivity. Data are expressed as percent release of incorporated [^{3}H]thymidine as compared to control counts (no anti-μ or anti-CD8). The mean values (± SEM) of four to five independent experiments are shown.

[b] Percent release ± SEM, n = 4–5.

using a [^{3}H]thymidine release assay. In principle, this assay measures the release of incorporated [^{3}H]thymidine, which reflects the amount of fragmented and solubilized DNA. As shown in TABLE 2, replicate experiments consistently show a high-level release of [^{3}H]thymidine incorporated into CD8:Igα and CD8:Igβ cells when stimulated with anti-CD8 mAb plus anti-Ig as cross-linker. By contrast, background levels in control cell lines typically do not exceed 5 percent. Soluble anti-CD8 mAb do not cause significant release of incorporated [^{3}H]thymidine, suggesting that stronger signaling is required for the occurrence of apoptosis. These results suggest that cytoplasmic tails of Igα and Igβ are functionally active domains; both of them are capable of transducing a signal independently.

ACKNOWLEDGMENTS

We thank Drs. M. Reth, K-M. Kim, and H. Flaswinkel for providing cDNA constructs.

The Kinase-dependent Function of Lck in T-Cell Activation Requires an Intact Site for Tyrosine Autophosphorylation[a]

HUA XU[c] AND DAN R. LITTMAN[b,d]

Departments of Microbiology and Immunology and of [b]Biochemistry and Biophysics and
Howard Hughes Medical Institute
University of California, San Francisco
San Francisco, California 94143-0414

INTRODUCTION

The lymphocyte-specific protein tyrosine kinase (PTK) p56*lck* (Lck) is a member of the Src family of cytoplasmic kinases that participate in many important signal transduction pathways in a variety of cell types.[1] Accumulating evidence indicates that Lck plays an essential role in both T-cell maturation and activation.[2] In mice lacking Lck expression, due to disruption of the gene, T-cell development was severely impaired.[3,4] Absence of Lck in mutant T-cell lines resulted in a defective response to stimulation with anti-T-cell receptor (TCR) antibody[5] or loss of a cytolytic response.[6] Lck is associated with the cytoplasmic regions of the T-cell surface glycoproteins CD4 and CD8 through its unique N-terminal domain.[7–11] This interaction has been shown to facilitate the coreceptor function of these molecules.[12,13] Lck bound to CD4 participates in antigen-specific signaling through both kinase-independent and -dependent functions. Kinase-independent functions appear to involve the Lck SH2 and SH3 domains,[14,15] which may be required to recruit CD4 to TCR complexes that are interacting with MHC class II molecules. In the absence of SH2 and SH3 function, the kinase domain of CD4-associated Lck can still contribute to the activation signal by way of its catalytic activity.[14] A kinase-dependent function of Lck may be

[a]This work was supported by NIH Grant AI23513 to D. R. Littman. H. Xu was a Howard Hughes Medical Institute Fellow of the Life Sciences Research Foundation. D. R. Littman is an Investigator of the Howard Hughes Medical Institute.

[c]Present address: Amgen, Inc., M.S. #8-1-A-236, Amgen Center, Thousand Oaks, CA 91320-1789.

[d]Address correspondence to D. R. Littman: Skirball Institute, New York University Medical Center, 540 First Avenue, New York, NY 10016.

observed even in the absence of a coreceptor, possibly due to interaction of Lck with other components of the TCR complex.[16,17] This interpretation is further supported by the finding that overexpression of an "activated" form of Lck in a CD4-negative T-cell hybridoma resulted in an enhanced response to antigen.[18] Taken together, these results indicate that the phosphotransferase activity of Lck directly participates in transducing the signals derived from the T-cell receptor and the CD4/CD8 coreceptors.

Studies on the regulation of Lck kinase activity have been guided by observations made with c-Src, the prototype of this family of tyrosine kinases. In c-Src, phosphorylation of two tyrosine residues, which are conserved in all members of the family, is involved in regulating kinase activity.[19,20] Substitution of the tyrosine located near the C-terminus (positions 527 and 505 in c-Src and Lck, respectively) resulted in activation of the kinase and revealed its oncogenic potential in transformation assays.[21–24] This is thought to reflect disruption of an inhibitory intramolecular interaction between the normally phosphorylated C-terminal tyrosine and the SH2 domain of the PTK.[25–27] In Lck, the phosphorylation state of Tyr-505 is thought to be positively and negatively regulated, respectively, by c-Src kinase (Csk) and by the protein tyrosine phosphatase CD45.[28–33] The importance of the second regulatory tyrosine residue, located within the kinase domain, is not as clear. This residue (Tyr-416 in c-Src and Tyr-394 in Lck) is the major *in vitro* autophosphorylation site, and its phosphorylation appears to have a positive effect on the function of Src family kinases.[23,34] Substitution of phenylalanine for Tyr-416 of v-Src resulted in little change in its *in vitro* kinase activity or its ability to transform cultured fibroblasts. However, the ability of the transformed cells to form tumors in animals was impaired.[35–37] In Lck, Tyr-394 is required for a constitutively activated mutant (Phe-505) to transform fibroblasts or to enhance the response to antigen stimulation in T cells.[38,39] Despite the functional defect, this double mutant molecule still augmented TCR-induced protein-tyrosine phosphorylation. As observed with c-Src, mutation of Tyr-394 in Lck had only a moderate effect on the baseline kinase activity (2- to 3-fold decrease).

We have recently established a system to study the mechanisms by which CD4-associated Lck functions in T cells.[14] A chimera between CD4 and Lck effectively replaced wild-type CD4 as coreceptor in a CD4-dependent, antigen-specific T-cell hybridoma. This enabled us to perform structure-function studies on Lck. Our data showed that the kinase activity of CD4-associated Lck contributes significantly to its biological function but that a kinase-independent function that requires an intact SH2 domain is also involved. Using this system, we have now examined the importance of the regulatory tyrosine residues (Tyr-394 and Tyr-505) of Lck in the control of kinase activity in T cells. We find that, in the context of normal regulation at Tyr-505, Tyr-394 is essential for the function of Lck kinase activity in antigen response as well as in the initiation of cellular protein tyrosine phosphorylation upon CD4 cross-linking. These observations indicate that the function of the Lck kinase during T-cell activation requires phosphorylation and dephosphorylation at both positive and negative regulatory tyrosines.

MATERIAL AND METHODS

Cell Lines

171.3 is a $CD4^-CD8^-$ T-cell hybridoma specific for a hen egg lysozyme peptide (residues 74-88) in association with the $I\text{-}A^b$ MHC class II molecule;[12] FT7.1, used as the antigen-presenting cell line, is derived from L cells and expresses $I\text{-}A^b$.[12] Culture conditions for various cell lines are as described.[14]

Construction of Mutant CD4/Lck Chimeras

Mutations at the regulatory tyrosine residues (Tyr-394 and Tyr-505) were first introduced into the murine Lck cDNA through oligonucleotide-directed mutagenesis. The intended mutations were verified by sequencing. The segment of the Lck cDNA containing each mutation was swapped into the CD4/Lck chimeric DNA. Chimeras containing mutations that eliminated kinase activity (Lys-273 to arginine) or the ability of the SH2 domain to bind tyrosine-phosphorylated proteins (Arg-154 to lysine) have been described.[14]

All the chimeric constructs were initially introduced into the pSM vector for expression in COS-7 cells.[10] They were subsequently cloned into retroviral vectors (pMV7 or DOL^-) for establishing packaging cell lines.

Expression of the Chimeric Proteins in T Cells

The different chimeric proteins were first tested in COS cells and then introduced into T cells by retrovirus-mediated gene transfer. The packaging cells were prepared as previously described[12,40] with the following modifications when the DOL^- vector was used. The culture supernatants harvested 72 hours after transient transfection of COS-7 cells with the construct and amphotrophic helper virus DNA were used to infect $\Psi 2$ cells; the infected cells were selected in medium containing 400 μg/mL G418 for approximately one week to establish stable lines. Infection and subsequent sorting of the T cells expressing the CD4/Lck chimeras at the surface were as previously described.[14]

To analyze cell-surface CD4 expression, 171.3-derived or transfected COS-7 cells were stained with FITC-conjugated anti-CD4 monoclonal antibody (mAb) GK1.5 (Caltag) and propidium iodide, which is excluded by live cells, and then analyzed on a FACScan.[10]

Stimulation of T Cells

For antigen stimulation, T cells were cultured with the lysozyme-derived peptide in the presence of FT7.1 cells for 24 hours, and the amount of IL-2 in the supernatant was then determined using the indicator cell line CTLL.[12]

Cross-linking of cell-surface molecules (CD4 and CD3) with soluble antibodies and subsequent immunoblotting of total cellular proteins were as described.[14] To stimulate T cells with immobilized antibodies, 10^5 171.3 derived cells were cocultured with 2 μL of latex beads (PolyScience) coupled to various mAbs according to the manufacturer's instructions. The culture supernatants were harvested 24 hours later for IL-2 assays.

Immunoprecipitation and Kinase Assays

Immunoprecipitation and kinase assays were carried out as described,[14] except that the *in vitro* kinase assays contained 2 μM ATP and 20 μCi [^{32}P]γ-ATP, and the reaction was carried out at 0 °C for 10 minutes. Quantitation of enolase phosphorylation was performed on a PhosphorImager (Molecular Dynamics).

RESULTS

Tyr-394 of Lck Is Essential for in Vivo *Kinase Function*

We previously showed that a chimeric molecule consisting of the extracellular and transmembrane domains of murine CD4 linked to the entire Lck sequence can fulfill the coreceptor function of wild-type CD4 in an antigen-specific CD4-dependent T-cell hybridoma.[14] We have now used this system to study the regulation of Lck kinase activity in T cells. We replaced the putative regulatory tyrosine residues, separately or in combination, with phenylalanine residues that cannot be phosphorylated (FIG. 1A). All of the constructs were first transfected into COS-7 cells and shown to encode transmembrane proteins possessing *in vitro* kinase activity (data not shown). They were then introduced by way of retrovirus-mediated gene transfer into the CD4-negative hybridoma 171.3, whose T-cell receptor reacts specifically with a hen egg lysozyme (HEL)-derived peptide antigen presented on the MHC class II molecule I-A^b.[12] The resulting cells were sorted for matching levels of surface CD4 expression, so that the responses of the different cell lines to antigen stimulation could be directly compared (FIG. 1B).

Cells expressing the mutant CD4/Lck chimeras were stimulated with the peptide antigen in the presence of antigen-presenting cells. T-cell activation was measured by the level of interleukin-2 (IL-2) secreted into the culture medium. A typical result is shown in FIG. 2A. Cells expressing a mutant CD4 (CD4T1) incapable of associating with endogenous Lck due to deletion of its cytoplasmic domain[10,12] served as negative controls and were largely unresponsive to antigen stimulation. Cells expressing the "wild-type" chimera (CtmLck) responded well, but replacement of Tyr-505, the negative regulatory site, with phenylalanine resulted in a consistently increased response to antigen (CtmLck505F in FIG. 2A and data not shown). By contrast, the antigenic response of cells expressing the chimera bearing a mutation at Tyr-394 was impaired; this phenotype was similar to that of cells expressing a kinase-negative mutant, resulting from the substitution of arginine for Lys-273 in the nucleotide-binding pocket of the kinase domain.[14] Cells expressing the double mutant (CtmLckDF) in which both Tyr-505 and Tyr-394 were substituted with phenylalanine had an intermediate phenotype between the two single mutants (FIG. 2A).

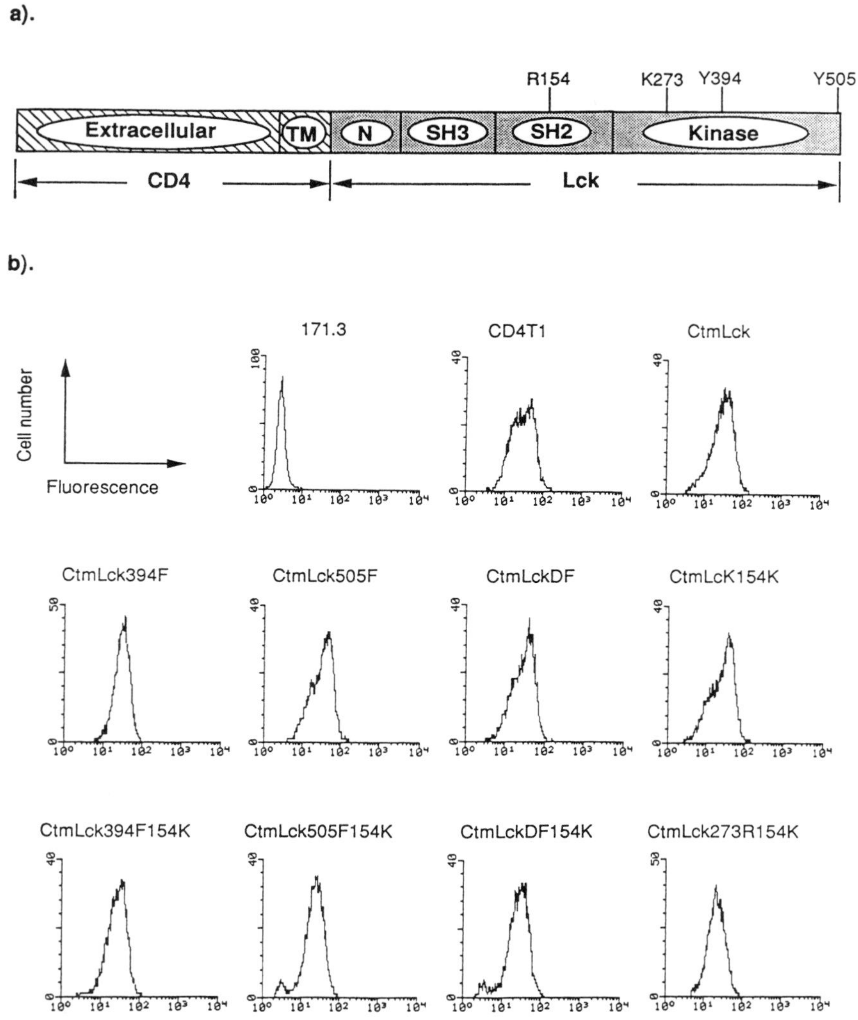

FIGURE 1. Expression of mutant CD4/Lck chimeras in T cells. a: Schematic drawing of the chimeric protein. The chimeras comprise the extracellular and transmembrane domains of CD4 and the entire Lck, as described in ref. 13. The residues within Lck that were mutated either alone or in combination are indicated. The drawing is not to scale. b: Surface CD4 expression in 171.3 cells. The various CD4/Lck DNAs were introduced into these cells by retrovirus-mediated gene transfer, and the infected cells were sorted as described in MATERIAL AND METHODS. The cells were stained with the anti-CD4 mAb GK1.5 coupled to fluorescein and analyzed on a FACScan. The parental 171.3 cells served as the negative control for CD4 expression. The CD4T1 cells express a mutant CD4 molecule with truncation of the cytoplasmic domain (described in refs. 10 and 11). The other cell lines were named according to the corresponding chimera (for example, the CtmLck394F line expresses the CD4/Lck chimera with a phenylalanine substitution at position 394 of Lck).

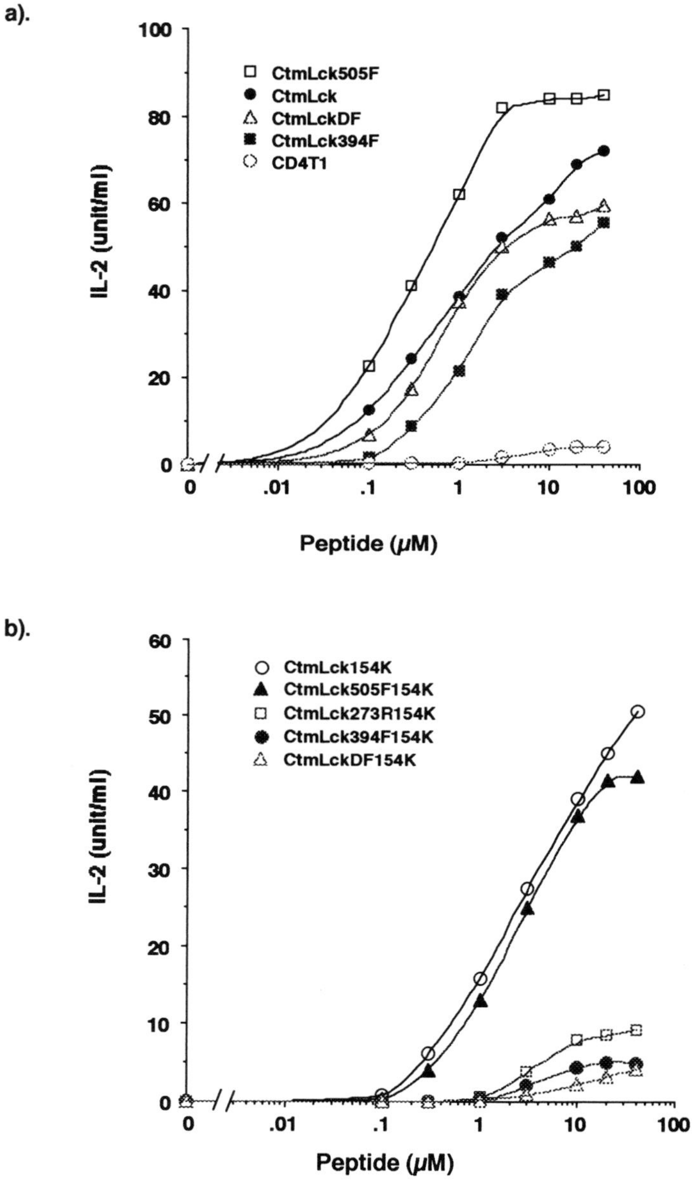

FIGURE 2. Effects of mutations of the regulatory tyrosine residues on the function of the CD4/Lck chimera. a: Responses of mutants having an intact SH2 domain. b: Responses of mutants that lack SH2 function. The 171.3-derived T cells were stimulated with various concentrations of the peptide antigen in the presence of antigen-presenting cells, and supernatants were analyzed for the level of IL-2 produced.

Our earlier work showed that the CD4/Lck chimera can function as a coreceptor in the absence of detectable kinase activity, but that it requires an intact SH2 domain.[14] To determine whether the biological activity of the Tyr-394 mutant chimera is contributed by the noncatalytic domains of Lck, we combined the tyrosine mutations with a mutation in the SH2 domain (Arg-154 changed to Lys) that abrogates association with phosphotyrosine-containing proteins. In the context of the SH2 mutation, the activating mutation at Tyr-505 had no effect (compare CtmLck154K with CtmLck505F154K). However, as observed with the kinase-negative mutant (CtmLck273R154K), chimeras having mutations at Tyr-394 (CtmLck394F154K and CtmLckDF154K) showed very little activity in the antigen-stimulation assays (FIG. 2B). In contrast to the Lys-273 substitution mutant devoid of catalytic activity, these chimeric proteins displayed substantial kinase activity when tested *in vitro* on acid-denatured enolase (FIG. 3A). Accompanying immunoblot analyses of the same immunoprecipitates confirmed that there were approximately equal amounts of the fusion proteins in each reaction (FIG. 3B). From quantitation of these results, we estimate that the specific kinase activity of the CtmLck394F154K mutant protein was about 50% that of the CtmLck505F154K chimera, yet it had dramatically lower biological activity in T cells (FIG. 2B).

Phosphorylation of Cellular Substrates upon Anti-CD4 Cross-linking Requires both Dephosphorylation at the Negative Regulatory Site (Tyr-505) and an Intact Autophosphorylation Site (Tyr-394)

To begin delineating the mechanisms underlying the phenotypes of the regulatory tyrosine mutants, we carried out a series of cross-linking experiments. Cells were stimulated with anti-CD4 monoclonal antibody plus polyclonal secondary antibodies, and the tyrosine phosphorylation status of intracellular proteins was analyzed by antiphosphotyrosine immunoblotting (FIG. 4). Cross-linking of the wild-type chimera resulted mainly in its own phosphorylation, whereas only a barely detectable signal was detected for the Tyr-394 single mutant (FIG. 4, lanes 1-4), a finding in agreement with previous reports that this residue is the major autophosphorylation site.[24,38] Strikingly, when the Tyr-505 mutant chimera was aggregated, there was heavy tyrosine phosphorylation not only of the chimeric protein itself, but also of multiple cellular proteins (FIG. 4, lanes 5 and 6). By contrast, no such phosphorylation events were detected when both of the regulatory tyrosine residues were substituted with phenylanaline (FIG. 4, lanes 7 and 8), indicating that the effect of the Tyr-394 mutation was dominant over that of the Tyr-505 mutation.

It is possible that the requirement of an intact autophosphorylation site for the activation of the kinase activity of Lck in T cells reflects the binding of a T cell-specific factor or factors to phosphorylated Tyr-394. To directly test this possibility, we transiently expressed the chimeric constructs in COS-7 cells. Total cell lysates were prepared, and tyrosine-phosphorylated proteins were detected by immunoblotting (FIG. 5). The activity of the tyrosine mutants in fibroblasts mirrored that in T cells, although antibody-mediated cross-linking was unnecessary in COS cells, probably due to the high-level expression of the chimeras. Expression of the wild-type protein (CtmLck) resulted in high-level autophosphorylation and in low-level phosphorylation

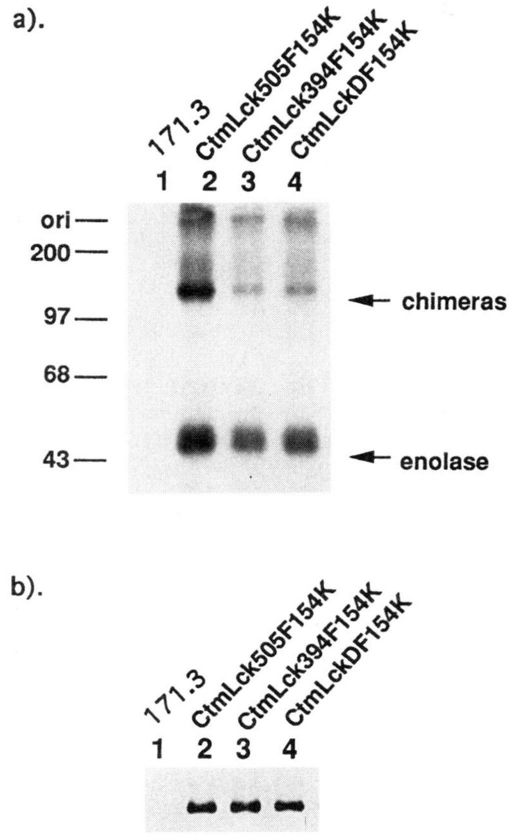

FIGURE 3. *In vitro* kinase activity of the mutant CD4/Lck chimeras. a: Chimeric protein was immunoprecipitated from the indicated cell lines, and part (30%) was incubated with $[^{32}P]\gamma$-ATP and enolase. The reaction products were analyzed by autoradiography following electrophoresis. Phosphorylation of enolase was linear under the conditions used. The positions of molecular weight markers are indicated on the left. b: Anti-CD4 immunoblot to determine the amount of chimeric protein present in the kinase reaction. 70% of each immunoprecipitate was used.

of some cellular proteins, suggesting that negative regulation of the kinase activity is not as effective in COS cells as in T cells. The CtmLck505F mutant displayed significantly increased activity in COS-7 cells, but chimeras containing a mutation at Tyr-394 (either alone or in combination with the Tyr-505 mutation) failed to significantly phosphorylate either themselves or cellular substrates, despite having high *in vitro* kinase activity towards exogenous substrate (FIG. 5A, lanes 3 and 5; and data not shown). These observations indicate that the defect of the Tyr-394 mutants is not limited to T cells. Additionally, phosphorylation of this tyrosine residue appears essential for the phosphorylation of cellular substrates.

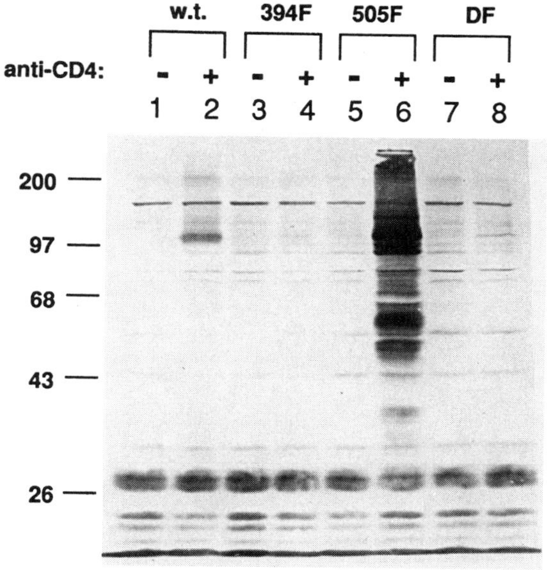

FIGURE 4. Effect of cross-linking the CD4/Lck chimeras on T-cell protein tyrosine phosphorylation. T cells expressing the wild-type or the indicated mutant chimeras were stimulated with the anti-CD4 mAb, GK1.5, followed by the secondary antibodies (for negative controls, GK1.5 was omitted) for 2 minutes at 37°C. Total cell lysates were prepared, and the proteins were separated by SDS-PAGE. Proteins phosphorylated on tyrosine residues were detected by immunoblotting with the antiphosphotyrosine mAb 4G10. Each lane represents total cellular protein from approximately 1.5×10^6 cells.

Phosphorylation of Cellular Substrates by Activated CD4-associated Lck Is Not Sufficient for IL-2 Production

As described above, antibody-induced aggregation of the CtmLck505F chimera led to tyrosine-phosphorylation of many cellular proteins in T cells. To determine whether this signal was sufficient for full activation of the T cells, we compared downstream events with those elicited by cross-linking of the TCR/CD3 complex. Cross-linking of either CD4 or CD3 in cells expressing CtmLck505F resulted in tyrosine phosphorylation of several common proteins, the most prominent of which are indicated in FIGURE 6A (lanes 2 and 3). Some additional proteins (55–60 kDa and > 100 kDa) were phosphorylated upon cross-linking of CD4 alone. When compared to cross-linking of CD3 or CD4 alone, co-cross-linking of these molecules resulted in a significant increase in the phosphorylation of multiple proteins (FIG. 6A, lane 4). We then determined whether activation of Lck alone was sufficient to induce full T-cell activation. As shown in FIGURE 6B, stimulation through the TCR/CD3 complex resulted in the expected production of IL-2. By contrast, no IL-2 was detected following aggregation of either the wild-type or the Phe-505 mutant chimera. Interest-

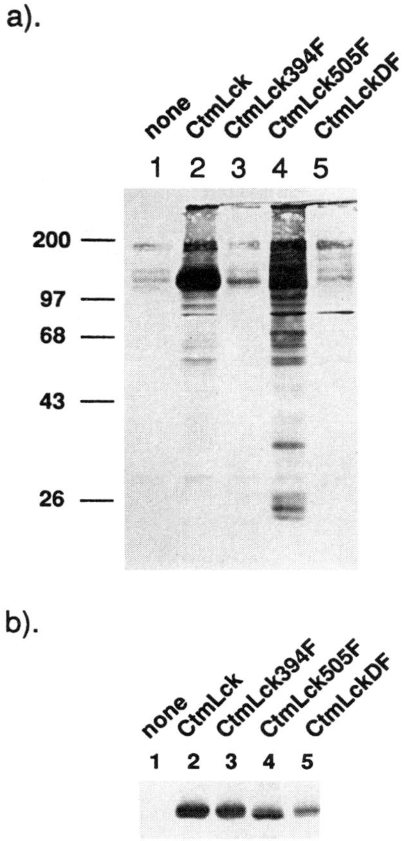

FIGURE 5. Protein tyrosine phosphorylation in COS cells expressing the CD4/Lck chimeras. a: The indicated constructs were transiently transfected into COS-7 cells, and total cell lysates were prepared after 48 hours and analyzed by immunoblotting with antiphosphotyrosine antibody. b: Immunoblot with anti-Lck antibodies to determine transfection efficiency. Protein from 10^5 cells was loaded in each lane.

ingly, cells expressing the activated chimera always produced a larger amount of IL-2 than cells expressing the wild-type fusion protein after cross-linking of TCR/CD3 or co-cross-linking of the chimeric protein with TCR/CD3. This is consistent with the notion that expression of a constitutively activated Src family kinase lowers the threshold of T-cell activation and enhances TCR-mediated signals.

DISCUSSION

Extensive studies on Src family kinases have identified two regulatory tyrosines that are conserved in all family members. The C-terminal tyrosine (Tyr-527 in c-Src

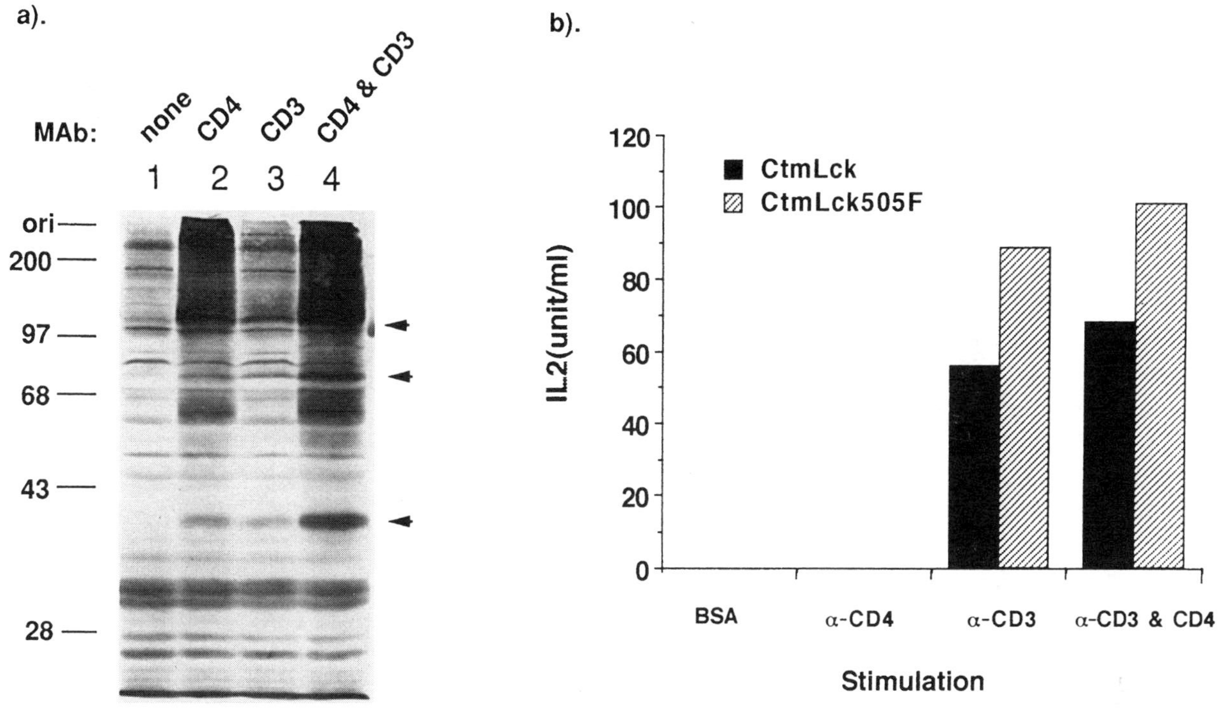

FIGURE 6. Signaling through the activated CD4/Lck chimeric protein is not sufficient for induction of IL-2 synthesis. a: Tyrosine phosphorylation of cellular proteins in CtmLck505F cells treated with immobilized anti-CD4, anti-CD3, or anti-CD4 plus anti-CD3 antibodies. Antiphosphotyrosine immunoblotting was as described in FIGURE. 4. Arrows indicate common bands induced by cross-linking with anti-CD4 or anti-CD3. The positions of molecular weight markers are shown on the left. b: IL-2 production in T cells stimulated with immobilized mAbs. Immobilized bovine serum albumin was used as a control.

and Tyr-505 in Lck) is involved in negative regulation of kinase activity and, possibly, of SH2/SH3 function, whereas the tyrosine at the major *in vitro* autophosophorylation site (Tyr-416 in Src and Tyr-394 in Lck) has been hypothesized to have a role in PTK effector function.[20] In this paper, we have explored the roles of these two tyrosines in Lck-mediated signal transduction during T-cell activation. We found that substitution of phenylalanine for Tyr-505 within a CD4/Lck chimeric molecule resulted in enhancement of the CD4-dependent response to antigen. Strikingly, mutation of Tyr-394 resulted in virtual ablation of the kinase-dependent response to antigen, despite little loss in its catalytic activity towards an artificial substrate. These results strongly support key roles for Tyr-394 and Tyr-505 in the regulation of Lck during TCR-mediated signal transduction in T cells.

Contributions of Tyr-394 and Tyr-505 to Lck Kinase Function

The mechanisms governing the kinase activity of CD4-associated Lck were analyzed in a T-cell hybridoma whose response to antigen is dependent upon the coreceptor. In an earlier study,[14] we found that the CD4/Lck chimeric molecule could effectively replace wild-type CD4 in this cell and that its activity was contributed to by both kinase-dependent and -independent functions of Lck. The latter function was postulated to be due to the ability of the Lck SH2 and, possibly, SH3 domains to interact with other proteins after a conformational change resulting from dephosphorylation of Tyr-505. Consistent with this notion, we show here that substitution of phenylalanine for Tyr-505 in the chimeric protein results in a small but consistently enhanced coreceptor function. This most likely reflects an increase in substrate accessibility, because there was no evidence of increased catalytic activity, and our results from the cross-linking experiments suggest that the mutant protein is in an "open" conformation.

By mutating the SH2 domain of Lck, we were able to focus exclusively on the kinase-dependent property of the chimeric protein. Mutation of Tyr-505 now had no effect on the response to antigen, probably because the interaction site for the phosphotyrosine was already destroyed by the mutation at residue 154. However, phenylalanine substitution for Tyr-394 ablated the kinase-dependent response to antigen. This effect was as dramatic as the substitution of arginine for Lys-273, which abolishes phosphotransferase activity. These results suggest that phosphorylation of Tyr-394 is an essential step for activation of Lck *in vivo*.

Analysis of protein tyrosine phosphorylation in T cells following antibody-mediated aggregation of the various CD4/Lck chimeras also supports the notion that Tyr-394 has a critical function. Only phosphorylation of the chimera itself was detected in cells expressing the wild-type protein, whereas mutation of Tyr-505 to phenylalanine resulted in tyrosine phosphorylation of not only the chimeric molecules but also of many cellular proteins. Transphosphorylation of the CD4/Lck chimera, mostly on Tyr-394, was thus observed upon its cross-linking in T cells regardless of whether tyrosine or phenylalanine was present at position 505. However, activation of a PTK pathway was only observed with the activated Phe-505 mutant protein, suggesting that Lck is normally in a "closed" conformation that is not accessible to substrates and that dephosphorylation at Tyr-505 is necessary to convert the kinase into an active form. Furthermore, the substitution of Tyr-394 with a phenylanaline residue renders the mutant chimera containing the activating Phe-505 mutation (CtmLckDF) incapable of phosphorylating cellular proteins upon anti-CD4-induced aggregation. These results indicate that Tyr-394 phosphorylation in addition to dephosphorylation at Tyr-505 is required for *in vivo* activation of the kinase activity.

Similar results were obtained in COS cells expressing the chimeric proteins, indicating that failure of the Phe-394 mutants to phosphorylate cellular substrates was not T cell-specific and was intrinsic to the mutation. Interestingly, when assayed *in vitro* using an artificial substrate, the Phe-394 mutant had kinase activity that was approximately 50% that of the control protein. In other published work,[38] mutants of the major autophosphorylation site of Src family kinases showed similar discrepancy in their ability to phosphorylate substrates *in vivo* versus *in vitro,* with tyrosine phosphorylation of cellular proteins greatly reduced despite only a moderate decrease in kinase activity toward a model substrate. Possible explanations for these observations are that phosphorylation of Tyr-394 or similar residues in other Src kinases increases the affinity of the kinases for their physiological substrates or alters the conformation of the kinases, allowing them to interact appropriately with these substrates. In the recently reported crystal structure of the insulin receptor kinase domain, the equivalent tyrosine in that molecule (Tyr-1,162) was shown to occupy the active site, blocking access of ATP and substrate.[41] It was proposed that transphosphorylation of the tyrosine may result in a conformational change that would allow efficient interaction with substrates.[41] A similar mechanism may be operational in the Src family kinases.

Regulation of the Phosphorylation State of Tyrosines in Lck

Numerous studies on Lck have focused on regulation of Tyr-505 phosphorylation, but little is known about regulation at the autophosphorylation site. The phosphorylation state of Tyr-505 is thought to be regulated by the opposing activities of Csk[42] and CD45. In a T-cell hybridoma, overexpression of Csk inhibited TCR-mediated T-cell activation, potentially by blocking activation of Lck, Fyn, and other Src family PTKs; accordingly, this inhibitory effect of Csk was overcome by expressing a mutant Fyn, whose C-terminal tyrosine was replaced with phenylalanine.[30] *In vitro* studies have also shown that Csk can preferentially phosphorylate this tyrosine in several Src family members.[28,43] In lymphocytes, the action of Csk is thought to be countered by the protein tyrosine phosphatase CD45, which activates Lck through removal of the phosphate group at Tyr-505.[31,32] It is not presently known whether Csk and CD45 are themselves regulated and, thereby, affect the phosphorylation state of Lck.

Several lines of evidence have suggested that phosphorylation of Tyr-394 is important for Lck effector function. Substitution of Tyr-394 with phenylalanine in an activated form of Lck (F505 mutant) decreased its transformation capacity in fibroblasts.[34,38] Additionally, it was demonstrated that Tyr-394 is required for activated Lck (Phe-505) to enhance T-cell responsiveness to antigen stimulation but unnecessary for augmentation of TCR-induced protein tyrosine phosphorylation. It is unclear from that study, however, if Tyr-394 is critical, inasmuch as, in the T-cell hybridoma studied, the wild-type protein (Tyr-505) behaved in the same way.[18,39] Consistent

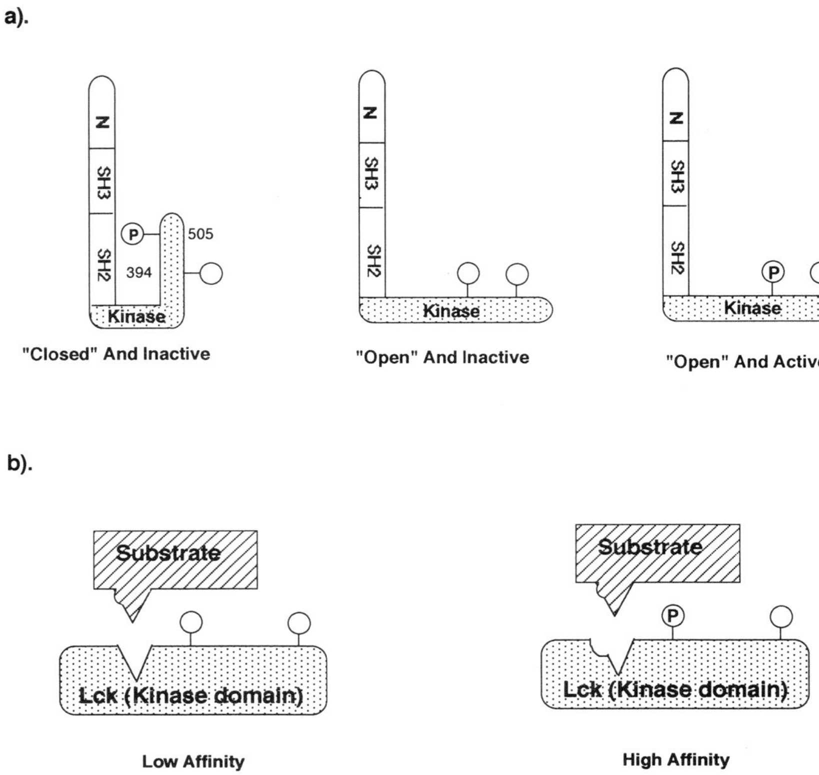

FIGURE 7. Postulated roles of the regulatory tyrosine residues in control of Lck function. a: Dephosphorylation of Tyr-505 renders the kinase domain accessible to substrates. Additionally, phosphorylation of Tyr-394 is required for the activation of the catalytic activity towards relevant protein substrates. b: A possible mechanism through which Tyr-394 positively regulates Lck is through a phosphorylation-induced conformational change in the kinase domain that results in increased substrate accessibility.

with the earlier study and the present results using T-cell hybridomas, it was recently shown that activation of a Fos promoter construct in BAF cells by activated Lck (F505) was also dependent on Tyr-394.[44]

The data presented herein demonstrate that an intact Tyr-394 is required for antigen-specific activation in the context of a form of Lck that can still be regulated at Tyr-505. The results are thus in agreement with the proposal[20] that phosphorylation/dephosphorylation at both the C-terminal regulatory tyrosine and the kinase domain autophosphorylation site are involved in regulating the function of Src family kinases in intact cells (FIG. 7). According to this model, in resting T cells, Lck is dephosphorylated on Tyr-394 and phosphorylated on Tyr-505 and consequently remains in a closed and inactive state. During T-cell activation, CD45-mediated dephosphorylation of Tyr-505 would disrupt the intramolecular interaction of the C-terminus with the noncatalytic domains (SH2 and SH3), converting the kinase to an open conformation. A second event, phosphorylation of Lck at Tyr-394, may then be required for efficient interaction of the catalytic domain with substrates, resulting in full Lck function. The temporal sequence of these events is not yet clear. Moreover, it is not yet known whether, during normal T-cell activation in response to antigen, phosphorylation of Tyr-394 in Lck is achieved through dimerization and transautophosphorylation, or through the activity of a separate PTK. It is possible, for example, that Tyr-394 is phosphorylated by other PTKs associated with the TCR complex, such as Fyn, ZAP-70, or Syk.[2]

Role of Lck in Downstream Events of T-Cell Activation

The extensive tyrosine phosphorylation of cellular proteins induced upon cross-linking of the activated CD4/Lck chimera (CtmLck505F) suggested that the pathway for T-cell activation may proceed through Lck. Previous studies have been divided on whether constitutively active Lck could elicit spontaneous production of IL-2 in T cells, possibly due to differences in cells used.[18,45] In our system, the patterns of tyrosine phosphorylation of cellular proteins were subtly different when cells were stimulated by cross-linking CtmLck505F or the TCR/CD3 complex. Only the latter stimulus resulted in full activation of the T cells, as monitored by IL-2 secretion. It is therefore likely that activation of the PTK pathway downstream of Lck must be supplemented by a second signal derived from the TCR for activation of factors involved in IL-2 gene transcription. The chimeric protein may need to gain access to substrates within the TCR/CD3 complex, which may occur only upon coaggregation of CD4 and TCR. Alternatively, activation of the kinase activity of CD4/Lck, alone, in the absence of other T-cell activation signals may deliver a negative signal. Consistent with this, cross-linking of CD4 has been previously shown to negatively regulate subsequent TCR-mediated T-cell growth.[46] Use of the chimeric proteins described here may facilitate identification of relevant substrates, whose phosphorylation may play key roles in positive or negative signaling in T cells.

SUMMARY

The cytoplasmic protein tyrosine kinase $p56^{lck}$ (Lck) has important signaling roles in T-cell development and activation. We have mutated the two known regulatory tyrosine residues of CD4-associated Lck and examined the effects on its kinase-dependent function in an antigen-specific CD4-dependent T-cell hybridoma. Substitution of phenylalanine for the negative regulatory tyrosine-505 within a CD4/Lck chimera resulted in a slightly increased response to antigen, whereas mutation of the major *in vitro* autophosphorylation site (tyrosine-394) completely abolished the kinase-dependent function of Lck. Even though its kinase activity was only slightly affected, the F394 mutant behaved similarly to a catalytically inactive chimeric protein. Cross-linking of the F505 mutant, but not of wild-type Lck or F394 mutants, resulted in tyrosine phosphorylation of multiple cellular proteins. Although the pattern of tyrosine phosphorylation resembled that observed upon T-cell receptor cross-linking, there was no induction of interleukin-2 synthesis upon cross-linking of the chimeric protein. These results suggest that the activity of the Lck kinase domain *in vivo* is controlled by dephosphorylation at the negative regulatory site and phosphorylation at the positive regulatory (autophosphorylation) site. Additionally, our data show that the specific kinase activity of Lck towards an artificial substrate need not correlate with its ability to phosphorylate cellular proteins or its biological function.

ACKNOWLEDGMENTS

We thank Paul Dazin for his expert assistance with FACS sorting and Kevin Boyd for the antibody-coupled latex beads. The antisera against murine CD4 and Lck were generous gifts from Alem Truneh and Joseph Bolen, respectively. We thank Craig Davis, Mark Hill, and Nigel Killeen for their helpful comments on the manuscript.

REFERENCES

1. Bolen, J. B., R. B. Rowley, C. Spana & A. Y. Tsgankov. 1992. The Scr family of tyrosine kinases in hemopoietic signal transduction. FASEB J. **6:** 3403-3409.
2. Weiss, A. & D. R. Littman. 1994. Signal transduction by lymphocyte antigen receptors. Cell **76:** 263-274.
3. Molina, T. J., K. Kishihara, D. P. Siderovski, W. van Ewijk, A. Narendran, E. Timms, A. Wakeham, C. J. Paige, K. U. Hartmann, A. Veillette, D. Davidson & T. W. Mak. 1992. Profound block in thymocyte development in mice lacking p56^{lck}. Nature **357:** 161-164.
4. Penninger, J., K. Kishihara, T. Molina, V. A. Wallace, E. Timms, S. M. Hedrick & T. W. Mak. 1993. Requirement for tyrosine kinase p56^{lck} for thymic development of transgenic gamma delta T cells. Science **260:** 358-361.
5. Straus, D. B. & A. Weiss. 1992. Genetic evidence for the involvement of the Lck tyrosine kinase in signal transduction through the T cell antigen receptor. Cell **70:** 585-593.
6. Karnitz, L., S. L. Sutor, T. Torigoe, J. C. Reed, M. P. Bell, D. J. Mckean, P. J. Leibson & R. T. Abraham. 1992. Effects of p56^{lck} deficiency on the growth and cytolytic effector function of an interleukin-2-dependent cytotoxic T-cell line. Mol. Cell. Biol. **12:** 4521-30.
7. Rudd, C. E., J. M. Trevillyan, J. D. Dasgupta, L. L. Wong & S. F. Schlossman. 1988. The CD4 receptor is complexed in detergent lysates to a protein-tyrosine kinase (pp58) from human T lymphocytes. Proc. Natl. Acad. Sci. USA **85:** 5190-5194.
8. Shaw, A. S., J. Chalupny, J. A. Whitney, C. Hammond, K. E. Amerein, P. Kavathas, B. M. Sefton & J. K. Rose. 1990. Short related sequences in the cytoplasmic domains of CD4 and CD8 mediate binding to the amino-terminal domain of the p56^{lck} tyrosine protein kinase. Mol. Cell. Biol. **10:** 1853-1862.
9. Shaw, A. S., K. E. Amerein, C. Hammond, D. F. Stern, B. M. Sefton & F. K. Rose. 1989. The Lck tyrosine protein kinase interacts with the cytoplasmic tail of the CD4 glycoprotein through its unique amino-terminal domain. Cell **59:** 627-636.
10. Turner, J. M., M. H. Brodsky, B. A. Irving, S. D. Levin, R. M. Perlmutter & D. R. Littman. 1990. Interaction of the unique N-terminal region of the tyrosine kinase p56^{lck} with cytoplasmic domains of CD4 and CD8 is mediated by cysteine motifs. Cell **60:** 755-765.

11. VEILLETTE, A., M. A. BOOKMAN, E. M. HORAK & J. B. BOLEN. 1988. The CD4 and CD8 T cell surface antigens are associated with the internal membrane tyrosine-protein kinase $p56^{lck}$. Cell **55:** 301-308.
12. GLAICHENHAUS, N., N. SHASTRI, D. R. LITTMAN & J. M. TURNER. 1991. Requirement for association of $p56^{lck}$ with CD4 in antigen-specific signal transduction in T cells. Cell **64:** 511-520.
13. ZAMOYSKA, R., P. DERHAM, S. D. GORMAN, P. VON HOEGEN, J. B. BOLEN, A. VEILLETTE & J. R. PARNES. 1990. Inability of CD8 alpha′ polypeptides to associate with $p56^{lck}$ correlates with impaired function *in vitro* and lack of expression *in vivo.* Nature **342:** 278-81.
14. XU, H. & D. R. LITTMAN. 1993. A kinase-independent function of Lck in potentiating antigen-specific T cell activation. Cell **74:** 633-643.
15. COLLINS, T. L. & S. J. BURAKOFF. 1993. Tyrosine kinase activity of CD4-associated $p56^{lck}$ may not be required for CD4-dependent T-cell activation. Proc. Natl. Acad. Sci. USA **90:** 11885-9.
16. KILLEEN, N. & D. R. LITTMAN. 1993. Helper T cell development in the absence of CD4-$p56^{lck}$ association. Nature **364:**729-732.
17. LEVIN, S. D., K. M. ABRAHAM, S. J. ANDERSON, K. A. FORBUSH & R. M. PERLMUTTER. 1993. The protein tyrosine kinase $p56^{lck}$ regulates thymocyte development independently of its interaction with CD4 and CD8 coreceptors. J. Exp. Med. **178:**245-55.
18. ABRAHAM, N., M. C. MICELI, J. R. PARNES & A. VEILLETTE. 1991. Enhancement of T-cell responsiveness by the lymphocyte-specific tyrosine protein kinase $p56^{lck}$. Nature **350:** 62-66.
19. COOPER, J. A. 1989. The *Src* family of protein tyrosine kinases. *In* Peptides and Protein Phosphorylation. B. Kemp & P. F. Alewood, Eds.: 85-113. CRC Press. Boca Raton, Florida.
20. COOPER, J. A. & B. HOWELL. 1993. The when and how of Src regulation. Cell **73:** 1051-1054.
21. AMREIN, K. E. & B. M. SEFTON. 1988. Mutation of a site of tyrosine phosphorylation in the lymphocyte-specific tyrosine protein kinase, $p56^{lck}$, reveals its oncogenic potential in fibroblasts. Proc. Natl. Acad. Sci. USA **85:** 4247-4251.
22. COOPER, J. A., K. L. GOULD, C. A. CARTWRIGHT & T. HUNTER. 1986. Tyr527 is phosphorylated in $pp60^{c\text{-}Src}$: implication for regulation. Science **231:** 1431-1434.
23. KMIECIK, T. E. & D. SHALLOWAY. 1987. Activation and suppression of $pp60^{c\text{-}Src}$ transforming ability by mutations of its primary sites of tyrosine phosphorylation. Cell **49:** 65-73.
24. MARTH, J. D., J. A. COOPER, C. S. KING, S. F. ZIEGLER, D. A. TINKER, R. W. OVERELL, E. G. KREBS & R. M. PERLMUTTER. 1988. Neoplastic transformation induced by an activated lymphocyte-specific protein tyrosine kinase ($pp56^{lck}$). Mol. Cell. Biol. **8:** 540-550.
25. MURPHY, S. M., M. BERGMAN & D. O. MORGAN. 1993. Suppression of c-Src activity by C-terminal Src kinase involves the C-Src SH2 and SH3 domains: analysis with *Saccharomyces cerevisiae.* Mol. Cell. Biol. **13:** 5290-5300.
26. SIEH, M., J. B. BOLEN & A. WEISS. 1993. CD45 specifically modulates binding of Lck to a phosphopeptide encompassing the negative regulatory tyrosine of Lck. EMBO J. **12:** 315-321.
27. SUPERTI-FURGA, G., S. FUMAGALLI, M. KOEGL, S. A. COURTNEIDGE & G. DRAETTA. 1993. Csk inhibition of c-Src activity requires both the SH2 and SH3 domains of Src. EMBO J. **12:** 2525-2634.
28. BERGMAN, M., T. MUSTELIN, C. OETKEN, J. PARTANEN, N. A. FLINT, K. E. AMREIN, M. AUTERO, P. BURN & K. ALITALO. 1992. The human $p50^{csk}$ tyrosine kinase phosphorylates $p56^{lck}$ at Tyr-505 and down regulates its catalytic activity. EMBO J. **11:** 2919-2924.
29. CAHIR MCFARLAND, E. D., T. R. HURLEY, J. T. PINGEL, B. M. SEFTON, A. SHAW & M. L. THOMAS. 1993. Correlation between Src family member regulation by the protein-

tyrosine-phosphatase CD45 and transmembrane signalling through the T-cell receptor. Proc. Natl. Acad. Sci. USA **90:** 1402-1406.

30. CHOW, L. M., M. FOURNEL, D. DAVIDSON & A. VEILLETTE. 1993. Negative regulation of T-cell receptor signalling by tyrosine protein kinase p50^{csk}. Nature **365:** 156-160.
31. MUSTELIN, T. & A. ALTMAN. 1990. Dephosphorylation and activation of the T cell tyrosine kinase pp56^{lck} by the leukocyte common antigen (CD45). Oncogene **5:** 809-813.
32. MUSTELIN, T., K. M. COGGESHALL & A. ALTMAN. 1989. Rapid activation of the T-cell tyrosine protein kinase pp56^{lck} by the CD45 phosphotyrosine phosphatase. Proc. Natl. Acad. Sci. USA **86:** 6302-6306.
33. OSTERGAARD, H. L., D. A. SHACKELFORD, T. R. HURLEY, P. JOHNSON, R. HYMAN, B. M. SEFTON & I. S. TROWBRIDGE. 1989. Expression of CD45 alters phosphorylation of the *Lck*-encoded tyrosine protein kinase in murine lymphoma T-cell lines. Proc. Natl. Acad. Sci. USA **86:** 8959-8963.
34. VEILLETTE, A. & M. FOURNEL. 1990. The CD4 associated tyrosine protein kinase p56^{lck} is positively regulated through its site of autophosphorylation. Oncogene **5:** 1455-1462.
35. CROSS, F. R. & H. HANAFUSA. 1983. Local mutagenesis of Rous Sacoma virus: the major sites of tyrosine and serine phosphorylation of p60^{Src} are dispensable for transformation. Cell **34:** 597-607.
36. SNYDER, M. A. & J. M. BISHOP. 1984. A mutation at the major phosphotyrosine in pp60$^{v\text{-}Src}$ alters oncogenic potential. Virology **136:** 375-386.
37. SNYDER, M. A., J. M. BISHOP, W. W. COLBY & A. D. LEVINSON. 1983. Phosphorylation of tyrosine-416 is not required for the transforming properties and kinase activity of pp60$^{v\text{-}Src}$. Cell **32:** 891-901.
38. ABRAHAM, N. & A. VEILLETTE. 1990. Activation of p56^{lck} through mutation of a regulatory carboxy-terminal tyrosine residue requires intact sites of autophosphorylation and myristylation. Mol. Cell. Biol. **10:** 5197-5206.
39. CARON, L., N. ABRAHAM, T. PAWSON & A. VEILLETTE. 1992. Structural requirement for enhancement of T-cell responsiveness by the lymphocyte-specific tyrosine protein kinase p56^{lck}. Mol. Cell. Biol. **12:** 2720-2729.
40. LANDAU, N. R. & D. R. LITTMAN. 1992. Packaging system for rapid production of murine leukemia virus vectors with variable tropism. J. Virol. **66:** 5110-5113.
41. HUBBARD, S. R., L. WEI, L. ELLIS & W. A. HENDRICKSON. 1994. Crystal structure of the tyrosine kinase domain of the human insulin receptor. Nature **372:** 746-754.
42. NADA, S., M. OKADA, A. MACAULEY, J. A. COOPER & H. NAKAGAWA. 1991. Cloning of a complementary DNA for a protein-tyrosine kinase that specifically phosphorylated a negative regulatory site of p60$^{c\text{-}Src}$. Nature **351:** 69-72.
43. OKADA, M., S. NADA, Y. YAMANASHI, T. YAMAMOTO & H. NAKAGAWA. 1991. CSK: a protein-tyrosine kinase involved in regulation of Src family kinase. J. Biol. Chem. **266:** 24249-24252.
44. SHIBUYA, H., K. KOHU, K. YAMADA, E. L. BARSOUMIAN, R. M. PERLMUTTER & T. TANIGUCHI. 1994. Functional dissection of p53lck, a protein tyrosine kinase which mediates interleukin-2-induced activation of the c-fos gene. Mol. Cell. Biol. **14:** 5812-5819.
45. LUO, K. & B. M. SEFTON. 1992. Activated *Lck* tyrosine protein kinase stimulates antigen-independent interleukin-2 production in T cells. Mol. Cell. Biol. **12:** 4724-4732.
46. HAUGHN, L., S. GRATTON, L. CARON, R. P. SEKALY, A. VEILLETTE & M. JULIUS. 1992. Association of tyrosine kinase p56^{lck} with CD4 inhibits the induction of growth through the $\alpha\beta$ T-cell receptor. Nature **358:** 328-331.

Coreceptors and Adapter Proteins in T-Cell Signaling

K. S. RAVICHANDRAN, J. C. PRATT,
S. SAWASDIKOSOL, H. Y. IRIE, AND
S. J. BURAKOFF

Division of Pediatric Oncology
Dana-Farber Cancer Institute and
the Department of Pediatrics
Harvard Medical School
44 Binney Street
Boston, Massachusetts 02115

Recognition of antigens by T cells involves the interaction between the antigen-specific T-cell receptor (TCR) and the antigenic peptide bound to the major histocompatibility complex (MHC) molecules on the surface of antigen-presenting cells (APC). Concurrent with the TCR, the antigen-MHC complexes are also recognized by the invariant CD4 or CD8 molecules expressed on mutually exclusive populations of mature T cells. A number of studies have demonstrated that CD4 and CD8 function as coreceptors and actively participate as adhesion and signaling molecules during T-cell recognition of antigen. Upon antigen engagement and ligation of the TCR on the cell surface (which can be mimicked by anti-TCR antibodies), a complex series of intracellular signals are transduced from the membrane, including the activation of protein tyrosine kinases and phosphatases, activation of protein kinase C, and release of intracellular calcium. A variety of intracellular proteins, including a recently defined class of adapter proteins, mediate the delivery of signals initiated by the TCR. This review will focus on how CD4 and CD8 coreceptors modulate the signals initiated by way of the TCR and the role of adapter proteins during T-cell signaling.

STRUCTURE AND FUNCTION OF CD4 AND CD8

CD4 is a 55-60 kDa monomeric glycoprotein expressed on ~60% of peripheral blood T lymphocytes. CD8 is expressed as either a homodimer of two α chains (38 kDa) or a heterodimer of an α and a β chain (28-30 kDa) on ~30% of the mature T-cell population.[1] There is a tight linkage between expression of CD4 or CD8 and the antigen recognition of the T cell with class II or class I MHC molecules, respectively. Although CD4 and CD8 were originally identified as phenotypic markers on T cells, several lines of evidence indicated an important role for CD4 and CD8 molecules in T-cell function: (1) Anti-CD4 or anti-CD8 antibodies inhibited a number of T lymphocyte effector functions *in vitro*. (2) Transfection of CD4 or CD8 into antigen-specific T-cell hybridomas (that did not express these molecules previously) rendered them responsive to antigen.[2–5] (3) In some cases, the existing minimal response of the T cell to antigen was enhanced by the coexpression of CD4 or CD8, as measured by proliferation, IL-2 production, or cytotoxicity.[4,6–10]

Subsequently, it was shown that CD4 or CD8 influences T-cell effector functions through their interaction with the same MHC molecules recognized by the antigen-specific TCR, and they have thus been termed coreceptors.[11–13] More recently, the CD4 and CD8 binding sites have been mapped to analogous regions in the nonpolymorphic domains of class II and class I MHC molecules.[14–16] These regions on MHC are distinct from regions that bind the antigenic peptide and interact with the TCR. When an MHC class I molecule carrying a point mutation in the CD8 binding region was expressed as a transgene in mice, it resulted in the lack of development of mature single $CD8^+$ T cells, providing further evidence that the site mapped *in vitro* was functionally relevant *in vivo*.[13]

An important role for CD4 and CD8 molecules in the development of T cells in the thymus has also been recognized. During thymic development, immature T cells go through stages when they express neither CD4 or CD8 (double negative), to a stage when they express both CD4 and CD8 (double positive), and subsequently mature into cells that express only CD4 or CD8, but not both (single positive). The interaction of CD4 or CD8 expressed on immature thymocytes with class II or class I MHC molecules appears to be essential for the generation of mature single positive cells. This is based upon the observation that injection of anti-CD4 or anti-class II MHC antibodies into mice blocks the development of mature $CD4^+$ T cells. Similarly, anti-CD8 or anti-class I MHC antibodies abrogated the maturation of $CD8^+$ T cells. Moreover, it has been shown that mice that have homozygous deletions of either class I or class II MHC genes lack single positive CD8 or CD4 T cells, respectively.[17–19] Thus, the interaction of CD4 or CD8 coreceptors with the MHC molecules plays a significant role in both immature and mature T-cell functions.[20–22]

CD4 and CD8 Coreceptors Function as both Adhesion and Signaling Molecules

Although the studies mentioned above indicated a role for CD4 and CD8 during T-cell recognition of antigen, it was not clear whether the coreceptors simply served to increase the strength of adhesion between the T cells and the antigen-presenting cell, transduced intracellular signals, or both.

Evidence for Coreceptor Involvement in Adhesion

A number of observations suggest a role for CD4 and CD8 in cell:cell adhesion.

(a) The ability of anti-CD4 or anti-CD8 antibodies to block T-cell function was inversely related to the avidity of the T-cell/APC interaction,[23] suggesting that when the avidity of the T-cell receptor (TCR) for antigen/MHC is low, the CD4:MHC class II interaction may become critical. This was demonstrated using a $CD4^+$ T-helper hybridoma that produced IL-2 upon stimulation with antigen plus MHC class II-bearing cells.[30,41] Moreover, although CD8 is present on class I MHC-specific T cells, when CD8 was expressed in a T-cell hybridoma specific for class II MHC, it was still able to increase IL-2 production, through its interaction with class I MHC on the APC (Ratnofsky *et al.*). These results, though suggesting a role in adhesion, are also consistent with a role for CD4 and CD8 in signaling.

(b) The loss of CD4-mediated enhancement of T-lymphocyte activation resulting from deletion of the cytoplasmic tail of CD4 (see below) can be overcome by expression of very high amounts of the CD4 deletion mutant.[24] This was also observed *in vivo*, when a cytoplasmic tail-deletion mutant of CD4, overexpressed as a transgene in mice lacking endogenous CD4, could rescue the functional development of $CD4^+$ T cells.[25] It is possible that the high level of coreceptor expression, by increasing the strength of adhesion between TCR and APC, may have simply augmented the signals initiated by the TCR alone.

(c) Direct binding between class II MHC molecules embedded in a lipid bilayer and CD4 could be demonstrated.[26] Similarly, a direct interaction between purified class I MHC molecules bound to plastic and CD8 could also be demonstrated.[27]

Taken together, these data suggest that CD4 and CD8 can increase the avidity of T cells for MHC, particularly in situations when the antigen is limiting or when the avidity of the TCR for antigen/MHC is low.

Evidence for a Signaling Role for the Coreceptors

Several observations suggested that CD4 and CD8 can also transduce intracellular signals during T-cell activation.

(a) Cross-linking CD4 or CD8 with the TCR/CD3 complex caused significantly greater activation than if the TCR/CD3 complex was cross-linked alone.[6–8,28–30] It is important to note that in these assays CD4 was physically joined to the TCR/CD3 complex. The negative effects of CD4 cross-linking alone, seen in some earlier reports, could be due to the sequestration of CD4 away from the TCR/CD3 complex rather than due to the generation of a negative signal by CD4.[6,7,30]

(b) The enhancement of T-lymphocyte activation resulting from expression of CD4 could be prevented by deletion of the cytoplasmic domain of CD4,[24] or by replacement of the cytoplasmic and transmembrane domains with a glycophosphatidylinositol anchor.[31] Because the glycophosphatidylinositol-linked CD4 mutant had the same avidity for MHC class II protein as the full-length CD4,[31] the enhancement of T-cell activation by ligation of CD4 with MHC class II molecules was dependent on the cytoplasmic domain of CD4 and not merely on the extracellular, ligand-binding domain.

(c) Both CD4 and CD8 interact with a cytoplasmic tyrosine kinase $p56^{lck}$ (Lck) (see below), and ligation of CD4 or CD8 molecules leads to activation of Lck and phosphorylation of several intracellular substrates.[32–34] Cross-linking of CD4 or CD8 with TCR leads to enhancement of tyrosine phosphorylation of a number of substrates as well as phosphorylation of proteins that are not induced by cross-linking either the TCR or the coreceptors alone.

(d) It has been demonstrated that interaction of CD8 on T cells with purified MHC class I molecules leads to phosphoinositide turnover and release of intracellular calcium.[35]

Taken together, these data suggest that both CD4 and CD8 play a role in transducing intracellular signals.

Interaction of CD4 or CD8 with the TCR/CD3 Complex

A number of observations suggest that CD4 and CD8 associate with the TCR/CD3 complex during T-cell activation and that this association may be critical for CD4- or CD8-mediated enhancement of T lymphocyte responsiveness. CD4 redistributes with the TCR/CD3 complex to the site of contact between the T cell and an APC bearing specific antigen (antigen that binds the TCR) but does not redistribute to the site of contact with an APC bearing non-specific antigen.[36] This indicates that CD4 redistribution is dependent on TCR recognition. CD4 or CD8 can be induced to comodulate from the surface with the TCR following stimulation with antibodies to the TCR/CD3 complex;[37] similarly, anti-TCR antibodies can induce CD4 or CD8 to cocap with the TCR/CD3 complex under activating conditions.[38] Interestingly, the ability of anti-TCR antibodies to cause CD4 cocapping correlates with the efficiency of antibody-mediated activation of the T lymphocyte.[38–40] Anti-TCR antibodies that induce CD4 cocapping with the TCR/CD3 complex also cause phosphorylation of the TCR-ζ chain much better than antibodies that do not induce CD4 cocapping.[40] Using fluorescence resonance energy transfer, Mittler *et al.*[41] demonstrated that CD4 coassociates with the TCR/CD3 complex within 10 minutes after antibody cross-linking of the TCR. This coassociation is dependent on the expression of the cytoplasmic domain of CD4, because CD4 deletion mutants failed to coassociate with the TCR/CD3 complex.[41] The coassociation of CD4 with the TCR/CD3 complex is specifically dependent on the association of the protein tyrosine kinase Lck with CD4 [ref 9, and see below].

Interaction of CD4 and CD8 with Lck

CD4 and CD8 associate noncovalently with the Src-related protein tyrosine kinase Lck,[32–34] suggesting that signaling by way of the coreceptors may be mediated, at least in part, by the activation of Lck. Lck is a 56 kDa protein composed of a unique amino-terminal domain, a Src-homology 2 (SH2) domain, a Src-homology 3 (SH3) domain, and a kinase domain. The CD4:Lck interaction occurs between the cytoplasmic domain of CD4 or CD8[42–45] and the N-terminal domain of Lck.[42] Cysteine residues 420 and 422 of human CD4[42] and residues 20 and 23 of Lck[46] are critical for this interaction. Cross-linking of CD4 or CD8 with monoclonal antibodies results in a rapid increase in the phosphorylation of Lck on both tyrosine and serine residues, an increase in Lck activity,[47,48] and an increase in tyrosine phosphorylation of a number of cellular proteins.[49,50] These data suggested that Lck may be involved in CD4- or CD8-mediated signal transduction. It has been estimated that ~50-70% of CD4 and ~25-40% of CD8 on the membrane associate with Lck.[48] Whether ligation of CD4 or CD8 leads to recruitment of more Lck, which may influence the magnitude of signals by way of the coreceptors, has not been addressed.

Association with Lck Is Necessary for Coreceptor Function via CD4 and CD8

Mutation of either of the two cysteine residues in the cytoplasmic domain of CD4 involved in Lck binding results in a loss of association between CD4 and Lck.[46]

Whereas expression of wild-type CD4 in a murine T-cell hybridoma resulted in enhanced antigen responsiveness,[4,51,52] expression of CD4 with mutations in the cysteines necessary for interaction with Lck did not enhance antigen responsiveness.[9,51] Similar results have been obtained for CD8. Interestingly, these CD4 mutants also failed to coassociate with the TCR/CD3 complex during T-cell stimulation, indicating that Lck is required to form a CD4- TCR/CD3 complex during T-cell activation.[9] Inasmuch as IL-2 production in response to TCR cross-linking with the mutant CD4 molecules was also diminished,[9] these data suggest that the loss in CD4 enhancement of antigen responsiveness is due largely to a lack of CD4/Lck-mediated signaling.

Signaling through CD4, CD8αα, and CD8αβ molecules

Initially, because both CD4 and CD8 molecules associate with Lck, it was believed that the CD4 and CD8 molecules would be involved in similar signaling pathways. However, Singer and colleagues observed that, in double positive thymocytes, engagement of CD4 by class II MHC or polyvalent anti-CD4 antibodies leads to inhibition of TCR expression and ζ chain phosphorylation, whereas CD8 engagement does not.[53] There are also other subtle differences between CD4 and CD8 that may point to differential signaling events during T-cell activation. For example, CD4 modulates from the cell surface upon ligation, whereas CD8 does not.[54] The avidity of CD8 for class I MHC molecules is increased upon TCR activation, although a similar phenomenon has not been observed for CD4.[55,56] In structural terms, CD4 is a monomer, whereas CD8 can exist as an $\alpha\alpha$ homodimer (capable of binding two Lck molecules) or as an $\alpha\beta$ heterodimer (with only the α chain binding to Lck[57]). Hence, it becomes necessary to understand the early signaling events mediated by CD4 and CD8 to learn more about their function as coreceptors.

Evidence for Differential Signaling by way of CD4 and CD8αα Homodimers

To better understand the early signaling events, a system was needed where both coreceptors would be expressed at comparable levels and had comparable coreceptor function as assessed by a late activation event such as interleukin-2 production. To address this, we chose to co-express CD4 and CD8 in a murine T-cell hybridoma and evaluate signals initiated through these two molecules. Initially, we generated transfectants (BYDP) that expressed high levels of human CD4 and CD8αα homodimers at the cell surface.[58] When the CD4 or CD8αα molecules were cross-linked with suboptimal levels of anti-TCR antibody, they both enhanced IL-2 production equivalently. Both CD4 and CD8αα bound equal amounts of Lck, and the associated Lck had comparable basal kinase activity.

When we examined the early signals initiated by the two molecules, several lines of evidence indicated that CD4 and CD8 molecules may initiate different intracellular signals: (1) Antibody cross-linking of CD4 resulted in fivefold greater activation of associated Lck compared to CD8 (as assessed by an *in vitro* kinase assay). (2) Using the sensitivity of the *in vitro* kinase assay combined with lysis under mild detergent conditions, several new phosphorylated bands were detected after TCR × CD4 cross-linking that were not seen after TCR × CD8 cross-linking. (3) Immunoblotting of cell lysates with antiphosphotyrosine antibody revealed that TCR × CD4 cross-linking resulted in significantly enhanced tyrosine phosphorylation of intracellular substrates compared to TCR × CD8 cross-linking. This was not due to a difference in the kinetics of activation between CD4 and CD8 as determined by a time course study. (4) A protein kinase C inhibitor, RO318220, significantly inhibited the enhancement of IL-2 production by CD8αα, although the enhancement by CD4 was far less affected.

These data suggested that CD4 and CD8αα may initiate different early signals and that CD4 and CD8αα may rely more on tyrosine kinase activation and protein kinase C-mediated signaling, respectively. *In vivo*, intraepithelial lymphocytes in the gut and natural killer cells express exclusively the CD8αα homodimer, and the above results may reflect CD8αα-mediated signaling in these cells. However, most thymocytes and peripheral T cells express mostly CD8αβ dimers, and signaling by these two forms of CD8 were therefore investigated.

CD8β Chain Influences Signaling by way of CD8α

Much of the earlier work demonstrated that CD8αα homodimers are sufficient to bind to MHC class I molecules and are sufficient for the enhancement of antigen responses by way of the TCR. The role of the β chain in CD8 signaling was largely ignored for several years because the CD8β chain is expressed on the surface only in the presence of the CD8α chain, and only the α chain interacts with Lck. However, recent reports indicate an important role for the CD8β chain in maturation of T cells in the thymus (reviewed in ref. 59). Three different studies noted a significant decrease in the number of peripheral $CD8^+$ T cells in mice bearing a homozygous disruption of the CD8β chain.[60–62] In these mice, thymocytes were arrested at the double-positive $CD4^+CD8^+$ stage, suggesting that the absence of CD8β affects positive selection of $CD8^+$ thymocytes. Besides these data, a cytoplasmic tail-deleted form of the CD8β chain, expressed as transgene in mice, acted as a dominant negative mutation and interfered with normal development of $CD8^+$ cells.[63] This suggested a potential signaling role for the cytoplasmic tail of the CD8β chain.

To address the contribution of the CD8β chain to CD8 signaling, we coexpressed CD8β along with CD8α and CD4 in the same murine T-cell hybridoma (BYDP) described above.[64] Several observations suggested that CD8β modifies the CD8α-associated Lck tyrosine kinase activity: (1) Cross-linking of CD8αβ molecules (using an antibody that specifically recognizes only CD8αβ heterodimers) showed an ~10-fold greater activation of Lck activity (assessed by an *in vitro* kinase assay) compared to cells expressing CD8αα alone. In fact, the activity observed after CD8αβ ligation was equivalent to that of CD4 cross-linking. Moreover, some of the phosphorylated substrates that were absent after TCR × CD8αα cross-linking were now observed after CD8αβ or CD4 cross-linking with the TCR. (2) In contrast to CD8αα cross-linking, cross-linking CD8αβ with the TCR led to enhanced tyrosine phosphorylation of intracellular substrates and was comparable to TCR × CD4 cross-linking. (3) Although CD8β is not directly associated with Lck, it stabilized the Lck interaction with CD8α, as determined by stable Lck association in harsher detergent lysis conditions.

Thus, the differential activity of Lck observed after CD8αα versus CD8αβ cross-linking may reflect the unique signaling potential of the CD8β chain and may, in part, account for the diminished ability of thymocytes to mature in CD8β-deficient mice.

ADAPTER PROTEINS

In the past few years, a new class of molecules, termed adapter proteins, have been shown to play a crucial role in intracellular signaling processes in a plethora of systems.[65] The adapter proteins, which mediate protein-protein interactions, have no apparent catalytic domain and are composed of one or more of SH2 and SH3 domains. SH2 and SH3 domains are modular domains that have been demonstrated to interact with specific phosphotyrosine-containing sequences and proline-rich sequences, respectively.[66,67] Examples of these adapter molecules include Grb2, Shc, Crk, Nck, and the p85 subunit of PI-3 kinase. Although these adapter proteins were originally described in the context of growth factor receptor signaling in fibroblasts, these proteins are widely expressed in all tissues and appear to interact with the same or analogous molecules in other cell types. The role of Shc, Grb2, Crk, and p85 in T-cell signaling is described below.

Shc and Grb2

Shc and Grb2 Help to Shuttle mSOS to the Membrane

Shc proteins are composed of a single SH2 domain, a glycine-proline rich collagen-homology domain (CH), and a unique N-terminal domain (N).[68] Two different isoforms of Shc (52 and 48 kDa), which differ in the length of the N domain, are expressed in T cells. Shc was first shown to function upstream of Ras proteins in PC12 pheochromocytoma cells.[69] The conversion of Ras from its inactive GDP-bound state to its active GTP-bound state occurs downstream of tyrosine kinase activation and is a crucial early event in signaling by way of many receptors, including the TCR.[70] Active GTP-bound Ras interacts with a serine-threonine protein kinase Raf and then, by way of a kinase cascade, leads to activation of MAP kinase (also called erk). It has been demonstrated, through the use of oncogenic and dominant negative inhibitors of Ras proteins and reporter constructs containing the IL-2 promoter, that activation of MAP kinases (through Ras) is essential for initiation of transcription from the IL-2 promoter.[71–73]

Shc has also been shown to be involved in Ras activation by a number of receptors, including tyrosine kinase receptors (such as the receptors for insulin, epidermal growth factor, and nerve growth factor) as well as receptors that do not possess intrinsic tyrosine-kinase activity but that signal through activation of cytoplasmic tyrosine-kinases (such as the TCR, B-cell receptor, and receptors for IL-2, IL-3, GM-CSF, and erythropoietin). We have previously demonstrated that Shc may play a role in coupling TCR activation to the Ras signaling pathway based on the following observations:[74]

(1) TCR cross-linking led to tyrosine-phosphorylation of Shc as early as 15 seconds, and the level of Shc phosphorylation was enhanced by cross-linking CD4/Lck with the TCR.

(2) Phosphorylated Shc subsequently interacted with Grb2. Grb2 is composed of a single SH2 domain and two SH3 domains. Whereas Grb2 interacts through its SH2 domain with phosphorylated Shc, it also interacts through its SH3 domains with a Ras guanine nucleotide exchange factor, mSOS. Thus upon T-cell activation, a complex between Shc, Grb2, and mSOS could be demonstrated.

(3) Several lines of evidence also demonstrated that Shc, through its own SH2 domain, interacts with the TCR-ζ chain. First, a GST-fusion protein encoding the SH2 domain of Shc precipitated the tyrosine-phosphorylated TCR-ζ chain when incubated with activated T-cell lysates. Second, a phosphopeptide, corresponding to the tyrosine-based motifs (TAMs) in the TCR-ζ chain, specifically precipitated Shc from T-cell lysates. Third, Shc was coimmunoprecipitated with the ζ chain only upon TCR activation. Moreover, subcellular fractionation (*i.e.*, membrane versus cytosolic) following TCR activation showed a movement of a fraction of Shc from the cytosol to the membrane following TCR activation (unpublished observations).

Inasmuch as Shc, Grb2, and mSOS are intracellular proteins and Ras is membrane-bound, this complex would have to be localized to the membrane, if Ras activation (through mSOS) were to be achieved. Our data that Shc interacts with the TCR-ζ chain suggested that the simultaneous interaction of Shc with the TCR, and with Grb2 and mSOS, may help to shuttle the Ras nucleotide exchange factor to the membrane, thereby leading to Ras activation (FIG. 1).

Shc Regulates the Grb2 Association with mSOS

In contrast to fibroblasts, where no changes in the levels of the Grb2:mSOS complex were observed following growth factor stimulation, stimulation by way of the T-cell receptor led to a three- to five-fold increase in the levels of the Grb2:mSOS complex.[75] This increased association was detected in leukemic T-cell lines as well as freshly isolated human peripheral blood lymphocytes. The enhanced Grb2:mSOS association occurred as early as 1 min after TCR cross-linking and was back to nearly basal levels by about 30 minutes. Several observations suggested that Shc may play a role in regulating the Grb2:mSOS association:

(1) A phosphopeptide that corresponds to the amino acid sequence surrounding Tyr317 of Shc (in the CH domain), which displaces Shc from Grb2, was able to inhibit the enhanced association between Grb2 and mSOS. Surprisingly, although the peptide can occupy the Grb2-SH2 domain and displace Shc from Grb2, the addition of the peptide to the lysates did not enhance the Grb2:mSOS interaction. Similar results were obtained with a 9-mer as well as a 16-mer Shc peptide. Thus it appeared that the simple occupancy of the Grb2-SH2 domain is not sufficient and that some structure of the native Shc molecule is necessary for this regulation.

(2) Addition of phosphorylated Shc to unactivated lysates was able to enhance the Grb2:mSOS association, essentially mimicking TCR activation. When fusion proteins encoding different regions of Shc were used, the region of Shc responsible for this regulation was narrowed to the 145 amino acid CH domain (which encompasses the Tyr317 site).

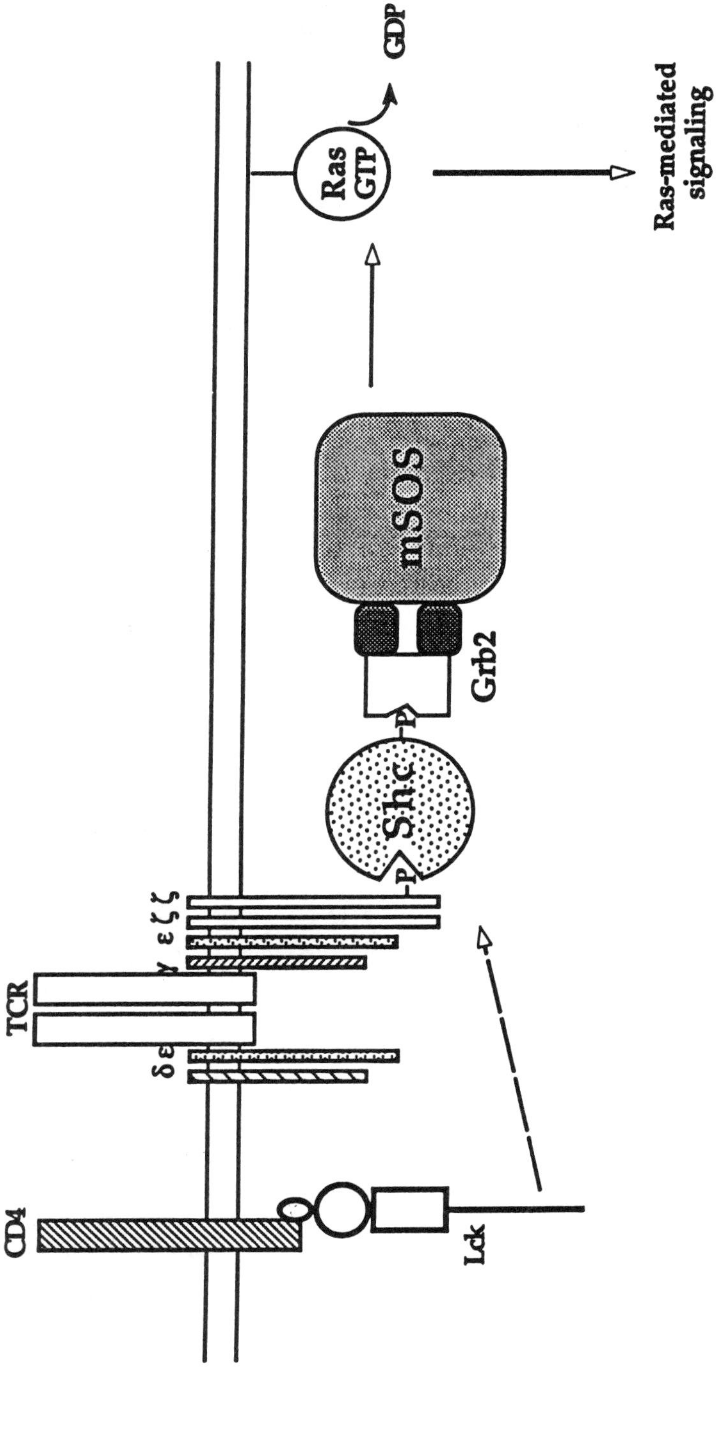

FIGURE 1. Model for Shc interaction with the TCR-ζ chain.

These data suggest that the interaction of tyrosine phosphorylated Shc with the Grb2-SH2 domain enhances the affinity of the Grb2-SH3 domains for mSOS and may serve as a mechanism for controlling the extent of Ras activation. In PC12 cells, the duration and magnitude of Ras activation seem to correlate with proliferation or differentiation signals initiated by epidermal growth factor (EGF) and nerve growth factor (NGF), respectively. Although, no such analogies have been observed so far in T cells, cross-linking of the TCR with CD4 resulted in a greater increase in the Grb2:mSOS complex compared to TCR cross-linking alone. These data suggest that, besides the simple translocation of the exchange factor mSOS to the membrane, coreceptors and adapter proteins may provide an additional level of control for modulating the extent of Ras activation, perhaps resulting in differing effects on T-cell function.

Crk

Originally identified as a part of a viral oncogene, three different Crk isoforms, Crk-I, Crk-II, and Crk-L, are expressed in T cells and most other cell types.[76] Whereas Crk-I is composed of a single SH2 domain and an SH3 domain, Crk-II and Crk-L contain an additional SH3 domain. A role for Crk in TCR-mediated signaling has been recently demonstrated.[77] Upon TCR activation, Crk proteins specifically interact with tyrosine-phosphorylated protein(s) migrating at 115-120 kDa (denoted p116). Although none of the three Crk isoforms was detectably phosphorylated upon TCR cross-linking, all three isoforms interacted, through their SH2 domain, with p116. It was also demonstrated that Crk interacted, by way of its SH3 domain, with another protein, C3G. C3G has been recently identified as another guanine nucleotide exchange factor for Ras, and the Crk:C3G interaction was found to be necessary for Ras activation in PC12 cells.[78,79] Whether the Crk-associated C3G functions in Ras activation concurrently with mSOS, or if they are nucleotide exchange factors operating under different activation conditions, awaits further investigation.

The precise identity of the p116 molecule remains uncertain at present; p116 (which has been observed by several groups as a prominant phosphoprotein(s) of 115-120 kDa in activated T cells) may represent a protein that runs as a diffuse band due to differential phosphorylation or may represent more than one protein with similar mobility on SDS-PAGE. At present, it is difficult to exclude either of these possibilities. Earlier studies indicated that p116 may be a T-cell isoform of the p130 Src substrate that was originally described in fibroblasts;[80,81] p130 has been recently cloned and shows nine motifs (YDXP, where X is any amino acid), that may be potential binding sites for the Crk-SH2 domain.[82] Moreover, p116 was recognized by a monoclonal antibody that was raised against the fibroblast p130. In our experiments, we find that p116 that is coimmunoprecipitated with Crk also cross-reacts with the anti-p130 antibody, 4F4. Interestingly, p116 has been reported to interact with the Fyn tyrosine kinase (through its SH2 domain), and Fyn has also been shown to directly associate with the members of the TCR complex. In fibroblasts, phosphorylated p130 has been localized to the plasma membrane under certain conditions.[82] It remains to be seen if a supramolecular complex of TCR:Fyn:p116:Crk:C3G exists on the membrane or, alternatively, if there are separate complexes that play independent roles in T-cell activation. One caveat to concluding that p116 is an isoform of p130 is that the anti-p130 antibody (which was originally raised to a phosphotyrosine-containing protein from v-Src-transformed cells) may cross-react with some tyrosine-phosphorylated proteins.[81] This conclusion, therefore, must await the cloning of p116 from T cells or a definitive identification of proteins of that apparent molecular weight.

Recently, p120-Cbl has been identified as a protein that is inducibly tyrosine-phosphorylated upon TCR cross-linking.[83] Cbl, which was also initially identified as a viral oncogene and subsequently cloned from mammalian cells, contains multiple tyrosine phosphorylation sites, one of which is within a YDXP motif.[84] We have also identified Cbl in Crk immunoprecipitates, and this interaction occurs through the Crk-SH2 domain (Sawasdikosol *et al.*, unpublished observations). Preliminary experiments, in which Cbl was precleared from cell lysates prior to Crk immunoprecipitations, suggest that Cbl may comprise much of the Crk-associated p116. Further experiments need to be performed to definitively identify the p116 associated with Crk.

p85 Subunit of PI-3 Kinases

Phosphatidylinositol 3′ kinase (PI-3 kinase) is composed of a 110 kDa catalytic subunit (p110) and an 85 kDa regulatory subunit (p85).[85] p85 has all the hallmarks of an adapter protein: it contains two SH2 domains and an SH3 domain, as well as two proline-rich sequences that can serve as binding sites for SH3 domains of other proteins. During signaling through PDGF or EGF, p85 interacts with the autophosphorylated receptor through its SH2 domain and thereby helps to shuttle the p110 subunit to the membrane.

It has been demonstrated that Lck associated with CD4 can interact by way of its SH3 domain with the proline-rich sequences in p85.[86,87] Cross-linking of CD4 either through antibodies or gp120 from human immunodeficiency virus-1 leads to an increase in PI-3 kinase activity associated with CD4. p85 also interacts with the SH3 domain of the Fyn tyrosine kinase. Besides its interaction with Lck and Fyn, we have also observed that p85 (by way of its SH2 domain) can interact with p120-Cbl (S. Sawasdikosol, S. J. Burakoff, unpublished observations). Because cross-linking of CD4 with the TCR leads to increased phosphorylation of Cbl and, in turn, greater association with p85, CD4 may regulate the extent of PI-3 kinase activation in multiple ways. Moreover, p85 has also been shown to interact (through its SH2 domain) with the costimulatory molecule, CD28. Thus p85 may function as an adapter in multiple ways during T-cell signaling. Although the precise role of the 3′ phosphorylated lipids in intracellular signaling remains unclear, recently p110 itself has been shown to serve as a direct downstream target of Ras in PC12 cells.

Coreceptors May Regulate Signaling through Adapter Proteins

Several observations suggest a cooperation between CD4 or CD8 coreceptors and adapter proteins during T-cell signaling: (1) Phosphorylation of Shc is enhanced upon cross-linking the TCR with CD4 or CD8. This leads to greater interaction of

Grb2 with Shc and also enhances the level of Grb2 bound to mSOS. Thus, coreceptors may regulate the extent of Ras activation through the level of Shc phosphorylation. Interestingly, cross-linking the TCR with CD8αβ heterodimers leads to greater phosphorylation of Shc compared to CD8αα homodimers and may, in turn, affect the extent of Ras activation. It is tempting to speculate that the extent of Ras activation regulated by way of the coreceptors may play a crucial role in thymic development of T cells, based on the observation that mice lacking the CD8β chain show a significantly diminished maturation of $CD4^{+}8^{+}$ thymocytes to single $CD8^{+}$ cells. (2) The level of phosphorylation on p116 is enhanced by cross-linking CD4 or CD8 with the TCR. Inasmuch as this, in turn, alters the levels of Crk associated with p116, coreceptors may alter the downstream signaling by way of Crk. (3) Coreceptors also influence interactions of p85 in several ways. Besides the association of CD4 or CD8 (through Lck) with p85, the coreceptors also regulate the level of phosphorylation on Cbl and alter the extent of p85 bound to Cbl.

CD4-Associated Lck: Another Adapter Protein?

Surprisingly, it has been demonstrated that the CD4-associated tyrosine kinase activity of Lck could be reduced or eliminated without a loss in CD4-dependent IL-2 production.[88,89] Expression of a chimeric receptor containing CD4 extracellular and transmembrane domains fused to a kinase-deleted Lck resulted in similar IL-2 production following stimulation with a CD4-dependent antigen as when a full-length, kinase-functional form of Lck was fused to the CD4.[89] Similarly, overexpression of a deletion mutant of Lck lacking the kinase domain resulted in greater than 95% inhibition of tyrosine kinase activity associated with CD4 but did not diminish IL-2 production in response to CD4-dependent antigen or to cross-linking with anti-CD4 and anti-TCR antibodies.[88] These data suggest that the kinase domain of Lck is not necessary for CD4 enhancement of IL-2 production, although the interaction of Lck with CD4 is essential for coreceptor activity.

In addition to the kinase domain, Lck also contains an SH2 domain and an SH3 domain. It is possible that CD4 coreceptor activity is mediated by either, or both, of these domains. A model that could reconcile these apparently paradoxical results is given below. The low level of Fyn associated with the TCR/CD3 complex[90] or free Lck (not associated with CD4) might initiate phosphorylation of TCR-ζ or CD3 chains, especially ε, to begin recruiting the tyrosine kinase ZAP-70 (which associates through its SH2 domains with phosphorylated ζ and ε chains). At this time, CD4-associated Lck may be recruited to the TCR/CD3 complex by its interaction with phosphorylated ZAP-70. Recently, it has been demonstrated that Lck, through its SH2 domain, can interact with phosphorylated ZAP-70.[91] At this time, Lck may bring other proteins into the TCR/CD3 complex (via its SH2 and SH3 domains) that are essential for CD4/Lck-mediated coreceptor activity. Such a model would be consistent with the finding that the kinase activity of Lck associated with CD4 is dispensable, whereas the association of Lck with CD4 is essential for the coreceptor activity of CD4. A similar scenario may exist for CD8, although it has not been analyzed. The precise molecular interactions and importance of the SH2 and SH3 domains of Lck in coreceptor activity await further investigation.

REFERENCES

1. PARNES, J. R. 1989. Molecular biology and function of CD4 and CD8. Adv. Immunol. **44:** 265-311.
2. MARRACK, P., R. ENDERS, R. SHIMONKEVITZ, A. ZLOTNIK *et al.* 1983. The major histocompatibility complex-restricted antigen receptor on T cells. II. Role of the L3T4 product. J. Exp. Med. **158:** 1077-1091.
3. GREENSTEIN, J. L., B. MALISSEN & S. J. BURAKOFF. 1985. Role of L3T4 in antigen-driven activation of a class I-specific T cell hybridoma. J. Exp. Med. **162:** 369-374.
4. SLECKMAN, B. P., A. PETERSON, W. K. JONES & J. A. FORAN *et al.* 1987. Expression and function of CD4 in a murine T-cell hybridoma. Nature **328:** 351-353.
5. RATNOFSKY, S. E., A. PETERSON, J. L. GREENSTEIN & S. J. BURAKOFF. 1987. Expression and function of CD8 in a murine T cell hybridoma. J. Exp. Med. **166:** 1747-1757.
6. ANDERSON, P., M.-L. BLUE, C. MORIMOTO & S. F. SCHLOSSMAN, 1987. Cross-linking of T3 (CD3) with T4 (CD4) enhances the proliferation of resting T lymphocytes. J. Immunol. **139:** 678-682.
7. EMMRICH, F., L. KANZ & K. EICHMANN. 1987. Cross-linking of the T cell receptor complex with the subset-specific differentiation antigen stimulates interleukin 2 receptor expression in human CD4 and CD8 T cells. Eur. J. Immunol. **17:** 529-534.
8. EICHMANN, K., J. I. JONSSON, I. FALK & R. EMMRICH. 1987. Effective activation of resting mouse T lymphocytes by cross-linking submitogenic concentrations of the T cell antigen receptor with either Lyt-2 or L3T4. Eur. J. Immunol. **17:** 643-650.
9. COLLINS, T. L., S. UNIYAL, J. SHIN, J. L. STROMINGER *et al.* 1992. $p56^{lck}$ association with CD4 is required for the interaction between CD4 and the TCR/CD3 complex and for optimal antigen stimulation. J. Immunol. **148** (7):2159-2162.
10. CHALUPNY, N. J., J. A. LEDBETTER & P. KAVATHAS. 1991. Association of CD8 with $p56^{lck}$ is required for early T cell signalling events. EMBO J. **10** (5): 1201-1207.
11. ALDRICH, C. J., R. E. HAMMER, S. JONES-YOUNGBLOOD, U. KOSZINOWSKI *et al.* 1991. Negative and positive selection of antigen-specific cytotoxic T lymphocytes affected by the α3 domain of MHC class I molecules. Nature **352:** 718-721.
12. INGOLD, A. L., C. LANDEL, C. KNALL, G. A. EVANS *et al.* 1991. Co-engagement of CD8 with the T cell receptor is required for negative selection. Nature **352:** 721-723.
13. KILLEEN, N., A. MORIARTY, H.-S. TEH & D. R. LITTMAN. 1992. Requirement for CD8-major histocompatibility complex class I interaction in positive and negative selection of developing T cells. J. Exp. Med **176:** 89-97.
14. KONIG, R., L.-Y. HUANG & R. N. GERMAIN. 1992. MHC class II interaction with CD4 mediated by a region analogous to the MHC class I binding site for CD8. Nature **356:** 796-799.
15. CAMMAROTA, G., A. SCHEIRLE, B. TAKACS, D. M. DORAN *et al.* 1992. Identification of a CD4 binding site on the β2 domain of HLA-DR molecules. Nature **356:** 799-801.
16. SALTER, R. D., R. J. BENJAMIN, P. K. WESLEY, S. E. BUXTON *et al.* 1990. A binding site for the T cell co-receptor CD8 on the α3 domain of HLA-A2. Nature **345:** 41-46.
17. COSGROVE, D., D. GRAY, A. DIERICH, J. KAUFMAN *et al.* 1991. Mice lacking MHC class II molecules. Cell **66:** 1051-1066..
18. GRUSBY, M. J., R. S. JOHNSON, V. E. PAPAIOANNOU & L. H. GLIMCHER. 1991. Depletion of $CD4^+$ T cells in major histocompatibility complex class II-deficient mice. Science **253:** 1417-1420.
19. FUNG-LEUNG, W.-P., M. W. SCHILHAM, A. RAHEMTULLA, T. M. KUNDIG *et al.* 1991. CD8 is needed for development of cytotoxic T cells but not helper T cells. Cell **65:** 443-449.
20. BILL, J. & E. PALMER. 1989. Positive selection of $CD4^+$ T cells mediated by MHC class II-bearing stromal cell in the thymic cortex. Nature **341:** 649-651.
21. BENOIST, C. & D. MATHIS. 1989. Positive selection of the T cell repertoire: where and when does it occur? Cell **58:** 1027-1033.

22. Berg, L. J., A. M. Pullen, B. Fazekas de St. Groth, D. Mathis *et al.* 1989. Antigen/MHC-specific T cells are preferentially exported from the thymus in the presence of their MHC ligand. Cell **58:** 1035-1046.
23. Biddison, W. E., P. E. Rao, M. A. Talle, G. Goldstein *et al.* 1984. Possible involvement of the T4 molecule in T cell recognition of class II HLA antigens. Evidence from studies of CTL-target cell binding. J. Exp. Med. **159:** 783-797.
24. Sleckman, B. P., A. Peterson, J. A. Foran, J. C. Gorga *et al.* 1988. Functional analysis of a cytoplasmic domain-deleted mutant of the CD4 molecule. J. Immunol. **141** (1): 49-54.
25. Killeen, N. & D. R. Littman. 1993. Helper T-cell development in the absence of CD4-$p56^{lck}$ association. Nature **364:** 729-732.
26. Doyle, C. & J. L. Strominger. 1987. Interaction between CD4 and class II MHC molecules mediates cell adhesion. Nature **330:** 256-259.
27. Rosenstein, Y., S. Ratnofsky, S. J. Burakoff & S. H. Herrmann. 1989. Direct evidence for binding of CD8 to HLA class I antigens. J. Exp. Med. **169:** 149-160.
28. Emmrich, F., U. Strittmatter & K. Eichmann. 1986. Synergism in the activation of human CD8 T cells by cross-linking the T-cell receptor complex with the CD8 differentiation antigen. Proc. Natl. Acad. Sci. USA **83:** 8298-8302.
29. Owens, T., B. Fazekas de St. Groth & J. F. A. P. Miller. 1987. Coaggregation of the T-cell receptor with CD4 and other T-cell surface molecules enhances T-cell activation. Proc. Natl. Acad. Sci. USA **84:** 9209-9213.
30. Walker, C., F. Bettens & W. J. Pickler. 1987. Activation of T cells by cross-linking an anti-CD3 antibody with a second anti-T cell antibody: mechanism and subset-specific activation. Eur. J. Immunol. **17:** 873-880.
31. Sleckman, B. P., Y. Rosenstein, V. E. Igras, J. L. Greenstein *et al.* 1991. Glycolipid-anchored form of CD4 increases intercellular adhesion but is unable to enhance T cell activation. J. Immunol. **147** (2): 428-431.
32. Veillette, A., M. A. Bookman, E. M. Horak & J. B. Bolen. 1988. The CD4 and CD8 T cell surface antigens are associated with the internal membrane tyrosine-protein kinase $p56^{lck}$. Cell **55:** 301-308.
33. Rudd, C. E., J. M. Trevillyan, J. D. Dasgupta, L. L. Wong *et al.* 1988. The CD4 receptor is complexed in detergent lysates to a protein-tyrosine kinase (pp58) from human T lymphocytes. Proc. Natl. Acad. Sci. USA **85:** 5190-5194.
34. Barber, E. K., J. D. Dasgupta, S. F. Schlossman, J. M. Trevillyan *et al.* 1989. The CD4 and CD8 antigens are coupled to a protein-tyrosine kinase ($p56^{lck}$) that phosphorylates the CD3 complex. Proc. Natl. Acad. Sci. USA **86:** 3277-3281.
35. O'Rourke, A. M. & M. F. Mescher. 1992. Cytotoxic T-lymphocyte activation involves a cascade of signalling and adhesion events. Nature **358:** 253-255.
36. Kupfer, A., S. J. Singer, C. A. Janeway & S. L. Swain. 1987. Coclustering of CD4 (L3T4) molecule with the T-cell receptor is induced by specific direct interaction of helper T cells and antigen-presenting cells. Proc. Natl. Acad. Sci. USA **84:** 5888-5892.
37. Saizawa, K., J. Rojo & C. A. Janeway. 1987. Evidence for a physical association of CD4 and the CD3:α:β T-cell receptor. Nature **328:** 260-263.
38. Rojo, J. M., K. Saizawa & C. A. Janeway. 1989. Physical association of CD4 and the T-cell receptor can be induced by anti-T-cell receptor antibodies. Proc. Natl. Acad. Sci. USA **86:** 3311-3315.
39. Portoles, P., J. Rojo, A. Golby, M. Bonneville *et al.* 1989. Monoclonal antibodies to murine CD3ε define distinct epitopes, one of which may interact with CD4 during T cell activation. J. Immunol. **142** (12): 4169-4175.
40. Dianzani, U., A. Shaw, B. K. Al-Ramadi, R. T. Kubo *et al.* 1992. Physical association of CD4 with the T cell receptor. J. Immunol. **148** (3): 678-688.
41. Mittler, R. S., S. J. Goldman, G. L. Spitalny & S. J. Burakoff 1989. T-cell receptor-CD4 physical association in a murine T-cell hybridoma: induction by antigen receptor ligation. Proc. Natl. Acad. Sci. USA **86:** 8531-8535.

42. Shaw, A. S., K. E. Amrein, C. Hammond, D. F. Stern *et al.* 1989. The lck tyrosine protein kinase interacts with the cytoplasmic tail of the CD4 glycoprotein through its unique amino-terminal domain. Cell **59:** 627-636.
43. Shaw, A. S., J. Chalupny, J. A. Whitney, C. Hammond *et al.* 1990. Short related sequences in the cytoplasmic domains of CD4 and CD8 mediate binding to the amino-terminal domain of the $p56^{lck}$ tyrosine protein kinase. Mol. Cell. Biol. **10** (5): 1853-1862.
44. Vega, M. A., M.-C. Kuo, A. C. Carrera & J. L. Strominger. 1990. Structural nature of the interaction between T lymphocyte surface molecule CD4 and the intracellular protein tyrosine kinase lck. Eur. J. Immunol. **20:** 453-456.
45. Veillette, A., B. P. Sleckman, S. Ratnofsky, J. B. Bolen *et al.* 1990. The cytoplasmic domain of CD4 is required for stable association with the lymphocyte-specific tyrosine kinase $p56^{lck}$. Eur. J. Immunol. **20:** 1397-1400.
46. Turner, J. M., M. H. Brodsky, B. A. Irving, S. D. Levin *et al.* 1990. Interaction of the unique N-terminal region of tyrosine kinase $p56^{lck}$ with cytoplasmic domains of CD4 and CD8 is mediated by cysteine motifs. Cell **60:** 755-756.
47. Marth, J. D., D. B. Lewis, M. P. Cooke, E. D. Mellins *et al.* 1989. Lymphocyte activation provokes modification of a lymphocyte-specific protein tyrosine kinase ($p56^{lck}$). J. Immunol **142** (7): 2430-2437.
48. Luo, K. & B. M. Sefton. 1990. Cross-linking of T-cell surface molecules CD4 and CD8 stimulates phosphorylation of the *lck* tyrosine protein kinase at the autophosphorylation site. Mol. Cell. Biol. **10** (10): 5305-5313.
49. Veillette, A., J. B. Bolen & M. A. Bookman. 1989. Alterations in tyrosine protein phosphorylation induced by antibody-mediated cross-linking of the CD4 receptor of T lymphocytes. Mol. Cell. Biol. **9** (10): 4441-4446.
50. Veillette, A., M. A. Bookman, E. M. Horak, L. E. Samelson *et al.* 1989. Signal transduction through the CD4 receptor involves the activation of the internal membrane tyrosine-protein kinase $p56^{lck}$. Nature **338:** 257-259.
51. Glaichenhaus, N., N. Shastri, D. R. Littman & J. M. Turner. 1991. Requirement for association of $p56^{lck}$ with CD4 in antigen-specific signal transduction in T cells. Cell **64:** 511-520.
52. Gay, D., P. Maddon, R. Sekaly, M. A. Talle *et al.* 1987. Functional interaction between human T-cell protein CD4 and the major histocompatibility complex HLA-DR antigen. Nature **328:** 626-629.
53. Nakayama, T., C. H. June, T. I. Munitz, M. Sheard *et al.* 1990. Inhibition of T cell receptor expression and function in immature $CD4^+CD8^+$ cells by CD4. Science **249:** 1558-1561.
54. Sleckman, B. P., M. Bigby, J. L. Greenstein, S. J. Burakoff *et al.* 1989. Requirements for modulation of the CD4 molecule in response to phorbol myristate acetate. J. Immunol. **142** (5): 1457-1462.
55. O'Rourke, A. M., J. Rogers & M. F. Mescher. 1990. Activated CD8 binding to class I protein mediated by the T-cell receptor results in signalling. Nature **346:** 187-189.
56. O'Rourke, A. M. & M. F. Mescher. 1993. The roles of CD8 in cytotoxic T lymphocyte function. Immunol. Today **14** (4): 183-188.
57. Zamoyska, R., P. Derham, S. D. Gorman, P. von Hoegen *et al.* 1989. Inability of $CD8\alpha'$ polypeptides to associate with $p56^{lck}$ correlates with impaired function *in vitro* and lack of expression *in vivo*. Nature **342:** 278-281.
58. Ravichandran, K. S., & S. J. Burakoff. 1994. Evidence for differential intracellular signaling via CD4 and CD8 molecules. J. Exp. Med. **179:** 727-732.
59. Zamoyska, R., 1994. The CD8 coreceptor revisited: one chain good, two chains better. Immunity **1:** 243-246.
60. Fung-Leung, W. P., T. M. Kundig, K. Ngo, J. Panakos *et al.* 1994. Reduced thymic maturation but normal effector function of $CD8^+$ cells in $CD8\beta$ gene-targeted mice. J. Exp. Med. **180:** 959.

61. Crooks, M. E. C. & D. R. Littman. 1994. Disruption of T lymphocyte positive and negative selection in mice lacking the CD8β chain. Immunity **1:** 277-285.
62. Nakayama, K.-I., K. Nakayama, I. Negishi, K. Kuida *et al.* 1994. Requirement for CD8 β chain in positive selection of CD8-lineage T cells. Science **263:** 1131-1133.
63. Itano, A., D. Cado, F. K. M. Chan & E. Robey. 1994. A role for the cytoplasmic tail of the β chain of CD8 in thymic selection. Immunity **1:** 287-290.
64. Irie, H. Y., K. S. Ravichandran & S. J. Burakoff. 1995. CD8β chain influences CD8α-associated Lck kinase activity. J. Exp. Med. **181:** 1267-1273.
65. Pawson, T. & G. D. Gish. 1992. SH2 and SH3 domains: From structure to function. Cell **71:** 359-362.
66. Songyang, Z., S. E. Shoelson, M. Chaudhuri, G. Gish *et al.* 1993. SH2 domains recognize specific phosphopeptide sequences. Cell **72:** 767-778.
67. Ren, R., B. J. Mayer, P. Cicchetti & D. Baltimore. 1993. Identification of a ten-aminoacid proline-rich SH3 binding site. Science **259:** 1157-1161.
68. Pelicci, G., L. Lanfrancone, F. Grignani, J. McGlade *et al.* 1992. A novel transforming protein (SHC) with an SH2 domain is implicated in mitogenic signal transduction. Cell **70:** 93-104.
69. Rozakis-Adcock, M., J. McGlade, G. Mbamalu, G. Pelicci *et al.* 1992. Association of the Shc and Grb2/Sem5 SH2-containing proteins is implicated in activation of the Ras pathway by tyrosine kinases. Nature **360:** 689-692.
70. Downward, J. 1992. Regulatory mechanisms for ras proteins. BioEssays **14:** 177-184.
71. Rayter, S., M. Woodrow, S. Lucas, D. Cantrell *et al.* 1992. p21ras mediates control of IL2 gene promoter function in T cell activation. EMBO J. **11:** 4549-4556.
72. Baldari, C., G. Macchia & J. Telford. 1992. Interleukin-2 promoter activation in T cell expressing activated Ha-ras. J. Biol. Chem. **267:** 4289-4291.
73. Izquierdo, M., S. Leevers, C. Marshall & D. Cantrell. 1993. $p21^{ras}$ couples the T cell antigen receptor to extracellular signal-regulated kinase 2 in T lymphocytes. J. Exp. Med. **178:** 1199-1208.
74. Ravichandran, K. S., K. K. Lee, Z. Songyang, L. C. Cantley *et al.* 1993. Interaction of Shc with ζ chain of the T cell receptor upon T cell activation. Science **262:** 902-905.
75. Ravichandran, K. S., U. Lorenz, S. E. Shoelson & S. J. Burakoff 1995. Interaction of Shc with Grb2 regulates the association of Grb2 with mSOS. Mol. Cell. Biol. **15:** 593-600.
76. Reichman, C. T., B. J. Mayer, S. Keshav & H. Hanafusa. 1992. The product of the cellular crk gene consists primarily of SH2 and SH3 regions. Cell Growth & Differen. **3:** 451-460.
77. Sawasdikosol, S., K. S. Ravichandran, J. H. Chang, K. K. Lee *et al.* 1995. Interaction of Crk with p116 upon T cell activation. J. Biol. Chem. In press.
78. Tanaka, S., T. Morishita, Y. Hashimoto, S. Hattori *et al.* 1994. C3G, a guanine nucleotide-releasing protein expressed ubiquitously, binds to the Src homology 3 domains of CRK and GRB2/ASH proteins. Proc. Natl. Acad. Sci. USA **91:** 3443-3447.
79. Matsuda, M., Y. Hashimoto, K. Muroya, H. Hasegawa *et al.* 1994. CRK protein binds to two guanine nucleotide-releasing proteins for the ras family and modulates nerve growth factor-induced activation of ras in PC12 cells. Mol. Cell. Biol. **14:** 5495-5500.
80. Da Silva, A. J., O. Janssen & C. E. Rudd. 1992. T cell receptor zeta/CD3-p59fyn(T)-associated p120/130 binds to the SH2 domain of p59fyn(T). J. Exp. Med. **178:** 2107-213.
81. Tsygankov, A. Y., C. Spana, R. B. Rowley, R. C. Penhallow *et al.* 1994. Activation-dependent tyrosine phosphorylation of Fyn-associated proteins in T lymphocytes. J. Biol. Chem. **269:** 7792-7800.
82. Sakai, R., A. Iwamatsu, N. Hirano, S. Ogawa *et al.* 1994. A novel signaling molecule, p 130, forms stable complexes *in vivo* with v-Crk and v-Src in a tyrosine phosphorylation-dependent manner. EMBO J. **13:** 3748-3756.

83. DONOVAN, J. A., R. L. WANGE, W. Y. LANGDON & L. E. SAMELSON. 1994. The protein product of the c-cbl protooncogene is the 120-kDa tyrosine-phosphorylated protein in jurkat cells activated via the T cell antigen receptor. J. Biol. Chem. **269:** 22921-22924.
84. BLAKE, T. J., M. SHAPIRO, H. C. MORSE III & W. Y. LANGDON. 1991. The sequences of the human and mouse c-cbl proto-oncogenes show v-cbl was generated by a large truncation encompassing a proline-rich domain and a leucine zipper-like motif. Oncogene **6:** 643-657.
85. CANTLEY, L. C., K. R. AUGER, C. CARPENTER, B. DUCKWORTH *et al.* 1991. Oncogenes and signal transduction. Cell **64:** 281-302.
86. PRASAD, K. V. S., O. JANSSEN, R. KAPELLER, M. RAAB *et al.* 1993. Src-homology 3 domain of protein kinase $p59^{fyn}$ mediates binding to phosphatidylinositol 3-kinase in T cells. Proc. Natl. Acad. Sci. USA **90:** 7366-7370.
87. PRASAD, K. V. S., R. KAPELLER, O. JANSSEN, H. RAPKE *et al.* 1993. Phosphatidylinositol (PI) 3-kinase and PI 4-kinase binding to the CD4-p56lck complex: the p56lck SH3 domain binds to PI 3-kinase but not PI 4-kinase. Mol. Cell. Biol **13:** 7708-7717.
88. COLLINS, T. L. & S. J. BURAKOFF 1993. Tyrosine kinase activity of CD4-associated $p56^{lck}$ may not be required for CD4-dependent T-cell activation. Proc. Natl. Acad. Sci. USA **90** (24): 11885-11889.
89. XU, H. & D. R. LITTMAN. 1993. A kinase-independent function of Lck in potentiating antigen-specific T cell activation. Cell **74:** 633-643.
90. SAMELSON, L. E., A. F. PHILLIPS, E. T. LUONG & R. D. KLAUSNER. 1990. Association of the fyn protein-tyrosine kinase with the T-cell antigen receptor. Proc. Natl. Acad. Sci. USA **87:** 4358-4362.
91. DUPLAY, P., M. THOME, F. HERVÉ & O. ACUTO. 1994. p56lck interacts via its src homology domain with the ZAP-70 kinase. J. Exp. Med. **179:** 1163-1172.

Symmetry of the Activation of Cyclin-dependent Kinases in Mitogen and Growth Factor–stimulated T Lymphocytes[a]

JAIME F. MODIANO, JOANNE DOMENICO, AGOTA SZEPESI, NAOHIRO TERADA, JOSEPH J. LUCAS, AND ERWIN W. GELFAND[b]

Division of Basic Sciences
Department of Pediatrics
National Jewish Center for Immunology and Respiratory Medicine
1400 Jackson Street
Denver, Colorado 80206

INTRODUCTION

Several interdependent biochemical pathways are activated following either stimulation of resting T cells through the antigen receptor or stimulation of activated T cells through the interleukin-2 (IL-2) receptor.[1–5] Many of the events that occur after the engagement of either of these receptors are qualitatively similar, such as the activation of nonreceptor tyrosine kinases,[3,5] mitogen-activated protein kinase (MAPK) pathways and ribosomal S6 kinases,[6,6a] and preexisting transcription factors, leading to the expression of specific growth-associated genes.[7,8] However, there are other events that are triggered either specifically by stimulation through the T-cell receptor, such as the hydrolysis of phosphoinositides with subsequent increases in the concentration of intracellular calcium and protein kinase C,[9–12] or specifically through IL-2 receptors, such as activation of members of the Janus (Jak) kinase family of enzymes.[13–16]

Our aims were to delineate those regulatory events that are associated with cell-cycle entry and early G1 phase versus those that are necessary for later cell-cycle progression. Of particular interest were the functions of the members of the cyclin-dependent kinase (CDK) family of proteins and their roles in the regulation of the *RB*-1 (retinoblastoma susceptibility) gene product ($p110^{Rb}$). The $p110^{Rb}$ protein is a tumor suppressor that acts by positively and negatively regulating the transcription of various growth-associated genes.[17–20] In normal resting and proliferating cells, the

[a] This work was supported by National Institutes of Health Grants POI-HL-36577 and AI-26490 (E. W. Gelfand) and by American Cancer Society Grant IM-746 (J. J. Lucas). J. F. Modiano is the recipient of Clinical Investigator Development Award HL-03130 from the National Institutes of Health.

[b] To whom correspondence should be addressed.

function of p110^{Rb} is controlled through the phosphorylation of numerous serine and threonine residues[19,21–24] by Rb-kinases that likely include one or several CDKs.[23–29] In resting (G0) cells, p110^{Rb} is hypophosphorylated.[19,21–23] The protein becomes phosphorylated during the G1 and S phases, and it remains so throughout the S and G2 phases.[21,24] Several CDKs, in association with their cyclin partner, bind directly to p110^{Rb} and phosphorylate it *in vitro*,[26,27,29–33] but it is not known which of the CDKs is responsible for these phosphorylation events *in vivo*.[24] We and others have established the hierarchy of expression of the known CDKs and their cyclin partners in T cells as they progress through the cell cycle. Based on the kinetics of expression of the CDKs and their cyclin partners, and *in vitro* specificity of the complexes, it appears that either Cdk2, Cdk4, Cdk6, or a combination thereof could function as the G1-phase Rb-kinase(s).

To establish the role of these CDKs and their importance in T-cell cycle progression, it was important to distinguish the events that regulate entry into the cell cycle and progression through the early stages of the G1 phase from those that control passage through the putative ''R'' point (at which the cells are irreversibly committed to undergo DNA synthesis), and the G1 to S transition. The entry and progression of lymphocytes through the cell cycle are dependent on the coordinated expression of a number of genes and proteins in a temporally defined sequence.[34] In particular, the production of IL-2 is regulated very closely at various levels, and it requires the sustained activation of the full complement of biochemical cascades that are initiated by antigenic or mitogenic stimulation.[12,35–37] We took advantage of this fact to separate the IL-2-independent events that lead to cell cycle entry from those that require IL-2 and regulate progression to the S phase. Our data show that T-cell activation stimulates a transient phosphorylation of p110^{Rb}, expression and activation of newly synthesized Cdk4 and Cdk6, and activation of preexisting Cdk2 in an IL-2-independent fashion. By comparison, the interaction of IL-2 with its receptor in activated T cells stimulates sustained phosphorylation of p110^{Rb} along with a second, greater phase of expression and activation of newly synthesized Cdk4, Cdk6, and Cdk2.

MATERIAL AND METHODS

Cell Cultures

Peripheral blood T cells were purified from plateletpheresis residues by Ficoll-Hypaque (1.077 g/mL) density gradient centrifugation followed by depletion of adherent cells from the peripheral blood mononuclear cells and E-rosetting on neuraminidase-treated sheep erythrocytes as described.[35,38] The phenotype of the resulting T-cell population was ≥96% CD3$^+$, ≤6% DR$^+$, ≤2% CD16$^+$, and <1% CD20$^+$ or CD14$^+$. The proliferative response of T cells to mitogens was determined by the incorporation of [^{3}H]thymidine into DNA, 48 hours after the onset of culture, and following precipitation with 10% trichloroacetic acid as described.[39]

Induction of Competence

Suboptimal stimuli can induce resting T cells to acquire a ''competent'' state. Specifically, competent T cells are those that exit G0 but stop their passage through the cell cycle at a defined point before entry into S phase. These cells have been termed competent because the addition of a progression signal (such as IL-2) will induce the cells to progress into the S phase and undergo cell division. Peripheral blood T cells were rested overnight in complete medium. These cells were rendered competent to proliferate, as described by a brief stimulation with 12,13-phorbol dibutyrate (PDB, 10 nM) and ionomycin (500 nM) followed by extensive washing or by suboptimal concentrations of phytohemagglutinin (PHA, 0.5 μg/mL) that do not stimulate production of IL-2.[35,40] The competent cells were divided into two groups. One group received anti-human recombinant IL-2 antibody (2 μg/mL), whereas the other received human recombinant IL-2 (25 nM) for the duration of culture (''progressing'' T cells). In each experiment, an equal number of T cells as those that were rendered competent were allowed to remain unstimulated (received vehicle), or were stimulated to proliferate by the addition of PDB and ionomycin throughout the culture period (''proliferating'' T cells).

Flow Cytometry

Analyses for all the experiments were done on an Epics Profile or an Epics 751 (Coulter Corporation, Hialeah, FL) flow cytometer. The expression of surface T-cell antigens, $p34^{cdc2}$, and cellular DNA content was analyzed by single or dual-wave flow cytometry, as described.[41,42]

Northern Blotting

Cytosolic RNA was isolated from resting, competent, progressing, or proliferating T cells, and the steady-state expression of IL-2, *cdk*4, *cdk*2, cyclin D2, and β_2-microglobulin mRNAs was assessed as described.[12,40]

Immunoblotting

The relative levels of the hypophosphorylated or of the phosphorylated species of Rb, and the accumulation of intracellular IL-2, Cdk4, Cdk2, and cyclin D2 were analyzed, as described, using enhanced chemiluminescense (ECL) or an alkaline phosphatase system.[12,40]

CDK Kinase Assays

The kinase activities of Cdk2, Cdk4, and Cdk6 were determined as described.[32,40,40a] Histone HI was used as a substrate for Cdk2, and $p56^{Rb}$, a truncated recombinant Rb protein, was used as a substrate for Cdk4 and Cdk6. One unit/min. of CDK-associated kinase activity was defined as the incorporation of 1 fmol of phosphate/minute into the substrate.

RESULTS

Progression of Competent T Cells through the Cell Cycle

We determined the cell-cycle status of resting, competent, progressing, or proliferating T cells, by assessing DNA synthesis, DNA content, expression of $p34^{cdc2}$, and expression of IL-2 receptors and transferrin receptors. Resting peripheral blood T cells comprised a population of cells that were predominantly (>96%) in the G0 phase of the cell cycle and lacked significant expression of surface receptors for IL-2 or transferrin (FIGURES 1A and 1B). These resting T cells could be rapidly induced to enter the cell cycle in a synchronous manner upon continuous stimulation by PDB and ionomycin, a condition that bypasses the requirement for antigen receptor engagement. These combined stimuli resulted in entry and progression through the cell cycle by most of the T cells (FIG. 1G). The expression of the α chain of the IL-2 receptor and of the transferrin receptor was markedly increased (up to 30-fold greater mean fluorescence intensity) in 80-95% of these cells, as compared to the unstimulated controls (FIG. 1H).

Stimulation of T cells by PDB and ionomycin for 20 minutes, followed by washing (induction of competence), failed to induce a brisk proliferative response: nearly all of these cells remained in the G0-G1 stages of the cell cycle as measured by DNA synthesis and by analysis of DNA content (FIG. 1C). However, the competent cells exhibited a measurable increase (15-25% change in mean fluorescence intensity) in the expression of surface IL-2 receptor α and transferrin receptors (FIG. 1D). This low level of expression of receptors for IL-2 and transferrin by competent T cells was functionally significant, because the addition of IL-2 (25 nM, a progression signal) at the onset of culture restored 50-90% of the proliferative response and allowed the cells to progress beyond the G1 phase ("progressing T cells;" FIG. 1E). As has been described previously,[36,43,44] IL-2 up-regulated further the surface expression of both its own receptors and transferrin receptors in the competent T cells (FIG. 1F).

Competent cells failed to progress through the G1 phase because they cannot sustain critical levels of IL-2 production, although they express adequate levels of IL-2 receptors.[35–38,45] FIGURES 2A and 2B show that resting T cells did not express IL-2 mRNA. These cells also did not produce any detectable IL-2 protein.[40] Conversely, when T cells were stimulated continuously by PDB and ionomycin, they expressed high levels of IL-2 mRNA after 4 hours (FIGURES 2A and 2B); this expression of IL-2 mRNA was sustained over 24 hours and resulted in high levels of IL-2 production. T cells that were rendered competent by a brief stimulation with PDB and ionomycin (FIG. 2A) or by stimulation with suboptimal concentrations of PHA (FIG. 2B) did not express IL-2 mRNA or produce detectable levels of IL-2 protein,[40] despite their up-regulation of receptors for IL-2 and transferrin. Also, the addition of exogenous IL-2 did not induce expression of the endogenous IL-2 gene in progressing T cells (FIG. 2A). The competent T cells resembled T cells that were treated with the immunosuppressants cyclosporin A (CsA) or FK506 before or upon activation,[34,46,47] inasmuch as CsA inhibits T-cell activation by interfering with the transcription of the IL-2 gene,[34,40,46,47] and some of the inhibitory effects of CsA can be rescued by the addition of exogenous IL-2.[40,48]

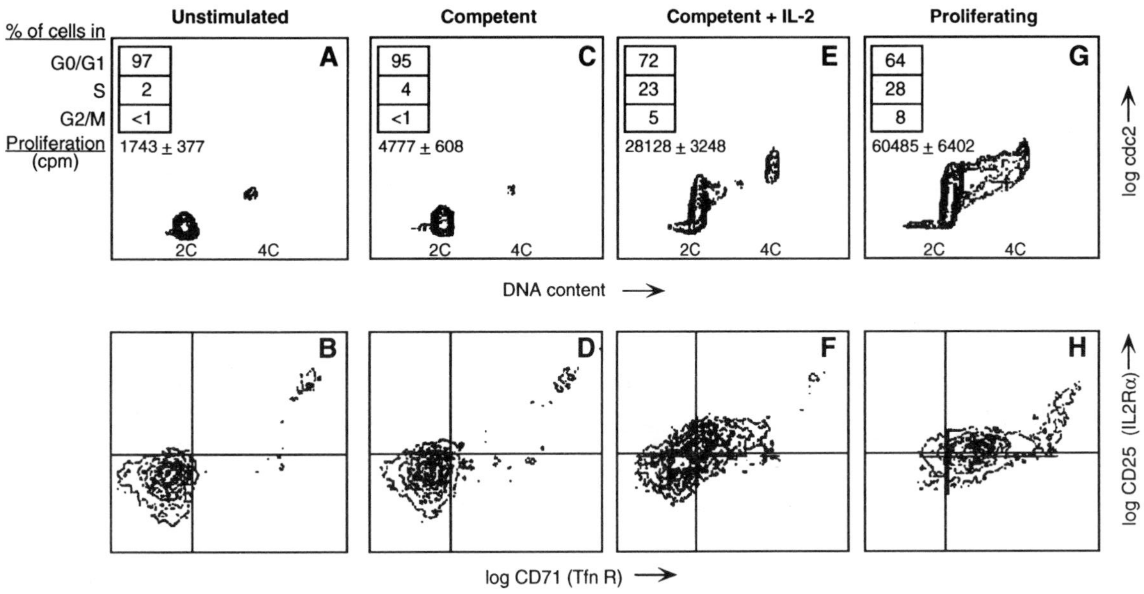

FIGURE 1. Entry and progression through the cell cycle by resting, competent, and proliferating T cells. The relative expression of $p34^{cdc2}$ and DNA content (**A**, **C**, **E**, **G**) or the surface expression of IL-2 receptor α (IL2Rα) and transferrin (Tfn R) receptors (**B**, **D**, **F**, **H**) was evaluated by dual-wave flow cytometry 40 hours after the onset of culture. Cell proliferation was determined 48 hours after the onset of culture by incorporation of $[^3H]$thymidine into DNA. Resting human peripheral blood T cells remained unstimulated throughout the culture period (**A**, **B**); were rendered "competent" (by a brief stimulation with PDB [10 nM] and ionomycin [500 nM] followed by extensive washing, [**C**, **D**]); were induced to a "progressing" state (by the addition of IL-2 to competent cells [**E**, **F**]); or were stimulated to proliferate by the addition of PDB and ionomycin throughout the culture period [**G**, **H**]). The flow cytometric results shown are representative of three independent experiments; the proliferation data reflect the mean ± standard error of 12 independent experiments.

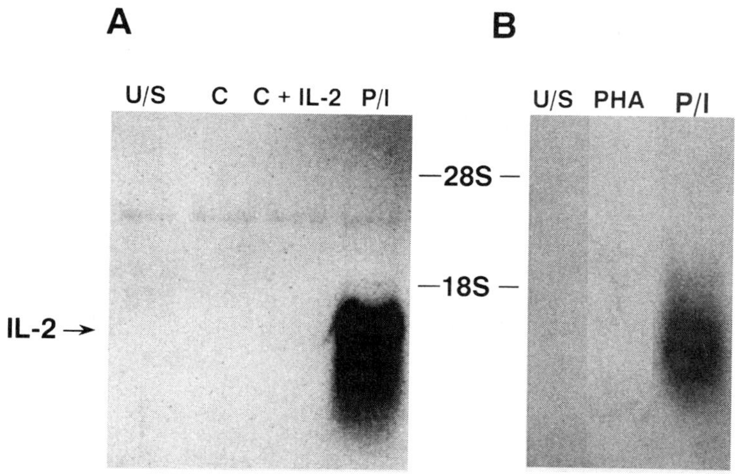

FIGURE 2. IL-2 gene expression in resting, competent, and progressing T cells. The expression of IL-2 mRNA was measured by Northern blotting in (**A**) resting T cells (unstimulated, U/S), T cells that were rendered competent by a brief stimulation with PDB and ionomycin as in FIG. 1 (C), T cells induced to a "progressing" state as in FIG. 1 (C + IL-2), or T cells stimulated with PDB and ionomycin throughout the culture period (P/I); or in (**B**) resting T cells (U/S), T cells that were rendered competent by stimulation with suboptimal concentrations of PHA (0.5 μg/mL), or T cells stimulated with PDB and ionomycin throughout the culture period (P/I). Cytosolic mRNA was isolated 4 hours after the onset of culture. The expression of β_2-microglobulin was used to ensure that the amounts of RNA present in each sample were approximately equivalent. The results shown are representative of four experiments.

Phosphorylation of Rb in the T-Cell Cycle

Several of the serine/threonine sites in the $p110^{Rb}$ have been shown to be phosphorylated differentially during the G1, S, and G2 phases of the cell cycle in proliferating T cells.[21] To determine the relative importance of these phosphorylation events during entry versus progression of T cells through the cell cycle, we examined the state of phosphorylation of $p110^{Rb}$ in resting, competent, and progressing T cells. In agreement with previous studies, FIGURE 3A shows that $p110^{Rb}$ was hypophosphorylated in resting T cells and that the continuous stimulation by PDB and ionomycin induced a sustained phosphorylation of $p110^{Rb}$ as the cells entered the cell cycle and progressed into the S phase. We could further resolve this sustained phosphorylation of $p110^{Rb}$ into at least two separate components. Induction of competence stimulated the first phase of $p110^{Rb}$ phosphorylation in an IL-2-independent fashion (FIGURES 3B and 3C). This phosphorylation staged peaked at ~15 hours and remained only as long as the cells were competent. The second phase of $p110^{Rb}$ phosphorylation followed stimulation of the competent cells by IL-2, and it remained as the cells entered the S phase (FIG. 3C). These findings confirm and extend the observations that the phosphorylation of $p110^{Rb}$ occurs at various stages of the cell cycle,[21,23,25] and also suggest that these phosphorylation reactions could be performed by different protein kinases.

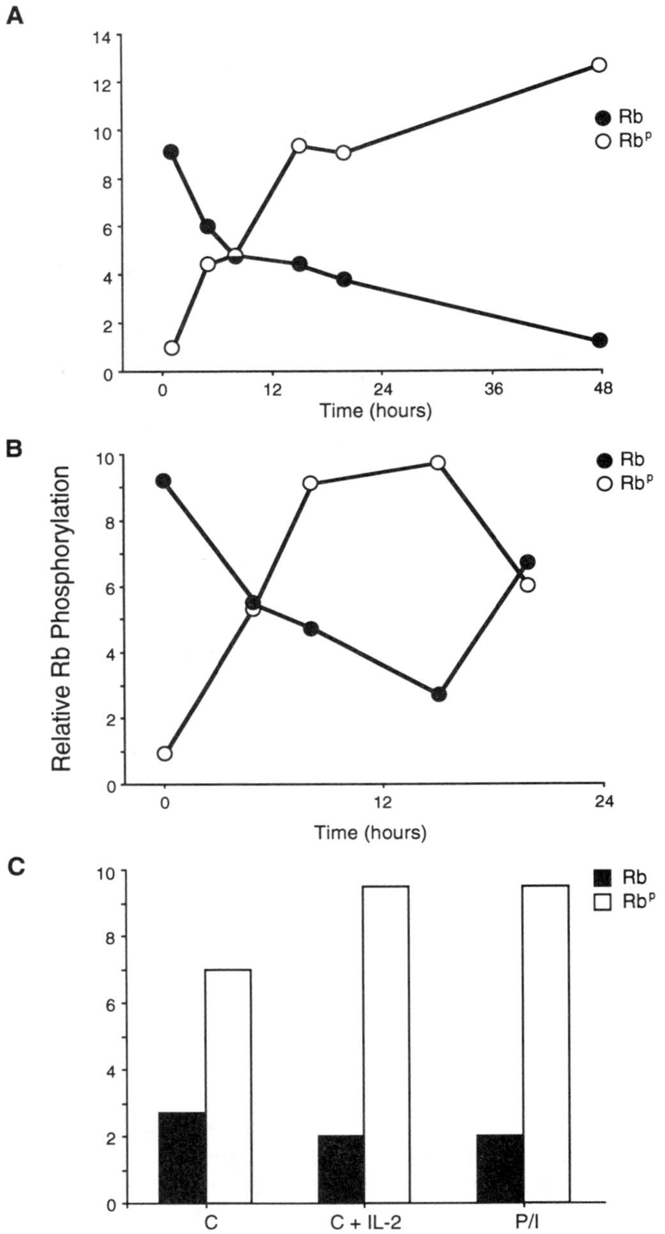

FIGURE 3. Phosphorylation of $p110^{Rb}$ in resting, competent, progressing, and proliferating T cells. The phosphorylation of $p110^{Rb}$ was assessed by immunoblotting (**A**) in T cells stimulated with PDB and ionomycin throughout the culture period, (**B**) in T cells rendered competent as in FIG. 1, or (**C**) in competent T cells (C), progressing T cells (C + IL-2), or T cells stimulated continuously with PDB and ionomycin (P/I) 12 hours after the onset of culture. The relative levels of hypophosphorylated $p110^{Rb}$ (closed circles and closed bars) and phosphorylated $p110^{Rb}$ (open circles and open bars) were quantified from the immunoblots by scanning and integration of the images into a Macintosh II Ci computer with the ScanAnalysis software; the amount of phosphorylated $p110^{Rb}$ in resting cells was assigned an arbitrary value of 1.0. Similar results were obtained in two additional experiments.

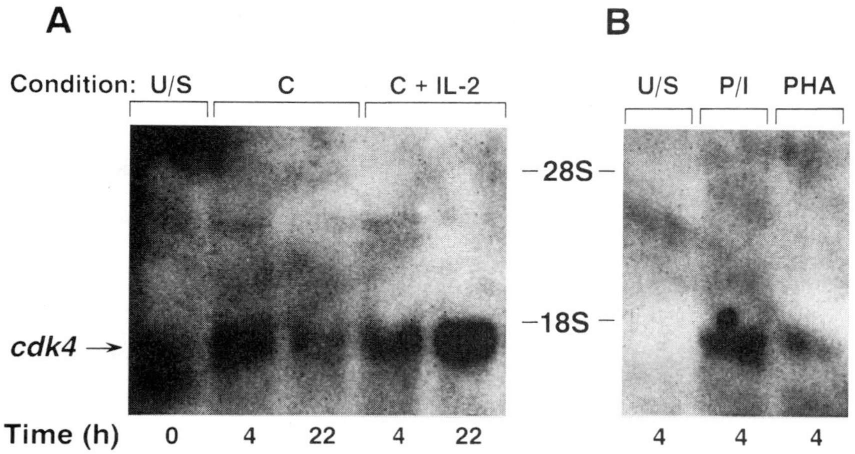

FIGURE 4. Expression of the *cdk*4 gene in resting, competent, and progressing T cells. **A**: The expression of *cdk*4 mRNA was measured by Northern blotting in resting T cells (unstimulated, U/S), T cells that were rendered competent by a brief stimulation with PDB and ionomycin as in FIG. 1 (C), or in progressing T cells (competent T cells stimulated with exogenous IL-2, [C + IL-2]). Cytosolic mRNA was isolated at the onset of culture (0 hours), or after 4 or 22 hours. **B**: The expression of *cdk*4 mRNA was measured in resting T cells (U/S), T cells stimulated with PDB and ionomycin throughout the culture period (P/I), or T cells that were rendered competent by stimulation with suboptimal concentrations of PHA (0.5 μg/mL). Cytosolic mRNA was isolated 4 hours after the onset of culture. The expression of β_2-microglobulin was used to ensure that the amounts of RNA present in each sample were approximately equivalent. The results shown are representative of three independent experiments.

Expression and Activity of Cyclin-dependent Kinases in Resting, Competent, and Progressing T Cells

Either Cdk2, Cdk4, or Cdk6 could have been responsible for the phosphorylation of $p110^{Rb}$ as T cells entered the cell cycle and progressed through the G1 phase. We examined the expression and activity of these CDKs in resting, competent, and progressing T cells. Resting T cells did not express messenger RNA for *cdk*2[40] or *cdk*4 (FIGURES 4A and 4B), but they contained detectable levels of Cdk2 and Cdk6 protein and lesser amounts of Cdk4 protein.[40,40a,49] Immunoprecipitates of these proteins from resting T cells had little or no kinase activity *in vitro*.[40,40a]

The continuous stimulation of resting T cells with PDB and ionomycin induced, in order of appearance, an increase in Cdk6, Cdk4, and Cdk2, along with their cyclin partners, cyclin D2, cyclin D3, cyclin E, and cyclin A.[40a,49] Accumulation of Cdk6 protein occurred within the first 2 to 3 hours in these cells and continued into the S phase. Immunoprecipitates of Cdk6 had Rb-kinase activity that was detectable 3 hours after stimulation and that peaked after 30 hours. Expression of *cdk*4 mRNA in the activated T cells was detectable after 4 hours (FIG. 4), and it continued to increase for at least 24 hours. The Cdk4 protein was detectable after 6 to 8 hours; at this time, Cdk4 immunoprecipitates also began to exhibit kinase activity *in vitro*

(FIG. 5). Both the expression and the kinase activity of Cdk4 continued to increase through the G1/S transition. The expression of *cdk*2 mRNA in activated T cells was detectable after approximately 10 hours; levels of Cdk2 protein above those detected in resting cells accumulated between 15 hours and the G1/S transition[40,49a] This accumulation of Cdk2 protein during the G1 phase was associated with increases in immunoprecipitable Cdk2 kinase activity prior to entry into the S phase that was entirely mediated by Cdk2/cyclin E complexes.

To further define the kinetics of expression and activation of the CDKs and the role of growth or progression factors, we examined these responses in competent T cells prior to IL-2 synthesis. Induction of competence by a brief exposure to PDB and ionomycin, or by stimulation with suboptimal concentrations of PHA, resulted in an IL-2-independent, CsA-resistant increase in the level of *cdk*4 mRNA expression that was detectable 3 to 4 hours after stimulation (FIGURES 4A and 4B) but that did not increase further for the next 20 hours. The accumulation of Cdk6 and Cdk4 proteins also increased 6 to 8 hours after the induction of competence,[40,40a] and immunoblot analyses of the Cdk4 and Cdk6 immunoprecipitates showed that a single species of cyclin D2, but no cyclin D3, was associated with Cdk4 in the active complexes recovered from the competent T cells (FIG. 5). There were concomitant increases in the activation of Cdk4 and Cdk6 kinases in competent T cells (FIG. 5) within 12 hours after induction of competence, and the kinase activity associated with these proteins continued to increase for at least 20 hours.

In contrast to the observations for *cdk*4, competent T cells did not express any *cdk*2 mRNA, nor was there an increase in the levels above preexisting Cdk2 protein. However, despite this absence of *cdk*2 gene expression and the lack of an increase in Cdk2 protein accumulation, induction of competence in T cells stimulated an increase in Cdk2 H1-kinase activity of ~50% of the activity seen in the proliferating cells 20 to 24 hours after stimulation.[40]

Stimulation of competent T cells by IL-2 did not affect the initial accumulation of Cdk6 or Cdk4 by these cells, but it induced a marked increase in mRNA expression and in the accumulation of these proteins after 22 hours (FIG. 4A). The expression of cyclin D2 mRNA and the accumulation of cyclin D2 protein in competent T cells seemed particularly sensitive to stimulation by IL-2, frequently increasing to levels greater than those seen in proliferating T cells. Furthermore, the faster mobility species of cyclin D2 was also present in Cdk4 and Cdk6 immunoprecipitates from these cells (FIG. 5). Stimulation of competent T cells with IL-2 restored the early phase of Cdk4 activity to levels comparable to those seen in T cells that were stimulated continuously by PDB and ionomycin (FIG. 5). Finally, the addition of IL-2 to competent T cells stimulated *de novo* expression of *cdk*2 mRNA and Cdk2 protein, as well as a further increase in the activity of Cdk2 kinase to levels similar to those in the T cells stimulated continuously by PDB and ionomycin.[40]

DISCUSSION

We compared the mitogen- or growth factor-stimulated changes in the expression and function of Cdk2, Cdk4, and Cdk6, and in their relation to p110[Rb] phosphorylation in normal human T cells. The data show that induction of competence in these

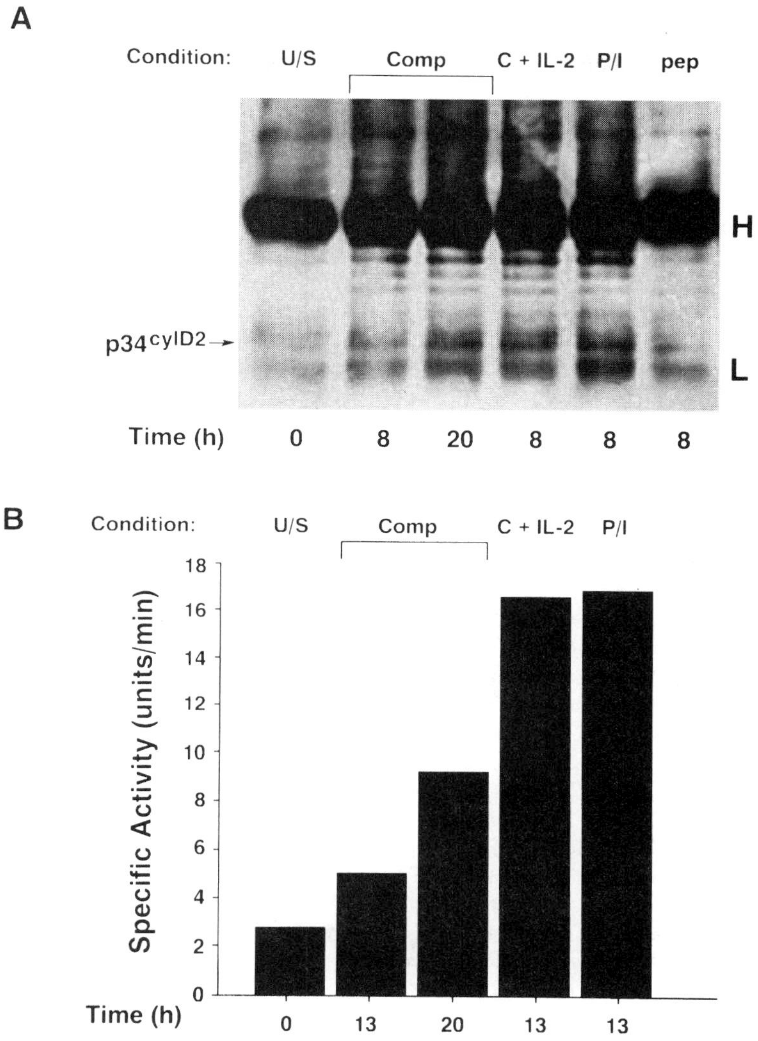

FIGURE 5. Association of cyclin D2 with Cdk4, and Cdk4 kinase activity in resting, competent, and progressing T cells. **A**: Immunoblotting was used to determine the presence of cyclin D2 in Cdk4 immunoprecipitates from resting T cells (unstimulated, U/S), competent T cells (Comp), progressing T cells (C + IL-2), or T cells stimulated continuously with PDB and ionomycin (P/I). The Cdk4 complexes were immunoprecipitated at the onset of culture (0 hours), or after 8 or 20 hours. The specificity of the association between Cdk4 and cyclin D2 was assessed by using a 5-fold excess of the immunogenic peptide used to prepare the Cdk4 antibody as a competitor in the immunoprecipitation step of parallel samples of those stimulated continuously with PDB and ionomycin for 8 hours (pep). H and L reflect the position of the heavy chain (~53 kDa) and light chain (~25 kDa), respectively, of the anti-Cdk4 antibody used for immunoprecipitation. Similar results were obtained in two additional experiments. **B**: The Cdk4 kinase activity was measured in Cdk4 immunoprecipitates from resting T cells (U/S), from competent T cells (Comp), from progressing T cells (C + IL-2), or from T cells stimulated continuously by PDB and ionomycin (P/I). Cells were lyzed at the indicated time, and the *in vitro* kinase activity of Cdk4 complexes immunoprecipitated from the cells was determined using a recombinant, truncated form of the Rb protein ($p56^{Rb}$) as a substrate. One unit/min was defined as the incorporation of 1 fmol of phosphate per minute into the substrate. The data shown are representative of four experiments.

cells stimulated a transient, IL-2-independent, and CsA-resistant phase of $p110^{Rb}$ phosphorylation that remained only as long as the cells were responsive to IL-2. These data support previous findings that showed specific phosphorylation of $p110^{Rb}$ during the G1 phase,[21,23] and suggest that these phosphorylation events may be required to maintain the competent state and responsiveness to a progression signal.

There were also significant increases in the levels of Cdk4 and Cdk6 proteins in competent T cells, along with their cyclin partner, cyclin D2. The increased amounts of Cdk4, Cdk6, and cyclin D2 were functionally significant, because there were concomitant increases in Cdk4- and Cdk6-associated kinase activities. Furthermore, despite the lack of an increase in the expression of Cdk2 in competent T cells, there was an increase in Cdk2-associated kinase activity in these cells. The data do not distinguish which of these proteins serves as the IL-2-independent Rb-kinase *in vivo* (see FIG. 6). Nevertheless, in our experiments, Cdk4 and Cdk2 demonstrated relatively greater levels of Rb-kinase activity than did Cdk6 (unpublished data). The data also suggest that the signals that stimulate the expression and activation of Cdk4 and Cdk6 are sufficient to activate Cdk2, and that these events may be causally related. Previous reports showed that TGFβ1 inhibited the expression of Cdk4 and the activation of Cdk2 in mink lung epithelial cells.[49] This inhibition of Cdk2 activity seemed to be related to inefficient association of the preexisting Cdk2 with cyclin E, rather than to a decrease in Cdk2 expression.[50] Additionally, in these cells, overexpression of Cdk4 restored the activity of Cdk2 and stimulated phosphorylation of Rb in the presence of TGFβ1, whereas overexpression of Cdk2 did not.[49] In T cells, TGFβ1 also has been shown to inhibit phosphorylation of Rb during the G1 phase.[51] Thus, at least a significant proportion of Rb phosphorylation in competent T cells may be carried out by Cdk4.

The addition of IL-2 to competent T cells stimulates progression through the cell cycle accompanied by the sustained phosphorylation of $p110^{Rb}$. It is possible that the IL-2-dependent patterns of Rb phosphorylation in the progressing T cells are not only quantitatively, but also qualitatively different from the profiles observed during the competence phase. IL-2 stimulated a further increase in the expression and activities of Cdk4 and Cdk6, along with newly synthesized and activated Cdk2. These increases in Cdk2 activity in the progressing T cells may be related to the observed increase in the amount of Cdk2 protein as the cells progress into the late G1 phase; to increased levels of Cdk4 (and Cdk6) activity, resulting in Cdk2 activation; to the dissociation of small molecular weight inhibitors, such as $p27^{Kip1}$ from the Cdk2 protein;[52] or to a combination thereof (FIG. 6). As was the case for competent T cells, our data do not distinguish which of these proteins (Cdk2, Cdk4, or Cdk6) is the major source of IL-2-dependent Rb-kinase *in vivo* in this phase of T-cell cycle progression. However, it is possible that the Cdk2 protein may play a more significant role at this stage.

Recent work describing the events that take place during T-cell cycle entry and progression indicates that resting T cells contain the molecular components necessary to form three potentially active CDK/cyclin complexes after mitogenic stimulation that induces $p110^{Rb}$ phosphorylation and entry into the cell cycle; that is, Cdk6/cyclin D2, Cdk4/cyclin D2, and Cdk2/cyclin E. Knowledge of the precise kinetics of activation of these kinases and their dependence on the various biochemical pathways initiated by mitogenic stimulation of T cells should lead to a better under-

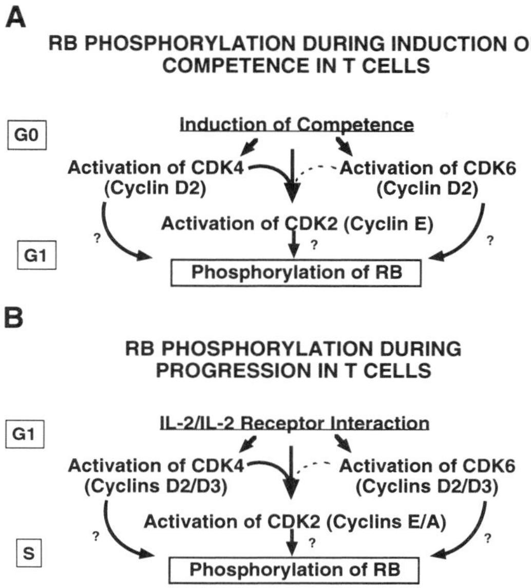

FIGURE 6. Activation of G1-phase cyclin-dependent kinases and phosphorylation of $p110^{Rb}$ in competent and progressing T cells. **A**: Signals that render normal human T cells competent to proliferate are sufficient to induce the expression of Cdk4, Cdk6, and cyclin D2; the association of Cdk4 with cyclin D2 and of Cdk6 with cyclin D2; and the activation of these complexes in an IL-2-independent fashion. Activation of the Cdk4 kinase (and possibly of the Cdk6 kinase) may be required for the association of preexisting Cdk2 and cyclin E in these cells, as well as for the activation of the resultant complex. One or all of these active Cdk/cyclin complexes may act as a functional G1-phase Rb kinase in T cells that is necessary for entry into the cell cycle and progression through the G1 phase. **B**: A progression signal such as IL-2 stimulates a second phase of Cdk4, Cdk6, and cyclin D2 expression, as well as expression of cyclin D3, cyclin A, and of new Cdk2 and cyclin E. Stimulation of IL-2 also results in further association of both Cdk4 and Cdk6 with cyclins D2 and D3, and of Cdk2 with cyclin E and cyclin A, and potentially, in the activation of all these complexes. One or several of these active Cdk/cyclin complexes may then act as the functional Rb kinases that are necessary for G1 to S transition and S phase progression in normal human T cells.

standing of the elusive link between early signaling events, and cell-cycle entry and progression in normal T cells.

SUMMARY

The entry of resting T cells into the G1 phase of the cell cycle after stimulation by mitogens is controlled by a series of biochemical events that are independent of growth factors. These events follow the initial signals stimulated through the engagement of the T-cell receptor and include activation of the cyclin-dependent kinases Cdk6, Cdk4, and Cdk2, as well as a transient phosphorylation of the retinoblastoma gene product ($p110^{Rb}$) by one or several of these proteins. A progression signal such as that delivered by interleukin-2 then induces a second phase of Cdk6, Cdk4, and Cdk2 activation, along with sustained phosphorylation of $p110^{Rb}$ in the activated T cells. This second signal is required to carry the cells into the S phase and beyond. Quantitative and qualitative differences in the expression and activity of these proteins may be critical to maintain the delicate balance that is necessary to ensure the normal progression of T cells through the cell cycle.

REFERENCES

1. GELFAND, E. W. 1989. Adv. Regul. Cell Growth **1:** 119-140.
2. WEISS, A., B. A. IRVING, L. K. TAN & G. A. KORETZKY. 1991. Semin. Immunol. **3:** 313-24.
3. WEISS, A. 1993. Cell **73:** 209-12.
4. TANIGUCHI, T. & Y. MINAMI. 1993. Cell **73:** 5-8.
5. TERADA, N., R. A. FRANKLIN, J. J. LUCAS & E. W. GELFAND. 1994. T-cell signaling and activation through the IL-2 receptor complex. *In* Growth Factors and Signal Transduction in Development. M. Nilsen-Hamilton, Ed. Wiley-Liss. New York.
6. FRANKLIN, R. A., A. TORDAI, H. PATEL, A. M. GARDNER, G. L. JOHNSON & E. W. GELFAND. 1994. J. Clin. Invest. **93:** 2134-2140.

6a. FRANKLIN, R. A., H. SAWAMI, N. TERADA, J. J. LUCAS & E. W. GELFAND. Activation of MAP2-K and $p90^{rsk}$ by IL-2 occurs independently of protein kinase C. Manuscript submitted for publication.

7. REED, J. C., J. D. ALPERS, P. C. NOWELL & R. G. HOOVER. 1986. Proc. Natl. Acad. Sci. USA **83:** 3982-3986.
8. SAWAMI, H., N. TERADA, R. A. FRANKLIN, H. OKAWA, T. UCHIYAMA, J. J. LUCAS & E. W. GELFAND. 1992. J. Cell. Physiol. **153:** 367-377.
9. MILLS, G. B., P. GIRARD, S. GRINSTEIN & E. W. GELFAND. 1988. Cell **55:** 91-100.
10. MILLS, G. B., D. J. STEWART, A. MELLORS & E. W. GELFAND. 1986. J. Immunol. **136:** 3019-3024.
11. MILLS, G. B., R. K. CHEUNG, S. GRINSTEIN & E. W. GELFAND. 1985. J. Immunol. **134:** 2431-2435.
12. MODIANO, J. F., R. KOLP, R. J. LAMB & P. C. NOWELL. 1991. J. Biol. Chem. **266:** 10552-10561.
13. JOHNSTON, J. A., M. KAWAMURA, R. A. KIRKEN, Y. Q. CHEN, T. B. BLAKE, K. SHIBUYA, J. R. ORTALDO, D. W. MCVICAR & J. J. O'SHEA. 1994. Nature **370:** 151-153.
14. WITTHUHN, B. A., O. SILVENNOIEN, O. MIURA, K. S. LAI, C. CWIK, E. T. LIU & J. N. IHLE. 1994. Nature **370:** 153-157.
15. MIYAZAKI, T., A. KAWAHARA, H. FUJII, Y. NAKAGAWA, Y. MINAMI, Z. J. LIU, I. OISHI, O. SILVENNOINEN, B. A. WITTHUHN, J. N. IHLE & T. TANIGUCHI. 1994. Science **266:** 1045-1047.
16. RUSSELL, S. M., J. A. JOHNSTON, M. NOGUCHI, M. KAWAMURA, C. M. BACON, M. FRIEDMANN, M. BERG, D. W. MCVICAR, B. A. WITTHUHN, O. SILVENNOINEN, A. S. GOLDMAN, F. C. SCHMALSTIEG, J. N. IHLE, J. J. O'SHEA & W. J. LEONARD. 1994. Science **266:** 1042-1045.
17. LEE, W. H., J. Y. SHEW, F. D. HONG, T. W. SERY, L. A. DONOSO, L. J. YOUNG, R. BOOKSTEIN & E. Y. LEE. 1987. Nature **329:** 642-645.
18. JACKS, T., A. FAZELI, E. M. SCHMITT, R. T. BRONSON, M. A. GOODELL & R. A. WEINBERG. 1992. Nature **359:** 295-300.
19. MITTNACHT, S., P. W. HINDS, S. F. DOWDY & R. A. WEINBERG. 1991. Quant. Biol. **56:** 197-209.
20. CHEN, L. I., T. NISHINAKA, K. KWAN, I. KITIBAYASHI, K. YOKOYAMA, Y. H. FU, S. GRUNWALD & R. CHIU. 1994. Mol. Cell. Biol. **14:** 4380-4389.

21. DeCaprio, J. A., Y. Furukawa, F. Ajchenbaum, J. D. Griffin & D. M. Livingston. 1992. Proc. Natl. Acad. Sci. USA **89:** 1795-1798.
22. Furukawa, Y., J. A. DeCaprio, A. Freedman, Y. Kanakura, M. Nakamura, T. J. Ernst, D. M. Livingston & J. D. Griffin. 1990. Proc. Natl. Acad. Sci. USA **87:** 2770-2774.
23. Terada, N., J. J. Lucas & E. W. Gelfand. 1991. J. Immunol. **147:** 698-704.
24. Mittnacht, S., J. A. Lees, D. Desai, E. Harlow, D. O. Morgan & R. A. Weinberg. 1994. EMBO J. **13:** 118-127.
25. Lucas, J. J., N. Terada, A. Szepesi & E. W. Gelfand. 1992. J. Immunol. **148:** 1804-1811.
26. Ewen, M. E. H. K. Sluss, C. J. Sherr, H. Matsushime, J. Kato & D. M. Livingston. 1993. Cell **73:** 487-497.
27. Kato, J., H. Matsushime, S. W. Hiebert, M. E. Ewen & C. J. Sherr. 1993. Genes & Dev. **7:** 331-342.
28. Dowdy, S. F., P. W. Hinds, K. Louie, S. I. Reed, A. Arnold & R. A. Weinberg. 1993. Cell **73:** 499-511.
29. Akiyama, T., T. Ohuchi & S. Sumida. 1992. Proc. Natl. Acad. Sci. USA **89:** 7900-7904.
30. Lees, E., B. Faha, V. Dulic, S. I. Reed & E. Harlow. 1992. Genes & Dev. **6:** 1874-1885.
31. Hu, Q. J., J. A. Lees, K. J. Buchkovich & E. Harlow. 1992. Mol. Cell. Biol. **12:** 971-980.
32. Meyerson, M. & E. Harlow. 1994. Mol. Cell. Biol. **14:** 2077-2086.
33. Matsushime, H., D. E. Quelle, S. A. Shurtleff, M. Shibuya, C. J. Sherr & J. Y. Kato. 1994. Mol. Cell. Biol. **14:** 2066-2076.
34. Crabtree, G. R. 1989. Science **243:** 355-361.
35. Kumagai, N., S. H. Benedict, G. B. Mills & E. W. Gelfand. 1987. J. Immunol. **139:** 1393-1399.
36. Kumagai, N., S. H. Benedict, G. B. Mills & E. W. Gelfand. 1988. J. Immunol. **140:** 37-43.
37. Or, R., H. Renz, N. Terada & E. W. Gelfand. 1992. Clin. Immunol. Immunopathol. **64:** 210-217.
38. Kern, J. A., J. C. Reed, R. P. Daniele & P. C. Nowell. 1986. J. Immunol. **137:** 764-769.
39. Modiano, J. F., E. Kelepouris, J. A. Kern & P. C. Nowell. 1988. J. Cell. Physiol. **135:** 451-458.
40. Modiano, J. F., J. Domenico, A. Szepesi, J. J. Lucas & E. W. Gelfand. 1994. J. Biol. Chem. **269:** 32972-32978.
40a. Lucas, J. J., A. Szepesi, J. F. Modiano, J. Domenico & E. W. Gelfand. 1995. Identification and regulation of the PLSTIRE-protein (CDK6) as a major cyclin D-associated CDK4 homologue in normal human T lymphocytes. J. Immunol. In press.
41. Takase, K., M. Sawai, K. Yamamoto, J. Yata, Y. Takasaki, H. Teraoka & K. Tsukada. 1992. Cell Growth & Differen. **3:** 515-521.
42. Takase, K., N. Terada, A. Szepesi, H. Teraoka, E. W. Gelfand & J. J. Lucas. 1994. Cell Growth & Differen. **5:** 1051-1059.
43. Smith, K. A. & D. A. Cantrell. 1985. Proc. Natl. Acad. Sci. USA **82:** 864-868.
44. Reed, J. C., W. C. Greene, R. G. Hoover & P. C. Nowell. 1985. J. Immunol. **135:** 2478-82.
45. Kumagai, N., S. H. Benedict, G. B. Mills & E. W. Gelfand. 1988. J. Immunol. **141:** 3747-52.
46. Emmel, E. A., C. L. Verweij, D. B. Durand, K. M. Higgins, E. Lacy & G. R. Crabtree. 1989. Science **246:** 1617-1620.
47. Terada, N., R. Or, K. Weinberg, J. Domenico, J. J. Lucas & E. W. Gelfand. 1992. J. Biol. Chem. **267:** 21207-21210.

48. GELFAND, E. W., R. K. CHEUNG & G. B. MILLS. 1987. J. Immunol. **138:** 1115-1120.
49. EWEN, M. E., H. K. SLUSS, L. L. WHITEHOUSE & D. M. LIVINGSTON. 1993. Cell **74:** 1009-1020.
49a. LUCAS, J. J., A. SZEPESI, J. DOMENICO, A. TORDAI, N. TERADA & E. W. GELFAND. 1995. Differential regulation of the synthesis and activity of the major cyclin-dependent kinases $p34^{cdc2}$, $p33^{cdk2}$, and $p34^{cdk4}$, during cell cycle entry and progression in normal human T lymphocytes. J. Cell. Physiol. In press.
50. KOFF, A., M. OHTSUKI, K. POLYAK, J. M. ROBERTS & J. MASSAGUE. 1993. Science **260:** 536-539.
51. AHUJA, S. S., F. PALIOGIANNI, H. YAMADA, J. E. BALOW & D. T. BOUMPAS. 1993. J. Immunol. **150:** 3109-3118.
52. FIRPO, E. J., A. KOFF, M. J. SOLOMON & J. M. ROBERTS. 1994. Mol. Cell. Biol. **14:** 4889-4901.

Molecular and Genetic Insights into T-Cell Antigen Receptor Signaling

ARTHUR WEISS, THERESA KADLECEK, MAKIO IWASHIMA,[a] ANDREW CHAN,[b] AND NICOLAI VAN OERS

Departments of Medicine and of Microbiology and Immunology
Howard Hughes Medical Institute
University of California, San Francisco
San Francisco, California 94143

INTRODUCTION

Advances in the study of the signal transduction mechanisms of receptors on T cells have led to the elucidation of the molecular basis of T-cell immune deficiency syndromes. Conversely, these clinical disorders have led to unexpected insights into the biology of these receptors. This manuscript will focus on our studies that examine the role of protein tyrosine kinases (PTKs) in T-cell antigen receptor (TCR) signal transduction. These studies contributed to the identification of the molecular basis of an unusual autosomal recessive form of the severe combined immunodeficiency syndrome (SCID).

Stimulation of the TCR induces the tyrosine phosphorylation of several cellular proteins, including subunits of the TCR (CD3 and ζ chains), phospholipase C $\gamma 1$ (PLC $\gamma 1$), the product of the protooncogene vav, and mitogen-activated protein kinases (MAP kinase).[1,2] Phosphorylation of these proteins contributes to a cascade of biochemical events that culminate in a cellular differentiation. For instance, phosphorylation of PLC $\gamma 1$ results in its activation.[3] The activation of PLC $\gamma 1$ leads to an increase in cytoplasmic free calcium and activation of protein kinase C, events that have been causally related to the induction of interleukin 2 gene transcription.[2] The tyrosine phosphorylation events induced by the TCR result from the activation of cytoplasmic PTKs by the TCR.

Considerable progress has been made in understanding how the TCR regulates PTKs. Unlike many growth factor receptors, the oligomeric TCR chains do not have PTK or protein tyrosine phosphate (PTPases) domains. Instead, the TCR uses sequence motifs contained in the cytoplasmic tails of the CD3 and ζ chains to interact with cytoplasmic PTKs.[2]

[a] Current address: Mitsubishi Kasei Industries of Life Sciences, 11 Minamiooya Machida, Tokyo, Japan 194.

[b] Current address: Department of Medicine, Howard Hughes Medical Institute, Washington University School of Medicine, St. Louis, MO.

THE TCR CD3 AND ζ CHAINS INTERACT WITH CYTOPLASMIC PTKS

The TCR, the product of at least six genes, consists of an antigen-binding subunit and subunits involved in signaling. The αβ heterodimer is the ligand-binding subunit that recognizes peptides bound to major histocompatibility complex molecules. The CD3 and ζ chains are the signal transduction subunits of the TCR complex. The signaling function of these chains has been established by using chimeric receptors in which the cytoplasmic domains of the CD3 ε or ζ chains have been linked to the extracellular and transmembrane domains of other transmembrane molecules.[4–6] Such studies have shown that the cytoplasmic domains of both the CD3 ε and ζ chains contain all of the sequence information necessary to couple the TCR to intracellular signaling machinery.

Mutagenesis studies of the cytoplasmic domains of the CD3 and ζ chains mapped the functional domains in the cytoplasmic sequences and explained the apparent redundancy in the function of these subunits.[6–9] These studies identified a functional sequence motif, first noted by Reth,[10] termed ARAM (for antigen recognition activation motif) in CD3 and ζ chains, which is sufficient to confer upon heterlogous receptors signal transduction function. The minimal ARAM sequence necessary for function is $YXXL(X)_{6\text{-}8}YXXL$. The ARAM is triplicated within the cytoplasmic domain of ζ and contained as a single copy in each of the CD3 chains. The ARAM motif is also found in the associated chains of other antigen receptors, including the B-cell antigen receptor and the mast cell IgE Fc receptor. In addition, ARAM sequences are present in the cytoplasmic domains of transmembrane proteins of two viruses involved in B-cell transfomation. Thus, all of these plasma membrane molecules appear to use a common mechanism to interact with intracellular signaling machinery.

Each of the antigen receptors listed above contain multiple ARAMs. The significance of this multiplicity of ARAMs in these receptors is not clear. The different ARAMs have been suggested to contain sufficient sequence disparity to interact with distinct intracellular proteins involved in signal transduction.[6,11] So, the presence of multiple ARAMs may serve to diversify signals initiated by a single receptor binding event. Alternatively, the incorporation of multiple copies of ARAMs within a receptor with a single binding subunit may represent a means of signal amplification. In support of this, three copies of a ζ ARAM linked to the transmembrane and extracellular domains of CD8 functioned more efficiently than a chimera with one copy of this motif.[8] In fact, the responses of the chimera with the triplicated ARAM were comparable to that of the wild-type ζ sequence, which contains three distinct ARAMs. Thus, the presence of multiple ARAMs within a single receptor can play a role in the signal amplification mechanism. This has the net effect of increasing the sensitivity of the TCR to antigen.

ARAMS INTERACT WITH MEMBERS OF TWO DISTINCT FAMILIES OF PTKS

At least two families of PTKs have been implicated in TCR signal transduction. Among Src family members, Fyn and Lck appear to play a role in TCR signaling.

Small amounts of Fyn have been coimmunoprecipitated with the TCR under mild conditions of cell solubilization.[12,13] Although disruption of the *fyn* gene in mice does not alter T-cell development, mature single positive ($CD4^+$ or $CD8^+$) thymocytes exhibit markedly diminished responses to TCR stimuli.[14,15] However, peripheral T cells have a much milder signaling defect and can respond to antigen. Stronger evidence exists for the involvement of another *src* family PTK, Lck. Lck is associated with the cytoplasmic tails of the coreceptors CD4 and CD8 through interactions involving cysteine residues contained both in the cytoplasmic domains of these coreceptors and the unique N-terminal domain of Lck.[16] Coligation of the TCR and CD4 coreceptor results in increased signal transduction initiated by the TCR.[17] Moreover, the TCR signals induced by antigen are potentiated by the engagement of the CD4 coreceptor, and these signals are enhanced by the interaction of Lck with CD4.[18] Cell lines deficient in Lck kinase function have a marked impairment in TCR signal transduction.[19,20] Moreover, disruption of the *lck* gene results in a profound arrest in thymocyte development at the early double positive ($CD4^+/CD8^+$) stage.[21] Thus, evidence exists that both of these PTKs play important roles in TCR signal transduction, though Lck appears to play a more central role.

The Syk and ZAP-70 PTKs are homologous cytoplasmic PTKs.[22,23] Unlike the Src family PTKs, Syk and ZAP-70 lack an N-terminal myristilation site, an SH3 domain, or a C-terminal negative regulatory site of tyrosine phosphorylation. Instead, Syk and ZAP-70 contain two tandem N-terminal SH2 domains and a C-terminal kinase domain. ZAP-70 is expressed only in T cells and natural killer cells,[23] whereas Syk is expressed more widely.[22] Syk is expressed at very low levels in peripheral T cells, at somewhat higher levels in the thymus, and at considerably higher levels in B cells.[22,24]

Both ZAP-70 and Syk are recruited to the tyrosine phosphorylated ζ and CD3 chains of the stimulated TCR in the Jurkat T-cell line.[25] ZAP-70 associates only with the tyrosine phosphorylated ARAMs of the CD3 and ζ chains.[8,23,24,26] Moreover, TCR stimulation also induces the tyrosine phosphorylation of both ZAP-70 and Syk.[23,24] Genetic and biochemical studies suggest that Lck and ZAP-70 interact with an ARAM in the TCR in a sequential manner (FIG. 1A).[27] Stimulation of the TCR induces the tyrosine phosphorylation of ARAMs by an Src family member. This is likely to involve Lck, or possibly Fyn. In the Lck-deficient J.CaM1 T-cell line, stimulation of the TCR fails to induce the phosphorylation of ζ or the CD3 chains. Moreover, in this cell, ZAP-70 is neither recruited to the TCR nor is it tyrosine phosphorylated.[27] Identical observations have been made in a CD45-deficient cell, J45.01, in which Lck is thought to be inactivated due to hyperphosphorylation at its C-terminal negative regulatory site (ref. 28 and unpublished data). These observations place Lck upstream of ZAP-70.

We have used a heterologous cell system to study these interactions in further detail. Cos 18 stably expresses a CD8/ζ chimera as a TCR surrogate.[23] Transfection of Lck or Fyn alone can induce a low level of tyrosine phosphorylation of the CD8/ζ chimera. In the presence of either of these Src family members, ZAP-70 or Syk associate with the CD8/ζ chimera.[23,24,27] This is also associated with the tyrosine phosphorylation of ZAP-70 or Syk as well as an increase in CD8/ζ phosphorylation. The kinase activity of Lck, but not ZAP-70, is required for the phosphorylation of CD8/ζ as well as for the recruitment and phosphorylation of ZAP-70.[27] These studies

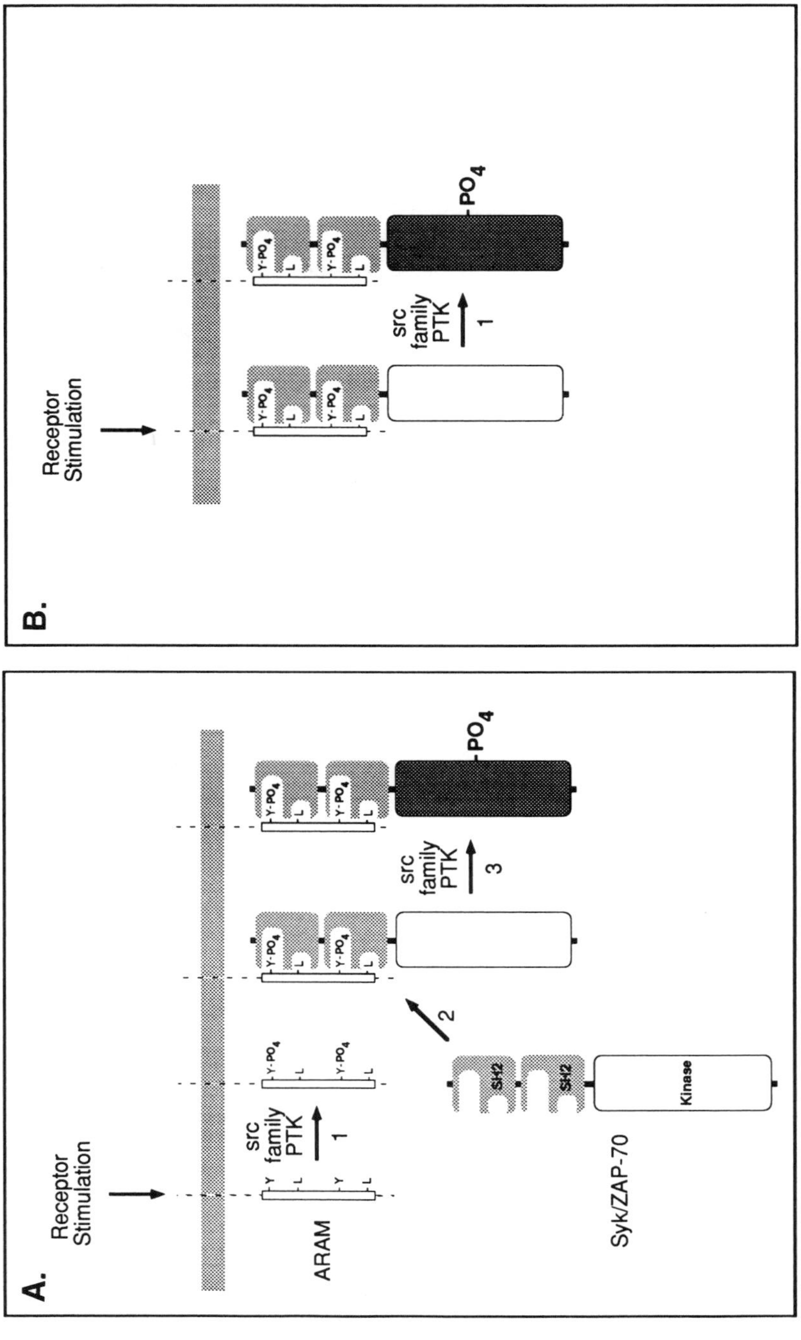

FIGURE 1. Model of the interactions of an ARAM contained in an antigen receptor with members of the Src or ZAP-70/Syk families of PTKs. A: T-cell lines and clones. B: Thymocytes and lymph node T cells.

provide strong evidence that Lck (or in some T cells, Fyn) plays a role that is upstream of ZAP-70 in initiating signal transduction by the TCR.

Both tyrosines within an ARAM are important for signal transduction and in the recruitment of ZAP-70.[6–8] Mutation of either tyrosine eliminates the function of the ARAM in signal transduction.[6,7] Thus, phosphorylation of both tyrosines is likely to be required for the function of the ARAM. This probably reflects the interaction of the ARAM with both of the SH2 domains of ZAP-70 or Syk. This is supported by studies with synthetic phosphopeptides whose sequences are based on a ζ chain ARAM. Only a doubly phosphorylated ARAM could form a stable complex with ZAP-70.[27] Moreover, the phosphotyrosine binding activity of both SH2 domains of ZAP-70 is required for this interact.[27,29] These observations offer an explanation for the requirement for two tyrosines within an ARAM for signal transduction function.

The sequential interactions of the TCR ARAMs with these two families of PTKs is an attractive one inasmuch as it incorporates into it the function of the coreceptors, CD4 and CD8. These coreceptors can enhance the sensitivity of the T cell to antigen.[30] They appear to do this by augmenting TCR signal transduction events,[17] presumably through mechanisms involving Lck. Thus, coreceptors not only function to bind the MHC molecules and contribute to the binding energetics of the TCR/major histocompatibility complex molecule/peptide, but to deliver and concentrate Lck in proximity with the ARAMs.

Once recruited to the TCR complex, the function of ZAP-70 is not so clear. In the Cos 18 system, a very large increase in cellular tyrosine phosphorylation is observed only if catalytically active ZAP-70 is cotransfected with either Lck or Fyn.[23,27] In further support of the critical function of ZAP-70 is the correlation between the sequence motifs responsible for ζ-chain signal transduction and the sequence requirements for ZAP-70 interactions.[8] The most compelling evidence that ZAP-70 plays a critical role in TCR signal transduction comes from observations of a relatively rare form of the SCID syndrome described below.

ZAP-70 IS CONSTITUTIVELY ASSOCIATED WITH THE TYROSINE PHOSPHORYLATED ζ CHAIN *IN VIVO*

In contrast to the studies performed with T-cell lines and clones, the signaling events that occur *in vivo* within thymocytes or lymph node T cells appear to differ. The ζ chain of the TCR is constitutively phosphorylated in thymocytes and lymph node T cells that are freshly isolated from mice.[31,32] Dephosphorylation of ζ occurs over hours after the cells are separated from the intact organ. This tyrosine phosphorylated form of ζ in thymocytes has been proposed to reflect the interaction of CD4 with class II MHC molecules *in vivo*.[31–34] Moreover, it has been suggested that the TCRs that contain tyrosine phosphorylated ζ in thymocytes may have impaired signal transduction function.[34] These observations differ from the model described above developed with cell lines in which tyrosine phosphorylation of ζ is involved in the initiation of signal transduction by the TCR.

To explore the differences in the status of phosphorylation of the ζ chain in thymocytes and cell lines and examine its impact on the association of ZAP-70, we have performed immunoprecipitation studies. These studies revealed that, unlike the situation in T-cell lines or clones, ZAP-70 is constitutively associated with phospho-ζ *in vivo*.[35] Stimulation of the TCR induces the tyrosine phosphorylation of CD3 chains and of ZAP-70 associated with phospho-ζ. Based on studies of mice deficient in CD4 or MHC expression, it appears that the constitutive phosphorylation of ζ and the constitutive association of ZAP-70 are independent of TCR or coreceptor engagement by MHC molecules *in vivo*. These studies suggest that in contrast to the sequential model of the interaction of ARAMs with PTKs shown in FIGURE 1A, the tyrosine phosphorylation of the ARAM and recruitment of ZAP-70 has already occurred *in vivo* (FIG. 1B). Whether this constitutive ζ phosphorylation and ZAP-70 association is dependent upon an Src PTK, such as Lck, is not clear. However, it is likely that the initial signaling events involved in TCR engagement *in vivo* are regulated differently than they are in cell lines and clones *in vitro*. The identification of molecules involved in regulating ζ chain phosphorylation and ZAP-70 association is under active investigation.

ZAP-70 MUTATIONS IN A SCID SYNDROME ASSOCIATED WITH A TCR SIGNAL TRANSDUCTION DEFECT

Most forms of SCID result in an absence or paucity of T and sometimes B cells. However, a minority of patients with SCID have normal or increased numbers of T cells that are functionally impaired. One group of these patients is uniquely characterized by the presence of $CD4^+$ T cells but are deficient in $CD8^+$ T cells in their peripheral blood.[36–38] The CD4 T cells fail to proliferate to TCR-dependent stimuli, such as phytohemagglutinin and anti-CD3 monoclonal antibodies, but do respond to the combination of calcium ionophore and phorbol ester, reagents that mimic the downstream events induced by the TCR. They also respond to exogenous interleukin 2. Natural killer cell cytolytic activity and B cell responses to membrane immunoglobulin stimulation are also preserved. These observations suggest that the CD4 T cells have a selective defect in TCR signal transduction. Indeed, analysis of freshly isolated cells or short-term T-cell lines revealed that stimulation of the TCR on these $CD4^+$ T cells failed to induce increases in tyrosine phosphoproteins or in cytoplasmic free calcium.[37,38]

To identify the molecular basis for this selective defect in TCR signal transduction, the expression of PTKs and PTPases implicated in TCR signal transduction was assessed. Normal levels of the CD45 PTPase were detected on these cells. Although normal levels of Lck, Fyn, and Syk protein could be detected by Western blot analysis, no ZAP-70 protein could be detected.[36–38] To date, five affected families have been studied in detail. Three of these families are Canadian Mennonites, and two mutant alleles have been identified in these patients. One is a missense mutation in the kinase domain; and the other represents a point mutation in an intron that is responsible for a new splice acceptor site.[36,38] ZAP-70 is an autosomal gene encoded on chromosome 2q12.[38] Therefore, it is not surprising that the patients are homozygotes or compound heterozygotes, inheriting mutant alleles from both parents who are heterozygous carriers. The patients in the other two families inherited mutant alleles homozygously and are the result of consanguinity. One patient's mutation represents a thirteen base pair deletion in the kinase domain.[37] The other is represented by the absence of detectable transcripts (E. Gelfand and A. Weiss, unpublished data). The transcripts that are expressed give rise to unstable proteins that are catalytically inactive when analyzed in the Cos cell system. These data suggest that the mutations in ZAP-70 account for this rare form of the SCID syndrome. They also demonstrate conclusively the critical role that ZAP-70 plays in signal transduction by the TCR in $CD4^+$ peripheral T cells.

Still unexplained, but perhaps providing unique and unexpected insights into T-cell biology, is the relative paucity of $CD8^+$ T cells in the periphery and the explanation for how $CD4^+$ T cells can develop in the absence of signal transduction by their TCRs. At least one study has shown that CD4 and CD8 antigens can be detected on cortical thymocytes.[36] However, only CD4 antigens were detected on medullary thymocytes from such patients. This suggests a block in positive selection of $CD8^+$ but not $CD4^+$ thymocytes. Both $CD4^+$ and $CD8^+$ cells express ZAP-70.[24] Thus, the differential expression of ZAP-70 does not explain the failure of $CD8^+$ cells to be selected. It is possible that the diminished association of Lck with CD8 compared to CD4 plays a role here.[39] Perhaps the CD4 lineage is favored under conditions of impaired signal transduction, that is, in the absence of ZAP-70. If peripheral $CD4^+$ cells cannot signal, how then are these cells positively selected in the thymus? One possible explanation is that Syk compensates for the loss of ZAP-70 in the thymus. Because Syk is expressed at higher levels in the thymus than in peripheral T cells,[24] it may play a somewhat redundant role in the thymus. However, Syk may not fully compensate for ZAP-70. Thus, the preferential selection of the CD4 lineage may reflect the bias imposed by the preferential association of Lck with this coreceptor.[39] Clearly, a more definitive explanation of the phenotype in these patients will await more detailed studies of the thymocytes from these patients as well as an available mouse model.

CONCLUSION

Studies of the basic mechanisms involved in TCR signal transduction reveal the complex, yet highly coordinated, interactions of the receptor with distinct PTKs. Such studies have contributed to our understanding of the roles of the CD3 and ζ subunits of the TCR. In addition, insights into the roles of at least two distinct families of cytoplasmic PTKs have been provided. These studies have led to a molecular understanding of a rare form of a SCID syndrome. Conversely, the SCID syndrome provides strong genetic evidence for the critical role that the ZAP-70 PTK plays in TCR signal transduction. Finally, the unusual phenotype of the T cells in this SCID syndrome has led to new questions regarding the roles of the ZAP-70 and Syk PTKs in T-cell development and function.

REFERENCES

1. Samelson, L. E. & R. D. Klausner. 1992. J. Biol. Chem. **267:** 24913-24916.
2. Weiss, A. & D. R. Littman. 1994. Cell **76:** 263-274.
3. Nishibe, S. *et al.* 1990. Science **250:** 1253-1256.
4. Irving, B. & A. Weiss. 1991. Cell **64:** 891-901.
5. Romeo, C. & B. Seed. 1991. Cell **64:** 1037-1046.

6. LETOURNEUR, F. & R. D. KLAUSNER. 1992. Science **255:** 79-82.
7. ROMEO, C., M. AMIOT & B. SEED. 1992. Cell **68:** 889-897.
8. IRVING, B. A., A. C. CHAN & A. WEISS. 1993. J. Exp. Med. **177:** 1093-1103.
9. WEGENER, A.-M. K. *et al.* 1992. Cell. **68:** 83-95.
10. RETH, M. 1989. Nature. **338:** 383-384.
11. CLARK, M. R. *et al.* 1992. Science **258:** 123-126.
12. SAMELSON, L. E., A. F. PHILLIPS, E. T. LUONG & R. D. KLAUSNER. 1990. Proc. Natl. Acad. Sci. USA **87:** 4358-4362.
13. SAROSI, G. P. *et al.* 1992. Int. Immunol. **4:** 1211-1217.
14. APPLEBY, M. W. *et al.* 1992. Cell **70:** 751-763.
15. STEIN, P. L., H.-M. LEE, S. RICH & P. SORIANO. 1992 Cell **70:** 741-750.
16. VEILLETTE, A., N. ABRAHAM, L. CARON & D. DAVIDSON. 1991. Semin. Immunol. **3:** 143-152.
17. LEDBETTER, J. A., L. K. GILLILAND & G. A. SCHIEVEN. 1990. Semin. Immunol. **2:** 99-106.
18. GLAICHENHAUS, N., N. SHASTRI, D. R. LITTMAN & J. M. TURNER. 1991. Cell **64:** 511-520.
19. STRAUS, D. & A. WEISS. 1992. Cell **70:** 585-593.
20. KARNITZ, L. *et al.* 1992. Mol. Cell. Biol. **12:** 4521-4530.
21. MOLINA, T. J. *et al.* 1992. Nature **357:** 161-164.
22. TANIGUCHI, T. *et al.* 1991. J. Biol. Chem. **266:** 15790-15796.
23. CHAN, A. C., M. IWASHIMA, C. W. TURCK & A. WEISS. 1992. Cell **71:** 649-662.
24. CHAN, A. C. *et al.* 1994. J. Immunol. **152:** 4758-4766.
25. CHAN, A.C., B. A. IRVING, J. D. FRASER & A. WEISS. 1991. Proc. Natl. Acad. Sci. USA **88:** 9166-9170.
26. WANGE, R. L., A.-N. T. KONG & L. E. SAMELSON. 1992. J. Biol. Chem. **267:** 11685-11688.
27. IWASHIMA, M., B. A. IRVING, N. S. C. VAN OERS, A. C. CHAN & A. WEISS. 1994. Science **263:** 1136-1139.
28. SIEH, M., J. B. BOLEN & A. WEISS. 1993. EMBO J. **12:** 315-322.
29. WANGE, R. L., S. N. MALEK, S. DESIDERIO & L. E. SAMELSON. 1993. J. Biol. Chem. **268:** 19797-19801.
30. JANEWAY, Jr., C. A. 1992. Annu. Rev. Immunol. **10:** 645-674.
31. NAKAYAMA, T., A. SINGER, E. D. HSI & L. E. SAMELSON. 1989. Nature **341:** 651-654.
32. VAN OERS, N. S. C. *et al.* 1993. Mol. Cell. Biol. **13:** 5771-5780.
33. NAKAYAMA, T. *et al.* 1991. Proc. Natl. Acad. Sci. USA **88:** 9949-9953.
34. NAKAYAMA, T. *et al.* 1993. Proc. Natl. Acad. Sci. USA **90:** 10534-10538.
35. VAN OERS, N. S. C., N. KILLEEN & A. WEISS. 1994. Immunity **1:** 675-685.
36. ARPAIA, E.l, M. SHAHAR, H. DADI, A. COHEN & C. M. ROIFMAN. 1994. Cell **76:** 947-958.
37. ELDER, M. E. *et al.* 1994. Science **264:** 1596-1599.
38. CHAN, A. C. *et al.* 1994. Science **264:** 1599-1601.
39. WIEST, D. L. *et al.* 1993. J. Exp. Med. **178:** 1701-1712.

Signal Transduction Mediated by the T-Cell Antigen Receptor

LAWRENCE E. SAMELSON, JERALD A. DONOVAN, NOAH ISAKOV, YASUO OTA, AND RONALD L. WANGE

Cell Biology and Metabolism Branch
National Institute of Child Health and Human Development
National Institutes of Health
Building 18T, Room 101
18 Library Dr. MSC 5430
Bethesda, Maryland 20892-5430

BACKGROUND

The function of the T-cell antigen receptor (TCR) is to bind to foreign antigenic peptides, themselves bound to cell-surface proteins encoded by the major histocompatibility complex (MHC).[1] This binding event, in turn, results in an activation of multiple biochemical pathways, and, ultimately, depending on the setting, leads to the activation, inactivation, or differentiation of the T cell.[2,3] The receptor consists of multiple proteins encoded by different genes (FIG. 1). The recognition subunits, α and β, in the majority of TCRs, are disulfide-linked immunoglobulin-like molecules each comprised of a variable and constant region.[4] The variable regions are in turn encoded by several genetic elements and together comprise the binding regions for interaction with the peptide-MHC ligand. These chains are associated with several nonpolymorphic molecules, the CD3 γ, δ, and ε chains and the TCRζ chains.[5] The CD3 chains are found within the TCR as dimers of γ-ε and δ-ε. The TCRζ chain and in some T cells the splice-variant eta chain, or the closely related Fc receptor gamma chain, are found as disulfide-linked dimers within the receptor.[6] Multiple studies have revealed that these nonpolymorphic molecules are required to ensure surface expression of the TCR. In addition, extensive results from the past two years demonstrate a critical role for an amino acid motif found in the cytoplasmic domains of each of the CD3 chains and in three copies in each TCRζ chain. These motifs have a consensus sequence YXX(L/I)$X_{(7-8)}$YXX(L/I) and are known as TAMs (tyrosine-based activation motifs),[7] ARAMs (antigen receptor activation motifs),[8] or ARH-1 (antigen receptor homology) motifs.[9] Chimeric molecules comprised of extracellular receptor domains fused to cytoplasmic domains containing one or more of these TAMs are capable of activating T cells. Intact TAMs are necessary and sufficient for the functional response. Proteins that interact with TAMs are discussed below.

We demonstrated in 1986 that activation of T cells by way of the TCR results in tyrosine phosphorylation of a receptor subunit, later shown to be the TCRζ chain.[10,11] The observation that the TCR is coupled to a protein tyrosine phosphorylation pathway has continued to stimulate our investigations to this date. Since that

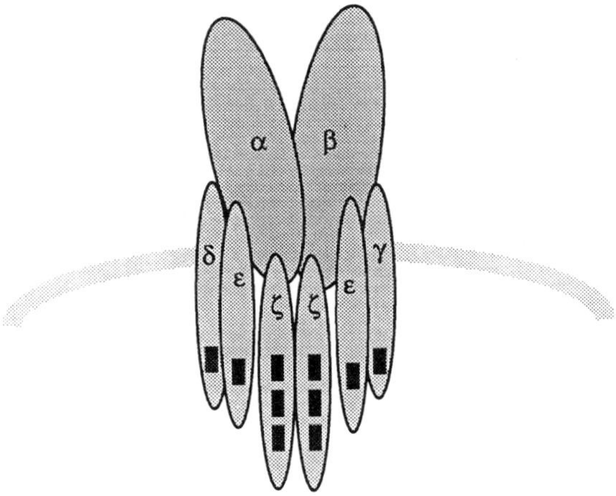

FIGURE 1. The T-cell antigen receptor (TCR) is composed of multiple chains arranged as dimers. The α and β chains bind the antigen-MHC complex, and the CD3γ, δ, and ε chains and the TCRζ chains are signaling components. The tyrosine-based activation motifs (TAMs) are depicted as rectangles.

time, we and others have provided data confirming the central role of this pathway in T-cell activation. Kinetic, pharmacologic, and genetic experiments have shown that induction of tyrosine phosphorylation is a rapid and required event in T-cell activation.[12–14] Biochemical and genetic studies have revealed that several nonreceptor protein tyrosine kinases (PTK) are either constitutively associated with the TCR[15] or the coreceptors CD4 and CD8,[16,17] or associate with the TCR upon TCR activation.[18] The manner in which they associate and are activated, and their regulation by other tyrosine kinases and tyrosine phosphates, remains an active area of research and is discussed below.

Much attention in signal transduction research recently has been directed at identifying PTKs and characterizing the biochemical events that are coupled to them. This is also true in the TCR system. It was clear for years that phospholipase C and consequently protein kinase C were activated through the TCR. The gamma isoform of PLC was one of the first demonstrated substrates of TCR-coupled tyrosine kinases, and its tyrosine phosphorylation is thought to regulate enzyme activity.[19] Pathways downstream of PLC are thus indirectly regulated by tyrosine phosphorylation. The ras pathway, as in fibroblast systems, has also been shown to be controlled by tyrosine phosphorylation in T cells.[20] These biochemical pathways are central to T-cell activation, but it is likely that we are only beginning to comprehend the complexity of pathways coupled to TCR-associated tyrosine kinases.

The goals in the field continue to be the characterization of the interaction of TCR and protein tyrosine kinases. The mechanism of kinase activation and regulation is another central question. Identification of substrates of these kinases, their function and the manner in which tyrosine phosphorylation regulates this function is an additional long-range objective. This manuscript reviews our approach to these topics over the past several years.

PROTEIN TYROSINE KINASES

Lck and Fyn

Identification of the kinase(s) responsible for tyrosine phosphorylation of the TCR and phosphorylation of other intracellular substrates has been a central concern since the identification of TCRζ chain phosphorylation in the mid-1980s. It was clear from analysis of the predicted amino acid sequence of all of the TCR subunits that, unlike many of the growth factor receptors, the PTK(s) would not be encoded within a receptor subunit. Instead, nonreceptor PTKs, for example, of the src family, were viewed as possible TCR-associated PTK candidates (FIG. 2). The T-cell specific PTK, Lck, was the first example of a nonreceptor PTK shown to be noncovalently coupled to integral membrane receptor molecules, CD4 and CD8.[16,17] Inasmuch as these so-called accessory molecules are intimately involved in the TCR response, it seemed likely that Lck could be involved in TCR-mediated activation. Numerous studies support this model. Mice engineered to be deficient in this enzyme demonstrate a profound defect in T-cell development and thymic structure.[21] The more direct evidence for Lck function in TCR-mediated activation comes from analysis of mutants of the Jurkat T-cell line, which fail to be activated through the TCR. The mutation was shown to be in the gene encoding Lck, and the cell line was reconstituted by transfection with wild-type Lck.[22] The current model for Lck involvement in TCR activation is that the kinase is brought into contact with TCR subunits when the TCR and either CD4 or CD8 bind separate faces of the same MHC molecule containing the antigenic peptide.[23] Activation of the kinase is thought to occur by several mechanisms, including CD45-mediated dephosphorylation of an inhibitory tyrosine phosphate at the Lck COOH-terminus and/or by the aggregation process, which might result in kinase-kinase cross-activation.[23]

For a number of reasons, we thought it unlikely that Lck was the only PTK involved in TCR function. Some T cells lack CD4 or CD8; certain T-cell lines are deficient in Lck; and early experiments failed to demonstrate Lck kinase activation upon TCR or CD3 engagement, though CD4 or CD8 cross-linking resulted in robust PTK activation.[24] We succeeded in demonstrating that, Fyn, another member of the src kinase family associates with the TCR directly.[15] We detected the interaction after immunoprecipitation and with an *in vitro* phosphorylation reaction. This interaction was resistant to digitonin but was disrupted when Triton X-100 was used for T-cell solubilization. Immunoprecipitation with antibodies binding the TCR specifically coprecipitated Fyn, and, reciprocally, immunoprecipitation with antibodies binding Fyn resulted in coprecipitation of the TCRζ and CD3 subunits. With both sets of immunoprecipitations, we detected additional proteins of 130, 120, and 60 kDa, suggesting that the Fyn kinase was part of a complex, and that this complex interacts with the TCR.

We performed several additional studies following these initial observations. We were concerned that the interaction of TCR and Fyn was simply an artifact of inadequate membrane solubilization with digitonin. To overcome this potential problem, we developed a cellular permeabilization and chemical cross-linking protocol to stabilize TCR-PTK interactions.[25] Cells were permeabilized with tetanolysin, and proteins were cross-linked with the water-soluble chemical cross-linker, 3,3′ dithiobis-

(sulfosuccinimidylpropionate) (DTSSP). Under these conditions, we found that solubilization with Triton X-100 did not disrupt the specific TCR/Fyn interaction. No Lck was detected bound to the TCR, and, again, the 130, 120, and 60 kDa proteins were detected in the same complex. We also used this cross-linking technique to demonstrate that Fyn could be cross-linked to the TCRζ chain and that only a small fraction of TCR (2-4%) bound this PTK.

A set of experiments was performed in collaboration with Andrey Shaw, to map the interaction of Fyn with TCR subunits.[26] A full-length Fyn cDNA was isolated, and a sequence encoding an epitopic tag was added. This cDNA was coexpressed in a vaccinia expression system with chimeric proteins comprised of the VSV G protein extracellular domain fused to cytoplasmic domains from the various CD3 and TCRζ subunits. Coexpression studies revealed that Fyn specifically bound the cytoplasmic domains of the CD3 epsilon and gamma chains as well as TCRζ and eta chains. Neither Src nor Lck demonstrated the same binding. This fact enabled Shaw's group to generate PTK chimeras and map the interaction by Fyn with TCRζ chain to the first ten residues of the kinase. This result is analogous to earlier mapping studies localizing the site of Lck interaction with CD4 to the amino-terminal section of Lck. More recent data from the Shaw laboratory indicates that Fyn interacts with TAM residues, though individual residues involved in binding cannot be mapped.[27] They also suggest that lipid moieties shown to be bound to the amino-terminal end of Fyn are likely to influence the interaction with TCR subunits.

ZAP-70

In the past three years our focus on T-cell tyrosine kinases has changed somewhat to a concentration on the ZAP-70 enzyme. Weiss' group had shown in the summer of 1991 that PTK activity and a 70 kDa tyrosine phosphorylated protein associated with the TCRζ chain in activated Jurkat T cells.[18] We had not observed these phenomena in our analysis of the murine T-cell line, 2B4. However, we were able to reproduce the reported findings in Jurkat, and our first series of studies were directed at a further characterization of this system.[28] Using *in vitro* kinase assays as a method of labeling protein, we found that an activated PTK and a 70 kDa protein bound to CD3 epsilon as well as to the TCRζ chain. We also observed that Fyn was coprecipitated with this activity complex. The most intensively labeled band in these immunoprecipitates was a 70 kDa tyrosine phosphorylated protein. We showed that this protein bound a ^{32}P-labeled, nonhydrolyzable, photoreactive analogue of ATP. This was the first experiment suggesting directly that the 70 kDa protein is itself a PTK.

Subsequently, Weiss' group cloned the cDNA encoding the 70 kDa protein and showed directly that it is a PTK, named by them, ZAP-70[29] (FIG. 2). In transfection studies, using COS cells, they also demonstrated that its association with TCRζ and its activation was dependent on the presence of an src family kinase, either Fyn or Lck. Analysis of the amino acid structure of ZAP-70 revealed an interesting feature. In addition to its kinase domain, this PTK and the closely related Syk PTK contains two adjacent SH2 domains. SH2 domains are well known to bind to proteins bearing phosphotyrosine residues and adjacent residues that define the specificity of interac-

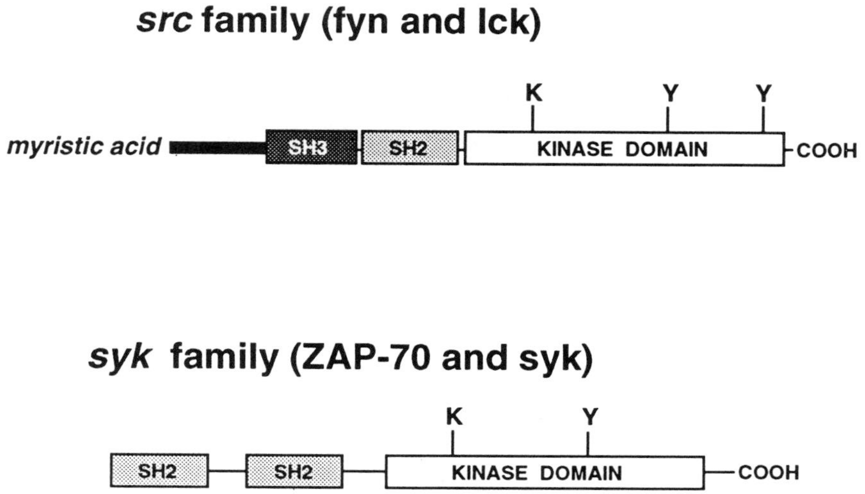

FIGURE 2. Structural motifs in two families of protein tyrosine kinases. Members of these families are involved in TCR-mediated signaling.

tion.[30,31] It is now clear that SH2-phosphotyrosine interactions determine protein-protein interactions involved in many signaling pathways.

In this period, a number of investigators, beginning with Reth,[32] had realized that the cytoplasmic domains of TCR subunits as well as analogous subunits found in the B cell and IgE receptors contained structures defined by the consensus sequence $YXX(L/I)X_{(7-8)}YXX(L/I)$. This motif is known alternatively as a TAM,[7] ARAM,[8] or ARH-1.[9] Studies with chimeric constructs containing TAMs or mutant TAMs have demonstrated that these structures are necessary and sufficient for T-cell activation. The old data from our laboratory demonstrating TCR-induced tyrosine phosphorylation of TCRζ, and newer evidence for similar phosphorylation of CD3 chains, can now be reinterpreted as being phosphorylations within the TAM motif.[10,33] Inasmuch as each TAM has two tyrosine residues and the ZAP kinase contains two SH2 domains, it appeared likely to us and others that induction of tyrosine phosphorylation on TCR TAMs generated the "ligand" to which ZAP-70 could bind.

To test the hypothesis that it is the tandem SH2 domains of ZAP-70 that bind to the receptor chains, we generated bacterial fusion proteins containing glutathione-*S*-transferase (GST) coupled to the tandem SH2 domains of ZAP-70 or the individual C- and N-terminal SH2 domains.[34] We obtained a number of GST fusion proteins containing SH2 domains from other kinases and signaling molecules. We found that the tandem ZAP-70 SH2, but not the individual SH2 domains, affinity purified a very limited number of tyrosine phosphorylated proteins from lysates of activated, but not unstimulated Jurkat T cells. This result was observed with lysates prepared from Jurkat cells stimulated either through the TCR or with the tyrosine phosphatase inhibitor, pervanadate. By contrast, fusion proteins composed of SH2 domains from a number of src kinase proteins, and from phospholipase C or PI3-kinase, bound a large number of tyrosine phosphorylated proteins. The proteins specifically bound to ZAP-70 SH2 domains were shown to be the TCRζ and CD3 epsilon chains. Our conclusion from this study was that the tandem SH2 domains of ZAP-70 are capable of a very specific and perhaps high-affinity interaction with the activated, phosphorylated TCR.

A model of sequential PTK activation has been proposed by Weiss and colleagues.[35] Our version of this model, incorporating work from that lab, ours, and others is depicted in (FIGURE 3). In the resting T cell, one finds Fyn constitutively associated with the TCR, whereas CD4 or CD8, depending on the subset of T cell, is associated with Lck. Engagement of the TCR leads to association of receptor and coreceptor molecules at the same ligand, a peptide-MHC complex. The src kinases become activated in this process because of proximity and kinase-kinase interaction and/or because of the additional presence of the tyrosine phosphatase, CD45 (not shown). This molecule activates Src-family kinases by dephosphorylation of inhibitory tyrosine residues.[36] Target substrates of Fyn and/or Lck are the TAMs of the TCR. The double tyrosine phosphorylation of a TCRζ TAM is depicted. The ZAP-70 kinase binds these phosphorylated TAMs and becomes activated. Substrates of all of these activated enzymes are then phosphorylated, the next stage in T-cell activation.

Current Studies

Our initial work with the tandem SH2 domains of the ZAP-70 kinases (GST-ZAP-$(SH2)_2$) described above, demonstrated that, when expressed as a fusion protein with GST, they are capable of binding the activated TCR. Our assumption had been that this interaction is a direct binding of the two SH2 domains to the tyrosine phosphorylated TAMs. To test this we, in collaboration with J. Watts and R. Aebersold, began by obtaining the tyrosine phosphorylated cytoplasmic domain of TCRζ (residues 52-164).[37] This polypeptide was prepared synthetically, tyrosine phosphorylated *in vitro* using recombinant Lck and ^{32}P-γ-ATP, and purified by HPLC. After phosphorylation and purification was performed, an analysis revealed that, on average, 2-3 tyrosine residues in this polypeptide were phosphorylated. This material was incubated with various GST fusion constructs, and we observed saturable binding to GST-ZAP-$(SH2)_2$. The single amino terminal SH2 domain bound, but with about 100-fold lower affinity, and we detected no binding to the C-terminal SH2 or to GST alone. These studies confirmed that the tandem SH2 domains were capable of direct binding to the TCR but were limited because the ligand was not well defined.

To overcome this problem we obtained synthetic peptides with the sequences of the three TCRζ TAMs and the one epsilon TAM (FIG. 4). These could be tyrosine phosphorylated as above and separated by HPLC into peaks shown by mass spectrometry to contain either doubly tyrosine phosphorylated or singly tyrosine phosphorylated TAMs (the latter, a pool of the two monophosphorylated species). Additional peptides were synthetically prepared with nonradioactive phosphate at both or either tyrosine of the TAM peptide. Studies using the radiolabeled TAMζ_3 peptide showed that binding to GST-ZAP-$(SH2)_2$ was saturable, the N-terminal SH2 domain bound with at least 200-fold lower affinity, and the C-terminal SH2 domain and GST alone did

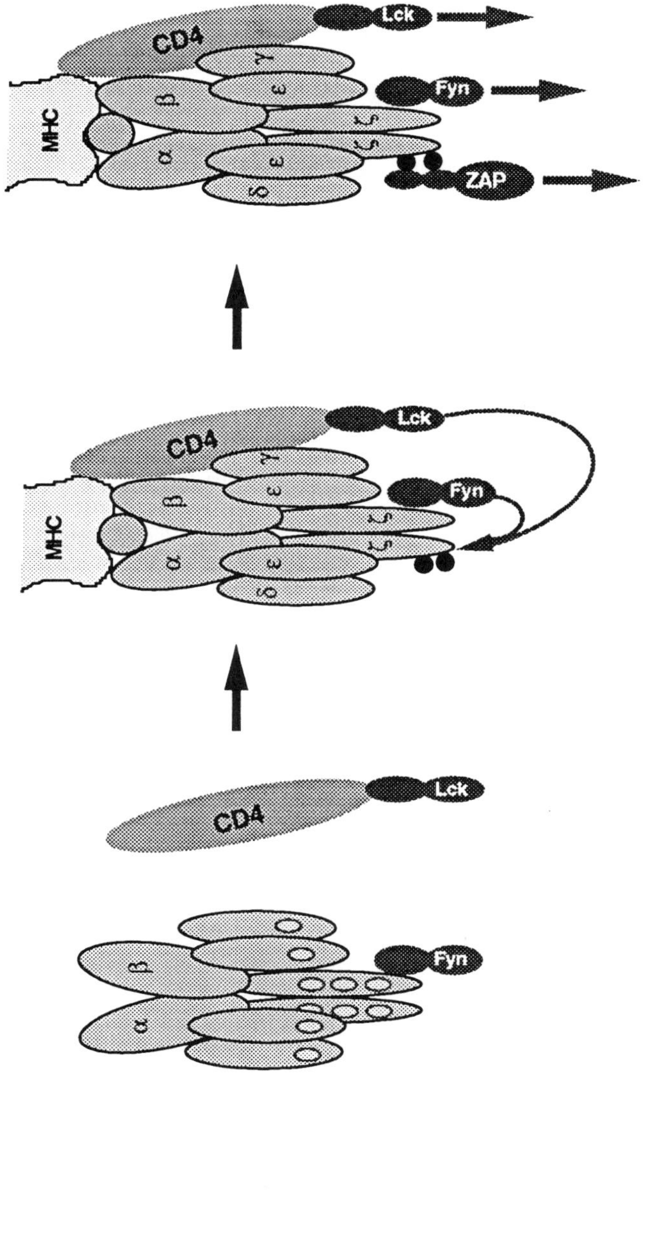

FIGURE 3. Sequential activation of protein tyrosine kinases following TCR engagement. The details of this model are described in the text. The peptide antigen is depicted as an oval between TCR and MHC in the second and third panels. Tyrosine phosphorylation of a single TAM is represented by two black balls. Arrows leading from the kinases represent their activation and phosphorylation of substrates.

TAMζ1	L	**Y**	N	E	**L**	N	K	G	R	R	E	E	-	**Y**	D	V	**L**	D
TAMζ2	L	**Y**	N	E	**L**	Q	K	D	K	M	A	E	A	**Y**	S	E	**I**	G
TAMζ3	L	**Y**	Q	G	**L**	S	T	A	T	K	D	T	-	**Y**	D	A	**L**	H
TAMε	D	**Y**	E	P	**I**	R	K	G	Q	R	D	L	-	**Y**	S	G	**L**	N

FIGURE 4. Amino acid sequence of synthetic peptides used in this study were derived from the human TCRζ or CD3ε chains. Conserved tyrosine (**Y**) and leucine/isoleucine (**L/I**) residues are indicated with bold letters.

not bind. Competition studies showed that unlabeled (pTyr)$_2$-TAMζ$_3$ blocked binding with half-maximal inhibition at 3–4 μM under the conditions used in these studies. TAMζ$_3$ peptides containing either of the single phosphotyrosines and non-phosphorylated peptides failed to block.

The TCR subunits contain multiple TAMs, three in each zeta chain, and one in each of the CD3 chains (γ,δ,ε). Both TCRζ and CD3ε chains are duplicated in each TCR, thereby increasing the number of TAMs per receptor. To test whether other TAMs could bind the ZAP-70 SH2 domains, we also prepared doubly tyrosine phosphorylated TAMζ$_1$, TAMζ$_2$, and TAMε peptides. These also demonstrated saturable binding. These reagents and the corresponding nonradioactive peptides were then used to determine the relative affinity of each peptide for the tandem ZAP SH2 domains. Regardless of which peptide was labeled, the hierarchy of competition was the same, TAMζ$_1$≥TAMζ$_2$>TAMε≥TAMζ$_3$. The relative apparent affinities of these peptides for the tandem SH2 domains, as determined by competition (IC$_{50}$), varied between 0.1–3.0 μM. These data show that the presence of the two SH2 domains, in tandem, results in high-affinity binding to the doubly phosphorylated TAMs. Because ZAP-70 has been shown to be activated only when bound to the TCR, this apparent cooperative binding between individual SH2 domains insures a high level of specificity. Kinase activation is, thus at least in part, regulated by assembly with the TCR. In addition, the work suggests that there is a specificity of TAM-ZAP-70 interaction. Though we saw binding of ZAP to four different TAMs, the affinity differences were such that one might predict a hierarchy of ZAP interaction *in vivo*. In addition, these specific differences suggest that other TAM-binding proteins might favor one TAM over another. Such binding preferences have been seen in the binding of two other proteins, SHC and PI-3-kinase, to the TCR.[38,39] Two explanations have been provided to account for the presence of so many TAMs in the TCR. Data support the idea that multiple TAMs result in signal amplification.[8] However, our data suggest that multiple TAMs are required because of specificity differences in the interaction of multiple proteins with different TAMs.

These binding studies served as a prelude to a number of experiments we performed designed to specifically inhibit the interaction of the TCR with ZAP-70 in cells and thereby block T-cell activation. Medicinal chemists in the NCI, Terry Burke and Peter Roller, had synthesized an analogue of phosphotyrosine, difluorophosphonomethyl phenylalanine (F$_2$Pmp), which is resistant to tyrosine phosphatases[40] (FIG. 5). In addition, they had developed techniques to incorporate this analogue into peptides.

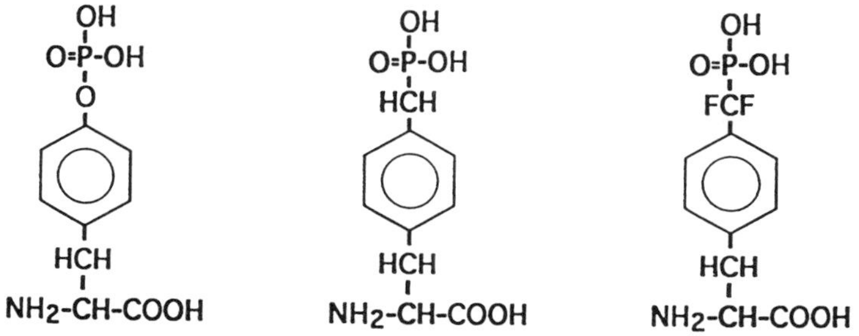

Phosphotyrosine **Phosphonomethyl Phenylalanine** **Difluorophosphonomethyl Phenylalanine**

FIGURE 5. A comparison of the structures of phosphotyrosine, phosphomethyl phenylalanine, and difluorophosphonomethyl phenylalanine (F_2Pmp).

In collaboration with a number of groups, they have shown that peptides made with this reagent bind to SH2 domains with affinities similar to that seen with peptides containing phosphotyrosine. These investigators then prepared the TAMζ_3 peptide with F_2Pmp in place of both tyrosines [(F_2Pmp)$_2$-TAMζ_3)], in place of either N- or C-terminal tyrosine [(F_2Pmp)$_N$-TAMζ_3 or (F_2Pmp)$_C$-TAMζ_3, respectively], or with a control peptide in which blocking groups remain on the F_2Pmp.

We first demonstrated that (F_2Pmp)$_2$-TAMζ_3 competes for binding to GST-ZAP-(SH2)$_2$ with the same affinity as the (pTyr)-TAMζ_3 peptide.[41] With this evidence for high-affinity binding, we thought it reasonable to introduce the (F_2Pmp)$_2$-TAMζ_3 reagent into T cells in an attempt to block activation. Jurkat T cells were permeabilized with tetanolysin, and (F_2Pmp)$_2$-TAMζ_3, or various control peptides, were introduced prior to T-cell activation with an anti-CD3 antibody. The TCR from permeabilized, activated T cells bound tyrosine phosphorylated ZAP-70 as had been shown previously with intact cells. Addition of (F_2Pmp)$_2$-TAMζ_3 prior to activation blocked association of ZAP with the TCR. Precipitation with antibodies to ZAP-70 showed that the blockade induced by the reagent resulted in an inhibition of tyrosine phosphorylation in the total ZAP-70 population. Singly F_2Pmp-substituted peptides and the control peptide with residual blocking groups failed to block association and tyrosine phosphorylation. Additional controls were performed that indicated that the inhibitor was working in the cells rather than during the immunoprecipitation.

Two additional experiments with the (F_2Pmp)$_2$-TAMζ_3 reagent were very informative. The first relied on our observation that ZAP-70 kinase activity can be measured *in vitro* using the cytoplasmic fragment of the band 3 protein as an exogenous substrate as had been previously shown for the related kinase, Syk.[42] In the *in vitro* kinase assay we observed (F_2Pmp)$_2$-TAMζ_3, but not a control peptide, blocked both phosphorylation of ZAP-70 (possibly autophosphorylation) and phosphorylation of the band 3 fragment. Activation of the enzyme thus seems to require assembly with the TCR. The consequence of the block in kinase assembly and activation was tested by standard antiphosphotyrosine blotting. In this experiment, in addition to the block in ZAP-70 tyrosine phosphorylation, we detected a decrease in phosphorylation of substrates of 120, 80, and 35–40 kDa. These proteins are likely candidates to be direct substrates of ZAP-70. The $(F_2Pmp)_2$-TAMζ_3 reagent serves as a prototype of reagents that might prove useful as immunosuppressive drugs.

SUBSTRATES

Background and Initial Results

Several years after demonstrating that the TCR is tyrosine phosphorylated upon engagement with antigen, we performed a number of studies in which intracellular substrates were detected by transferring T-cell lysates from one- and two-dimensional gels followed by immunoblotting with antiphosphotyrosine antibodies. This work included experiments indicating that tyrosine phosphorylation is a very rapid response to TCR engagement that precedes activation of phospholipase C.[12] Subsequently we demonstrated that pharmacologic inhibition of tyrosine kinases with herbimycin resulted in a blockade of T-cell activation.[14] Inhibition of calcium elevation, PI turnover, and lymphokine production were observed following incubation with herbimycin. Additionally we showed that substrate patterns differed slightly depending on the mode of T-cell activation.[43] In all of this work, the increase in tyrosine phosphorylation on intracellular proteins served merely as a marker of the activation of this pathway. We became very interested in identifying what proteins we were detecting on our immunoblots. Identification and characterization of tyrosine kinase substrates have become a significant part of our work.

To purify T-cell PTK substrates, we began by screening different T-cell populations to find sources of abundant tyrosine phosphorylated proteins. We chose the MRL/lpr mouse strain as the most convenient source of T-cell protein. This strain, now known to carry a mutation in the Fas gene,[44] demonstrates massive lymphoproliferation. Absence of Fas results in a failure of lymphocyte apoptosis in the lymph nodes and spleen, with accumulation of T cells. One can harvest as many as 5×10^9 T cells from one mouse. We had also shown previously that these cells contain activated tyrosine kinases and that their TCRs contain constitutively tyrosine phosphorylated TCRζ chains.[45] More recently it was shown that the Fyn PTK is overexpressed in these cells.[46] Cell suspensions were made from lymph nodes and spleens and were incubated with phenylarsine oxide (PAO), a potent inhibitor of T-cell tyrosine phosphatases.[47] After lysis in the presence of additional tyrosine phosphatase inhibitors, we detected an enormous increase in the level of protein tyrosine phosphorylation. With this optimized protocol, immunoaffinity purification with the 4G10 antiphosphotyrosine monoclonal antibody resulted in the specific purification of microgram quantities of several proteins.

These proteins were transferred to nylon membranes, and in collaboration with W. Burgess, subjected to trypsin digestion, peptide purification, and amino acid sequence determination. Amino acid sequence from peptides isolated from the most prominent band, an 81 kDa protein, identified it as ezrin.[48] This molecule is a member of a family of proteins related to erythrocyte band 4.1.[49] Members of this family are cytosolic proteins that interact with the cytoskeleton. In particular, the band 4.1 protein acts as a linker between cytoskeleton and plasma membrane molecules. It had been previously demonstrated that ezrin is tyrosine phosphorylated in A431 cells upon treatment with epidermal growth factor (EGF).[50] In those studies, EGF enhanced the redistribution of ezrin from cytosol to the cytoskeleton. We performed a series of experiments demonstrating that T-cell ezrin was cytosolic and did not shift to insoluble fractions upon TCR engagement or PAO treatment, though PAO did induce a small fraction to shift to the membrane.[48] The function of this molecule in lymphocytes as well as in the EGF system remains to be determined. For us, what was most important at the time was that our use of PAO-treated MRL/lpr lymphocytes and the immunoaffinity purification with antiphosphotyrosine antibodies led us to a known tyrosine kinase substrate. The result gave us the expectation that other substrates could be purified and identified by this approach.

VCP

In our studies of human T lymphocytes, we had noticed a prominent tyrosine kinase substrate of approximately 100 kDa. A band of this size was detected by Ponceau S stain after antiphosphotyrosine immunoaffinity purification, as described above. Amino acid determination revealed that peptides isolated after tryptic digestion of the isolated protein had sequences identical to that of the valosin-containing protein (VCP) isolated from a porcine cDNA library. VCP was first identified as a result of attempts to clone cDNA encoding a putative peptide hormone called valosin. The earlier cloning studies revealed that the 90 kDa VCP was likely to be a ubiquitous cytosolic protein without any of the characteristics of a peptide hormone precursor.[51] For our purposes, the existence of the porcine sequence simplified our cloning of the murine sequence. Analysis of the full murine sequence showed it to be 99.6% identical to the porcine, differing by three conservative amino acid differences out of 806 amino acids.[52]

Our initial studies confirmed that VCP is expressed in T cells and is a tyrosine kinase substrate in lymphocytes. Antisera were generated by immunizing with peptides generated from the VCP sequence. These reagents were used to confirm the expression of VCP in T cells. To estimate the stoichiometry of VCP tyrosine phosphorylation, lysates of activated T cells were subjected to immunoprecipitation with antiphosphotyrosine antibodies. The fraction of VCP isolated and detected by immunoblotting with anti-VCP antibodies was 3-5% of total. To prove conclusively that VCP is a tyrosine kinase substrate in T cells, we cotransfected BW5147 thymoma cells with cDNA encoding (1) a chimeric receptor protein for activating the cells (IL-2 receptor external and transmembrane domains fused to the CD3ε cytoplasmic tail, TTε), (2) murine VCP with a C-terminal epitopic tag (VCPmyc), and (3) a neomycin selection marker. A stable cell line expressing surface TTε and VCPmyc was established (BW125.1C) and was activated by cross-linking the chimeric receptor molecule with biotinylated anti-IL-2 receptor and avidin. Immunoprecipitation with anti-myc antibodies followed by antiphosphotyrosine immunoblotting revealed a greater than 10-fold increase in tyrosine phosphorylation on the VCPmyc protein.

Computer analysis of the VCP protein sequence proved to be of great interest. VCP contains duplicated domains of approximately 220 amino acids containing a nucleotide binding site consensus sequence. An increasing number of proteins are now known to contain one or two of these domains.[53] Proteins with one such domain include several transcription factors and a component of the proteasome. Proteins with two such domains, besides VCP, include the sec18/NSF protein involved in intracellular vesicular transport, p97 ATPase from *X. laevis*, and cdc48p from *S. cerevisiae*. The latter protein was identified in mutant yeast cells that, at the restrictive temperature (*i.e.*, below 16°C), are disrupted in mitosis with abnormal spindles and microtubule disruption.[53] A few features are common to those members of this family that have been adequately analyzed. Several of the proteins are found to be homooligomers, with NSF a tetramer[54] and p97ATPase a hexamer.[55] Additionally, these two molecules have been shown to have ATPase activity.

We continued our analysis of VCP in several ways.[56] The site of tyrosine phosphorylation in VCP was mapped by using a GST-VCP fusion protein as a substrate and MRL/lpr membranes as a source of tyrosine kinases. GST-VCP was tyrosine phosphorylated *in vitro* and digested with trypsin, and two-dimensional phosphopeptide maps were performed. The sequences of potential phosphopeptides were analyzed to determine appropriate secondary protease digestions of the initial peptides. When these redigestions were performed we were led to the presumptive identification of two tyrosine residues in the C-terminus of VCP as sites of phosphorylation. We confirmed this identification by site-directed mutagenesis of these two tyrosines in GST-VCP. We found that the majority of phosphorylation occurs on tyrosine 805 with small amounts detected on Tyr 796. We then performed labeling experiments in the BW125.1C T cells, described above. The phosphopeptide maps obtained from this experiment showed that the same phosphopeptides found after *in vitro* phosphorylation of VCP were identified in VCP isolated after TCR engagement in intact cells.

To determine whether VCP is an oligomer, we subjected cytosol of the T-cell transfectant to sucrose gradient centrifugation. The majority of immunoblottable VCP comigrated with the 19.2S marker, thyroglobulin. This migration and apparent size is proof of the oligomeric structure of VCP and is consistent with a hexameric structure. Similar results were obtained on a sizing column. To determine whether VCP is also an ATPase, we expressed the tagged protein in bacteria as a GST fusion protein and proved that this too has oligomeric structure. We then incubated bacterial VCP with [^{32}P]γ-ATP and showed that free phosphate was generated. The amount of ATP hydrolyzed increased linearly over time and was dependent on the concentration of VCP. Finally, we showed that the tyrosine phosphorylation of VCP has no detectable effect on the ATPase activity of the protein. Preliminary results suggest that the sites of tyrosine phosphorylation are important. Mutant VCP, in which the two tyrosine phosphorylation sites are mutated to phenylalanine, was expressed in HeLa cells using an inducible promoter system. Induction of mutant VCP, but not wild-type VCP results in a decrease in the rate of cell growth. The mechanism for this phenomenon is under investigation.

p120^{cbl}-The 120 kDa Substrate

In earlier studies we noted a prominent 120 kDa tyrosine kinase substrate that was rapidly phosphorylated following TCR engagement.[12] We have recently identified this substrate as the protein product of the c-*cbl* protooncogene, p120^{cbl}.[57] The limited amount of information that is available on this protein is intriguing and suggests that it may be important in T-cell function. The c-*cbl* protooncogene has had only limited characterization after the discovery of the viral oncogene v-*cbl*. The v-*cbl* form is transforming in early B-cell, myeloid lineage cells and fibroblasts.[58] The product of the v-*cbl* oncogene is a 100 kDa fusion protein composed of a retroviral gag protein and a truncated form of p120^{cbl}, which lacks the C-terminal 60% of residues encoded by the protooncogene.

Data base searches fail to reveal any protein with significant similarity to p120^{cbl}. However, the deduced amino acid sequence contains a number of interesting features.[59] The proteins encoded by both oncogene and protooncogene contain a putative nuclear localization signal and a region rich in basic amino acids. Features present in p120^{cbl} but lost in the truncation that lead to the v-*cbl* product include a stretch of about 200 amino acids that contain 23% proline, a RING motif related to zinc finger, and a putative zinc finger. Many of these features might suggest that the protein is a transcription factor. However, whereas the v-*cbl* product has been localized to the nucleus, p120^{cbl} is exclusively cytoplasmic.[60] Another oncogenic form of cbl is found in the 70Z pre-B-cell line. The cbl-encoded protein in this cell is similar to p120^{cbl}, except for a 17 amino acid internal deletion at a position just distal to where the v-*cbl* truncation occurs. Interestingly, this oncogenic form is cytoplasmic and tyrosine phosphorylated.[61] After learning of some of these results, we obtain antibodies generated against p120^{cbl} constructs to immunoprecipitate and immunoblot p120^{cbl} from Jurkat cells. Initial experiments demonstrated that p120^{cbl} is tyrosine phosphorylated in Jurkat T cells in response to TCR engagement or after treatment of the cells with the tyrosine phosphatase inhibitor, pervanadate. Immunodepletion experiments revealed that all of the prominent 120 kDa tyrosine phosphoprotein observed after TCR ligation is p120^{cbl}. Maximal tyrosine phosphorylation could be detected at 1 minute.

In the past year, several papers have been published showing that a 116-120 kDa protein can be purified from T cells through its association with proteins bearing SH3 domains.[62–64] SH3 domains, like the earlier-defined SH2 domains, can be found in a number of effector and linker molecules. Their structure has been defined by NMR, and they, like SH2 domains, interact with proteins by way of a defined sequence specificity. The SH3 domains bind proteins rich in proline, and in one study, in which a degenerate peptide approach was used to define binding specificity, the sequence PXXP was found to be a minimal consensus binding motif.[65] The 116-120 kDa proteins, mentioned above, were among a number of proteins isolated from lysates of activated T cells by way of interaction with the linker protein Grb2, a molecule containing an SH2 domain flanked by two SH3 domains. This set of proteins was detected by antiphosphotyrosine immunoblotting and includes proteins of 120, 75, and 36. Grb2 is known to be involved in coupling growth factor receptor tyrosine kinases to SOS, a protein involved in ras activation by virtue of its ability to load ras with GTP.[66,67] Thus the impetus for surveying T-cell lysates for Grb2-binding proteins arises from a desire to understand how the TCR is coupled to the ras activation pathway. We addressed the question of whether p120^{cbl} could be the 120 kDa protein observed in these Grb2 experiments by using GST fusion proteins composed of full-length Grb2. To determine whether Grb2 SH3 domains were responsible for binding, we also used individual C-terminal or N-terminal Grb2 SH3 domains coupled to GST. We found that the full-length, and, to a lesser degree, the N-terminal SH3 domain of Grb2, but not the C-terminal SH3 domain, bound p120^{cbl}. The interaction with this domain is decreased upon T-cell activation. Similar interactions occur in T cells.

[**Note Added In Proof:** The sequence defined as TAM in the text is by consensus renamed ITAM for immune receptor tyrosine-based activation motif.[68]]

REFERENCES

1. Schwartz, R. H. 1985. Annu. Rev. Immunol. **3:** 237.
2. Samelson, L. E. & R. D. Klausner. 1992. J. Biol. Chem. **267:** 24913.
3. Weiss, A. & D. R. Littman. 1994. Cell **76:** 263.
4. Kronenberg, M., G. Siu, L. E. Good & N. Shastri. 1986. Annu. Rev. Immunol. **4:** 529.
5. Ashwell, J. D. & R. D. Klausner. 1990. Annu. Rev. Immunol. **8:** 139.
6. Weissman, A. M. 1994. Chem. Immunol. **59:** 1.
7. Letourneur, F. & R. D. Klausner. 1992. Science **255:** 79.
8. Irving, B. A., A. C. Chan & A. Weiss. 1993. J. Exp. Med. **177:** 1093.
9. Clark, M. R., K. S. Campbell, A. Kazlauskas, S. A. Johnson, M. Hertz, T. A. Potter, C. Pleiman & J. C. Cambier. 1992. Science **258:** 123.
10. Samelson, L. E., M. D. Patel, A. M. Weissman, J. B. Harford & R. D. Klausner. 1986. Cell **46:** 1083.
11. Baniyash, M., P. Garcia-Morales, E. Luong, L. E. Samelson & R. D. Klausner. 1988. J. Biol. Chem. **263:** 18225.
12. June, C. H., M. C. Fletcher, J. A. Ledbetter & L. E. Samelson. 1990. J. Immunol. **144:** 1591.
13. Mustelin, T., K. M. Coggeshall, N. Isakov & A. Altman. 1990. Science **247:** 1584.
14. June, C. H., M. C. Fletcher, J. A. Ledbetter, G. L. Schieven, J. N. Siegal, A. F. Phillips & L. E. Samelson. 1990. Proc. Natl. Acad. Sci. USA **87:** 7722.
15. Samelson, L. E., A. F. Phillips, E. T. Luong & R. D. Klausner. 1990. Proc. Natl. Acad. Sci. USA **87:** 4358.
16. Veillette, A., M. A. Bookman, E. M. Horak & J. B. Bolen. 1988. Cell **55:** 301.
17. Rudd, C. E., J. M. Trevillyan, J. D. Dasgupta, L. I. Wong & S. F. Schlossman. 1988. Proc. Natl. Acad. Sci. USA **85:** 5190.
18. Chan, A. C., B. A. Irving, J. D. Fraser & A. Weiss. 1991. Proc. Natl. Acad. Sci. USA **88:** 9166.
19. Weiss, A., G. Koretzky, R. C. Schatzman & T. Kadlecek. 1991. Proc. Natl. Acad. Sci. USA **88:** 5484.
20. Izquierdo, M., J. Downward, J. D. Graves & D. A. Cantrell. 1992. Mol. Cell. Biol. **12:** 3305.
21. Molina, T. J., K. Kishihara, D. P. Siderovski, W. van Ewijk, A. Narendran, E. Timms, A. Wakeham, C. A. Paige, K.-U. Hartmann, A. Veillette, D. Davidson & T. W. Mak. 1992. Nature **357:** 161.
22. Straus, D. B. & A. Weiss. 1992. Cell **70:** 585.
23. Peri, K. G. & A. Veillette. 1994. Chem. Immunol. **59:** 19.
24. Veillette, A., M. A. Bookman, E. M. Horak, L. E. Samelson & J. B. Bolen. 1989. Nature **338:** 257.

25. SAROSI, G. A., P. M. THOMAS, M. EGERTON, A. F. PHILLIPS, K. W. KIM, E. BONVINI & L. E. SAMELSON. 1992. Int. Immunol. **4:** 121.
26. TIMSON GAUEN, L. K., A. N. KONG, L. E. SAMELSON & A. S. SHAW. 1992. Mol. Cell. Biol. **12:** 5438.
27. GAUEN, L. K., Y. ZHU, F. LETOURNEUR, Q. HU, J. B. BOLEN, L. A. MATIS, R. D. KLAUSNER & A. S. SHAW. 1994. Mol. Cell. Biol. **14:** 3729.
28. WANGE, R. L., A. N. KONG & L. E. SAMELSON. 1992. J. Biol. Chem. **267:** 11685.
29. CHAN, A. C., M. IWASHIMA, C. W. TURCK & A. WEISS. 1992. Cell **71:** 649.
30. KOCH, C. A., D. ANDERSON, M. F. MORAN, C. ELLIS & T. PAWSON. 1991. Science **252:** 669.
31. SONGYANG, Z., S. E. SHOELSON, M. CHAUDHURI, G. GISH, T. PAWSON, W. G. HASER, F. KING, T. ROBERTS, S. RATNOFSKY, R. J. LECHLEIDER, B. G. NEEL, R. B. BIRGE, J. E. FAJARDO, M. M. CHOU, H. HANAFUSA, B. SCHAFFHAUSEN & L. C. CANTLEY. 1993. Cell **72:** 767.
32. RETH, M. 1989. Nature **338:** 383.
33. QIAN, D., I. GRISWOLD-PRENNER, M. R. ROSNER & F. W. FITCH. 1993. J. Biol. Chem. **268:** 4488.
34. WANGE, R. L., S. N. MALEK, S. DESIDERIO & L. E. SAMELSON. 1993. J. Biol. Chem. **268:** 19797.
35. IWASHIMA, M., B. A. IRVING, N. S. VAN OERS, A. C. CHAN & A. WEISS. 1994. Science **263:** 1136.
36. MCFARLAND, E. C., E. FLORES, R. J. MATTHEWS & M. L. THOMAS. 1994. Chem. Immunol. **59:** 40.
37. ISAKOV, N., R. L. WANGE, W. H. BURGESS, J. D. WATTS, R. AEBERSOLD & L. E. SAMELSON. 1994. J. Exp. Med. **181:** 375.
38. RAVICHANDRAN, K. S., K. K. LEE, L. C. SONGYANG, L. C. CANTLEY, P. BURN & S. J. BURAKOFF. 1993. Science **262:** 902.
39. EXLEY, M., L. VARTICOVSKY, M. PETER, J. SANCHO & C. TERHORST. 1994. J. Biol. Chem. **269:** 15140.
40. BURKE, Jr., T. R., M. S. SMYTH, A. OTAKA, M. NOMIZU, P. P. ROLLER, G. WOLF, R. CASE & S. S. SHOELSON. 1994. Biochemistry **33:** 6490.
41. WANGE, R. L., N. ISAKOV, T. R. BURKE, Jr., A. OTAKA, P. P. ROLLER, J. D. WATTS, R. AEBERSOLD & L. E. SAMELSON. 1995. J. Biol. Chem. **270:** 944.
42. HARRISON, M. L., C. C. ISAACSON, D. L. BURG, R. L. GEAHLEN & P. S. LOW. 1994. J. Biol. Chem. **269:** 955.
43. HSI, E. D., J. N. SIEGEL, Y. MINAMI, E. T. LUONG, R. D. KLAUSNER & L. E. SAMELSON. 1989. J. Biol. Chem. **264:** 10836.
44. WATANABE-FUKUNAGA, R., C. J. BRANNAN, N. COPELAND, N. A. JENKINS & S. NAGATA. 1992. Nature **356:** 314.
45. SAMELSON, L. E., W. F. DAVIDSON, H. C. MORSE III & R. D. KLAUSNER. 1986. Nature **324:** 6741.
46. KATAGIRI, T., K. URAKAWA, Y. YAMANASHI, K. SEMBA, T. TAKAHASHI, K. TOYOSHIMA, T. TAMAMOTO & K. KANO. 1989. Proc. Natl. Acad. Sci. USA **86:** 10064.
47. GARCIA-MORALES, P., Y. MINAMI, E. LUONG, R. D. KLAUSNER & L. E. SAMELSON. 1990. Proc. Natl. Acad. Sci. USA **87:** 9255.
48. EGERTON, M., W. H. BURGESS, D. CHEN, B. J. DRUKER, A. BRETSCHER & L. E. SAMELSON. 1992. J. Immunol. **149:** 1847.
49. GOULD, K. L., A. BRETSCHER, F. S. ESCH & T. HUNTER. 1989. EMBO J. **8:** 4133.
50. BRETSCHER, A. 1989. J. Cell. Biol. **108:** 921.
51. KOLLER, K. J. & M. J. BROWNSTEIN. 1987. Nature **325:** 542.
52. EGERTON, M., O. R. ASHE, D. CHEN, B. J. DRUKER, W. H. BURGESS & L. E. SAMELSON. 1992. EMBO J. **11:** 3533.
53. FRÖHLICH, K.-U., H.-W. FRIES, M. RÜDIGER, R. ERDMANN, D. BOTSTEIN & D. MECKE. 1991. J. Cell Biol. **114:** 443.

54. Block, M. R., B. S. Glick, C. A. Wilcox, F. T. Wieland & J. E. Rothman. 1988. Proc. Natl. Acad. Sci. USA **85:** 7852.
55. Peters, J.-M., M. J. Walsh & W. W. Franke. 1990. EMBO J. **9:** 1757.
56. Egerton, M. & L. E. Samelson. 1994. J. Biol. Chem. **269:** 11435.
57. Donovan, J. A., R. L. Wange, W. A. Langdon & L. E. Samelson. 1994. J. Biol. Chem. **269:** 22921.
58. Langdon, W. A., J. W. Hartley, S. P. Klinken, S. K. Ruscetti & H. C. Morse III. 1989. Proc. Natl. Acad. Sci. USA **86:** 1168.
59. Blake, T. J., M. Shapiro, H. C. Morse III & W. Y. Langdon. 1991. Oncogene **6:** 653.
60. Langdon, W. Y., K. G. Heath & T. J. Blake. 1992. Curr. Top. Microbiol. Immunol. **182:** 467.
61. Andoniou, C. E., C. B. F. Thien & W. Y. Langdon. 1994. EMBO J. **13:** 4515.
62. Motto, D. G., S. E. Ross, J. K. Jackman, Q. Sun, A. L. Olson, P. R. Findell & G. A. Koretzky. 1994. J. Biol. Chem. **269:** 21608.
63. Buday, L., S. E. Egan, P. Rodriguez Viciana, D. A. Cantrell & J. Downward. 1994. J. Biol. Chem. **269:** 9019.
64. Reif, K., L. Buday, J. Downward & D. A. Cantrell. 1994. J. Biol. Chem. **269:** 14081.
65. Yu, H., J. K. Chen, S. Feng, D. C. Dalgarno, A. W. Brauer & S. L. Schreiber. 1994. Cell **76:** 933.
66. Lowenstein, E. J., R. J. Daly, A. G. Batzer, W. Li, B. Margolis, R. Lammers, A. Ullrich, D. Skolnick, D. Bar-Sagi & J. Schlessinger. 1992. Cell **70:** 431.
67. Egan, S. E., B. W. Giddings, M. W. Brooks, L. Buday, A. M. Sizeland & R. A. Weinberg. 1993. Nature **363:** 45.
68. Cambier, J. C. 1995. Immunol. Today **16:** 110.

Genetic Dissection of the Transducing Subunits of the T-Cell Antigen Receptor

BERNARD MALISSEN, GRACE KU,
MIRJAM HERMANS, ERIC VIVIER, AND
MARIE MALISSEN

Centre d'Immunologie INSERM-CNRS de Marseille-Luminy
Case 906
13288 Marseille Cedex 9, France

INTRODUCTION

The specific recognition of antigens by T cells and its ensuing conversion into intracellular signals is accomplished by the T-cell antigen receptor (TCR)-CD3 complex. Triggering of this complex on resting mature T lymphocytes leads to the activation of hundreds of genes among which are those coding for cytokines and cytokine receptors. Upon interaction with their specific receptors, the resulting cytokines induce the clonal expansion and differentiation events that constitute the hallmarks of T-cell activation.[1,2] Under physiological circumstances, engagement of the TCR alone is not sufficient to induce the full genetic program leading to the production of activated T cells capable of carrying out cytotoxic or helper functions. The coincident triggering of additional T-cell surface molecules, denoted as "costimulatory receptors," is required to sustain the priming signals resulting from TCR engagement. For instance, the interaction of the CD28 molecule on T cells with the B7-1 ligand on antigen-presenting cells generates signals that synergize with those originating from the TCR to increase the activity of a few intracellular effector molecules endowed with the capacity to integrate distinct signals. In T cells, the protein kinase JNK and the NF-AT transcriptional complex are located at points of convergence of the TCR- and CD28-operated signaling pathways and therefore constitute two of these signal coincidence detectors.[3,4]

The TCR-CD3 complex consists of a clonally variable, antigen-binding TCR $\alpha\beta$ (or TCR $\gamma\delta$) heterodimer that is noncovalently associated with a group of invariant nonantigen-binding polypeptides, denoted CD3-γ, CD3-δ, CD3-ε, CD3-ζ, and CD3-η (FIG. 1). All these subunits have to be assembled in a coordinated fashion inside the endoplasmic reticulum for efficient transport to the cell surface. The stoichiometry and spatial arrangement of the TCR-CD3 subunits found within cell surface-expressed complexes are not known. Recent studies have suggested that each TCR-CD3 complex contains only one TCR α chain, one TCR β chain, and two CD3-ε chains.[5] However, the stoichiometry of the CD3-δ, CD3-γ, and CD3-ζ/η chains still remains to be determined.

The cytoplasmic segments of the various CD3 chains are responsible for coupling the antigen-binding unit to intracellular signaling pathways. Their signal transduction

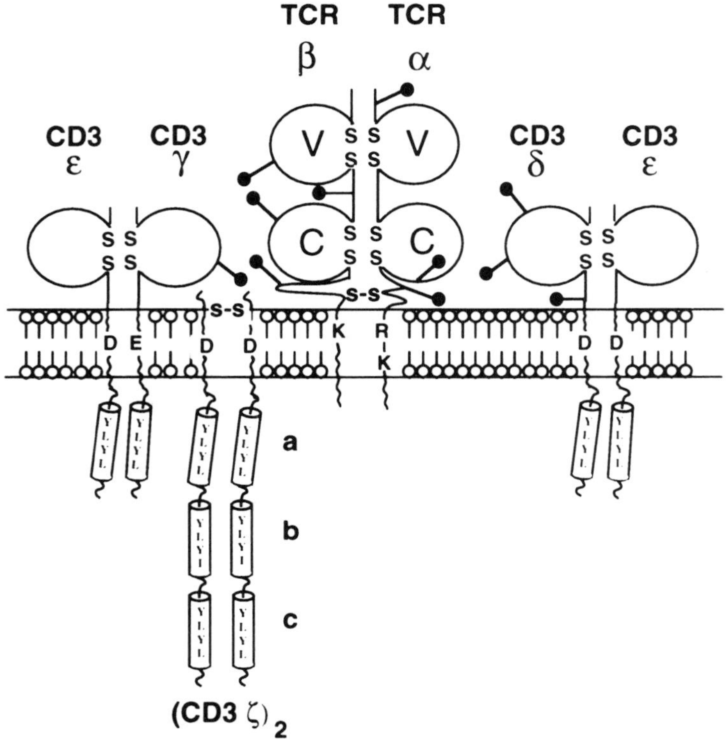

FIGURE 1. Subunit composition of the TCR-CD3 complex found on the majority of mature T lymphocytes. The TCRα and TCRβ gene products are expressed within the complex as disulfide-linked heterodimers, possess short cytoplasmic tails (4 to 5 residues), and contain clonally variable regions that determine the antigenic specificity of the T cell. The evolutionary related CD3-γ, CD3-δ, and CD3-ε subunits display immunoglobulin-like extracellular domains and are expressed as noncovalently associated γε and δε pairs. The ζ and η subunits contain only 9 residues in their extracellular domains and share significant structural homology with the γ chain of the high-affinity IgE receptor (FcεRI). The CD3-ζ and CD3-η polypeptides derive from the same gene transcripts by way of an alternative splicing event, and as a consequence differ only at their carboxy termini. When coexpressed in a single T cell, the CD3-ζ, CD3-η, and FcεRIγ polypeptides can combine to form various disulfide-linked homo- and heterodimers (*e.g.*, ζζ, ζγ, ζη . . .), each capable of associating with a core made of the TCR αβ, CD3 γε, and CD3 δε pairs. The cytoplasmic domains of the CD3 subunits are considerably larger (44 to 113 residues) than those of the TCR chains and are responsible for coupling the TCR heterodimer to the intracellular signaling machinery. The recurrent functional domains found in each of the CD3 subunits are shown as cylinders.

capability has been attributed to the presence of a recurrent functional domain of ~ 20 amino acids.[6] Common to each copy of this domain is a pair of Tyr-X-X-Leu/Ile sequences (where X corresponds to a variable residue) separated by seven or eight variable residues. This consensus sequence, hereafter referred to as the $(YXXL/I)_2$ motif, is expressed as a single copy in the cytoplasmic tail of the CD3-δ, CD3-γ, and CD3-ε subunits. The CD3-ζ polypeptide appears to be unique among this set of signaling devices, in the sense that it displays three concatenated copies of the $(YXXL/I)_2$ motif (denoted as ζa, ζb, and ζc in FIG. 1). Cross-linking of chimeric molecules composed of the extracellular and transmembrane parts of the CD4, CD8α, CD16, or CD25 molecules and of a single copy of the $(YXXL/I)_2$ motif suffices to elicit most of the early and late activation events that normally occur when antigen receptors are stimulated.

THE MODULAR ARCHITECTURE OF THE TCR-CD3 COMPLEX

The analysis of TCR-CD3 complexes harboring functionally impaired CD3-ζ subunits has led to the view that the TCR-CD3 complexes are normally composed of two parallel signal transduction modules made of the γδε and ζ chains, respectively.[7] Although direct evidence is still lacking, it is likely that each of the CD3 polypeptides can individually act as an autonomous transducer once expressed in the context of a whole TCR-CD3 complex. Therefore, such modular architecture should permit the generation of various TCR isoforms made of distinct CD3 subunit combinations. Moreover, assuming that individual CD3 subunits have unique functional properties (see below), such combinatorial diversity should allow T cells to assemble TCR-CD3 isoforms with different transducing characteristics. In support of this view, the analysis of mouse gut intraepithelial lymphocytes has revealed that the γ chain of the high-affinity IgE receptor (FcεRIγ, see legend of FIG. 1) could be incorporated into TCR-CD3 complexes in lieu of CD3-ζ and CD3-η.[8–11] Furthermore, other unique combinations of TCR/CD3 subunits appear to be expressed with a temporally restricted pattern. For instance, in contrast to mature T cells, pre-T cells can assemble TCR-CD3 complexes devoid of the TCRα chain.[12] Therefore, changes in the subunit composition of the TCR-CD3 complex may occur both during T-cell development and among mature T-cell subpopulations and be responsible for coupling antigen recognition to distinct signaling cassettes.

ROLE OF THE $(YXXL/I)_2$ MOTIFS IN THE INITIATION OF T-CELL ACTIVATION

The precise mechanisms by which the oligomerization of the various $(YXXL/I)_2$ motifs present in the CD3 subunits generates intracellular signals remain unknown. However, as an immediate result of antigen binding, the tyrosine residues found in each YXXL/I sequence become phosphorylated,[13,14] and act subsequently as "docking" sites for the src-homology 2 (SH2) domains found in certain intracellular adaptor and effector molecules. For instance, a 70 kDa protein tyrosine kinase, termed ZAP-70, has been shown to specifically associate with the tyrosine phosphorylated CD3-ζ and CD3-ε polypeptides. ZAP-70 differs from the src-family kinases in that it possesses two N-terminal SH2 domains and lacks a myristylation signal or other recognizable membrane localization motifs (FIG. 2). Further experiments have suggested that the high-affinity binding of ZAP-70 to ζ or ε is direct and requires that its two N-terminal SH2 domains bind in a coordinated fashion to the pair of phosphorylated YXXL/I sequences found within each activated motif.[15,16] Neither of the ZAP-70 SH2 domains expressed singly as a glutathione S-transferase (GST) fusion protein could bind to the phosphorylated ζ or ε subunits.[16,17] By contrast, the single SH2 domain of the Shc adaptor protein appears capable of interacting directly with at least one of the phosphorylated YXXL/I sequences found in the activated ζ chains.[18] Therefore, once phosphorylated, the two YXXL/I sequences present within a given motif appear capable of being recognized either in a concerted mode by intracellular effector/adaptor proteins containing tandem SH2 domains (*e.g.*, ZAP-70, syk) or in an independent mode by effector/adaptor proteins containing single SH2 domain (*e.g.*, Shc).

In the resting state, the unphosphorylated motifs appears to be constitutively associated with at least one resident tyrosine kinase belonging to the Src-family kinases (FIG. 2). This association appears independent of the YXXL/I sequences, and, most likely, determined by the amino acids that immediately flank them.[19] Upon receptor cross-linking, these resident tyrosine kinases may be responsible for the phosphorylation of the various $(YXXL/I)_2$ motifs.

It had been originally hypothesized that all the phosphorylated $(YXXL/I)_2$ motifs have the same binding specificity (*i.e.*, are capable of docking to the very same set of intracellular effector/adaptor proteins). Accordingly, the presence of several redundant $(YXXL/I)_2$ motifs in a given antigen receptor (*e.g.*, $\geq$ 10 for the TCR, $\geq$ 4 for the B-cell antigen receptor, and $\geq$ 3 for the high-affinity IgE receptor) may have merely evolved as a way of amplifying the intracellular signal(s) resulting from ligand binding. Alternatively, this multiplicity of redundant motifs may constitute a potential way of tuning the activation signals proportionally to the strength of the antigenic stimulation. However, recent results suggest that the various $(YXXL/I)_2$ motifs found in antigen receptors are nonredundant, each being capable of association with distinct subsets of SH2-containing proteins, and thus responsible for the coupling to a unique signaling pathway.[20,21]

GENETIC DISSECTION OF THE PRE-TCR/CD3 COMPLEX

Germline gene targeting experiments have already highlighted the contribution of a few genes to the cellular and molecular processes that govern mouse intrathymic T-cell development. These experiments have also suggested the existence of developmental control points (checkpoints) that ensure that T cells do not complete their intrathymic differentiation program if they have failed to assemble functional TCR genes or if they express TCR αβ combinations with inappropriate specificities (*i.e.*, TCRs that are self-reactive, unable to cooperate with the set of coexpressed self-MHC molecules, or displayed with a mismatched CD4/CD8 coreceptor molecule). These ''proofreading'' mechanisms probably operate by way of the triggering of the pre-TCR and TCR complexes that are sequentially expressed during T-cell development (FIG. 3). By activating intracellular effectors, the pre-TCR and TCR complexes

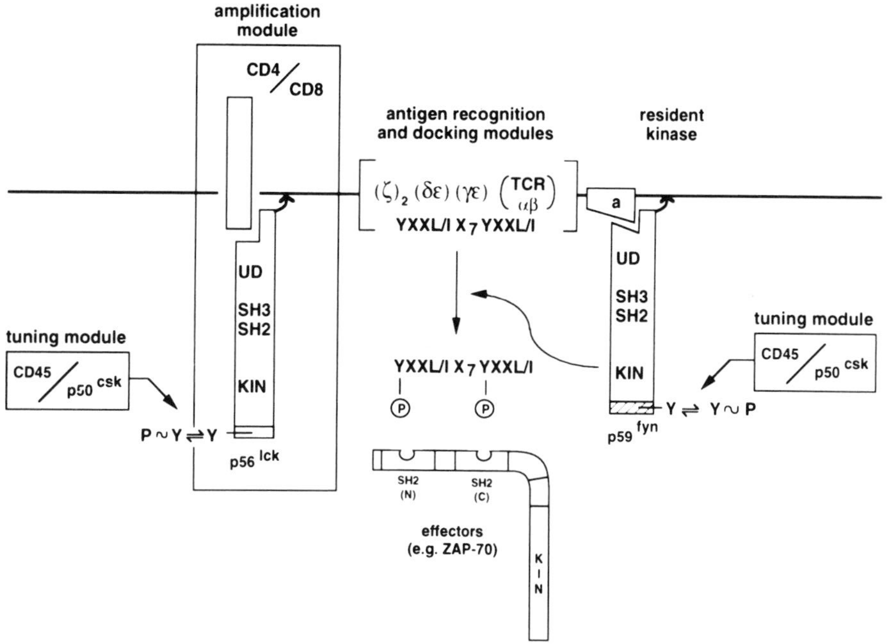

FIGURE 2. A model for the initiation of T-cell activation. TCR stimulation, with or without engagement of the CD4/CD8 amplification module, results in the phosphorylation of the $(YXXL/I)_2$ motifs by a small amount of prebound src family protein tyrosine kinase (*e.g.*, $p59^{fyn}$ or $p56^{lck}$) and permits the translocation of ZAP-70 to the plasma membrane in proximity to its substrates. In the case of the TCR-CD3 complex, it appears to be tyrosine kinases (*e.g.*, ZAP-70 or syk) rather than other downstream effectors/adapters (such as GTPase activating proteins, PI 3-kinase, or phospholipase C) that bind to the phosphorylated $(YXXL/I)_2$ motifs. Following ligand binding the TCR-CD3 complex may become physically associated with a signal amplification module made of the CD4 : $p56^{lck}$ or CD8 : $p56^{lck}$ complexes. As a result, $p56^{lck}$ may contribute to the phosphorylation of the CD3 subunits and the subsequent recruitment of ZAP-70. According to this view, one of the roles played by CD4 or CD8 is to increase the probability of bringing lck molecules into close proximity with the CD3 subunits. Alternatively, the CD3 subunits may not be the direct targets of the lck molecules associated with the CD4/CD8 coreceptors, and other molecules, such as ZAP-70 or fyn, may constitute their physiological substrates and account for their role in antigen-driven T-cell activation.

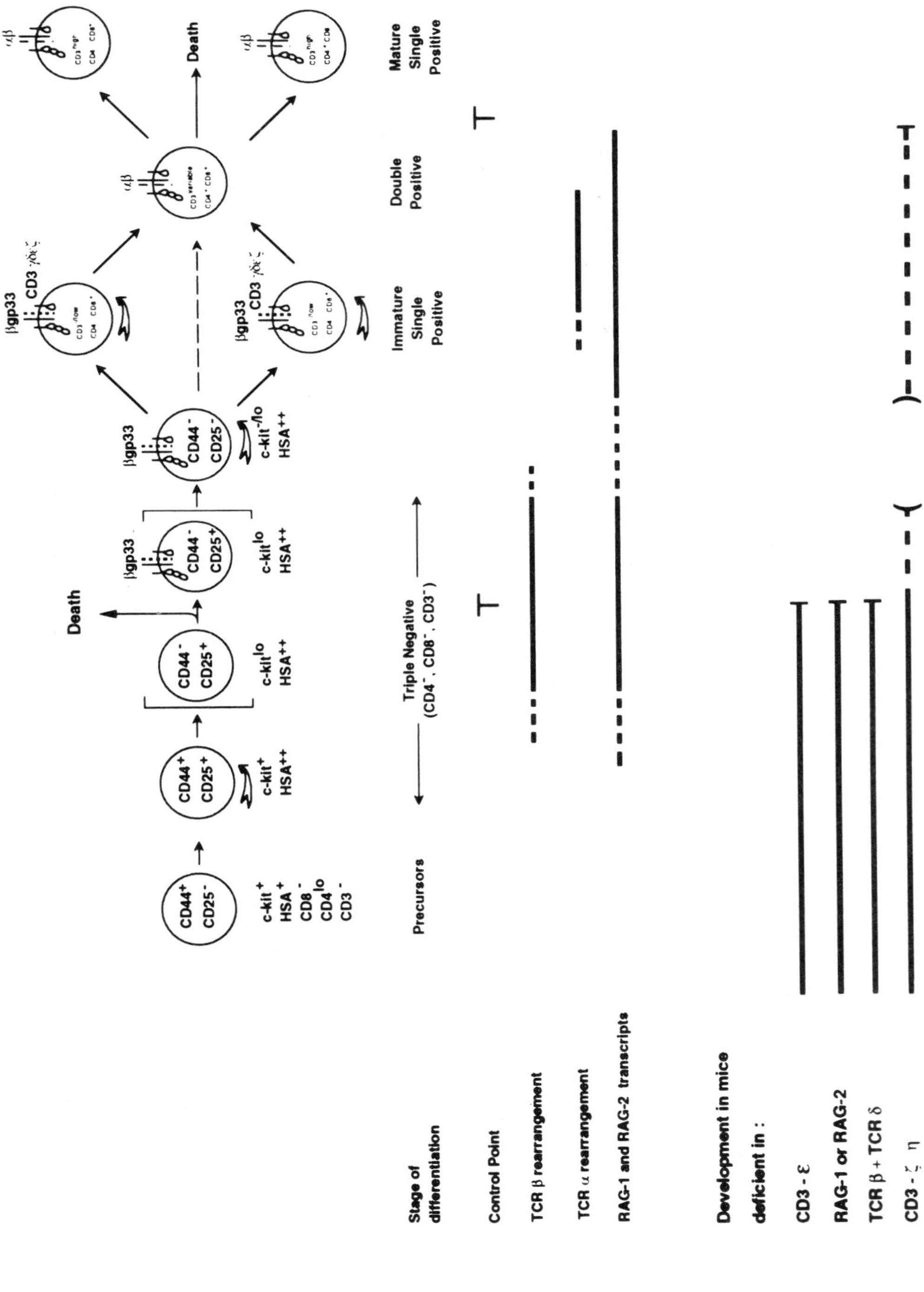

Stage of differentiation
Precursors
Triple Negative (CD4⁻, CD8⁻, CD3⁻)
Immature Single Positive
Double Positive
Mature Single Positive
CD44⁺ CD25⁻
c-kit⁺ HSA⁺ CD8⁻ CD4^lo CD3⁻
CD44⁺ CD25⁺
c-kit⁺ HSA⁺⁺
CD44⁻ CD25⁺
c-kit^lo HSA⁺⁺
Death
βgp33
CD44⁻ CD25⁻
c-kit⁻/lo HSA⁺⁺
βgp33 CD3 γδεζ
αβ
Death
Control Point
TCR β rearrangement
TCR α rearrangement
RAG-1 and RAG-2 transcripts
Development in mice deficient in :
CD3 - ε
RAG-1 or RAG-2
TCR β + TCR δ
CD3 - ζ η

may modulate a gene expression pattern and drive the maturing T cells through a precise developmental sequence.

Mice deficient in V(D)J recombination (*e.g.*, RAG-$1^{-/-}$ or RAG-$2^{-/-}$ mice, FIG. 3) have very small thymuses in which T cells fail to develop beyond the $CD44^-$ $CD25^+$ triple negative stage. Introduction of a productively rearranged TCR β transgene partially relieves this developmental blockade and drives the thymocytes up to the double-positive stage. However, further maturation to the single-positive stage was still blocked in these TCR β reconstituted mice. By contrast, the introduction of a productively rearranged TCR α transgene into the RAG-$1^{-/-}$ background was unable to induce a similar transition from the double-negative to the double-positive stage. For a long time, it was not clear in which way TCR β polypeptides could drive early T-cell development in the complete absence of TCR α polypeptides. However, a "TCR α-less" molecular complex denoted as the pre-TCR complex has been recently identified on early fetal thymocytes and on transformed pre-T-cell lines.[22] On these cells, the TCR β chain forms an 80 kDa disulfide-linked heterodimer with a glycosylated protein of 33 kDa. This protein has been denoted gp33, and its corresponding gene has been cloned and shown to encode a transmembrane protein, the expression of which is limited to immature thymocytes. gp33 mRNA are found in thymocytes from RAG-$2^{-/-}$ mice and therefore did not require V(D)J recombination to be expressed. Aside from the TCR β chain and gp33 components, the CD3-ε, –γ, and –δ polypeptides appear to participate in this complex, whereas CD3-ζ has been found to be only loosely associated with it.

The pre-TCR is thought to function as a sensing device capable of checking for the assembly of productively rearranged TCRβ chain genes. However, it is still not clear whether it has to be expressed at the cell surface and bind to an intrathymic ligand or whether its intracellular display is sufficient to drive the cells beyond the $CD44^-$ $CD25^-$ developmental stage. Recent analysis of CD3-ε-deficient mice indicates that pre-T-cell development up to the $CD44^-$ $CD25^+$ triple-negative stage does not depend on the expression of the CD3-ε gene. However, the expression of CD3-ε is absolutely required for development beyond the $CD44^-$ $CD25^-$ triple-negative cells. Therefore, these data suggest that CD3–ε constitutes an essential component of the pre-TCR complex.

The thymuses of mice deficient in the CD-ζ chain gene vary considerably in size from animal to animal (2.4×10^7 cells on average) and contain cells that are arrested

FIGURE 3. The intrathymic differentiation of αβ T cells is sequentially controlled by the pre-TCR and TCR complexes. Precursors entering the thymus carry their TCR loci in germline configuration and may develop along the γδ or αβ T-cell lineages. Upon commitment to the αβ lineage, immature $CD3^-$ $CD4^-$ $CD8^-$ ("triple negative") thymocytes differentiate into $CD4^+$ $CD8^+$ ("double positive") cells, a small percentage of which mature into either $CD4^+$ $CD8^-$ or $CD4^-$ $CD8^+$ ("mature single positive") cells that correspond to the end stage of the intrathymic differentiation pathway. Rearrangements at the TCR β locus occur at the $CD44^-$ $CD25^+$ stage and precede rearrangement at the TCR α locus. Phenotypic transitions accompanied by strong cellular proliferation are denoted by →. Also shown are the stages at which development in mice is arrested as a result of inactivation of the following genes : RAG-1 or RAG-2, TCR β + TCR δ, CD3-ζ/η, and CD3-ε. The inactivation of the gene encoding CD3-ζ/η reduces the number of double-positive thymocytes.

at the double-positive stage (FIG. 3). Based on the expression of the CD4 and CD8 molecules, as well as on total cell number counts, the CD3-ζ mutation appears to block thymocyte differentiation at an earlier stage than the mutation in TCR α, and does not affect the progression to the double-positive stage as much as the TCR β mutation. Therefore, the CD3-ζ mutation does not fit, in a straightforward way, into the current two-control point scheme of T-cell development (FIG. 3). A detailed examination of the composition of the double-negative subsets found in ζ-deficient mice and of their transition to the double-positive stage show that these thymocytes "jump" over the $CD44^-$ $CD25^-$ developmental stage by making a direct transition from the penultimate $CD44^-$ $CD25^+$ double-negative stage to the double-positive stage.[23] Moreover, 10 to 15% of them still express CD25 at their surface.[8] This unique pattern of progression toward the double-positive stage could be due to the fact that the "ζ-less" pre-TCR complexes trigger only a few cell divisions prior to the expression of CD4 and CD8. In wild-type adult thymuses, following the engagement of the pre-TCR and the silencing of the CD25 gene, a high rate of cell divisions would result in the rapid dilution of the residual CD25 mRNA and protein pools and in the disappearance of CD25 from the cell surface prior to the acquisition of high levels of CD4 and CD8. By contrast, in the CD3-ζ deficient mice, due to the induction of fewer cell divisions by the partial pre-TCR complex, CD25 expression would persist for a longer time at the surface of the maturing cells. Taken together, these data suggest that the ζ chain associated with the pre-TCR is dispensable for inducing the transition to the double-positive stage but might have a specific role in inducing the proliferation of the corresponding "transitional" thymocytes. The pre-TCR/CD3 complex interacts with some intracellular effectors ($p56^{lck}$?) independently of the CD4 and CD8 coreceptors and generates signals that lead to phenotype shifts that include CD4 and CD8 coreceptor expression, cellular expansion, as well as the induction of TCR α rearrangements.

REFERENCES

1. BEADLING, C. & K. A. SMITH. 1993. Tactics for the isolation of interleukin-2-induced immediate-early genes. Semin. Immunol. **5:** 365–373.
2. KRÄMER, S., C. MAMALAKI, I. HORAK, A. SCHIMPL, D. KIOUSSIS & T. HÜNIG. 1994. Thymic selection and peptide-induced activation of T cell receptor-transgenic CD8 T cells in interleukin-2-deficient mice. Eur. J. Immunol. **24:** 2317–2322.
3. SU, B., E. JACINTO, M. HIBI, T. KALLUNKI, M. KARIN & Y. BEN-NERIAH. 1994. JNK is involved in signal integration during costimulation of T lymphocytes. Cell **77:** 727–736.
4. CRABTREE, G. R. & N. A. CLIPSTONE. 1994. Signal transmission between the plasma membrane and nucleus of T lymphocytes. Annu. Rev. Biochem. **63:** 1045–1083.
5. PUNT, J. A., J. L. ROBERTS, K. P. KEARSE & A. SINGER. 1994. Stoichiometry of the T cell antigen receptor complex : each TCR/CD3 complex contains one TCRα, one TCRβ, and two CD3-ε chains. J. Exp. Med. **180:** 587–593.
6. MALISSEN, B. & A.-M. SCHMITT-VERHULST. 1993. Transmembrane signalling through the T cell receptor-CD3 complex. Curr. Opinion Immunol. **5:** 324–333.
7. WEGENER, A. M. K., F. LETOURNEUR, A. HOEVELER, T. BROCKER, F. LUTON & B. MALISSEN. 1992. The T cell receptor/CD3 complex is composed of at least two autonomous transduction modules. Cell **68:** 83–95.

8. MALISSEN, M., A. GILLET, B. ROCHA, J. TRUCY, E. VIVIER, C. BOYER, F. KÖNTGEN, N. BRUN, G. MAZZA, E. SPANOPOULOU *et al.* 1993. T cell development in mice lacking the CD3-ζ/η gene. EMBO J. **12:** 4347-4355.
9. LIU, C. P., R. UEDA, J. SHE, J. SANCHO, B. WANG, G. WEDDELL, J. LORING, C. KURAHA, E. C. DUDLEY, A. HAYDAY *et al.* 1993. Abnormal T cell development in CD3-$\zeta^{-/-}$ mutant mice and identification of a novel T cell population in the intestine. EMBO J. **12:** 4863-4875.
10. OHNO, H., S. ONO, N. HIRAYAMA, S. SHIMADA & T. SAITO. 1994. Preferential usage of the Fc receptor γ chain in the T cell antigen receptor complex by γ/δ T cells localized in epithelia. J. Exp. Med. **179:** 365-369.
11. GUY-GRAND, D., B. ROCHA, P. MINTZ, M. MALASSIS-SERIS, F. SELZ, B. MALISSEN & P. VASSALLI. 1994. Different use of T cell receptor transducing modules in two populations of gut intraepithelial lymphocytes are related to distinct pathways of T cell differentiation. J. Exp. Med. **180:** 673-679.
12. GROETTRUP, M. & H. VON BOEHMER. 1993. A role for a pre-T cell receptor in T-cell development. Immunol. Today **14:** 610-614.
13. QIAN, D., I. GRISWOLD-PRENNER, M. R. ROSNER & F. W. FITCH. 1993. Multiple components of the T cell antigen receptor complex become tyrosine-phosphorylated upon activation. J. Biol. Chem. **268:** 4488-4493.
14. SANCHO, J., R. FRANCO, T. CHATILA, C. HALL & C. TERHORST. 1993. The T cell receptor associated CD3-ε protein is phosphorylated upon T cell activation in the two tyrosine residues of a conserved signal transduction motif. Eur. J. Immunol. **23:** 1636-1642.
15. IWASHIMA, M., B. A. IRVING, N. S. C. VAN OERS, A. C. CHAN & A. WEISS. 1994. Sequential interactions of the TCR with two distinct cytoplasmic tyrosine kinases. Science. **263:** 1136-1139.
16. KOYASU, S., A. G. D. TSE, P. MOINGEON, R. E. HUSSEY, A. MILDONIAN, J. HANNISIAN, L. K. CLAYTON & E. L. REINHERZ. 1994. Delineation of a T cell activation motif required for binding of protein tyrosine kinases containing tandem SH2 domains. Proc. Natl. Acad. Sci. USA **91:** 6693-6697.
17. WANGE, R. L., S. N. MALEK, S. DESIDERIO & L. E. SAMELSON. 1993. Tandem SH2 domains of ZAP-70 bind to the T cell antigen receptor ζ and CD3-ε from activated Jurkat cells. J. Biol. Chem. **268:** 19797-19801.
18. RAVICHANDRAN, K. S., K. K. LEE, Z. SONGYAND, L. C. CANTLEY, P. BURN & S. J. BURAKOFF. 1993. Interaction of Shc with the ζ chain of the T cell receptor upon T cell activation. Science **262:** 902-905.
19. TIMSON-GAUEN, L. K., Y. ZHU, F. LETOURNEUR, Q. HU, J. B. BOLEN, L. A. MATIS, R. D. KLAUSNER & A. S. SHAW. 1994. Interactions of $p59^{fyn}$ and ZAP-70 with T cell receptor activation motifs: defining the nature of a signalling motif. Mol. Cell. Biol. **14:** 3729-3741.
20. JOUVIN, M. H., M. ADAMCZEWSKI, R. NUMEROF, O. LETOURNEUR, A. VALLÉ & J.-P. KINET. 1994. Differential control of the tyrosine kinase lyn and syk by the two signaling chains of the high affinity immunoglobulin E receptor. J. Biol. Chem. **269:** 5918-5925.
21. CHOQUET, D., G. KU, S. CASSARD, B. MALISSEN, H. KORN, W. H. FRIDMAN & C. BONNEROT. 1994. Different patterns of calcium signaling triggered through two components of the B lymphocyte antigen receptor. J. Biol. Chem. **269:** 6491-6497.
22. GROETTRUP, M., K. UNGEWISS, O. AZOGUI, R. PALACIOS, M. J. OWEN, A. C. HAYDAY & H. VON BOEHMER. 1993. A novel disulfide-linked heterodimer on pre-T cells consists of the T cell receptor β chain and a 33 kd glycoprotein. Cell **75:** 283-294.
23. CROMPTON, T., M. MOORE, H. R. MACDONALD & B. MALISSEN. 1994. Double-negative thymocyte subsets in CD3-ζ chain-deficient mice: absence of HSA^+ $CD44^-$ $CD25^-$ cells. Eur. J. Immunol. **24:** 1903-1907.

The Cyclosporin-sensitive Transcription Factor NFATp Is Expressed in Several Classes of Cells in the Immune System[a]

DON Z. WANG, PATRICIA G. McCAFFREY,[b] AND ANJANA RAO[c]

Committee on Immunology and Department of Pathology
Harvard Medical School and
Division of Cellular and Molecular Biology
Dana Farber Cancer Institute
Boston, Massachusetts 02115

INTRODUCTION

The transcription factor NFATp is a preexisting subunit of NF-AT, the nuclear factor of activated T cells.[1,2] NFATp has been implicated in the induction of several cytokine genes during the immune response.[1] NFATp is a target for the immunosuppressive drugs cyclosporin A (CsA) and FK506, which inhibit a calcium- and calcineurin-dependent signaling pathway in activated T cells.[1–4] In resting T cells, NFATp resides in the cytoplasm and is fully phosphorylated; following stimulation, it rapidly becomes dephosphorylated by way of a calcineurin-dependent pathway and then translocates to the nucleus.[5] In the nucleus, NFATp cooperates with members of the Fos and Jun family of transcription factors at composite regulatory elements in the promoter/enhancer regions of several cytokine genes, including the interleukin 2 (IL2) gene, and is thereby thought to mediate cytokine gene transcription.[1,6–10]

The molecular cloning of NFATp[11] and the related protein NFATc[12] identified these proteins as members of a novel family of transcription factors.[13,14] The DNA-binding domains of the two proteins are highly homologous (~75% identity) and show a weak degree of sequence similarity to Rel family proteins.[13,15] Although the similarity to Rel is slight (~17% overall identity), certain sequence motifs appear to be conserved.[13,15] Site-directed mutagenesis and chemical modification studies indicate that at least one of these sequence motifs subserves a similar DNA-binding

[a] This work was supported by NIH Grants CA42471 and GM46227, and a Grant from Hoffman-La Roche, Inc. (to A. Rao). P. G. McCaffrey was a Special Fellow of the Leukemia Society of America. D. Z. Wang is a graduate student supported by the Division of Medical Sciences of Harvard University.

[b] Present address: Vertex Pharmaceuticals, Inc., 40 Allston Street, Cambridge, MA 02139.

[c] Address correspondence to Dr. Anjana Rao, B465, Dana-Farber Cancer Institute, 44 Binney Street, Boston, MA 02115.

function in NFAT- and Rel-family proteins.[15] In addition, there are several functional resemblances between NFATp and NFκB:[13,14] NFATp binds to and is functionally active at certain κB sites;[16–18] both NFATp and NFκB can cooperate functionally with transcription factors of the basic region-leucine zipper family;[11,13] and both NFATp and NFκB are cytoplasmic proteins that translocate to the nucleus in activated cells.[5,19]

In the present study, the expression of NFATp in various cell lines and tissues was examined using specific antibodies. The results indicate that NFATp has a relatively narrow distribution in adult tissues. NFATp is present in lymphoid organs (thymus and spleen) but not in cardiac or skeletal muscle, small intestine, liver, kidney, or fat tissue. NFATp is not present in fibroblasts or neutrophils but is expressed in cell lines and primary cells representing various cell types within the immune system, including T cells, B cells, NK cells, mast cells, monocytes, and macrophages. We discuss the possibility that NFATp participates in calcium-dependent, CsA-sensitive cytokine gene transcription in several classes of immune system cells during the immune response.

MATERIAL AND METHODS

Cell Lines, Primary Cells, and Tissues

The murine T-cell line EL4, the human T-cell line JURKAT, the mouse B-cell lines A20 and TA3, the mouse monocyte/macrophage cell lines P338D1 and RAW264, the mouse mast cell line P815, the rat basophilic leukemia cell line (RBL.2H3, hereafter called RBL) (a mast cell model), and the mouse L-cell fibroblast cell line were cultured in DMEM containing 10% FCS, 10 mM HEPES, 2 mM L-glutamine, 50 units/mL penicillin, and 50 μg/mL streptomycin. All cell lines except TA3[20] are available from the American Type Culture Collection, Rockville, MD. The IL2-dependent Thl T-cell clone Ar-5 was cultured as previously described.[21,22]

B cells and T cells from mouse spleen, prepared by complement-dependent cytolysis,[23] were kindly given by H. Wortis and M. Teutsch of Tufts Medical School, with purity of >95% for B220$^+$ B cells and >80% for Thy1.2$^+$ T cells. Murine bone marrow-derived mast cells (BMMC)[24] were cultured from murine bone marrow by M. Gurish of Harvard Medical School, with >95% purity. Human NK blasts and T blasts[25] were from J. Ritz and M. Robertson of the Dana Farber Cancer Institute and are >99% pure. Human neutrophils (>98% pure, with <2% and <1% contamination of eosinophils and basophils, respectively) were from R. Jack of Harvard Medical School.[26] Human granulocytes (neutrophils, >99% pure) and monocytes (>95% pure) were a kind gift of R. Briggs of Vanderbilt University.[27]

Tissue samples from brain, heart, lung, small intestine, skeletal muscle, kidney, liver, spleen, and thymus were obtained from BALB/c mice and washed once in ice-cold PBS.

Antisera to NFATp

An antiserum to a recombinant fragment of NFATp (NFATpXS)[11] was used for immunoprecipitation. Antipeptide antisera generated against peptides in the N-terminal region of NFATp (67.1 peptide) and in the DNA binding domain (48 peptide) were used for Western analysis after affinity purification over peptide-conjugated Affigel columns (BioRad). The sequence of the 67.1 peptide (AISSPSGLAYPDDVL-DYGL) is not conserved in NFATc. The 48 peptide (SAHELPMVERQDMDSCLVY) differs in sequence from NFATc at 8 of 19 positions, and we have not tested whether antisera to peptide 48 crossreact with NFATc. To investigate the tissue distribution of the three alternatively spliced isoforms of NFATp, polyclonal rabbit antisera were generated against peptides corresponding to the three different C-termini of the three isoforms, none of which are represented in NFATc (isoform A, SMDQNQTPSPHWQRHKEVASPGWI; isoform B, DDELIDTHLSWIQNIL; isoform C, DDVNEIIRKEFSGPPSRNQT).

SDS Lysis of Cultured Cells and Tissues

For SDS lysis of cultured cells, the cells were washed once in cold PBS-0.1% BSA, resuspended in screw-capped Sarstedt microcentrifuge tubes (1×10^8/mL) in suspension buffer (10 mM Tris pH 8.0, 50 mM NaCl, 1 mM EDTA) containing protease and phosphatase inhibitors (100 μg/mL soybean trypsin inhibitor, 100 μg/mL aprotinin, 25 μM leupeptin, 10 mM iodoacetamide, 2 mM phenylmethylsulfonyl-fluoride (PMSF), 20 mM sodium pyrophosphate, 50 mM sodium fluoride, 1 mM sodium orthovanadate), and solubilized by adding an equal volume of 10% SDS. The lysates were immediately boiled for 5 min at 100°C, passed through a 26 gauge needle to shear DNA, and then boiled for another 10 min and centrifuged for 10 minutes. The protein concentrations of the supernatants were determined by the method of Lowry. Lysates were aliquoted, quick frozen, and stored at −20°C.

For SDS lysis of tissue samples, each sample (~100 μg) was homogenized using a 1 mL Dounce homogenizer in 300 μL cell-suspension buffer containing protease and phosphatase inhibitors, immediately lysed by adding 300 μL of 10% SDS, and treated as described above.

Immunoprecipitation of NFATp

For immunoprecipitation of NFATp from cell lysates, cells were harvested, washed once in cold PBS-0.1% BSA, and resuspended (2×10^7 cells/mL) in cold RIPA buffer (50 mM Tris pH 8.0; 150 mM NaCl; 1% Triton X-100; 1% deoxycholic acid; 0.1% SDS; 5 mM EDTA), containing the protease and phosphatase inhibitors listed above. The lysates were passed through 26 gauge needles five times to shear DNA and centrifuged in microcentrifuge tubes for 10 minutes. The supernatants (500 μL) were incubated with anti-NFATp antisera (1-2 μg of affinity-purified antibodies or 1-10 μL of unfractionated antiserum) overnight and then immunoprecipitated with protein A-Sepharose for 1 hour. The beads were washed four times in 1 mL of RIPA buffer containing protease inhibitors (2 mM PMSF, 10 μg/mL aprotinin, 25 μM leupeptin), washed twice in 1 mL TBS (10 mM Tris, pH 8.0; 150 mM NaCl) (all procedures done at 4°C), and boiled in Laemmli-reducing sample buffer.

For treatment of immunoprecipitates with calcineurin, the washed immunoprecipitates (~7×10^6 cells/lane) were treated at 30°C for 10 min with 150 nM calcineurin and 333 nM calmodulin (both from Sigma) in a buffer containing 1.5 mM $MnCl_2$, 0.5 mM EDTA, and 15 mM 2-mercaptoethanol, in the absence or presence of the phosphatase inhibitor sodium pyrophosphate (20 mM).

Western Analysis

Cell lysates (50-200 μg) or immunoprecipitated proteins were fractionated by electrophoresis on 6% SDS-polyacrylamide gels for 4 h at 25 mA constant current. The proteins were transferred to nitrocellulose.[28] The membranes were blocked in 100 mL TBS buffer plus 2% gelatin for 2 h and washed in 100 mL TBST (TBS + 0.05% Tween 20 + 0.02% gelatin) twice for 15 min and three more times for 5 minutes. The blots were then incubated at room temperature with anti-NFATp antibodies (1-2 μg/mL affinity-purified antipeptide antibodies, anti-67.1, or anti-48; 1:500 dilution of unfractionated antiserum to NFATp isoform B; or 1:1000 dilution of unfractionated antiserum to NFATp isoform C) for 3 h, washed in TBST (twice for 15 min, four more times for 5 min), incubated again with goat-anti-rabbit horseradish peroxidase-conjugated (Fab)2 fragment (Cappel) for 60 min, and washed extensively in TBST (twice for 15 min, nine times for 5 min). The specificity of staining with antipeptide antisera was confirmed by preincubating the antisera with their cognate peptides before incubation with the blot. Immunoreactive bands were detected using the ECL system (Amersham Life Science). The antiserum to isoform B is relatively weak, was used at a dilution of 1:500, and thus detects several nonspecific bans (see left panel of FIG. 5). The apparent molecular weights of the immunoreactive bands were estimated using prestained high-range molecular weight markers from BioRad.

RESULTS

NFATp Is Present in Several Classes of Immune System Cells

Several lymphoid and nonlymphoid cell lines were tested for expression of NFATp. Cells were lysed in RIPA buffer, and the whole-cell extracts were immunoprecipitated with an antiserum to recombinant NFATp. After electrophoresis on SDS gel and transfer to nitrocellulose, the membrane was blotted with an affinity-purified antibody to NFATp (anti-67.1). As shown in FIGURE 1A, the immunoprecipitate from the murine T-cell clone Ar-5 showed a major immunoreactive band migrating with an apparent molecular mass of ~138 kDa (lane 1). This band was specific in that it was not seen if the immune serum had been blocked by preincubation with the specific peptide 67.1 (lane 8). A similar band of ~145 kDa was seen in the human T-cell line JURKAT (lane 4). The difference in migration of the immunoreactive bands seen in lane 1 and other lanes is likely due to the fourfold difference in the amount of protein loaded. In other cell lines, such as the murine T-cell line EL4 (lane 3), the murine B-cell lines A20 and TA3 (lanes 2, 7), and the rat basophilic cell line RBL (lane 6), a slower migrating band (~125 kDa) and a smear of proteins extending up to ~145 kDa were also detected. The immunoreactive proteins migrating between ~125 and ~145 kDa were specific inasmuch as they were not seen after peptide block (lanes 9-11, 13, 14). No immunoreactive bands were detected in L cells (lane 5), supporting our previous observation that NFATp is not expressed in L cells (9, 32). The nonspecific 97 kDa band seen in all samples probably represents reaction of the goat-anti-rabbit second antibody used in the Western blotting procedure with the immunoprecipitating antibodies.

In subsequent experiments, we used a protocol in which cells were resuspended in buffer containing protease and phosphatase inhibitors and lysed immediately by boiling in SDS. Under these conditions, we detected a major band of ~145 kDa in Ar-5 T cells (FIG. 1B, lanes 4, 8), which was also observed in JURKAT and RBL cells (lanes 6, 7) and in the murine monocyte/macrophage cell line RAW264 (lane 10). By contrast, the ~125 kDa band was the major band detected in TA3 and EL4 cells (lanes 3, 5) and in the murine mast cell line P815 (lane 11), whereas A20 had both bands and the intermediate smear (lane 2). No NFATp was detected in L cells (lanes 1, 12) or in another murine monocyte/macrophage cell line P338D1 (lanes 9). The varying intensities of the ~125-145 kDa bands in different lanes are likely to reflect differences in the abundance of NFATp in different cell lines. Cell lines of similar lineages showed significant variations as well (A20 vs. TA3, and P338D1 vs. RAW264). The 90 kDa band seen in the JURKAT lane is also detected in some other human cell lines (see FIG. 2), and its nature is currently unknown.

One explanation for the detection of immunoreactive bands with different apparent molecular masses is that these bands represent differentially phosphorylated forms of NFATp. When NFATp immunoprecipitates from Ar-5 cells were treated *in vitro* with calcineurin and calmodulin, the apparent molecular weight of the specifically immunoprecipitated NFATp was decreased by 10-20 kDa (FIG. 1C, compare lanes 2 and 3). This shift was due to dephosphorylation and not proteolysis, because it was completely blocked by the phosphatase inhibitor pyrophosphate (lane 4). Thus the differentially migrating forms of NFATp seen in different cell lines may at least partly reflect differences in the phosphorylation status of the protein, arising either from intrinsic differences among different cell types or from variable degrees of dephosphorylation during cell lysis.

NFATp Is Present in Several Primary Cells of the Immune System

Because transformed cell lines might aberrantly express NFATp, we directly examined NFATp expression in primary cells of the immune system. As shown in FIGURE 2, the ~125-145 kDa bands were detected in murine splenic B and T cells (lanes 1, 2), human T and NK blasts (lanes 5, 6), murine bone marrow-derived mast cells (BMMC, lane 9), and human monocytes (lane 10). By contrast, no immunoreactive proteins were detected in two different preparations of human granulocytes (neutrophils) (lanes 7, 8).

NFATp Is Present in Spleen and Thymus but Not in Other Tissues

We next examined the expression of NFATp in several murine tissues. When Western blots of SDS lysates were probed with anti-67.1 serum, the spleen and thymus showed two sets of closely spaced doublets migrating at ~125 and ~145 kDa (FIG. 3, lanes 9, 10). No immunoreactive bands were seen in lysates of whole brain,

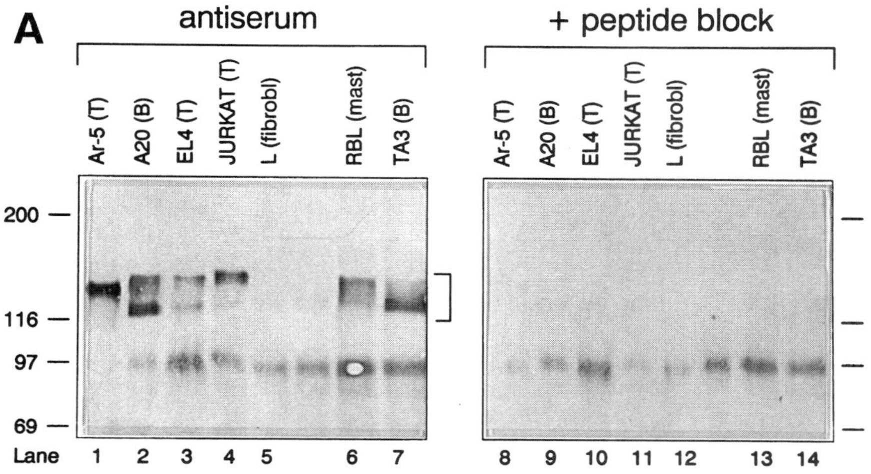

FIGURE 1. NFATp is expressed in cell lines representing several classes of immune system cells. **A**: RIPA extracts of Ar-5 cells (5×10^6) and other cell lines (2×10^7) were immunoprecipitated with an antiserum to recombinant NFATp, and Western blots were probed with anti-67.1 antiserum in the absence (lanes 1-7) or presence (lanes 8-14) of specific peptide (67.1). The bracket denotes the position of NFATp. **B**: Western blots of SDS lysates of Ar-5 cells (50 μg) and other cell lines (200 μg) were probed with anti-67.1 antiserum. **C**: Dephosphorylated NFATp migrates with lower apparent molecular weight. Immunoprecipitates of Ar-5 cells were treated with calcineurin in the presence or absence of the phosphatase inhibitor pyrophosphate and fractionated on an SDS gel. The Western blot was probed with anti-67.1 serum.

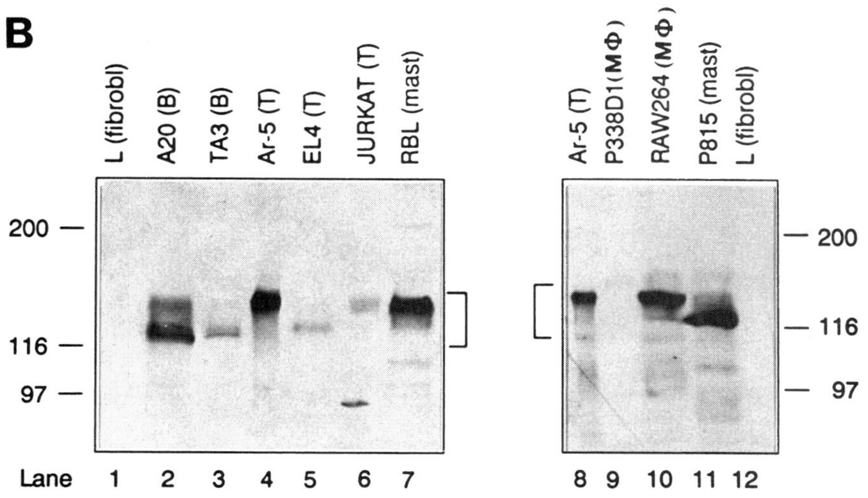

FIGURE 1B.

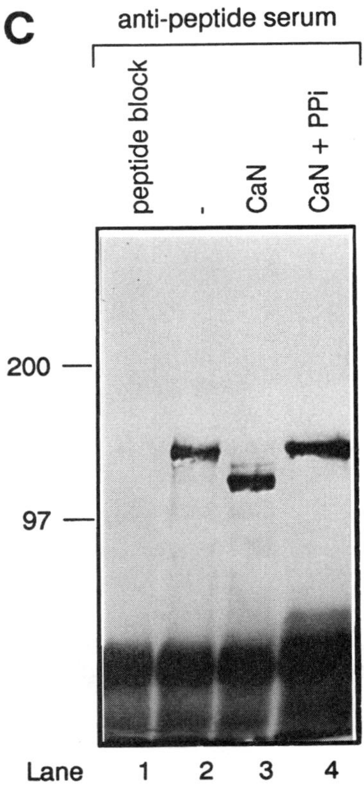

heart, small intestine, lung, skeletal muscle, or fat (lanes 1-4, and data not shown). Although kidney and liver showed some immunoreactive bands (lanes 5, 6), these bands migrated with different apparent molecular weights than those from lymphoid tissues. Moreover, they showed no reactivity with a second antipeptide antibody (anti-48) (lanes 12, 13) that recognized NFATp in Ar-5 cells (lane 14). We therefore conclude that NFATp is selectively expressed in lymphoid tissues.

Confirmation of Primary Structure Predicted from Analysis of cDNA Clones

Analysis of cDNA clones predicts the existence of at least three isoforms of NFATp, related by alternative splicing at the C-terminus.[11,29] The longest isoform (isoform A) is 1064 amino acids long, yielding a predicted molecular mass of 115 kDa.[29] The two shorter isoforms (isoforms B and C) are 923 and 927 amino acids long, respectively, with predicted molecular masses of 100 kDa.[29] To confirm the predicted primary structure of the three NFATp isoforms, we immunoprecipitated

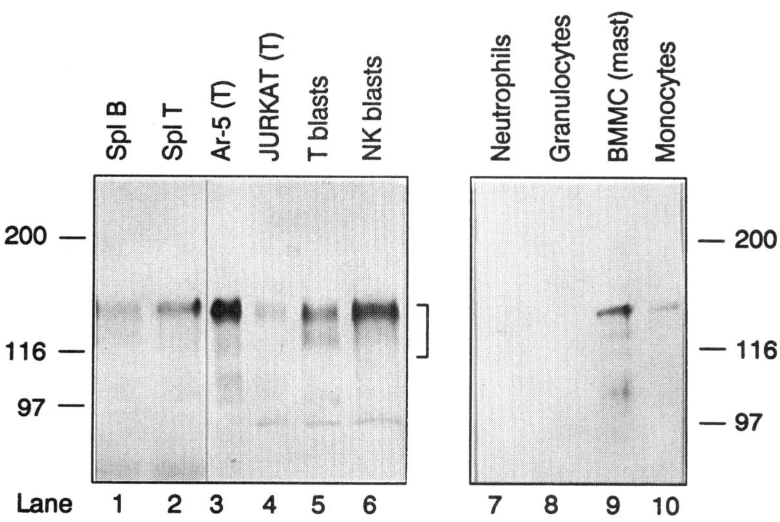

FIGURE 2. Most primary cells of the immune system express NFATp. Western blots of SDS lysates of several primary immune system cells (of either human or mouse origin, as indicated in MATERIAL AND METHODS) were probed with anti-67.1 serum. The lanes contained 100 μg (left panel) or 200 μg (right panel) of protein per lane, except for Ar-5 cells (50 μg).

NFATp with antisera to specific peptides or fragments of the protein and tested the precipitates by Western analysis for reactivity with antisera to different peptides (FIG. 4).

As a first step, we asked whether isoforms B and C were expressed as proteins in T and B cells and whether both isoforms contained the DNA-binding domain (FIG. 4, left panel). Lysates from the untransformed murine T-cell clone Ar-5 and the transformed murine B-cell line A20 were immunoprecipitated with antiserum to NFATpXS, an internal 464-amino acid fragment of NFATp that contains the DNA-binding domain.[11] When Western blots of the immunoprecipitates were probed with antisera to isoforms B and C, both T cells and B cells were seen to express both isoforms. The major immunoreactive band in T-cell lysates (lanes 2, 4) migrated at the position of fully phosphorylated NFATp (as detected by the antiserum to the N-terminal 67.1 peptide; see FIGURES 1-3), whereas the major band in B-cell lysates (lanes 1, 3) migrated at the position of dephosphorylated NFATp. Because SDS lysates of A20 cells contain both phosphorylated and dephosphorylated forms of NFATp (FIG. 1 A and B, lanes 2; FIG. 3, lane 11), this result is likely to reflect dephosphorylation of NFATp in the B-cell lysates during the extended periods required for immunoprecipitation. The antiserum to isoform A showed no reactivity with proteins immunoprecipitated from either T-cell or B-cell lysates (data not shown). However, the antiserum does recognize recombinant isoform A expressed in cos cells (data not shown), indicating that isoform A is either not expressed in T or B cells or is expressed at concentrations too low to be detected by this technique.

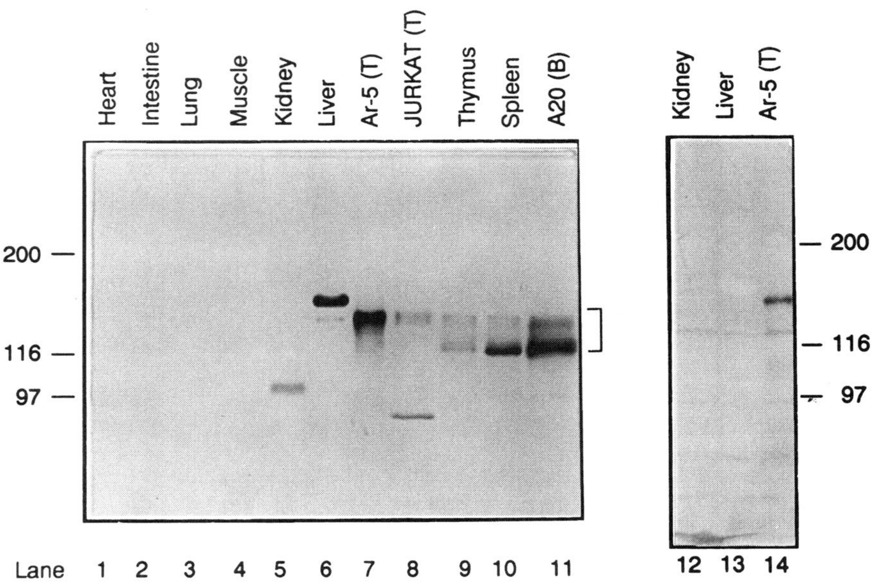

FIGURE 3. NFATp is selectively expressed in lymphoid tissues. Left panel: SDS lysates of several tissue samples were loaded at 200 μg per lane (except for Ar-5, 50 μg) and probed with anti-67.1 serum. Right panel: A Western blot of SDS lysate of kidney and liver (200 μg each) and Ar-5 cells (50 μg) was probed with anti-48 serum.

We next asked whether isoforms B and C contained the N-terminal 67.1 peptide (FIG. 4, right panel). B-cell lysates were immunoprecipitated with antisera to the N-terminal 67.1 peptide (lane 1); to the C-terminal peptides of isoforms A, B, and C (lanes 2-4); or to the internal fragment NFATpXS (lane 5), and immunoblotted with the antiserum to 67.1. This analysis indicated that both isoforms B and C contained the N-terminal 67.1 peptide (lanes 3, 4). The antiserum to isoform A did not precipitate any proteins immunoreactive with the 67.1 antiserum (lane 2), again suggesting that isoform A is expressed at undetectably low concentrations.

Expression of NFATp Isoforms in Cell Lines

A panel of cell lines expressing NFATp, and representing several different classes of immune system cells, were tested for expression of the major B and C isoforms of NFATp. As shown in FIGURE 5, both isoforms were detected in each of the cell lines (Ar-5, RBL, JURKAT, RAW264, and P815) examined. Together, these results demonstrate that of the three C-terminally spliced isoforms of NFATp detected by cDNA cloning, the two shorter isoforms (B and C) are clearly expressed as proteins in several cell types. Both isoforms are likely to encode proteins that are functional in DNA-binding and transcriptional activity, inasmuch as they contain the central

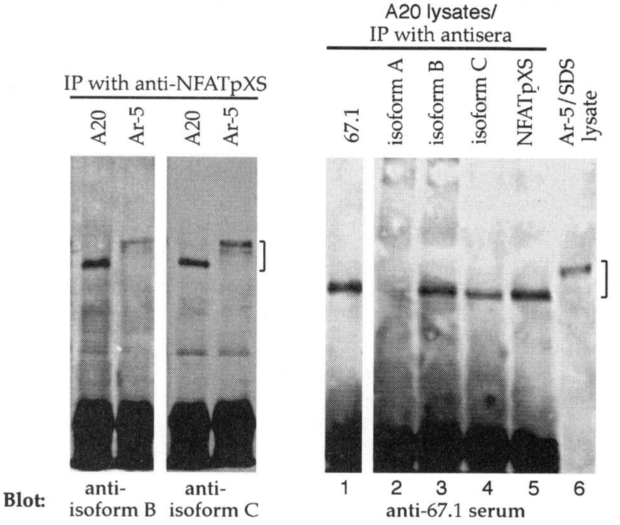

FIGURE 4. NFATp isoforms B and C are expressed in T and B cells and contain the N-terminal 67.1 peptide. Left panel: RIPA extracts of Ar-5 T cells and A20 B cells were immunoprecipitated (IP) with antiserum to NFATpXS, and Western blots were probed with antisera to the C-terminal peptides of isoforms B and C. Right panel: RIPA extracts of A20 B cells were immunoprecipitated with the indicated antisera, and Western blots were probed with anti-67.1 serum.

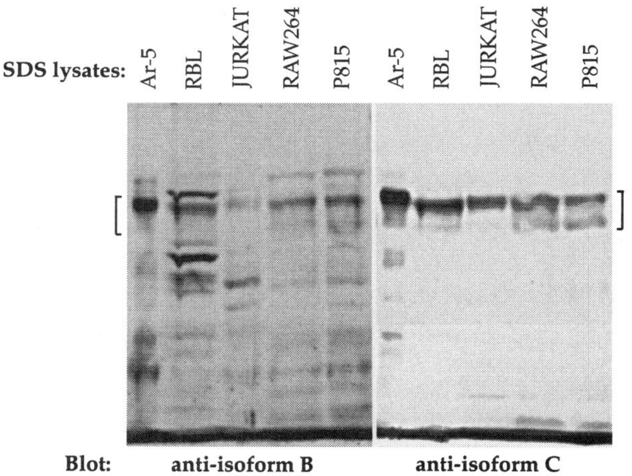

FIGURE 5. NFATp isoforms B and C are expressed in cell lines representing several classes of immune system cells. Western blots of SDS lysates of the indicated cell lines were probed with antisera to the C-terminal peptides of isoforms B and C. The bracket indicates the position of NFATp.

region of NFATp, which includes the DNA-binding domain, as well as the N-terminal region, which includes peptide 67.1 and contains a transactivation domain.[29]

DISCUSSION

In this study, we demonstrate that the cyclosporin-sensitive transcription factor NFATp is not a ubiquitously expressed protein, but rather shows a relatively limited distribution in adult tissues. Western analysis with antisera against an N-terminal peptide indicates that NFATp is selectively expressed in lymphoid tissues (spleen and thymus). By contrast, NFATp was not detected in nonlymphoid tissues, including skeletal and cardiac muscle, liver, lung, kidney, small intestine, and fat. Similarly, NFATp was not detected in SDS lysates of whole adult brain, although a more careful analysis indicated that NFATp was expressed in certain specific subregions of the murine nervous system.[30] Among the cell types represented in spleen and thymus, NFATp was detected not only in T cells, but also in B cells, NK cells, mast cells, monocytes, and macrophages. NFATp has also been detected in an endothelial cell line derived from human umbilical vein (D. Z. Wang, unpublished results).

By RNAse protection analysis, mRNA for NFATp has reportedly been detected in a variety of tissues including heart and brain.[12] These results do not necessarily conflict with the Western analysis results described in this work: it is possible that mRNA abundance does not match protein abundance in these tissues, or that the proteins expressed are alternatively spliced forms of NFATp that lack regions near both the N- and C-termini of the protein.

Our results are consistent with previous studies showing that NFATp or closely related proteins are expressed in all subsets of T-cell clones tested ($CD4^+$ vs. $CD8^+$ T cells, Th1 vs. Th2),[31,32] as well as in B cells[33–36] and NK cells (Aramburu and Perussia, personal communication). As previously suggested, NFATp may play a role in the calcium-dependent, CsA-sensitive transcription of several cytokine genes in stimulated T cells.[1] In fact, sequence elements capable of binding NFATp have been described in the regulatory regions of several cytokine genes, including the TNFα, IL4, and GM-CSF promoters and the GM-CSF/IL3 intergenic enhancer, and expression of each of these cytokines in strongly inhibited by CsA and FK506.[1,17,18,31,37–40] Nevertheless, it is clear that cytokine production during the immune response is not restricted to T cells:[41,42] for instance, mast cells produce IL1, IL3/GM-CSF, IL4, IL6, and TNFα; NK cells produce IFNγ and TNFα; macrophages produce GM-CSF and TNFα; and B cells produce TNFα. Because each of these cell types bears a surface receptor capable of initiating calcium mobilization and cytokine expression when cross-linked,[43] it is conceivable that NFATp may play a role in Ca^{2+}-dependent cytokine gene transcription in these diverse cell types, perhaps by mechanisms involving calcineurin and similar to those of T cells.[5] In support of this hypothesis, the induction of TNFα mRNA in B cells stimulated by surface Ig cross-linking is sensitive to CsA,[44] as is the induction of IL4 and IL3/GM-CSF in mast cells after stimulation by PMA and ionomycin,[45] or the induction of IL1 and IL6 in BMMC stimulated through FceRI.[46] Further experiments are needed to establish a role for NFATp in regulating cytokine gene transcription in different cell types during the immune response.

ACKNOWLEDGMENTS

We greatly appreciate the generosity of H. Wortis, M. Teutsch, M. Gurish, R. Jack, R. Briggs, J. Ritz, and M. Robertson for giving us murine and human primary lymphoid cells. We also thank our colleagues C. Loh, A. W. Ho, K. T. Y. Shaw, and J. Jain for their kind help.

REFERENCES

1. Rao, A. 1994. Immunol. Today **15:** 274-281.
2. Crabtree, G. R. & N. A. Clipstone. 1994. Annu. Rev. Biochem. **63:** 1045-1083.
3. Liu, J. 1993. Immunol. Today **14:** 290-295.
4. McCaffrey, P. G., B. A. Perrino, T. R. Soderling & A. Rao. 1993. J. Biol. Chem. **268:** 3747-3752.
5. Shaw, K.T.-Y., A. M. Ho, A. Raghavan, A. Rao & P. G. Hogan. In preparation.
6. Jain, J., P. G. McCaffrey, V. E. Valge-Archer & A. Rao. 1992. Nature **356:** 801-804.
7. Jain, J., P. G. McCaffrey, Z. Miner, T. K. Kerppola, J. N. Lambert, G. L. Verdine, T. Curran & A. Rao. 1993. Nature **365:** 352-355.
8. Jain, J., Z. Miner & A. Rao. 1993. J. Immunol. **151:** 837-848.
9. Northrop, J. P., K. S. Ullman & G. R. Crabtree. 1993. J. Biol. Chem. **268:** 2917-2923.
10. Boise, L. H., B. Petryniak, X. Mao, C. H. June, C.-Y. Wang, T. Lindsten, R. Bravo, K. Kovary, J. M. Leiden & C. G. Thompson. 1993. Mol. Cell. Biol. **13:** 1911-1919.
11. McCaffrey, P. G., C. Luo, T. K. Kerppola, J. Jain, T. M. Badalian, A. M. Ho, E. Burgeon, W. S. Lane, J. N. Lambert, T. Curran, G. L. Verdine, A. Rao & P. G. Hogan. 1993. Science **262:** 750-754.
12. Northrop, J. P., S. N. Ho, D. J. Thomas, L. Chen, L. Timmerman, G. P. Nolan, A. Admon & G. R. Crabtree. 1994. Nature **369:** 497-502.
13. Nolan, G. P. 1994. Cell **77:** 795-798.
14. Rao, A. 1995. J. Leukocyte Biol. **57:** 536-542.
15. Jain, J., E. Burgeon, T. M. Badalian, P. G. Hogan & A. Rao. 1995. J. Biol. Chem. **270:** 4138-4145.
16. McCaffrey, P. G., J. Jain, C. Jamieson, R. Sen & A. Rao. 1992. J. Biol. Chem. **267:** 1864-1871.
17. Goldfeld, A. E., P. G. McCaffrey, J. L. Strominger & A. Rao. 1993. J. Exp. Med. **178:** 1365-1379.
18. McCaffrey, P. G., A. E. Goldfeld & A. Rao. 1994. J. Biol. Chem. **269:** 30445-30450.
19. Beg, A. & A. S. Baldwin, Jr. 1993. Genes & Dev. **7:** 2064-2070.
20. Glimcher, L. H., T. Hamano, R. Asofsky, D. H. Sachs, M. Pierres, L. E. Samelson, S. O. Sharrow & W. E. Paul. 1983. J. Immunol. **130:** 2287-2294.
21. Rao, A., S. J. Faas & H. Cantor. 1984. J. Exp. Med. **159:** 479-494.
22. Wong, J. G. P. & A. Rao. 1990. J. Biol. Chem. **275:** 4685-4693.
23. Ying-zi, C., E. Rabin & H. H. Wortis. 1991. Int. Immunol. **3:** 467-476.
24. Razin, E., J. N. Ihle, D. Seldin, J. M. Mencia-huerta, H. R. Katz, P. A. LeBlanc, A. Hein, J. P. Caulfield, K. F. Austen & R. L. Stevens. 1984. J. Immunol. **132:** 1479-1485.
25. Robertson, M. J., T. J. Manley, C. Donahue, H. Levine & J. Ritz. 1993. J. Immunol. **150:** 1705-1714.
26. Neuman, E., J. W. Huleatt & R. M. Jack. 1990. J. Immunol. **145:** 3325-3332.
27. Briggs, R., L. Dworkin, J. Briggs, E. Dessypris, J. Stein & J. Lian. 1994. J. Cell Biochem. **54:** 198-206.

28. Towbin, H., T. Staehelin & J. Gordon. 1979. Proc. Natl. Acad. Sci. USA **76:** 4350-4354.
29. Luo, C., E. Burgeon, J. Carew, T. Badalian, P. G. Hogan & A. Rao. In preparation.
30. Ho, A. M., J. Jain, A. Rao, & P. G. Hogan. 1994. J. Biol. Chem. **269:** 28181-28186.
31. Rooney, J. W., M. R. Hodge, P. G. McCaffrey, A. Rao & L. H. Glimcher. 1994. EMBO J. **13:** 625-633.
32. Jain, J., E. A. Nalefski, P. G. McCaffrey, R. S. Johnson, B. M. Spiegelman, V. Papaioannou & A. Rao. 1994. Mol. Cell. Biol. **14:** 1566-1574.
33. Brabletz, T., I. Pietrowski & E. Serfling. 1991. Nucleic Acids. Res. **19:** 61-67.
34. Yaseen, N. R., A. L. Maizel, F. Wang & S. Sharma. 1993. J. Biol. Chem. **268:** 14285-14293.
35. Venkataraman, L., D. A. Francis, Z. Wang, J. Liu, T. Rothstein & R. Sen. 1994. Immunity **1:** 189-196.
36. Choi, M. S. K., R. D. Brines, M. J. Holman & G. G. B. Klaus. 1994. Immunity **1:** 179-187.
37. Masuda, E. S., H. Tokumitsu, A. Tsuboi, J. Shlomai, P. Hung, K.-I. Arai & N. Arai. 1993. Mol. Cell. Biol. **13:** 7399-7407.
38. Cockerill, P. N., M. F. Shannon, A. G. Bert, G. R. Ryan & M. A. Vadas. 1993. Proc. Natl. Acad. Sci. USA **90:** 2466-2470.
39. Szabo, S. J., J. S. Gold, T. L. Murphy & K. M. Murphy. 1993. Mol. Cell. Biol. **13:** 4793-4805.
40. Chuvpilo, S., C. Schomberg, R. Gerwig, A. Heinfling, R. Reeves, F. Grummt & E. Serfling. 1993. Nucleic Acids Res. **21:** 5694-5704.
41. Arai, K.-I., F. Lee, A. Miyajima, S. Miyatake, N. Arai & T. Yokota. 1990. Annu. Rev. Biochem. **59:** 783-836.
42. Paul, W. E. & R. E. Seder. 1994. Cell **76:** 253-262.
43. Weiss, A. & D. R. Littman. 1994. Cell **76:** 263-274.
44. Goldfeld, A. E., E. K. Flemington, V. A. Boussiotis, C. M. Theodos, R. G. Titus, J. L. Strominger & S. H. Speck. 1992. Proc. Natl. Acad. Sci. USA **89:** 12198-12201.
45. Hatfield, S. & N. W. Roehm. 1992. Pharmacol. Exp. Ther. **260:** 680-687.
46. Kaye, R. E., D. A. Fruman, B. E. Bierer, M. W. Albers, L. D. Zydowsky, S. I. Ho, Y.-J. Jin, M. C. Castells, S. L. Schreiber, C. T. Walsh, S. J. Burakoff, K. F. Austen & H. R. Katz. 1992. Proc. Natl. Acad. Sci. USA **89:** 8542-8546.

Signal Transduction by the B-Cell Antigen Receptor

ANTHONY L. DEFRANCO,[a–c]
JAMES D. RICHARDS,[a,c] JONATHAN H. BLUM,[b,c]
TRACY L. STEVENS,[a,c] DEBBIE A. LAW,[a,c]
VIVIEN W.-F. CHAN,[b,c] SANDIP K. DATTA,[a,c]
SHAUN P. FOY,[d] SHARON L. HOURIHANE,[d]
MICHAEL R. GOLD,[e] AND LINDA MATSUUCHI [d]

[a]*Department of Microbiology and Immunology*
[b]*Department of Biochemistry and Biophysics*
[c]*G. W. Hooper Foundation*
University of California, San Francisco
San Francisco, California 94143-0552
[d]*Department of Zoology*
[e]*Department of Microbiology and Immunology*
University of British Columbia
Vancouver, British Columbia

The proper functioning of the immune system requires that B lymphocytes respond and make antibodies to foreign antigens but not to self-components. This self/nonself-discrimination is known to be a complicated process, but a key element is the recognition of antigen by membrane forms of immunoglobulin displayed on the surface of the B cell. Signaling by this B-cell antigen receptor (BCR) can promote B-cell activation or B-cell inactivation, depending on the context.[1]

The BCR is composed of membrane immunoglobulin (mIg) complexed with a heterodimer of two transmembrane polypeptides, Ig-α and Ig-β. These related polypeptides each have one N-terminal extracellular Ig-like domain, a single transmembrane domain, and a moderate size cytoplasmic domain (61 and 48 amino acids, respectively).[2] Based on the fact that an mIg unit has two transmembrane heavy chains, it is generally presumed that one mIg unit combines with two Ig-α/Ig-β heterodimers in the BCR complex. As described by Michael Reth in this volume, additional polypeptides associated with mIgM or with other isotypes of mIg (mIgD, mIgG, etc.) have recently been identified.[3,4] The roles of these new proteins are not known at this time. By contrast, Ig-α and Ig-β have been demonstrated to play a role in intracellular trafficking of the BCR and in signal transduction. Some of our studies on these issues are described below.

BCR ASSEMBLY AND TRAFFICKING

The IgM isoform of mIg requires Ig-α and Ig-β to come to the cell surface. In nonlymphoid cells, which lack expression of these B-cell specific gene products, mIgM expression results in localization in the endoplasmic reticulum (ER).[5] Transfection of expression vectors for Ig-α and Ig-β, as well as Ig heavy and light chains into nonlymphoid cells, results in formation of complete BCR complexes and cell-surface expression.[6,7] Thus, a likely hypothesis is that some property of the mIgM molecule and of the Ig-α/Ig-β heterodimer leads to retention of unassembled components in the ER, and this retention is relieved by complex formation, allowing transit of assembled BCRs to the cell surface.

Both the ER retention function and the ability to assemble with Ig-α/Ig-β depend primarily on the μ heavy chain transmembrane domain. For example, replacement of the 41 C-terminal amino acids of the μ chain (representing the extracellular spacer region, the transmembrane region, and the cytoplasmic tail) with the corresponding region from MHC class I or CD8 molecules results in chimeric mIgM molecules that can go to the cell surface in the absence of Ig-α and Ig-β.[5] Moreover, expression of such chimeras in B-cell lines results in cell-surface expression in the absence of association with Ig-α/Ig-β.[8,9] Thus, the C-terminal region of the μ heavy chain is necessary for ER retention of unassembled mIgM molecules and for complex formation with Ig-α/Ig-β heterodimers.

The transmembrane domains of each of the heavy chains are unusual in that they have a high fraction of hydrophilic amino acid residues, and many of these hydrophilic amino acids are conserved in most of the different heavy chains.[2] For the μ heavy chain transmembrane domain, 10 out of 26 residues have side chains with hydroxyl groups. This transmembrane domain has two polar patches with four out of five and five out of six adjacent hydroxylated residues. Clustered mutation (to hydrophobic residues) of either N-terminal[5] or C-terminal[10] of these regions abrogates the ER retention function of the transmembrane domain. In addition, we have found that mutation of two adjacent residues in the C-terminal polar patch (YS to VV) allowed cell-surface expression in the nonlymphoid AtT20 cells.[11] Similar results were recently reported with the same mutation introduced into the human μ chain transmembrane domain.[12] These results suggest that the two polar patches are involved in the interaction between mIgM and one or more proteins that are localized to the ER, and might therefore keep mIgM in the ER until it can assemble with Ig-α/Ig-β. This ER retention protein may be calnexin, as calnexin has been shown to bind to the μ heavy chain and to bind less well to the human μ chain with the YS to VV mutation.[12]

The transmembrane region of μ heavy chain is also important for mIgM's assembly with Ig-α/Ig-β. This possibility was originally suggested by the observation that the BCR complex is only stable in very mild detergents such as digitonin and CHAPS, and disrupted by harsher nonionic detergents such as Triton X-100 and NP-40.[2] Moreover, as mentioned above, chimeric proteins with the C-terminal region of μ chain replaced with the equivalent region of other proteins do not assemble with Ig-α/Ig-β.[8,9] In addition, the YS to VV mutant form of murine mIgM assembles poorly with Ig-α and Ig-β.[11] Five- to tenfold less Ig-α is associated with this mutant mIgM compared to wild-type mIgM. The associated molecules do not exhibit greatly increased dissociation from mIgM during immunoprecipitation, suggesting that their binding energy is not greatly affected. Thus, it is possible that the decreased assembly with Ig-α/Ig-β is a result of more rapid exit from the ER of unassembled mIgM molecules, rather than a direct effect on assembly of Ig-α/Ig-β with mIgM. Alternatively, the YS to VV mutant could affect mIgM interactions with both the ER retention factor and Ig-α/Ig-β. This second possibility is favored by experiments in which the human μ chain was mutated in this way and then introduced into a murine β-cell line. In this context, the human YS to VV mutant μ heavy chain exhibited a complete defect in assembly with Ig-α/Ig-β.[13] This result suggests that human mIgM has a somewhat weaker interaction with murine Ig-α/Ig-β than does murine mIgM, and that the YS to VV mutation further weakens this interaction, resulting in the failure of mutated human mIgM to form BCR complexes with Ig-α/Ig-β.

In contrast to the importance of the transmembrane domain, the three-amino acid-long μ chain cytoplasmic domain does not seem to be critical for BCR assembly and function. Point mutations of this domain, as long as they retain its positively charged character, do not affect receptor function.[9,14] If the cytoplasmic domain is deleted, however, then the μ chain is converted to a glycosyl-phosphatidylinositol-linked protein, a reaction that includes enzymatic removal of the transmembrane domain.[9,15] This form of mIgM is not associated with Ig-α/Ig-β,[9,16] further demonstrating the role of the μ chain transmembrane region in BCR assembly. Thus, it appears to be the transmembrane domain of μ that is required for the assembly of functional BCR complexes.

SIGNAL TRANSDUCTION FUNCTION OF THE BCR

Stimulation of the BCR leads to activation of multiple intracellular protein tyrosine kinases, which are essential for the biological effects of this receptor.[17] Targets of BCR-induced tyrosine phosphorylation include known signaling components such as phospholipases Cγ1 and γ2, phosphatidylinositol 3-kinase, Shc, Vav, and rasGAP.[18]

The signaling function of the BCR is mediated by the cytoplasmic domains of Ig-α and Ig-β. For example, as mentioned above, chimeric proteins in which the C-terminal region of the μ heavy chain has been replaced by equivalent regions of other proteins fail to assemble with Ig-α and Ig-β. Such chimeric proteins also fail to initiate signaling reactions such as tyrosine phosphorylation of cytoplasmic proteins or elevation of intracellular free calcium.[9,19,20] Similarly, the YS to VV mutation of the transmembrane domain described above exhibits decreased assembly with Ig-α and Ig-β and correspondingly decreased signaling ability.[11] The equivalent mutation of human μ heavy chain abolishes its interaction with Ig-α and Ig-β and signaling, as well.[13,14,16] Similarly, association of mIgM with Ig-α/Ig-β is lost upon deletion of the cytoplasmic tail of μ chain, as is signaling function.[9,14,16] Thus, association of mIgM with the accessory proteins Ig-α/Ig-β is closely correlated with signaling ability.

SIGNALING ABILITY OF THE IG-α AND IG-β CYTOPLASMIC DOMAINS

The cytoplasmic domains of Ig-α and Ig-β are 61 and 48 amino acids long, respectively. Each contains one copy of a sequence motif found in a number of components of the T-cell antigen receptor and of Fc receptors, recently renamed the ITAM (for immunoreceptor tyrosine-based activation motif).[21] This motif contains two YXXL/I sequences (Y: tyrosine, L: leucine, I: isoleucine, X: any amino acid) with a characteristic spacing of 6–8 residues. The cross-linking of chimeric proteins with ITAMs and surrounding sequences from the cytoplasmic domains from T-cell antigen receptor ζ or CD3 ε chains triggers signaling reactions characteristic of the intact T-cell antigen receptor (reviewed in ref. 22). Similarly, in our experiments, chimeric proteins containing the ITAMs and local surrounding sequences of Ig-α or Ig-β are fully capable of inducing all of the BCR signaling events examined.[23] Similar experiments have been performed in a number of laboratories.[22] In general, chimeras containing all or part of the cytoplasmic domains of Ig-α are fully active, whereas chimeras containing the cytoplasmic domain of Ig-β have a more variable activity, ranging from being as good as the Ig-α chimeras to having only partial signaling ability. The reasons for these discrepancies are not yet apparent. Perhaps the most significant variables in these experiments are the cell lines expressing the chimeras, which may vary in their expression level for signaling components (particularly the tyrosine kinases). According to this hypothesis, the Ig-α chimeras may be inherently more active than the Ig-β chimeras under conditions of limiting amounts of a particular signaling component. Clearly this issue warrants further experimentation.

The fact that Ig-α and Ig-β chimeras are equally able, at least in some contexts, to activate the signaling reactions examined suggest that ITAMs function in the initiation of tyrosine kinase activation rather than in the selection of targets for tyrosine phosphorylation. If the latter were the case, then Ig-α and Ig-β would be expected to bind to different signaling targets and thus to exhibit complementary function. On the contrary, we have not seen significant association of the signaling targets of tyrosine phosphorylation either with the intact BCR or with the ITAM-containing chimeric proteins (data not shown). Cross-linking the Ig-α and Ig-β-containing chimeras did induce their tyrosine phosphorylation and moreover induced the binding of Syk and Lyn.[23] Lyn has one SH2 domain and Syk has two SH2 domains, so we propose that phosphorylation of the ITAM tyrosines promotes binding of these tyrosine kinases via their SH2 domains. Another tyrosine kinase, Fyn was also associated with the chimeras, although this association was present in unstimulated cells as well. The amount of Fyn bound was estimated as being about one percent of total cellular Fyn.[23]

These and related observations in T cells have led to the following model of ITAM action in BCR signaling. Cross-linking of the BCR clusters the cytoplasmic domains of Ig-α and Ig-β. This leads to phosphorylation of ITAM tyrosines, either by Src-family tyrosine kinases prebound to the BCR (as seen with Fyn) or by Src-family tyrosine kinases generally present in the membrane. The newly phosphorylated ITAMs are good binding sites for the tyrosine kinases Syk and Lyn. Once bound, these protein tyrosine kinases phosphorylate adjacent ITAM tyrosines (a step that requires cross-linking), providing a positive feedback loop leading to extensive ITAM phosphorylation and to clustering of tyrosine kinase molecules. During this process, Syk becomes tyrosine phosphorylated strongly[24] (Richards *et al.*, manuscript in preparation), suggesting that is a substrate for either Lyn, Fyn, or itself. This phosphorylation may activate Syk, as phosphorylation is a common mechanism by which protein kinases are activated.[25] Indeed, this possibility is suggested by an experiment of Kolanus *et al.*[26] in which they made a chimeric transmembrane protein that includes Syk attached to the cytoplasmic domains. Cross-linking this chimeric protein triggers phosphoinositide breakdown and cytolytic effector function in T cells. Moreover, these observations suggest that Syk is the effector kinase responsible for phosphorylating at least some of the downstream signaling targets.

This model of BCR signaling suggests that the *src*-family tyrosine kinases Lyn and Fyn initiate the signaling by phosphorylating the clustered cytoplasmic domains of Ig-α and Ig-β, and that subsequent events in the antigen receptor signaling cascade require Syk recruitment and action. Interestingly, cross-linking BCR molecules expressed in nonlymphoid AtT20 cells results in strong tyrosine phosphorylation of Ig-α and Ig-β cytoplasmic domains but deficient signaling ability with regard to downstream targets.[7] These cells express Fyn but do not express Syk. Transfection of these cells with a Syk expression vector restores some downstream signaling reactions, such as phosphorylation of Shc and activation of MAP kinase (Richards *et al.*, in preparation). Other signaling events, including phosphoinositide hydrolysis and elevation of intracellular free calcium are still not activated or weakly activated. This could reflect the absence of a B cell-specific adapter protein needed to direct phosphorylation of these signaling targets, or it could reflect a dominant inhibitory influence expressed in the AtT20 cells. These possibilities are currently under investigation. The restoration of partial signaling ability in AtT20 cells required the Syk tyrosine kinase activity, as it was not seen when these cells were transfected with a mutant form of Syk in which the ATP-binding function was compromised. These observations support the idea that Syk is responsible for phosphorylating at least some downstream signaling targets.

SUMMARY

The antigen receptor of B lymphocytes (BCR) plays important roles in recognition of foreign antigens and self-components to allow the immune system to make appropriate antibody responses. The BCR is a complex between membrane immunoglobulin and the Ig-α and Ig-β heterodimer. Site-directed mutagenesis experiments have shown that the μ heavy chain transmembrane domain plays a key role in the association of mIgM with Ig-α/Ig-β. In the absence of complex formation, mIgM is retained in the endoplasmic reticulum, and this function is also specified by the μ chain transmembrane domain. The ability of various mutant mIgM molecules to associate with Ig-α/Ig-β correlates well with their ability to induce signal transduction reactions such as protein tyrosine phosphorylation and phosphoinositide breakdown. Thus, the signaling ability of the BCR appears to reside in the Ig-α/Ig-β heterodimer. The cytoplasmic domains of Ig-α and Ig-β each contain an ITAM sequence, which is defined by its limited homology with subunits of the T-cell antigen receptor and of Fc receptors. Moreover, chimeric proteins containing these ITAMs and surrounding sequences from the cytoplasmic domains of Ig-α or Ig-β exhibit signaling function characteristics of the intact BCR. The Ig-α and Ig-β chimeras are each capable of inducing all of the BCR signaling events tested and thus represent redundant functions. Cross-linking these chimeras leads to their phosphorylation and to binding of the intracellular tyrosine kinases Lyn and Syk. The BCR expressed in the nonlymphoid AtT20 cells, which express the Src-family tyrosine kinase Fyn but not Syk, was not able to trigger vigorous signaling reactions. Introduction of the active form of Syk into these cells restored some signaling events. These results are consistent with a model in which the ITAMs act to initiate the BCR signaling reactions by binding and activating tyrosine kinases.

REFERENCES

1. DeFranco, A. L. 1993. Structure and function of the B cell antigen receptor. Annu. Rev. Cell Biol. **9:** 377-410.
2. Reth, M. 1992. Antigen receptors on B lymphocytes. Annu. Rev. Immunol. **10:** 97-121.
3. Terashima, M., K.-M. Kim, T. Adachi, P. J. Nielsen, M. Reth, G. Kohler & M. C. Lamers. 1994. The IgM antigen receptor of B lymphocytes is associated with prohibitin and prohibitin-related protein. EMBO J. **13:** 3782-3792.
4. Kim, K.-M., T. Adachi, P. J. Nielsen, M. Terashima, M. C. Lamers, G. Kohler & M. Reth. 1994. Two new proteins preferentially associated with membrane immunoglobulin D. EMBO J. **13:** 3793-3800.
5. Williams, G. T., A. R. Venkitaraman, D. J. Gilmore & M. S. Neuberger. 1990. The sequence of the mu transmembrane segment determines the tissue specificity of the transport of immunoglobulin M to the cell surface. J. Exp. Med. **171:** 947-952.
6. Venkitaraman, A. R., G. T. Williams, P. Dariavach & M. S. Neuberger. 1991. The B cell antigen receptor of the five immunoglobulin classes. Nature **352:** 777-781.
7. Matsuuchi, L., M. R. Gold, A. Travis, R. Grosschedl, A. L. DeFranco & R. B. Kelly. 1992. The membrane IgM-associated proteins MB-1 and Ig-β are sufficient to promote surface expression of a partially functional B-cell antigen receptor in a nonlymphoid cell line. Proc. Natl. Acad. Sci. USA **89:** 3404-3408.
8. Hombach, J., T. Tsubata, L. Leclercq, H. Stappert & M. Reth. 1990. Molecular components of the B-cell antigen receptor complex of the IgM class. Nature **343:** 760-762.
9. Blum, J. H., T. L. Stevens & A. L. DeFranco. 1993. Role of the μ immunoglobulin heavy chain transmembrane and cytoplasmic domains in B cell antigen receptor expression and signal transduction. J. Biol. Chem. **268:** 27238-27247.
10. Cherayil, B. J., K. MacDonald, G. L. Waneck & S. Pillai. 1993. Surface transport and internalization of the membrane IgM chain in the absence of the Mb-1 and B29 proteins. J. Immunol. **151:** 11-19.
11. Stevens, T. L., J. B. Blum, S. P. Foy, L. Matsuuchi & A. L. DeFranco. 1994. A mutation of the μ transmembrane that disrupts endoplasmic reticulum retention: effects on association with accessory proteins and signal transduction. J. Immunol. **152:** 4397-4406.
12. Grupp, S. A., R. N. Mitchell, K. L. Schreiber, D. J. McKean & A. K. Abbas. 1995. Molecular mechanisms that control expression of the B lymphocyte antigen receptor complex. J. Exp. Med. **181:** 161-168.
13. Grupp, S. A., K. Campbell, R. N. Mitchell, J. C. Cambier & A. K. Abbas. 1993. Signaling-defective mutants of the B lymphocyte antigen receptor fail to associate with Ig-α and Ig-β/γ. J. Biol. Chem. **268:** 25776-25779.
14. Shaw, A. C., R. N. Mitchell, Y. K. Weaver, J. Campos-Torres, A. K. Abbas & P. Leder. 1990. Mutations of immunoglobulin transmembrane and cytoplasmic domains: Effects on intracellular signaling and antigen presentation. Cell **63:** 381-392.
15. Mitchell, R. N., A. C. Shaw, Y. K. Weaver, P. Leder & A. K. Abbas. 1991. Cytoplasmic tail deletion converts membrane immunoglobulin to a phosphatidylinositol-linked form lacking signaling and efficient antigen internalization functions. J. Biol. Chem. **266:** 8856-8860.
16. Sanchez, M., Z. Misulovin, A. L. Burkhardt, S. Mahajan, T. Costa, R. Franke, J. B. Bolen & M. Nussenzweig. 1993. Signal transduction by immunoglobulin is mediated through Ig-α and Ig-β. J. Exp. Med. **178:** 1049-1055.
17. DeFranco, A. L. 1995. Transmembrane signaling by antigen receptors of B and T lymphocytes. Cur. Opinion Cell. Biol. **7:** 163-175.

18. DEFRANCO, A. L. 1994. Signaling pathways activated by protein tyrosine phosphorylation in lymphocytes. Curr. Opinion Immunol. **6:** 364-371.
19. WEBB, C. F., C. NAKAI & P. W. TUCKER. 1989. Immunoglobulin receptor signalling depends on the carboxyl terminus but not the heavy-chain class. Proc. Natl. Acad. Sci. USA **86:** 1977-1981.
20. DUBOIS, P. M., J. STEPINSKI, J. URBAIN & C. H. SIBLEY. 1992. Role of the transmembrane and cytoplasmic domains of surface IgM in endocytosis and signal transduction. Eur. J. Immunol. **22:** 851-857.
21. CAMBIER, J. C. 1995. New nomenclature for the Reth motif (or ARH1/TAM/ARAM/YXXL). Immunol. Today **16:** 110.
22. SEFTON, B. M. & J. A. TADDIE. 1994. Role of tyrosine kinases in lymphocyte activation. Curr. Opinion Immunol. **6:** 372-379.
23. LAW, D. A., V. W. F. CHAN, S. K. DATTA & A. L. DEFRANCO. 1993. B-cell antigen receptor motifs have redundant signalling capabilities and bind the tyrosine kinases PTK72, Lyn and Fyn. Curr. Biol. **3:** 645-657.
24. HUTCHCROFT, J. E., M. L. HARRISON & R. L. GEAHLEN. 1991. B lymphocyte activation is accompanied by phosphorylation of a 72-kDa protein-tyrosine kinase. J. Biol. Chem. **266:** 14846-14849.
25. MORGAN, D. O. & H. L. DEBONDT. 1993. Protein kinase regulation: insights from crystal structure analysis. Curr. Opinion Cell. Biol. **6:** 239-246.
26. KOLANUS, W., C. ROMEO & B. SEED. 1993. T cell activation by clustered tyrosine kinases. Cell **74:** 171-183.

Interaction of Shc with Grb2 Regulates the Grb2 Association with mSOS

K. S. RAVICHANDRAN, U. LORENZ,
S. E. SHOELSON, AND S. J. BURAKOFF

Division of Pediatric Oncology
Dana-Farber Cancer Institute and Department of Pediatrics
Harvard Medical School
Boston, Massachusetts 02115

Stimulation of growth-factor receptors, T-cell or B-cell antigen receptors, and many cytokine receptors activates p21^{ras} proteins.[1] The conversion of Ras from its inactive GDP-bound state to its active GTP-bound state, which occurs downstream of tyrosine kinase activation, involves several intracellular signaling proteins.[2] Adapter proteins, which have no apparent catalytic domain, but contain one or more Src-homology 2 (SH2) and Src-homology 3 (SH3) domains,[3] play a key role in linking many receptors to Ras activation. SH2 domains bind to specific phosphotyrosine-containing sequences, whereas SH3 domains bind to proline-rich sequences.

The adapter protein Shc has been implicated in Ras signaling through many receptors, including the T-cell antigen receptor (TCR), the interleukin-2 receptor, the interleukin-3 receptor, and the erythropoietin receptor, as well as the receptors for epidermal growth factor, nerve growth factor, and insulin. Shc is composed of a single SH2 domain, a glycine/proline-rich collagen homology (CH) domain, and a unique N-terminal domain.[4] Once of the mechanisms of TCR-mediated Ras activation involves tyrosine phosphorylation of Shc.[5] Tyrosine-phosphorylated Shc subsequently interacts with Grb2 (through the Grb2-SH2 domain), and Grb2, in turn, (by way of its SH3 domains) interacts with the Ras GTP/GDP exchange factor, mSOS. Shc also interacts through its own SH2 domain with the tyrosine-phosphorylated TCR-ζ chain. Thus, the simultaneous interaction of Shc with the TCR, and with Grb2 and mSOS, may help to shuttle the Ras nucleotide exchange factor to the membrane, thereby leading to Ras activation.

Although it was shown that growth-factor stimulation in fibroblasts did not alter the levels of Grb2 associated with mSOS,[2,6,7] we observed that in T-cell lines as well as in freshly isolated human T cells, TCR stimulation leads to a significant increase in the levels of Grb2 associated with mSOS. The increased Grb2:mSOS association occurred within 1 min after TCR stimulation and returned to basal levels within 30 minutes. The increased Grb2 binding to mSOS still occurred by way of the Grb2-SH3 domains and was regulated through the SH2 domain of Grb2. The following observations suggested a role for Shc in regulating the Grb2:mSOS association: (1) a phosphopeptide corresponding to the sequence surrounding Tyr^{317} of Shc, which displaces Shc, from the Grb2-SH2 domain, abolished the enhanced association between Grb2 and mSOS, whereas a control peptide had no effect; (2) although the Shc peptide by itself is unable to increase the Grb2:mSOS association, addition of phosphorylated Shc to unactivated T-cell lysates was sufficient to enhance the interaction of Grb2 with mSOS. Furthermore, using fusion proteins encoding different domains of Shc, we identified that the CH domain of Shc (which includes the Tyr^{317} site) can mediate this effect. Thus, the Shc-mediated regulation of the Grb2:mSOS association may provide a means for controlling the extent of Ras activation following receptor stimulation.

REFERENCES

1. Downward, J. 1992. BioEssays **14:** 177-184.
2. Schlessinger, J. 1993. Trends Biochem. Sci. **18:** 273-275.
3. Pawson, T. & G. D. Gish. 1992. Cell **71:** 359-362.
4. Pelicci, G. *et al.* 1992. Cell **70:** 93-104.
5. Ravichandran, K. S. *et al.* 1993. Science **262:** 902-905.
6. Li, N. *et al.* 1993. Nature **363:** 85-87.
7. Rozakis-Adcock, M. *et al.* 1993. Nature **363:** 83-85.

A 115 kDa Tyrosine Phosphorylated Protein Associates with Grb-2 in Activated Jurkat Cells[a]

HERMAN MEISNER AND MICHAEL P. CZECH

Program in Molecular Medicine and
Department of Biochemistry and Molecular Biology
University of Massachusetts Medical School
373 Plantation Street
Worcester, Massachusetts 01655

T-cell receptor cross-linking produces a rapid activation of Ras as a result of either activation of the guanine nucleotide exchange proteins Sos-1 and Sos-2 or suppression of the GTPase-activating protein (GAP).[1,2] A fraction of Sos is bound stably to a ubiquitous adaptor, Grb2, by a proline-*Src* homology (SH) 3 interaction. Grb2, a protein composed of one SH2 and two SH3 domains, links receptor and nonreceptor protein tyrosine kinases to Ras pathways by complexing with Sos.[1] In addition, Grb2 binds stably to a major tyrosine phosphorylated protein of 115–120 kDa in T cells. A protein of similar size also binds to SH3 and SH2 groups of Lck, Fyn, Src, p85, and Grb2 *in vitro*.[3,4] Recently, a 120 kDa tyrosine phosphorylated protein that binds to the Grb2 N-SH3 domain in T cells was identified as the product of the protooncogene c-*cbl*.[5] *Cbl* is a cytosol and cytoskeletal associated protein containing multiple phosphotyrosine (pY) residues in the N Terminus and an equal number of proline-rich motifs in the C terminus. The viral oncogene product is a truncated form of the native protein and is found in the nucleus where it presumably acts as a transcription factor. We confirmed that the pY115 protein is *Cbl* by precipitating lysates from resting and stimulated Jurkat cells with *Cbl* or Grb2 antisera and blotting against anti-c-*Cbl* or the anti-pY antibody 4G10. Both antisera precipitate a protein of the same size that is recognized by *Cbl* antiserum. The Grb2-*Cbl* complex is constitutive, because activation results in a greater pY content but not amount of *Cbl* bound to Grb2. In this work, we sought to characterize the role of Grb2 and p85 SH2/SH3 groups in binding to p115/*Cbl* after T-lymphocyte activation and the extent to which Sos is associated with this protein.

We tested the affinity of p115/*Cbl* in Jurkat cells for the SH2 and SH3 domains of Grb2, using a Grb2-GST fusion protein, or fusion proteins with a mutation in the N-SH3 binding domain (P49L) or the SH2 domain (R86K). After fully activating Jurkat cells by cross-linking CD3 and CD4 receptors, tyrosine-phosphorylated p115/*Cbl* in cell lysates binds equally to Grb2-GST, P49L, and R86K mutants at every concentration between 0.2 and 10 μg protein. Interestingly, in lysates of Jurkat cells that were partially activated with 10 μg OKT3/mL, *Cbl* exhibited a different affinity

[a] This work was supported in part by the JDF International Program Project, Grant #892004.

profile. Although wild-type Grb2 bound *Cbl* from these cells with a similar dose-response curve, the affinity of p115/*Cbl* for the P49L construct was greatly reduced. Binding of a significant amount of *Cbl* to this mutant Grb2 is only seen at high levels of fusion protein. Thus, in the fully activated state, *Cbl* can bind with high affinity to either the SH2 or SH3 domain of Grb2; however, in the partially tyrosine phosphorylated state (OKT3 only), the N-terminal SH3 domain is required to bind *Cbl*.

Several reports have noted that p85 SH2 domains bind to a 115-120 kDa protein in hematopoietic cells. We therefore examined whether a fusion protein containing both SH2 domains of p85α as well as the Grb2-GST binds *in vitro* to a common 115 kDa protein. Cell lysates were first mixed with 10 μg of p85SH2-GST or Grb2-GST, and, after centrifugation, the respective supernatants were bound to either Grb2-GST or p85SH2-GST. Both fusion proteins were found to bind the same pY115 band protein identified as *Cbl*.

To test whether Sos and *Cbl* may be associated in Jurkat cells, we prepared antibodies against fusion proteins containing the catalytic and proline-rich C terminus of mouse Sos-1. These antisera immunoprecipitate Sos-1 but not Sos-2 from Jurkat cells. When cell lysates were mixed with these antibodies and the resolved immune complexes probed with *Cbl* antisera, a p115 band was not present in activated cell lysates. When these blots were probed with Grb2, both Sos antibodies immunoprecipitated a small fraction of Grb2, and the stoichiometry of the Grb2-Sos complex was unaffected by activation. We conclude that complexes containing Sos/Grb2 and Cbl/Grb2 are formed in activated T cells.

REFERENCES

1. Schlessinger, J. 1993. TIBS **18:** 273-278.
2. Downward, J., J. Graves, P. Warne, S. Rayter & D. Cantrell. 1990. Nature (London) **346:** 719-723.
3. Motto, D., S. Ross, J. Jackman, Q. Sun, A. Olson, P. Findell & G. Koretzky. 1994. J. Biol. Chem. **269:** 21608-21613.
4. Exley, M., L. Varticovski, M. Peter, J. Sancho & C. Terhorst. 1994. J. Biol. Chem. **269:** 15140-15146.
5. Donovan, J., R. Wange, W. Langdon & L. Samelson. 1994. J. Biol. Chem. **269:** 22921-22924.

Csk Associates with the TCR ζ and ε Chains through Its SH2 Domain

THORUNN RAFNAR, JONATHAN P. SCHNECK, MARY E. BRUMMET, DAVID G. MARSH, AND BRANIMIR ČATIPOVIĆ

Division of Clinical Immunology
Johns Hopkins Asthma and Allergy Center
Johns Hopkins University School of Medicine
Baltimore, Maryland 21224

C-terminal Src kinase (Csk) down-regulates the activity of Src family kinases by phosphorylating their C-terminal tyrosine.[1] Overexpression of Csk in T-cell hybridomas results in decreased phosphorylation upon TCR stimulation and reduced IL-2 production.[2] Presumably this effect is due to down-regulation of the Src family kinases Lck and Fyn that are activated upon TCR stimulation. However, Csk is a cytosolic kinase, whereas the Src kinases are attached to the cell membrane. Until the present studies, stable association between Src kinases and Csk had not been detected; therefore, it was not known how Csk was targeted to Lck and Fyn.

To analyze the T-cell proteins with which Csk interacts, we produced recombinant fusion proteins consisting of the SH2, SH3, or SH2/SH3 domains of Csk, attached to glutathione S-transferase (GST). The recombinant proteins were produced in *E. coli* and purified on glutathione-coated agarose beads. The beads were incubated with lysates of stimulated or unstimulated Jurkat cells, and the bound proteins were analyzed by Western blotting using an antiphosphotyrosine antibody (FIG. 1). Several tyrosine-phosphorylated proteins bound to the Csk-SH2-GST and Csk-SH2/3-GST fusion proteins, whereas no phosphoproteins bound to either Csk-SH3-GST or GST alone. The two major proteins that bound to the SH2 domain of Csk had molecular masses of 60 kDa (p60) and 21 kDa (p21); in addition, four proteins with molecular masses 100 kDa, 70 kDa, 52 kDa, and 25 kDa were detected after activation. p60, p21, and p25 were shown to bind directly to the SH2 domain, whereas the binding of the three other proteins was indirect, that is, was lost upon denaturation of the proteins by boiling in 1% SDS. p60 was constitutively phosphorylated on tyrosine, but p21, as well as the four minor bands, bound only after activation of the Jurkat cells.

The 70 kDa protein was identified by Western blotting as the protein tyrosine kinase ZAP-70. Because ZAP-70 is known to associate with the phosphorylated ε and ζ chains of the TCR, we investigated whether p21 and p25 might be components of the TCR. Jurkat cells were stimulated for 1 minute with OKT3 antibody, lysed, and immunoprecipitated with antibodies to either ε or ζ (FIG. 2). After radiolabeling of the ε and ζ chains with [^{32}Pγ]ATP *in vitro*, the precipitates were boiled in 1% SDS to release the TCR chains, and the diluted supernatants were reprecipitated with Csk-SH2-GST. Csk-SH2 bound the radiolabeled ε and ζ chains as efficiently as a

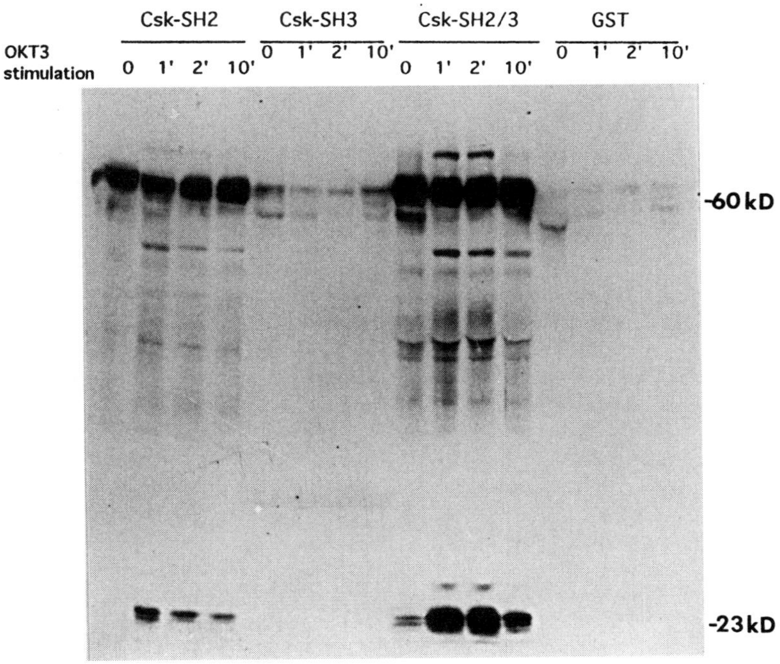

FIGURE 1. Binding of tyrosine-phosphorylated proteins to Csk fusion proteins. Lysates of Jurkat cells, unstimulated and stimulated with monoclonal antibody OKT3 for 1 min, 2 min, and 10 min were incubated with the Csk-SH2-GST, Csk-SH3-GST, Csk-SH2/3-GST, or GST bound to glutathione-agarose beads. The adsorbates were separated by SDS-PAGE on 8% acrylamide gel, transferred to nitrocellulose, and probed with the antiphosphotyrosine monoclonal antibody 4G10.

ZAP-70-$(SH2)_2$-GST-fusion protein. Additional experiments, including surface iodination of Jurkat cells, followed by Csk-SH2 binding and diagonal electrophoresis, and comigration of Csk-SH2-bound material with the ε and ζ chains confirmed that Csk binds to ε and ζ by way of its SH2 domain. We propose that Csk binds to the phosphorylated ε and ζ chains upon activation and that this binding may cause Csk to become activated and phosphorylate Lck and Fyn.

It has been shown that, upon activation, the ARAM sequences in the ε and ζ chains of the TCR become phosphorylated on two tyrosines within each motif.[3] At least two proteins, ZAP-70 and Shc, bind to the phosphorylated ARAMs, acting as downstream effectors in the signal-transduction cascade. We used phosphopeptides based on the ε ARAM sequence to test whether they would inhibit the binding of Csk-SH2 to ε and ζ. One of the peptides, in which the second tyrosine was phosphorylated, inhibited the binding significantly. Thus, our data suggest that the balance between Csk, Shc, and ZAP-70 may be an important factor in regulating TCR signaling.

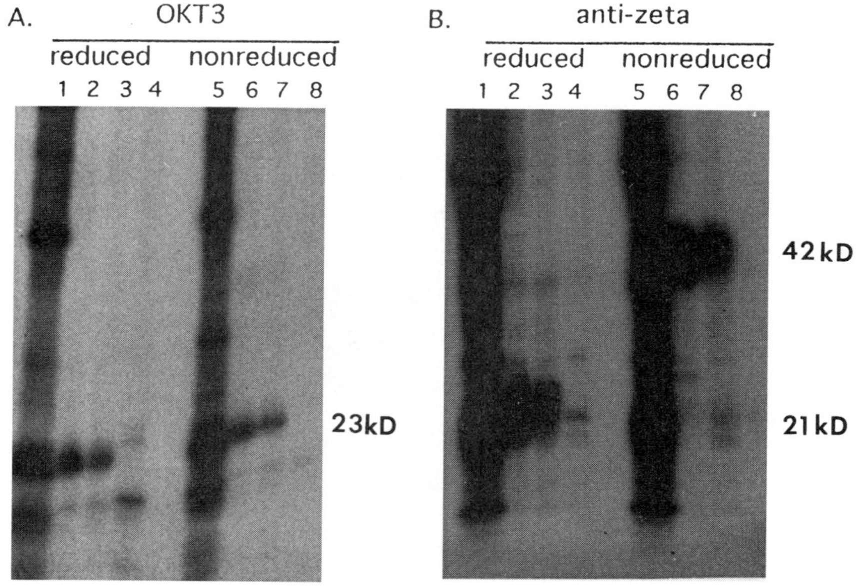

FIGURE 2. Binding of Csk-SH2 domain to radiolabeled ε and ζ chains of TCR complex. **A**: The ε chain was immunoprecipitated with monoclonal antibody OKT3 from lysates of stimulated Jurkat cells. The complex was subjected to an *in vitro* kinase assay in the presence of [^{32}Pγ]ATP. One-tenth of the immunoprecipitate was analyzed directly (lanes 1 and 5), whereas the rest was boiled in 1% SDS, diluted ten times, and incubated with Csk-SH2-GST (lanes 2 and 6), ZAP-70-$(SH2)_2$-GST (lanes 3 and 7), or GST beads (lanes 4 and 8). The bound proteins were separated by SDS-PAGE, either under reducing (lanes 1-4) or nonreducing conditions (lanes 5-8). The gel was dried and exposed to X-ray film. **B**: Same as in **A** except that the immunoprecipitation was performed using anti-ζ serum.

REFERENCES

1. Cooper, J. A. & B. Howell. 1993. Cell **73:** 1051-1054.
2. Chow, L. M. L., M. Fournel, D. Davidson & A. Veillette. 1993. Nature **365:** 156-160.
3. Chan, A. C., D. M. Desai & A. Weiss. 1994. Annu. Rev. Immunol. **12:** 555-592.

Fc Receptor Stimulation of PI 3-Kinase in NK Cells Is Associated with Protein Kinase C–independent Granule Release and Cell-mediated Cytotoxicity

JOY D. BONNEMA, LARRY M. KARNITZ, RENEE A. SCHOON, ROBERT T. ABRAHAM, AND PAUL J. LEIBSON[a]

Department of Immunology
Mayo Clinic and Foundation
Rochester, Minnesota 55905

INTRODUCTION

Although diverse signaling events are initiated by stimulation of multichain immune recognition receptors on lymphocytes, it remains unclear as to which specific signal transduction pathways are functionally linked to granule exocytosis and cellular cytotoxicity. In the case of natural killer (NK) cells, it has been presumed that the rapid activation of protein kinase C (PKC) enables them to mediate antibody-dependent cellular cytotoxicity (ADCC) and "natural" cytotoxicity toward tumor cells. However, using cloned human NK cells and specific inhibitors of PKC (GF109203X) and phosphatidylinositol 3-kinase (wortmannin), we determined that Fc receptor stimulation triggers granule release and ADCC through a PKC-independent, but phosphatidylinositol 3-kinase (PI 3-kinase)-sensitive pathway. By contrast, natural cytotoxic activity initiated after recognition of the tumor cell line K562 was regulated by way of a PI 3-kinase-independent, but PKC-dependent pathway. Taken together, these results suggest that (1) PI 3-kinase activation induced by FcR ligation is functionally coupled to granule exocytosis and ADCC; and (2) the signaling pathways involved in ADCC versus natural cytotoxicity are distinct.

MATERIAL AND METHODS

Material and methods are as previously described.[1]

RESULTS AND DISCUSSION

NK cells mediate cytotoxicity against susceptible tumor targets and virus-infected cells by either ADCC or direct cell-mediated cytotoxicity.[2] A crucial event in the

[a] To whom correspondence should be addressed.

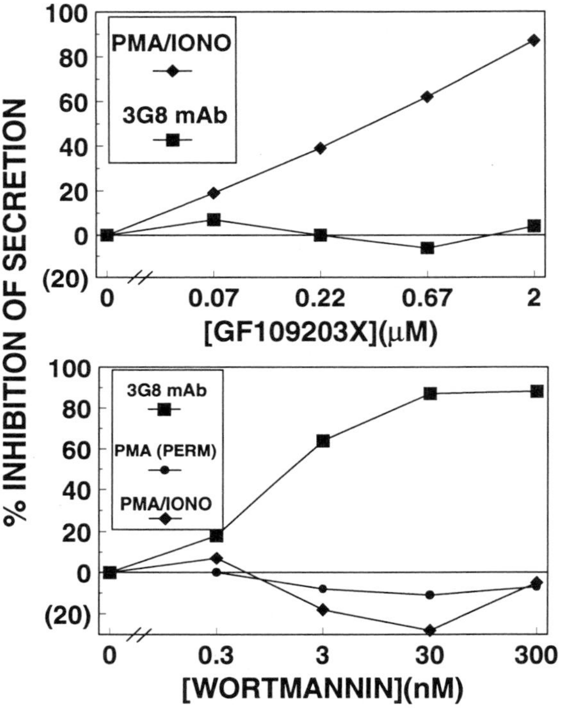

FIGURE 1. Selective inhibition of FcR-induced secretion by wortmannin. Cloned $CD16^{+}$, $CD3^{-}$ human NK cells were preincubated for 15 min at 37°C with the indicated concentrations of either GF109203X (top) or wortmannin (bottom), left intact or permeabilized (PERM), stimulated with the appropriate stimulus, and assayed for percent hexosaminidase released. (Bonnema *et al.*[1] With permission from the Rockefeller University Press.)

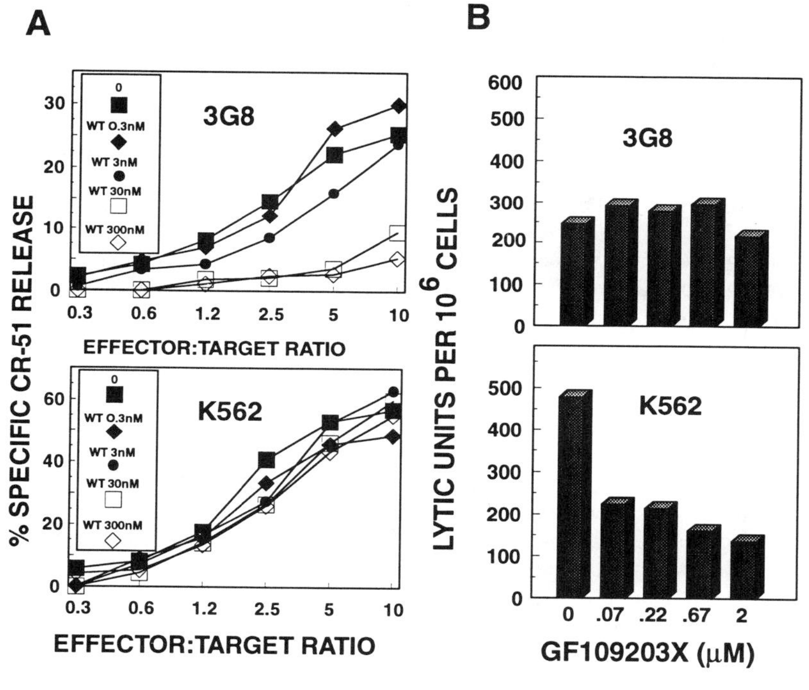

FIGURE 2. FcR-dependent cytotoxicity and direct antitumor cytotoxicity are regulated by separate signaling pathways. Cloned NK cells were preincubated for 15 min at 37°C with the indicated concentrations of either wortmannin (WT) (A) or GF109203X (B). Drug-treated cells were then incubated for 1 h with either ^{51}Cr-labeled 3G8 (anti-FcγRIII) hybridoma cells (top) or ^{51}Cr-labeled K562 tumor cells (bottom). (Bonnema *et al.*[1] With permission from the Rockefeller University Press.)

development of either of these modes of killing is the receptor-stimulated secretion of granule-derived proteins.[3] It remains unclear, however, which signals generated after FcR ligation regulate granule exocytosis from NK cells. Because direct pharmacologic activation of PKC can induce secretion from NK cells,[4] we first evaluated whether PKC plays a similar role after FcR stimulation. The fluorescence assay for hexosaminidase, previously shown to be a sensitive method for measuring granule exocytosis from NK cells,[4] was used to quantitate secretion. Cloned $CD16^+$, $CD3^-$ human NK cells were first exposed to varying concentrations of the selective PKC inhibitor GF109203X and then stimulated with either 3G8 monoclonal antibody (anti-FcγRIII)-coated polystyrene beads or a combination of the PKC-activating phorbol ester, PMA, and the calcium ionophore, ionomycin. As shown in the top panel of FIGURE 1, pretreatment of NK cells with concentrations of GF109203X that fully inhibited PMA/ionomycin-induced secretion did not alter FcR-induced granule release. These results suggest that although PKC activation can result in granule exocytosis from NK cells, FcR-stimulated granule release is regulated by PKC-independent signaling pathways.

The novel finding that FcR-induced granule exocytosis is not regulated by PKC prompted us to investigate whether PI 3-kinase, a potential modulator of vesicular transport,[5,6] might influence secretion from NK cells. PI 3-kinase has previously been shown to be activated following FcR ligation in NK cells and has been shown to be inhibited by nanomolar concentrations of the fungal metabolite, wortmannin.[6,7] Pretreatment of NK cells with wortmannin potently inhibited FcR-induced secretion in a concentration-dependent manner with an IC_{50} of ~2 nM (FIG. 1, bottom). It is important to note that although wortmannin blocked FcR-induced granule release from NK cells, PKC-dependent release from either intact cells stimulated with PMA/ionomycin, or permeabilized cells stimulated with PMA in buffer containing high free Ca^{2+} concentration, was not inhibited by wortmannin (FIG. 1, bottom). These data, together with the differential effects of GF109203X on NK cell secretion (FIG. 1, top), suggest that FcR-induced secretion is regulated by a PI 3-kinase-dependent, PKC-independent pathway. Conversely, pharmacologic activation by phorbol esters induces secretion by way of a separate PKC-dependent, PI 3-kinase-independent pathway.

Because granule exocytosis is believed to deliver the "lethal hit" during ADCC as well as during direct, antitumor cell-mediated killing, we performed parallel analysis evaluating the potential roles of PI 3-kinase and PKC in regulating FcγRIII-dependent versus direct cell-mediated killing. Results from these investigations indicated that whereas FcR-dependent killing of the 3G8 (anti-FcRγRIII) hybridoma cells was wortmannin-sensitive (FIG. 2A, top), and GF109203X-resistant (FIG. 2B, top), killing of the tumor line K562 was wortmannin-resistant (FIG. 2A, bottom) and GF109203X-sensitive (FIG. 2B, bottom). These data suggest that although FcR-initiated granule release and ADCC proceed through PI 3-kinase dependent pathways, direct recognition of susceptible tumor targets initiates granule release and killing through PKC-dependent pathways.

REFERENCES

1. BONNEMA, J. D., L. M. KARNITZ, R. A. SCHOON, R. T. ABRAHAM & P. J. LEIBSON. 1994. J. Exp. Med. **180:** 1427.
2. TRINCHIERI, G. 1989. Biology of natural killer cells. Adv. Immunol. **47:** 187.
3. HENKART, P. A., M. P. HAYES & J. W. SHIVER. 1993. The granule exocytosis model for lymphocyte cytotoxicity and its relevance to target cell DNA breakdown. *In* Cytotoxic Cells: Recognition, Effector Function, Generation, and Methods. M. V. Sitkovsky & P. A. Henkart, Eds.: 153-165. Birkhaeuser Boston, Inc., Cambridge, MA.
4. TING, A. T., R. A. SCHOON, R. T. ABRAHAM & P. J. LEIBSON. 1992. J. Biol. Chem. **267:** 23957.
5. SCHU, P. V., K. TAKEGAWA, M. J. FRY, J. H. STACK, M. D. WATERFIELD & S. D. EMR. 1993. Science (Wash D.C.). **260:** 88.
6. YANO, H., S. NAKANISHI, K. KIMURA, N. HANAI, Y. SAITOH, Y. FUKUI, Y. NONOMURA & Y. MATSUDA. 1993. J. Biol. Chem. **268:** 25846.
7. KANAKARAJ, P., B. DUCKWORTH, L. AZZONI, M. KAMOUN, L. C. CANTLEY & B. PERUSSIA. 1994. J. Exp. Med. **179:** 551.

Potentiation of B-Cell Antigen Receptor-mediated Signal Transduction by the Heterologous *src* Family Protein Tyrosine Kinase, *src*

JIEJIAN LIN, ANNE L. BURKHARDT,[a]
JOSEPH B. BOLEN,[a] AND LOUIS B. JUSTEMENT

Department of Microbiology and Immunology
The University of Texas Medical Branch
Galveston, Texas 77555

[a]*Department of Molecular Biology*
Bristol-Myers Squibb Pharmaceutical Research Institute
Princeton, New Jersey 08354

B-cell activation in response to antigen-receptor (AgR) ligation is dependent on the activation of one or more *src* family protein-tyrosine kinases (PTK), including *lyn*, *fyn*, and *blk*.[1] Very little is known concerning the relative importance of these PTK for B-cell activation at the present time. Based on previously reported findings, we demonstrated that association of the above PTK with the AgR may involve a competitive binding process.[2] However, the basis for this has not yet been delineated. If in fact the different *src* family PTK bind competitively, then it is possible that their physical association with the AgR is controlled by their relative level of expression within the B cell. Moreover, it is possible that the association of these PTK with the AgR complex in the B cell is mediated by conserved structural elements that are common to members of the *src* family. If this is the case, then it is logical to propose that overexpression of a heterologous *src* family PTK might result in its association with the AgR complex. Whether such an association would be mediated by a direct interaction between the specific PTK and either Igα or Igβ is not known.[3] Nevertheless, it should be possible to determine whether the interaction provides a functional link between kinase and AgR by monitoring phosphorylation and activation of the PTK in response to AgR cross-linking. Currently, it is not known whether the different *src* family PTK exhibit distinct substrate specificities and thus perform unique functions during B-cell activation. If the different *src* family PTK do not exhibit unique substrate specificities, and overexpression of a heterologous PTK leads to its association with the AgR and its activation upon AgR cross-linking, then it should be possible to observe alterations in the phosphorylation of certain endogenous substrates in the B cell.

In order to address these questions, we have transfected the B-cell lymphoma K46-17μmλ (K46) with a cDNA construct encoding the PTK *src*, which is not expressed in either the K46 cell line or normal splenic B cells. After isolation of stable transfectants, we were able to demonstrate that *src* associates with the AgR complex in resting B cells based on coprecipitation of the Igα/Igβ heterodimer from radiolabeled AgR immune-complex material using anti-*src* antibody. Cross-linking of the AgR on *src* transfectants using anti-IgM antibody resulted in tyrosine phosphorylation of *src in vivo* within 1 to 5 min, suggesting that the kinase is associated with the AgR in a functional manner and that it becomes activated in response to ligand binding. Subsequently, experiments were performed to compare the phosphorylation of specific substrates in response to cross-linking of the AgR on parental K46 and *src*-expressing transfectants. The B cell-specific glycoprotein, CD22, was observed to become hyperphosphorylated on tyrosine to a much greater extent (~5- to 8-fold) in *src* transfectants than in parental K46 cells. These observations suggest that heterologous *src* family PTK are able to associate with the AgR and become activated in response to AgR cross-linking. These data further suggest that there may be a common mechanism involved in mediating the association of distinct *src* family PTK with the B-cell AgR complex. Experiments are currently underway to examine this question in further detail.

REFERENCES

1. Pleiman, C. M., D. D'Ambrosio & J. C. Cambier. 1994. The B-cell antigen receptor complex: structure and signal transduction. Immunol. Today **15**(9): 393-399.
2. Lin, J. & L. B. Justement. 1992. The MB-1/B29 heterodimer couples the B cell antigen receptor to multiple *src* family protein tyrosine kinases. J. Immunol. **149**(5): 1548-1555.
3. Clark, M. R., S. A. Johnson & J. C. Cambier. 1994. Analysis of Ig-α-tyrosine kinase interaction reveals two levels of binding specificity and tyrosine phosphorylated Ig-α stimulation of Fyn activity. EMBO J. **13**(8): 1911-1919.

Regulation of Human Natural Killer-Cell Lytic Activity by Serine/ Threonine Phosphatases and Kinases

ANIL BAJPAI[a] AND ZACHARIE BRAHMI[a,b]

Department of [a]Medicine and [b]Microbiology/Immunology
Indiana University Medical Center
Indianapolis, Indiana 46202

The combined activities of phosphoprotein phosphatases and kinases regulate the very early events in the activation of cytolytic lymphocytes. The role of tyrosine kinases and phosphotyrosine phosphatase (CD45) in the regulation of lytic functions mediated by natural killer (NK)-cell and cytolytic T lymphocytes (CTL) is well established.[1,2] The potential role of ser/thr phosphatases in the regulation of lymphocyte lytic activity is not well documented. Recently we have shown that both CD45 and ser/thr phosphatases are important in the regulation of lytic activity of CTL and NK cells.[2] We hypothesized that ser/thr phosphatases may be acting by regulating cellular kinases associated with the lytic activities of lymphocytes. Here, we provide evidence for the identification of two ser/thr kinases regulated by calyculin A-sensitive phosphatases that appear to play a direct role in tumor cell lysis by cytolytic lymphocytes.

YT-INDY, an NK-like subline derived from original human non-ATL thymic lymphoma,[3] was used. Cells were maintained in complete medium, and chromium release assays for cytotoxicity measurement were done as described previously.[2] Kinases were identified by an *in vitro* kinase renaturation assay of blotted protein kinases.[4]

Incubation of YT-INDY cells with two ser/thr phosphatase inhibitors, calyculin A and okadaic acid, for 30 min showed a dose-dependent inhibition of cytolytic activity only with calyculin A (FIG. 1a). Okadaic acid essentially had no effect up to 1 μM, whereas caliculyn A inhibited >98% lytic activity at 20 nM. The effect of calyculin A and okadaic acid on YT-INDY cells was similar to what we observed with fresh NK cells.[2] The coordinated regulation of phosphoprotein phosphatases and phosphokinases has been documented in lymphocyte activation. In this study, we observed several protein bands with kinase activity in calyculin A- and okadaic acid-treated cells. The two protein bands observed at 78 kDa and 60 kDa had strong kinase activity in calyculin A-treated cells. The 78 kDa protein kinase that had elevated kinase activity in calyculin A-treated cells was not specific to calyculin A treatment, as it was also present with basal activity in untreated control and okadaic acid-treated cells. However, the 60 kDa-associated kinase activity was specific for

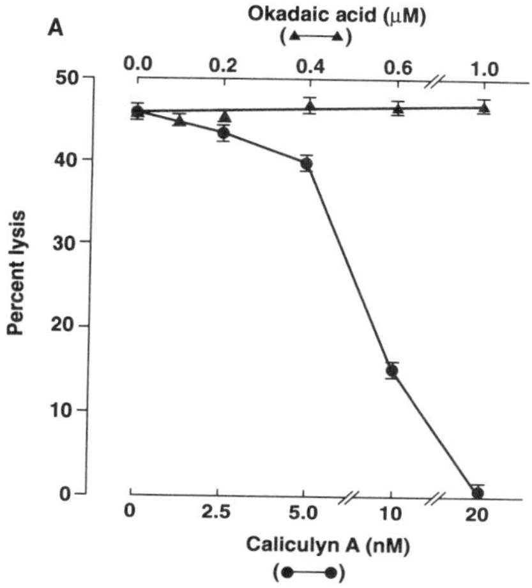

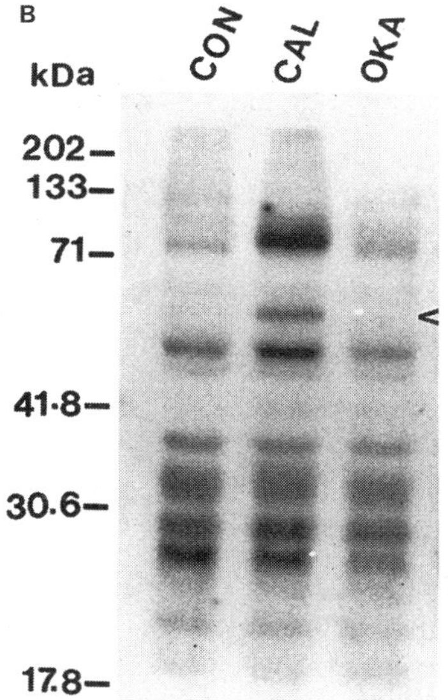

FIGURE 1. Effect of calyculin A and okadaic acid on YT-INDY cell-mediated cytotoxicity and cellular kinases. **A**: YT-INDY cells were treated with calyculin A and okadaic acid for 30 min at 37°C. After incubation, cells were washed and used for a chromium release assay using ^{51}Cr-labeled Raji cells as targets. **B**: YT-INDY cells were treated with 20 nM calyculin A and 1 μM okadaic acid for 30 min at 37°C. After incubation, cells were washed, lysed, and assayed for kinase activity. CON: control; CAL: calyculin A-treated cells; and OKA: okadaic acid-treated cells. Arrow indicates 60 kDa kinase.

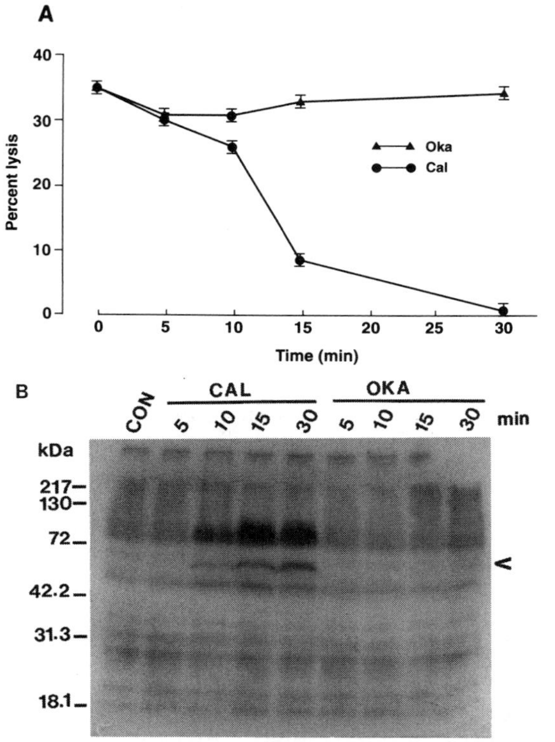

FIGURE 2. Kinetics of calyculin A and okadaic acid-mediated regulation of YT-INDY cell-mediated cytotoxicity and appearance of kinases. YT cells were treated with 20 nM calyculin A and 1 μM okadaic acid for indicated times at 37°C. After incubation, cells were washed and (**A**) used for cytotoxicity assay and (**B**) for kinase assay. CON: control; CAL: calyculin A; and OKA: okadaic acid. Arrow indicates 60 kDa kinase.

calyculin A treatment (FIG. 1b). Okadaic acid had no effect on the kinase activity associated with the 60 kDa protein.

To confirm the association of the 60 kDa kinase activity with calyculin A-mediated inhibition of the lytic activity of YT-INDY cells, we examined the kinetics of the loss of lytic activity and the appearance of 60 kDa kinase. We observed a direct parallel time course between these two events (FIG. 2). Calyculin A inhibited 14%, 26%, 77%, and >98% of lytic activity mediated by YT-INDY after 5, 10, 15, and 30 min incubation, respectively (FIG. 2a). Calyculin A treatment of YT-INDY cells led to a gradual increase in 60 kDa kinase activity that started appearing as early as 5 min and peaked after 30 min of incubation. Okadaic acid had no significant effect either on the lytic activity or the appearance of 60 kDa kinase in YT-INDY cells. However, treatment of YT-INDY cells with calyculin A induced the 78 kDa associated kinase activity, which appears to be present with basal activities in untreated control and okadaic acid-treated cells.

These data further support the differential role of calyculin A and okadaic acid on the lytic activity of NK cells and implicate the association of the appearance of 60 kDa kinase with the calyculin A-mediated inhibition of NK cell-mediated tumor cell lysis.

REFERENCES

1. Perlmutter, R. M., D. L. Steven, M. W. Appleby, S. J. Anderson & J. Alberola-Ila. 1993. Regulation of lymphocyte function by protein phosphorylation. Annu. Rev. Immunol. **11:** 451-499.
2. Bajpai, A. & Z. Brahmi. 1994. Target cell-induced inactivation of cytolytic lymphocytes: role and regulation of CD45 and calyculin A-inhibited phosphatases in response to IL-2. J. Biol. Chem. **269:** 18864-18869.
3. Yodoi, J., K. Teshigawara, T. Nikaido, K. Fukui, T. Noma, T. Honjo, M. Takigawa, M. Sasaki, N. Minato, M. Tsudo, T. Uchiyama & M. Maeda. 1985. TCGF (IL-2)-receptor inducing factor(s): 1. Regulation of IL-2 receptor on a natural killer-like cell line (YT cells). J. Immunol. **134:** 1623-1630.
4. Ferrell Jr., J. E. & G. S. Martin. 1991. Assessing activities of blotted protein kinases. Methods Enzymol. **200:** 430-435.

Mechanisms of Enhanced Nuclear Translocation of the Transcription Factors c-Rel and NF-κB by CD28 Costimulation in Human T Lymphocytes[a]

JENN-HAUNG LAI, GYÖRGYI HORVATH, YONGQIN LI, AND TSE-HUA TAN[b]

Department of Microbiology and Immunology
Baylor College of Medicine
Room M929
Houston, Texas 77030

CD28 is a pivotal T-cell molecule that plays important roles in T-cell activation, HIV-1 pathogenesis, and the induction and maintenance of peripheral tolerance and autoimmunity. The interaction of B7-1 and B7-2 on the antigen-presenting cells with CD28 on the T cells provides a costimulatory signal that is critical in the activation of T-cell responses.[1] Blocking CD28 function can inhibit T-cell activation with the antigen. It can also drive the T cells into a state of long-lasting antigen-specific unresponsiveness (anergy). The CD28-mediated signal transduction pathway is distinct from the T-cell receptor/CD3 signaling pathway. A characteristic of CD28 signaling is that it is resistant to the immunosuppressant cyclosporin A (CsA) but sensitive to the immunosuppressant rapamycin. CD28 costimulation, in cooperation with another stimulus, stimulates the secretion of multiple lymphokines. The induction of lymphokine gene expression by CD28 costimulation is mediated by both transcriptional and posttranscriptional mechanisms. CD28 costimulation activates lymphokine gene transcription through a unique motif, CD28RE,[2] which is conserved among several lymphokine promoters. CD28 costimulation also activates transcription from the HIV-1 LTR and the IL-2Rα promoter by way of the κB elements. This suggests that the κB enhancer-binding proteins, namely the Rel family of transcription factors, are involved in the CD28 signaling pathway.

The Rel family of transcription factors includes c-Rel, RelA (p65), RelB, NFKB1 (p50 and precursor p105), and NFKB2 (p52 and precursor p100). They share a 300-aa conserved domain, called the Rel domain. In resting cells, Rel family proteins are

[a] This work was supported by National Institutes of Health Grant RO1 GM-49875 from the National Institute of General Medical Sciences.

[b] Address for correspondence: Tse-Hua Tan, Ph.D., Department of Microbiology and Immunology, Baylor College of Medicine, M-929, One Baylor Plaza, Houston, Texas 77030-3498.

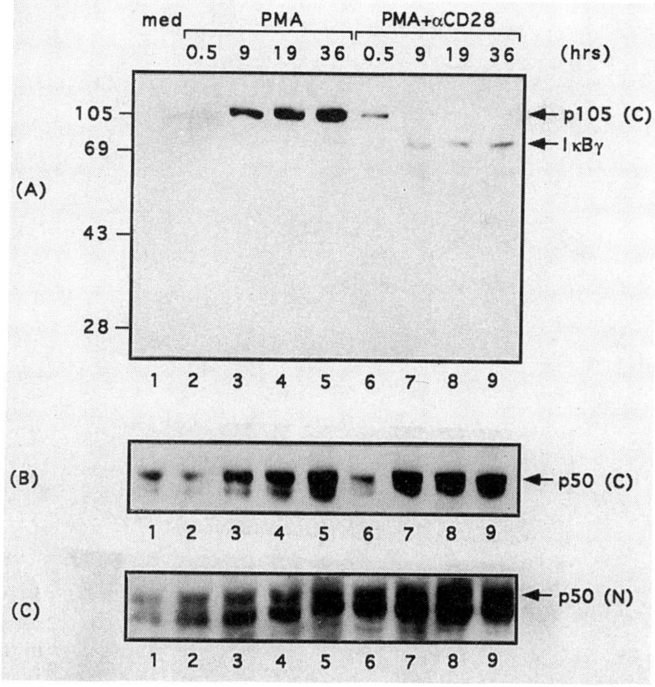

FIGURE 1. CD28 signaling caused enhanced processing of the precursor NFKB1 p105 to p50 and enhanced nuclear translocation of p50. Human peripheral blood T cells were purified from buffy coats by negative selection. 2×10^7 T cells were stimulated with phorbol myristate acetate (PMA; 5 ng/mL) in the presence or absence of anti-CD28 mAb ascites (1:2,000 dilution) for varying periods of time. The nitrocellulose filter containing either cytoplasmic or nuclear extracts was subjected to immunoblotting with rabbit antisera raised against the peptide sequence derived from either the carboxyl terminus (A) or the amino terminus (B and C) of NFKB1 p105. C, cytoplasmic extracts; N, nuclear extracts. med, medium.

sequestered in the cytoplasm by their physical association with IκB proteins or the ankyrin-containing cytoplasmic proteins, NFKB1 (p105) and NFKB2 (p100). Stimulation of the cells with mitogenic stimuli leads to rapid but transient down-regulation of the IκBα. This causes the translocation of Rel family proteins into the nucleus and subsequent activation of gene transcription. Our previous results showed that c-Rel, p65, and p50 are part of the components of the CD28-responsive complex (CD28RC).[3]

RESULTS AND DISCUSSION

We found that CD28 signaling enhances and accelerates the translocation of cytoplasmic c-Rel and NF-κB into the nucleus in both human primary T cells[4] and

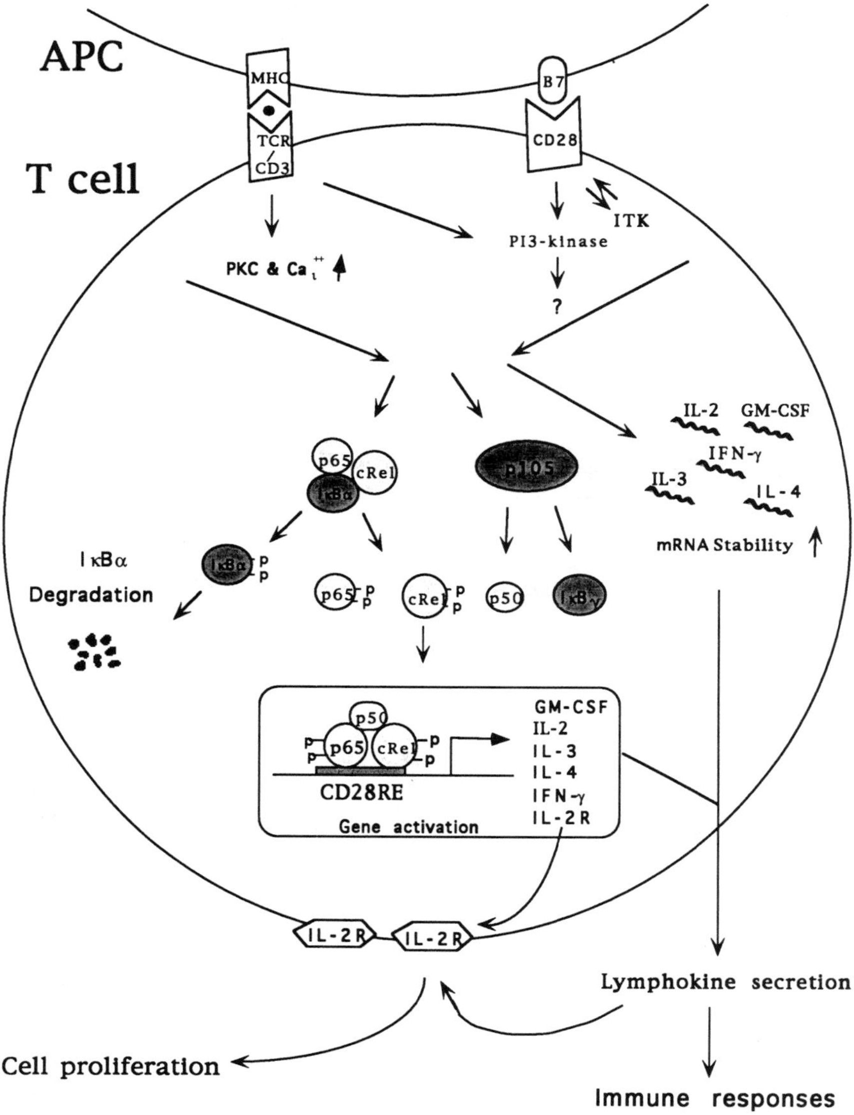

FIGURE 2. A schematic description of the sequence of events that leads to the enhanced nuclear translocation of c-Rel and NF-κB by CD28 costimulation in human T lymphocytes.

Jurkat T cells.[5] Specifically, the enhanced translocation of c-Rel and NF-κB by CD28 signaling is relatively resistant to CsA but is sensitive to rapamycin.[5] Our data also show that CD28 signaling, distinct from other stimuli, such as PMA, IL-1, and TNFα, causes a sustained down-regulation of the inhibitor IκBα,[5] which is also prevented by rapamycin but not by CsA.[5] The counter regulation of IκBα by CD28 signaling and rapamycin observed in Jurkat T cells is also reproducible in primary T cells.[5] By contrast, the PMA plus ionomycin-mediated down-regulation of IκBα was prevented by CsA but not by rapamycin.[5] Our data suggest that IκBα is the downstream target of both CD28 signaling and rapamycin.

We also examined the effects of CD28 signaling on two other c-Rel inhibitors, NFKB1 p105 and NFKB2 p100. We found that CD28 signaling does not down-regulate the levels of NFKB2 p100.[5] Interestingly, when antibody against the N-terminus of NFKB1 p105 was used, we invariably detected a slight down-regulation of p105 levels by Western blotting analysis.[4,5] Consistent with this result, we also observed decreased levels of the phosphorylated p105, detected by immunoprecipitation assays, after CD28 costimulation.[4] To further substantiate this observation, we used the antibody raised against the C-terminus of p105, which can detect one of the processed products of p105, the 70-kDa IκBγ. Interestingly, CD28 signaling resulted in the appearance of IκBγ in the cytoplasm (FIG. 1A). These results suggest that CD28 signaling can also enhance the processing of the precursor p105 to p50, which results in the induction of nuclear levels of p50 (FIG. 1C). However, we found that both c-Rel and p65 (but not p50) bind and synergistically activate the CD28RE enhancer activity in cotransfection assays.[6] Because the Jurkat T cells we used in the transfection assays contain high basal levels of p50, we suggest that moderate levels of p50 are required for CD28-mediated gene activation, and excess levels of p50 will inhibit the CD28RE enhancer activity. In conclusion, our data indicate that IκBα and p105 are the downstream targets of CD28 signaling. A sustained down-regulation of IκBα and an enhanced processing of p105 by CD28 costimulation led to an accelerated and sustained nuclear translocation of c-Rel and NF-κB (p50 and p65), which may, in turn, have activated lymphokine gene transcription (FIG. 2).

REFERENCES

1. JUNE, C. H., J. A. BLUESTONE, L. M. NADLER & C. B. THOMPSON. 1994. The B7 and CD28 receptor families. Immunol. Today **15:** 321-331.
2. FRASER, J. D., B. A. IRVING, G. R. CRABTREE & A. WEISS. 1991. Regulation of interleukin-2 gene enhancer activity by the T cell accessory molecule CD28. Science **251:** 313-316.
3. GHOSH, P., T.-H. TAN, N. RICE & H. YOUNG. 1993. The IL-2 CD28-responsive complex contains at least three members of the NF-κB family: c-Rel, p50 and p65. Proc. Natl. Acad. Sci. USA **90:** 1696-1700.
4. BRYAN, R. G., Y. LI, J.-H. LAI, M. VAN, N. R. RICE, R. R. RICH & T.-H. TAN. 1994. The effect of CD28 signal transduction on c-Rel in human peripheral blood T cells. Mol. Cell. Biol. **14:** 7933-7942.
5. LAI, J.-H. & T.-H. TAN. 1994. CD28 signaling causes a sustained down-regulation of IκBα which can be prevented by the immunosuppressant rapamycin. J. Biol. Chem. **269:** 30073-30076.
6. LAI, J.-H., G. HORVATH, J. SUBLESKI, P. GHOSH, J. BRUDER & T.-H. TAN. 1995. RelA is a potent transcriptional activator of the CD28 response element within the IL-2 promoter. Mol. Cell. Biol. **15:** 4260-4271.

Cytokine Signal Transduction through a Homo- or Heterodimer of gp130

TADAMITSU KISHIMOTO,[a] TAKASHI TANAKA,[a]
KANJI YOSHIDA,[b] SHIZUO AKIRA,[b] AND
TETSUYA TAGA[b]

[a]*Department of Medicine III*
Osaka University Medical School
2-2 Yamada-oka, Suita
Osaka 565, Japan
and
[b]*Institute for Molecular and Cellular Biology*
Osaka University
1-3 Yamada-oka, Suita
Osaka 565, Japan

INTRODUCTION

One of the characteristic features of cytokines is their functional pleiotropy and redundancy. One cytokine shows a wide variety of biological functions on various tissues and cells, and different cytokines often show similar activities on certain cells. This functional pleiotropy and redundancy can now be explained on the basis of molecular structure of the cytokine receptor system.[1–3] The cytokine receptor system usually consists of two polypeptide chains, that is, a ligand-specific receptor and a common signal transducer. This unique system was originally identified in the IL-6 receptor system, which consists of an 80 kDa IL-6 receptor (IL-6R) and a 130 kDa signal transducer (gp130).[4] Later, gp130 turned out to be used as a signal transducer also for several other multifunctional cytokines, such as LIF (leukemia inhibitory factor), OM (oncostatin M), CNTF (ciliary neurotrophic factor), and IL-11[1] (FIG. 1). This principle has been shown to be applied not only to IL-6-related cytokines but also to most other cytokine receptor systems. In the hemopoietic system, IL-3, IL-5, and granulocyte macrophage colony stimulating factor (GM-CSF) share a common β component as a signal transducer,[2] and in the lymphoid system, IL-2, IL-4, IL-7, IL-9, and IL-15 share the IL-2Rγ chain as a common γ component[3] (FIG. 2). These results could explain the functional redundancy of cytokines.

INITIAL STEP OF THE SIGNAL TRANSDUCTION

Interaction of IL-6 with the IL-6 receptor induces interaction and homodimerization of gp130. The dimer could be dissociated into monomers only by reduction, showing the formation of disulfide linkage between two monomers.[5] After stimulation

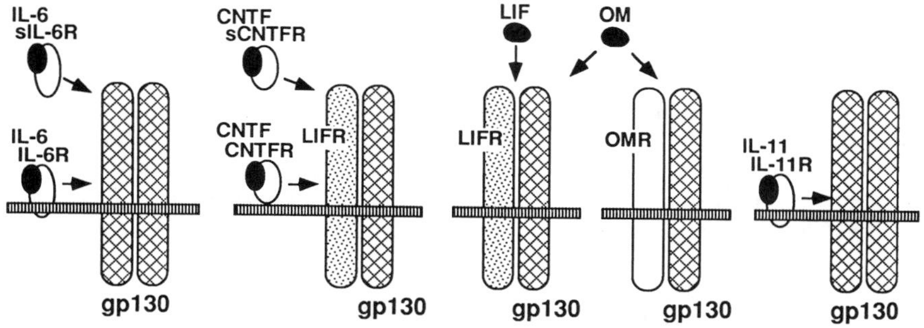

FIGURE 1. Schematic model of receptor complexes sharing gp130 as a critical signal-transducing receptor component. IL-6 stimulation induces homodimerization of gp130, whereas stimulation of LIF, OM, and CNTF induces heterodimerization of gp130 and LIFR. The IL-11 receptor complex is demonstrated to use gp130 as a signal transducer, but its precise composition needs to be clarified.

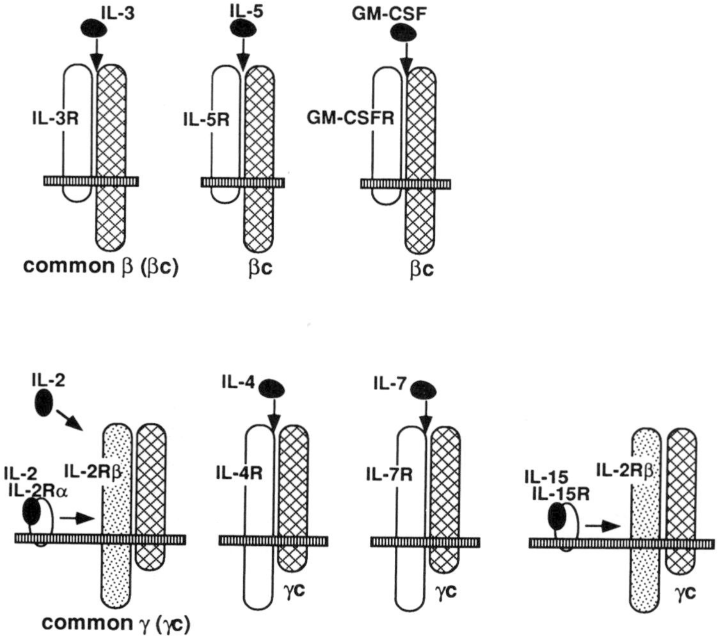

FIGURE 2. Depiction of receptor systems sharing a common β (βc) or a common γ (γc). In the latter receptor system, the IL-2Rβ chain is also shared in the IL-15 receptor complex where its α chain remains to be molecularly cloned.

and homodimerization, tyrosine-specific phosphorylation of gp130 was observed.[5–7] Because gp130 does not possess a tyrosine kinase in the cytoplasmic domain, the result indicates the interaction of gp130 with a cytoplasmic tyrosine kinase. Several cytokine receptors, such as erythropoietin and interferons, have been shown to interact with a family of JAK kinases.[8] The study with anti-JAK1, JAK2, and TYK2 antibodies confirmed the interaction and activation of JAK kinases with a homodimer of gp130.[9–11]

Most of the cytokine receptor family members possess two highly conserved amino acid stretches in their cytoplasmic regions, called box1 and box2. In box1, there exists a proline-X-proline motif[12] (FIG. 3). It was shown that the presence of box1 is necessary and sufficient for the activation of JAK kinases.[9,11] When the mutation was introduced into the proline-X-proline motif, the JAK kinase activation was abrogated.

MOLECULAR CLONING OF APRF/STAT3

IL-6 induces the acute-phase gene expression in liver cells. Thus, in order to study the down-stream substrate of JAK kinases, hepatocytes were employed. Ten minutes after *in vivo* injection of IL-6, nuclear extracts of murine hepatocytes showed the activity necessary to bind with the type II IL-6 response element in the gel shift assay.[13–15] This binding activity was completely blocked by antiphosphotyrosine antibody, indicating that a certain molecule was activated to bind with the promoter of the acute-phase genes after tyrosine-specific phosphorylation. The molecule was called APRF (acute-phase response factor) and purified to a single moiety with oligonucleotide-conjugated beads. According to partial amino acid sequences of the protein, the cDNA encoding APRF was cloned.[14] The nucleotide and deduced amino acid sequences revealed that this molecule conserved SH2 and SH3 domains and showed 55% overall homology with ISGF3p91 protein (STAT1; signal transducer and activator of transcription 1). Approximately at the same time, STAT3 was cloned by the homology with STAT1, and the result showed that STAT3 is identical with APRF.[16] Both APRF and STAT3 were shown to be tyrosine-specifically phosphorylated by IL-6 stimulation.

Not only IL-6, but also LIF, OM, and CNTF could activate APRF/STAT3 and could induce its tyrosine-specific phosphorylation in a hepatoma cell line, HepG2, as well as in a myeloid leukemia cell line, M1 (FIG. 4). The result showed that a signal through gp130 induces tyrosine-specific phosphorylation of APRF/STAT3, which is most likely a downstream substrate of JAK kinases in the gp130 signaling pathway.

IL-6 activated the APRF/STAT3 protein and to a much lesser extent the p91/STAT1 protein in a hepatoma cell line. By contrast, γ-interferon (γ-IFN) induced tyrosine-specific phosphorylation of p91/STAT1α but not APRF/STAT3. Thus, both gp130 and the γ-IFN receptor could activate the same JAK kinases but induce a different combination of the members of the STAT family.[14–17] This could be a clue for explaining the mechanism of how each cytokine could transduce its specific signal, leading to specific gene expression. The result also suggests the specific interaction of each receptor with its own set of the STAT family members. Coexpression of gp130, JAK kinase, and APRF/STAT3 in Cos cells confirmed the specific

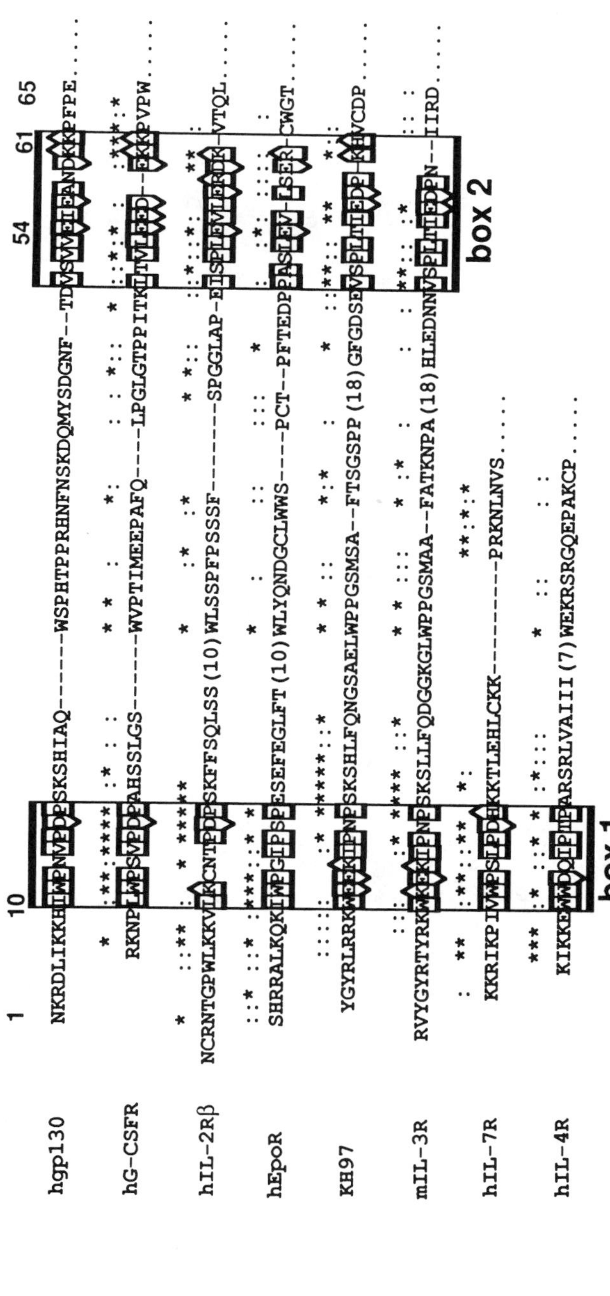

FIGURE 3. Highly conserved amino acid stretches, box 1 and box 2, in the cytoplasmic regions of the members of the cytokine receptor family. Membrane proximal cytoplasmic amino acid sequences of representative members of the cytokine receptor family are aligned. In box 1 and box 2, amino acids of the same nature (*i.e.*, hydrophobicity and charge) are indicated by the same symbols.

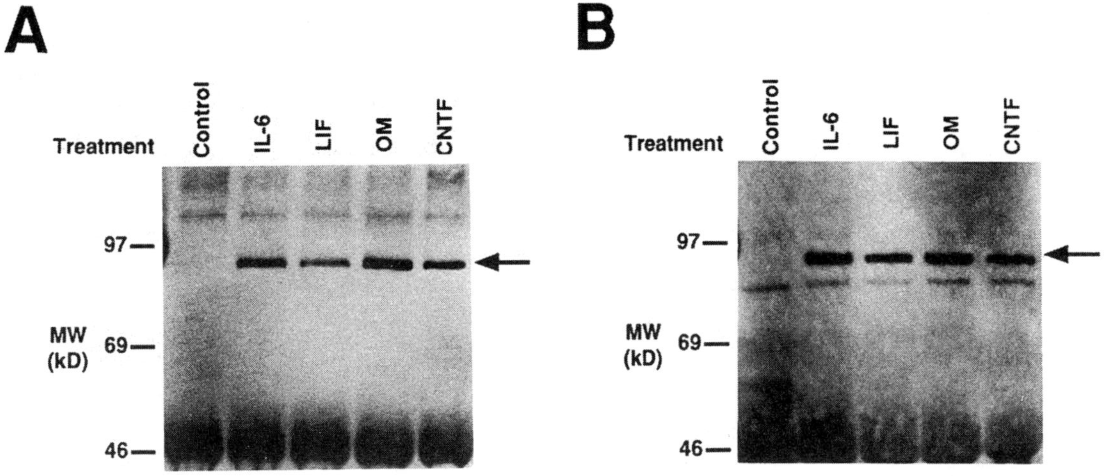

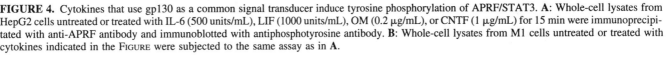

FIGURE 4. Cytokines that use gp130 as a common signal transducer induce tyrosine phosphorylation of APRF/STAT3. **A**: Whole-cell lysates from HepG2 cells untreated or treated with IL-6 (500 units/mL), LIF (1000 units/mL), OM (0.2 μg/mL), or CNTF (1 μg/mL) for 15 min were immunoprecipitated with anti-APRF antibody and immunoblotted with antiphosphotyrosine antibody. **B**: Whole-cell lysates from M1 cells untreated or treated with cytokines indicated in the FIGURE were subjected to the same assay as in **A**.

interaction of gp130 and APRF only when JAK kinase is present (M. Narazaki *et al.*, unpublished observation).

A TRANSCRIPTION FACTOR, NF-IL6

NF-IL6 was originally identified as a nuclear factor binding to a 14 bp palindromic sequence within an IL-1-responsive element in the human IL-6 gene.[18] The gene encoding NF-IL6 was cloned, and the result revealed that NF-IL6 is a transcription factor with a leucine zipper structure and highly homologous with C/EBP, the first nuclear factor proposed to contain the leucine zipper structure.

Expression of NF-IL6 and C/EBP is different, that is, C/EBP is constitutively expressed in hepatocytes and adipocytes, but NF-IL6 is not detectable in normal tissues and inducible by the stimulation with LPS (lipopolysaccharide), IL-1, TNF (tumor necrosis factor), or IL-6 in hepatocytes and monocytes.[19] NF-IL6 is a transcription factor not only for the expression of inflammatory cytokines but also for the expression of the IL-6-inducible genes, such as the acute-phase genes in hepatocytes. The type I IL-6 responsive element in the various acute-phase genes includes the recognition sequence of NF-IL6.

NF-IL6 is a phosphoprotein. Phosphoamino acid analysis showed that phosphorylation occurred on the serine and threonine residues. By transient expression of a series of site-directed mutants of NF-IL6 and subsequent phosphopeptide mapping, three phosphorylated residues were identified. Ser-231 and Thr-235, both located within the serine-rich domain adjoining the bZip domain, and Ser-325 located within the leucine zipper.[20] Closer inspection of NF-IL6 phosphorylation sites revealed the presence of the consensus sequence for MAP kinase in the region immediately surrounding Thr-235 (SSPPGTPSP). When vectors expressing NF-IL6 and/or oncogenic $p21^{ras}$ were cotransfected with an IL-6 promoter-luciferase gene reporter construct, simultaneous expression of both NF-IL6 and oncogenic $p21^{ras}$ resulted in a dramatic synergistic stimulation of the reporter gene. Two-dimensional phosphopeptide mapping showed that oncogenic $p21^{ras}$ expression dramatically augmented the phosphorylation on Thr-235 of NF-IL6. Furthermore, substitution of Ala for Thr-235 resulted in the loss of ras-dependent activation of NF-IL6. These results demonstrated that ras-dependent activation of NF-IL6 is mediated through phosphorylation of Thr-235 by MAP kinases.

As summarized in FIG. 5, there are two pathways in the signal transduction through a homodimer of gp130, that is, (1) activation of JAK kinase followed by tyrosine-specific phosphorylation of APRF/STAT3 and its translocation into nuclei, and (2) activation of unidentified tyrosine kinase followed by the activation of the Ras-MAP kinase cascade, whose ultimate substrate is NF-IL6.

PREPARATION OF NF-IL6$^{-/-}$ MICE BY GENE TARGETING

To know the specific regulatory role of NF-IL6 *in vivo*, NF-IL6$^{-/-}$ mice were generated by gene targeting in embryonic stem cells.[21] The NF-IL6 targeting vector was constructed by inserting a neomycin resistance gene into the bZip domain of

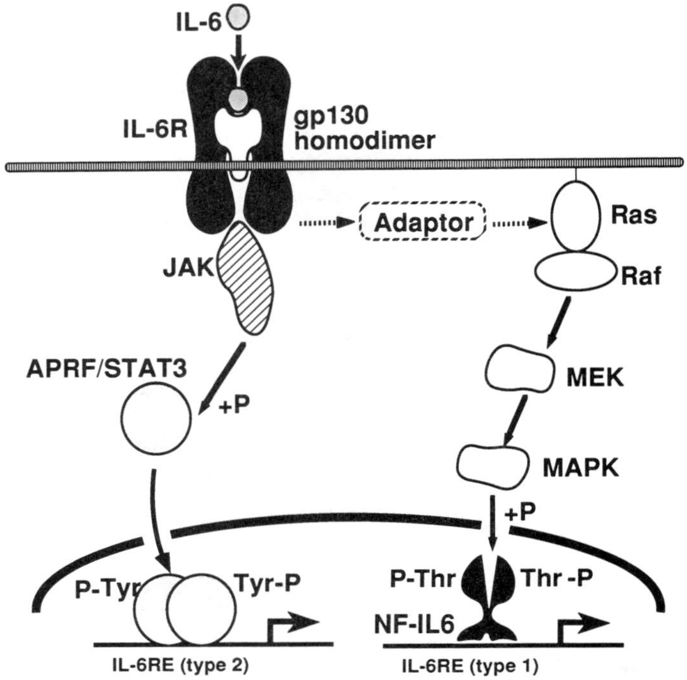

FIGURE 5. Schematic model of the two pathways from gp130 to the target gene expression. See text for the details.

the NF-IL6 gene. Two lines of mice carrying the mutation at the NF-IL6 locus have been generated from independently isolated ES cell lines. Homozygous mutant mice were obtained from heterozygous parents at a frequency lower than the expected Mendelian ratio, suggesting prenatal mortality. Born mice, however, appeared normal and presented no significant increase in mortality rate under specific pathogen-free (SPF) conditions. However, NF-IL6$^{-/-}$ mice presented a high susceptibility to *Listeria* infection. When NF-IL6$^{-/-}$ and control mice were infected intraperitoneally with 5 $\times$ 10^2 colony-forming units (CFU) (50% lethal dose of wild mice is 1 $\times$ 10^6 CFU) of *L. monocytogenes*, all NF-IL6$^{-/-}$ mice died within 5 days after challenge with *Listeria* (FIG. 6). Histopathological examination showed the multiple foci of microabscesses in the liver and spleen of NF-IL6$^{-/-}$ mice. Electromicroscopic observation revealed escape of a large number of the pathogens from phagosome to cytoplasm in activated macrophages from NF-IL6$^{-/-}$ mice. These results demonstrate an essential role of NF-IL6 in intracellular killing of *Listeria* by macrophages.

A currently established concept about the killing mechanism of intracellular bacteria by macrophages is as follows: γ-interferon and TNF produced by T cells or NK cells activate macrophages, and the activated macrophages produce nitric oxide (NO), which kills bacteria inside phagosomes.[22] Thus, we studied the production of γ-IFN and TNF after infection of *Listeria* and NO production in maximally

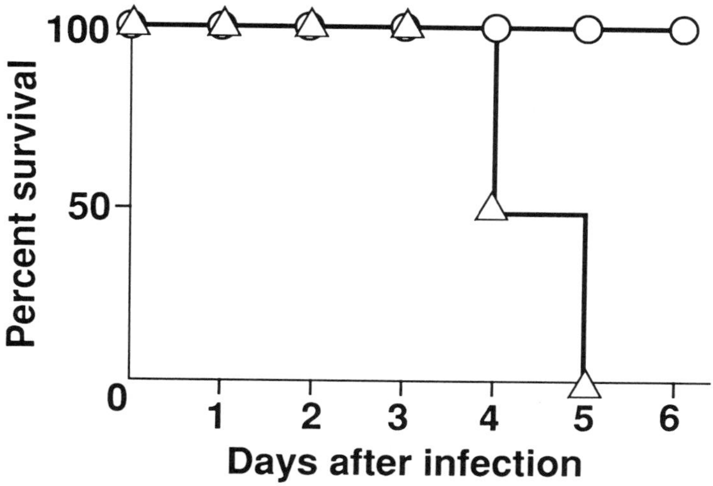

FIGURE 6. Survival of mice infected with *L. monocytogenes*. Wild-type and NF-IL6$^{-/-}$ mice were injected intraperitonealy with 5 × 10^2 CFU of *L. monocytogenes* (n = 6) (50% lethal dose for wild-type mice was 1 × 10^6 CFU). Percent survival of wild-type (O) and NF-IL6$^{-/-}$ mice (△) is shown.

activated macrophages. Expression of γ-IFN and TNF was comparable between wild-type and NF-IL6$^{-/-}$ mice. Moreover, NO production in activated macrophages from NF-IL6$^{-/-}$ mice was not impaired. These results clearly demonstrated that a NO-independent bactericidal mechanism is present and that a molecule(s) whose expression is controlled by NF-IL6 is indispensable for bactericidal activity of macrophages.[21]

NF-IL6 was originally identified as a transcription factor for inflammatory cytokines including IL-6, IL-8, IL-1, and TNF, and for the acute-phase genes. Interestingly, however, all these cytokine gene expressions were not impaired. Acute-phase gene expression was also intact in the knockout mice, suggesting the presence of the other transcription factor(s), which may redundantly control the expression of these genes. One of the candidates may be a transcription factor(s) belonging to a C/EBP family, such as NF-IL6β or Ig/EBP.

PREPARATION OF GP130$^{-/-}$ MICE

As described, gp130 is shared by the receptors of IL-6, LIF, OM, CNTF, and IL-11 as a common cytokine signal transducer. In order to know the *in vivo* function of gp130, an attempt was made to prepare gp130$^{-/-}$ mice by gene targeting. The gp130 targeting vector was constructed by inserting a neomycin resistance gene into the second exon of the gp130 gene at the 3′ proximity of the translational initiation codon. Three mutant ES clones were established, and chimeric mice and heterozygote mice were prepared. To obtain homozygous mutant mice, matings between heterozygous parents were carried out. However, no gp130$^{-/-}$ homozygous progenitors were obtained, indicating that the gp130 deficiency is lethal. Careful examination of embryos could detect the presence of gp130$^{-/-}$ fetuses in the early embryonic stages, but no newborn mice without gp130 expression were obtained, showing that gp130 is essential for the development of embryos.

Histological study revealed abnormal cardiac development in gp130 embryos, in which extremely thin ventricular walls were observed. The result suggested that signals through gp130 are essential for the development of cardiac muscle cells. Indeed, gp130 and APRF/STAT3 are abundantly expressed in cardiac muscle cells, although IL-6 receptor expression is very low or undetectable, suggesting a presence of a natural ligand for gp130 in cardiac muscle cells. Recently, cytokine, called cardiotrophin-1 (CT-1), which induces cardiac cell hypertrophy, has been isolated. CT-1 seems to belong to the IL-6/LIF/CNTF family and may be a natural ligand of gp130 in the heart.

The number of mononuclear cells in fetal livers of gp130$^{-/-}$ mice was much lower than that in wild mice, indicating impairment of homopoietic development in fetal liver. Indeed, a CFU-S assay with fetal liver cells revealed that no or very few colonies were detected in recipient spleens when fetal liver cells of gp130$^{-/-}$ mice were transferred. The result confirmed that signals through gp130 are essential for normal development of hemopoietic lineage cells.

CONCLUSION

Most of the cytokines and their receptors involved in immune regulations and hematopoiesis have been molecularly identified. The characteristic features of cytokines, their functional pleiotropy and redundancy, may now be explained on the basis of the molecular structures of cytokine receptors. Most cytokine receptors consist of multiple chains, that is, private ligand-binding receptors specific for each cytokine and public signal transducers common to several cytokines. Homo- or heterodimerization of receptor molecules by ligand binding triggers the association and activation of associated cytoplasmic tyrosine kinases of the JAK family, for example, JAK2 in the case of cytokines, including erythropoietin, granulocyte-colony stimulating factor, IL-3, granulocyte-macrophage colony stimulating factor, IL-6, and growth hormone. This is followed by tyrosine-specific phosphorylation and activation of the member of the STAT family, for example, STAT3/APRF for IL-6-related cytokines. Alternatively, or in addition, a transcription factor(s) is activated by the induction of the Ras-MAP kinase cascade. Therefore, a general outline of the process of signal transduction from cytokine receptors to the activation of their target genes has been figured out, and the identity of most of the important players in this process has been determined. Targeted disruption of the genes encoding the molecules involved in the IL-6 signaling pathway revealed the *in vivo* function of these molecules. NF-IL6 was shown to be the transcription factor essential for bactericidal activity of macrophages. Signals through gp130 were shown to be essential for cardiogenesis and hemopoiesis.

REFERENCES

1. Kishimoto, T., T. Taga & S. Akira. 1994. Cytokine signal transduction. Cell **76:** 253-262.

2. MIYAJIMA, A., T. KITAMURA, N. HARADA, T. YOKOTA & K.-I. ARAI. 1992. Cytokine receptors and signal transduction. Annu. Rev. Immunol. **10:** 295-331.
3. TANIGUCHI, T. & Y. MINAMI. 1993. The IL-2/IL-2 receptor system: a current overview. Cell **73:** 5-8.
4. TAGA, T., M. HIBI, Y. HIRATA, K. YAMASAKI, K. YASUKAWA, T. MATSUDA, T. HIRANO & T. KISHIMOTO. 1989. Interleukin-6 triggers the association of its receptor with a possible signal transducer, gp130. Cell **58:** 573-581.
5. MURAKAMI, M., M. HIBI, N. NAKAGAWA, T. NAKAGAWA, K. YASUKAWA, K. YAMANISHI, T. TAGA & T. KISHIMOTO. 1993. IL-6-induced homodimerization of gp130 and associated activation of a tyrosine kinase. Science **260:** 1808-1810.
6. TAGA, T., M. NARAZAKI, K. YASUKAWA, T. SAITO, D. MIKI, M. HAMAGUCHI, S. DAVIS, M. SHOYAB, G. D. YANCOPOULOS & T. KISHIMOTO. 1992. Functional inhibition of hematopoietic and neurotrophic cytokines by blocking the interleukin-6 signal transducer gp130. Proc. Natl. Acad. Sci. USA **89:** 10998-11001.
7. DAVIS, S., T. H. ALDRICH, N. STAHL, L. PAN, T. TAGA, T. KISHIMOTO, N. Y. IP & G. D. YANCOPOULOS. 1993. LIFRβ and gp130 as heterodimerizing signal transducers of the tripartite CNTF receptor. Science **260:** 1805-1808.
8. ZIEMIECKI, A., A. G. HARPUR & A. F. WILKS. 1994. JAK protein tyrosine kinases: their role in cytokine signalling. Trends Cell Biol. **4:** 207-212.
9. NARAZAKI, M., B. A. WITTHUHN, K. YOSHIDA, O. SILVENNOINEN, K. YASUKAWA, J. N. IHLE, T. KISHIMOTO & T. TAGA. 1994. Activation of JAK2 kinase mediated by the interleukin 6 signal transducer gp130. Proc. Natl. Acad. Sci. USA **91:** 2285-2289.
10. LÜTTICKEN, C., U. M. WEGENKA, J. YUAN, J. BUSCHMANN, C. SCHINDLER, A. ZIEMIECKI, A. G. HARPUR, A. F. WILKS, K. YASUKAWA, T. TAGA, T. KISHIMOTO, G. BARBIERI, S. PELLEGRINI, M. SENDTNER, P. C. HEINRICH & F. HORN. 1994. Association of transcription factor APRF and protein kinase Jak1 with the interleukin-6 signal transducer gp130. Science **263:** 89-92.
11. STAHL, N., T. G. BOULTON, T. FARRUGGELLA, N. Y. IP, S. DAVIS, B. A. WITTHUHN, F. W. QUELLE, O. SILVENNOINEN, G. BARBIERI, S. PELLEGRINI, J. N. IHLE & G. D. YANCOPOULOS. 1994. Association and activation of Jak-Tyk kinases by CNTF-LIF-OSM-IL-6 β receptor components. Science **263:** 92-95.
12. MURAKAMI, M., M. NARAZAKI, M. HIBI, H. YAWATA, K. YASUKAWA, M. HAMAGUCHI, T. TAGA & T. KISHIMOTO. 1991. Critical cytoplasmic region of the IL-6 signal transducer, gp130 is conserved in the cytokine receptor family. Proc. Natl. Acad. Sci. USA **88:** 11349-11353.
13. WEGENKA, U. M., J. BUSCHMANN, C. LÜTTICKEN, P. C. HEINRICH & F. HORN. 1993. Acute-phase response elements are rapidly activated by interleukin-6 at the posttranslation level. Mol. Cell. Biol. **13:** 276-288.
14. AKIRA, S., Y. NISHIO, M. INOUE, X-J. WANG, S. WEI, T. MATSUSAKA, K. YOSHIDA, T. SUDO, M. NARUTO & T. KISHIMOTO. 1994. Molecular cloning of APRF, a novel IFN-stimulated gene factor 3 p91-related transcription factor involved in the gp130-mediated signaling pathway. Cell **77:** 63-71.
15. WEGENKA, U. M., C. LÜTTICKEN, J. BUSCHMANN, J. YUAN, F. LOTTSPEICH, W. MÜLLER-ESTERL, C. SCHINDLER, E. ROEB, P. C. HEINRICH & F. HORN. 1994. The interleukin-6-activated acute-phase response factor is antigenically and functionally related to members of the signal transducer and activator of transcription (STAT) family. Mol. Cell. Biol. **14:** 3186-3196.
16. ZHON, Z., Z. WEN & J. E. DARNELL JR. 1994. Stat3: a STAT family member activated by tyrosine phosphorylation in response to epidermal growth factor and interleukin-6. Science **264:** 95-98.
17. FELDMAN, G. M., E. F. PETRICOIN III, M. DAVID, A. C. LARNER & D. S. FINBLOOM. 1994. Cytokines that associate with the signal transducer gp130 activate the interferon-induced transcription factor p91 by tyrosine phosphorylation. J. Biol. Chem. **269:** 10747-10752.

18. Akira, S., H. Isshiki, T. Sugita, O. Tanabe, S. Kinoshita, Y. Nishio, T. Nakajima, T. Hirano & T. Kishimoto. 1990. A nuclear factor for IL-6 expression (NF-IL6) is a member of a C/EBP family. EMBO J. **9:** 1897-1906.
19. Akira, S. & T. Kishimoto. 1992. IL-6 and NF-IL6 in acute-phase response and viral infection. Immunol. Rev. **127:** 25-50.
20. Nakajima, T., S. Kinoshita, T. Sasagawa, K. Sasaki, M. Naruto, T. Kishimoto & S. Akira. 1993. Phosphorylation at threonine-235 by a ras-dependent mitogen-activated protein kinase cascade is essential for transcription factor NF-IL6. Proc. Natl. Acad. Sci. USA **90:** 2207-2211.
21. Tanaka, T., S. Akira, K. Yoshida, M. Umemoto, Y. Yoneda, N. Shirafuji, H. Fujiwara, S. Suematsu, N. Yoshida & T. Kishimoto. 1994. Targeted disruption of the NF-IL6 gene discloses its essential role in bacteria killing and tumor cytotoxicity by macrophages. Cell. In press.
22. Nathan, C. F. & J. M. Hibbs. 1991. Role of nitric oxide synthesis in macrophage antimicrobial activity. Curr. Opinion Immunol. **3:** 65-70.

IL-2 Signaling Involves Recruitment and Activation of Multiple Protein Tyrosine Kinases by the IL-2 Receptor[a]

TADATSUGU TANIGUCHI,[b] TADAAKI MIYAZAKI,
YASUHIRO MINAMI, ATSUO KAWAHARA,
HODAKA FUJII,[b] YOKO NAKAGAWA,
MASANORI HATAKEYAMA,[c] AND ZHAO-JUN LIU[b]

Institute for Molecular and Cellular Biology
Osaka University
Yamadaoka 1-3, Suita-shi, Osaka 565, Japan

INTRODUCTION

The IL-2 system has been extensively studied in the context of clonal proliferation of T cells, and it has become a paradigm of how interleukins and other soluble mediators, collectively termed cytokines, function in the development and regulation of the immune system. In fact, the proliferation of matured, resting T cells is initiated by way of a process of signal transduction, wherein the specific interaction of the antigen/MHC molecule and the T-cell antigen receptor complex (TCR) triggers the induction of IL-2 and its homologous receptor (IL-2R). IL-2 is also known to affect proliferation, differentiation, and activation of other lymphoid cell types, which include (immature) thymocytes, B cells, lymphokine-activated killer (LAK) and natural killer (NK) cells. Even nonlymphoid cells, such as monocytes, oligodendroglial cells, and embryonic fibroblasts express IL-2R, and they also respond to IL-2, albeit in diverse ways. Collectively, these observations suggest that IL-2R is capable of transmitting signals in a variety of cell types but that the nature of the signals may not be the same in all cell types.[1-4]

Among growth factor receptors, the IL-2R offers a unique example; it is a multichain complex that consists of at least three distinct membrane components,

[a] Our work was supported by a Grant-in-Aid for Special Project Research, Cancer Bioscience, from the Ministry of Education, Science, and Culture of Japan.

[b] Present address: Department of Immunology, Faculty of Medicine, University of Tokyo, Hongo 7-3-1, Bunkyo-ku Tokyo 113, Japan.

[c] Present address: Department of Viral Oncology, Cancer Institute, Japanese Foundation for Cancer Research, Tokyo 170, Japan.

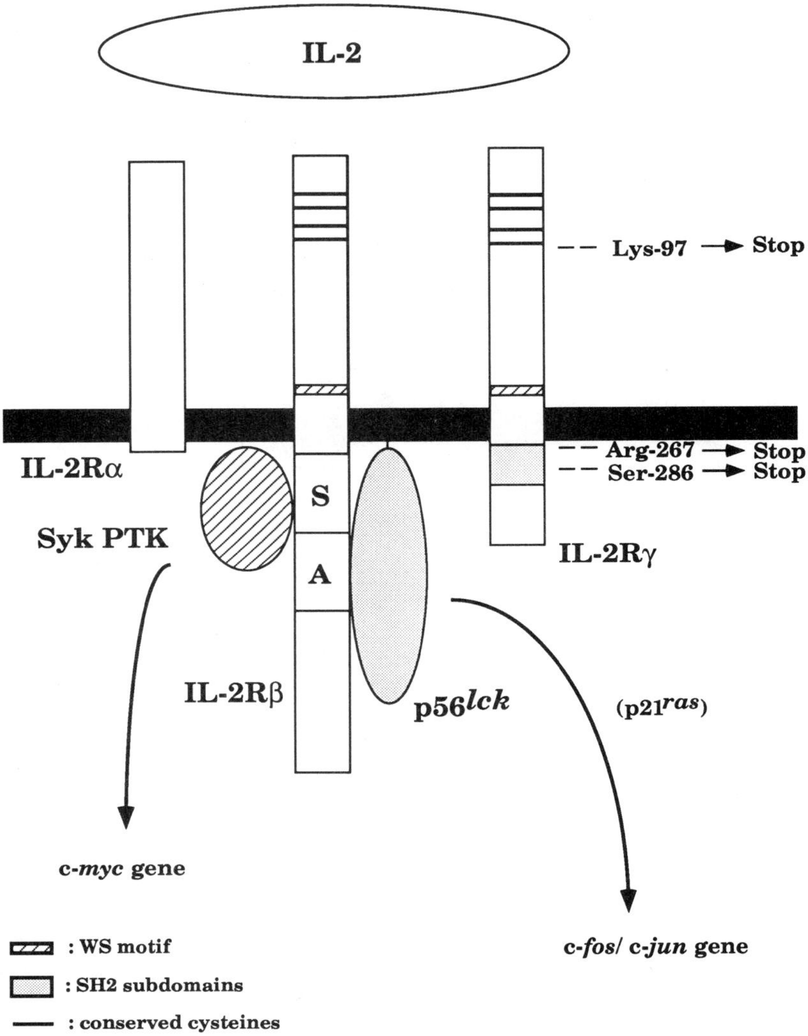

FIGURE 1. Diagram of the IL-2 receptor subunits and the interaction of $p56^{lck}$ and Syk PTK. S, serine-rich region of the IL-2Rβ chain; A, acidic region of the IL-2Rβ chain. The positions of non-sense point mutations within the IL-2Rγ gene of the XSCID patients are indicated.

IL-2R α chain (IL-2Rα), β chain (IL-2Rβ), and γ chains (IL-2Rγ) (FIG. 1). Previous studies have demonstrated that both IL-2Rβ and IL-2Rγ are required to transmit the IL-2 signal to the cell interior.[5–8] When the human IL-2Rβ cDNA was introduced into the murine IL-3-dependent pro-B cell line, BAF-B03, which constitutively expresses endogenous IL-2Rα and IL-2Rγ, these cells proliferated in response to IL-2. Further expression studies with deletion mutant IL-2Rβ cDNAs revealed that a restricted cytoplasmic region of IL-2Rβ, designated the "serine-rich" region (S region), is essential for the mitotic response to IL-2 (FIG. 1).[5] It has also been shown that the signal(s) elicited from the S region leads to the induction of the c-*myc* gene, whose expression is critical for the cell-cycle progression.[9] The role of the cytoplasmic region of IL-2Rγ in IL-2 signaling has also recently been demonstrated. In fact, ectopic expression of a mutant IL-2Rγ lacking most of its cytoplasmic region inhibits the ability of the intact IL-2R to transmit the IL-2 proliferative signal in BAF-B03 cells.[6] Furthermore, it was shown that ligand(s)- or antibody-induced heterodimerization of the cytoplasmic domains of IL-2Rβ and IL-2Rγ triggers cellular proliferation, indicating that functional cooperation between the cytoplasmic domains of IL-2Rβ and IL-2Rγ is critical to IL-2 signaling.[7,8] The human IL-2Rγ gene maps to the locus closely linked to X-linked severe combined immunodeficiency (XSCID), and nonsense mutations were found within the IL-2Rγ gene of XSCID patients (FIG. 1).[10]

It is particularly important that members of the new cytokine receptor family, which includes IL-2Rβ and IL-2Rγ, all lack the intrinsic protein tyrosine kinase (PTK) domain that is the hallmark of other growth factor receptors, such as EGF, PDGF, and CSF-1 receptors. Hence receptors of this family might use a mechanism(s) distinct from the "classical type" growth factor receptors in transmitting proliferative signals. Although lacking intrinsic kinase activity, the IL-2R couples ligand binding to induction of tyrosine phosphorylation of cellular substrates, including IL-2Rβ and IL-2Rγ. In this article we will summarize our results on the identification and function of PTKs, which are associated with and activated by the IL-2R complex.

RESULTS AND DISCUSSION

The p56^{lck} *and Other* src*-family PTKs*

As the first of the series of the functional intermolecular associations between the new family of cytokine receptors and nonreceptor type PTKs, we and others have demonstrated that the IL-2Rβ chain interacts with the nonreceptor type PTK, p56^{lck}, both physically and functionally. In fact, IL-2Rβ coprecipitates with p56^{lck} in an NK cell line, YT, and IL-2 stimulation of peripheral blood lymphocytes results in the p56^{lck} PTK activation. cDNA coexpression studies in COS cells and in BAF/B03 have revealed that the interaction of the two molecules occurs between a cytoplasmic region of IL-2Rβ, which is characterized for its abundance of acidic amino acids, termed "acidic region" (FIG. 1), and a region spanning the aminoterminal half of the p56^{lck} PTK domain.[11] In fact, this molecular interaction appears to be indispensable for the IL-2-induced p56^{lck} PTK activation.[12] Interestingly, in BAF/B0 cells, in which p56^{lck} is not expressed, another *src*-family PTK, p58^{fyn}, is activated by interacting with IL-2Rβ in an analogous manner, suggesting a certain redundancy in the IL-2-induced *src*-family PTK activation. In addition, the PTK activation appears to be linked to the p21^{ras} activation, and this IL-2Rβ-*src*-family PTK-p21^{ras} pathway seems to be linked to the activation of nuclear protooncogenes, *c-jun* and *c-fos*.[9,13] These findings suggest that other cytokine receptors may also use similar mechanisms. On the other hand, an IL-2Rβ mutant lacking the "acidic" region (therefore overt PTK activation is not observed) is still capable of delivering the proliferative signal(s) in BAF/B0 cells. Hence this raises the issue of whether the p56^{lck} PTK pathway is involved in any way in IL-2-induced cell proliferation.

Expression studies with mutant IL-2Rβ and epidermal growth factor receptor (EGFR) cDNAs in BAF-B03 cells revealed the presence of another IL-2R-linked pathway leading to the induction of the nuclear proto-oncogene c-*myc* (see below). This pathway is activated by the wild-type and the A-mutant but not by the S-mutant IL-2Rβ chains. EGFR stimulation results in the activation of the c-*fos*/c-*jun* pathway but not the c-*myc* pathway, and the c-*myc* gene must be ectopically expressed in order for the EGF-stimulated cells to progress beyond the S-phase of the cell cycle.[9] Hence c-*myc* is a target of the IL-2R mediated signal, which is required for the progression of the hematopoietic cell cycle. Constitutive expression of c-*myc*, however, does not cause IL-2 (or IL-3)-independent growth of BAF-B03 cells.

More recently, we observed that an activated form of p56^{lck}, carrying a phenylalanine substitution at position 505 (p56*lck* F505), can cooperate with either c-*myc* or *bcl*-2 in the induction of BAF cell proliferation, providing strong evidence supporting a role for p56^{lck} in transmitting mitogenic signals (T. Miyazaki, unpublished data).

The Jak Family PTKs

Recently, it has been shown that IL-2 and IL-4 induce the tyrosine phosphorylation and activation of the *Janus* kinases, Jak1 and Jak3, which suggests that, in addition to the *src*-family PTKs, these kinases might also associate with the IL-2R and participate in IL-2 signaling.[14,15]

To determine if Jaks associate with the IL-2R subunits, CD4 chimeric receptors were used that could be detected with the same monoclonal antibody to human CD4 (anti-CD4). Extracts of COS cells cotransfected with each chimeric receptor cDNA and one of the Jak cDNAs were immunoprecipitated with anti-CD4 and subjected either to immunoblot analysis with antisera to the respective Jaks or to an *in vitro* kinase assay. Both assays showed that Jak1 could be coimmunoprecipitated with CD4β, but not with CD4γ or CD4γM1, whereas Jak3 could be coprecipitated with CD4γ, but not with CD4β or CD4γM1. By contrast, the association of Jak2 with any of the CD4-IL-2R chimeras was marginal, if at all. Thus, the cytoplasmic domains of IL-2Rβ and IL-2Rγ possess regions capable of selectively associating with Jak1 and Jak3, and additional lymphoid-specific proteins are not required.[16]

We next examined the association sites within the respective cytoplasmic domains of the IL-2R subunits with the respective Jaks, by using the similar assay system. It was found that the cytoplasmic S region of IL-2Rβ is required for its association with Jak1, whereas the COOH-terminal 48 amino acids of IL-2Rγ are necessary for its association with Jak3 (FIG. 2). Importantly, these two regions are known to participate in IL-2-induced cell proliferation. It is also worth pointing out that the COOH-terminal region of IL-2Rγ is deleted by non-sense mutations in many XSCID patients[10] (see FIG. 1). The utilization of IL-2Rγ by a number of receptors explains the activation of Jak3 by IL-4, IL-7, and IL-9 and predicts its activation in response to IL-13 or IL-15.

We also investigated the importance of these associations in the activation of Jaks by IL-2. We examined whether Jaks were activated after IL-2 stimulation in a

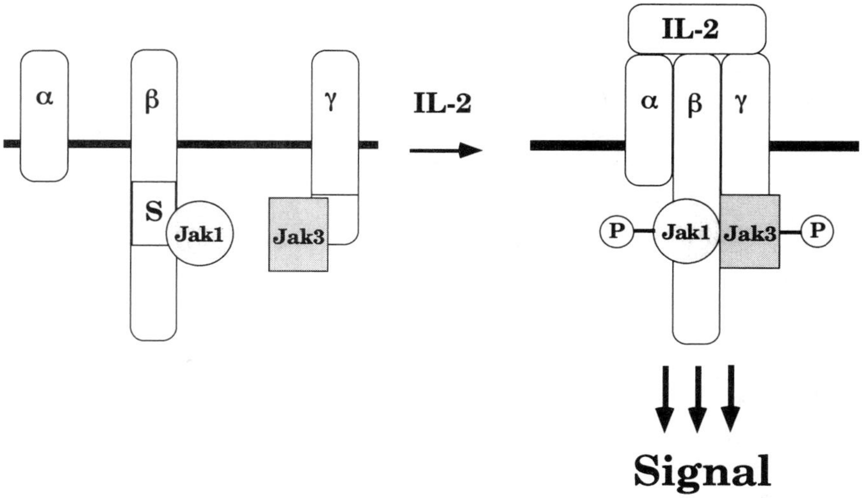

FIGURE 2. Selective association and activation of Jak1 and Jak3 with the IL-2Rα and β chains. For the details, see text and ref. 16.

BAF-B03-derived cell line, FWT-2, which expresses both human IL-2Rβ and human IL-2Rγ. Stimulation of FWT-2 cells by IL-2 induced tyrosine phosphorylation of endogenous Jak1 and Jak3, but not of Jak2. Furthermore, an *in vitro* kinase assay also revealed that Jak1 and Jak3, but not Jak2, were activated after IL-2 stimulation. Thus, IL-2 induced the selective activation of Jak1 and Jak3, and this activation correlated with the ability of the Jaks to associate with IL-2R.[16]

In order to test the role of Jak3 in an IL-2 response, the Jak3 cDNA was stably expressed in the NIH 3T3 fibloblast-derived cell line, 3T3αβγ. This cell line expresses the high-affinity IL-2R complex by the stable introduction and expression of cDNAs for the three IL-2R subunits, but does not respond to IL-2. The 3T3αβγ cells express Jak1 and Jak2, but not Jak3. When the Jak3 cDNA was stably expressed in this cell line, it was found that Jak3-expressing clones all responded to IL-2 by thymidine incorporation at levels that were comparable to serum stimulation.[16]

IL-2 stimulation induces the association of the cytoplasmic domains of IL-2Rβ and IL-2Rγ, an event that is required for biological responses. Our results suggest that this may be necessary to bring Jak1 and Jak3 into close proximity and promote cross-activation (FIG. 2). Our reconstituted experiments with fibroblasts for the first time provide functional evidence for the involvement of Jaks in the growth signal transmission by cytokines.

The Syk Family PTKs

Among the PTKs expressed in lymphocytes, the Syk family PTks have been extensively studied in the context of antigen-induced lymphocyte activation.[17,18] In view of the involvement of both the Syk family and *src*-family PTKs in antigen-induced lymphocyte signaling, we examined the possibility of whether the Syk PTK is also involved in IL-2 signaling along with *src*-family PTKs. In fact, initial observations with porcine PBLs suggested the potential involvement of Syk PTK in the IL-2 system.[19] Unlike the *src*-family PTKs, the Syk family PTKs, Syk, and ZAP-70 PTKs contain two SH2 domains and no N-terminal myristylation site. Syk and ZAP-70 PTKs have been shown to be coupled with several cell-surface receptors, including the T and B cell antigen receptors, members of the immunoglobulin (Ig)G and IgE Fc receptor families, and the granulocyte colony-stimulating factor (G-CSF) receptor. The Syk and ZAP-70 PTKs have distinct tissue distributions: Syk PTK is expressed in T cells, B cells, myeloid cells, polymorphonuclear neutrophils, and platelets, whereas ZAP-70 PTK expression is restricted to T and natural killer cells. Inasmuch as IL-2 can act on a variety of hematopoietic cells, an interesting possibility is that Syk PTK may be involved in IL-2 signaling.[17,18,20]

We examined whether or not Syk PTK is physically or functionally coupled with IL-2R. Transient coexpression of Syk PTK and various CD4/IL-2R chain chimeras in COS cells revealed that Syk PTK associates constitutively with IL-2Rβ by way of the cytoplasmic S-region of IL-2Rβ, the region required for both IL-2 induced expression of c-*myc* gene and cellular proliferation. Further confirmation that the S region of IL-2Rβ is required for the association with Syk PTK was provided by analyzing the association of Syk PTK with IL-2Rβ in the BAF-B03-derived cell lines, each expressing the IL-2Rβ mutant cDNAs. It is noteworthy that the cytoplasmic S region of IL-2Rβ does not contain the sequence Tyr-X-X-Leu-X(6-8)-Tyr-X-X-Leu (variously called the antigen-recognition activation motif (ARAM), the tyrosine-based activation motif (TAM), or the antigen receptor homology 1 (ARH1) motif), a potential recognition site for Syk/ZAP-70 family PTKs.[17,18,21,22] The critical role of ARAM (TAM or ARH1) in the association of receptors with ZAP-70 PTK has been well established. The association of tyrosine-phosphorylated ARAM (TAM or ARH1) in CD3 or the ζ subunits of TCR with the SH2 domains of ZAP-70 PTK recruits ZAP-70 PTK to the activated TCR and results in the tyrosine phosphorylation of ZAP-70 PTK. Thus, our results may represent a novel example of an intermolecular association between an Syk/ZAP-70 family PTK and a cell-surface receptor(s). We have found that Syk PTK physically associates with IL-2Rβ and is activated following IL-2 stimulation.

In addition, antibody-mediated clustering of Syk PTK in BAF-B03 cells resulted in the induction of the c-*myc* but not the c-*fos* gene. Collectively, this structure-function study of IL-2Rβ in BAF-B03 cells supports the idea that Syk PTK plays a role in mediating IL-2-induced expression of the c-*myc* gene and subsequent cellular proliferation (Y. Minami, manuscript in preparation).

CONCLUSIONS AND PERSPECTIVES

The IL-2/IL-2R system has been extensively studied in the context of the regulation of the immune response and has also become a paradigm of how cytokines transmit proliferative signals in hematopoietic cells. Previous studies have demonstrated the importance of the cooperation between IL-2Rβ and IL-2Rγ cytoplasmic domains in

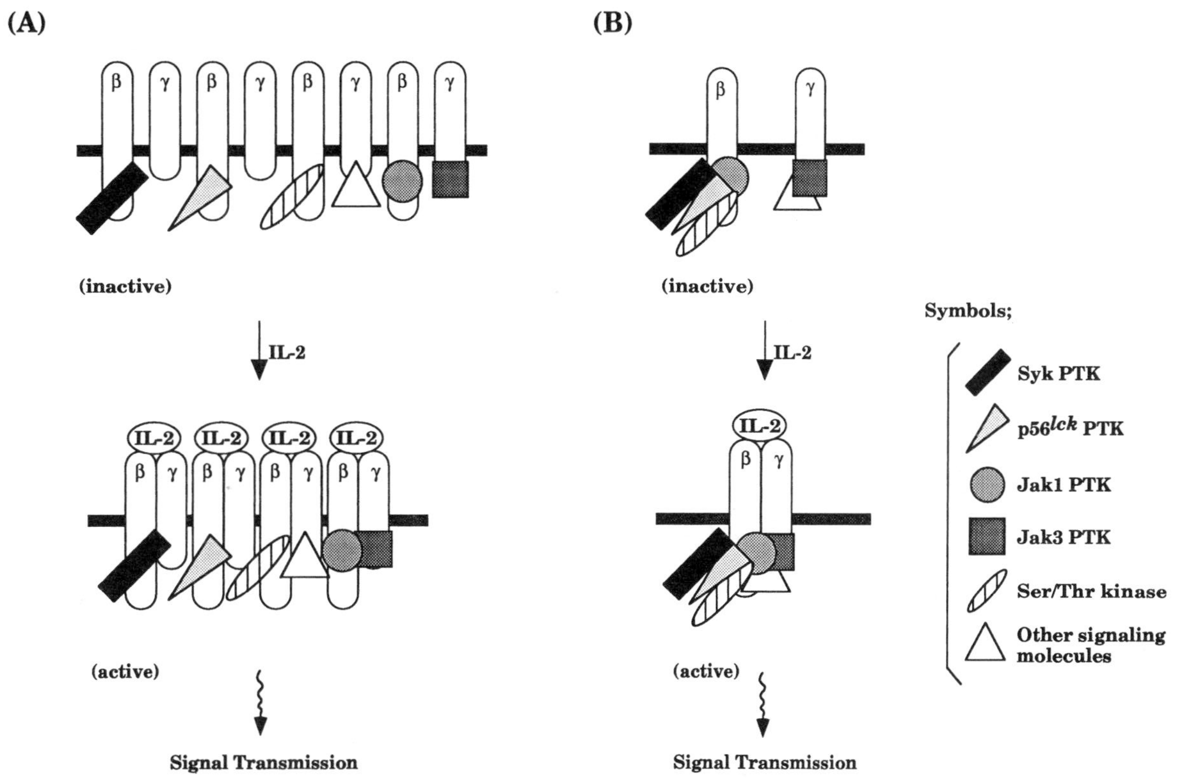

FIGURE 3. How multiple protein kinases function in concert. Left panel (**A**): each IL-2Rβ (and possibly IL-2Rγ) recruits distinct kinase molecules. Upon IL-2 stimulation, these molecules function in concert. Right panel (**B**): the IL-2Rβ chain can simultaneously recruit multiple kinases that are activated upon IL-2 stimulation. Whether or not IL-2 binding to IL-2R triggers clustering has not been experimentally demonstrated. For convenience, IL-2Rα has been omitted, and the exact association sites for the respective kinases have been ignored. Other signaling molecules that have been reported to possibly couple with IL-2R are not included, *i.e.*, phosphatidylinositol 3-kinase, Raf-1 kinase, and glycosylphosphatidylinositol (GPI)-specific phospholipase (reviewed in ref. 4).

generating proliferative signals. The series of our studies have revealed the interaction of multiple nonreceptor PTKs with the IL-2R complex; that is, p56lck, Syk PTK, and Jak1/Jak3. Furthermore, it was shown that a serine/threonine kinase, distinct from the previously identified serine/threonine kinases, also associates with IL-2Rβ (Y. Minami and T. Taniguchi, unpublished observation).

These previous and present findings raise the interesting issue of how a given cytokine receptor may function to activate many distinct signaling molecules. At present it is not clear whether there are many distinct subpopulations of IL-2Rβ (or IL-2Rγ), each of which associates with the respective signaling molecules or if the subpopulation of cellular IL-2Rβ (or IL-2Rγ) that associates with one signaling molecule can simultaneously associate with other signaling molecules (FIG. 3). In either case, our results collectively suggest that the concerted action of signaling molecules may be required for full activation of downstream signaling pathways. This phenomenon may be common among cytokine receptors, which otherwise lack any intrinsic PTK motifs. Instead, the cytoplasmic domains of cytokine receptors have structurally evolved to recruit multiple signaling molecules, including PTKs, and have used them to trigger the full-scale activation of multiple downstream signaling pathways leading to cellular proliferation and/or differentiation. It will be important to study how these kinases cooperate with each other and to identify molecules involved in the downstream signaling cascades.

SUMMARY

The IL-2 receptor (IL-2R) consists of three subunits, the IL-2Rα, IL-2Rβ, and IL-2Rγ chains, the last of which is also used in the receptors for IL-4, IL-7, IL-9, IL-13, and IL-15. The IL-2-induced proliferative signals emanate from the cytoplasmic domains of IL-2Rβ and IL-2Rγ, but the nature and function of the signaling molecules that transmit these signals are not fully understood. Here we summarize our current understanding of the mechanisms by which IL-2R transmit signals by using multiple protein kinases. In fact, at least four protein tyrosine kinases (PTKs) are physically associated with IL-2R: $p56^{lck}$ (and its members), Syk PTK, and the *Janus* kinases, Jak1 and Jak3. cDNA expression studies revealed that the activation of these PTKs is critical for IL-2-induced proliferative signal transmission. Our findings indicate that a unique property of the IL-2R cytoplasmic domains is to recruit a variety of signaling molecules, which may suggest a mechanism by which these PTKs and other signaling molecules function in concert.

ACKNOWLEDGMENTS

The results summarized in this article have been obtained in collaboration with a number of groups, which include those of Drs. R. M. Perlmutter (University of Washington), J. N. Ihle (St. Jude Children's Research Hospital), and H. Yamamura (Kobe University).

REFERENCES

1. SMITH, K. A. 1988. Interleukin-2: Inception, impact, and implications. Science **240:** 1169-1176.
2. WALDMANN, T. A. 1989. The multi-subunit interleukin receptor. Annu. Rev. Biochem. **58:** 875-911.
3. TANIGUCHI, T. & Y. MINAMI. 1993. The IL-2/IL-2 receptor system: A current overview. Cell **73:** 5-8.
4. MINAMI, Y., T. KONO, T. MIYAZAKI & T. TANIGUCHI. 1993. The IL-2 receptor complex: Its structure, function, and target genes. Annu. Rev. Immunol. **11:** 245-267.
5. HATAKEYAMA, M., H. MORI, T. DOI & T. TANIGUCHI. 1989. A restricted cytoplasmic region of IL-2 receptor β chain is essential for growth signal transduction but not for ligand binding and internalization. Cell **59:** 837-845.
6. KAWAHARA, A., Y. MINAMI & T. TANIGUCHI. 1994. Evidence for the critical role of the cytoplasmic region of the IL-2 receptor γ chain in IL-2, IL-4 and IL-7 signalling. Mol. Cell. Biol. **14:** 5433-5440.
7. NAKAMURA, Y., S. M. RUSSELL, S. A. MESS, M. FRIEDMANN, M. ERDOS, C. FRANCOIS, Y. JACQUES, S. ADELSTEIN & W. J. LEONARD. 1994. Heterodimerization of the IL-2 receptor β- and γ-chain cytoplasmic domains is required for signalling. Nature **369:** 330-333.
8. NELSON, B. H., J. D. LORD, & P. D. GREENBERG. 1994. Cytoplasmic domains of the interleukin-2 receptor β and γ chains mediate the signal for T-cell proliferation. Nature **369:** 333-336.
9. SHIBUYA, H., M. YONEYAMA, J. NINOMIYA-TSUJI, K. MATSUMOTO & T. TANIGUCHI. 1992. IL-2 and EGF receptors stimulate the hematopoietic cell cycle via different signaling pathways: Demonstration of a novel role for c-*myc*. Cell **70:** 57-67.
10. NOGUCHI, M., H. YI, H. M. ROSENBLATT, A. H. FILIPOVICH, S. ADELSTEIN, W. S. MODI, O. W. MCBRIDE & W. J. LEONARD. 1993. Interleukin-2 receptor γ chain mutation results in X-linked severe combined immunodeficiency in humans. Cell **73:** 147-157.
11. HATAKEYAMA, M., T. KONO, N. KOBAYASHI, A. KAWAHARA, S. D. LEVIN, R. M. PERLMUTTER & T. TANIGUCHI. 1991. Interaction of the IL-2 receptor with the *src*-family kinase p56*lck*: Identification of novel intermolecular association. Science **252:** 1523-1528.
12. MINAMI, Y., T. KONO, K. YAMADA, N. KOBAYASHI, A. KAWAHARA, R. M. PERLMUTTER & T. TANIGUCHI. 1993. Association of p56*lck* with IL-2 receptor β chain is critical for the IL-2-induced activation of p56*lck*. EMBO J. **12:** 759-768.
13. SATOH, T., Y. MINAMI, T. KONO, K. YAMADA, A. KAWAHARA, T. TANIGUCHI & Y. KAZIRO. 1992. Interleukin 2-induced activation of ras requires two domains of interleukin 2 receptor β subunit, the essential region for growth stimulation and *lck*-binding domain. J. Biol. Chem. **267:** 25423-25427.
14. JOHNSTON, J. A., M. KAWAMURA, R. A. KIRKEN, Y-Q. CHEN, T. B. BLAKE, K. SHIBUYA, J. R. ORTALDO, D. W. MCVICAR & J. J. O'SHEA. 1994. Phosphorylation and activation of the Jak-3 Janus kinase in response to interleukin-2. Nature **370:** 151-153.
15. WITTHUHN, B. A., O. SILVENNOINEN, O. MIURA, K. S. LAI, C. CWIK, E. T. LIU & J. N. IHLE. 1994. Involvement of the Jak-3 Janus kinase in signalling by interleukins 2 and 4 in lymphoid and myeloid cells. Nature **370:** 153-157.
16. MIYAZAKI, T., A. KAWAHARA, H. FUJII, Y. NAKAGAWA, Y. MINAMI, Z-J. LIU, I. OISHI, O. SILVENNOINEN, B. A. WITTHUHN, J. N. IHLE & T. TANIGUCHI. 1994. IL-2 receptor subunits selectively associate with and functionally activate Jak1 and Jak3. Science In press.
17. SAMELSON, L. E. & R. D. KLAUSNER. 1992. Tyrosine kinases and tyrosine-based activation motifs: Current research on activation via the T cell antigen receptor. J. Biol. Chem. **267:** 24913-24916.
18. WEISS, A. & D. R. LITTMAN. 1994. Signal transduction by lymphocyte antigen receptors. Cell **76:** 263-274.

19. QIN, S., T. INAZU, C. YANG, K. SADA, T. TANIGUCHI & H. YAMAMURA. 1994. Interleukin 2 mediates p72*syk* activation in peripheral blood lymphocytes. FEBS Lett. **345:** 233-236.
20. ASAHI, M., T. TANIGUCHI, E. HASHIMOTO, T. INAZU, H. MAEDA & H. YAMAMURA. 1993. Activation of protein-tyrosine kinase p72*syk* with concanavalin A in polymorphonuclear neutrophils. J. Biol. Chem. **268:** 23334-23338.
21. CAMBIER, J. C. 1992. Signal transduction by T-and B-cell antigen receptors: Converging structures and concepts. Curr. Opinion Immunol. **4:** 257-264.
22. CHAN, A. C., D. M. DESAI & A. WEISS. 1994. The role of protein tyrosine kinases and protein tyrosine phosphatases in T cell antigen receptor signal transduction. Annu. Rev. Immunol. **12:** 555-592.

Costimulation Requirement for AP-1 and NF-κB Transcription Factor Activation in T Cells

STEFFEN JUNG, AVRAHAM YARON,
IRIT ALKALAY, ADA HATZUBAI,
AYELET AVRAHAM, AND YINON BEN-NERIAH

Lautenberg Center for General and Tumor Immunology
The Hebrew University
Hadassah Medical School
POB 12272
Jerusalem 91120 Israel

INTRODUCTION

Activation of resting T cells, resulting in proliferation and cytokine secretion, requires two distinct signals from antigen-presenting cells (APC).[1] One signal is generated by the engagement of the clonotypic T-cell receptor (TCR) by antigenic peptides in the context of MHC molecules and is transduced by the CD3 complex. The second, the costimulatory signal, arises from the interaction of an invariant APC surface protein, a member of the B7 family (B7-1, B7-2) with members of the CD28 receptor family (CD28, CTLA4) present on the T cell.[2] Delivery of a single stimulus (TCR) not only fails to activate the cells but drives them into a state of persistent unresponsiveness (anergy).[3] Whereas the elements involved in the transduction of the TCR signal have been defined,[4] the biochemical character of the CD28 signal remains unknown. Excessive engagement of either TCR or CD28 alone does not lead to activation of resting T cells. Also, as opposed to the TCR signal, the CD28 signal is at least partially Ca^{2+}-independent and cyclosporin A (CsA)-resistant.[5] Therefore, the costimulatory signal is believed to be fundamentally different from the TCR one. Independently operating signals could account for the broad, continuous range of IL-2 production seen in cytotoxic and humoral immune responses.[6] Yet the question remains: Where and how are the two signals integrated so as to result in IL-2 gene activation? In view of the modular nature of the IL-2 promoter/enhancer, some previous models favored signal integration at the level of the IL-2 promoter, by way of the joint action of individually activated transcription factors.[7] Inasmuch as full activation of each IL-2 transcription factor has been shown to require costimulation, it would appear that signal integration occurs at some common step, upstream of the IL-2 promoter.

With regard to composition, the transactivating complexes acting on the IL-2 promoter can be divided into two groups: complexes either representing or containing members of the NF-κB/Rel family (NF-κB, CD28RC)[8–10] and complexes related to or associated with AP-1 (AP-1, NF-AT, NF-IL2).[11,12] In-depth knowledge of the regulation of NF-κB and AP-1 is thus likely to provide insights into T-cell activation, particularly with respect to signal integration.

MATERIAL AND METHODS

Cell Culture and Activating Agents

Jurkat cells were cultured in RPMI 1640, supplemented with 10% fetal calf serum. Activation was carried out by a 10-minute cell incubation in the presence of phorbol myristate acetate (PMA) and A23187 Ca^{2+} ionophore at concentrations of 5 to 10 ng/mL and 1 μM, respectively. Anti-CD3 (OKT3) and anti-CD28 (9.3) antibodies were used at concentrations of 5 and 3 μg/mL, respectively. Tosyl-phe-chloromethylketone (TPCK) was added to the cultures at a concentration of 25 μM for 30 min, followed by the addition of PMA in 0.2% DMSO or DMSO alone for an additional 10 minutes.

Transfection Experiments

Jurkat T cells were transfected with 1 μg of luciferase constructs driven by the IL-2 promoter (IL-2-Luc) or the HIV long terminal repeat (LTR) promoter (HIV-LTR-Luc). Using the Biorad Genepulser (250 V, 960 μFd), transfection efficiency was about 15%, as determined by flow cytometry after cotransfection of a surface marker. One day posttransfection, cells were subjected to the various stimuli for 10 minutes. Six hours later, the cells were lysed, and luciferase activity was determined with the Promega luciferase assay system. For IκB cotransfection, 2 μg pCMV-MAD3 was added to each assay.

Protein Kinase Assays

JNK Assay

The solid-phase JNK assay was essentially performed as described.[13] In brief, cell extracts were incubated for 1 h at 4°C with glutathione S-transferase (GST)-c-Jun fusion proteins immobilized on glutathione-SH agarose beads, followed by extensive washing. Bound JNK was detected by the addition of [γ-^{32}P]ATP. After a 10-min incubation at 30°C, the reaction was terminated, the products were resolved by SDS-PAGE, and the incorporation of [^{32}P]phosphate into the GST-c-Jun protein was visualized by autoradiography.

MAP Kinase Assay

MAP kinase assays of cytoplasmic cell extracts were performed with polyclonal rabbit antiserum as described.[14] The kinases were immunoprecipitated with protein-A-sepharose beads (Pharmacia) and washed. The immune complexes were incubated

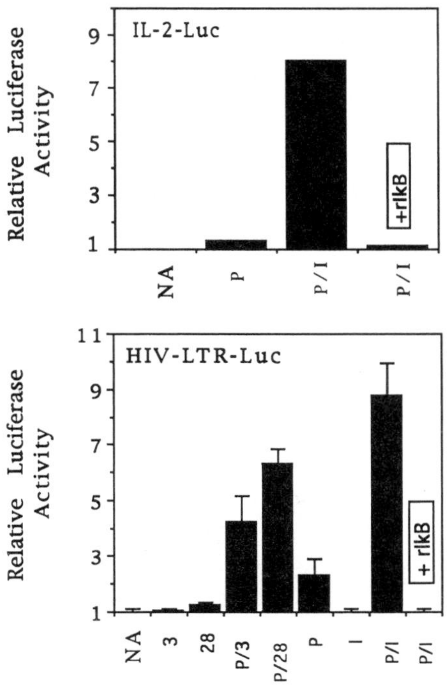

FIGURE 1. Effects of various stimulating agents on Jurkat T cells transiently transfected with IL-2 and HIV-LTR promoter-driven luciferase reporter genes (IL-2-Luc, HIV-LTR-Luc). The histograms depict relative luciferase activity as determined by a luminometer. NA, nonactivated; 3, anti-CD3; 28, anti-CD28; P, PMA; I, ionophore; rIκB, exogenous, recombinant IκB.

for 20 min with myelin basic protein (MBP), a MAP kinase substrate, in buffer containing [γ-^{32}P]ATP. Following SDS-PAGE, labeled MBP was autoradiographically visualized.

RESULTS AND DISCUSSION

Costimulation-dependent Phosphorylation of IκB

The costimulatory signal delivered by CD28, in conjunction with the TCR/CD3 stimulus, reportedly activates IL-2 promoter-driven reporter genes and the native IL-2 gene in cell lines and primary T cells, respectively.[9–11] Activation of the IL-2 promoter has been shown to be associated with stimulation of members of the NF-κB/Rel transcription factor family. In order to further characterize the effect of the costimulatory signal on NF-κB-dependent promoter activity, we carried out transfection studies with the Jurkat T-cell line. Luciferase-based reporter constructs controlled by the IL-2 promoter or the HIV-LTR promoter were used for this purpose. The HIV-LTR promoter contains two κB sites and is dependent on NF-κB for transcriptional activity.[15] Costimulation, as mimicked by suboptimal PMA concentrations in combination with the engagement of one receptor or Ca^{2+} ionophore,[5,7] leads to a three- to tenfold increase in luciferase activity (FIG. 1), as compared to that following a single stimulus.

NF-κB activation is achieved by the dissociation of the NF-κB/Rel factor from an inhibitory protein, IκB, which retains the latent transcription factor in the cytoplasm.[16] As overexpression of IκB has been shown to block NF-κB activation,[17] we tested the effect of exogenous IκB on the costimulation of the two reporter genes. Cotransfection of the IκB expression vector with the reporter constructs abolished the costimulation-dependent activation of both the IL-2 and the HIV-LTR promoters (FIG. 1), proving the essential role of NF-κB in the induction process.

Dissociation of NF-κB from the inhibitor is thought to depend on the proteolytic degradation of IκB.[18,19] Yet, the initial event observed in T-cell activation is a costimulation-dependent phosphorylation of the inhibitor.[20] Although the need for IκB modification prior to its proteolytic inactivation has not been established, recent studies with protease and kinase inhibitors suggest that the two events are linked.[20] Integration of the two stimuli provided by TCR/CDR and CD28 engagement thus seems to result in the activation of an uncharacterized IκB kinase, which through modification and inactivation of the inhibitor, leads to NF-κB activation.

Costimulation-dependent Activation of Jun N-terminal Kinase, JNK

Aside from NF-κB, the IL-2 promoter is also largely controlled by AP-1 and the related transcription factors, NF-AT and NF-IL2, which contain AP-1 components.[11,12] Rapid responses in eukaryotic cells, such as the activation of AP-1 and NF-κB in T cells, do not require *de novo* protein synthesis, suggesting the involvement of posttranslational modifications. Phosphorylation of c-Jun (one of the two AP-1 components) leads to a pronounced increase in its transactivating activity.[13] The kinase responsible for the phosphorylation of the N-terminal transactivation domain of c-Jun (JNK) was recently identified.[21] It is a member of the family of mitogen-activated protein (MAP) kinases.[21] Given the intimate involvement of AP-1 in T-cell activation, the characterization of JNK prompted experiments addressing the role of JNK activation in costimulation of T lymphocytes.[14] Whereas other members of the MAP kinase group, ERK1 and ERK2, were shown to be fully activated by a single stimulus, namely PMA or TCR/CDR engagement, JNK activation requires costimulation by CD28. The synergistic activation of JNK is tightly correlated with the induction of IL-2 production and has proved to be sensitive to CsA. Thus, JNK seems to be involved in the integration of the independent signaling pathways leading to T-cell activation. To date, JNK is the first characterized protein kinase capable of integrating distinct signals in T cells.

Relation of JNK to the Putative IκB Kinase

The putative IκB kinase and JNK share several features. Both enzymes are ubiquitously expressed but unique in their activation requirements in T cells. Whereas a single stimulus is sufficient to activate the two kinases in non-T cells, their activation in T cells is dependent on a costimulatory signal.[14,20] Furthermore, the kinetics of activation of the two kinases are similar, maximal activity being reached 5 min after stimulation.[14,20] These overlapping properties would seemingly indicate that the two kinases are but one.

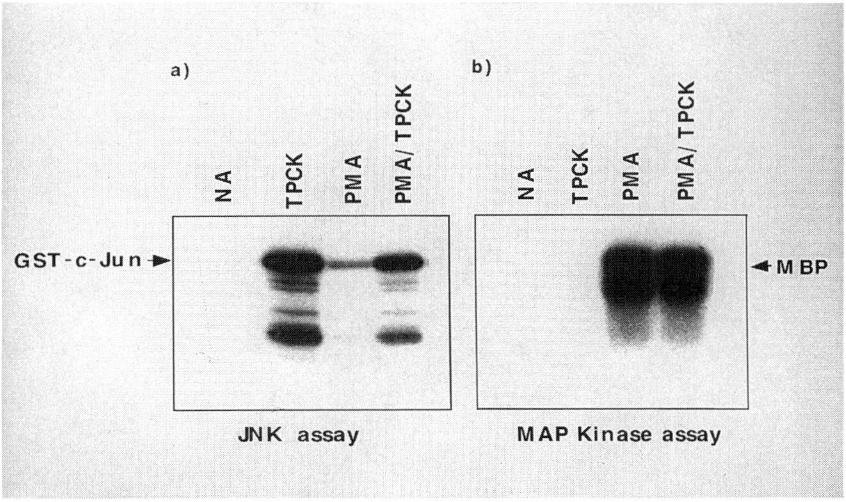

FIGURE 2. Effects of TPCK on activation status of JNK (a) and MAP kinases (b). [^{32}P]Phosphorylation of GST-c-Jun by JNK is visualized after SDS-PAGE of extracts of nonactivated cells (NA), or cells treated with TPCK, PMA, or a combination of the two agents. The faster moving bands are degradation products of GST-c-Jun. MAP kinase activity was determined by the ability of extracts to label the substrate myelin basic protein (MBP) with [^{32}P].

If IκB is indeed a substrate of JNK, reagents blocking IκB phosphorylation, and thereby NF-κB activation, would also be expected to interfere with JNK activation. To test this assumption, we analyzed the activation status of JNK in cells activated in the presence of TPCK, an inhibitor of the inducible IκB modification.[20] TPCK proved to be a powerful inducer of JNK activity, even in the absence of any additional signal (FIG. 2a). Although the drug's mode of action remains unknown, the effect must be JNK-specific, as TPCK failed to affect other members of the MAP kinase family (FIG. 2b). Yet, as argued above, the activation of JNK by TPCK is inconsistent with its putative role in the NF-κB/IκB system.

Further evidence that JNK and the IκB kinase are distinct molecules comes from the results of in-gel kinase assays,[13] recently carried out in our laboratory (unpublished results). We found that the kinases capable of phosphorylating recombinant IκB in the in-gel kinase assay had molecular masses distinct from that of JNK.

SUMMARY

The transcriptional activity of the IL-2 promoter requires T-cell costimulation delivered by the TCR and the auxiliary receptor CD28. Several transcription factors participate in IL-2 promoter activation, among which are AP-1-like factors and NF-κB. Protein phosphorylation has an important role in the regulation of these two factors: (1) it induces the transactivating capacity of the AP-1 protein c-Jun; and (2)

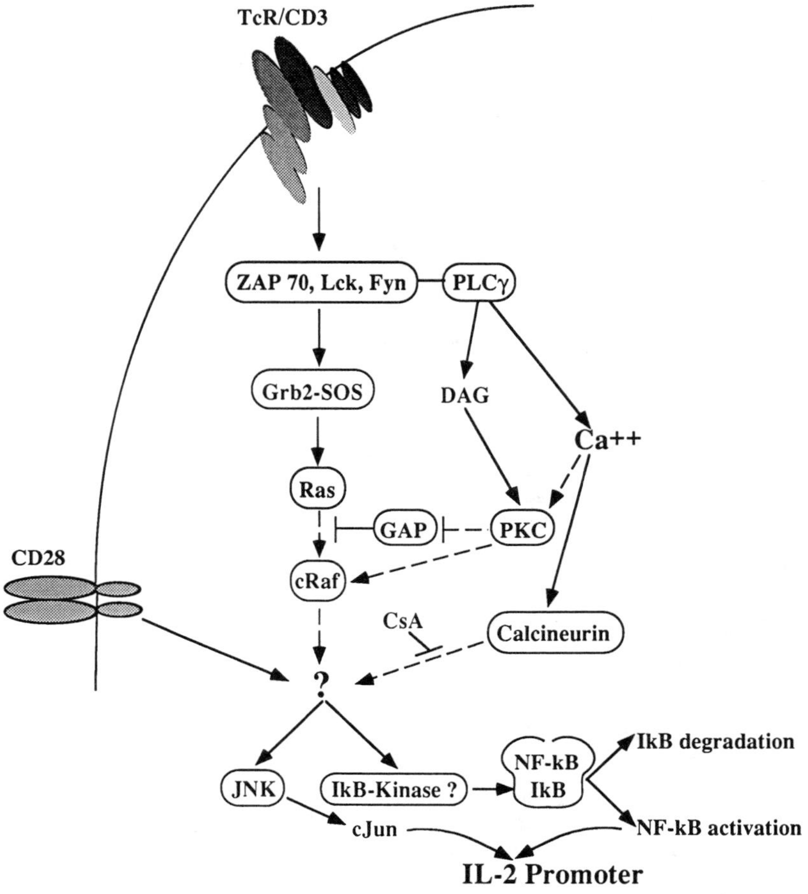

FIGURE 3. A central signaling pathway in T cells.

it is involved in the release of the cytoplasmic inhibitor, IκB, from NF-κB, allowing translocation of the latter into the nucleus. We have recently shown that both phosphorylation processes require T-cell costimulation. Furthermore, in activated T cells, the kinetics of the two phosphorylation events are essentially similar. According to our results, however, the kinases responsible for the two processes are distinct entities. Whereas TPCK inhibits phosphorylation of IκB and, consequently, activation of NF-κB, it markedly enhances the activity of JNK, the MAP kinase-related kinase that phosphorylates the transactivation domain of c-Jun. We, therefore, propose the activation scheme presented in FIGURE 3 for T-cell costimulation. Costimulation results in the activation of a signaling pathway that leads to the simultaneous induction of the two transcription factors, AP-1 and NF-κB. Integration of the signals generated by TCR and CD28 engagement occurs along this pathway, which then bifurcates to induce IκB phosphorylation and NF-κB activation on the one hand, and JNK activation and c-Jun phosphorylation on the other. We are currently engaged in defining where the two signals integrate along the AP-1/NF-κB pathway.

REFERENCES

1. BRETSCHER, P. & M. COHN. 1970. A theory of self-nonself discrimination. Science **169:** 1042-1049.
2. JUNE, C. H., J. A. BLUESTONE, L. M. NADLER & C. B. THOMPSON. 1994. The B7 and CD28 receptor families. Immunol. Today **15:** 321-331.
3. SCHWARTZ, R. H. 1990. A cell culture model for T lymphocyte clonal anergy. Science **248:** 1349-1355.
4. WEISS, A. & D. LITTMAN. 1994. Signal transduction by lymphocyte antigen receptors. Cell **76:** 263-274.
5. LINSLEY, P. S. & J. A. LEDBETTER. 1993. The role of the CD28 receptor during T cell responses to antigen. Annu. Rev. Immunol. **11:** 191-212.
6. UMLAUF, S. W., B. BEVERLY, S.-M. KANG, K. BRORSON, A.-C. TRAN & R. H. SCHWARTZ. 1993. Molecular regulation of the IL-2 gene: rheostatic control of the immune system. Immunol. Rev. **133:** 177-197.
7. CRABTREE, G. R. 1989. Contingent genetic regulatory events in T lymphocyte activation. Science **243:** 355-361.
8. HOYOS, B., D. W. BALLARD, E. BOEHNLEIN, M. SIEKEVITZ & W. C. GREENE. 1989. Kappa B-specific DNA binding proteins: role in the regulation of human interleukin-2 gene expression. Science **244:** 457-460.
9. GOSH, P., T.-H. TAN, N. R. RICE, A. SICA & H. A. YOUNG. 1993. The interleukin 2 CD28-responsive complex contains at least three members of the NF κB family: c-Rel, p50 and p65. Proc. Natl. Acad. Sci. USA **90:** 1696-1700.
10. VERWEIJ, C. L., M. GEERTS & L. A. AARDEN. 1991. Activation of interleukin-2 gene transcription via T-cell surface molecule CD28 is mediated through an NF-κB-like response element. J. Biol. Chem. **266:** 14179-14182.
11. ULLMAN, K., J. NORTHTRUP, A. ADMON & G. R. CRABTREE. 1993. Jun family members are controlled by a calcium-regulated, cyclosporin A-sensitive signaling pathway in activated T-lymphocytes. Genes & Dev. **7:** 188-196.
12. JAIN, J., P. G. MCCAFFREY, V. E. VALGE-ARCHER & A. RAO. 1992. Nuclear factor of activated T cells contains Fos and Jun. Nature **356:** 801-804.
13. HIBI, M., A. LIN, T. SMEAL, A. MINDEN & M. KARIN. 1993. Identification of an oncoprotein- and UV responsive protein kinase that binds and potentiates the c-Jun activation domain. Genes & Dev. **7:** 2135-2148.
14. SU, B., E. JACINTO, M. HIBI, T. KALLUNKI, M. KARIN & Y. BEN-NERIAH. 1994. JNK mediates signal integration during costimulation of T lymphocytes. Cell **77:** 727-736.
15. NABEL, G. & D. BALTIMORE. 1987. An inducible transcription factor activates the expression of human immunodeficiency virus in T cells. Nature **326:** 711-713.
16. BAEUERLE, P. & D. BALTIMORE. 1988. IκB: a specific inhibitor of the NF-κB transcription factor. Science **242:** 540-546.
17. BEG, A. A., S. M. RUBEN, R. I. SCHEINMAN, S. HASKILL, C. A. ROSEN & A. J. BALDWIN. 1992. IκB interacts with the nuclear localization sequences of the subunits of NF κB: a mechanism for cytoplasmic retention. Genes & Dev. **6:** 1899-1913.
18. HENKEL, T., T. MACHLEIDT, I. ALKALAY, M. KRONKE, Y. BEN-NERIAH & P. A. BAEUERLE. 1993. Rapid proteolysis of IκB is necessary for activation of transcription factor NF κB. Nature **365:** 182-185.
19. MELLITS, K. H., R. T. HAY & S. GOODBOURN. 1993. Proteolytic degradation of MAD3 (IκB) and enhanced processing of the NF κB precursor p105 are obligatory steps in the activation of NF κB. Nucleic Acids Res. **21:** 5059-5066.

20. Alkalay, I., A. Yaron, A. Hatzubai, S. Jung, A. Avraham, O. Gerlitz, I. Pashut-Lavon & Y. Ben-Neriah. 1995. *In vivo* stimulation of IκB phosphorylation is not sufficient to activate NF-κB. Mol. Cell. Biol. **15:** 1294.
21. Derijard, B., M. Hibi, I. Wu, T. Barrett, B. Su, T. Deng, M. Karin & R. Davis. 1994. JNK: a protein kinase stimulated by UV light and Ha-Ras that binds and phosphorylates the c-Jun activation domain. Cell **76:** 1025-1037.

Structural Aspects of Cytokine/ Receptor Interactions

NICOS A. NICOLA

The Cooperative Research Centre
for Cellular Growth Factors and the
Walter and Eliza Hall Institute of Medical Research
PO Royal Melbourne Hospital
Parkville, VIC 3050 Australia

INTRODUCTION

The cytokines are a group of proteins, usually produced at local sites of inflammation, that serve to regulate the production, migration, and functional activation of hemopoietic cells. As a result of their site of action, cytokines have a relatively robust structure, usually stabilized by disulfide bonds and glycosylation. Their sphere of influence, however, is limited by a number of mechanisms, including immobilization on the cell-surface matrix and receptor-mediated internalization and destruction, as well as inhibitors of cytokine action in the circulation. Perhaps because they are localized in their actions, many cytokines appear to be used for quite different biological responses at different sites and at different times in the body (cytokine pleiotropy). On the other hand, most biological functions can be effected by several different cytokines that may be produced under different circumstances as a result of the use of individual gene promoters (cytokine redundancy).

Cytokine pleiotropy and redundancy are beginning to be understood from a study of the organization of their cellular receptors. All cytokines appear to elicit intracellular responses by converting an extracellular receptor aggregation event into an intracellular aggregation of receptor cytoplasmic domains that present novel recognition surfaces to intracellular signaling molecules. Cytokine redundancy may arise from the use of common receptor subunits, whereas pleiotropy may arise from the distribution of receptor subunits on different cells or by the divergence of multiple signaling pathways from one receptor subunit.

The pleiotropy and redundancy of cytokine action provide a flexible and subtle means of biological control of inflammatory responses but provide the clinician with many problems in the use of cytokines to manage human disease. How are unwanted biological effects due to the systemic spread of cytokines to be controlled? How are biological effects due to multiple cytokines to be mimicked or inhibited? The multiple interaction surfaces between cytokines, and their various receptor subunits, and their functional consequences are beginning to be understood in molecular detail and will be the basis in the future for the design of cytokine mimics and antagonists of improved specificity. In this review, the state of knowledge of the interaction surfaces of a few cytokine/receptor systems will be described.

STRUCTURES OF THE CYTOKINES

The cytokines and growth factors can be classified into several groups of structurally conserved families, including those displaying the cystine knot, β-trefoil, β-jellyroll, and 4-α-helical bundle motifs.[1–3] Other subclassifications based on biological function, regulation, gene structure, and receptor usage have also been described.[2,3] Because the cytokine/receptor pair can be thought of as a coevolving structure where complementary interacting structures must be maintained, it is of some interest to compare the structures of related cytokines and their receptors in order to try and define common recognition motifs. One of the largest cytokine/receptor families comprises the 4-α-helical bundle cytokines and their hemopoietin domain receptors, which will be discussed in this review.

The cytokines known or thought to take up 4-α-helical bundle structures include growth hormone (GH), prolactin (PRL), the interleukins (IL), the colony-stimulating factors (G-CSF, GM-CSF, M-CSF), the interferons, erythropoietin (Epo), thrombopoietin, stem cell factor, flk-2 ligand, leukemia inhibitory factor (LIF), oncostatin-M (OSM), and ciliary neurotrophic factor. All except the interferons, IL-10, M-CSF, SCF, and flk-2 ligand bind to hemopoietin domain receptors. This large family of cytokines is characterized by a similar topological organization of four main α-helices, with the A and D helices forming one antiparallel helical pair and the B and C helices forming another. The A and B helices and the C and D helices are parallel, so that this arrangement has been termed "up-up-down-down." The structure requires long overhand loops between the A and B helices and the C and D helices in order to allow for helical chain reversal.

The 4-α-helical cytokines can be divided into two major subgroups based on overall chain length, helical segment length, structures in the long overhand loops, and the overall shape of the molecule. The long-chain cytokines (GH, PRL, Epo, G-CSF, IL-6, IL-11, IL-12, LIF, OSM, CNTF) have larger helices, usually contain additional short helices in the overhand loops, and have the shape of elongated cylinders. The short-chain cytokines (IL-2, 3, 4, 5, 6, 7, 13, GM-CSF) have shorter helices, contain one short antiparallel β-strand in each of the two overhand loops, and have the shape of an oblate ellipsoid.[1]

All of the 4-α-helical cytokines interact with at least two receptor subunits in order to elicit biological responses. In some cases (GH, PRL, Epo, G-CSF), this interaction is with two identical receptor subunits (homodimerization), whereas in others (IL-6, IL-11, LIF, OSM, CNTF, IL-2, 3, 4, 5, 7, 9, 13, GM-CSF) two or more different (but usually structurally related) receptor subunits are aggregated (heterodimerization). Thus each cytokine must present two separate interaction surfaces, one for each receptor subunit. These two interaction surfaces have been mapped in some detail for growth hormone by mutagenesis studies and by direct visualization of the complex of GH with two identical receptor subunits.[4] Some surprises have emerged from this study. First, two quite separate regions of the growth hormone molecule (site I and site II) interact with essentially the same regions on each of the two receptor molecules. Second, although the ligand/receptor contact areas are quite large (1230 $Å^2$ or 31 residues for site I, and 900 $Å^2$ or 20 residues for site II), the residues that make significant contributions to the binding energy are far fewer (*e.g.*, 6 or 7 residues for site I account for 80-90% of the binding energy[5]). Third, the binding mechanism is sequential with GH binding to the first receptor chain at site I and only then binding to the second receptor chain at site II.[4] This may result, in part, because the site II interactions are weaker than those at site I and therefore additional stabilizing interactions between the two receptor chains (site III) are required to form a stable trimeric complex. Alternatively, or additionally, binding of GH to the first receptor chain at site I may alter the conformation of GH, the receptor, or both, to expose new interaction surfaces for site II. Such a conformational change in GH has been proposed from a comparison of the structure of human growth hormone in the receptor complex compared to the structure of porcine GH in the absence of receptor.[4]

The site I contacts on GH are clustered on the D-helix, the A-helix, and the short helical segments in the A-B loop, with most of the energetically important contributions arising from residues on the exposed face of the D-helix. Site II contacts are located on the opposite face of the 4-α-helical bundle and involve residues at the N-terminus, the beginning of the A-helix, and a few residues on the C-helix (FIG. 1).

As mentioned above, the binding sites on each receptor subunit for site I or site II of GH are essentially the same. Each receptor subunit consists of two β-barrels, each containing seven antiparallel β-strands (A1-G1, A2-G2). These subdomains are homologous to fibronectin-III repeats and are arranged nearly perpendicular to each other. The GH contacts occur at the interface of these two subdomains and primarily involve residues in the loop regions between the A1-B1 and E1-F1 β-strands of the first subdomain and the B2-C2 and F2-G2 β-strands of the second subdomain. In addition, residues in the G1 β-strand and in the linker region between the two subdomains are involved in GH contacts. The orientation of all of these residues are similar but not identical in site I compared to site II contacts. In particular, the F2-G2 contacts used in site I are not used in site II (FIG. 1).

THE GM-CSF, IL-3, IL-5 RECEPTOR SYSTEMS

The structure of the GH/GH receptor complex has served as a paradigm for other 4-α-helical bundle cytokines interacting with hemopoietin domain receptors. However, many other cytokine receptor systems do not involve receptor homodimerization but rather receptor heterodimerization or heterotrimerization. For example GM-CSF, IL-3, and IL-5 each interact with unique receptor α-chains (equivalent to site I interactions in GH) and a common receptor β-chain (equivalent to site II interactions). Although these receptor systems use two distinct receptor subunits, each subunit is nevertheless a member of the hemopoietin receptor superfamily and contains at least one domain that has strong structural similarity to the GH receptor binding domain. They are likely to use a similar binding interface to that used by the GH receptor, and the fact that the common β-chain can recognize site II interaction residues on each of three quite different cytokines further reinforces the flexibility in design of the ligand-binding surface of hemopoietin domain receptors already noted in GH/GH receptor interactions. By further analogy with the GH/GH receptor system, the common β-chain of IL-3, IL-5, and GM-CSF receptors does not interact with any of these cytokines unless they have bound to their respective α-chains first.

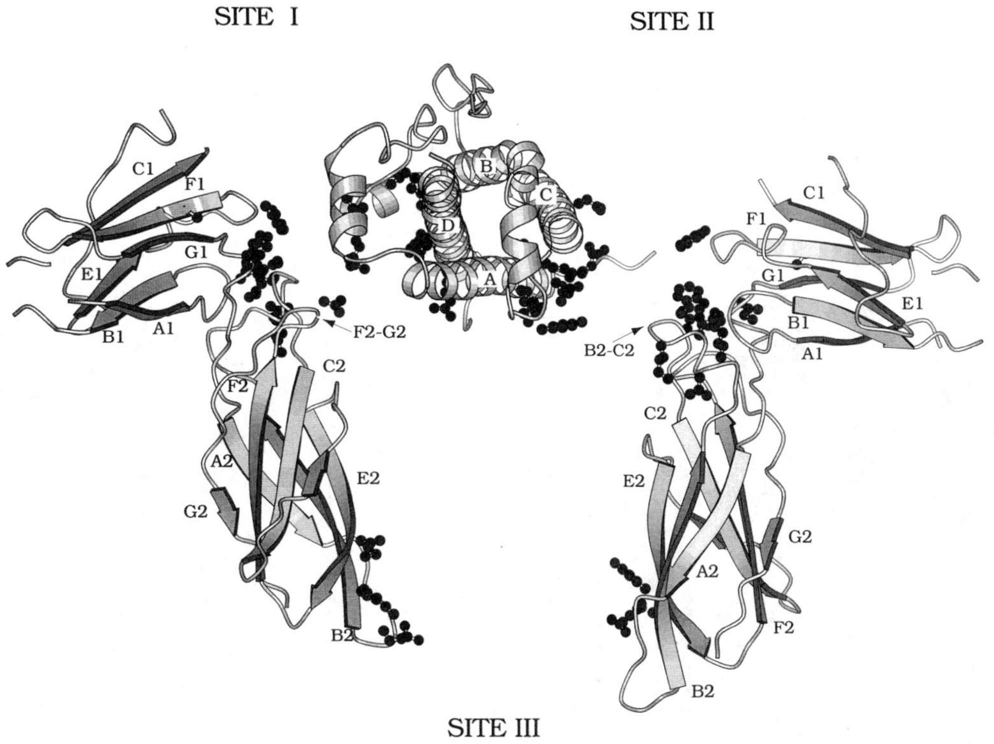

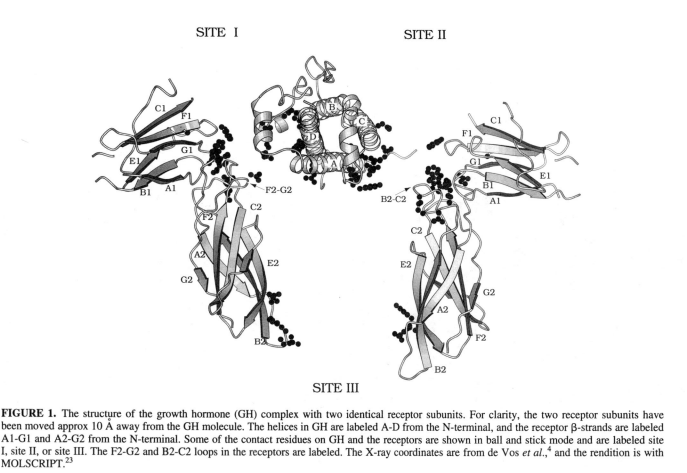

FIGURE 1. The structure of the growth hormone (GH) complex with two identical receptor subunits. For clarity, the two receptor subunits have been moved approx 10 Å away from the GH molecule. The helices in GH are labeled A-D from the N-terminal, and the receptor β-strands are labeled A1-G1 and A2-G2 from the N-terminal. Some of the contact residues on GH and the receptors are shown in ball and stick mode and are labeled site I, site II, or site III. The F2-G2 and B2-C2 loops in the receptors are labeled. The X-ray coordinates are from de Vos *et al.*,[4] and the rendition is with MOLSCRIPT.[23]

This again implies that receptor-receptor contacts between the α-chains and the common β-chains, which stabilize these interactions, are compatible in the three different combinations, revealing another level of flexibility in protein-protein interactions.

In the GM-CSF/receptor system, a very high degree of correspondence in ligand/receptor recognition interfaces has been observed with those described for the GH/receptor interaction. Mutagenesis studies have implicated the D-helix, and the A- and C-helices as involved in receptor binding, although initial studies did not distinguish between binding to the α- or β-receptor subunits.[6–8] More recent studies using interspecies chimeras of GM-CSF and selective receptor binding or bioassays have identified the regions of the GM-CSF molecule that interact with the β-chain of the receptor.[9] These consist of residues on the exposed face of the N-terminal half of the A-helix, and mutagenesis studies have revealed a critical role for glutamate 21, inasmuch as charge-reversal mutants completely abrogated high-affinity binding to the α-β receptor complex without affecting the low-affinity binding of human GM-CSF to isolated α-receptor subunits.[10,11]

We reasoned that this homology in the binding interface of human GM-CSF for the β-chain and the site II interaction surface of GH for the second GH-receptor subunit might extend to homologous interaction surfaces on the β-chain receptor for GM-CSF and the second GH-receptor subunit. Previous studies on the homologous murine IL-3 β-chain had revealed that the binding site for IL-3 was within the membrane proximal hemopoietin domain rather than the N-terminal one and included residues that, by sequence alignment, would correspond to the B2-C2 loop of the GH receptor.[12] Sequence alignments of the human GM-CSF receptor β-chain with GH receptor, βIL-3, and the IL-2 receptor β-chain allowed us to define the presumptive B2-C2 loop in the GM-CSF receptor β-chain, and this served as the target for mutagenesis studies. The charge reversal mutants of glutamate 21 in human GM-CSF had strongly suggested that it formed a stabilizing salt bridge with a positively charged residue on the β-chain, so we concentrated on mutating each of the positively charged residues in the presumptive B2-C2 loop to three different amino acids. We found that all mutations of histidine 367 of the β-chain completely abrogated high-affinity binding of GM-CSF, but mutation of all adjacent residues in this loop were without effect.[13] These data suggest that a stabilizing salt bridge between glutamate 21 on GM-CSF and histidine 367 on the β-chain of the receptor is a major determinant in converting the low-affinity interaction of GM-CSF with the α-chain of the receptor into a high-affinity interaction when the β-chain is also present. The data also suggest that the structural elements involved in ligand/receptor interactions may be highly conserved in some members of the hemopoietin receptor family.

There is considerable evidence that the direct interaction of GM-CSF with the β-chain of the receptor serves to stabilize ligand binding and dramatically reduce the kinetic off rate, but this interaction does not seem to be required for the ligand-induced heterodimerization of receptor α- and β-chains. Rather, this appears to result from ligand-induced conformational changes in the α-chain that expose an interaction surface for the β-chain. When the human GM-CSF receptor α-chain was expressed in mouse cells coexpressing the mouse β-chain, there was no detectable high-affinity conversion of human GM-CSF binding, yet human GM-CSF could elicit proliferative signals in such cells with apparent full efficiency (on a per receptor occupied basis).[14,15]

Similarly, mutant human β-chains involving changes in Histidine 367, when coexpressed with human receptor α-chains, were capable of proliferative signaling in response to human GM-CSF in mouse CTLL cells.[13] Because both α-chains and β-chains are required for proliferative signaling,[16] this suggests that receptor aggregation is independent of high-affinity GM-CSF binding and relies primarily on receptor/receptor contacts of the type described for the GH/receptor complex (site III, FIG. 1).

THE LIF/OSM/IL-6/IL-11/CNTF RECEPTOR FAMILY

As for the GM-CSF/IL-3/IL-5 receptor system, this family of receptors is also characterized by a common receptor β-chain, termed gp130.[17] IL-6, IL-11, and CNTF each have a ligand-specific, low-affinity α-chain that belongs to the hemopoietin family of receptors.[17] The α-chain for the LIF receptor is a low-affinity LIF-binding protein, but it also forms a third receptor chain of the CNTF receptor and forms a part of the OSM receptor. This arrangement serves to blur the distinction between α- and β-chains of hemopoietin receptor complexes, but all components of this family of receptors are structurally related in their ligand-binding hemopoietin domains.

Several of these receptors, including gp130, LIF receptor α-chain, IL-6, and CNTF receptor α-chains are produced in soluble form, and particularly high levels of soluble LIF receptor α-chain (2-300 μg/mL) are found in mouse serum.[18] Soluble LIF receptor α-chain and gp130 appear to act as pure cytokine antagonists and, given the pleiotropic effects of this family of cytokines, may serve to prevent the systemic spread of locally produced cytokines. However, soluble IL-6, CNTF, and possibly IL-11 receptor α-chains can still associate with and activate cell-bound gp130 in the presence of the appropriate ligand,[17] so that they could potentially act as cytokine agonists or even generate cell responses in cells that do not normally express the appropriate receptor α-chain.

This family of cytokines belongs to the long-chain 4-α-helical bundle structural subfamily as confirmed by the recent X-ray structural determination of mouse LIF.[19] We had noted that, although mouse LIF and human LIF were highly homologous in amino acid sequence (78%) and probably structure, they displayed only a one-way interspecies cross-reactivity in their receptor binding and biological activities, namely both species of LIF could bind to the mouse LIF receptor, but only human LIF could bind to the human LIF receptor.[20] Somewhat surprisingly, human LIF bound to the mouse LIF receptor α-chain with a much higher affinity (K_D = 10 pM) than did mouse LIF (K_D = 1-2 nM), yet human LIF bound to its own receptor α-chain with a typically low affinity (K_D = 0.5-1 nM). We took advantage of these observations to generate a large series of human/mouse LIF chimeras and test their binding to human or mouse LIF receptor α-chains in order to determine, on the one hand, the amino acid residues responsible for binding to the human LIF receptor α-chain and, on the other hand, the residues responsible for the unusual high-affinity binding to the mouse LIF receptor α-chain.[21,22] All chimeras were purified, accurately quantitated, and tested for structural integrity by their capacity to bind to the mouse LIF receptor and stimulate mouse cells (M1 cells for myeloid differentiation). The structural compatibility of mouse and human LIF was evidenced by the fact that all chimeras had full biological activity in the mouse cell bioassay.

Quantitation of the receptor-binding assays revealed some interesting findings. (1) Almost all chimeras that contained human LIF amino acid residues showed simple additive effects in all assays. That is, the cumulative effect of all critical amino acid substitutions identified was essentially the sum of all of the individual contributions to binding energy. This again suggests that major conformational differences between mouse and human LIF are unlikely and that the binding energy does not arise from cooperative interactions between contact residues. (2) There was a near perfect correlation between the capacity of each chimera to engender binding affinity for the human LIF receptor α-chain and an increase in the binding affinity to the mouse LIF receptor α-chain. This strongly suggests that the interaction surface on human LIF recognized by the human LIF receptor α-chain is the same surface that provides the additional interactional energy on the mouse LIF receptor α-chain to generate high-affinity binding. It implies that the receptor interaction surface on LIF has changed during the course of evolution from mouse to humans and, in turn, that it recognizes a new site on the human LIF receptor. Nevertheless, human LIF has retained its vestigial mouse LIF receptor binding site at the same time as it has acquired the new receptor binding site. (3) Six amino acids were identified in human LIF that together accounted for nearly 90% of the binding energy to the human LIF receptor α-chain. This is very similar to the number of energetically important contacts identified in site I of GH, and by analogy with that system, it is likely that a much larger number of residues on hLIF contact the receptor. However, in contrast to the GH/receptor system (and to the GM-CSF/receptor system), the important contact residues were located primarily near helix-loop boundaries at the N-terminus of the C-helix (H112, S113), the D-helix (V155, K158), the B-C loop (S107), and the short helix in the A-B loop (D57). Moreover, rather than forming a patch in the center of the interaction surface, as in GH, the residues in LIF form a ring around the periphery of the presumptive interaction surface at one end of the molecule (FIG. 2). If laid on top of each other, the LIF receptor interaction surface is at the top of the molecule, whereas the GH site I interaction surface is on the side (FIG. 2). Based on these observations, it may not be as straightforward as in the case of GM-CSF to predict the corresponding contact residues in the LIF receptor.

CONCLUSIONS

Most cytokine/receptor activation systems begin by the cytokine-induced aggregation of the extracellular domains of two or more structurally related receptor subunits. At least three important protein-protein interaction surfaces have been identified: (1) between the cytokine and the ligand-specific receptor subunit that is usually a low-affinity reaction that initiates the chain of events; (2) between the cytokine and a second receptor subunit that generates high-affinity binding and ensures that the cytokine will be stably associated with the receptor complex (this increases the sensitivity of cells to the cytokine and increases the probability of signal transduction); (3) a cytokine-induced interaction between the receptor subunits that is primarily responsible for receptor aggregation and hence signal transduction. Targeting these different interaction surfaces may provide quite different therapeutic interventions. In the first case, cytokine-specific reagents may be produced. In the second case,

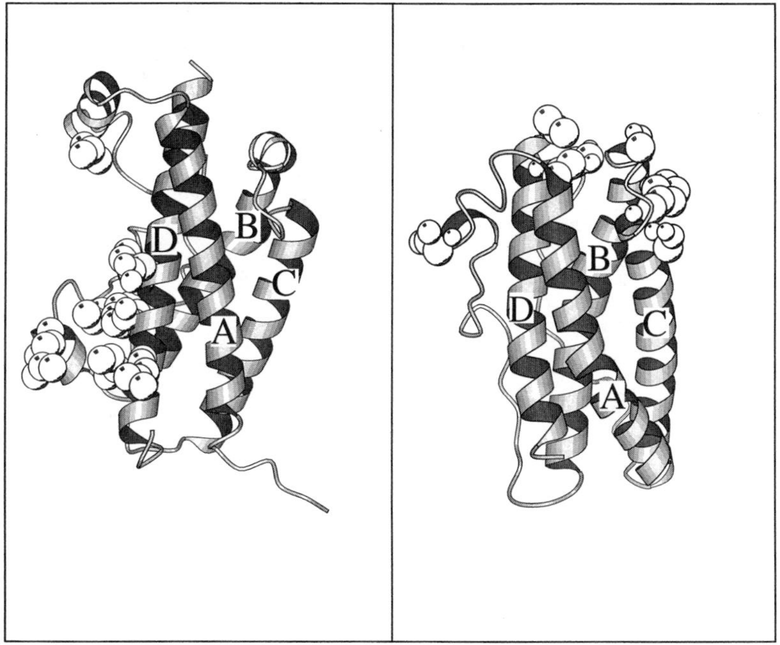

FIGURE 2. The energetically important contact residues involved in growth hormone (GH) (left) and leukemia inhibitory factor (LIF) (right) site I receptor contacts. Contact residues are shown in CPK mode (spheres), and the four main helices are labeled A-D. The GH structure is from de Vos *et al.*,[4] and the LIF structure is from Robinson *et al.*[19] The rendition is with MOLSCRIPT.[23]

cytokine-specific or generic reagents may be produced, inasmuch as the second receptor subunit may be used by several functionally related cytokines. In the last case, generic inhibitors of receptor activation may be produced. Because the energetically important interaction residues in all of these cases are quite small, it may be possible to design cytokine agonists and antagonists with improved bioavailability, biolocalization, or pharmacokinetic properties and overcome some of the current problems associated with cytokine therapy.

REFERENCES

1. SPRANG, S. R. & J. F. BAZAN. 1993. Cytokine structural taxonomy and mechanisms of receptor engagement. Curr. Opinion Struct. Biol. **3:** 815-27.
2. BOULAY, J. & W. E. PAUL. 1993. Hemapoietin sub-family classification based on size, gene organization and sequence homology. Curr. Biol. **3:** 573-581.
3. NICOLA, N. A. 1994. Introduction to the cytokines. *In* Guidebook to Cytokines and Their Receptors. N. A. Nicola, Ed. Oxford University Press. Oxford, UK.
4. DE VOS, A. M., M. ULTSCH & A. A. KOSSIAKOFF. 1992. Human growth hormone and extracellular domain of its receptor: crystal structure of the complex. Science **255:** 306-12.
5. CUNNINGHAM, B. C. & J. A. WELLS. 1993. Comparison of a structural and functional epitope. J. Mol. Biol. **234:** 554-63.
6. KAUSHANSKY, K., S. G. SHOEMAKER, S. ALFARO & C. BROWN. 1989. Hematopoietic activity of granulocyte/macrophage colony-stimulating factor is dependent upon two distinct regions of the molecule: functional analysis based upon the activities of interspecies hybrid growth factors. Proc. Natl. Acad. Sci. USA **86(4):** 1213-7.
7. SHANAFELT, A. B. & R. A. KASTELEIN. 1989. Identification of critical regions in mouse granulocyte-macrophage colony-stimulating factor by scanning-deletion analysis. Proc. Natl. Acad. Sci. USA **86(13):** 4872-6.
8. SHANAFELT, A. B., K. E. JOHNSON & R. A. KASTELEIN. 1991. Identification of critical amino acid residues in human and mouse granulocyte-macrophage colony-stimulating factor and their involvement in species specificity. J. Biol. Chem. **266(21):** 13804-10.
9. SHANAFELT, A. B., A. MIYAJIMA, T. KITAMURA & R. A. KASTELEIN. 1991. The amino-terminal helix of GM-CSF and IL-5 governs high affinity binding to their receptors. EMBO J. **10(13):** 4105-112.
10. SHANAFELT, A. B. & R. A. KASTELEIN. 1992. High affinity ligand binding is not essential for granulocyte-macrophage colony-stimulating factor receptor activation. J. Biol. Chem. **267(35):** 25466-72.
11. LOPEZ, A. F., M. F. SHANNON, T. HERCUS, N. A. NICOLA, B. CAMBARERI, M. DOTTORE, M. J. LAYTON, L. EGLINTON & M. A. VADAS. 1992. Residue 21 of human granulocyte-macrophage colony-stimulating factor is critical for biological activity and for high but not low affinity binding. EMBO J. **11(3):** 909-16.
12. WANG, H.-M., T. OGOROCHI, K.-i. ARAI & A. MIYAJIMA. 1992. Structure of mouse interleukin 3 (IL-3) binding protein (AIC2A). Amino acid residues critical for IL-3 binding. J. Biol. Chem. **267(2):** 979-83.
13. LOCK, P., D. METCALF & N. A. NICOLA. 1994. Histidine 367 of the human common beta chain of the receptor is critical for high affinity binding of human granulocyte macrophage colony stimulating factor. Proc. Natl. Acad. Sci. USA **91(1):** 252-256.
14. METCALF, D., N. A. NICOLA, D. P. GEARING & N. M. GOUGH. 1990. Low-affinity placenta-derived receptors for human granulocyte-macrophage colony-stimulating factor can deliver a proliferative signal to murine hemopoietic cells. Proc. Natl. Acad. Sci. USA **87(12):** 4670-4.
15. KITAMURA, T., K. HAYASHIDA, K. SAKAMAKI, T. YOKOTA, K. ARAI & A. MIYAJIMA. 1991. Reconstitution of functional receptors for human granulocyte/macrophage colony-

stimulating factor (GM-CSF): evidence that the protein encoded by the AIC2B cDNA is a subunit of the murine GM-CSF receptor. Proc. Natl. Acad. Sci. USA **88(12):** 5082-6.

16. Sakamaki, K., I. Miyajima, T. Kitamura & A. Miyajima. 1992. Critical cytoplasmic domains of the common beta subunit of the human GM-CSF, IL-3 and IL-5 receptors for growth signal transduction and tyrosine phosphorylation. EMBO J. **11(10):** 3541-9.
17. Kishimoto, T., T. Taga & S. Akira. 1994. Cytokine signal transduction. Cell **76:** 253-262.
18. Layton, M. J., B. A. Cross, D. Metcalf, L. D. Ward, R. J. Simpson & N. A. Nicola. 1992. A major binding protein for leukemia inhibitory factor in normal mouse serum: Identification as a soluble form of the cellular receptor. Proc. Natl. Acad. Sci. USA **89:** 8616-20.
19. Robinson, R. C., L. M. Grey, D. Staunton, H. Vankelecom, A. B. Vernallis, J.-F. Moreau, D. I. Stuart, J. K. Heath & E. Y. Jones. 1994. The crystal structure and biological function of leukemia inhibitory factor: Implications for receptor binding. Cell **77:** 1101-16.
20. Layton, M. J., P. Lock, D. Metcalf & N. A. Nicola. 1994. Cross-species receptor binding characteristics of human and mouse leukemia inhibitory factor suggest a complex binding interaction. J. Biol. Chem. **269:** 17048-55.
21. Owczarek, C. M., M. J. Layton, D. Metcalf, P. Lock, T. A. Willson, N. M. Gough & N. A. Nicola. 1993. Inter-species chimaeras of leukaemia inhibitory factor define a major human receptor binding determinant. EMBO J. **12:** 3487-95.
22. Layton, M. J., C. M. Owczarek, D. Metcalf, R. L. Clark, D. K. Smith, H. R. Treutlein & N. A. Nicola. 1994. Conversion of the biological specificity of murine to human leukemia inhibitory factor by replacing 6 amino acid residues. J. Biol. Chem. **269:** 29891-96.
23. Kraulis, P. J. 1991. Molscript: a program to produce both detailed and schematic plots of protein structure. J. Appl. Cryst. **24:** 946-50.

Cell Growth Signal Transduction Is Quantal[a]

KENDALL A. SMITH

Division of Immunology
Department of Medicine
The New York Hospital–Cornell Medical Center
525 East 68th Street
New York, New York 10021

INTRODUCTION

The meeting entitled Receptor Activation by Antigens, Cytokines, Hormones, and Growth Factors was one of the best meetings I have attended in my scientific career. It was remarkable because it brought together so many top scientists from so many separate fields. Even more important, most of these people participated throughout the meeting. As a result, the formal presentations were first-rate, and the discussions were sparkling. Many of the perplexing issues that I had been preoccupied with for many years regarding interleukin 2 (IL-2) signaling were finally resolved as a consequence of this meeting. Therefore, I have elected to summarize my current concepts of how the IL-2R operates, instead of detailing my oral presentation at the meeting, which concerned isolation and characterization of cytokine response (CR) genes.

THE CELL CYCLE AND VARIABILITY

The variability of cell-cycle times among individual cells of a genetically homogenous population has been a perplexing, but fundamental, property of all cells studied thus far. Indeed, from the earliest experiments performed with bacteria, to subsequent studies on yeast, protozoa, avian, and mammalian cells, all cell populations behave similarly with respect to division rate: individual cell-cycle times follow a normal distribution when examined as a function of division rate (reviewed in ref. 1). Thus, some cells divide very rapidly, whereas others divide slowly, with most distributed about the mean.[1,2] Kinetic studies of the cell cycle indicate that most of the variability in transit times occurs in the prereplicative phase (G_1) of the cell cycle, inasmuch as the replicative phases (*i.e.*, S, G_2/M) remain constant.[1]

Various theories were proposed in the 1960s and 1970s to account for the variability of cell growth rates, including two fundamentally opposite models. The deterministic model, proposed by Koch and Schaechter,[3] is based upon the assumption that cells are

[a]Work by the author cited in this review was supported by Grants 5R01-AI-32031 and 5R01-A1-32122 from the National Institute of Allergy and Infectious Diseases, National Institutes of Health.

functionally different, even when they may be genetically clonal, and that cell-cycle variability arises from the cumulative effects of many small differences. By comparison, the probabilistic model originally proposed by Burns and Tannock,[4] and later popularized by Smith and Martin,[5] is based on the assumption that cells initiating their cycle are functionally identical, and that the transition of cells from an indeterminant, resting phase to a determinant proliferative phase, is a matter of chance, quite independent of other events or properties. Therefore, both models predict ultimate failure in trying to understand this individual variability further. The deterministic theory states that there are so many differences between individual cells that it will be extremely difficult to decipher all of them, and the probabilistic theory states that there are no differences between cells, rather cell-cycle entry is a stochastic process, impossible to predict.

Particularly noteworthy is the observation that cell-cycle duration is not genetically determined, so that it is not passed on at division, and there is a poor correlation between the cell-cycle times of mother and daughter cells.[1] One consequence of this phenomenon is that there is no selection for more rapid transit times, and a cell population maintains the same mean cell-cycle time over many generations. However, cycle times of sister cells do show a positive correlation, indicting that some property important for the determination of cycle times is inherited, only to disappear with subsequent cycles.

Spudich and Koshland[6] noted a similar individual variability of bacterial cells responding to chemotactic stimuli. Even though the cells studied were the progeny of a single cell and therefore genetically homogeneous, they each showed different response times to chemotactic stimuli, this despite all environmental conditions being identical. To account for these findings, it was proposed that there may be some critical molecules present in small, limiting numbers, and that the individual variation of the concentrations of these molecules leads to the behavioral differences observed. As early as 1945, Delbrück[7] also proposed variations in small numbers of molecules as a possible cause of the variability of bacteriophage burst size. The same argument was proffered to explain the rapid asynchronization of prokaryotic and eukaryotic cell cultures.[6] Thus, if one crucial step leading to growth and cell division is controlled by a small number of molecules, then a Poisson distribution of these molecules would guarantee that cell cycles would quickly get out of phase, as some cells would divide more rapidly than others. It follows that such a nongenetic individuality of a critical molecule required for signaling cell-cycle progression could be a way in which the variability of cell-cycle times could be understood.

These issues were discussed lucidly by Pardee *et al.*[1] in trying to decipher the causes underlying the variability of mammalian cell-cycle times. By 1979 it was felt that "what is needed now is a comparison to test the most fundamental difference between deterministic and probabilistic models, namely, that the former assumes that individual cells in a population differ in some fundamental ways related to rate of cycle transit, and the latter assumes that cells are identical in this regard. We need to know whether any critical cell properties [molecules] can predict cycle times."

THE LIGAND-RECEPTOR MODEL FOR T-CELL GROWTH

Ten years ago, in studies focused on the critical determinants of T lymphocyte (T cell) cycle progression, our experiments indicated that neither the deterministic nor the probabilistic models explain the variability of T-cell growth. Instead, a new cell-growth model (termed the ligand-receptor model) was formulated that fully accounted for the variables that determine T-cell cycle progression.[8] By comparison to the deterministic model, which attempted to explain the variability by cumulative effects of many small differences between cells, we found that only three parameters determine T-cell cycle progression: the concentration of the lymphocyte growth factor, interleukin 2 (IL-2); the IL-2 receptor (R) density on each cell; and the duration of the IL-2R interaction. These experiments and conclusions were possible because we had painstakingly developed crucial reagents, including synchronized IL-2R$^+$ T cells, homogeneous immunoaffinity-purified IL-2, and monoclonal antibodies reactive with both IL-2 and the IL-2Rs. By contrast, investigators studying fibroblast cell-cycle progression at the time had for the most part only employed serum as the growth stimulus, so that the growth of the cells could not be related to the concentration of a single hormone molecule. In addition, quantitative and qualitative assessments of any growth factor receptors were precluded.

Particularly noteworthy were the findings that IL-2R expression by individual cells within a population follows a gaussian distribution (*i.e.*, log–normal distribution). Those familiar with the analysis of cell-surface molecules in immunology will recognize that all cell-surface markers are distributed in this fashion, even among cloned cell populations. Also, from the very first experiments describing IL-2-promoted cellular proliferation, a symmetrically sigmoid log-dose response curve was found. Subsequently, when radiolabeled IL-2 was generated, the concentrations of IL-2 necessary to bind to IL-2 receptors were found to be coincident with those that promote T-cell cycle progression. Among other things, these data indicated that the rate-limiting step in IL-2-stimulated T-cell growth occurs at the interaction of IL-2 with its receptor.

In a synchronized cell population, the rate of entry into the cell cycle was found to be directly related to the concentration of IL-2. Thus, at a receptor saturating IL-2 concentration ($\geq$ 100 pM), the G_1 lag time was 12 hours, whereas at a half-saturating concentration (10 pM), the G_1 lag time was nearly double (~ 20 hours). These findings were reminiscent of those obtained with murine embryonic fibroblasts (3T3 cells) exposed to various serum concentrations. If limiting the ligand resulted in prolonged cell-cycle transit times, then limiting the receptor should do likewise. The separation of a synchronized T-cell population into two subpopulations on the basis of receptor density, using the fluorescence-activated cell sorter (FACS), yielded the predicted results. The subset with the highest density of IL-2Rs transited the cell cycle more rapidly than the subset with lower IL-2R density. It follows that the IL-2 log-dose response curve is explicable by the log-normal distribution of the IL-2Rs: those cells with the highest density of IL-2Rs would be expected to respond to the lowest IL-2 concentrations and vice versa.

All of these observations are consistent with the idea that the number of IL-2R interactions ultimately dictates the cell-growth response. There is some finite number of interactions that must occur before the cell makes the quantal, irrevocable decision to replicate DNA. It follows that the cell has some mechanism of counting the number of IL-2R interactions, and a certain threshold exists that must be surpassed for G_1 progression to S phase. This threshold phenomenon can also be observed by IL-2 washout experiments. Even though IL-2R binding comes to steady state within 10 minutes, at least 5 hours (300 minutes) of continuous IL-2 exposure is required before any of the cells within the population progress to S phase. This indicates that the signals generated by the IL-2Rs present initially are insufficient to surpass the threshold to promote DNA replication.

The IL-2 T-cell system provided both a quantitative and qualitative model that accounted completely for the growth characteristics of T cells. Thus the prediction by Spudich and Koshland[6] was correct: the ligand-receptor interaction is the critical step in signaling cell growth, there being no evident downstream rate-limiting reaction. Accordingly, the cumulative number of receptors occupied ultimately determines the response, and in the case of IL-2Rs, the initial density expressed by activated T cells appears to be insufficient to signal cell-cycle progression. Inasmuch as activated T cells express only ~ 1500 mean sites per cell, this number of receptors is insufficient to signal cell-cycle progression, even when fully occupied.

At the time that these experiments were performed, it was unclear why so much time was required to reach the threshold of activation, although this phenomenon was well-known, having first been elucidated by Pardee[9] in serum-stimulated 3T3 cell-cycle progression. Pardee noted that there is some critical point in the mid-G_1 interval, which he termed the "restriction point," that must be surpassed before the cells are irrevocably committed to proceed to S phase. It is intuitively logical that the greatest control over a process as complicated and as important as cell division would reside in a small number of receptor molecules that initate the signal by communicating to the interior of the cell through a quantal, all or none, mechanism. However, it was not entirely evident exactly how the receptors accomplished this feat. Thus, more information was necessary as to the structure and function of growth-factor receptors.

RECEPTOR FUNCTION: BINDING AND SIGNALING

The concept of receptors was first introduced by Langely in 1878[10] to explain his findings on the antagonistic effects of atropine on pilocarpine-induced salivary gland secretion in cats. Almost three decades later,[11] he elaborated on his findings and defined a "receptive substance" as "the recipient of stimuli from drugs or hormones, which function to transfer these stimuli to the effector organs thereby eliciting a response." In addition, subsequent experiments revealed that most receptors exhibit very high affinity and show stereo-specificity for their ligands, thereby allowing the signal to be generated by very low ligand concentrations.

The first quantitative explanation of receptor function was developed by Clark,[12] who proposed in 1926 that ligands and receptors obey the law of mass action. As applied to hormone receptors, Clark assumed that the magnitude of the biologic response would be directly proportional to the number of receptors occupied, and this idea served as the basis for the occupancy theory of receptor function. Although this treatment provided a rational chemical explanation for the affinity of binding a ligand by its receptor, it did not address in a quantitative manner the other characteristic of a true receptor, the ability of a ligand to elicit a response subsequent to binding. In addition, over the next few decades investigators noted several phenomena that were not explained by the simple occupancy theory. For example, different ligands binding to the same receptor were found to differ in their "intrinsic activities" to elicit a response. Thus, partial agonists and antagonists were recognized. As well, 100% receptor occupancy was not always found necessary to elicit a maximum response, and the concept of "spare receptors" was formulated.

To account for these phenomena and in an attempt to render receptor signaling quantifiable in chemical terms, Paton[13] proposed the rate theory of hormone receptor action. Like Clark, Paton assumed that ligand binding obeys the law of mass action. However, Paton made the additional assumption that each association of a ligand to a receptor produces a "quantum" of excitation. As well, he assumed that the number of excitations per unit time determined the response observed. Accordingly, ligands could differ in how rapidly they associate and dissociate from a receptor, and therefore, the most powerful agonists would have the fastest dissociation rate, producing more quanta of excitation per time unit. Conversely, a partial agonist or an antagonist would have correspondingly slower dissociation rates. When the IL-2 receptor system was examined by rate of binding experiments, it was intriguing to discover that the very high affinity of IL-2 binding is created by a very fast rate of association ($t_{1/2} \leq$ 2 s) and a slow rate of dissociation ($t_{1/2}$ = 45 min).[14] Therefore the rate theory of Paton did not seem to be operating, nor did the occupancy theory seem to fit either, in that simple binding to the receptors did not elicit a response.

One additional observation necessary to begin to understand how the IL-2R system really functions related to receptor metabolism, that is, receptor synthesis and turnover rates.[15] In the absence of IL-2, the receptors disappear from the cell surface relatively slowly, with a $t_{1/2}$ equal to 150 minutes. However, when the cells are exposed to IL-2, the disappearance rate accelerates 10-fold, so that the $t_{1/2}$ equals 15 minutes. Because the synthesis rate of new receptors is not influenced appreciably by IL-2 binding, the consequences of the IL-2-promoted accelerated internalization of high affinity IL-2Rs is a readjustment downward in the number of Rs expressed, and a 50%-60% decrease in the number of detectable high affinity Rs occurs within 1-2 hours. Therefore, in the presence of IL-2 at steady state, it is possible to calculate the appearance rate of new receptors on the cell surface: given the $t_{1/2}$ of R disappearance, the endocytosis rate constant

$$k = \frac{\ln 2}{15} = 4.67 \times 10^{-2} \text{ min}^{-1},$$

and the rate of receptor synthesis

$$\begin{aligned} R &= k \times R \text{ at steady state} \\ &= 4.67 \times 10^{-2} \text{ min}^{-1} \times 750 \text{ R/cell} \\ &= 35 \text{ R/cell/min.} \end{aligned}$$

Inasmuch as at least 5 hours (300 min) are necessary to reach the threshold of activation, approximately 10,500 Rs must be occupied by ligand.

All of these data are consistent with the concept proposed by Bourne[16] that a receptor has two distinct features, a switch and a timer. As discussed recently by De Meyts,[17] "The switch functions to transform the imprint signal received from the ligand into a change in the three dimensional shape of the signal generator domain, thereby activating it. The timer determines how long the signal generator domain will remain 'ON' in response to reception of a signal by the detector domain."

Similarly, Kuhl distinguishes between *allosteric* changes as activating factors in macromolecules, and *allochronic* changes in which the timing and duration of the changes play an equally important regulatory role.[18]

With respect to mitogenic receptors, there is now a growing consensus that the switch and timing functions are both regulated by ligand occupancy, and are related, but independent, functions of the assembly of the functioning signal transduction units. First elucidated by Schlessinger[19] and Williams[20] in studies with the family of mitogenic receptors that contain tyrosine kinase activity within their cytoplasmic domains (*e.g.*, EGF, PDGF), activation of the receptor is mediated by ligand-induced dimerization. This receptor dimerization brings the autophosphorylation sites of one receptor chain in contact with the active site of the kinase domain of the other chain, leading to transphosphorylation. These findings dispelled the notion that signaling involved some ligand-promoted communication between the extracellular and cytoplasmic domains of a single receptor chain. Now, it is realized that there are several ways in which ligand binding causes dimerization of independent receptor chains:[17] (1) A monomeric ligand causes a conformational change in the receptor upon binding that enhances the affinity of one receptor molecule for a second receptor molecule: this seems to be operative for the EGF receptor. (2) A dimeric ligand, (*e.g.* PDGF and colony, stimulating factor – 1) binds to two receptor chains together simultaneously. (3) The receptor is a covalent disulfide-linked dimer that binds a monovalent ligand, thereby inducing close proximity and site interactions of the cytoplasmic domains. This is the case for the insulin receptor group. (4) A monovalent ligand has two asymmetric binding sites for monovalent receptor chains. Binding to both sites stabilizes a dimeric receptor complex. The growth hormone receptor functions in this manner. (5) A monovalent ligand binds to a noncovalently linked heterodimeric receptor, promoting the association of two distinct cytoplasmic domains, both of which are necessary for signaling. This is operative in several receptors of the cytokine family (*e.g.*, IL-2, IL-4, and IL-7). (6) A monovalent ligand binds to a single chain that induces its association with a separate, signaling chain, which then dimerizes with another ligand-binding chain/signaling pair. This appears to be involved in IL-6 and related cytokine family members.

With regard to IL-2 and its receptor, kinetic binding studies revealed a cooperative binding model,[14] whereby the IL-2R α chain contributes a very rapid association rate ($k = 10^7\ M^{-1}\ s^{-1}$, $t_{1/2} = 1.7$ s), and, together, the β, γ heterodimer contributes a slow dissociation rate ($k' = 10^{-4}\ s^{-1}$, $t_{1/2} = 45$ min). Thus, from the ratio of the rate constants, a very high affinity ligand-receptor interaction is created:

$$K_d = \frac{k'}{k} = \frac{10^{-4}\ s^{-1}}{10^7\ M^{-1}\ s^{-1}} = 10^{-11}M.$$

Inasmuch as the dissociation rate is 3-fold slower than the endocytosis rate (*i.e.*, $t_{1/2} = 45$ min vs. 15 min), once the ligand is bound to the receptor on the cell surface, it continues to signal until the ligand-receptor complex is endocytozed and degraded.

Recent studies[21] are consistent with an allosteric switch model for the β and γ chains of the IL-2R, in that the β chain is associated with one specific tyrosine kinase, termed JAK-1, whereas the γ chain is associated with a related, but distinct kinase, JAK-3. Thus, ligand-receptor occupancy promotes the close approximation of JAK-1 and JAK-3, and, presumably, their transphosphorylation and activation.

The slow dissociation rate of the ligand ensures that this allosteric ''switch'' will be left ''On,'' as long as the receptor remains on the cell surface.

Now, it only remains left to relate the variability of cell-cycle times and the threshold phenomenon with the apparent ability of the cell to count the number of ligand-receptor occupations upon the formation of a stable, allosteric complex that is ON. Also, it is necessary to understand the importance of the small number of IL-2 receptors expressed by normal T cells. Clues can be surmised from recent studies on insulin mitogenicity. The mitogenic potency of a series of insulin analogues was found to be inversely related to the dissociation rate, rather than the affinity of receptor binding.[17] Therefore, the longer the ligand is bound, the longer the receptor is ON, generating second messenger molecules. It follows that changes in the ligand or the receptor that accelerate dissociation would lead to partial agonists or antagonists.

A further implication of this model is that the rate-limiting step in the mitogenic response is dictated by the number of receptors. Consequently, we would expect that mutations resulting in the overexpression of receptors should potentiate the mitogenic response to a point ultimately determined by more downstream molecular events. Actually, Wells and co-workers[22] have already reported data consistent with this idea. Truncation of the C-terminus of the EGF receptor results in the loss of ligand-promoted acceleration of receptor endocytosis. This mutation results in cells with ~ 2-fold more EGF receptors than wild-type receptors transfected into NIH-3T3-derived NR6 cells. Most notably, when exposed to varying concentrations of EGF, the cells with more receptors were more sensitive to the mitogenic effects of EGF, such that there was an approximate 8-fold shift to the left in the EGF log-dose response curve. Because the mitogenic response is determined by a finite number of ligand-receptor interactions, it follows that cells with greater numbers of receptors will achieve the requisite number at lower ligand concentrations.

The practical implications of this notion are of potentially great importance, not only for our understanding of the regulation of normal cell growth, but for malignancy as well. If a particular mutation results in the overexpression of a growth-factor receptor, the requisite number of receptors occupied might be triggered by very low ligand concentrations, inasmuch as ligand-receptor interactions obey the law of mass action. Actually, this phenomenon has already been described by the artificial overexpression of normal EGF receptors.[23,24] In addition, because mitogenic signaling is a quantal phenomenon, the abnormal overexpression of any of the molecules involved in the biochemical pathway would be expected to result in a transformed cell. Of course, there are already numerous examples of cellular protooncogenes that function in signaling normal cell cycle progression that are known to mediate malignant transformation.

These considerations illuminate even further the importance of identifying the individual molecules responsible for receptor-promoted signal transduction of the mitogenic response. With regard to the IL-2 receptor, at least two pathways have been identified thus far that are thought to be important for ultimately signaling the G_1 progression to S phase: the JAK-STAT pathway[21] and the Ras-Raf-MAPK pathway.[25] Both of these pathways end in the activation of specific transcription factors, which are responsible for stimulating the expression of specific immediate/early target genes, the products of which then promote G_1 progression. In this regard, it is noteworthy that we have recently developed methods to enrich for, and select, ligand-induced immediate/early gene expression.[26,27] However, it is important now to move beyond the simple qualitative identification of signaling molecules and target genes and begin to quantitate the concentrations of molecules and rates of the reactions necessary to promote cell-cycle progression. Only then will we have a true picture of the function of both the switch and the timer in receptor-mediated mitogenesis.

REFERENCES

1. PARDEE, A. B., B.-Z. SHILO & A. L. KOCH. 1979. Variability of the cell cycle. *In* Hormones and Cell Culture. G. H. Sato & R. Ross, Eds.: p. 373. Cold Spring Harbor Laboratory, Cold Spring Harbor, NY.
2. COOK, J. R. & B. COOK. 1962. Effect of nutrients on the variation of individual generation times. Exp. Cell Res. **28:** 524.
3. KOCH, A. & M. SCHAECHTER. 1962. A model for statistics of the cell division process. J. Gen. Microbiol. **29:** 435.
4. BURNS, V. W. & I. F. TANNOCK. 1970. On the existence of a G_0 phase in the cell cycle. Cell Tissue Kinet. **3:** 321.
5. SMITH, J. A. & L. MARTIN. 1973. Do cells cycle? Proc. Natl. Acad. Sci. USA **70:** 1263.
6. SPUDICH, J. L. & D. E. KOSHLAND. 1976. Non-genetic individuality: chance in the single cell. Nature **262:** 467-471.
7. DELBRÜCK, M. 1945. The burst size distribution in the growth of bacterial viruses (bacteriophages). J. Bacteriol. **50:** 131-135.
8. CANTRELL, D. A. & K. A. SMITH. 1984. The interleukin-2 T-cell system: A new cell growth model. Science **224:** 1312-1316.
9. PARDEE, A. B. *et al.* 1974. A restriction point for control of normal animal cell proliferation. Proc. Natl. Acad. Sci. USA **71:** 1286.
10. LANGLEY, M. A. 1878. On the physiology of the salivary secretion. J. Physiol. (Lond.) **1:** 339-369.
11. LANGLEY, M. A. 1905. On the reaction of cells and of nerve-endings to certain poisons, chiefly as regards the reaction of striated muscle to nicotine and to curari. J. Physiol. (Lond.) **33:** 374-413.
12. CLARK, A. J. 1926. The antagonism of acetyl choline by atropine. J. Physiol. (Lond.) **61:** 547-556.
13. PATON, W. D. M. 1960. A theory of drug action based on the rate of drug-receptor combination. Proc. R. Soc. Lond. Ser. B Biol. Sci. **154:** 21-69.
14. WANG, H-M. & K. A. SMITH. 1987. The interleukin 2 receptor: Functional consequences of its bimolecular structure. J. Exp. Med. **166:** 1055-1069.
15. SMITH, K. A. 1989. The interleukin 2 receptor. Annu. Rev. Cell Biol. **5:** 397-425.
16. BOURNE, H. 1992. G proteins as signaling machines. Prog. Endocrinol. 9th Congr. Endocrinol. Nice, France. Parthenon Pub., London.
17. DE MEYTS, P. 1994. The structural basis of insulin and insulin-like growth factor-1 receptor binding and negative co-operativity, and its relevance to mitogenic versus metabolic signalling. Diabetologia **37** [Suppl 2]: S135-S148.
18. KÜHL, P. W. 1994. Excess-substrate inhibition in enzymology and high-dose inhibition in pharmacology: A re-interpretation. Biochem. J. **298:** 171.
19. SCHLESSINGER, J. 1988. Signal transduction by allosteric receptor oligomerization. Trends Biochem. Sci. **13:** 443-447.
20. WILLIAMS, L. T. 1989. Signal transduction by the platelet-derived growth factor receptor. Science **243:** 1564-1570.
21. JOHNSTON, J. A., M. KAWAMURA, R. A. KIRKEN, Y-Q CHEN, T. B. BLAKE, K. SHIBUYA, J. R. ORIALDO, D. W. MCVLCAR & J. J. O'SHEA. 1994. Phosphorylation and activation of the Jak-3 kinase in response to interleukin-2. Nature **370:** 151-157.

22. Wells, A., J. B. Welsh, C. S. Lazar, H. S. Wiley, G. N. Gill & M. G. Rosenfeld. 1990. Ligand-induced transformation by a noninternalizing epidermal growth factor receptor. Science **247:** 962-964.
23. di Fiore, P. P., J. H. Pierce, T. P. Fleming, R. Hazan, A. Ullrich, C. R. King, J. Schlessinger & S. A. Aaronson. 1987. Overexpression of the human EGF receptor confers an EGF-dependent transformed phenotype to NIH 3T3 cells. Cell **51:** 1063-1070.
24. Velu, T. J., L. Beguinot, W. C. Vass, M. C. Willingham, G. T. Merlino, I. Pastan & D. R. Lowy. 1987. Epidermal growth factor-dependent transformation by a human EGF receptor proto-oncogene. Science **238:** 1408-1410.
25. Zmuidzinas, A., H. J. Mamon, T. M. Roberts & K. A. Smith. 1991. Interleukin-2 triggered raf-1 expression, phosphorylation and associated kinase activity increase through G_1 and S in CD3 stimulated primary human T cells. Mol. Cell. Biol. **11:** 2794-2803.
26. Beadling, C., K. W. Johnson & K. A. Smith. 1993. Isolation of interleukin-2-stimulated immediate-early genes. Proc. Natl. Acad. Sci. USA **90:** 2719-2723.
27. Beadling, C. & K. A. Smith. 1994. In search of cytokine-response genes. Immunol. Today **15:** 197-199.

TNFα Is an Effective Therapeutic Target for Rheumatoid Arthritis[a]

MARC FELDMANN, FIONULA M. BRENNAN,
MICHAEL J. ELLIOTT, RICHARD O. WILLIAMS, AND
RAVINDER N. MAINI

Kennedy Institute of Rheumatology
Sunley Division
Lurgan Avenue
Hammersmith, London, W6 8LW

INTRODUCTION

Although there has not been much recent progress in defining the etiology of rheumatoid arthritis (RA) since it became apparent that approximately 80% of patients with definite RA expressed the "shared epitope" on the DRβ1 chain of DR4 (DW4, 14, or 15) or DR1,[1] there has been considerable progress on details of the pathogenesis, based on the analysis of the cytokine expression and regulation in human rheumatoid synovium cultures. In this chapter we will review the work from our own and other laboratories that has highlighted the importance of the cytokine network in disease pathogenesis and has permitted the definition of a new therapeutic target, TNFα. We expect that these results will provide an additional impetus to define the molecular mechanisms that lead to TNFα production, and by which TNF signals using its p55 and p75 receptors.

RESULTS AND DISCUSSION

Cytokine Regulation in Local Autoimmune Tissue Site

Cytokines are chiefly short range molecules with a wide range of biological properties. In a disease such as rheumatoid arthritis, which is localized to joints, the relevant molecular actions are those in the tissue site. This has prompted us to analyze the cytokine expression, regulation, and interactions in the rheumatoid synovium.

Our first set of observations was made possible by the cloning of cDNAs for cytokines IFNγ, IL-2, TNFα, and IL-1 in the 1983–1985 period, and subsequently we developed a sensitive slot-blotting approach to assess local mRNA production using small biopsies or operative specimens.[2,3] This clearly demonstrated abundant local synthesis of mRNA and was the first evidence for high-rate production of

[a] This work would not have been possible without the support of a number of granting agents, chiefly the Arthritis and Rheumatism Council, but with major contributions from the Nuffield Foundation; Medical Research Council; Wellcome Trust; Centocor, Inc.; and ICI plc.

cytokines. This was reproducible, irrespective of duration of disease or therapy, and implied that there was continuous cytokine synthesis in a local site of immunity and inflammation, unlike the transient cytokine synthesis traditionally noted using peripheral blood mononuclear cells (PBMC) stimulated extrinsically by mitogens.

Documentation of the chronic synthesis of cytokine mRNA was provided by culturing dissociated rheumatoid synovial cells (a mixture mainly of T lymphocytes and macrophages, with fewer fibroblasts, plasma cells, endothelium and dendritic cells). In the absence of extrinsic stimulation, these cells reassociated and produced abundant cytokine mRNA (*e.g.* IL-1, TNF, and IL-6).

However, there was a need to verify that the cytokine mRNA was translated effectively into protein. This was readily shown for TNFα, IL-1 (α and β), IL-6, and GM-CSF,[3–6] but the quantities of IL-2, lymphotoxin, and IFNγ detectable were less than expected from the level of mRNA.[7] The reasons for this discrepancy are not clear, and a variety of hypotheses has been proposed. Firestein and Zvaifler have provocatively proposed that, in late RA, T-cell activity is not important.[8] We have always been more conservative in interpreting this data and have considered the fact that T cells, trafficking to sites of their target and delivering directly to their target by cell contact, need not produce the large quantities of cytokines released by macrophages for controlling their extracellular environment. Other hypotheses to explain the low levels of T-cell cytokines include absorption and consumption by target cells, membrane expression, and the presence of cytokine inhibitors and of inhibitory cytokines. Despite the reproducibility of Firestein's findings that macrophage cytokine levels are high and that T-cell cytokine levels are low, there is increasing evidence that the latter cytokines are detectable at the protein level in signaling quantities.

Based on the emerging evidence that IL-1 was important in joint cartilage and bone destruction,[9,10] we investigated the regulation of its synthesis in the rheumatoid synovium, using *in vitro*-dissociated, but unstimulated synovial cultures. With neutralizing antisera to TNFα, it was shown that within 3 days the production of IL-1 was markedly diminished in all RA cultures. This was our first insight into the major importance of TNFα in the disease process.[4] It was soon followed by the demonstration that GM-CSF production was also regulated by TNFα.[6]

A Hierarchy of Cytokine Interactions

By studying the effects of anti-TNFα on the expression of other cytokines, it has been shown that TNFα also regulates IL-6, IL-8, and IL-10 expression in RA cultures. This work is most easily interpretable in terms of a hierarchy, with TNFα at the top, resembling an electrical circuit in series. If so, blocking TNFα at the head of the series (or cascade) should diminish the production of all other proinflammatory cytokines, and hence reduce disease activity. This view was in contrast to one usually expressed, that if many proinflammatory cytokines were present in the RA joint, then blocking one was unlikely to be of therapeutic benefit. On that basis, it was commonly believed that cytokines would not be good therapeutic targets.

Animal Models of Rheumatoid Arthritis Are Ameliorated by Anti-TNF

Our analysis *in vitro* had highlighted the importance of TNFα and suggested that it was a possible single molecular target in this multifactorial disease. To validate this concept, we injected with monoclonal anti-TNF antibody DBA/1 mice that had been induced to develop destructive arthritis by the injection of bovine collagen type II several weeks previously. A key point to emphasize was that treatment was begun *after* disease onset, mimicking a clinical situation. This is because research in the many papers on animal models does not discriminate between disease prevention, which is not relevant in the clinical situation and treatment. Our results, and those reported independently by others were clear-cut. Anti-TNF had a marked effect, not only on the inflammatory aspect of collagen-induced arthritis (the footpad swelling), but also on disease progression (the number of affected limbs). Additionally and most critically, in view of our long-term ambition to make a real impact on the disease process, the treatment decreased joint erosions, which histologically resemble the erosions of human RA.[11,12]

Other animal models of arthritis have supported our conclusion that TNFα is of major importance. Thus Kollias and colleagues[13] have generated transgenic mice, over-expressing human TNFα due to replacement of its UT-rich 3′-untranslated region by that of β globin. Surprisingly, as TNFα is disregulated in many tissues, there is only one major pathology, an erosive arthritis, which can be prevented with anti-TNFα antibody.

Clinical Trials of Chimeric Anti-TNFα

The work summarized above provided the rationale for testing the concept we had developed that TNFα was of major or pivotal importance in the pathogenesis of RA. We found a partner, Centocor Inc., who had developed a high-affinity chimeric (human IgG1, K mouse Fv) anti-TNFα antibody, termed cA2, from a murine antibody, generated in Dr. Jan Vilcek's laboratory in New York. It had been produced and developed for clinical trials of sepsis, but were able to convince Dr. J. Woody of Centocor to test it in RA patients who had failed existing therapy.

The results of the first phase I/II "open-label" (nonplacebo-controlled, non-blinded) clinical trial were highly encouraging.[14] Not only was there improvement in symptoms like pain and morning stiffness, which are major complaints of the patients (but difficult to quantitate accurately, as they are entirely subjective), but it was also possible to document improvements in clinical signs, such as joint swelling and joint tenderness, which are easier to quantitate with a trained observer. Our enthusiasm was really kindled, however, when the results of laboratory blood tests indicated that inflammatory markers, such as C-reactive protein and/or erythrocyte sedimentation rate (ESR) in RA patients, had been markedly diminished also by the anti-TNFα antibody. The degree of improvement and its duration (median 12 weeks) were most impressive for an antibody course of 2-4 infusions over a two-week interval (FIG. 1).

Since this trial, we have performed two further clinical trials. One was to retreat 7 of the original 20 open-label patients, after relapse, for up to three further courses, this time with half the previous dose. In all instances, there was marked benefit in subjective, semi-objective, and serum biochemistry.[15] However, the "gold standard" of clinical research is the randomized placebo-controlled double-blind trial. We have recently completed such a multicenter trial, together with three other teams in Europe,

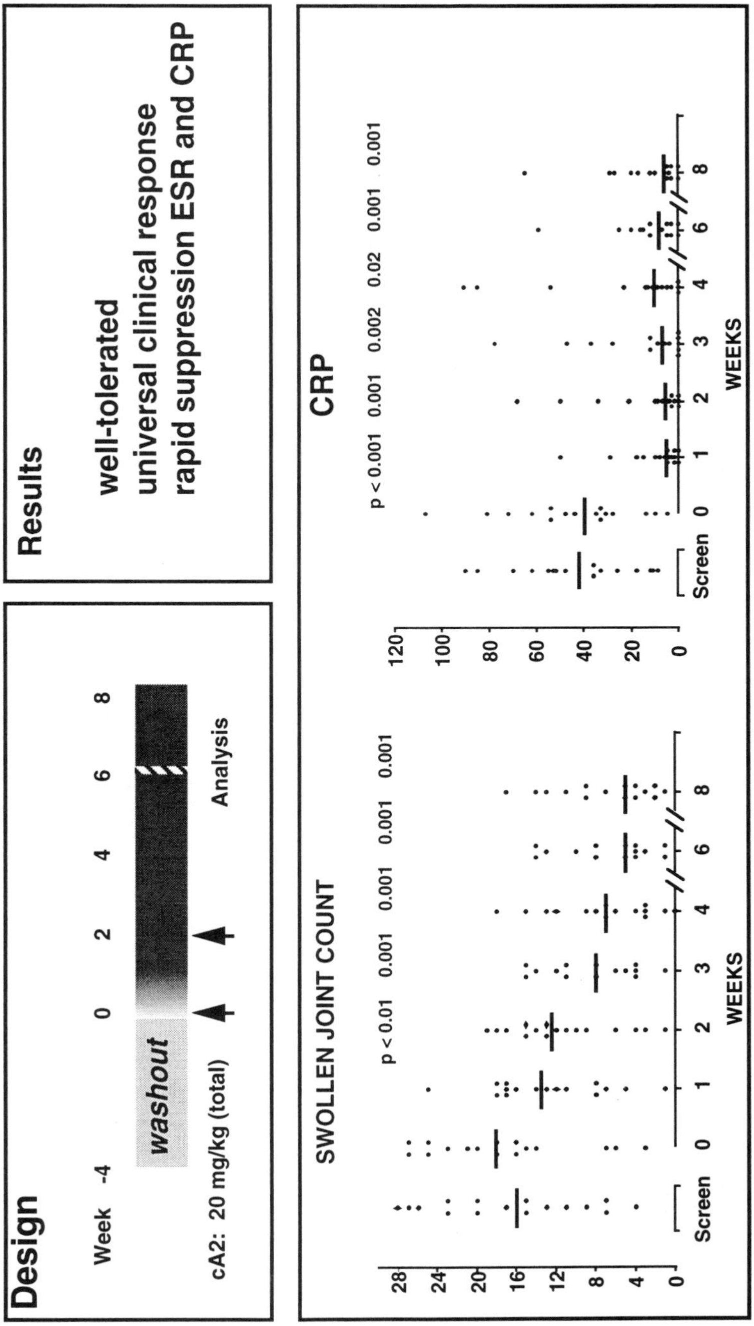

FIGURE 1. Open-label treatment with cA2 in RA. CRP, C reactive protein.

led by Professor J. Kalden in Erlangen, Professor J. Smolen in Vienna, and Professor F. Breedveld in Leiden. Seventy-three patients were enrolled in a randomized manner into three groups: a placebo (0.1% human serum albumin), 1 mg/kg, and 10 mg/kg of cA2, administered as a single intravenous infusion. Both investigators and patients were ignorant of which group they were in (blinded) until all the results were compiled and the case report forms completed for all patients.

The results[16] have formally proven that the blockade of TNFα by cA2 causes a marked clinical benefit, which is mirrored by all serum indices of activity. Only 8% of the placebo-treated patients met the preset criterion of improvement, the 20% "Paulus" compound score. By contrast, at 1 mg/kg, 44% of patients met these criteria, and a striking 80% of the 10 mg/kg cA2-treated patients improved. It is of interest that this is the first successful randomized placebo-controlled trial of a new bioengineered drug to show a clear-cut result, with a majority of patients improving with treatment. Surprisingly few other biological agents have been reported to be effective in placebo-controlled trials, and among these are two trials with anti-CD4 antibody, which were negative.[17,18] Two others reported clinically unimpressive results with IL-2 diphtheria toxin[19] and interferon-γ.[20]

Conclusions

It is now clear that effective TNFα blockade, exemplified by the use of cA2, is effective in most RA patients. Thus, although diseases with complex and unknown etiologies are often considered to be heterogeneous (*i.e.*, not all patients are DR4/DR1), our data suggest that in all RA patients the disease is critically dependent on excess TNFα, as judged by the degree, duration, and reproducibility of the clinical response. This pathway is thus an effective target for therapy.

Support for our conclusions has not been slow: we are aware, from meeting reports, of at least two other anti-TNF reagents that have been successfully used in RA. Celltech/Bayer has reported benefit in a trial led by Professors Isenberg and Panayi in London, using their anti-TNF antibody, and Dr. H. Fenner, representing Hoffman-La Roche, has reported efficacy with their IgG-TNF receptor fusion protein. Clearly a new area of targeted therapeutic research is opening up, and there are real prospects for patients to benefit.

SUMMARY

Rheumatoid arthritis is the most common of a number of diseases in which inflammation and tissue destruction is driven by an autoimmune process. Current therapy is inadequate, and this has prompted major research efforts, both in academia and industry, to understand more about the pathogenesis, and hence provide the rationale for new therapeutic strategies. Here we review our studies of cytokine expression and regulation in rheumatoid joints, which has culminated in demonstrating that TNFα blockade, using a chimeric (human IgG_1/K, mouse Fv) anti-TNFα antibody, cA2, markedly ameliorates arthritis. This defines a therapeutic target for rheumatoid arthritis.

ACKNOWLEDGMENTS

We appreciate the assistance of numerous colleagues over the years and the very generous gifts of cytokine reagents, chiefly from Industry, mainly from Genentech, Immunex, DNAX, Sandoz, Roche, Syntex, Ajinomoto, as well as from the laboratories of Professors T. Kishimoto and T. Taniguchi.

REFERENCES

1. Gregerson, P. K., J. Silver & R. J. Winchester. 1987. The shared epitope hypothesis: an approach to understanding the molecular genetics of susceptibility to rheumatoid arthritis. Arthritis Rheum. **30:** 1205–1213.
2. Buchan, G., K. Barrett, T. Fujita, T. Taniguchi, R. Maini & M. Feldmann. 1988. Detection of activated T cell products in the rheumatoid joint using cDNA probes to interleukin-2 (IL-2) IL-2 receptor and IFN-γ. Clin. Exp. Immunol. **71:** 295–301.
3. Buchan, G., K. Barrett, M. Turner, D. Chantry, R. N. Maini & M. Feldmann. 1988. Interleukin-1 and tumour necrosis factor mRNA expression in rheumatoid arthritis: prolonged production of IL-1α. Clin. Exp. Immunol. **73:** 449–455.
4. Brennan, F. M., D. Chantry, A. Jackson, R. Maini & M. Feldmann. 1989. Inhibitory effect of TNFα antibodies on synovial cell interleukin-1 production in rheumatoid arthritis. Lancet **2:** 244–247.
5. Hirano, T., T. Matsuda, M. Turner, N. Miyasaka, G. Buchan, B. Tang, K. Sato, M. Shimizu, R. Maini, M. Feldmann & T. Kishimoto. 1988. Excessive production of interleukin 6/B cell stimulatory factor-2 in rheumatoid arthritis. Eur. J. Immunol. **18:** 1797–1801.
6. Haworth, C., F. M. Brennan, D. Chantry, M. Turner, R. N. Maini & M. Feldmann. 1991. Expression of granulocyte-macrophage colony-stimulating factor in rheumatoid arthritis: regulation by tumor necrosis factor-α. Eur. J. Immunol. **21:** 2575–2579.
7. Brennan, F. M., D. Chantry, A. M. Jackson, R. N. Maini & M. Feldmann. 1989. Cytokine production in culture by cells isolated from the synovial membrane. J. Autoimmunity **2** (Suppl): 177–178.
8. Firestein, G. S. Z. N. 1990. How important are T cells in chronic rheumatoid synovitis? Arthritis Rheum. **33:** 768–773.
9. Pettipher, E. R., G. A. Higgs & B. Henderson. 1986. Interleukin 1 induces leukocyte infiltration and cartilage proteoglycan degradation in the synovial joint. Proc. Natl. Acad. Sci. USA **83:** 8749–8753.
10. Gowen, M., D. D. Wood, E. J. Ihrie, M. K. B. Mcguire & R. G. Russell. 1983. An interleukin-1 like factor stimulates bone resorption *in vitro*. Nature **306:** 378–380.
11. Williams, R. O., M. Feldmann & R. N. Maini. 1992. Anti-tumor necrosis factor ameliorates joint disease in murine collagen-induced arthritis. Proc. Natl. Acad. Sci. USA **89:** 9784–9788.
12. Thorbecke, G. J., R. Shah, C. H. Leu, A. P. Kuruvilla, A. M. Hardison & M. A. Palladino. 1992. Involvement of endogenous tumour necrosis factor α and transforming growth factor β during induction of collagen type II arthritis in mice. Curr. Opinions Immunol. **4:** 754–759.
13. Keffer, J., L. Probert, H. Cazlaris, S. Georgopoulos, E. Kaslaris, D. Kiossis & G. Kollias. 1991. Transgenic mice expressing human tumour necrosis factor: a predictive genetic model of arthritis. EMBO J. **10:** 4025–4031.
14. Elliott, M. J., R. N. Maini, M. Feldmann, A. Long-Fox, P. Charles, P. Katsikis, F. M. Brennan, J. Walker, H. Bijl, J. Ghrayeb & J. Woody. 1993. Treatment of rheumatoid arthritis with chimeric monoclonal antibodies to TNFα. Arthritis Rheum. **36:** 1681–1690.

15. ELLIOTT, M. J., R. N. MAINI, M. FELDMANN, A. LONG-FOX, P. CHARLES, H. BIJL & J. N. WOODY. 1994. Repeated therapy with a monoclonal antibody to tumour necrosis factor α (cA2) in patients with rheumatoid arthritis. Lancet **344:** 1125-1127.
16. ELLIOTT, M. J., R. N. MAINI, M. FELDMANN, J. R. KALDEN, C. ANTONI, J. S. SMOLEN, B. LEEB, F. C. BREEDVELD, J. D. MACFARLANE, H. BIJL & J. WOODY. 1994. Randomized double blind comparison of a chimaeric monoclonal antibody to tumour necrosis factor α (cA2) versus placebo in rheumatoid arthritis. Lancet **344:** 1105-1110.
17. MORELAND, L., P. PRATT & M. MAYES. 1993. Minimal efficacy of a depleting chimaeric anti-CD4 (cM-T412) in treatment of patients with refractory rheumatoid arthritis (RA) receiving concomitant methotrexate (MTX). Arthritis Rheum. **36:** 39.
18. VAN DER LUBBE, P. A., B. A. DIJKMANS & H. M. MARKUSSE. 1994. Lack of clinical effect of CD4 monoclonal antibody therapy in early rheumatoid arthritis: a placebo-controlled trial. Arthritis Rheum. In press.
19. MORELAND, L. W., K. W. SEWELL & W. F. SULLIVAN. 1993. Double-blind placebo-controlled phase II trial of diphtheria-interleukin-2 fusion toxin (DAB486IL-2) in patients with refractory rheumatoid arthritis (RA). Arthritis Rheum. **36:** 39.
20. HOFSCHNEIDER, P. H., U. WINTER, E. M. LEMMEL, B. BROLZ, W. GAUS, S. SCHMID, H. J. ORBERT & C. STETTER. 1992. Double-blind controlled phase III multicentre clinical trial with interferon gamma in rheumatoid arthritis. Rheumatol. Int. **12:** 185.

Investigation of Ligand Binding to Members of the Cytokine Receptor Family within a Microbial System

K. H. YOUNG AND B. A. OZENBERGER

American Cyanamid Company
Agricultural Research Division
Molecular and Cellular Biology Group
P. O. Box 400
Princeton, New Jersey 08543

Peptide ligand binding can induce a cellular response by specific interaction with a single receptor or by inducing specific receptor oligomerization. Single transmembrane domain receptors, such as those of the hematopoietic receptor superfamily, initiate signal transduction through such protein-protein interactions.[1] A microbial system for the functional expression of peptide ligands and corresponding receptors of this family would be invaluable for investigation of multiple protein associations that occur during receptor activation. The two-hybrid system[2] has been used in studies of intracellular protein interactions but is considered to be exclusionary of protein interactions that normally occur extracellularly.[3] We describe the expression of mammalian peptide hormones and extracellular domains of their cognate receptors in yeast. Reversible and specific ligand-receptor interactions were functionally integrated with an easily scorable yeast reporter system to create a novel biological forum for the investigation of these intricate protein associations.

Complementary DNA sequences encoding porcine growth hormone (GH)[4] or the extracellular region of the rat GH receptor (GHR)[4] were cloned into Gal4 fusion protein expression plasmids.[5] A yeast strain containing a UAS_{GAL}-*HIS3* reporter gene was transformed with both fusion plasmids or with a single fusion construct plus the opposing vector containing no heterologous DNA. All strains grew on nonselective medium (FIG. 1a). Only the strain containing both hybrid proteins grew on selective medium (FIG. 1b), suggesting that GH and GHR can mediate the reconstitution of Gal4 activity in an interaction suggestive of ligand-receptor binding. To substantiate this finding, the system was expanded to add a third plasmid that expresses "free ligand." Concurrent expression of nonfusion GH peptide with the GH and GHR fusions reversed the growth phenotype (FIG. 1b), suggesting *in vivo* competition and illustrating the reversibility of the ligand-receptor interaction. Similar results were observed in an analogous system established for prolactin (PRL)[4] and the PRL receptor (PRLR).[4] When PRL was expressed concurrently with the GH and GHR fusions, however, the growth phenotype was not reversed, demonstrating specificity of the GH-GHR fusion protein interaction.

The demonstration of ligand-receptor binding enabled the investigation of more complex protein interactions. In an alternative application of this technology, ligand-dependent receptor dimerization was examined by expressing two receptor extracellu-

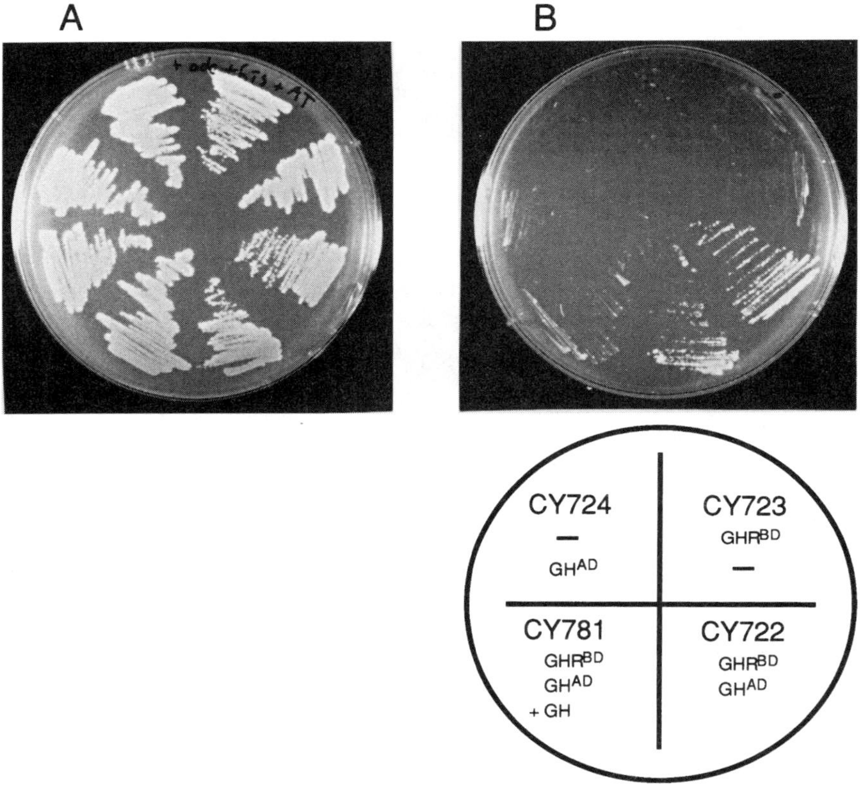

FIGURE 1. Growth assays of strains expressing GHR and GH fusion proteins. The position and the expressed heterologous fusion proteins of each strain are indicated in the diagram. Superscript denotes fusion partner as either the activation domain (AD) or DNA binding domain (BD) of the yeast Gal4 transcriptional activation protein. Ligand without superscript denotes expression as nonfusion protein. A dash denotes the opposing vector containing no heterologous cDNA. Two independent isolates of each strain were streaked on synthetic medium supplemented with 0.1 mM adenine, 60 mM 3-amino-triazole, and 0.3 mM histidine (nonselective, plate A), or the same medium lacking histidine (selective, plate B) and incubated at 30°C for 3 or 5 days, respectively.

lar domains as fusion proteins and the ligand as a nonfusion protein. The cDNAs encoding the mitogenic peptide vascular endothelial growth factor (VEGF)[4] and its receptor Flk1/KDR,[4] a type III tyrosine kinase receptor, were cloned into appropriate vectors. Yeast strains that expressed any single KDR fusion plasmid and VEGF, or the two KDR plasmids in the absence of VEGF, failed to grow or grew poorly on selective medium. Only the strain that expressed both fusion proteins plus VEGF exhibited enhanced growth on selective medium (Fig. 2). Similar results have been

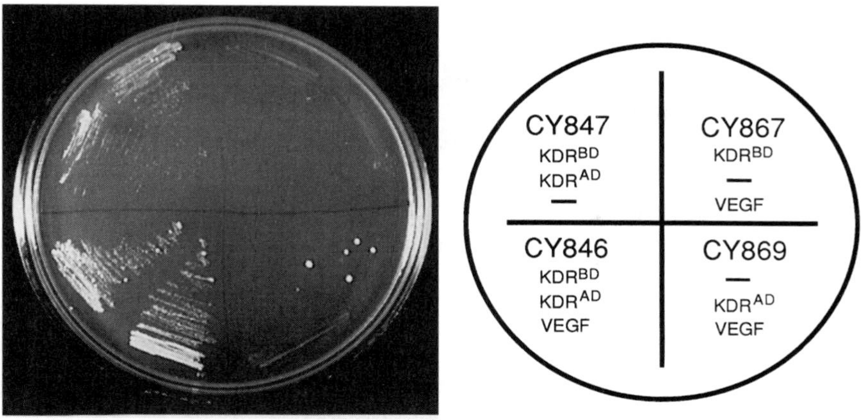

FIGURE 2. Growth assays of strains expressing KDR fusion proteins and VEGF. The position and expressed heterologous proteins of each strain are shown in the diagram. Superscript denotes fusion partner as either the activation domain (AD) or DNA binding domain (BD) of the yeast Gal4 transcriptional activation protein. Ligand without superscript denotes expression as nonfusion protein. A dash denotes the opposing vector containing no heterologous cDNA. Two independent isolates of each strain were streaked on selective medium, as described in FIG. 1, and incubated at 30°C for 5 days.

observed for GH and GHR. These data demonstrate that ligand-induced receptor dimerization can also be investigated in this microbial system.

The novel findings presented in this report demonstrate that our variation on the two-hybrid microbial expression system can be used to examine protein-protein interactions that normally occur at the extracellular surface. This technology introduces intriguing new approaches for rapid genetic dissection of ligand-receptor interactions.

REFERENCES

1. SATO, N. & A. MIYAJIMA. 1994. Curr. Opinion Biol. **6:** 174-179.
2. CHEIN, C.-T., P. L. BARTEL, R. STERNGLANZ & S. FIELDS. 1991. Proc. Natl. Acad. Sci. USA **88:** 9578-9582.
3. FIELDS, S. & R. STERNGLANZ. 1994. Trends Genet. **10:** 286-292.
4. All sequences available in GENBANK.
5. DURFEE, T., K. BECHERER, P.-L. CHEN, S.-H. YEH, Y. YANG, A. E. KILBURN, W.-H. LEE & S. E. ELLEDGE. 1993. Genes & Dev. **7:** 555-569.

The Proline-rich Motif Is Necessary but Not Sufficient for Prolactin Receptor Signal Transduction

KEVIN D. O'NEAL,[a,c] LI-YUAN YU-LEE,[b] AND WILLIAM T. SHEARER[a,b]

[a]*Departments of Allergy/Immunology Texas Children's Hospital* and

[b]*Microbiology and Immunology Baylor College of Medicine Houston, Texas 77030*

The prolactin receptor (PRL-R) is a member of the cytokine/hematopoietin receptor superfamily, which is involved in the growth and differentiation of hematopoietic cells. The superfamily was originally defined by the conservation of two extracellular amino acid (aa) sequence motifs among members, including two Cys pairs and the five aa Trp-Ser-X-Trp-Ser (X is any aa) motif.[1] Although a homology region containing box 1 and box 2 had been described for some of the cytokine receptors,[2] no intracellular motif shared among all family members had been characterized. By allowing similar aa to be grouped together (aliphatic (Al)=Val, Leu, Ile; aromatic (Ar)=Phe, Trp), the first intracellular consensus sequence, called the proline-rich motif (PRM), was identified.[3] Located near the membrane, the PRM is an eight aa motif that overlaps with box 1 and has the consensus sequence Al-Ar-Pro_1-X-Al-Pro_2-X-Pro_3. In the PRL-R, the PRM sequence is ^{243}Ile-Phe-Pro-Pro-Val-Pro-Gly-Pro^{250}. Mutagenesis or deletion of the PRM in several cytokine receptors has demonstrated that the PRM is critical for receptor function. For example, mutagenesis of Pro_2 and Pro_3 to Ser in gp130, the signal transducer for IL-6, abolished growth signaling and IL-6 induced gp130 tyrosine phosphorylation.[2] To further examine the role of the PRM in the PRL-R, double Ala substitutions were introduced into the PRM of the Nb2 PRL-R, and the mutant receptors were analyzed for their ability to transduce a signal to the PRL-inducible immediate-early gene interferon regulatory factor-1 (IRF-1).[4]

Three different double Ala mutants were generated by overlap-extension PCR and verified by sequencing (FIG. 1). Ala was substituted for Phe and Val (Nb2-FV), Pro_1 and Pro_2 (Nb2-P_1P_2), and Pro_2 and Pro_3 (Nb2-P_2P_3). Mutant or wild-type (wt) Nb2 PRL-Rs in the expression vector pECE were transiently cotransfected by electroporation into an IL-3-dependent premyeloid cell line, FDC-P1, with a reporter con-

[c]Address correspondence to Dr. Kevin D. O'Neal, Texas Children's Hospital, Department of Allergy/Immunology, MC 1-3291, 6621 Fannin St., Houston, TX 77030.

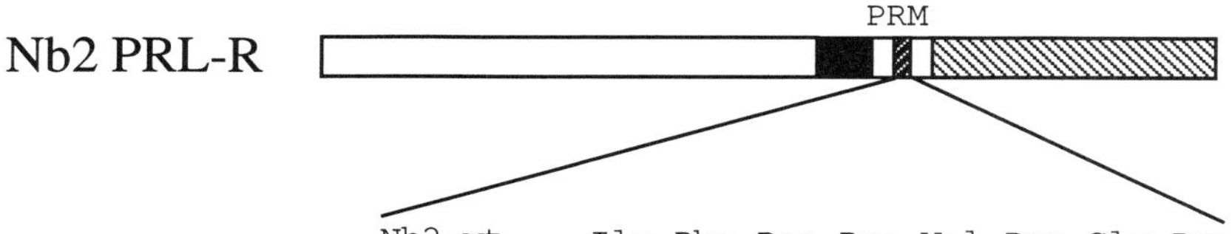

Nb2-wt	Ile-Phe-Pro-Pro-Val-Pro-Gly-Pro
Nb2-FV	Ile-**Ala**-Pro-Pro-**Ala**-Pro-Gly-Pro
Nb2-P_1P_2	Ile-Phe-**Ala**-Pro-Val-**Ala**-Gly-Pro
Nb2-P_2P_3	Ile-Phe-Pro-Pro-Val-**Ala**-Gly-**Ala**

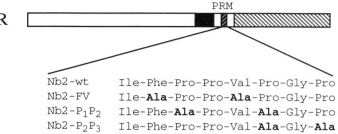

FIGURE 1. Double alanine substitutions in the PRM of the Nb2 PRL-R. From left to right in the receptor drawing are the extracellular domain (in white), the transmembrane region (in black), the PRM (small hatched box, labeled) within the first 27 aa of the cytoplasmic domain (in white), and the remaining cytoplasmic domain (cross-hatched) that is alternatively spliced. The aa sequences of the wild type (wt) and each of the three PRM mutants are shown with the Ala substitutions in boldface.

struct, containing the IRF-1 promoter (IRF-1-CAT) and treated with or without PRL for 40 h, as previously described.[5] An average of three separate experiments (± SEM) showed that all three mutant receptors were significantly impaired in their ability to activate the IRF-1 promoter when compared with the wild-type receptor: Nb2-wt (100%), Nb2-FV (34 ± 5%), Nb2-P_1P_2 (14 ± 3%), and Nb2-P_2P_3 (4 ± 2%) (FIG. 2).

These results indicate that aa residues within the PRM, both prolines and nonprolines, are important for Nb2 PRL-R signal transduction. The most severe mutant, Nb2-P_2P_3, essentially abolished receptor function, similar to the effect seen in the analogously mutated gp130.[2] Experiments are in progress to verify that the observed signaling defects is not due to an absence of expression on the cell surface.

The three forms of the PRL-R (short, Nb2, and long) differ only in the length and sequence of the cytoplasmic domain. Interestingly, all three forms contain the conserved PRM, but only the Nb2 and long PRL-Rs are able to activate IRF-1-CAT.[5] Taken together, these results suggest that the PRM is necessary but not sufficient for PRL-R signal transduction. Other regions of the cytoplasmic domain appear to be important to act in concert with the PRM for PRL-R function. The biochemical role of the PRM is unknown. It may serve as a structural motif or perhaps as a binding site for an SH3 domain-containing protein. A peptide corresponding to the PRL-R PRM was synthesized and displayed two conformers in equilibrium when analyzed by reverse-phase HPLC, most likely as the results of the *cis-trans* isomerization of one of the X-Pro peptide bonds.[6] In combination with modeling studies, these findings suggest a novel function for the PRM in cytokine receptors. The PRM may represent an endogenous ligand for an immunophilin, whose peptidly-prolyl *cis-trans* isomerase activity could isomerize the PRM and alter the conformation and possibly the activity state of the receptor. Further experiments are needed to test these hypotheses.

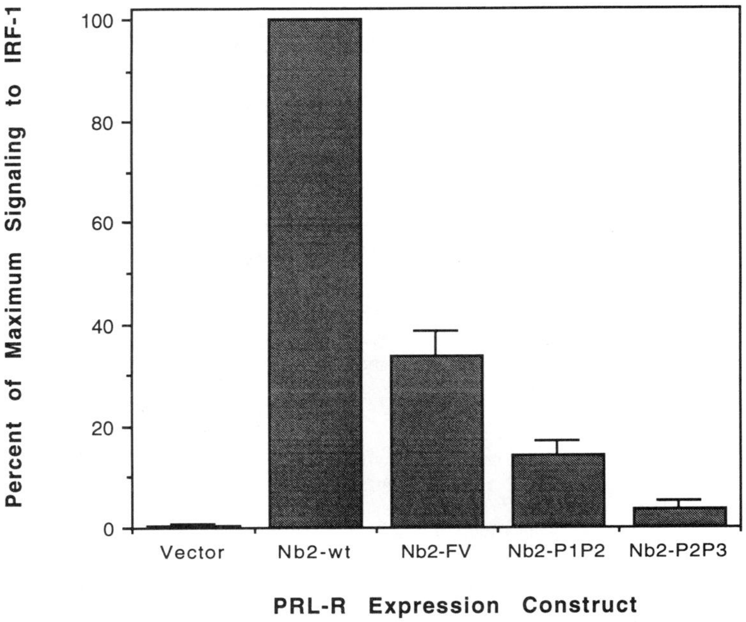

FIGURE 2. Relative signal transduction of wild-type and PRM mutant Nb2 PRL-Rs. The fold induction of the wt Nb2 PRL-R IRF-1-CAT in the presence of PRL was determined as described,[5] and was defined as 100%. The ratio of the fold PRL induction of the mutant receptors divided by the wt receptor gives the relative signaling capability, shown as a percentage. The values are provided in the text, and the control vector (with no PRL-R insert) was 0.4 ± 0.2% of maximum.

REFERENCES

1. Bazan, J. F. 1989. Biochem. Biophys. Res. Comm. **164:** 788-795.
2. Murakami, M., M. Narazaki, M. Hibi, H. Yawata, K. Yasukawa, M. Hamaguchi, T. Taga & T. Kishimoto. 1991. Proc. Natl. Acad. Sci. USA **88:** 11349-11353.
3. O'Neal, K. D. & L.-y. Yu-Lee. 1993. Lymphokine Cytokine Res. **12:** 309-312.
4. Yu-Lee, L.-y., J. A. Hrachovy, A. M. Stevens & L. A. Schwarz. 1990. Mol. Cell. Biol. **10:** 3087-3094.
5. O'Neal, K. D. & L.-y. Yu-Lee. 1994. J. Biol. Chem. **269:** 26076-26082.
6. O'Neal, K. D., M. V. Chari, C. H. McDonald, R. G. Cook, L.-y. Yu-Lee, J. D. Morrisett & W. T. Shearer. In preparation.

Activation of Multiple Protein Kinases by Interleukin-1[a]

J. E. DUNFORD, N. R. CORBETT, C. L. VARLEY,
C. H. DAWSON,[b] B. L. BROWN, AND
P. R. M. DOBSON[b]

Department of Human Metabolism and Clinical Biochemistry
[b]Institute for Cancer Studies
Sheffield University Medical School
Sheffield, S10 2RX, United Kingdom

Despite extensive research on the signaling mechanisms of interleukin-1 (IL-1), there is little consensus as to its mode of action.[1] It has been suggested that lipid second messengers, such as diradylglycerol and ceramide, are involved, which may activate kinase cascades.[2–4] Here we have investigated the role of protein kinases in IL-1 signal transduction in both the EL4 mouse thymoma and T47D human breast cancer cell lines, and the potential involvement of a lipid activator of kinases in T47D cells. Briefly, protein extracts were resolved on an SDS-polyacrylamide gel containing myelin basic protein (MBP) as an enzyme substrate. After denaturation and renaturation of the proteins, gels were incubated with [γ-^{32}P]ATP and the bands of kinase activity detected by autoradiography. It is clear that different kinases are activated by IL-1 in the two cell types. A double band of kinase activity at 42/44 kDa was observed to be rapidly and transiently increased by IL-1 in T47D cells (FIG. 1), but not in EL4 cells unless stimulated with phorbol myristate acetate (PMA) (TABLE 1). Western blotting analysis using antimitogen activated protein (MAP) kinase antibody also revealed a double band at 42/44 kDa. As MBP is an artificial substrate for MAP kinase, it is likely that these bands of kinase activity are the p42/44 MAP kinases, ERK2 and ERK1. Other proteins with apparent molecular masses of 140, 105, 90, and 59 kDa were also observed. The activity of the 59 kDa kinase was found to be rapidly and transiently increased in response to IL-1, and the 140 kDa was found to be activated over a longer time course, although this was not statistically significant in every experiment. All kinases were activated strongly by PMA (TABLE 1). Interestingly, IL-4, like IL-1, activated kinases using MBP as substrate with apparent molecular masses of 42, 44, and 59 kDa as well as a 54 kDa protein. The effects were additive in cells stimulated with both cytokines simultaneously. It is reported that IL-1 activates a kinase that phosphorylates threonine T669 of the epidermal growth factor receptor.[5] Our work shows activation of this kinase in both EL4 cells and T47D cells, regardless of MAP kinase activity, and indicates that a separate kinase is responsible for this phosphorylation.

[a]We thank the Yorkshire Cancer Research Campaign and the Arthritis and Rheumatism Council for financial support.

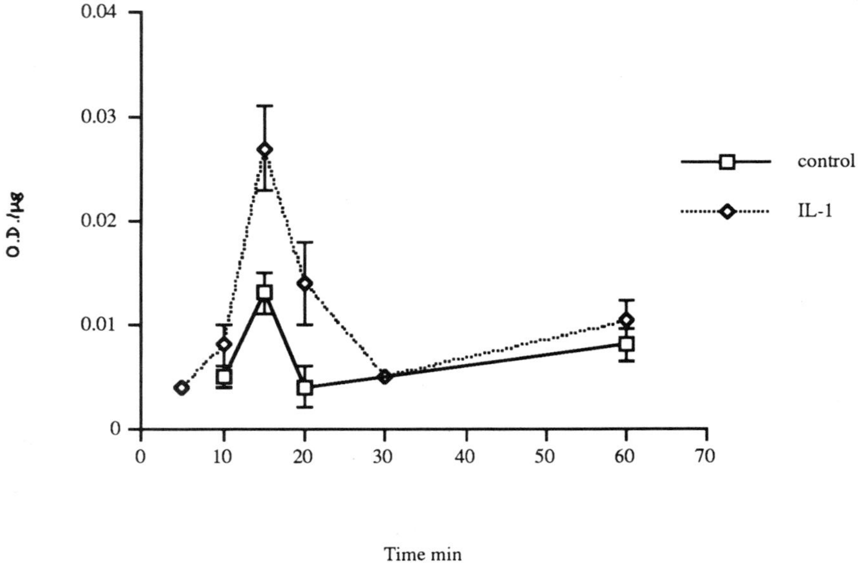

FIGURE 1. A time course of p42/44 MBP kinase activity.

To investigate lipid-activation of kinases, protein extracts from unstimulated cells were resolved on MBP gels that were then incubated simultaneously with [γ-^{32}P]ATP and either purified lipids or lipid extracts from IL-1-stimulated cells. The stabilized analogue of ceramide, C_6-ceramide, activated kinases with molecular masses of 95, 55, and 50 kDa. Kinases of similar molecular masses were activated by lipid extracts from IL-1-stimulated cells, providing circumstantial evidence for a kinase cascade initiated by the release of ceramide in T47D cells (TABLE 1).

In conclusion, multiple kinases, some of which may be affected by lipid mediators, are apparently activated by IL-1, and different kinase cascades may be important for IL-1 signaling in different cell types.

TABLE 1.

(a) Increase in Kinase Activity in EL4 Cells (fold stimulation)

Kinase	Cell Treatment	
	PMA (100 nM)	IL-1 (60 pM)
p42/44	8.1	1
p59	6.9	3.0
p90	7.1	1
p140	4.0	2.3

(b) Increase in Kinase Activity in T47D Cells (fold stimulation)

Kinase	Cell Treatment	
	PMA (100 nM)	IL-1 (60 pM)
p42/44	4.6	3.2
p59	1.3	2
p90	2.7	1.1
p140	1.8	1.5

(c) Increase in Kinase Activity due to Lipid Activation (fold stimulation)

Kinase	Kinase Treatment	
	ceramide (20 μM)	lipid extract
p50	1.9	1.7
p55	1.9	2.2
p90	1.8	1.6

REFERENCES

1. DOWER, S. K., J. E. SIMMS, D. P. CERRETTI & T. A. BIRD. 1992. *In* Interleukins: Molecular Biology and Immunology. T. Kishimoto, Ed.: Vol. 51: 33-64. Chemical Immunology. Karger. Basel.
2. DOBSON, P. R. M. & B. L. BROWN. 1989. *In* Advances in Second Messengers and Protein Phosphorylation Research. P. Greengard & A. Robison, Eds.: Vol. 24:170-175. Raven Press. New York.
3. DOBSON, P. R. M., C. P. PLESTED, D. R. JONES & B. L. BROWN. 1990. *In* Molecular and Cellular Biology of Cytokines. J. J. Oppenheim *et al.*, Eds.: 209-214.
4. KOLESNIK, R. & D. W. GOLDE. 1994. Cell **77:** 325-328.
5. BIRD, T. A., P. R. SLEATH, P. C. DEROOS, S. K. DOWER & G. D. VIRCA. 1991. J. Biol. Chem. **33:** 22661-22670.

IL-8 Signal Transduction in Human Neutrophils

CINDY KNALL,[a,f] G. SCOTT WORTHEN,[b,c] ANNE METTE BUHL,[a,e] AND GARY L. JOHNSON[a,d]

[a]*Division of Basic Sciences and*
[b]*Department of Medicine*
National Jewish Center for Immunology and Respiratory Medicine
Departments of [c]*Medicine and* [d]*Pharmacology*
University of Colorado Health Sciences Center
Denver, Colorado

[e]*Division of Biostructural Chemistry*
Department of Chemistry
University of Aarhus
Aarhus, Denmark

Receptors for the chemokine subfamily of cytokines, such as IL-8, are seven transmembrane-spanning, G protein-coupled receptors.[1] This is in contrast to the other cytokine receptors that are tyrosine-kinase coupled.[2] Although IL-8 induces a variety of cellular functions in polymorphonuclear (PMN) leukocytes, including chemoattraction and the respiratory burst,[3] the signal transduction pathway(s) triggering these activities remains undefined. Therefore, we have set out to analyze the signaling pathways activated by IL-8.

The mitogen-activated protein kinase (MAPK) pathway is triggered by a variety of cell-surface receptors through a series of sequential protein phosphorylation events that cause the activation of Raf, MAPK/ERK kinase (MEK), and MAPK.[4] We initially investigated the activation of MAPK and Raf in human PMNs upon IL-8 stimulation. Treatment of PMNs with IL-8 induced a time-dependent activation of MAPK (FIG. 1a). Furthermore, the level of MAPK activation was similar to that stimulated by C5a, a classic chemoattractant (data not shown). However, the duration of MAPK activation was prolonged in response to IL-8 as compared to that reported for C5a.[5] IL-8 also activated both Raf-1 (FIG. 1b) and B-Raf (FIG. 1c) in PMNs, although to a lesser extent than C5a. Similar to C5a,[5] IL-8 activated B-Raf for a longer period of time than Raf-1. Therefore, the MAPK activation may reflect a combination of both Raf-1 and B-Raf activation of MEK, with B-Raf activity accounting for the majority of the later MAPK activation.

Because Ras can activate Raf, we also looked for activation of the guanine nucleotide exchange activity of Ras upon IL-8 stimulation. Unlike C5a,[5] IL-8 did

[f]Address for correspondence: National Jewish Center for Immunology and Respiratory Medicine, 1400 Jackson Street, Denver, CO 80206.

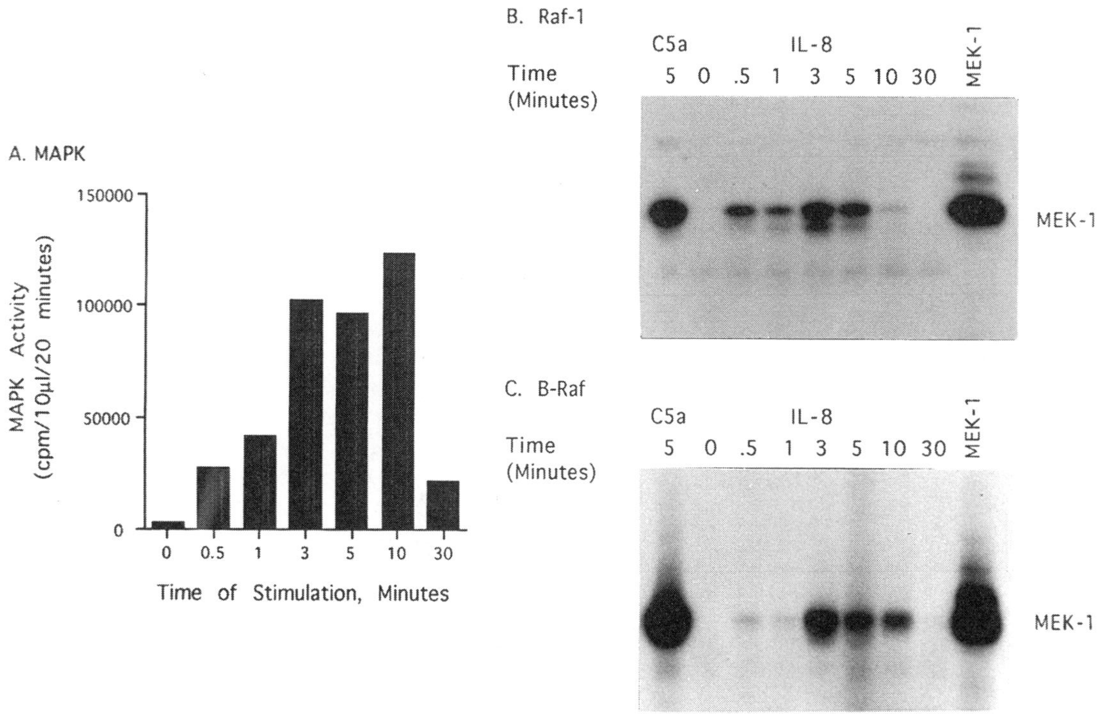

FIGURE 1. IL-8 stimulates MAPK, Raf-1, and B-Raf. Human PMNs that had been preincubated with protease inhibitors were stimulated with a final concentration of 25 nM IL-8 (recombinant, monocyte-derived) for various times and then assayed for either (A) MAPK, (B) Raf-1, or (C) B-Raf activity.[5] MAPK activity was determined by passing cell lysates (6×10^7 cell equivalents) over DEAE-Sephacel, eluting MAPK with 0.5 M NaCl, and then assaying for kinase activity against an epidermal growth factor receptor–derived peptide substrate. Raf activity was measured by immunoprecipitation of Raf-1 or B-Raf from cell lysates (4×10^7 cell equivalents) with anti-Raf-1 or anti-Raf-B antibodies and then assaying for kinase activity against a kinase-deficient MEK-1 substrate. Wild-type MEK-1 was used as a control for the *in vitro* kinase reaction.

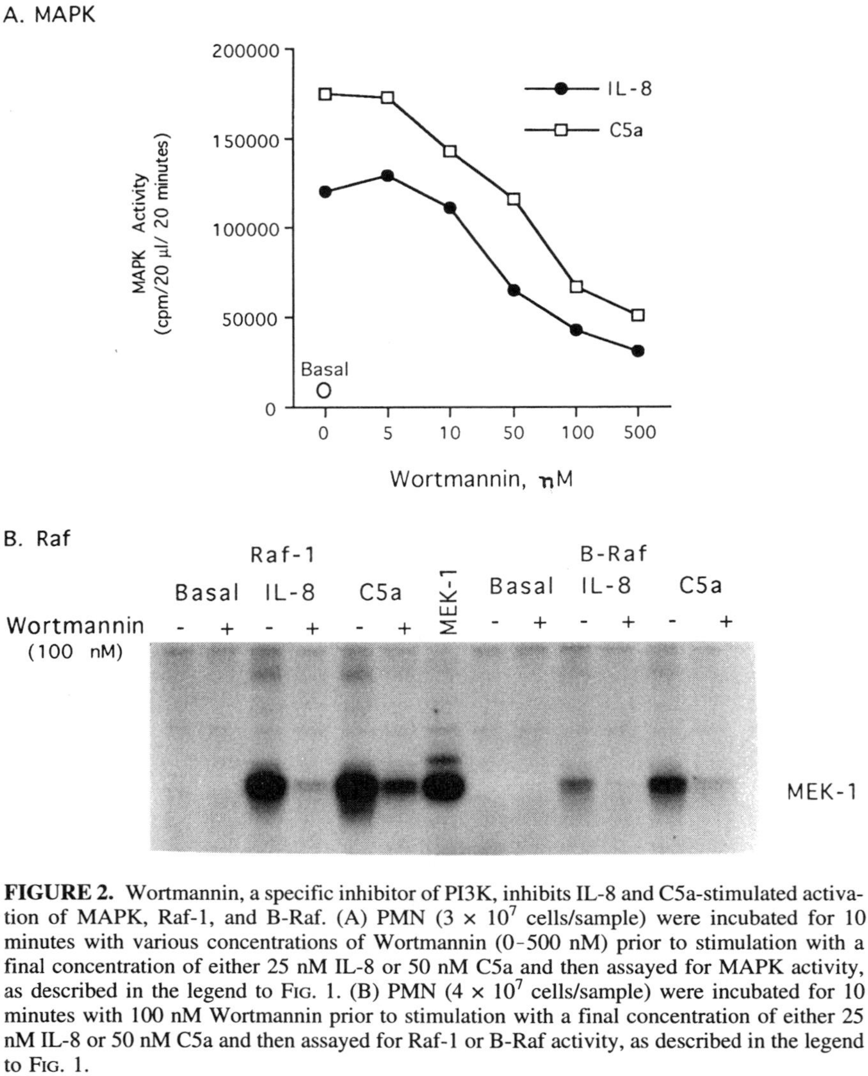

FIGURE 2. Wortmannin, a specific inhibitor of PI3K, inhibits IL-8 and C5a-stimulated activation of MAPK, Raf-1, and B-Raf. (A) PMN (3×10^7 cells/sample) were incubated for 10 minutes with various concentrations of Wortmannin (0–500 nM) prior to stimulation with a final concentration of either 25 nM IL-8 or 50 nM C5a and then assayed for MAPK activity, as described in the legend to FIG. 1. (B) PMN (4×10^7 cells/sample) were incubated for 10 minutes with 100 nM Wortmannin prior to stimulation with a final concentration of either 25 nM IL-8 or 50 nM C5a and then assayed for Raf-1 or B-Raf activity, as described in the legend to FIG. 1.

not stimulate any detectable guanine nucleotide exchange on Ras (data not shown). This may reflect an inability to detect a low level of exchange triggered by IL-8, inasmuch as the level of exchange induced by C5a was only twofold over background.[5] Alternatively, IL-8 may activate Raf by way of a PKC-dependent event.[4]

Previous reports have demonstrated the activation of phospho-inositide-3-kinase (PI3K) in PMNs stimulated by formylated Met-Leu-Phe,[6] a classic chemoattractant, and the requirement of PI3K activity for the respiratory burst.[7] We, therefore, tested the effect of Wortmannin, a specific inhibitor of PI3K,[7] on IL-8 and C5a-stimulated activation of MAPK and Raf. Surprisingly, Wortmannin inhibited both IL-8 and C5a-stimulated activation of MAPK (FIG. 2a), and Raf-1 and B-Raf (FIG. 2b). Furthermore, this inhibition was not directed at MAPK or Raf directly, inasmuch as addition of the inhibitor directly to the *in vitro* kinase reactions had no effect on the enzymatic activity of either of these molecules (data not shown). The data thus suggest that PI3K contributes to the activation of the Raf/MAPK pathway in PMNs stimulated by either classic chemoattractants or chemokines.

These data indicate that, like C5a, IL-8 activates the MAPK pathway through a Raf-mediated event. However, IL-8-stimulated activation of MAPK is prolonged, which may represent a contribution of both Raf-1 and B-Raf to the activation of MEK and subsequent activation of MAPK. Additionally, it appears that PI3K contributes to the activation of both Raf-1 and B-Raf, and the subsequent activation of MAPK. How this occurs is still unknown. Also, how this pathway activates or contributes to the activation of the various cellular functions of PMNs remains to be determined.

REFERENCES

1. MURPHY, P. M. 1994. Annu. Rev. Immunol. **12:** 593-633.
2. KISHIMOTO, T., T. TAGA & S. AKIRA. 1994. Cell **76:**253-262.
3. OPPENHEIM, J. J., C. O. C. ZACHARIAE, N. MUKAIDA & K. MATSUSHIMA. 1991. Annu. Rev. Immunol. **9:** 617-648.
4. BLUMER, K. J. & G. L. JOHNSON. 1994. TIBS **19:** 236-240.
5. BUHL, A. M., N. AVDI, G. S. WORTHEN & G. L. JOHNSON. 1994. Proc. Natl. Acad. Sci. USA. In press.
6. STEPHENS, L., T. JACKSON & P. T. HAWKINS. 1993. J. Biol. Chem. **268:** 17162-17172.
7. ARCARO, A. & M. P. WYMANN. 1993. Biochem. J. **296:** 297-301.

IL-8 Induces Calcium Mobilization in Interleukin-2-activated Natural Killer Cells Independently of Inositol 1,4,5 Trisphosphate[a]

ALA AL-AOUKATY, ADEL GIAID, AND
AZZAM A. MAGHAZACHI[b]

Northeastern Ontario Regional Cancer Centre
Department of Chemistry-Biochemistry
Laurentian University
Sudbury, Ontario, P3E 5J1
and
Department of Medicine
Faculty of Medicine
University of Ottawa
Ottawa, Ontario K1N 6N5
and
Montreal General Hospital
Montreal, Quebec, Canada H3G 1A4

IL-8 is a proinflammatory cytokine secreted by several cell types.[1] This cytokine is pleiotropic and induces the *in vitro* chemotaxis of neutrophils and T cells.[2] It belongs to the C-X-C branch of chemotactic cytokines, "chemokines."[3] IL-8-induced lymphocyte chemotaxis was inhibited by pertussis toxin (PT).[4] Similarly, PT inhibited IL-8-induced Ca^{2+} mobilization [5] and IL-8 augmentation of neutrophil growth inhibitory activity for *Candida albicans*.[6] These results suggest that IL-8 receptors are linked to guanine nucleotide binding (G) proteins. Molecular cloning confirmed the above contention. Receptors for IL-8 were cloned by several groups and found to be members of the seven transmembrane spanning domain family of receptors,[7–9] which characteristically bind G proteins. Along these lines, we have investigated whether IL-8 induces the locomotion of the antitumor effectors and the interleukin-2-activated natural killer (IANK) cells, and reported that IL-8 induces the random motility of human IANK cells which was mediated by G proteins, and in particular, G_o.[10]

Here we demonstrate that IL-8 induces the mobilization of intracellular Ca^{2+} upon incubation with either intact or streptolysin O (SLO)-permeabilized IANK cells.

[a] This work was supported by a Grant from the Northern Cancer Research Foundation. Dr. A. A. Maghazachi is a Career Scientist of the Ontario Cancer Treatment and Research Foundation. Dr. A. Giaid is supported by the Heart and Stroke Foundation of Canada.

[b] Address correspondence to Dr. A. A. Maghazachi, N.E.O.R.C.C., 41 Ramsey Lake Road, Sudbury, Ontario P3E 5J1, Canada.

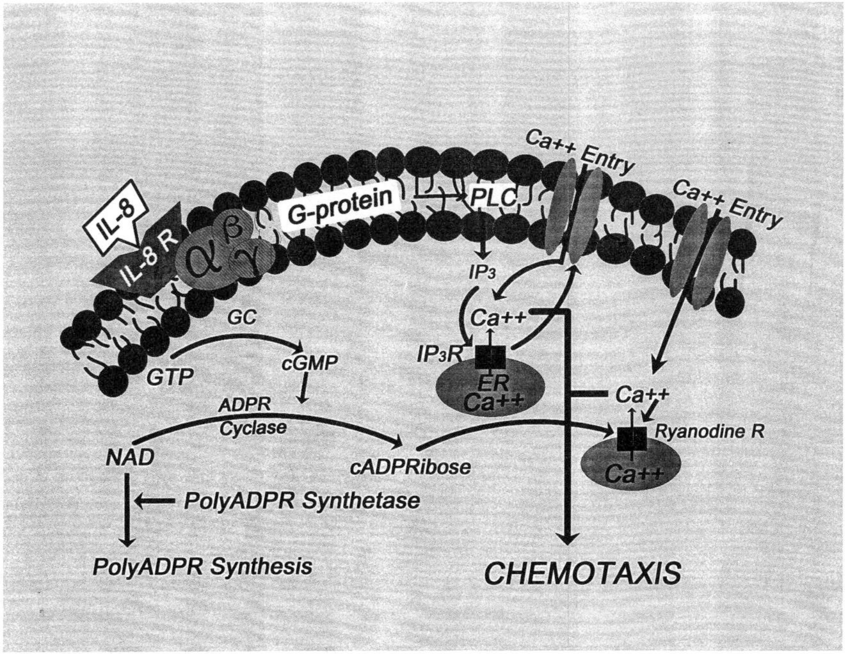

FIGURE 1. Upon binding of IL-8 to its putative receptors on IANK cells, it activates the G_s protein, which in turn may activate the guanylyl cyclase (GC), resulting in the conversion of GTP into cGMP. cGMP may activate the adenosine diphosphate ribosyl cyclase (ADPR-cyclase), which converts nicotinamide adenine dinucleotide (NAD) into cADP ribose. The latter may activate the ryanodine receptors, resulting in the efflux of calcium. Furthermore, IL-8 inhibits the synthesis of poly-ADPR synthetase. Also shown is a more familiar pathway of the accumulation of intracellular Ca^{2+} in mammalian cells. The latter pathway operates after the activation of G_q protein, which stimulates phospholipase C, resulting in the accumulation of Ins1, 4, 5 P_3 and the recruitment of calcium from the endoplasmic reticulum. This pathway may operate in IANK cells; however, it is unlikely that IL-8 activates it.

Furthermore, incorporation of anti-α_s virtually completely inhibited IL-8-induced $(Ca^{2+})_i$ mobilization in SLO-permeabilized IANK cells. These results suggest that IL-8 receptors in IANK cells may be linked to G_s protein, which mediates the signal necessary to mobilize $(Ca^{2+})_i$.

The ability of IL-8 to induce the mobilization of intracellular Ca^{2+} suggests that this cytokine may induce the accumulation of Ins1, 4, 5 P_3. However, incubation of IANK cells with IL-8 did not induce the accumulation of Ins1, 4, 5 P_3. In addition to the Ins1, 4, 5 P_3/Ins1, 4, 5 P_3 receptors (R), another pathway of stimulating the accumulation of intracellular Ca^{2+} is operative. This pathway uses cyclic adenosine diphosphate ribose (cADPR), which stimulates the ryanodine receptors (RR) to induce the efflux of Ca^{2+} from intracellular sources.[11] ADP-ribosyl cyclase, which catalyzes the conversion of cADPR from nicotinamide dinucleotide (NAD), shows about 70% homology to the CD38 B-cell marker,[12] suggesting that the cADPR/RR pathway may be functional in nonexcitable blood cells. Interestingly, cADPR is antagonized by poly-ADP-ribosyl synthetase, which uses NAD to form poly-ADP ribose.[13] Hence, depletion of NAD by poly-ADP-ribosyl synthetase results in the reduction of cADPR synthesis and, consequently, in the inhibition of the accumulation of intracellular Ca^{2+} by the cADPR/RR pathway.[14] Conversely, inhibition of the poly-ADP-ribosyl synthetase activity may result in the availability of NAD for ADP-ribosyl cyclase, which converts NAD into cADPR, resulting in the mobilization of $(Ca^{2+})_i$ through the cADPR/RR receptors.

To this end, we examined whether IL-8 affects the synthesis of poly-ADP-ribosyl synthetase and showed that IL-8 reduces the incorporation of $[^3H]$NAD in IANK cells. These results suggest that IL-8 may be using the cADPR/RR as an alternative pathway to the Ins1, 4, 5 P_3/Ins1, 4, 5 P_3R for the accumulation of intracellular calcium in IANK cells. Because the cADP ribose pathway is activated by cyclic GMP,[14] we have measured the level of cyclic GMP in IANK cells upon incubation with IL-8. Our results showed that IL-8 rapidly induces the accumulation of endogenous cyclic GMP in these cells, providing further support to the conclusion that IL-8 may induce the mobilization of $(Ca^{2+})_i$ through the activation and synthesis of cyclic ADP ribose.

FIGURE 1 shows the proposed signaling pathway of IL-8 in IANK cells. The possible coupling of G proteins, and in particular G_s to the cADPR/RR pathway in IANK cells, is highly important and is very exciting. G_s, which was found to be the most important G protein in mediating the cytolytic activity of IANK cells,[15] may play a major role in various biological functions of these cells.

REFERENCES

1. SMYTH, M. J., C. O. C. ZACHARIAE, Y. NORIHISA, J. R. ORTALDO, A. HISHINUMA & K. MATSUSHIMA. 1991. J. Immunol. **146:** 3815-3823.
2. LARSEN, C. G., A. O. ANDERSON, E. APPELLA, J. J. OPPENHEIM & K. MATSUSHIMA. 1989. Science **243:** 1464-1466.
3. SCHALL, T. J. 1991. Cytokine **3:** 165-183.
4. BACON, K. B. & R. D. R. CAMP. 1990. Biochem. Biophys. Res. Commun. **169:** 1099-1104.
5. DEWALD, B., B. THELEN & M. BAGGIOLINI. 1988. J. Biol. Chem. **263:** 16179-16184.

6. Djeu, J. Y., K. Matsushima, J. J. Oppenheim, K. Shiotsuki & D. K. Blanchard. 1990. J. Immunol. **144:** 2205-2210.
7. Murphy, P. M. & H. L. Tiffany. 1991. Science **253:** 1280-1283.
8. Holmes, W. E., J. Lee, W. J. Kuang, C. G. Rice & W. I. Wood. 1991. Science **253:** 1278-1280.
9. Cerretti, D. P., C. J. Kozlosky, T. Vanden Bos, N. Nelson, D. P. Gearing & M. P. Beckmann. 1993. Mol. Immunol. **30:** 359-367.
10. Sebok, K., D. Woodside, A. Al-Aoukaty, A. D. Ho, S. Gluck & A. A. Maghazachi. 1993. J. Immunol. **150:** 1524-1534.
11. Putney, J. W. 1993. Science **262:** 676-678.
12. Howard, M., J. C. Grimaldi, J. F. Bazan, F. E. Lund, L. Santos-Argumedo, R. M. E. Parkhouse, T. F. Walseth & H. C. Lee. 1993. Science **262:** 1056-1059.
13. Takasawa, S., K. Nata, H. Yonekuira & H. Okomoto. 1993. Science **259:** 370-373.
14. Gallione, A., A. White, N. Willmott, M. Turner, B. V. L. Potter & S. P. Watson. 1993. Nature **365:** 456-459.
15. Maghazachi, A. A. & A. Al-Aoukaty. 1994. J. Biol. Chem. **269:** 6796-6802.

Interleukin-11 Induces Tyrosine Phosphorylation, and c-jun and c-fos mRNA Expression in Human K562 and U937 Cells

SAMUEL E. ADUNYAH, GLENDORA C. SPENCER, ROLAND S. COOPER, JUAN A. RIVERO, AND KARAMBA CEESAY

Department of Biochemistry
Room 2113
Meharry Medical College
Nashville, Tennessee 37208

IL-11 is a multifunctional cytokine derived from bone marrow stromal cells. The biologic effects of IL-11 include regulation of lymphohematopoietic stem cell proliferation and differentiation. IL-11 also regulates megakaryocyte and B-lymphocyte maturation and activation of hepatocyte acute phase protein synthesis and adipogenesis.[1,2] It exhibits strong potentials as a clinical tool for treatment of myelosuppresive syndrome associated with cancer chemotherapy and bone marrow transplantation.[3] However, the mechanism of IL-11 is not clearly known.

To learn more about its signal transduction mechanism, we investigated whether IL-11 regulates protein phosphorylation and stimulates c-jun and c-fos mRNA expression during myeloid and erythroid proliferation. Human K562 and U937 leukemic cells were cultured in RPMI-1640 medium with 10% FBS in 5% CO_2. Cells were preincubated with vanadate (5 mM) for 30 min prior to treatment with IL-11 for various times. Monoclonal antibodies to c-fes, c-raf, and MAPK were used to immunoprecipiate these proteins from cell lysates and submitted to Western blotting with antiphosphotyrosine antibody (RC20). Total RNA was isolated from cells, size fractionated by gel electrophoresis, and hybridized with DNA probes of c-jun, c-fos, and tubulin, or analyzed by dot blotting.

The results in FIGURE 1 show that treatment or K562 cells with IL-11 (10 ng/mL) resulted in a rapid enhanced tyrosine phosphorylation of c-fes (within 20 s) and c-raf (within 30 s), followed by a rapid return to the basal level. Also, IL-11 induced a 3- to 4-fold stimulation of tyrosine phosphorylation of MAPK within 0.5 to 60 minutes. These results suggest that rapid stimulation of tyrosine phosphorylation of c-fes, raf-1, and MAPK occurs in response to IL-11. Similar results were obtained in U937 cells (data not shown). Furthermore, about 2- to 3-fold activation of protein kinase activities of c-fes, c-raf-1, and MAPK was detected within 2 to 5 min in response to IL-11 (data not shown). Treatment of cells with IL-11 for 2 h resulted in a dose-dependent stimulation of increases in c-jun and c-fos mRNA expression, with maximum effect attained at 5 ng/mL of IL-11 (FIG. 2). Further evidence indicated

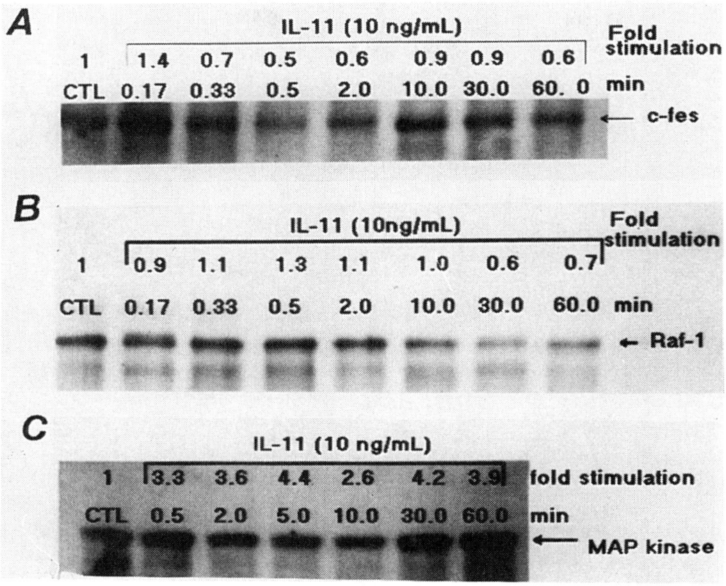

FIGURE 1. Time course of stimulation of tyrosine phosphorylation of c-fes, raf-1, and MAPK by IL-11. K562 (5×10^6) cells were stimulated with IL-11 (10 ng/mL), and cell lysates were analyzed for tyrosine phosphorylation. CTL = control (untreated).

that posttranscriptional regulation of c-jun mRNA expression occurs in response to IL-11. Finally, we detected about a 2-fold stimulation of cell growth and several fold increase in the levels of cdc2p34 and PCNA proteins within 6 to 16 h in IL-11-treated cells (data not shown)

In summary, these observations suggest that c-fes kinase, c-raf-1 kinase, MAPK, c-fos, and c-jun play important roles in the mechanism of regulation of growth and differentiation by IL-11.

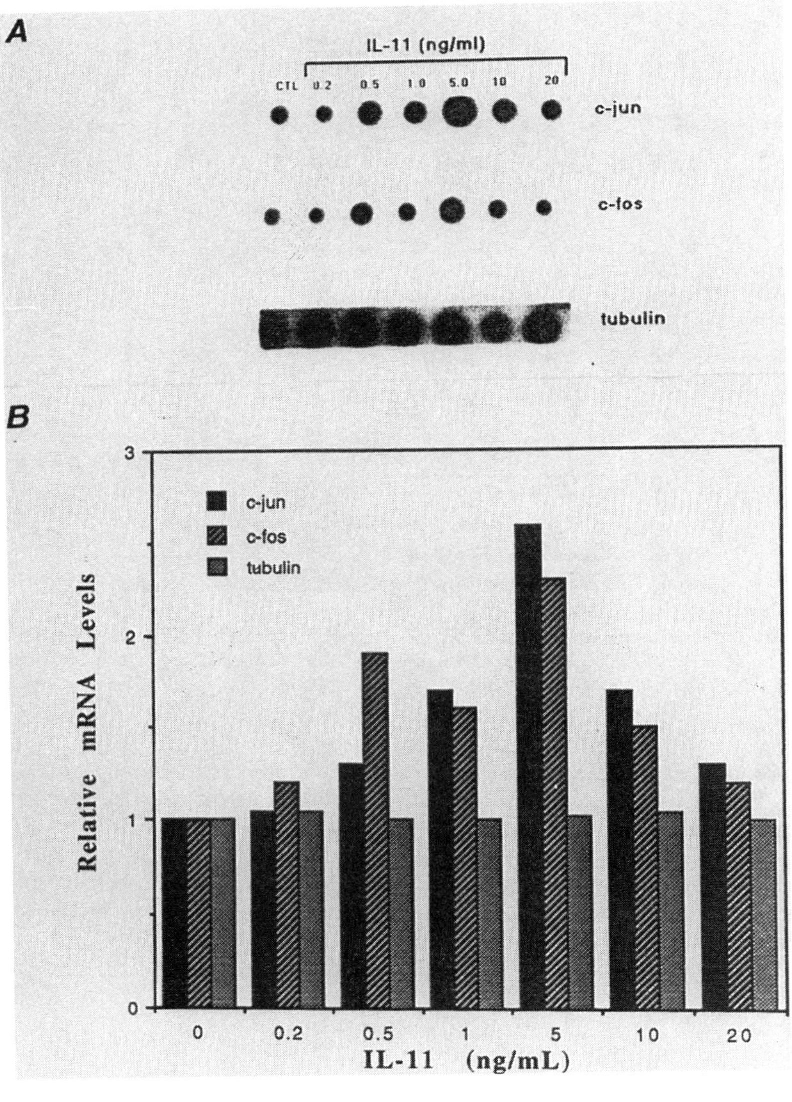

FIGURE 2. Dose-dependence of IL-11-induced c-fos and c-jun mRNA expression. K562 cells were treated with various doses of IL-11 for 2 hours. Total RNA was analyzed for mRNA expression by Northern blot hybridization. Panel A: autoradiograph; panel B: bar graphs of densitometric analysis of dot blots from two experiments.

REFERENCES

1. YIN, T. *et al.* 1992. J. Exp. Med. **175:** 211-216.
2. BRUNO, E. *et al.* 1991. Exp. Hematol. (NY) **19:** 378-383.
3. LEONARD, Jr., P. *et al.* 1994. Blood **83:** 1499-1506.

Analysis of the Interaction between Two TGF-β-binding Proteins and Three TGF-β Isoforms Using Surface Plasmon Resonance

M. D. O'CONNOR-McCOURT, P. SEGARINI,[a]
S. GROTHE, M. L.-S. TSANG,[b] AND
J. A. WEATHERBEE[b]

Biotechnology Research Institute
National Research Council Canada
Montréal, Québec, Canada

[a]Celtrix Pharmaceuticals
Santa Clara, California

[b]R&D Systems
Minneapolis, Minnesota

Transforming growth factor-β (TGF-β) is a peptide factor that stimulates the accumulation of extracellular matrix in cell cultures and is implicated in the pathological accumulation of matrix in fibrotic diseases.[1] The recent cloning of several cell-surface receptors and binding proteins for TGF-β represents a step towards the development of TGF-β antagonists that may be useful as a new approach for the treatment of these diseases. Soluble versions of the receptors and binding proteins can bind TGF-β and may prevent its interaction with cell-surface receptors, thereby neutralizing TGF-β activity. Indeed, it has been shown that the soluble extracellular domains of the type II[2] and type III[3] TGF-β receptors and β1-latency-associated peptide (LAP, the N-terminal remnant of the TGF-β1 precursor)[4] are able to neutralize the proliferation-inhibiting activity of TGF-β *in vitro* and that the TGF-β-binding protein, decorin, is able to antagonize the action of TGF-β in a fibrotic disease model.[5] As a step towards developing TGF-β antagonists, we have used a surface plasmon resonance (SPR)-based biosensor system (BIAcore™) to analyze the interaction between the three mammalian isoforms of TGF-β (TGF-β1, -β2, -β3) and two binding proteins; the extracellular domain of the type II TGF-β receptor (RII-ED),[2] and β1-LAP.[4] In order to adapt this system to measuring TGF-β interactions, TGF-β1, -β2, or -β3 was immobilized to the dextran layer on the surface of a gold sensor chip in the system's microflow cell. As the binding proteins flow past, they interact with the immobilized TGF-β, which results in an increased mass at the sensor chip surface. The resulting refractive index changes are measured as changes in the SPR angle and are plotted in the form of sensorgrams (resonance units [RUs] versus time). In this way, the progress of association and dissociation can be followed in real time. It can be seen in FIGURE 1 that the two binding proteins exhibit very different specificities; that is, the RII-ED associates only with TGF-β1 and -β3, whereas β1-

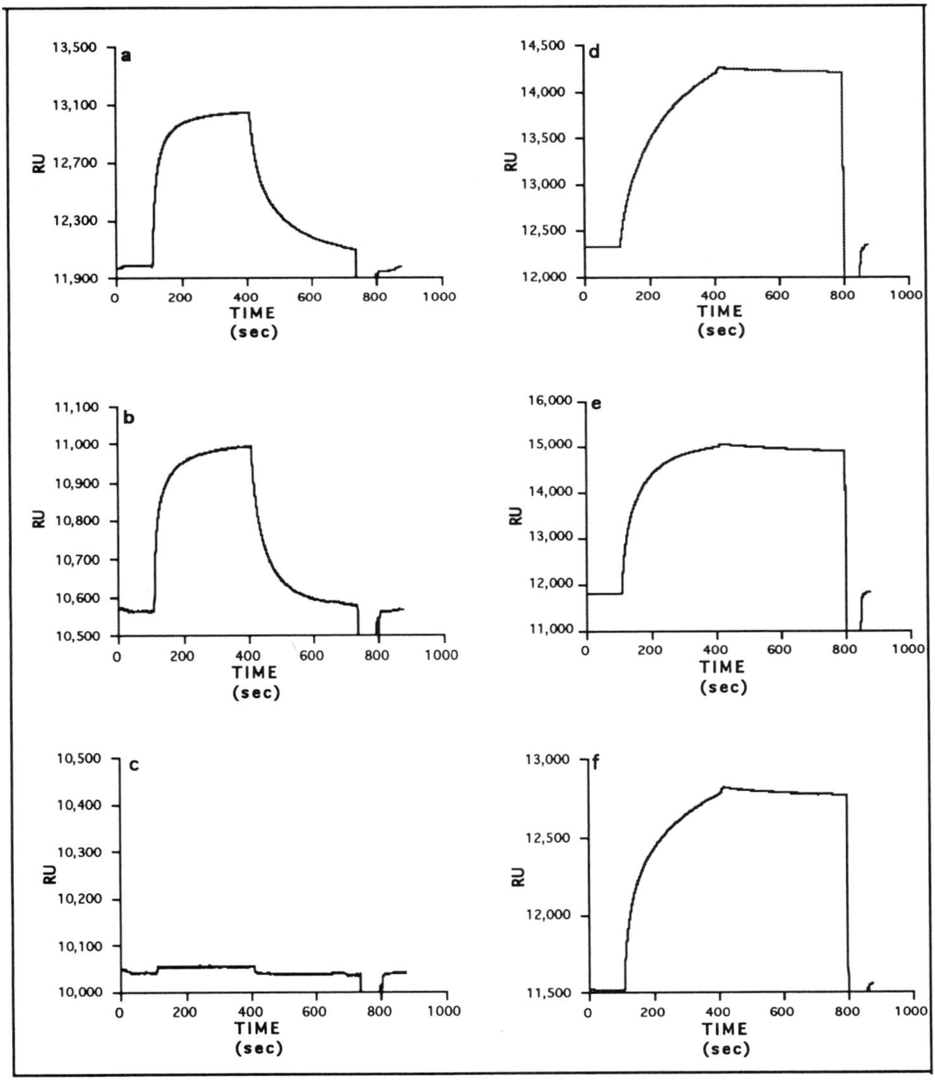

(Legend on p. 302.)

FIGURE 1. Sensorgrams (resonance units [RU] versus time) of the interaction of RII-ED and β1-LAP with TGF-β1, -β2, and -β3. A commercially available SPR-based biosensor system (BIAcore^{TM}, Pharmacia Biosensor, Uppsala, Sweden) was used. Recombinant TGF-β1 (a, d), -β2 (c, f), or -β3 (b, e) were immobilized onto CM5 sensor chips using the standard amine-coupling approach. Specifically, the surface was first activated with *N*-hydroxysuccinimide (NHS) and *N*-ethyl-*N'*-(dimethylaminopropyl) carbodiimide (EDC); a solution of TGF-β (between 3-12 μg/mL in 10 mM HAc, pH 4.5) was then injected over the modified surface; an injection of ethanolamine followed (to inactivate the remaining NHS groups). Immobilization was done at a flow rate of 5 μL/min with an injection volume for NHS/EDC, TGF-β, and ethanolamine of 30 μL. The amount of TGF-β immobilized was between 700-2500 RUs. All sensorgrams were run at a flow rate of 5 μL min^{-1} at 24°C in HBS (10 mM HEPES, 150 mM NaCl, 3.4 mM EDTA, pH 7.5). RII-ED (a, b, c) and β1-LAP (d, e, f) were injected at a concentration of 533 nM in a volume of 25 μL. Regeneration of the surface was performed by injecting 5 μL of 10 mM HCl. It can be seen that RII-ED interacts with TGF-β1 (a) and TGF-β3 (b) but not with TGF-β2 (c). By contrast, β1-LAP interacts with all three isoforms: TGF-β1 (d), TGF-β3 (e), and TGF-β2 (f).

LAP associates with all three TGF-β isoforms. This result is consistent with the isoform specificity of their neutralizing activities.[2,4] Apparent rate constants were obtained by varying the concentrations of RII-ED and β1-LAP and using linearized transformations of the primary data. RII-ED has a relatively fast apparent k_{on} ($1.8 \times 10^{+5}$ M^{-1} s^{-1}) and k_{off} ($2.2 \times 10^{-2} s^{-1}$), whereas β1-LAP has a slower apparent k_{on} ($4.3 \times 10^{+4} M^{-1} s^{-1}$) and k_{off} ($3.3 \times 10^{-4} s^{-1}$). The calculated equilibrium dissociation constants (K_d) for RII-ED and β1-LAP are approximately 120 nM and 8 nM, respectively. It should be mentioned that the slower off-rate of β1-LAP may be due to a classical avidity effect, inasmuch as it is a disulfide-linked dimer binding to an immobilized ligand. We are currently assessing which binding parameter (K_d, k_{on}, or k_{off}) is predictive of the neutralizing potency of these TGF-β binding proteins.

REFERENCES

1. Border, W. A. & N. A. Noble. 1994. N. Engl. J. Med. **331**(19): 1286-1292.
2. Tsang, M. L.-S., L. Zhou, B.-L. Zheng, J. Wenker, G. Fransen, J. Humphrey, J. M. Smith, M. D. O'Connor-McCourt, R. Lucas & J. A. Weatherbee. 1995. Cytokine. In press.
3. López-Casillas, F., H. M. Payne, J. L. Andres & J. Massagué. 1994. J. Cell Biol. **124**(4): 557-568.
4. Miller, D. M., Y. Ogawa, K. K. Iwata, P. ten Dijke, A. F. Purchio, M. S. Soloff & L. E. Gentry. 1992. Mol. Endocrinol. **6:** 694-702.
5. Border, W. A., N. A. Noble, T. Yamamoto, J. R. Harper, Y. Yamaguchi, M. D. Pierschbacher & E. Ruoslahti. 1992. Nature **360:** 361-364.

The Stress-activated Protein Kinases

A Novel ERK Subfamily Responsive to Cellular Stress and Inflammatory Cytokines

JOHN M. KYRIAKIS, JAMES R. WOODGETT,[a] AND JOSEPH AVRUCH[b]

Diabetes Research Laboratory
Massachusetts General Hospital East
149 13th Street
Charlestown, Massachusetts 02129

[a]Division of Cell and Molecular Biology
Ontario Cancer Institute
Princess Margaret Hospital
500 Sherbourne Street
Toronto, Ontario, M4X 1K9, Canada

THE RAS-MAP KINASE PATHWAY, AN UNUSUALLY BROAD REGULATORY NETWORK, IS A PARADIGM FOR SEVERAL CELLULAR SIGNALING MECHANISMS

A large and compelling body of biochemical and genetic evidence supports the view that the major effector pathway used by receptor tyrosine kinases and seven transmembrane receptors to specify cell differentiation or division involves a cascade of protein kinases recruited by the small GTPase Ras.[1] When charged with GTP, Ras binds directly to an aminoterminal noncatalytic domain on the c-Raf-1 protooncogene protein (Ser/Thr) kinase,[2] bringing the kinase to the plasma membrane; there Raf is activated by a poorly characterized process[3,4] that probably requires ATP and involves the 14.3.3 proteins[5] and other unidentified protein kinases.[6] Activated Raf then phosphorylates the MAP kinase kinase (MAPKK or MEK),[7] its only known substrate, which in turn catalyzes the phosphorylation of the p42/p44 MAP kinases (also known as ERK2 and ERK1, respectively) at a tyrosine and a threonine residue in the motif TEY, located in subdomain VIII of the MAPK catalytic domain;[8] the p42/p44 MAPKs are thereby activated and serve as the major multifunctional regulators in this pathway. The remarkably wide control the p42/p44 MAP kinases exert over cell function, specifying not only cell fate, but many aspects of the functions of mature, fully differentiated cells, is attributable to the ability of the MAP kinases to regulate several other classes of signal transducing proteins, such as other protein kinases, at least one phospholipase A_2, and a variety of transcriptional regulatory proteins.[9] Together

[b]To whom all correspondence should be addressed.

with the MAP kinases, these elements control the activity and abundance of proteins crucial to virtually every cellular function.

For example, the RSK enzyme is another multifunctional protein kinase whose activity is completely dependent on phosphorylation by p42/44 MAP kinase. RSK, however, has a radically different substrate specificity from the MAP kinases. Whereas p42/44 MAP kinase exhibits an apparently absolute requirement for a proline residue immediately carboxyterminal to the serine or threonine residue, RSK recognizes best the sequence (R)(R) RXXSX. The multifunctional nature of RSK is nicely illustrated by its role in the regulation of glycogen synthase, wherein it contributes to the dephosphorylation and activation of synthase by phosphorylating and inactivating glycogen synthase kinase-3 (GSK-3),[10,11] the most potent negative regulator of synthase, and concomitantly (at least in skeletal muscle) augmenting the activity of protein phosphatase-1, through a site-specific phosphorylation of the protein that binds phosphatase-1 to the glycogen particle.[12] Strong evidence implicates the MAP kinases in, for example, the regulation of specific gene expression, protein synthesis translational initiation, and microtubular rearrangement during cell division. Thus, it is clear that the ras-MAP kinase cascade has evolved as a versatile and powerful strategy for cell regulation in response to extracellular stimuli.

PROTEIN KINASE CASCADES ARE A COMMON REGULATORY MOTIF

The MAP kinase pathway was not the first protein kinase cascade to be recognized; in fact, the first well-characterized protein kinase, skeletal muscle phosphorylase b kinase,[13] requires phosphorylation by the cAMP-dependent protein kinase for activation in response to β-adrenergic stimulation.[14] Moreover, work on the regulation of cholesterol synthesis led to the discovery of the enzyme now known as the AMP-activated protein kinase,[15] a crucial negative regulator of *de novo* cholesterol and fatty acid biosynthesis, which is activated allosterically by AMP, and through phosphorylation by another Ser/Thr protein kinase. However, although the cAMP-dependent protein kinase may have as broad a range of substrates as the p42/p44 MAP kinases, and a considerable impact on regulated gene expression through its control of the CREB family of transcription factors, the essentially universal role of the MAP kinase cascade as the dominant effector arm of the ras protooncogene, which, in turn, is indispensable to the ability of receptor and cellular tyrosine kinases to direct cell fate, has served to create extraordinary interest in the MAP kinase pathway.

Although understanding of the regulatory relationships operative in the ras-MAP kinase pathway came entirely from work carried out in mammalian systems, the existence and broad utility of a similar protein kinase cascade had been previously and independently established by work in the lower eukaryotes, *Saccharomyces cerevisiae* and *Schizosaccharomyces pombe*. There it was shown that mating pheromones, which inform the cell to withdraw from the mitotic cycle and prepare for conjugation, convey a signal from a surface receptor through a heterotrimeric GTP binding protein to a cascade of protein kinases that proved to include elements with considerable structural homology with the mammalian p42/p44 MAP kinases (*KSSI* and *FUS3* in *S. cerevisiae*, $spk1^{+}$ in *S. pombe*) and their activator, MAPKK/MEK

(*STE7* in *S. cerevisiae*, *byr1* in *S. pombe*).[16–18] Subsequent work in *S. cerevisiae* demonstrated that architecturally related signaling cascades, consisting of kinases structurally homologus to those of the mating factor pathway, and thus to the mammalian MAPK pathway, also mediate the cellular response to several other extracellular stimuli.[18] Thus, the ability of *S. cerevisiae* to augment the synthesis of glycerol in response to hypertonic stress requires the function of the *HOG1* gene product, which encodes a protein kinase homologous to p42/p44 MAP kinase.[19] Normal cell wall construction by *S. cerevisiae* requires the activity of *MPK1*, a MAPK homologue distinct from *HOG1* and *KSS1/FUS3*. Each of these yeast MAP kinase homologues is regulated by a distinct dual specificity kinase homologous to mammalian MAPKK/MEK (*PBS2*, in the case of *HOG1*, and the *MKK1* and *MKK2* gene products in the case of *MPK1*). In turn, each yeast MAPK/MEK homologue is regulated by yet another Ser/Thr kinase. Although the *PBS2* activator has not yet been identified, the *STE7* and *byr1* gene products require activation by, respectively, the *STE11* and *byr2* gene products, while the product of the *BCK1* gene activates the *MKK1/MKK2* gene products.[18] Interestingly, aside from the features shared by all protein kinases, *STE11*, *byr2*, and *BCK1* bear little resemblance to Raf-1, the MAPK/MEK activator. However, six enzymes, referred to as MEK-kinases (MEKKs), have been cloned from mammalian cDNA libraries thus far.[20,21] When overexpressed in cells, one of these MEKKs (MEKK1) can activate the mammalian MAPK-specific MEKs *in situ*;[20] and endogenous MEKK1 immunoprecipitated from cells can activate MEK1 *in vitro*, and is itself activated *in situ* in what appears to be a Ras-dependent manner.[22] The activation of MEK by MEKK1 *in situ* is relatively modest, however, and MEKK overexpressed as a constitutively active kinase is less able to activate MEK than is endogenously expressed Raf-1.[20] Thus, the MEKKs may in fact more predominantly regulate pathways distinct from the MAPK pathway.

The existence of multiple, parallel protein kinase cascades in lower eukaryotes implied strongly that a similar multiplicity would be found in mammalian cells. This report summarizes work leading to the discovery of the stress-activated protein kinases (SAPKs), a family of proline-directed kinases that represents the MAPK homologues in a new protein kinase cascade that is similar in architecture to the MAPK pathway, but completely distinct in regulation and cellular function.

DISCOVERY, PURIFICATION, AND INITIAL CHARACTERIZATION OF A CYCLOHEXIMIDE-ACTIVATED P54 MAP-2 KINASE

The p70 S6 kinase is the major stimulus-regulated phosphorylating activity directed toward 40S ribosomal subunits in mammalian cells.[23] The enzyme requires multiple, independent upstream inputs for activation that are largely distinct from those acting on the p42/p44 MAP kinases and RSK. The p70 kinase is potently activated during hepatic regeneration in the rat; however, treatment of rats with the protein synthesis inhibitor cycloheximide gives an even greater increase in p70 activity *in situ*.[23] The p70 is most closely related in amino acid sequence to RSK and, like RSK, is known to be activated by Ser/Thr phosphorylation. Inasmuch as the primary activators of RSK *in situ* are the p42/p44 MAP kinases,[23–25] we questioned whether

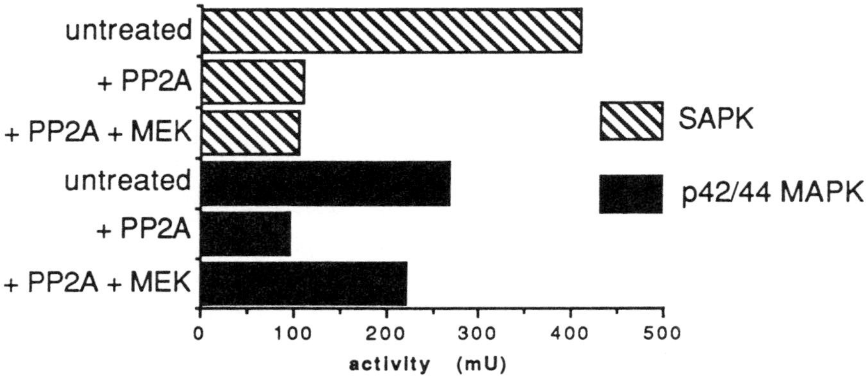

FIGURE 1. The MAPK-specific MEK does not reactivate SAPKs inactivated with phosphatase 2A. Purified MAPKs (a mixture of p42 and p44 MAPKs) and SAPK (purified from rat liver) were normalized to near identical MAP2 phosphorylating activity and inactivated with phosphatase 2A. After inhibition of the phosphatase with okadaic acid, the inactivated kinases were treated with purified MEK (from insulin-stimulated H4 hepatoma cells).

cycloheximide activated an enzyme related to, yet distinct from, the p42/p44 MAP kinases that might participate in the cycloheximide-induced activation of the p70 *in situ*. In fact, cycloheximide treatment of rats *in vivo* causes a very large increase in hepatic MAP-2 phosphorylating activity; purification of one major component of this activity to homogeneity yielded a 54-kDa polypeptide from which kinase activity could be renatured after SDS-PAGE and which was clearly distinct in size from the p42/p44 MAP kinases.[26] A comparison of the regulation and substrate specificity of the p42 and p54 kinases was then undertaken. A hallmark of the regulation of the p42 MAP kinase is the requirement for phosphorylation at both a tyrosine and a threonine residue for activity; treatment of the p42 with either a Ser/Thr-specific phosphatase or a tyrosine-specific phosphatase will inactivate the kinase.[27] The p54 kinase exhibited exactly the same susceptibility to protein phosphatases and, like p42, was inactivated specifically by either phosphatase 1 or 2A, as well as by a recombinant protein tyrosine phosphatase 1b.[28] Significantly, however, the phosphatase-inactivated p54 enzyme was completely resistant to reactivation by preparations of MEK that gave full reactivation of the p42 MAP kinase (FIG. 1). On this basis, we concluded that p54 was activated by dual specificity phosphorylation, analogous to that mediating p42 activation; however, p54 employed an immediate upstream activator entirely distinct from that regulating p42 MAP kinase.

As regards substrate specificities, the characteristic feature of the p42/p44 MAP kinases is the absolute requirement for a proline residue immediately carboxyterminal to the Ser/Thr residue.[29] Employing synthetic peptides based on the amino acid sequence surrounding Thr-669 in the EGF receptor, the ability of p42 and the p54 kinases to phosphorylate wild-type and variant sequences was compared.[29] The two enzymes phosphorylated a 16-residue wild-type peptide containing the sequence PLT669P at comparable rates and exclusively at Thr-669; replacement of P670 with

Ala led to a nearly complete loss of peptide phosphorylation by both kinases. Thus p54, like p42, is a proline-directed kinase. Nevertheless, it was immediately evident that the substrate specificities of the two kinases were easily distinguishable. Whereas replacement of P667 with ala inhibits peptide phosphorylation by p42 nearly as fully as replacement of P670, loss of P667 has no effect, and may even augment peptide phosphorylation by p54.[29] In addition, a set of basic peptides modeled on the site phosphorylated in myelin basic protein by p42 were not significantly phosphorylated by p54 under conditions in which they were well phosphorylated by p42.[29]

The differing specificities of these two proline-directed kinases were strongly evident in comparing their ability to phosphorylate proteins. When matched for identical MAP-2 phosphorylating activity, p54 phosphorylated myelin basic protein at 5-10% the rate seen with p42 MAP kinase and did not phosphorylate at all phosphatase-inactivated *Xenopus* S6 kinase II polypeptide (*Xenopus* RSK) under conditions where p42 gave substantial phosphorylation and reactivation of the same RSK polypeptide.[26]

THE c-JUN PROTOONCOGENE IS PHOSPHORYLATED IN ITS N-TERMINAL TRANSACTIVATION DOMAIN BY THE p54 KINASE

In contrast to RSK, the c-Jun protooncogene proved to be an excellent substrate for the p54 kinase, which phosphorylated recombinant c-Jun *in vitro* at 5-10 times the rate catalyzed by the p42/p44 MAP kinase.[30] The c-Jun protooncogene is a 43-kDa transcriptional regulatory protein of the bZip superfamily that contains a basic DNA binding domain followed by a leucine zipper in its carboxyterminal half and a transcriptional activation domain in its aminoterminal half.[31] c-Jun is a crucial component of the AP-1 DNA binding activity, where it is usually found in a heterodimer with the c-Fos protooncogene. c-Jun can also form homodimers or dimerize with other bZip family members, such as ATF-2. In contrast to c-Fos, whose abundance in unstimulated cells is very low, and whose activity is critically dependent on activation of gene transcription, considerable c-Jun polypeptide is found in unstimulated cells, but in an inactive form.[31] Thus, although c-*jun* gene expression is strongly autoactivated by AP-1, a prior activation of the c-Jun protein is necessary.[31] This is accomplished by two independently regulated sets of phosphorylation events that are directed to different regions of the c-Jun polypeptide. Just aminoterminal to the DNA binding domain is a cluster of three Ser/Thr residues that can be phosphorylated *in vitro* by GSK-3 and casein kinase-2; such phosphorylation of recombinant c-Jun essentially abolishes its DNA binding activity.[32,33] These sites are phosphorylated in resting cells and undergo dephosphorylation, concomitant with the appearance of stimulus-dependent AP-1 binding activity.[32] In addition, c-Jun contains two other phosphorylation sites in the aminoterminal transactivation domain, at ser 63 and 73. Mutation of these two residues to Leu or Ala completely eliminates c-Jun transactivating ability *in situ*.[30] The p54 kinase phosphorylates Ser-63 and Ser-73 selectively *in vitro* at a rate substantially faster than that catalyzed by p42 MAP kinase.[30] The rapid and selective phosphorylation of these important regulatory sites by p54 *in vitro* not only provided a useful substrate for p54 kinase assay, but marked p54 as a strong candidate regulator of c-Jun *in situ*.

MOLECULAR STRUCTURE OF THE p54/p46 KINASES REVEALS A SUBFAMILY OF THE ERKS

Using amino acid sequence of tryptic peptides derived from the purified rat liver p54 kinase polypeptide, a partial c-DNA encoding p54 was generated by polymerase chain reaction (PCR) using as template first strand c-DNA produced from rat liver mRNA.[34] Screening of a rat brain and human Hep G2 cell c-DNA libraries yielded eight independent classes of p54-related sequences, which appear to represent the products of three genes whose expression is further diversified by alternative splicing see (FIGURES 2 and 3). The p54αI sequence corresponds to the polypeptide purified from rat liver; p54αII is identical to aI except for the substitution of a 15 amino acid segment into subdomain IX of the catalytic domain, replacing p54αI residues 216 to 231. The p54β and p54γ sequences are 90% and 88% identical to p54αI in the catalytic domain. Each of the genes (α, β, and γ) is expressed as two molecular mass species due to the alternate insertion of 5 nucleotides (CACAG) within codon 377. This is just prior to the C-terminal region, but outside of the protein kinase catalytic domain, resulting in the generation of proteins of 54 or 46 kDa with identical catalytic domains (see FIG. 3). We thus refer to these proteins as p54α1, p46α1, etcetera.[34] Immunoblotting of cell extracts with antisera that cross-react with all eight gene products reveals expression of both 54 and 46 kDa forms in all tissues and cell lines tested to date. Northern blot analysis, however, indicates that whereas the α and γ isoforms are ubiquitously expressed, the β isoforms are most highly expressed in neural tissue.

Comparison of the amino acid sequence of the p54 kinases to other protein kinases indicates clearly that the p54 enzymes are members of the ERK family; the amino acid sequence through the catalytic domain of the p54s each exhibits about 40-45% identity to all known ERKs.[34] Importantly, the sequence of p54s is no more similar to that of the mammalian ERKs than to any of the homologues from lower eukaryotes. Moreover, the p54s are all nearly equal in sequence similarity to the various yeast ERKs (FIG. 4). Finally, the sequence in subdomain VIII of the p54 catalytic domain, corresponding to the site of activating phosphorylation of the p42/p44 MAPKs, is TPY in each of the p54 sequences, distinct from the TEY in mammalian ERK 1 and 2, and different from the corresponding sequences in all of the known yeast ERK structures.[34] On the basis of this comparison, we conclude that the p54/p46 kinases represent an entirely new ERK subfamily, whose counterpart in lower eukaryotes has not yet been detected.

REGULATION OF THE p54/p46 KINASES BY CELLULAR STRESS AND INFLAMMATORY CYTOKINES

A polyclonal antiserum raised to p54β, expressed as a prokaryotic recombinant, was reactive with each of the p54/p46 kinases, but not with ERK 1/2. The activity of SAPKs, immunoprecipitated from extracts of mammalian cells and assayed as c-Jun N-terminal kinase, was examined in response to a variety of stimuli, in comparison to the ERK 1 and 2 activities, and assayed as MBP kinase after chromatographic separation of these same extracts.[34] Treatment of NIH3T3 or HT29 cells with polypep-

```
MSDSKSDGQFYSVQVADSTFTVLKRYQQLKPIGSGAQGIVCAAFDTVLGINVAVKKLSRPFQ p54αI
:: :: : :::::.:.::::::::::::.:::::::::::::::.:.:: .:::.:::::::::
MSKSKVDNQFYSVEVGDSTFTVLKRYQNLKPIGSGAQGIVCAAYDAVLDRNVAIKKLSRPFQ p54β
::.:: ::.:::::..:::::::::::::::::::::::::::::::.:.:::::::::::::
MSRSKRDNNFYSVEIADSTFTVLKRYQNLKPIGSGAQGIVCAAYDAILERNVAIKKLSRPFQ p54γ

NQTHAKRAYRELVLLKCVNHKNIISLLNVFTPQKTLEEFQDVYLVMELMDANLCQVIHMELD p54αI
:::::::::::::: ::::::::: ::::::::.::::::::::.:::::::::::::: ::::
NQTHAKRAYRELVLMKCVNHKNIISLLNVFTPQKTLEEFQDVYLVMELMDANLCQVIQMELD p54β
::::::::::::::::::::::: ::::::::.::::::::::.::::::::::::::::::::
NQTHAKRAYRELVLMKCVNHKNIIGLLNVFTPQKSLEEFQDVYIVMELMDANLCQVIQMELD p54γ

HERMSYLLYQMLCGIKHLHSAGIIHRDLKPSNIVVKSDCTLKILDFGLARTACTNFMMTPYV p54αI
:::::::::::::::::::::::::::::::::::::::::::::::::::: : ::::::::
HERMSYLLYQMLCGIKHLHSAGIIHRDLKPSNIVVKSDCTLKILDFGLARTAGTSFMMTPYV p54β
::::::::::::::::::::::::::::::::::::::::::::::::::::::::::::::::
HERMSYLLYQMLCGIKHLHSAGIIHRDLKPSNIVVKSDCTLKILDFGLARTAGTSFMMTPYV p54γ

                                        GELVKGCVIFQGTDH p54α2
                                        .: :   : : : :
VTRYYRAPEVILGMGYKENVDIWSVGCIMAEMVLHKVLFPGRDYIDQWNKVIEQLGTPSAEF p54αI
::::::::::::::::::::::.::::::.:::: ::.:::::::::::::::::::::  ::
VTRYYRAPEVILGMGYKENVDIWSVGCIMGEMVRHKILFPGRDYIDQWNKVIEQLGTPCPEF p54β
::::::::::::::::::::::.:::::::::::  ::::::::::::::::::::::::::::
VTRYYRAPEVILGMGYKENVDLWSVGCIMGEMVCLKILFPGRDYIDQWNKVIEQLGTPCPEF p54γ

MKKLQPTVRNYVENRPKYPGIKFEELFPDWIFPSESERDKIKTSQARDLLSKMLVIDPDKRI p54αI
::::::::: ::::::: :. :: :::: .::..:: .:.:.:::::::::::::::::: :::
MKKLQPTVRNYVENRPKYAGLTFPKLFPDSLFPADSEHNKLKASQARDLLSKMLVIDPAKRI p54β
::::::::: ::::::::: .: ::::: ::::::::::::::::::::::::::::: .:::
MKKLQPTVRTYVENRPKYAGYSFEKLFPDVLFPADSEHNKLKASQARDLLSKMLVIDASKRI p54γ

SVDEALRHPYITVWYDPAEAEAPPPQIYDAQLEEREHAIEEWKELIYKEVMDWEERSKNGV- p54αI
:::.:: :::: :::::.: ::::::::: ::.::::.:::::::::::::::. ::..::::
SVDDALQHPYINVWYDPAEVEAPPPQIYDKQLDEREHTIEEWKELIYKEVMNSEEKTKNGVV p54β
:::.::::::::::::::.:.::::: : ::::::::::::::::::::::::. ::.:::::.
SVDEALQHPYINVWYDPSEAEAPPPKIPDKQLDEREHTIEEWKELIYKEVMDLEERTKNGVI p54γ

KDQPS--DAAV-SSKATPSQSSSINDISSMSTEHTLASDTDSSLDASTGPLEGCR COOH p54αI
: :::   ::: :: . :  :::.:::::::::. :::::::::.::.:::  ::
KGQPSPSGAAVNSSESLP-PSSSVNDISSMSTDQTLASDTDSSLEASAGPLGCCR COOH p54β
.::::: ::::       :  : ::::  ::::: ::::::::::: ::::::::
RGQPSPLGAAVINGSQHPVSSPSVNDSMSMSTDPTLASDTDSSLEAAAGPLGCCR COOH p54γ
       |
       |  differential splice
       |
KDQP--SAHMQQ COOH p46αI
: ::  ::  ::
KGQPSPSAQVQQ COOH p46β
.::::: :::::
RGQPSPLAQVQQ COOH p46γ
```

FIGURE 2. Primary structures of the known SAPK isoforms. Amino acid identities are indicated by (:), similarities by (.). Tryptic peptide sequences derived from p54 MAP kinase purified from the livers of cycloheximide-treated rats are underlined. Only the distinct sequence of the αII splice variant is shown. Likewise, only the C-terminal 14-16 residues of the 46 kDa isoforms are included. Other sequences are identical to the 54 kDa products.

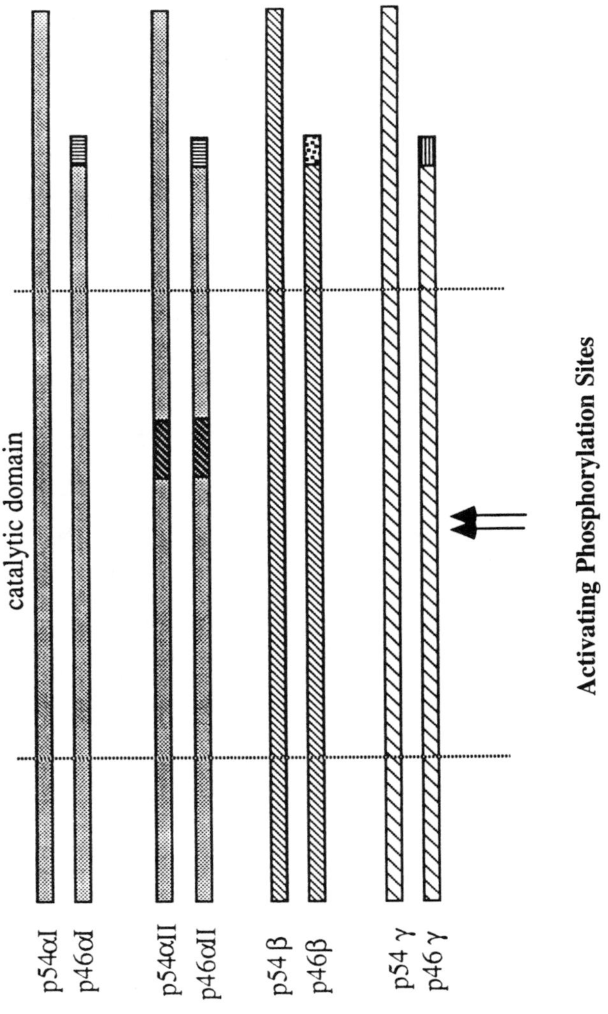

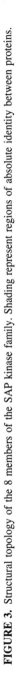

FIGURE 3. Structural topology of the 8 members of the SAP kinase family. Shading represent regions of absolute identity between proteins.

	p54α	p54β	p54γ	Slt-2	p44*mapk*	Hog-1	Kss-1	Fus-3	Spk-1
p54α	-								
p54β	90	-							
p54γ	88	95	-						
Slt-2	45	44	44	-					
p44*mapk*	44	43	43	49	-				
Hog-1	43	42	44	46	49	-			
Kss-1	42	41	42	52	52	50	-		
Fus-3	42	42	42	51	52	51	61	-	
Spk-1	43	42	43	51	56	52	57	58	-

FIGURE 4. Sequence homology comparison of the SAPKs and various yeast and mammalian MAPKs. The comparison is for the catalytic domains of the kinases shown. Figures are percent identity.

tide growth factors, such as FGF or EGF, or with active phorbol esters gave 5- to 10-fold increases in the activity of ERK 1/2, but stimulated p54/p46 activity < 2-fold. By contrast, addition of cycloheximide to cultured cells gave 4-to 8-fold activation of the p54/p46 kinases, similar to that observed *in vivo*, and also stimulated ERK 1/2 activity approximately 2- to 3-fold. The translational inhibitors emetine and, especially, anisomycin, also gave potent and relatively selective activation of the p54/p46, whereas puromycin and the RNA synthesis inhibitor, actinomycin D, were without effect. An even more robust and selective activation of the p54/p46 kinases was elicited by a brief heat-shock treatment (*e.g.*, 42° for 15 min) (FIG. 5).[34]

Translational inhibition and heat stress alter many aspects of cellular function; outcomes common to both include, for example, the denaturation of intracellular proteins and increased protein degradation. Establishing that such perturbations could couple to a specific signal transduction pathway required a more selective stimulus. To this end, we employed tunicamycin, an agent that inhibits N-linked glycosylation of membrane and secretory proteins. This inhibition leads to the accumulation, misfolding, and increased degradation of a subset of cellular proteins, but only within the lumen of the endoplasmic reticulum and Golgi. Treatment of a variety of cells with tunicamycin gave a 10- to 30-fold activation of the p54/p46 kinases, with only 2-to 3-fold activation of ERK 1/2 (ref. 34 and FIG. 6). These findings led us to speculate that the p54/p46 kinases were an important part of the cell's response to stress, and to examine their response to other forms of cellular stress.

Exposure of cells to ionizing radiation damages a variety of cellular constituents through the generation of free radicals. In response, the cell alters the pattern of gene expression (in part through an increase in AP-1 activity) and interrupts the cell

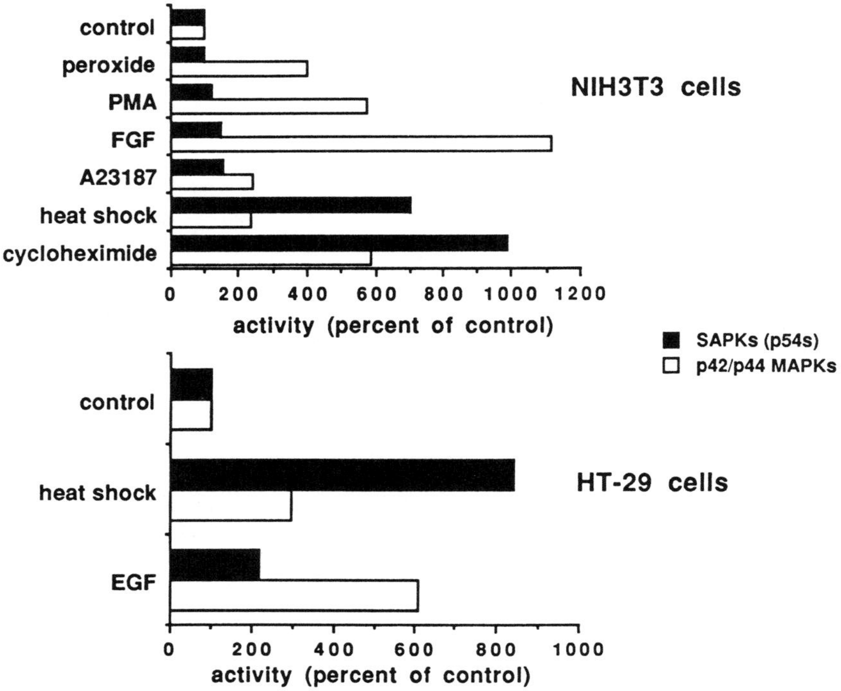

FIGURE 5. SAPKs are predominantly activated by cellular stress, whereas MAPKs are activated more strongly by mitogens. NIH3T3 cells or HT-29 cells were treated with the agonists shown, and extracts were then assayed for SAPKs (closed bars) or p42/p44 MAPKs (open bars).

division cycle through the recruitment of a set of "checkpoint" controls. The activity of the p54/p46 kinases is increased in a potent and selective fashion by exposure of cells to UV-C radiation.[34,35] Metabolic stress was induced in cell culture by causing ATP depletion through the inhibition of cellular energy metabolism with CN^- and 2-deoxyglucose. No increase in SAPK activity occurred during inhibition of substrate metabolism; however, release of this inhibition was followed within seconds by a 20- to 40-fold increase in p54/p46 kinase activity that was sustained; ERK 1/2 transiently increased 2-fold on restoration of metabolism.[36] A comparable response is seen *in vivo*; a brief occlusion of one renal artery in the rat followed by restoration of blood flow led to a marked and selective increase in p54/p46 activity during reperfusion.[36] This effect was not observed in the contralateral kidney. The unilaterality of this response argues against a humoral mechanism for p54/p46 activation, and the association of activation with reperfusion indicates that the initiating stimulus is not energy depletion (as is true for the AMP-activated kinase) but rather some alteration peculiar to the reperfused cell, such as the increase in free radicals known to occur in this state.[36]

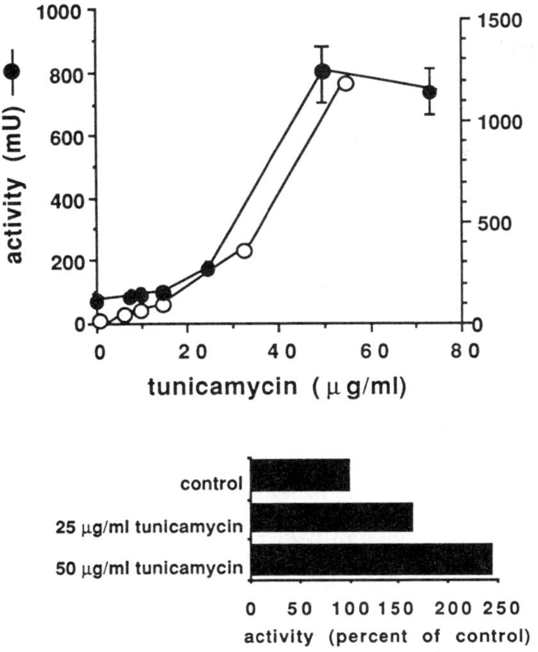

FIGURE 6. Selective activation of SAPK and GST-c-Jun-associated kinase activity by tunicamycin. HT-29 cells were treated with tunicamycin at the concentrations shown, and extracts were assayed for SAPKs (closed circles), GST-c-Jun-associated kinase (open circles), and p42/p44 MAPKs (bar graphs).

The perturbations described above represent different forms of cellular stress. In mammalian organisms, systemic stress, such as that caused by infection with microorganisms, and other encounters with proinflammatory agents and potent nonself antigens, elicits a coordinated interorgan response that is orchestrated primarily by glucocorticoids and an array of cytokines, such as tumor necrosis factor-α (TNF-α) and interleukin-1β (IL-1β). In an initial survey of cell lines known to be responsive to these cytokines, we found that the p54/p46 kinases were potently activated by TNF-α or IL-β, whereas in general they showed little (*i.e.*, < 2-fold) or no activation in response to Ras-linked growth factors or active phorbol esters.[34] The response of the MAPKs to these cytokines was variable; in the CCD-18Co cell line, for example, TNF-α activated the p42/p44 MAPKs as strongly as the p54/p46 kinases (about 8- to 10-fold), whereas in the EL4 thymoma cell line, IL-1β elicited a 4- to 5-fold activation of the p54/p46 kinases, but with no increase in p42/p44 MAPK activity (FIG.) 7. Even in those lines where TNF-α increased p42/p44 MAPK activity, the activation of these enzymes caused by Ras-linked growth factors and phorbol esters was much greater. An exception was observed in freshly isolated hepatocytes; TNF-α did give a greater activation of the p54/p46 kinases than of the p42/p44 MAPKs,

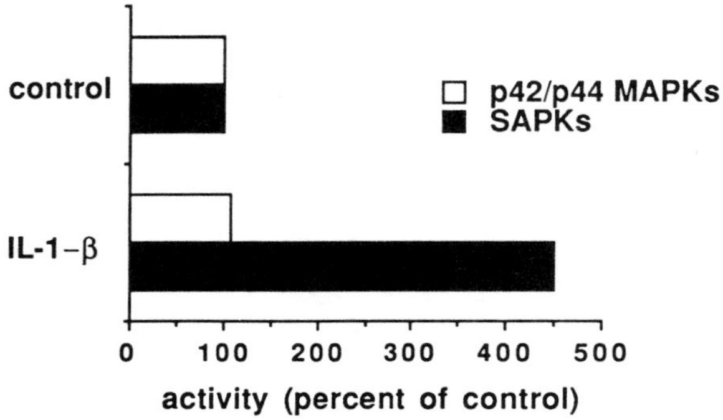

FIGURE 7. IL-1-β is a more potent SAPK and MAPK agonist. EL-4 murine thymoma cells were treated with IL-1-β and extracts assayed for SAPK (closed bars) or MAPK (open bars) activity.

but EGF gave a similar activation of both subfamilies.[34] Thus, potent activation of the p54/p46 kinases is seen as a consistent part of the response to the inflammatory cytokines, whereas activation of the p42/p44 MAPKs is less reliably observed; reciprocally, activation of RTKs consistently provides a strong activation of the p42/p44 MAPKs but only occasionally gives substantial (*e.g.*, > 2-fold) activation of the p54/p46 kinases.

On the basis of the potent activation of the p54/p46 kinases by a variety of cellular stresses and inflammatory cytokines, and their usually weak activation by RTK-linked growth factors, despite their clearcut structural homology to the p42/p44 MAPKs, we have chosen to refer to the subfamily of p54/p46 kinases as the *s*tress-*a*ctivated *p*rotein *k*inases (SAPKs) rather than *m*itogen-*a*ctivated *p*rotein *k*inases (MAPKs).[34] We envision these as two independently regulated subfamilies within the ERK family of the protein kinase superfamily.

COUPLING OF THE p54/p46 SAPKs TO CELL-SURFACE RECEPTORS

The architecture of the protein kinase cascades, known to regulate the mammalian p42/p44 MAPKs[1] and the various ERK isoforms identified in lower eukaryotes,[16–19,21] as well as the susceptibility of purified hepatic SAPK-p54α1 to deactivation by both ser/thr and tyr-specific phosphatases, points strongly to the likelihood that the SAPKs will be regulated by a homologous protein kinase cascade. The novel regulation of the SAPKs *in situ* and the resistance of SAPK-p54 α1 to reactivation by a MEK capable of reactivating p42/p44 MAPK, together with the unique amino acid sequence (*i.e.*, TPY) found in catalytic subdomain VIII in each SAPK isoform recovered thus far,[34] all indicate that the SAPK activator will be a kinase homologous to, but structurally and functionally distinct from, MEK1 and MEK2, the enzymes that regulate the p42/p44 MAPKs. We have obtained preliminary evidence for the existence of such a specific SAPK activator, which we have tentatively named SEK1; moreover, SEK1 itself appears to be regulated by ser/thr phosphorylation by a protein kinase distinct from c-Raf-1, the dominant activator of MEK1/2. The further elaboration of these elements will reveal the molecular structure of the second protein kinase cascade in mammalian cells involving an ERK subfamily. The likelihood is high that additional homologous cascades in mammalian cells remain to be discovered. Recently, three groups reported the isolation of a mammalian homologue of the yeast HOG1 protein kinase.[19,37–39] This enzyme, which is activated by treatment of cell lines with TNF-α, IL-1β, arsenite,[38,39] hyperosmolarity, or lipopolysaccharide (LPS) from gram-negative bacteria,[37] contains the sequence TGY in its catalytic subdomain VIII.[37,38]

Assuming that, as with the MAPKs, a two-tiered kinase cassette will be the regulatory motif immediately upstream of the SAPKs, the more interesting question becomes the identity of the SEK activator(s) and the nature of their coupling to the receptors. The ability of TNF-α/IL-1β to strongly activate the SAPKs in cell lines where RTK-linked growth factors, known to activate Ras, prove incapable of activating the SAPKs, indicates that Ras activation *per se* is insufficient to activate the SAPKs. We already know that oncogenic Ras and Raf do not activate the SAPKs *in situ* (ref. 40 and unpublished data). If the conservation between yeast and mammalian signaling mechanisms remains true, one would expect that members of the MEKK family would be the strongest candidates for SEK activators. TNF-α activation of SAPKs may proceed entirely through a Ras-independent pathway or may require, in addition to Ras, the input of some TNF-α receptor-specific element.

A body of evidence indicates that ceramide, the product of sphingomyelin hydrolysis, serves as an intermediate signal in at least some of the cellular responses to TNF-α.[41,42] In HepG2 cells, TNF-α activates SAPK activity approximately 15-fold. In comparison, treatment of the cells with a bacterial sphingomyelinase, added extracellularly, gave 5-fold activation of SAPK, whereas similar incubations with bacterial PL-C, PL-D, and PL-A_2 gave no SAPK activation.[34] Thus, sphingomyelin hydrolysis mediated by TNF-α may provide one of the signals upstream of the SAPKs. Much further work will be necessary to understand the multiplicity of signal transduction elements that couple receptors and cellular stress stimuli to the activation of the SAPKs.

THE SUBSTRATE SPECIFICITY, NATURAL SUBSTRATES, AND CELLULAR ROLE OF THE SAPKS

As described above, the c-Jun oncoprotein serves as a high-affinity substrate for the SAPKs *in vitro*, undergoing rapid and selective phosphorylation at Ser-63 and Ser-73 in the aminoterminal transactivation domain. Substantial evidence indicates that the SAPKs are among the dominant cellular c-Jun N-terminal kinases. Essentially all of the perturbations described above as activating SAPK activity also increase cellular AP-1 activity and/or c-Jun gene expression. In addition, we find that immunodepletion of extracts, prepared from TNFα or anisomycin-treated cells, with polyclonal antiserum to SAPK removes at least 70% of the c-Jun N-terminal kinase activity.[34]

Kraft and colleagues[43] observed that when extracts from phorbol ester-stimulated cells were passed over a column containing an immobilized GST-Jun fusion protein,

protein-kinase activity capable of phosphorylating c-Jun Ser-63/73 was adsorbed selectively. This technique has proved useful in the identification and characterization of c-Jun N-terminal kinases. When these c-Jun-associated kinases are visualized after SDS-PAGE and renaturation by in-gel assay of Jun phosphorylation, prominent ^{32}P-labeled bands at 40-46 kDa and 55 kDa, as well as several less intensely ^{32}P-labeled bands at higher molecular weight, are detected.[44] Although we have not made an exhaustive comparison, these patterns appear similar, but not identical, to the immunoblots of SAPKs immunoprecipitated from cell extracts. Moreover, a comparison of the relative amount of Jun-associated kinase activity and immunoprecipitated SAPK activity recovered from the same cell extracts shows an excellent correlation using extracts prepared from cells treated with protein synthesis inhibitors, heat shock, UV light, and TNF-α, suggesting that the SAPKs are the major Jun-associated kinases activated in response to these stimuli.[34] Nevertheless, we have encountered circumstances where immunodepletion of SAPKs removes only a small fraction of the Jun-associated kinase activity, such as in extracts from rat kidney made after one hour of reperfusion, or in cultured cells stimulated by selected other agonists.[36] Thus it is certain that c-Jun N-terminal kinases differing from the SAPKs in structure and regulation will be uncovered. Moreover, the designation of the SAPKs as Jun N-terminal kinases (JNKs), as suggested by Dérijard *et al.*,[35] overlooks the fact that avid substrates for the SAPKs other than c-Jun, such as ATF-2, are already known,[45] and evidence pointing to a role for the SAPKs in the regulation of a number of other transcriptional regulatory proteins is accumulating rapidly.

ATF-2 is another bZip protein, most homologous with CREB, that binds DNA as a homodimer; it also binds DNA in a heterodimeric association with CREB and other CREB family members, as well as with c-Jun.[46] Dimers of ATF-2/c-Jun bind preferentially to DNA sequences typical of cAMP-response elements, rather than to the classical phorbol ester response element that binds Jun/Jun and Jun/Fos dimers.[46] Hoeffler and co-workers observed that recombinant ATF-2 expressed in insect cells bound cognate DNA avidly but exhibited very poor DNA-binding activity when expressed in *E. coli*. Treatment of the baculoviral recombinant ATF-2 with phosphatase greatly diminished DNA binding. Phosphorylation of prokaryotic recombinant ATF-2 with p54 SAPK *in vitro* increased substantially ATF-2's ability to bind DNA, whereas phosphorylation by p42 MAPK to a comparable extent was perhaps 10% as effective, and the cAMP-dependent protein kinase was ineffective.[45] Thus, p54 catalyzed phosphorylation of ATF-2 *in vitro* appears to nullify an intramolecular domain inhibitory to DNA binding.

In addition to the requirement for a proline residue immediately carboxyterminal to the site of phosphorylation,[23] and the moderating influence of other amino acid residues immediately adjacent, a third critical determinant of the substrate specificity of the SAPKs has emerged recently from the ongoing characterization of Jun phosphorylation. The v-Jun oncogenic polypeptide contains a number of mutations and deletions as compared with the parent c-Jun polypeptide.[31] Interestingly, although Ser 63 and Ser 73 are both present in v-Jun, neither residue is phosphorylated *in situ*.[43,44,47] In addition, an immobilized GST-v-Jun fusion protein does not adsorb Jun N-terminal kinase activity from cell extracts. The v-Jun polypeptide contains a deletion corresponding to c-Jun residues 31-60, the so-called "delta domain." Deletion of c-Jun residues 31-46 abolishes completely the ability of GST-c-Jun to bind the SAPKs as well as the *in vitro* phosphorylation of Ser 63/73 by SAPKs; deletion of more carboxyterminal residues in the delta domain diminishes but does not abolish SAPK binding to, and phosphorylation of, c-Jun. Thus, the delta domain encompasses a binding site specific for the SAPKs, and perhaps other Jun N-terminal kinases.[33] This finding explains the relatively poor kinetic constants exhibited by SAPKs in their phosphorylation of synthetic peptides, corresponding to sequences surrounding putative sites of *in situ* substrate phosphorylation (e.g., K_ms > 0.2 mM); similar low-affinity peptide phosphorylation is also exhibited by the p42/44 MAPKs. It appears likely that the specific interaction between the SAPKs (and probably the MAPKs, as well) and their natural substrates involves the high-affinity binding of the kinase to a site on the substrate remote in the primary sequence from the actual site of phosphorylation, or perhaps even to a site located on another polypeptide with which the SAPK substrate is tightly associated. Interestingly, GST-c-Jun binds nonactivated and activated SAPK with comparably high affinity; the binding of SAPK to phospho-Jun has not yet been assessed. The requirement for a high-affinity bindings site to enable efficient SAPK phosphorylation of potential substrates may explain the rather remarkable ability of polylysine to expand the range of polypeptide phosphorylation catalyzed by SAPK *in vitro*;[23] polylysine may (artifactually) enable binding of SAPK to the numerous polypeptides in cell extracts that contain SP/TP residues in a favorable amino acid sequence context. The elucidation of the structural elements on SAPK and the c-Jun delta domain critical for high-affinity SAPK binding is of considerable intrinsic interest, and may also allow the facile detection of other native SAPK substrates. The concept of targeting subunits for protein (Ser/Thr) phosphatases, and for certain protein kinases (*e.g.*, the cAMP-dependent protein kinase), has gained considerable support in recent years;[48] the existence of discrete kinase-binding domains on the substrate itself represents a rather minor variation on this idea.

Despite the growing list of potential SAPK substrates, and the increasing insight into the determination of SAPK substrate specificity, the major roles of the SAPK pathway in cell regulation are entirely unknown. The cellular response to the varied stimuli that elicit SAPK activation includes proliferation as well as the induction of a delay in cell division, cellular repair as well as apoptosis; alteration in cellular phenotype (*i.e.*, cell differentiation) as well as the execution of preprogramed cell functions (*e.g.*, secretion). The contribution of the SAPKs to these cellular responses is presently obscure. This situation is likely to change in the very near future, as the kind of constitutively active and dominant inhibitory kinase mutants become available for the SAPK pathway, corresponding to those that have proved so useful in unraveling the ras-MAPK pathway.

REFERENCES

1. Avruch, J. A., X.-f. Zhang & J. M. Kyriakis. 1994. Trends Biochem. Sci. **19:** 279-283.
2. Zhang, X.-f., J. Settleman, J. M. Kyriakis, E. Takeuchi-Suzuki, S. J. Elledge, M. S. Marshall, J. T. Bruder, U. R. Rapp & J. Avruch. 1993. Nature **364:** 308-313.
3. Leevers, S. J., H. F. Paterson & C. J. Marshall. 1994. Nature **369:** 411-414.
4. Stokoe, D., S. G. Macdonald, K. Cadwallader, M. Symons & J. F. Hancock. 1994. Science **264:** 1463-1467.

5. Irie, K., Y. Gotoh, B. M. Yashar, B. Errede, E. Nishida & K. Matsumoto. 1994. Science **265:** 1716-1719.
6. Dent, P. & T. W. Sturgill. 1994. Proc. Natl. Acad. Sci. USA **91:** 9544-9548.
7. Kyriakis, J. M., H. App, X.-f. Zhang, P. Banerjee, D. L. Brautigan, U. R. Rapp & J. Avruch. 1992. Nature **358:** 417-421.
8. Ahn, N. G., R. Seger & E. G. Krebs. 1992. Curr. Opinion Cell. Biol. **4:** 992-999.
9. Davis, R. J. 1993. J. Biol. Chem. **268:** 14553-14556.
10. Sutherland, C., I. A. Leighton & P. Cohen. 1993. Biochem. J. **296:** 15-19.
11. Woodgett, J. R. Unpublished observations.
12. Dent, P., A. Lavoinne, S. Nakielny, F. B. Caudwell, P. Watt & P. Cohen. 1990. Nature **348:** 302-308.
13. Krebs, E. G., D. J. Graves & E. H. Fischer. 1959. J. Biol. Chem. **234:** 2867-2873.
14. Walsh, D. A., J. P. Perkin & E. G. Krebs. 1968. J. Biol. Chem. **243:** 3763-3765.
15. Carter, J. M., J. G. Gillespie & D. G. Hardie. 1994. Curr. Biol. **4:** 315-324.
16. Elion, E. A., P. L. Grisafi & G. R. Fink. 1990. Cell **60:** 649-664.
17. Courchesne, W. E., R. Kunisawa & J. A. Thorner. 1989. Cell **58:** 1107-1119.
18. Errede, B. & D. E. Levin. 1993. Curr. Opinion Cell. Biol. **5:** 254-260.
19. Brewster, J. L., T. de Valoir, N. D. Dwyer, E. Winter & M. C. Gustin. 1993. Science **259:** 1760-1763.
20. Lange-Carter, C. A., C. M. Pleiman, A. M. Gardner, K. J. Blumer & G. L. Johnson. 1993. Science **260:** 315-319.
21. Blumer, K. J. & G. L. Johnson. 1994. Trends Biochem. Sci. **19:** 236-240.
22. Lange-Carter, C. A. & G. L. Johnson. 1994. Science **265:** 1458-1461.
23. Kyriakis, J. M. & J. Avruch. 1994. S6 kinases and MAP kinases: sequential intermediates in insulin/mitogen-activated protein kinase cascades. *In* Protein Kinases: Frontiers in Molecular Biology. J. R. Woodgett, Ed. Oxford University Press. Oxford, UK.
24. Sturgill, T. W., L. B. Ray, E. Erikson & J. L. Maller. 1988. Nature **334:** 715-718.
25. Ahn, N. G. & E. G. Krebs. 1990. J. Biol. Chem. **265:** 11495-11501.
26. Kyriakis, J. M. & J. Avruch. 1990. J. Biol. Chem. **265:** 17355-17363.
27. Anderson, N. G., J. L. Maller, N. K. Tonks & T. W. Sturgill. 1990. Nature **343:** 651-653.
28. Kyriakis, J. M., D. L. Brautigan, T. S. Ingebritsen & J. Avruch. 1991. J. Biol. Chem. **266:** 10043-10046.
29. Mukhopadhyay, N. K., D. J. Price, J. M. Kyriakis, S. Pelech, J. Sanghera & J. Avruch. 1992. J. Biol. Chem. **267:** 3325-3335.
30. Pulverer, B. J., J. M. Kryiakis, J. Avruch, E. Nikolakaki & J. R. Woodgett. 1991. Nature **353:** 670-673.
31. Woodgett, J. R. 1990. Semin. Cancer Biol. **1:** 389-397.
32. Boyle, W. J., T. Smeal, L. H. K. Defize, P. Angel, J. R. Woodgett, M. Karin & T. Hunter. 1991. Cell **64:** 573-584.
33. Lin, A., J. Frost, T. Deng, T. Smeal, N. Al-lawi, U. Kikkawa, T. Hunter, D. Brenner & M. Karin. 1992. Cell **70:** 777-789.
34. Kyriakis, J. M., P. Banerjee, E. Nikolakaki, T. Dai, E. A. Rubie, M. F. Ahmad, J. Avruch & J. R. Woodgett. 1994. Nature **369:** 156-160.
35. Dérijard, B., M. Hibi, I.-H. Wu, T. Barett, B. Su, T. Deng, M. Karin & R. J. Davis. 1994. Cell **76:** 1025-1037.
36. Pombo, C. M., J. V. Bonventre, J. Avruch, J. R. Woodgett, J. M. Kyriakis & T. Force. 1994. J. Biol. Chem. **269:** 26546-26551.
37. Han, J., J.-D. Lee, L. Bibbs & R. J. Ulevitch. 1994. Science **265:** 808-811.
38. Rouse, J., P. Cohen, S. Trigon, M. Morange, A. Alonzo-Llamazares, D. Zamanillo, T. Hunt & A. Nebreda. 1994. Cell **78:** 1027-1037.
39. Freshney, N. W., L. Rawlinson, F. Guesdon, E. Jones, S. Cowley, J. Hsuan & J. Saklatvala. 1994. Cell **78:** 1039-1049.
40. Sun, H., N. K. Tonks & D. Bar-Sagi. 1994. Science **266:** 285-288.

41. Hannun, Y. A. 1994. J. Biol. Chem. **269:** 3125-3128.
42. Kolesnick, R. & D. W. Golde. 1994. Cell **77:** 325-328.
43. Adler, V., C. C. Franklin & A. S. Kraft. 1992. Proc. Natl. Acad. Sci. USA **89:** 5341-5345.
44. Hibi, M., A. Lin, T. Smeal, A. Minden & M. Karin. 1993. Genes & Dev. **7:** 2135-2148.
45. Abdel-Hafiz, H. A.-M., L. E. Heasley, J. M. Kyriakis, J. Avruch, D. J. Kroll, G. L. Johnson & J. P. Hoeffler. 1992. Mol. Endocrinol. **6:** 2079-2089.
46. Habener, J. F. 1990. Mol Endocrinol. **4:** 1087-1094.
47. Dai, T. & J. R. Woodgett. In preparation.
48. Hubbard, M. J. & P. Cohen. 1993. Trends Biochem. Sci. **18:** 172-177.

Protein Serine/Threonine Kinases of the MAPK Cascade

J. D. GRAVES, J. S. CAMPBELL, AND E. G. KREBS

Department of Pharmacology
Box 357280
University of Washington
Seattle, Washington 98195-7280

INTRODUCTION

The covalent modification of cellular proteins by phosphorylation is a prominent feature of pathways by which the cell-surface receptors of eukaryotes transduce extracellular signals. Many cellular processes are known to be modulated in some manner by phosphorylation/dephosphorylation mechanisms; substrates for protein kinases include metabolic enzymes, transcription factors, cytoskeleton proteins, cell-cycle regulators, and other protein kinases. In this respect, a number of protein kinase "cascades" have been recognized, in which a series of protein kinases phosphorylate and regulate one another in a sequential fashion. One such protein kinase cascade, known as the mitogen-activated protein kinase (MAPK) cascade, is activated as an early event in the response of cells to a wide variety of stimuli.[1–3] Signals from diverse receptors, including receptor protein tyrosine kinases, nonreceptor protein tyrosine kinases, cytokine receptors, and heterotrimeric G protein-coupled receptors have all been reported to result in activation of the MAPKs. Stimulation of this pathway has been observed during growth factor-induced DNA synthesis, differentiation, secretion, and metabolism.[3,4] The MAPK pathway has been highly conserved during eukaryotic evolution. Genetic analysis of signal transduction pathways in *Schizosaccharomyces pombe* and *Saccharomyces cerevisiae,* as well as in *Drosophila melanogaster* and *Caenorhabditis elegans,* have revealed that homologous pathways function in cellular differentiative pathways and in response to external stimuli.[5–7] As a result of these observations, the MAPK pathway has been suggested to play a critical role in the transduction of diverse receptor-generated signals from the membrane to the cytoplasm and nucleus.

IDENTIFICATION OF THE MAPK CASCADE

The discovery that many hormone and growth factor receptors possess intrinsic protein tyrosine kinase activity provoked an intensive effort by a number of laboratories to understand the biochemical mechanisms by which signals derived from these receptors are transmitted. Two broad approaches have been employed: looking downstream from activated receptors, and upstream from hormone-regulated cellular

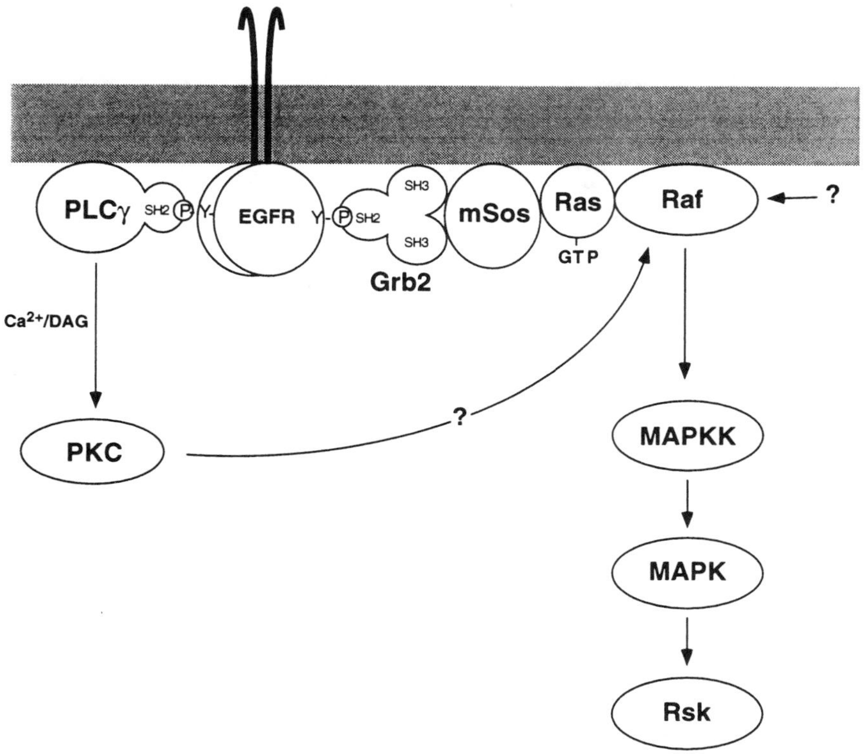

FIGURE 1. Receptor tyrosine kinase coupling to the MAPK cascade. Schematic representation of the mechanism by which the EGF receptor couples to the MAPK cascade. See text for further explanation. DAG = diacylglycerol.

events. Collectively, these efforts have been rewarded by the elucidation of many of the components that comprise the MAPK cascade (see FIGS. 1 and 2).

THE UPSTREAM APPROACH

Studies of the physiological responses of cells to insulin, whose receptor is a tyrosine kinase, provided the first clues that serine/threonine kinases may play an important role in mitogenic signal transduction. Treatment of rat liver or 3T3-L1 cells with insulin resulted in an increase in the serine/threonine phosphate content of a number of proteins, including glycogen synthase, ribosomal protein S6, and the insulin receptor itself. Although the effect of S6 phosphorylation on protein synthesis remains unclear, this reaction served as a convenient starting point from which investigators could work upstream toward the receptor.

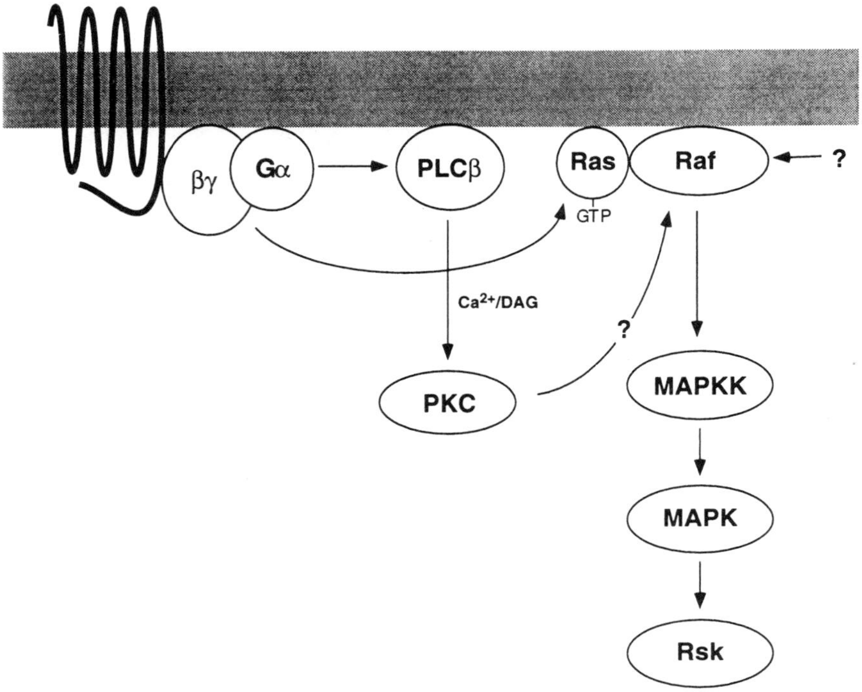

FIGURE 2. G protein-coupled receptor activation of the MAPK cascade. Schematic representation of the mechanism by which G protein-coupled receptors activate the MAPK cascade. See text for further explanation.

Using 40S ribosomes as a source of ribosomal protein S6, a growth factor-stimulated S6 kinase activity was detected that was regulated by protein phosphorylation.[1,8] Initially, a partially purified microtubule-associated protein 2 (MAP-2) protein kinase obtained from insulin-stimulated 3T3-L1 cells[9] was found to phosphorylate and reactivate a homogeneous phosphatase-treated S6 kinase (S6 kinase II or $p90^{rsk}$; Rsk) obtained from *Xenopus laevis* eggs.[10] The reactivation of S6 kinase was subsequently confirmed by Cobb and co-workers,[11] using insulin-stimulated S6 kinase isolated from rat liver and MAP-2 kinase isolated from cells overexpressing the insulin receptor. At about this time, the name of the MAP-2 kinases was changed to mitogen-activated protein (MAP) kinases. Meanwhile, Ahn and co-workers demonstrated that an enzyme that catalyzed the phosphorylation of a synthetic peptide modeled after the S6 protein phosphorylation sites could be phosphorylated and activated by fractions containing MAPK.[12,13] The vertebrate MAPK family (also referred to as extracellular signal-regulated kinases or ERKs) have molecular masses in the range of 40-44 kDa and 54-64 kDa. The best characterized are $pp44^{MAPK}$

(ERK1) and p42^{MAPK} (ERK2), whereas another isozyme, p62 ERK3, has also been studied in some detail.[14] MAPKs are often referred to as ''proline-directed'' protein kinases because the most stringent consensus sequence for their substrate recognition is proline-leucine-serine/threonine-proline.[4]

Subsequent studies showed that MAPK could be inactivated by either a protein serine/threonine[10–12] or tyrosine phosphatase,[15] indicating that phosphorylation on both tyrosine and serine/threonine were required for activity. These findings generated considerable excitement because it was thought that MAPK might serve as a substrate for and be activated by a tyrosine kinase, constituting the long sought for convergence point between serine/threonine and tyrosine phosphorylation pathways. Two partially purified fractions from epidermal growth factor (EGF)-stimulated cells were found to activate MAPK from unstimulated cells or MAPK that had been inactivated by phosphatase treatment.[13] The ability of both recombinant ERK2 and purified ERK1 to slowly undergo autophosphorylation on both tyrosine and threonine residues[16] raised the possibility that MAPK activators might be proteins that enhanced the rate of autophosphorylation of MAPK. It was subsequently shown that the MAPK activators were indeed MAP kinase kinases (MAPKKs).[17,18] Activation of ERKs 1 and 2 requires phosphorylation on both threonine and tyrosine residues within a conserved TEY motif within kinase subdomain VIII. MAPKK phosphorylates both of these residues and belongs to a class of dual specificity kinases that can phosphorylate both tyrosine and serine or threonine residues in proteins. The structural basis for the dual specificity of MAPKK is not understood, and a detailed characterization of any member of this class of kinases has yet to be described. MAPKK displays a very high degree of specificity for MAPK, no other substrates have been identified, and MAPK will not recognize denatured MAPK or peptides derived from MAPK.[18] Two isoforms of MAPKK have been purified from cells in culture (45 and 46 kDa), and three peaks of activity have been identified in rabbit muscle.[18,19] Two forms of rat and human MAPKK have been cloned, referred to as MAPKK-1a and MAPKK-2, as well as a human splice variant, MAPKK-1b, which lacks 26 amino acids in subdomain V.[20,21] MAPKK is also referred to as mitogen-activated or ERK-activated kinase (MEK).[20] There have been at least a dozen reports of the cloning of this and other isoforms from a variety of cell types.

MAPKK was quickly recognized to be regulated by phosphorylation.[22] Interestingly, the enzyme was inactivated by PP2A but not by protein tyrosine phosphatases. This result indicated that there was still at least one more upstream serine/threonine kinase, a MAPKK kinase (MAPKKK), between the cell-surface receptor and MAPKK. The technique that had allowed identification of MAPKK and MAPK in this laboratory failed to reveal any potential MAPKKK activities in growth factor-treated cells. However, several groups provided evidence that the serine/threonine kinase, Raf-1, possesses MAPKKK activity.[23–25] Biochemically, Raf-1 has been shown to have a highly selective substrate specificity, greatly preferring MAPKK over other substrates.[26] Recent studies have established that B-Raf, a 95-kDa a serine threonine kinase related to Raf-1, can also function as a MAPKKK.[27,28] In contrast to Raf-1, which is ubiquitously expressed, B-Raf is predominantly expressed in neuronal tissues and testes. It is unclear whether A-Raf shares the ability to function as a MAPKKK with its close homologues. Evidence exists to suggest that Raf-1 and

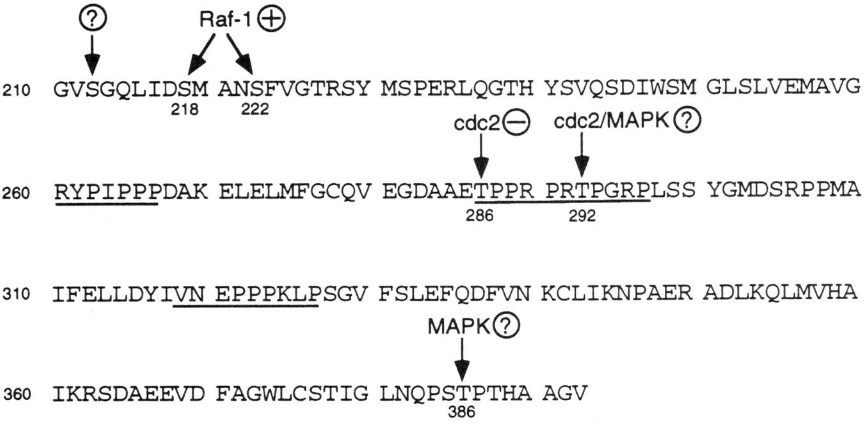

FIGURE 3. Regulatory sites of phosphorylation within human MAPKK-1a. Serines 218 and 222 have been identified as sites of phosphorylation and activation by MAPKKKs Raf-1 both *in vitro* and *in vivo*. Threonines 286 and 292 have been shown to be phosphorylated by cdc2 kinase and to inhibit MAPKK activity *in vitro*. MAPKK-2 lacks both of these sites. Thr386 is also phosphorylated by MAPK without affecting its activity *in vitro*. Proline-rich regions are underlined. See text for further explanation.

B-Raf may not be the only physiologically significant MAPKKK present in mammalian cells.[27,29,30] Several MAPKKK activities, found in high molecular weight complexes, have been identified in extracts from *Xenopus* oocytes and PC-12 cells.[31–33] Although these activities may constitute novel MAPKKKs, it is also possible that these high molecular weight activities contain one of the MAPKKKs described above. In this respect, Raf-1 has been identified to exist in a multisubunit complex with hsp90 and a 50 kDa protein in CHO cells and to associate with a protein, termed 14-3-3.[34,35]

Studies in *Xenopus* oocytes have revealed that c-Mos, a serine/threonine kinase, which had been previously identified as a component of cytostatic factor (CSF), possesses MAPKKK activity. Injection of an active c-Mos fusion protein into oocytes or incubation with cell-free extracts from oocytes or eggs results in activation of MAPK.[17,36] Furthermore, activated c-Mos-fusion protein can phosphorylate and reactivate purified, phosphatase-inactivated rabbit muscle MAPKK *in vitro.*[17] Expression of c-Mos in NIH-3T3 fibroblasts has also been observed to result in activation of MAPK.[37] Despite the accumulated evidence to suggest that c-Mos can activate the MAPK cascade *in vitro* and may function as a *Xenopus* MAPKKK *in vivo,* the role of c-Mos in events unrelated to fertilization remains to be determined.

Alignment of MAPKK sequences to those within subdomains VII and VIII of other protein kinases suggested that serines 218 and 222 might occupy regulatory positions (see FIG. 3). The systematic mutation of serine residues within this region to alanine, rendering the mutated residue unable to be phosphorylated, has revealed that phosphorylation of both serine 218 and 222 is required for full activation of MAPKK by growth factors *in vivo* or Raf-1 *in vitro.*[38–40] Phosphorylation of either serine 218 or 222 appears to be sufficient for partial activation, although phosphorylation of serine 222 may be required for activation *in vivo.*[39–41] Mutation of these residues to aspartate or glutamate, negatively charged residues thought to mimic the effect of phosphorylation, has revealed that mutation of each residue singly rendered MAPKK more easily activable whereas mutation of both residues renders the enzyme constitutively active.[40,42,43] Mutation of serine 212 to alanine has been reported to result in activation rather than inhibition of MAPKK.[40] The role of this residue in MAPKK regulation remains to be fully elucidated.

THE DOWNSTREAM APPROACH

Meanwhile, other researchers were employing a downstream approach in an attempt to unravel the mechanisms by which growth factor receptors transduce their signals. In addition, a considerable effort was being made towards an understanding of the role of protein phosphorylation in oncogenesis. Particular emphasis was placed on the relationship between tyrosine kinase oncogenes, such as v-src and erb-b; serine/threonine kinase oncogenes, such as v-raf, and the GTP-binding protein oncogene encoded by v-ras. From a combination of these studies, it became apparent that products of the ras protooncogenes played a central role in the regulation of cellular growth and differentiation.[44] The general mechanisms by which growth factor receptor tyrosine kinases, such as the EGF and EGF-epidermal growth factor (PDGF)-platelet-derived growth factor PDGF receptors, couple to downstream components are now understood and have been reviewed elsewhere.[45,46] Specific phosphotyrosine residues in activated growth factor receptors and certain of their substrates serve as binding sites for intracellular proteins containing src-homology 2 (SH2) domains and/or SH3 domains.[46,47] A significant breakthrough in understanding how receptors regulate Ras was made when Grb2, an adapter protein consisting of an SH2 domain and two SH3 domains, was found to recruit an activator of Ras to the activated EGF receptor.[48] Grb2 binds to the EGF receptor by way of its SH2 domain and to mSos (son of sevenless), a guanine nucleotide exchange factor for Ras, by way of its two SH3 domains. Sos then activates Ras by accelerating the exchange of GDP for GTP.[48]

The piece of the puzzle that remained to be solved concerned the relationship between Ras and the MAPKKK activity of Raf-1. This final link between those working downstream and those working upstream was established using the yeast two-hybrid system and by the finding that active, GTP-bound Ras interacts with Raf-1.[49,50] The excitement generated by these results was only slightly tempered by the fact that the association of Raf-1 with activated Ras was not sufficient to activate the kinase activity of Raf *in vitro.*[44] The contribution of the Ras-Raf association to Raf activation has been revealed by an elegant series of experiments in Chris Marshall's laboratory.[51] By attaching the lipid modification (CAAX) motif of Ras to the Raf carboxy-terminus, Raf was constitutively targeting to the membrane. Both the basal and growth factor-stimulated specific activities of Raf-CAAX were found to be significantly elevated over endogenous Raf. Furthermore, the kinase activity of Raf-CAAX was not suppressed by inhibitory Ras mutants, suggesting that Ras serves to recruit Raf to the membrane where something happens that results in its activation. The precise nature of the signal or signals that activate membrane-localized Raf remains unclear.[44]

Compared to what is known about the coupling of receptor tyrosine kinases to the MAPK cascade, our understanding of how G protein-coupled receptors regulate the cascade is incomplete. However, evidence exists to suggest that common themes may exist between pathways using G proteins and those mediated by tyrosine kinases. Receptors coupled to G_q and G_i, such as those for thrombin, acetylcholine, and α-adrenergic agonists, can activate the MAPK cascade.[52,53] Both Ras-dependent and -independent mechanisms have been implicated in the regulation of MAPK by these receptors. One G protein-mediated Ras-independent pathway is likely to involve activation of PLCβ isozymes by GTP-bound α_q and βγ subunits leading to activation of PKC.[54] Ras-dependent pathways have been implicated in activation of the MAPK pathway mediated both by α_{i2} subunits and by free βγ dimers.[55,56] Precisely how α subunits and βγ dimers might regulate Ras is currently the subject of investigation.

In summary, we now possess a relatively complete picture of the pathway by which receptors activate the sequential phosphorylation events that constitute the MAPK cascade (see FIGS. 1 and 2). What we do not have are the answers to many important questions concerning the regulation and function of the cascade. These questions are fundamental to our understanding of signal transduction and are being actively pursued by many laboratories around the world.

MULTIPLE MAPK PATHWAYS IN YEAST AND MAMMALS

When the MAPK family (ERKs 1, 2, and 3) was first cloned, striking sequence homology was noticed with the *S. cerevisiae* protein kinases FUS3, KSS1, MPK1, and HOG1, and *S. pombe* Spk-1.[57,58] In *S. cerevisiae,* HOG1 functions in a pathway that protects the cell from hypertonic conditions, FUS3 and KSS1 function in pathways mediating the transcriptional effects of mating pheromone, whereas MPK1 is required for the maintenance of cell-wall structure against isotonic conditions.[5] When MAPKK was sequenced, it was found to be homologous to the yeast kinases, STE7, HOG4(PBS2), MKK1, MKK2, and byr1[20,21] (and reviewed in refs. 5, 6) that had been shown to be "upstream" of the yeast MAPK homologues. Interestingly, no yeast kinases homologous to Raf-1 have been found upstream from MAPKK and MAPK homologues in yeast; epistatic studies demonstrated that the *S. cerevisiae* STE11 kinase functions upstream of STE7→FUS3 and byr2 functions upstream of Byr1→Spk1 in *S. pombe* (See FIG. 3). Upstream in the *S. cerevisiae* pheromone-mating pathway, the seven transmembrane pheromone receptor couples to the STE20 kinase by way of G protein βγ dimers. The product of the STE5 gene, which interacts with multiple components of this pathway, is required for transmission of the STE20-derived signal to STE11→STE7→FUS3. In *S. pombe,* G protein α subunits and the Ras homologue ras1 function upstream of byr2 in the pheromone-mating pathway.

Genetic analysis has demonstrated that each of the yeast MAPK homologues functions in a separate pathway and has specific MAPKK and MAPKKKs that are required for the appropriate phenotype. Elimination of a protein kinase in one pathway does not compromise the function of another. For example, *S. cerevisiae* lacking HOG1 do not respond to osmotic stress but retain their ability to respond to mating pheromone.[59] This high degree of specificity appears, at least in part, to be a function of the substrate specificity of kinases within these pathways. Thus, efficient suppression of conjugal defects caused by STE11 deletion was only achieved when both byr2 and byr1 were expressed.[60] Similarly, ERK2 only partially restored pheromone response defects caused by deletion of spk1, and rescue of byr2 or byr1 defects required coexpression of both MAPKK1 and Raf-1.[60,61] Although FUS3 and KSS1 appear to be redundant in their ability to activate the transcription factor STE12, only FUS3 can phosphorylate FAR1, an event that is required to trigger G_1 cell-cycle arrest.[62,63] In this respect, mammalian ERK1 and ERK2 appear to be largely redundant in terms of both regulation, substrate specificity, and function. This apparent redundancy has led to the suggestion that mammalian cells, like yeast, may contain multiple functionally distinct MAPK pathways. In this respect, a mammalian homologue of STE11 and byr2, termed MEK kinase (MEKK), has been cloned.[64] Overexpression of MEKK in COS cells elevated basal MAPK activity but only slightly enhanced EGF-induced MAPK activity.[64] More recently, MEKK activity has been shown to be dependent on Ras and to complement a BCK1 mutant in *S. cerevisiae.*[28,65] Collectively, these results suggest that MEKK may function in a Ras-dependent kinase cascade distinct from the MAPK cascade.

The recent cloning of several protein kinases from mammalian cells that are related to conventional MAPKs and yeast MAPKs would appear to confirm this suggestion. Kyriakis and co-workers originally described a 54 kDa kinase that was phosphorylated on serine/threonine and tyrosine in liver extracts from cycloheximide-treated rats.[66] Purification and cloning revealed a family of five closely related (88–90% identity) stress-activated protein kinases (SAPKα1, α2, β, and γ).[67] Independently, Karin's laboratory was pursuing a 46 kDa kinase that was activated by UV radiation, bound to the amino-terminal of c-Jun, and then phosphorylated the transcription factor at serines 63 and 73.[68] This group recently reported the cloning of what they refer to as c-Jun kinases (JNK1 and JNK2).[69] JNK2 is identical to SAPKβ, whereas the smaller 46 kDa JNK1 differs by a truncation at the carboxyterminus.[69] The SAPK/JNKs are only 40–45% sequence identical to the MAPKs but have the sequence TPY within kinase subdomain VIII, corresponding to the TEY phosphorylation loop motif in MAPKs.[67,69] However, the known MAPKKs do not phosphorylate and activate the SAPKs, suggesting that they are likely to have their own SAPKKs.[67] Like MAPKs, the SAPK/JNKs are proline-directed kinases, preferring substrate sequences containing a serine/threonine-proline consensus. Although little is currently known about SAPK-substrate specificities, evidence exists to suggest that they may differ from ERKs in some significant respects. For example, SAPKs do not appear to phosphorylate and activate Rsk.[67] Unlike MAPKs, which have been implicated in the phosphorylation of c-Jun in the carboxyterminus (serine 243), SAPKs phosphorylate Jun within their aminoterminal transactivation domains.[68,70] Most of the information concerning the relative specificities of the MAPKs and SAPKs are derived from *in vitro* studies. Great care must be taken to determine which kinases are responsible for these phosphorylations *in vivo.*

The SAPK/JNKs, which are weakly stimulated by mitogenic stimuli, are responsive to the inflammatory cytokine tumor necrosis factor (TNF)-α and to cellular stress that is associated with protein misfolding, such as heat shock, UV radiation, and inhibition of protein glycosylation or translational elongation.[67,69,71] In addition, hypertonic conditions have been shown to activate JNK1 in mammalian cells.[72] Despite the relatively weak homology, SAPK/JNKs have been suggested to be mammalian homologues of *S. cerevisiae* HOG1, which functions in an osmosensing pathway. Support for this hypothesis was provided by the ability of JNK1 to replace HOG1, in a manner dependent on PBS2 function.[72] However, another group has reported the existence of a mammalian kinase that is also regulated by heat shock and osmotic stress.[73] This kinase, termed reactivating kinase (RK), was identified by virtue of its ability to phosphorylate MAPKAP kinase-2, previously considered to be a MAPK substrate.[74] Interestingly, RK was recognized by antisera raised against the *Xenopus* Mpk2 kinase, a close relative of HOG1. Meanwhile, a MAPK relative, which is regulated by hypertonic shock and lipopolysaccharide, was cloned from rat lymphocytes.[75] This 38 kDa kinase is most closely related to *S. cerevisiae* HOG1 and *Xenopus* Mpk2 (86% identity), may be identical to RK, and probably represents the true mammalian HOG1 homologue.

The considerable overlap observed between stimuli that activate MAPK, SAPK, and p38 raises important questions concerning the relationship between these pathways. For example, MAPK is weakly responsive to heat shock and stress in some cells.[67] In common with MAPK, JNK1 is stimulated by activated Ras and is responsive, abeit weakly, to phorbol esters and mitogens.[67,69,71] One recent paper has reported that EGF rapidly and transiently activates 45 and 55 kDa kinases thought to be SAPKs.[76] Like the SAPK/JNKs, p38 is activated by osmotic stress and appears to be weakly activable by phorbol esters.[75] The biochemical basis for this promiscuity will require the characterization of upstream components. One answer might lie in the essential difference between genetic analyses that demonstrate a functional requirement and biochemical studies that might reveal relationships between components in parallel pathways. Unfortunately, biochemical experiments that might reveal the extent of cross-talk between MAPK pathways in yeast have not been performed. Genetic analyses of the role of the MAPK cascade in mammalian cellular responses must also be interpreted with this distinction in mind.

REGULATION OF THE MAPK CASCADE

One aspect of the MAPK cascade that has only recently received attention is its inactivation.[77] The significant differences in kinetics of MAPK activation and deactivation that have been observed with different stimuli are likely to reflect the multiplicity of activation and inactivation mechanisms. For example, in PC12 cells, EGF-stimulated MAPK activity is rapid and transient, but the response to NGF is rapid and persists for an hour or more.[78,79] A biphasic response, comprising a rapid and transient initial peak followed by a second prolonged activation phase, has been observed in lung fibroblasts in response to thrombin.[80] Inasmuch as MAPK activity is likely to be determined by the relative activities and/or concentrations of upstream activators and MAPK phosphatases, attenuation of the MAPK response might occur at multiple points within the cascade. For example, down-regulation of receptor expression by internalization may serve to terminate the receptor-derived signal. In addition, dephosphorylation of the growth-factor receptor at critical tyrosine residues may result in dissociation of SH2-containing signaling molecules, thereby terminating the signal.[46] Although this is an attractive model, no tyrosine protein phosphatase specific for this reaction has been identified. Candidates include the SH2-containing PTPases (PTP.2C, Syp, etc.) and the transmembrane PTPases, such as RPTPα, LAR, and CD45.[81,82] With G protein-coupled receptors, desensitization as a result of recruitment of kinases, such as βARK, might play a role in signal termination.[83]

Another level at which control may be exerted over the temporal activity of the MAPK cascade is that of Ras. The activation state of Ras is determined by interplay between factors that activate Ras by stimulating the exchange of bound GDP for GTP (guanine nucleotide-releasing factors or GRFs, such as mSos and the rasGRF family) and by the activities of factors that inactivate Ras by accelerating its intrinsic capacity to hydrolyze GTP (GTPase-activating proteins or GAPs such as rasGAP and neurofibromin).[84] Mechanisms by which the activities of these regulators of Ras are controlled are beginning to emerge.[84] Another possibility concerns the competition between Ras targets for Ras.GTP. Multiple targets have been identified for Ras.GTP, including both potential regulators (rasGAP and neurofibromin) and effectors (Raf-1, A-Raf, p110 PI(3)K, and rhoGDS; reviewed in ref. 85. For example, the interaction between Raf and Ras at the membrane might be expected to protect Ras from the action of RasGAPs and lead to accumulation of Ras in the GTP-bound state. Such competition might also explain why Ras targets, such as rasGAP and PI(3)K, have been placed both upstream and downstream of Ras by different experimental criteria. Significantly, levels of Ras.GTP observed upon stimulation of PC12 cells with EGF or NGF closely resemble the temporal profiles of EGF- and NGF-induced MAPK activity.[86] Ras.GTP levels may, therefore, be an important determinant of MAPK activation.

PHOSPHATASES AND THE MAPK CASCADE

Specific phosphatases involved in dephosphorylating and inactivating Raf-1, MAPKK, MAPK, and Rsk have yet to be unequivocally identified. However, the serine/threonine protein phosphatases 2A (PP2A) and 1 (PP1) dephosphorylate and inactivate Raf-1, MAPKK, MAPK, and p90^{rsk} *in vitro*.[81,82] Because phosphorylation of MAPK on both serine/threonine and tyrosine residues is required for activity, phosphatases specific for phosphoserine/threonine, phosphotyrosine, or dual specificity phosphatases may be physiologically relevant to its inactivation. Much recent work on phosphatases has focused on dual specificity phosphatases that may inactivate MAPK by dephosphorylating both residues within the TEY motif. Multiple *in vitro* candidates have been identified, including a family that is homologous to the vaccinia virus-encoded phosphatase (VH1).[87–89] This family includes the product of the immediate early gene 3CH134, also referred to as MAPK phosphatase 1 (MKP-1).[87,90] The human 3CH134 homologue was independently cloned as the product of the CL100 stress-induced gene.[91,92] Another member of this family, PAC-1, has been identified as a T cell-specific inducible MAPK phosphatase localized to the nucleus.[93]

MKP-1/CL100 is rapidly and transiently induced by a variety of stimuli, including growth factors, cellular stress, and elevators of cAMP.[87,90] The MKP-1 mRNA is detectable within minutes, accumulating for 1-2 hours, and protein synthesis begins by about an hour. The protein is very sensitive to degradation, having a half-life of 40 minutes.[87,90] Several *in vitro* studies have demonstrated the ability of MKP-1/CL100 to dephosphorylate MAPK on both threonine and tyrosine.[90,94] Overexpression of MKP-1 in COS cells blocked MAPK phosphorylation and activation in response to serum, activated Ras, or activated Raf-1. Under the same conditions, a catalytically inactive mutant of MKP-1 (cysteine 251→serine) was found to enhance MAPK phosphorylation but block MAPK activity by forming a tight complex with phosphorylated MAPK.[90] Furthermore, treatment of NIH 3T3 cells with cycloheximide, an inhibitor of protein synthesis, blocked MKP-1 induction and prolonged MAPK activation in response to serum.[90]

Despite these results, questions remain concerning the physiological relevance of MKP-1. Because of the relatively late induction of MKP-1, it is unlikely that the phosphatase could account for the rapid inactivation of MAPK that is observed in many cells. In a direct attempt to determine the contribution of MKP-1, a recent study employed a combination of actinomycin D and cycloheximide to inhibit transcription and translation of MKP-1 in PC12 cells.[95] Under these conditions, when no MKP-1 mRNA or protein was detectable, the kinetics of MAPK activation and inactivation were observed to be the same as in untreated cells. These results suggest that although MKP-1 may play a role in attenuating the MAPK response at later time points, distinct mechanisms are likely to be responsible for the rapid inactivation of MAPK. One possibility is that other phosphatases, including both serine/threonine and tyrosine phosphatases may be involved. For example, the serine/threonine phosphatase PP2A can inactivate MAPK *in vitro* and may also function in this role *in vivo.*[96] Other candidate MAPK phosphatases include the ERK1 phosphatase activity from PC-12 cells[97] and the *Xenopus* MAPK tyrosine phosphatase.[98]

One physiological circumstance in which an inducible MAPK phosphatase could play a role has been identified in the MM14 mouse myoblast cell line (Campbell *et al.*, in press). In these cells, basic fibroblast growth factor (bFGF) stimulates proliferation and represses differentiation. Removal of bFGF from the culture medium initiates the muscle differentiation program resulting in the phenotypic differentiation of myoblasts to myocytes. After withdrawal from serum and bFGF for 3 hours, bFGF stimulated MAPKK activity, but MAPK activity was not detected. After 10 hours of mitogen withdrawal, MAPKK and MAPK activities were all stimulated by bFGF treatment. The inability of bFGF to stimulate MAPK after 3 hours of withdrawal was found to correlate with the induction of a MAPK phosphatase activity detected in cell extracts. This dephosphorylating activity was observed to diminish during the commitment to terminal differentiation. The identity of this phosphatase and its relationship to the MPK1/CL100 family is being investigated.

FEEDBACK REGULATION OF THE CASCADE

A number of observations suggest that the kinases of the MAPK cascade itself may be involved in the inactivation of their own pathway through negative feedback loops. Although feedback regulation of the MAPK cascade is speculative, such inhibition mechanisms are not uncommon in metabolic and signaling pathways. A feedback loop in the MAPK cascade could result from upstream components becoming substrates for the activated "downstream" kinases. These phosphorylation events could either alter specific activities measured *in vitro* or influence protein-protein interactions critical for signal transmission. Several components in the MAPK cascade become hyperphosphorylated upon activation of the cascade. Several upstream components of the MAPK cascade appear to be phosphorylated by MAPK, including the EGF receptor,[99] Raf-1,[100–102] and MAPKK.[32,103] Both mSos and the insulin receptor substrate 1 (IRS-1) appear to be phosphorylated by a kinase(s) from the cascade that has yet to be unequivocally identified.[104,105] In this respect, Errede and co-workers have demonstrated that the *S. cerevisiae* MAPK homologue, FUS3, down-regulates its own phosphorylation pathway.[106] However, MAPK phosphorylation has no apparent effect on the enzymatic activity of components of the pathways that have been studied.

An example of this paradox is provided by MAPK phosphorylation of MAPKK, originally studied in *Xenopus* oocytes.[32] MAPKK1 isoforms have three potential phosphorylation sites for proline-directed kinases within the proline-rich regions in the carboxy terminus: threonines 286, 292, and 386 (see FIG. 4). Threonines 292 and 286 have been reported to be phosphorylated by cdc2 kinase, resulting in an inhibition of MAPKK activity measured *in vitro*.[107] A recent study has demonstrated that recombinant MAPK phosphorylates purified MAPKK at theonines 292 and 386.[103] However, phosphorylation by MAPK did not alter basal MAPKK activity or the phosphorylation and activation of MAPKK by Raf-1 as measured *in vitro*. Considering the importance that has been demonstrated for protein-protein interactions between components of the MAPK cascade, an alternative possibility may exist. The residues phosphorylated by MAPK are located within a region of MAPKK that contains multiple clustered proline residues. Short sequences containing multiple proline residues, with the consensus PPXP, have been shown to mediate interactions with target proteins containing SH3 domains.[108] Thus, it is possible that phosphorylation within a proline-rich domain affects the association between MAPKK and another protein or proteins important for its regulation or function. A MAPK site within the proline-rich C-terminal of Sos has also been shown to be phosphorylated by MAPK *in vitro*, but the consequences of this event for Sos activity or association with Grb2 or Ras.GDP are unknown.[109]

CROSS-TALK WITH OTHER PROTEIN KINASE PATHWAYS

It is becoming increasingly apparent that growth factor signal transduction pathways are subject to an elaborate network of positive and negative cross-regulatory inputs from other pathways. The hierarchical organization of the MAPK cascade makes it a particularly good target for such "cross-talk." In this manner, multiple physiological processes, such as cell-cycle progression, cytoskeleton rearrangements, and cellular metabolism, may be coordinated to ensure that the cell commits to an appropriate functional outcome. Such interactions also provide the possibility of cell-specific responses. Bifurcation of signaling pathways begins at the level of the

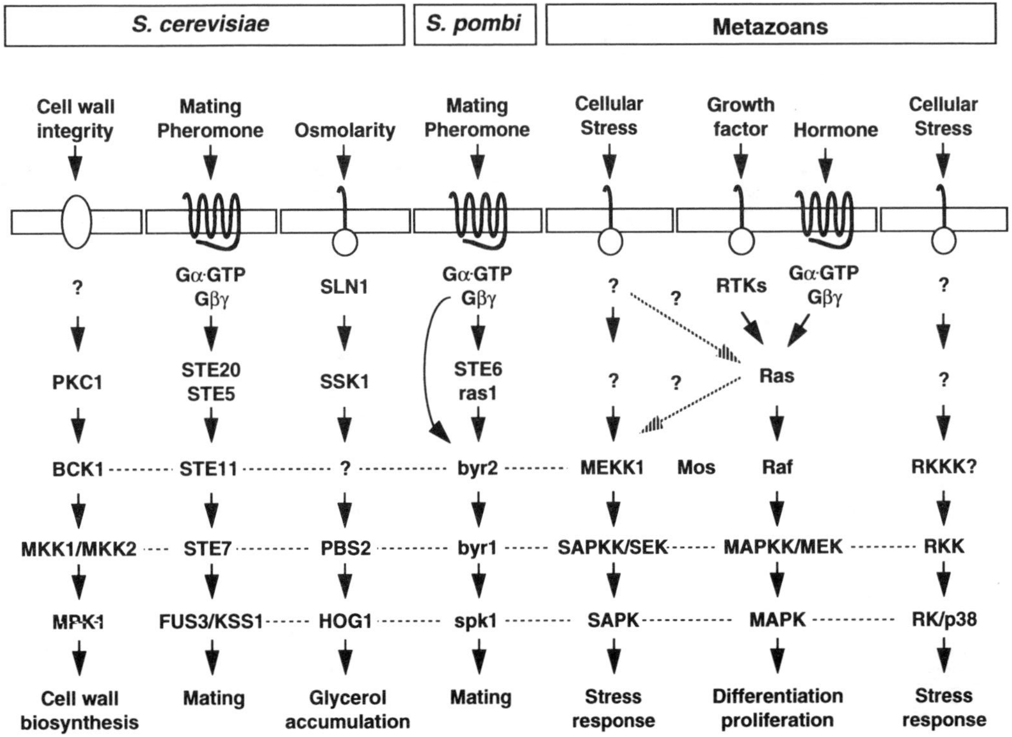

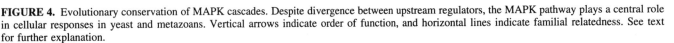

FIGURE 4. Evolutionary conservation of MAPK cascades. Despite divergence between upstream regulators, the MAPK pathway plays a central role in cellular responses in yeast and metazoans. Vertical arrows indicate order of function, and horizontal lines indicate familial relatedness. See text for further explanation.

receptors themselves, where, for example, activated receptor tyrosine kinases bind multiple signaling effectors through their SH2 domains. Integration of multiple signal transduction pathways may occur at the level of substrates, where either multiple inputs or sequential phosphorylation is required for activation. One example is seen in the sequential phosphorylation of glycogen synthase, first by casein kinase II (CK-II), and then by GSK-3.[110] Similar phosphorylations may occur involving transcription factors that are phosphorylated by multiple kinases, such as c-Jun, $p62^{TCF}$, and c-Myc.[11]

CYCLIC AMP AND THE MAPK CASCADE

Several laboratories have reported that elevation of cAMP levels, through the activation of cAMP-dependent protein kinase (PKA), inhibits activation of the MAP kinase cascade in response to stimulation of growth factor receptor tyrosine kinases, G-protein-coupled receptors, and other activators.[112–116] This work showed that growth factor stimulation of Raf-1, MAPKK, and MAPK activities were inhibited, but proximal events, such as receptor tyrosine kinase autophosphorylation, phosphatidylinositol metabolism, and Ras activation, were unaffected by elevation of intracellular cAMP or addition of cAMP analogues to cells.[113,114,116] Although these results localized the effect to a point between Ras and Raf, the identity of the relevant PKA target or targets remains unclear. Because Raf-1 has been shown to be phosphorylated by PKA, both *in vitro* and *in vivo,* one target has been suggested to be Raf itself. An early study reported that phosphorylation of Raf-1 by PKA, within its amino-terminal Ras-interaction domain, resulted in a reduced affinity of Raf-1 for Ras.GTP.[117] However, such a mechanism would not explain the ability of cAMP agonists to antagonize transformation by v-Raf, a process that is independent of Ras. An explanation may be provided by the observation that PKA also phosphorylates Raf-1 within the kinase domain, resulting in inhibition of Raf-1 activity.[118] Interestingly, c-Mos and B-Raf have also been shown to be inhibited by PKA activity.[27,119] Another potential PKA target is Rap1a, a small GTP-binding protein of the Ras family.[84] Rap1a was first identified by virtue of its ability to revert v-Ras transformation. This effect is thought to be a consequence of the similarity between the effector domains of Rap1a and Ras, which allow Rap1a to sequester effectors of Ras action. However, whether Rap1a functions *in vivo* as an endogenous antagonist of Ras action and mediates the inhibitory effects of cAMP remains unclear.

An intriguing aspect of the biological effects of cAMP concerns the fact that although cAMP agonists have been shown to inhibit growth in many cells, they display mitogenic effects in others.[120] Consistent with this finding is that elevation of cAMP can activate the MAPK cascade in these cells.[54,121] For example, treatment of pheochromocytoma (PC12) cells with cAMP agonists or cAMP analogues was found to synergize with either phorbol esters, EGF, or nerve growth factor (NGF) for the induction of MAPKK and MAPK activity.[27,121] NGF-stimulated neurite outgrowth was significantly enhanced by cAMP agonists in these cells.[122] In common with cells in which cAMP has been observed to antagonize MAPK activation, growth-factor stimulation of Raf-1, B-Raf, and MEKK are inhibited by cAMP in PC12 cells.[27,28] These results would appear to suggest the presence of a growth factor-regulated MAPKKK activity in PC12 cells, distinct from Raf-1 and B-Raf, that is not inhibited by cAMP. Elucidation of the mechanism by which cAMP antagonizes activation of the MAPK cascade in the majority of cell types will probably require an understanding of the stimulatory effects of cAMP in other cells.

PKC AND THE MAPK CASCADE

Phorbol ester treatment of cells, most likely through the activation of protein kinase C (PKC), stimulates the MAPK cascade. In fibroblasts, phorbol ester stimulation of the MAPK pathway occurs through both Ras-dependent and -independent mechanisms.[104] Expression of a dominant negative mutant of Ras did not block phorbol ester stimulation of ERK2 phosphorylation, consistent with two independent pathways leading to the activation of MAPK in these cells. Several studies have suggested that Raf-1 might be the site of action of PKC. Stimulation of fibroblasts with PMA activates Raf-1 hyperphosphorylation and histone kinase activity.[123] Consistent with a role for Raf-1 downstream of PKC, expression of dominant-negative Raf-1 mutants blocks phorbol ester-induced transcription of an AP-1-driven reporter gene construct.[124] Various studies indicate Raf-1 may be a direct target for PKC although the effects of PKC's phosphorylation are controversial. PKCα was shown to phosphorylate a synthetic peptide modeled after a putative Raf autophosphorylation site and to activate Raf-1 *in vitro*.[125] By contrast, MacDonald and co-workers demonstrated that although PKC (α, β, and γ) could phosphorylate Raf-1, this did not alter Raf-1's activity, as assessed by its ability to phosphorylate MAPKK.[126] Since translocation of Raf to the membrane by activated Ras has been identified as a critical step in the activation of Raf by growth factors, it is possible that Raf translocation also plays an important role in PKC-mediated activation of Raf.

FUNCTION OF THE MAPK CASCADE

Despite the significant progress that has been made in the molecular characterization of the MAPK pathway, its precise role in cellular responses remains unclear. In this respect, MAPK has been implicated in the phosphorylation and regulation of other kinases, such as Rsk;[10] phosphorylation of the transcription factors c-Myc, c-Jun, and p62^{TCF}/Elk-1[70,127,128] regulation of cytosolic phospholipase A_2;[129] phosphorylation of cytoskeletal components, such as microtubule-associated protein 2 (MAP-2);[6] and phosphorylation of the translational initiation inhibitor PHAS-1[4,130] (see FIG. 5). Thus, in response to a wide variety of stimuli, the MAPK cascade has been suggested to function in the regulation of such diverse cellular responses as growth, differentiation, morphology, and metabolism.

THE MAPK CASCADE IN CELL GROWTH AND DIFFERENTIATION

Genetic analyses in *D. melanogaster* and *C. elegans* have revealed that a MAPK cascade plays an essential role in cellular differentiation.[131,132] Results from the

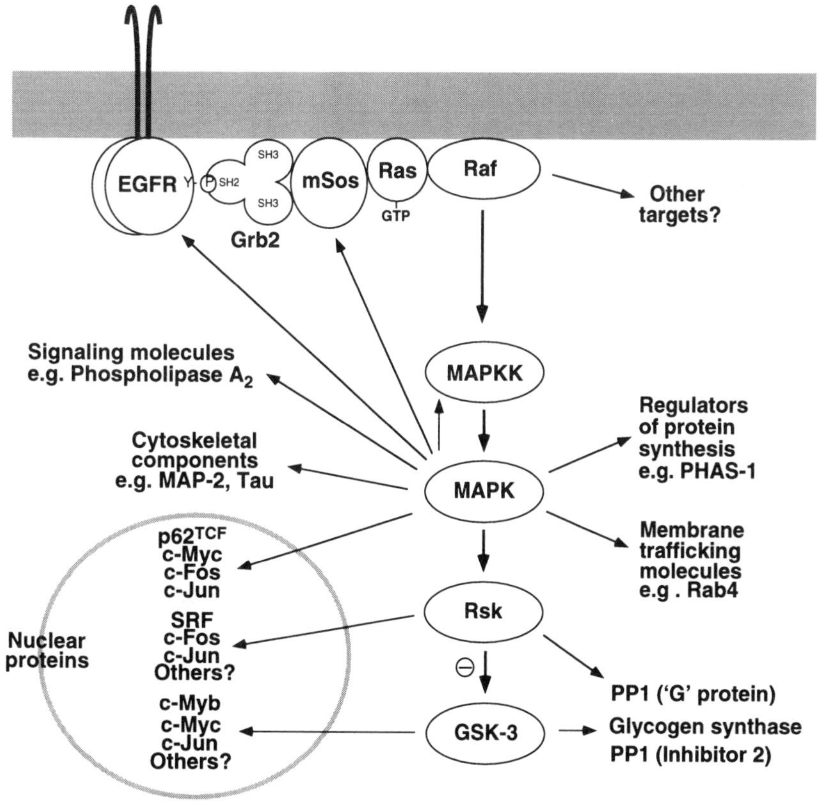

FIGURE 5. Major substrates of the MAPK cascade. MAPK has been implicated in the phosphorylation and regulation of diverse substrates, including other kinases, transcription factors, signaling molecules, cytoskeletal components, and regulators of protein synthesis. See text for further explanation.

biochemical analysis of mammalian cells are not as clear. One particularly intriguing aspect is that MAPK is activated by growth factors that promote both growth and differentiation. For example, in the rat pheochromocytoma cell line, PC12, both NGF and EGF induce MAPK activation, but NGF causes the cells to differentiate, whereas EGF elicits a mitogenic response. An explanation for this apparent paradox may be provided by the observation that NGF treatment results in a sustained activation of the MAPK pathway although the response to EGF is transient[78,79] (reviewed in ref. 3). Although this suggests that quantitative differences in MAPK activation may be responsible for determining signaling specificity and cellular outcome, the possibility that qualitative differences may also contribute cannot be excluded. In the developing *Xenopus* embryo, both fibroblast growth factor (FGF) and activin promote mesoderm induction in explanted animal caps, but only FGF activates the MAPK cascade, suggesting that both MAPK-dependent and -independent pathways to differentiation exist in *Xenopus* embryos.[133] Therefore, another possibility is that a signal derived from the MAPK cascade is insufficient to determine whether a cell will grow or differentiate and that MAPK activation is interpreted in the context of growth, differentiation, or cell-specific signals.

Activation of the MAPK pathway by both signals to enter and exit the cell cycle suggests that it may be a factor in the decision by cells whether to initiate the cycle and divide, or to arrest the cell cycle and differentiate. In vertebrates, MAPK activity has been reported to fluctuate during the cell cycle with peaks of MAPK activity seen at the G_1/S and G_2/M borders, suggesting that MAPK functions to regulate both entry into the cell cycle and subsequent progression through the cycle.[134] Expression of dominant negative and antisense ERKs has established that MAPK activity is necessary for growth factor-induced gene transcription and proliferation in fibroblasts.[135] Although these studies established the necessity of MAPK activation for proliferation, they did not address the question of whether MAPK activation was sufficient to induce cell growth or differentiation. In an attempt to answer this question, several groups have employed mutants of MAPKK that are either interfering or constitutively active.[40,42,43,136] Interfering MAPKK mutants were found to inhibit growth factor-induced proliferation and to revert Ras or Src transformation in NIH 3T3 cells.[40,136] Importantly, expression of constitutively active mutants in these cells resulted in transformation as determined by formation of foci, growth in soft agar, and tumorigenicity in nude mice.[43,136] Furthermore, expression of the phosphatase MKP-1 blocked DNA synthesis induced by activated Ras in rat embryo fibroblasts.[137] However, in PC12 cells, interfering MAPKK mutants were observed to block NGF-induced proliferation, whereas expression of a constitutively active mutant induced differentiation.[136]

Although these results suggest that MAPKK activity, and therefore MAPK activity, is necessary and sufficient for both differentiation and proliferation, activation of the MAPK cascade is only one of a number of bifurcating signaling events associated with the activation of growth factor receptors. Multiple potential effectors of Ras action have also been identified, including rasGAP, Raf, and PI(3) kinase, raising the possibility that other pathways may contribute to growth and differentiative signaling.[85] For example, the SAPK/JNK pathway has also been reported to be stimulated by activated Ras and to be sufficient for induction of the immediate early genes Fos and Jun.[69,76] These results draw into question the role of MAPK as the primary effector of nuclear events associated with mitogenic signaling pathways. Further research employing a combination of genetic and biochemical approaches will be required to determine the precise role of the MAPK cascade in cell growth and differentiation.

THE MAPK CASCADE AND THE CYTOSKELETON

Growing evidence exists to link components of the MAPK cascade to the microtubule and actin cytoskeletons. For example, the microtubule-associated proteins (MAPs) are associated with the actin cytoskeleton *in vivo* and act as MAPK substrates *in vitro*[138,139] (reviewed in ref. 4). Reconstitution experiments *in vitro* suggest that MAPK phosphorylation of MAPs may play a role in regulation of microtubule dynamics.[140,141] Furthermore, MAPK and MAPKK have been found in association with the microtubule and actin cytoskeletons in NIH/3T3 and Rat2 cells, and MAPK has been identified in microtubule organizing centers (refs. 142 and 143, and Reszka, in press). MAPKK mutants profoundly affect cell morphology, suggesting that MAPKK and MAPK may function in rearrangement of the microtubular network during the interphase-metaphase transition of the cell cycle.[40,42,43,136] A role for the MAPK cascade in regulating these events is consistent with the observation of an increase in MAPK activity during the G_2 to M transition.[134]

THE MAPK CASCADE AND METABOLISM

In addition to the phosphorylation and regulation of substrates involved with growth and differentiation, the MAPK cascade has also been implicated in the control of cellular metabolism associated with energy production, and biosynthetic and degradative events. For example, a direct role for MAPK in regulating protein synthesis has been suggested by the identification of the regulator of translation, PHAS-1, as a MAPK substrate.[130] PHAS-1 was identified in Rat adipocytes as a protein phosphorylated in response to insulin. By binding to eIF-4E, the mRNA cap-binding protein, PHAS-1, which is a component of the eIF-4F complex, inhibits the translation of capped mRNA. Increased phosphorylation of PHAS-1 in response to insulin is associated with decreased binding to eIF-4E. Recent results have demonstrated that PHAS-1 is phosphorylated and regulated by MAPK, suggesting a mechanism by which insulin might release eIF-4 from inhibition by PHAS-1 to stimulate eIF-4F activity.[130] Because PHAS-1 is expressed in a wide variety of tissues, these results suggest a general mechanism by which hormones and growth factors that activate the MAPK pathway might regulate protein synthesis.

A second metabolic pathway that has been investigated with respect to its regulation by the MAPK cascade is glycogen metabolism. Here it has been shown that glycogen synthase kinase-3 (GSK-3) and protein phosphatase-1 (PP-1), key enzymes in the coordinate regulation of glycogen synthase (GS), appear to be regulated by Rsk.[144–146] Although this might suggest a possible pathway to explain insulin stimulation of glycogen synthesis, insulin-induced MAPK activity was not found to correlate with its metabolic actions.[147,148] Rsk has been shown to phosphorylate and inactivate GSK-3 *in vitro,* correlating with the inhibition of GSK-3 activity observed in response to growth factors.[149] Work in this laboratory has directly demonstrated a role for MAPK/Rsk in inhibition of GSK-3 by showing that, in cells expressing an inhibitory mutant of MAPKK, growth factor inhibition of GSK-3 is blocked (Eldar-Finkelman, in press). A major question that remains is whether the MAPK cascade/Rsk is sufficient to induce these changes. In this respect, the phosphorylation and inactivation of GS by GSK-3 requires prior phosphorylation of GS by CK-II.[150] Another example of hierarchical phosphorylation by CK-II and GSK-3 is the phosphorylation of inhibitor 2 of the Mg.ATP-dependent form of PP-1,[151] resulting in activation of the phosphatase. The possibility that CK-II may contribute to the regulation of metabolic and growth or differentiative processes by the MAPK cascade remains to be investigated. Future studies will likely reveal numerous additional interactions between metabolism and the MAPK cascade.

REFERENCES

1. Cobb, M. H., T. G. Boulton & D. J. Robbins. 1991. Cell Regul. **2:** 965-978.
2. Blenis, J. 1993. Proc. Natl. Acad. Sci. USA **90:** 5889-5892.
3. Chao, M. V. 1992. Cell **68:** 995-997.
4. Davis, R. J. 1993. J. Biol. Chem. **268:** 14553-14556.
5. Errede, B. & D. E. Levin. 1993. Curr. Opinion Cell. Biol. **5:** 254-260.
6. Neiman, A. M. 1993. TIG **9:** 390-394.
7. Perrimon, N. 1993. Cell **74:** 219-222.
8. Novak-Hofer, I. & G. Thomas. 1984. **259:** 5995-6000.
9. Ray, L. B. & T. W. Sturgill. 1987. Proc. Natl. Acad. Sci. USA **84:** 1502-1506.
10. Sturgill, T. W., L. B. Ray, E. Erikson & J. L. Maller. 1988. Nature (Lond.) **334:** 715-718.
11. Gregory, J. S., T. G. Boulton, B. C. Sang & M. H. Cobb. 1989. J. Biol. Chem. **264:** 18397-18401.
12. Ahn, N. G. & E. G. Krebs. 1990. J. Biol. Chem. **265:** 11495-11501.
13. Ahn, N. G., R. Seger, R. L. Bratlien, C. D. Diltz, N. K. Tonks & E. G. Krebs. 1991. J. Biol. Chem. **266:** 4220-4227.
14. Robbins, D. J., E. Zhen, M. Cheng, S. Xu, C. A. Vanderbilt, D. Ebert, C. Garcia, A. Dang & M. H. Cobb. 1993. J. Am. Soc. Nephrology **4:** 1104-1110.
15. Anderson, N. G., J. L. Maller, N. K. Tonks & T. W. Sturgill. 1990. Nature **343:** 651-653.
16. Seger, R., N. G. Ahn, T. G. Boulton, G. D. Yancopoulos, N. Panayotatos, E. Radziejewska, L. Ericsson, R. L. Bratlien, M. H. Cobb & E. G. Krebs. 1991. Proc. Natl. Acad. Sci. USA **88:** 6142-6146.
17. Posada, J., N. Yew, N. G. Ahn, G. F. Vande Woude & J. A. Cooper. 1993. Mol. Cell. Biol. **13:** 2546-2553.
18. Seger, R., N. G. Ahn, J. Posada, E. S. Munar, A. M. Jensen, J. A. Cooper, M. H. Cobb & E. G. Krebs. 1992. J. Biol. Chem. **267:** 14373-14381.
19. Matsuda, S., H. Kosako, K. Takenaka, K. Moriyama, H. Sakai, T. Akiyama, Y. Gotoh & E. Nishida. 1992. EMBO J. **11:** 973-982.
20. Crews, C. M., A. Alessandrini & R. L. Erikson. 1992. Science **258:** 478-480.

21. SEGER, R., D. SEGER, F. J. LOZEMAN, N. G. AHN, L. M. GRAVES, J. S. CAMPBELL, L. ERICSSON, M. HARRYLOCK, A. M. JENSEN & E. G. KREBS. 1992. J. Biol. Chem. **267:** 25628-25631.
22. GÓMEZ, N. & P. COHEN. 1991. Nature **353:** 170-173.
23. KYRIAKIS, J. M., H. APP, X. F. ZHANG, P. BANERJEE, D. L. BRAUTIGAN, U. R. RAPP, & J. AVRUCH. 1992. Nature **358:** 417-421.
24. DENT, P., W. HASER, T. A. HAYSTEAD, L. A. VINCENT, T. M. ROBERTS & T. W. STURGILL. 1992. Science **257:** 1404-1407.
25. HOWE, L. R., S. J. LEEVERS, N. GÓMEZ, S. NAKIELNY, P. COHEN & C. J. MARSHALL. 1992. Cell **71:** 335-342.
26. FORCE, T., J. V. BONVENTRE, G. HEIDECKER, U. RAPP, J. AVRUCH & J. M. KYRIAKIS. 1994. Proc. Natl. Acad. Sci. USA **91:** 1270-1274.
27. VAILLANCOURT, R. R., A. M. GARDNER & G. L. JOHNSON. 1994. Mol. Cell. Biol. **14:** 6522-6530.
28. LANGE CARTER, C. A. & G. L. JOHNSON. 1994. Science **265:** 1458-1461.
29. KIZAKA KONDOH, S. & H. OKAYAMA. 1993. FEBS Lett. **336:** 255-258.
30. HAYSTEAD, C. M., P. GREGORY, A. SHIRAZI, P. FADDEN, C. MOSSE, P. DENT & T. A. HAYSTEAD. 1994. J. Biol. Chem. **269:** 12804-12808.
31. ITOH, T., K. KAIBUCHI, T. MASUDA, T. YAMAMOTO, Y. MATSUURA, A. MAEDA, K. SHIMIZU & Y. TAKAI. 1993. Proc. Natl. Acad. Sci. USA **90:** 975-979.
32. MATSUDA, S., Y. GOTOH & E. NISHIDA. 1993. J. Biol. Chem. **268:** 3277-3281.
33. GOMEZ, N., S. TRAVERSE & P. COHEN. 1992. FEBS Lett. **314:** 461-465.
34. WARTMANN, M. & R. J. DAVIS. 1994. J. Biol. Chem. **269:** 6695-6701.
35. FREED, E., M. SYMONS, S. G. MACDONALD, F. MCCORMICK & R. RUGGIERI. 1994. Science **265:** 1713-1716.
36. NEBREDA, A. R. & T. HUNT. 1993. EMBO J. **12:** 1979-1986.
37. NEBREDA, A. R., C. HILL, N. GOMEZ, P. COHEN & T. HUNT. 1993. FEBS Lett. **333:** 183-187.
38. ZHENG, C. F. & K. L. GUAN. 1994. EMBO J. **13:** 1123-1131.
39. ALESSI, D. R., Y. SAITO, D. G. CAMPBELL, P. COHEN, G. SITHANANDAM, U. RAPP, A. ASHWORTH, C. J. MARSHALL & S. COWLEY. 1994. EMBO J. **13:** 1610-1619.
40. SEGER, R., D. SEGER, A. A. RESZKA, E. S. MUNAR, H. ELDAR-FINKELMAN, G. DOBROWOLSKA, A. M. JENSEN, J. S. CAMPBELL, E. H. FISHCER & E. G. KREBS. 1994. J. Biol. Chem. **269:** 25699-25709.
41. GOTOH, Y., S. MATSUDA, K. TAKENAKA, S. HATTORI, A. IWAMATSU, M. ISHIKAWA, H. KOSAKO & E. NISHIDA. 1994. Oncogene **9:** 1891-1898.
42. HUANG, W. & R. L. ERIKSON. 1994. Proc. Natl. Acad. Sci. USA **91:** 8960-8963.
43. MANSOUR, S. J., W. T. MATTEN, A. S. HERMANN, J. M. CANDIA, S. RONG, K. FUKASAWA, G. F. VANDE WOUDE & N. G. AHN. 1994. Science **265:** 966-970.
44. AVRUCH, J., X. ZHANG & J. M. KYRIAKIS. 1994. Trends Biochem. Sci. **18:** 279-283.
45. EGAN, S. E. & R. A. WEINBERG. 1993. Nature **365:** 781-783.
46. SCHLESSINGER, J. 1994. Trends Biochem. Sci. **18:** 273-275.
47. PAWSON, T. & G. D. GISH. 1992. Cell **71:** 359-362.
48. BUDAY, L. & J. DOWNWARD. 1993. Cell **73:** 611-620.
49. VOJTEK, A. B., S. M. HOLLENBERG & J. A. COOPER. 1993. Cell **74:** 205-214.
50. ZHANG, X. F., J. SETTLEMAN, J. M. KYRIAKIS, E. TAKEUCHI SUZUKI, S. J. ELLEDGE, M. S. MARSHALL, J. T. BRUDER, U. R. RAPP & J. AVRUCH. 1993. Nature **364:** 308-313.
51. LEEVERS, S. J., H. F. PATERSON & C. J. MARSHALL. 1994. Nature **369:** 411-420.
52. BLUMER, K. J. & G. L. JOHNSON. 1994. Trends Biochem. Sci. **19:** 236-240.
53. COOK, S. & F. MCCORMICK. 1994. Nature **369:** 361-362.
54. FAURE, M., T. A. VOYNO-YASENETSKAYA & H. R. BOURNE. 1994. J. Biol. Chem. **268:** 7851-7854.

55. Johnson, G. L., A. M. Gardner, C. Lange-Carter, N. X. Qian, M. Russell & S. Winitz. 1994. J. Cell Biochem. **54:** 415-422.
56. Crespo, P., N. Xu, W. F. Simonds & J. S. Gutkind. 1994. Nature **369:** 418-420.
57. Boulton, T. G. & M. H. Cobb. 1991. Cell Regul. **2:** 357-371.
58. Boulton, T. G., S. H. Nye, D. J. Robbins, N. Y. Ip, E. Radziejewska, S. D. Morgenbesser, R. A. Depinho, N. Panayotatos, M. H. Cobb & G. D. Yancopoulos. 1991. Cell **65:** 663-675.
59. Brewster, J. L., T. de Valoir, N. D. Dwyer, E. Winter, M. C. Gustin, L. M. Ballou, H. Luther, G. Thomas, T. G. Boulton, G. D. Yancopoulos, J. S. Gregory, C. Slaughter, C. Moomaw, J. Hsu & M. H. Cobb. 1990. Science **249:** 64-67.
60. Neiman, A. M., B. J. Stevenson, H. P. Xu, G. F. J. Sprague, I. Herskowitz, M. Wigler & S. Marcus. 1993. Mol. Biol. Cell **4:** 107-120.
61. Hughes, D. A., A. Ashworth & C. J. Marshall. 1993. Nature **364:** 349-352.
62. Elion, E. A., J. A. Brill & G. R. Fink. 1991. Proc. Natl. Acad. Sci. USA **88:** 9392-9396.
63. Peter, M., A. Gartner, J. Horecka, G. Ammerer & I. Herskowitz. 1993. Cell **73:** 747-760.
64. Lange Carter, C. A., C. M. Pleiman, A. M. Gardner, K. J. Blumer & G. L. Johnson. 1993. Science **260:** 315-319.
65. Blumer, K. J., G. L. Johnson & C. A. Lange Carter. 1994. Proc. Natl. Acad. Sci. USA **91:** 4925-4929.
66. Kyriakis, J. M. & J. Avruch. 1990. J. Biol. Chem. **265:** 17355-17363.
67. Kyriakis, J. M., P. Banerjee, E. Nikolakaki, T. Dai, E. A. Rubie, M. F. Ahmad, J. Avruch & J. R. Woodgett. 1994. Nature **369:** 156-160.
68. Hibi, M., A. Lin, T. Smeal, A. Minden & M. Karin. 1994. Genes & Dev. **7:** 2135-2148.
69. D'erijard, B., M. Hibi, I. H. Wu, T. Barrett, B. Su, T. Deng, M. Karin & R. J. Davis. 1994. Cell **76:** 1025-1037.
70. Baker, S. J., T. K. Kerppola, D. Luk, M. T. Vandenberg, D. R. Marshak, T. Curran & C. Abate. 1992. Mol. Cell. Biol. **12:** 4694-4705.
71. Su, B., E. Jacinto, M. Hibi, T. Kallunki, M. Karin & Y. Ben Neriah. 1994. Cell **77:** 727-736.
72. Galcheva-Gargova, Z., B. Derijard, I-H. Wu & R. J. Davis. 1994. Science **265:** 806-808.
73. Rouse, J., P. Cohen, S. Trigon, M. Morange, A. Alonso-Llamazares, D. Zamanillo, T. Hunt & A. R. Nebreda. 1994. Cell **78:** 1027-1037.
74. Stokoe, D., D. G. Campbell, S. Nakielny, H. Hidaka, S. J. Leevers, C. Marshall & P. Cohen. 1992. EMBO J. **11:** 3985-3994.
75. Han, J., J.-D. Lee, L Bibbs & R. J. Ulevitch. 1994. Science **265:** 808-810.
76. Cano, E., C. A. Hazzalin & L. C. Mahadevan. 1994. Mol. Cell. Biol. **14:** 7352-7362.
77. Nebreda, A. R. 1994. Trends Biochem. Sci. **19:** 1-2.
78. Traverse, S., N. Gomez, H. Paterson, C. Marshall & P. Cohen. 1992. Biochem. J. **288:** 351-355.
79. Nguyen, T. T., J. C. Scimeca, C. Filloux, P. Peraldi, J. L. Carpentier & E. van Obberghen. 1993. J. Biol. Chem. **268:** 9803-9810.
80. Kahan, C., K. Seuwen, S. Meloche & J. Pouyssegur. 1992. J. Biol. Chem. **19:** 13369-13375.
81. Charbonneau, H. & N. K. Tonks. 1992. Annu. Rev. Cell Biol. **8:** 463-493.
82. Fischer, E. H., N. K. Tonks, H. Charbonneau, M. F. Cicirelli, D. E. Cool, C. D. Diltz, E. G. Krebs & K. A. Walsh. 1990. *In* Advances in Second Messenger and Cyclic Nucleotide Research. Y. Nishizuka, Ed.: 273-279. Raven Press. New York.
83. Inglese, J., N. J. Freedman, W. J. Koch & R. J. Lefkowitz. 1993. J. Biol. Chem. **268:** 23735-23738.
84. Boguski, M. S. & F. Mccormick. 1993. Nature **366:** 643-654.

85. FEIG, L. A. & B. SCHAFFHAUSEN. 1994. Nature **370:** 508-509.
86. MUROYA, K., S. HATTORI & S. NAKAMURA. 1992. Oncogene **7:** 277-281.
87. CHARLES, C. H., A. S. ABLER & L. F. LAU. 1992. Oncogene **7:** 187-190.
88. ISHIBASHI, T., D. P. BOTTARO, A. CHAN, T. MIKI & S. A. AARONSON. 1992. Proc. Natl. Acad. Sci. USA **89:** 12170-12174.
89. ZHENG, C. F. & K. L. GUAN. 1993. J. Biol. Chem. **268:** 11435-11439.
90. SUN, H., C. H. CHARLES, L. F. LAU & N. K. TONKS. 1993. Cell **75:** 487-493.
91. ALESSI, D. R., C. SMYTHE & S. M. KEYSE. 1993. Oncogene **8:** 2015-2020.
92. KWAK, S. P., D. J. HAKES, K. J. MARTELL & J. E. DIXON. 1994. J. Biol. Chem. **269:** 3596-3604.
93. WARD, Y., S. GUPTA, P. JENSEN, M. WARTMANN, R. J. DAVIS & K. KELLY. 1994. Nature **367:** 651-654.
94. CHARLES, C. H., H. SUN, L. F. LAU & N. K. TONKS. 1993. Proc. Natl. Acad. Sci. USA **90:** 5292-5296.
95. WU, J., L. F. LAU & T. W. STURGILL. 1994. FEBS Lett. **353:** 9-12.
96. FROST, J. A., A. S. ALBERTS, E. SONTAG, K. GUAN, M. C. MUMBY & J. R. FERAMISCO. 1994. Mol. Cell. Biol. **14:** 6244-6252.
97. PERALDI, P., J. C. SCIMECA, C. FILLOUX & E. VAN OBBERGHEN. 1994. Endocrinology **132:** 2578-2585.
98. SARCEVIC, B., E. ERIKSON & J. L. MALLER. 1994. J. Biol. Chem. **268:** 25075-25083.
99. TAKISHIMA, K., I. GRISWOLD PRENNER, T. INGEBRITSEN & M. R. ROSNER. 1991. Proc. Natl. Acad. Sci. USA **88:** 2520-2524.
100. ANDERSON, N. G., P. LI, L. A. MARSDEN, N. WILLIAMS, T. M. ROBERTS & T. W. STURGILL. 1991. Biochem. J. **277:** 573-577.
101. KYRIAKIS, J. M., T. L. FORCE, U. R. RAPP, J. V. BONVENTRE & J. AVRUCH. 1993. J. Biol. Chem. **268:** 16009-16019.
102. LEE, R. M., M. H. COBB & P. J. BLACKSHEAR. 1992. J. Biol. Chem. **267:** 1088-1092.
103. MANSOUR, S. J., K. A. RESING, J. M. CANDI, A. S. HERMANN, J. W. GLOOR, K. R. HERSKIND, M. WARTMANN, R. J. DAVIS & N. G. AHN. 1994. J. Biochem. **116:** 304-314.
104. BURGERING, B. M., A. M. DE VRIES SMITS, R. H. MEDEMA, P. C. VAN WEEREN, L. G. TERTOOLEN & J. L. BOS. 1993. Mol. Cell. Biol. **13:** 7248-7256.
105. SUN, X. J., M. MIRALPEIX, M. G. J. MYERS, E. M. GLASHEEN, J. M. BACKER, C. R. KAHN & M. F. WHITE. 1992. J. Biol. Chem. **267:** 22662-22672.
106. ERREDE, B., A. GARTNER, Z. ZHOU, K. NASMYTH & G. AMMERER. 1993. Nature **362:** 261-264.
107. ROSSOMANDO, A. J., P. DENT, T. W. STURGILL & D. R. MARSHAK. 1994. Mol. Cell. Biol. **14:** 1594-1602.
108. REN, R., B. J. MAYER, P. CICCHETTI & D. BALTIMORE. 1993. Science **259:** 1157-1161.
109. CHERNIACK, A. D., J. K. KLARLUND & M. P. CZECH. 1994. J. Biol. Chem. **269:** 4717-4720.
110. ROACH, P. J. 1991. J. Biol. Chem. **266:** 14139-14142.
111. HUNTER, T. & M. KARIN. 1992. Cell **70:** 375-387.
112. GRAVES, L. M., K. E. BORNFELDT, E. W. RAINES, B. C. POTTS, S. G. MACDONALD, R. ROSS & E. G. KREBS. 1993. Proc. Natl. Acad. Sci. USA **90:** 10300-10304.
113. BURGERING, B. M., G. J. PRONK, P. C. VAN WEEREN, P. CHARDIN & J. L. BOS. 1993. EMBO J. **12:** 4211-4220.
114. COOK, S. J. & F. MCCORMICK. 1993. Science **262:** 1069-1072.
115. SEVETSON, B. R., X. KONG & J. C. J. LAWRENCE. 1993. Proc. Natl. Acad. Sci. USA **90:** 10305-10309.
116. WU, J., P. DENT, T. JELINEK, A. WOLFMAN, M. J. WEBER & T. W. STURGILL. 1993. Science **262:** 1065-1069.
117. WU, J., J. K. HARRISON, P. DENT, K. R. LYNCH, M. J. WEBER & T. W. STURGILL. 1993. Mol. Cell. Biol. **13:** 4539-4548.

118. Hafner, S. A., H. S. Adler, H. Mischak, P. Janosch, G. Heidecker, A. Wolfman, S. Pippig, M. Lohse, M. Ueffing & W. Kolch. 1994. Mol. Cell. Biol. **14:** 6696-6703.
119. Daar, I., N. Yew & G. F. Vande Woude. 1993. J. Cell Biol. **120:** 1197-1202.
120. Boynton, A. L. & J. F. Whitfield. 1983. *In* Advances in Cyclic Nucleotide Research. P. Greengard & G. A. Robison, Eds.: 194-295. Raven Press. New York.
121. Frodin, M., P. Peraldi & E. van Obberghen. 1994. J. Biol. Chem. **269:** 6207-6214.
122. Young, S. W., M. Dickens, J. M. Tavar'e, S. Winitz, M. Russell, N. X. Qian, A. Gardner, L. Dwyer & G. L. Johnson. 1993. FEBS Lett. **268:** 19196-19199.
123. Morrison, D. K., D. R. Kaplan, U. Rapp & T. M. Roberts. 1988. Proc. Natl. Acad. Sci. USA **85:** 8855-8859.
124. Bruder, J. T., G. Heidecker & U. R. Rapp. 1992. Genes & Dev. **6:** 545-556.
125. Kolch, W., G. Heidecker, G. Kochs, R. Hummel, H. Vahidi, H. Mischak, G. Finkenzeller, D. Marm'e & U. R. Rapp. 1993. Nature **364:** 249-252.
126. MacDonald, S. G., C. M. Crews, L. Wu, J. Driller, R. Clark, R. L. Erikson & F. McCormick. 1993. Mol. Cell. Biol. **13:** 6615-6620.
127. Seth, A., F. A. Gonzalez, S. Gupta, D. L. Raden & R. J. Davis. 1992. J. Biol. Chem. **267:** 24796-24804.
128. Marais, R., J. Wynne & R. Treisman. 1993. Cell **73:** 381-393.
129. Lin, L. L., M. Wartmann, A. Y. Lin, J. L. Knopf, A. Seth & R. J. Davis. 1993. Cell **72:** 269-278.
130. Lin, T., X. Kong, T. A. J. Haystead, A. Pause, G. Belsham, N. Sonenberg & J. C. Lawrence, Jr. 1994. Science **266:** 653-656.
131. Biggs III, W. H., K. H. Zavitz, B. Dickson, A. van der Straten, D. Brunner, E. Hafen & S. L. Zipursky. 1994. EMBO J. **13:** 1628-1635.
132. Wu, Y. & M. Han. 1994. Genes & Dev. **8:** 147-159.
133. Graves, L. M., J. L. Northrop, B. C. Potts, E. G. Krebs & D. Kimelman. 1994. Proc. Natl. Acad. Sci. USA **91:** 1662-1666.
134. Tamemoto, H., T. Kadowaki, K. Tobe, K. Ueki, T. Izumi, Y. Chatani, M. Kohno, M. Kasuga, Y. Yazaki & Y. Akanuma. 1992. J. Biol. Chem. **267:** 20293-20297.
135. Pages, G., P. Lenormand, G. L'allemain, J. C. Chambard, S. Meloche & J. Pouyss'egur. 1993. Proc. Natl. Acad. Sci. USA **90:** 8319-8323.
136. Cowley, S., H. Paterson, P. Kemp & C. J. Marshall. 1994. Cell **77:** 841-852.
137. Sun, H., N. K. Tonks & D. Bar-Sagi. 1994. Science **266:** 285-288.
138. Asai, D. J., W. C. Thompson, L. Wilson, C. F. Dresden, H. Schulman & D. L. Purich. 1985. Proc. Natl. Acad. Sci. USA **82:** 1434-1438.
139. Cross, D., C. Vial & R. B. Maccioni. 1993. J. Cell Sci. 105: 51-60.
140. Jameson, L. & M. Caplow. 1981. Proc. Natl. Acad. Sci. USA **78:** 3413-3417.
141. Drechsel, D. N., A. A. Hyman, M. H. Cobb & M. W. Kirschner. 1992. Mol. Biol. Cell **3:** 1141-1154.
142. Verlhac, M. H., H. de Pennart, B. Maro, M. H. Cobb & H. J. Clarke. 1993. Dev. Biol. **158:** 330-340.
143. Verlhac, M.-H., J. Z. Kubiak, H. J. Clarke & B. Maro. 1994. Development **120:** 1017-1025.
144. Sutherland, C. & P. Cohen. 1994. FEBS Lett. **338:** 37-42.
145. Sutherland, C., D. G. Campbell & P. Cohen. 1993. Eur. J. Biochem. **212:** 581-588.
146. Vandenheede, J. R., S. D. Yang, J. Goris & W. Merlevede. 1980. J. Biol. Chem. **255:** 11768-11774.
147. Robinson, L. J., Z. F. Razzack, J. C. J. Lawrence & D. E. James. 1993. J. Biol. Chem. **268:** 26422-26427.

148. PANG, L., D. F. LAZAR, D. E. MOLLER, J. S. FLIER & A. R. SALTIEL. 1993. Biochem. Biophys. Res. Commun. **196:** 301-310.
149. SAITO, Y., J. R. VANDENHEEDE & P. COHEN. 1994. Biochem. J. **303:** 27-31.
150. ZHANG, W., A. A. DEPAOLI ROACH & P. J. ROACH. 1993. Arch. Biochem. Biophys. **304:** 219-225.
151. PARK, I. K., P. ROACH, J. BONDOR, S. P. FOX & A. A. DEPAOLI ROACH. 1994. J. Biol. Chem. **269:** 944-954.

Structural Aspects of Receptor Dimerization

c-Kit as an Example[a]

JANNA M. BLECHMAN AND YOSEF YARDEN[b]

Department of Chemical Immunology
The Weizmann Institute of Science
Rehovot 76100, Israel

Signal transduction across biological membranes ubiquitously involves a receptor oligomerization step. Although the necessity of this process was functionally demonstrated in many ligand-receptor systems, its structural basis is largely unknown. Here we describe the results of our work on the stem cell factor receptor (c-Kit). These studies indicate that an intrinsic dimerization site exists on the extracellular portion of the receptor. Unlike the ligand-binding site, the distinct putative dimerization site is a relatively small membrane proximal site that may involve a contiguous protein sequence. Mapping of the dimerization site enabled its mutagenesis and allowed closer examination of the relationships between receptor dimerization and signal transduction. Evidently, receptor dimerization and ligand-binding affinities are intimately coupled, implying causative relationships between ligand binding and dimer formation. As expected, ligand-bound monomeric receptors were found to be defective in both short- and long-term signaling.

RECEPTOR OLIGOMERIZATION IN SIGNAL TRANSDUCTION

Cell growth and differentiation are controlled by soluble and membrane-bound polypeptide factors that bind to specific cell-surface receptors, transmitting biochemical signals across the cell membrane and finally regulating the transcription of specific genes. Receptor molecules are highly selective for their ligands, and they transduce a wide range of signals, resulting in cell maintenance, mitogenesis, migration, and differentiation. The effects of a set of growth factors are mediated by high-affinity receptors with intrinsic tyrosine-kinase activity (RTKs). The RTKs are distinguished by a common structure represented by an extracellular (EC) glycosylated domain, a single transmembrane (TM) region, and an intracellular (IC) tyrosine kinase domain. The extracellular portions of the receptors are designed on the basis of combinations of several main structural modules, such as immunoglobulin (Ig)-like domains, fibronectin type III-like domains (FN III), cysteine-rich domains, and others.[1] Ligand

[a] This work was supported by Grants from Baxter Pharmaceutical Co., the Israel Cancer Research Fund, and the Forchheimer Center for Molecular Genetics.

[b] Send correspondence to Dr. Yosef Yarden.

binding to an RTK results in the activation of the catalytic function and consequent autophosphorylation of specific tyrosine residues of the cytoplasmic domain. Phosphotyrosine residues serve as high-affinity docking sites for cytoplasmic messengers, containing Src-homology 2 (SH2) domains.[2] The multiplicity of receptor phosphorylation sites combined with the ability of SH2-containing cytoplasmic proteins to couple to many signaling pathways provides the biochemical basis for both the specificity of ligand-mediated signaling and for the high degree of similarity in the activation pattern of RTKs. It is now well established that receptor dimerization or oligomerization is involved in the activation process that follows binding of many growth factors, cytokines, and hormones to their cognate receptors. Thus, EGF,[3] growth hormone,[4] erythropoietin,[5] and multivalent aggregates that contain IgG or IgE[6] cause rapid receptor clustering that appears to be essential for transmembrane signaling. Besides members of the insulin-receptor family that preexist as inactive dimers, other RTKs undergo rapid, but reversible, dimerization of occupied receptors, followed by a concomitant increase in autophosphorylation.[7–9]

Ligands for RTKs are known to exist both as monomeric molecules (EGF, NDF, FGF), or covalently (PDGF, CSF-1) and noncovalently (SCF) associated dimers. Both RTKs and their ligands are known to form families of structurally related molecules, providing the basis for the formation of heterodimeric complexes. Heterodimerization of different isoforms of the platelet-derived growth factor receptors (PDGF-Rs) or fibroblast growth factor receptors (FGFRs) underlies the regulation of signaling, either through receptor-expression levels[10] or by different ligands,[11] thereby increasing the diversity of signal transduction in response to extracellular signals. The involvement of an oligomerization step in receptor signaling is supported by the phenomenon of dominance of certain mutant alleles of receptor tyrosine kinases in heterozygous animals. This may be explained by the formation of dimeric complexes between wild-type and mutant receptors, which prevent proper signaling.[12]

An oligomerization-activation mechanism is essential also for signaling by cytokine receptors that lack an intrinsic enzymatic activity.[13] Cytoplasmic tyrosine kinases of the JAK family, which are normally responsible for transmission of cytokine-induced signals, form inactive complexes with the β subunit of cytokine receptors, implying that dimerization is essential also for activation of intracellular tyrosine kinases. Likewise, an equilibrium between association and dissociation of trimeric complexes of G proteins underlies the activation cycle of various effectors.[14] The choice between the formation of homo- and heterodimeric complexes and their dissociation to monomers may activate/inactivate families of intracellular regulatory molecules and transcription factors. Thus, overexpression of c-Myc and Max synergizes to cause malignant transformation, whereas overexpression of Max alone leads to growth inhibition.[15,16] Similarly, transformation of chicken embryo fibroblasts correlates with Jun homodimerization,[17] whereas heterodimers of c-Jun and JunB repress gene expression.[18]

Despite the wealth of information on the functional role of receptor dimerization, the structural basis of receptor-receptor interactions remains largely unknown. Two potential mechanisms that underlie receptor activation may be considered: an intramolecular model, involving activation of the cytoplasmic portion through allosteric alterations in the EC domain; and an intermolecular mechanism that is supported by observations of ligand-induced receptor dimers. Ligand binding to soluble EC domains of the receptors for insulin[19] and EGF[20] induces conformational changes, supporting the role for an intramolecular mechanism. Depending on ligand structure, either monomeric or dimeric, the formation of receptor dimers may be attributed either to allosteric alterations in the ligand-binding portion (monomeric ligand), or to stabilization of receptor dimers by ligand bivalency (dimeric ligands). For example, polyvalent organization of various cytokines may stabilize the formation of heterodimeric or trimeric receptor complexes.[4,21] However, it was proposed that even small ligands, such as EGF, may bivalently bind to their receptors.[22] Induction of high-affinity EGF binding to heterodimers between ErbB-1 (EGF-receptor) and ErbB-2 may indirectly support the asymmetric binding of a monomeric ligand to both counterparts of the complex through more than one binding site.[23]

Ligand binding to its cognate receptor in several systems showed some degree of redundancy. Thus, PDGF-Rα homodimer displays two binding sites characterized by high and low affinity to PDGF-AA.[11] The 1 : 1 or 1 : 2 stoichiometry within the ligand-receptor complex of human growth hormone (hGH) and its receptor may depend on glycosylation of the receptor or on its attachment to the cell membrane.[24] These data suggest the involvement of protein domains that are distinct from the ligand-binding portions, in stabilization of ligand-receptor complexes. This prediction is supported by reports on ligand-independent receptor autophosphorylation at high-receptor density that may be explained by spontaneous formation of receptor dimers.[11,25]

STRUCTURAL MOTIFS INVOLVED IN RECEPTOR DIMERIZATION

Structural analysis of the hGH-receptor complex revealed the existence of two overlapping receptor-binding sites on the ligand and a substantial receptor-receptor interface area adjacent to the membrane.[26] Segregation of amino acid residues in IL-6R, that are required for ligand binding and signal transduction by way of trimerization with gp130, was also observed.[27] Similar to the limited number of amino acids that are involved in antigen-antibody contact area, the hormone-binding epitopes of the hGH receptor depend on only a few strong interactions.[27a] FIGURE 1 schematically depicts the structures of several receptors and their presumed sites of interactions. Soluble domains of the receptors for the dimeric PDGF and SCF molecules were shown to undergo ligand-induced association[28,29] (FIG. 1A). The EC portion of hGHR and IL-6R is involved in the formation of homo- and heterooligomeric complexes, respectively, that are stabilized by the polyvalent monomeric ligands[4,27] (FIG. 1B). Heterodimerization of the β and γ subunits of IL-2R is also mediated by the EC portion of each protein[30] (FIG. 1C). Interestingly, the γ-subunit is able to associate with different β-chains,[13] but the formation of the IL-2R complex is induced by monovalent binding of IL-2 to the α-component.[30]

In addition to EC-mediated associations, deletions or insertions within the TM domain indicate the importance of the TM stretch for ErbB-2,[31] CD8,[32] or c-Ros[33] function (FIG. 1C). Substitution of a valine residue for glutamic acid or a glutamine at position 664 of ErbB-2 activated the kinase[34] and correlated with increased receptor dimerization.[35,36] Likewise, a three amino acid-long insertion within the TM domain

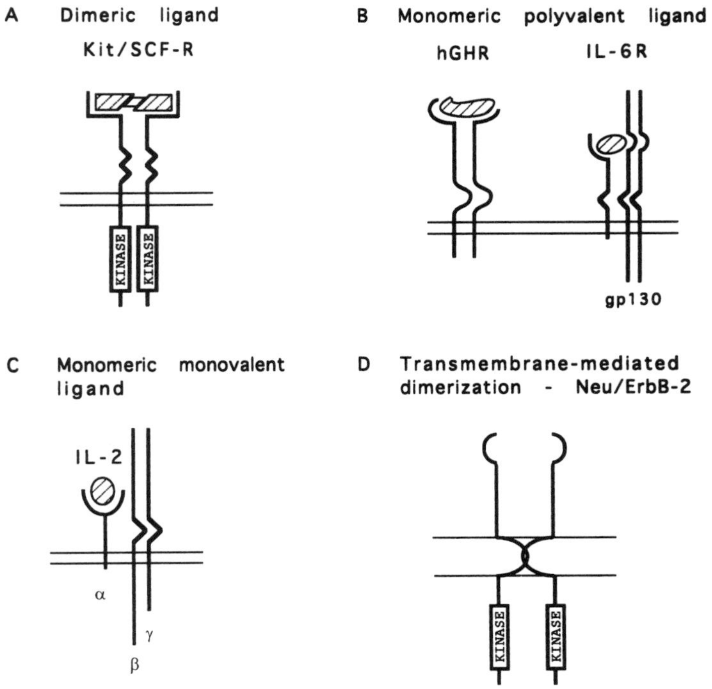

FIGURE 1. Representation of protein domains that are involved in ligand-induced receptor oligomerization. A: Dimeric complex of the Kit tyrosine kinase receptor that is mediated by the binding of dimeric SCF and an intrinsic membrane proximal portion of Kit. B: Homo- and heterooligomeric complexes of cytokine receptors for GH or IL-6 that are stabilized by binding of polyvalent monomeric ligands. C: A heterooligomeric complex of the IL-2 receptor that is induced by a monovalent monomeric ligand and is mediated by membrane proximal portions of receptor subunits. D: A dimeric complex of the oncogenic Neu/ErbB-2 protein that is induced by a single mutation at the TM domain.

provided a positive effect on the biological activity of v-Ros.[33] The TM domain of the insulin receptor (IR) was shown to have a passive role in insulin-dependent signaling,[37] but its substitution with the mutated TM domain of ErbB-2 conferred ligand-independent activation.[38] The increased dimer stabilization by the mutated TM domain of ErbB-2 may be explained by its α-helical conformation and the induction of an extra interdomain hydrogen bond.[31] Thus, domains that are located distally to the catalytic or regulatory domains appear to be involved in receptor dimerization. Consistent with this, truncations of the extracellular regions of certain receptors resulted in the generation of constitutively active kinases: deletions of the EC domains confer tumorigenicity to the oncogenic receptors encoded by v-*erbB,* v-*fms,* v-*kit,* and v-*ros,*[39] and the transforming potential of the IR was remarkably enhanced by deletion of most of the EC sequence.[40] The studies on IR,[40] Ros,[33] or

Kit[41] revealed that sequences immediately upstream to the transmembrane domains of these RTKs impose negative constraints on the transforming potential of the receptors. Alteration of the membrane-proximal region of the β-chain of cytokine receptors also resulted in ligand-independent activation of signaling.[42] These reports strengthen the role of the EC domain in the regulation of receptor function following ligand binding. A list of mutations in EC regions that affect receptor function are presented in TABLE 1. Some mutations promote ligand-independent activation through receptor dimerization as has been shown for Ret,[43] EGFR,[44] or EPO-R.[45] Mutations in IR,[46] hGHR,[47] and in IL-6R[27] apparently cause alterations in protein conformation and result in loss of dimer formation and signaling. Two-point mutations in the EC domain of murine CSF-1R, in contrast to its human[48] and feline[49] homologues, caused tumorigenicity, but only in combination with a truncation at the C-terminal region.[50] Removal of the EC portion was not sufficient for Met/HGF receptor activation, which required the presence of an extra sequence known to promote spontaneous dimerization.[51]

c-KIT AS A PROTOTYPIC RECEPTOR

Unlike many other RTKs, whose physiological roles are still unknown, the action of c-Kit *in vivo* is relatively well characterized.[52] The c-*kit* proto-oncogene is allelic with the White spotting (W) locus, whose mutations affect melanogenesis, erythropoiesis, and gametogenesis. Kit displays structural similarity to other RTKs, and especially to the receptors for the PDGFs and to the macrophage growth factor (CSF-1) receptor. This group of receptors contains five Ig-like domains in the EC portion that are connected through a single transmembrane stretch to a bisected cytoplasmic tyrosine kinase domain. The ligand of Kit, the stem-cell factor (SCF), is a noncovalently held dimer that is encoded by the Steel locus at mouse chromosome 10. The phenotypes of SCF-defective mice are in many aspects indistinguishable from the landmarks displayed by W-mutant mice.[53] On the cellular level SCF affects cell proliferation, migration, survival, and, to some extent, cell differentiation and was shown to act synergistically with other growth factors.[52]

Following SCF binding, Kit undergoes extensive dimerization and a subsequent activation of its catalytic function.[29] Several SH2-containing proteins were shown to associate with Kit upon ligand binding: Kit kinase strongly associates with phosphatidylinositol 3′-kinase (PI3K) through interaction with the kinase insert domain.[54] Prevention of PI3-K binding by Tyr 719 substitution for Phe abolished SCF-induced PI3-K activation.[55] The interaction with phospholipase Cγ (PLCγ) is less efficient and does not result in a significant change in the production of phophatidylinositol in SCF-stimulated fibroblasts.[52] SCF activation results in an increase in the ratio of GTP/GDP bound to Ras.[56] However, modification of ras-GAP was demonstrated only in cells that overexpress this protein.[57] Although the association of Kit with the Vav protein is relatively weak, Vav could be phosphorylated by an activated Kit.[58] Two serine/threonine kinases are included in the Kit-induced intracellular cascades; these are Raf-1 and the mitogen-activated protein kinase (MAPK).[56,59] In comparison with other RTKs, Kit promotes weaker interactions with its intracellular substrates, and this correlates with the weak mitogenic potential of SCF, which is synergized

TABLE 1. EC Domain Mutations That Affect Signaling but Not Ligand Binding

Receptor Family	Receptor Protein	Ref. no.	Type of EC Domain	Type of Mutations	Functional Changes	Proposed Structural Basis
Cytokine receptors	hGHR	47	FN III type	Asp^{152} for His	Abolished signaling.	Altered conformation, preventing homodimerization.
	IL-6R	27		Mutations of seven amino acids in the membrane-proximal region.		Prevent IL-6R contact with gp130.
	β chain for IL-3R, IL-5R, GM-SCFR	42		Duplication of 37 membrane proximal amino acids.	Ligand-independent activation.	Spontaneous homodimerization of β chain.
	EPO-R	45		Arg^{129}, Glu^{132}, or Glu^{133} substitution for Cys.		Formation of stable S-S linked homodimers.
Tyrosine-kinase receptors	Ret	43	Cysteine-rich	Cys^{634} for Tyr, Arg, or Trp.		
	ErbB-1	44		Cys-$(Ala)_3$ insertion between amino acids 618 and 619.		
	ErbB-2	79		Ala^{653} for Cys.	No significant effect.	
	IR	46	α-chain	Phe^{382} for Val.	Impaired receptor phosphorylation.	Impaired α-β heterodimerization.
	CSF-IR	48 49	Ig-like	Substitutions of Leu^{301} (human) and Leu^{301} + Ala^{374} (feline).	Ligand-independent activation.	Conformational changes in the membrane-proximal Ig-like domains.
	Kit	41		4 amino acid deletion in the 5th Ig-like domain.	Increased level of receptor phosphorylation.	

by other growth factors. The study of the synergy of SCF with other lymphokines on the promotion of proliferation and differentiation of hematopoietic progenitor cells revealed that JAK2 can physically associate with Kit, and JAK2 undergoes phosphorylation in response to SCF binding.[60] Thus, Kit appears to mediate signaling by way of several alternative pathways.

Although under physiological conditions SCF exists as a noncovalently associated dimer, two lines of evidence indicate that dimer formation by Kit is independent of ligand bivalency: First, no transition to monomeric receptors was observed at high SCF concentration, although this is expected, in analogy to antibody-antigen interactions.[29] Second, heterodimers between human (h) and murine (m) Kit molecules could be forced by either human (h) or rodent (r) SCF, although mKit is not recognized by hSCF.[61] These results attributed the stabilization of dimers to direct receptor-receptor interactions. In addition, an increased affinity of hKit for rSCF supports the possibility that ligand bivalency contributes to the stability of the final complex of two receptors and one dimeric SCF, in analogy to the model of hGH-hGHR complex.[4] Therefore, we proposed a sequential model of Kit activation: monovalent SCF binding exposes a putative receptor site that facilitates rapid dimerization of the receptors, and this is further stabilized by binding of the second arm of the dimeric SCF molecule.[29] We assumed that the putative dimerization site resides on the ectodomain because a truncated EC portion of Kit retained ligand-induced dimer formation,[62] in analogy to the EC portion of the PDGF-R.[28] The EC domain of Kit includes five Ig-like domains, characterized by a primary sequence of 70-100 residues with an essential disulfide bond spanning 40-60 amino acids. As in PDGF-R and in CSF-1R, the fourth Ig homology unit lacks the cysteine-mediated intradomain bond, probably resulting in an increased degree of domain flexibility. In order to map structural determinants of Kit that are involved in ligand binding and in receptor-receptor interactions, we employed three experimental approaches: (1) monoclonal antibodies (mAbs) were generated against the ectodomain of Kit and tested for inhibition of either ligand binding or receptor dimerization; (2) progressive truncation mutants of the soluble ectodomain of Kit were generated and analyzed; and (3) chimeric human-mouse Kit proteins, as well as deletion mutants of the full-length Kit protein, were analyzed for their ligand binding and dimerization potentials.

Mapping of the Ligand-Binding Site

Certain mAbs that were raised against the EC portion of human Kit fully inhibited ligand binding, induced SCF dissociation from preformed SCF-Kit complexes, and prevented ligand-mediated receptor dimerization and autophosphorylation.[63] The antibodies affected also long-term Kit activation, namely, SCF-induced proliferation of MO7e human megakaryoblastic cells. Two inhibitory mAbs, whose epitopes do not overlap, were able to recognize soluble truncated fragments of Kit that contained two or three amino-terminal Ig-like domains. Consistent with this observation, these mutants of the Kit extracellular portion were able to bind SCF.[63] However, deletion of the third Ig-like module significantly reduced the affinity of the soluble protein to SCF. Further analysis of the role played by individual Ig-like loops was performed by constructing chimeric mouse-human Kit proteins.[61] Although replacement of the second Ig-like unit of mKit with the corresponding portion of hKit was the only single-domain substitution that allowed the murine receptor to bind the human ligand, the third human unit enhanced ligand binding when combined with the second Ig loop. Interestingly, this effect was mediated largely by a decrease in the rate of SCF dissociation, supporting a stabilizing role for the third Ig-like module.[61] Additional analyses implied that the first Ig unit negatively affects ligand binding, and, unlike in hKit, the third subdomain of mKit, rather than the second, is the predominant motif of the binding cleft. In conclusion, the three N-terminal Ig-like units of Kit are involved in the formation of the binding cleft.

Similarly, the three N-terminal Ig-like domains of PDGF-R are also sufficient for PDGF-AA binding, and the second domain has the major role.[64] Interestingly, analogous to the negative effect of the most amino-terminal Ig-like domain of Kit, the N-terminal Ig-like domains of FGFR[10] and cytokine receptors for IL-6[27] and G-CSF[65] are not involved in ligand binding. However, as in the case of Kit, a role of noncontiguous receptor portions in ligand recognition has been demonstrated also for the EGF[66] and for the insulin[67] binding sites.

Mapping of the Dimerization Site

Despite the fact that a soluble Kit protein composed of the three N-terminal Ig-like domains fully retained ligand-binding ability, it totally lost the ligand-induced dimerization function.[63] This implied that the ability of the Kit ectodomain to undergo SCF-induced dimerization depends on a region within the two membrane-proximal Ig-like domains and their intervening sequences. In order to characterize the putative dimerization site we concentrated on two mAbs, denoted K27 and K69, that exerted only a limited effect on ligand binding to Kit, but could inhibit SCF-induced dimer formation (FIG. 2A). A similar effect was displayed by a monovalent fragment of mAb K27, implying that antibody bivalency is not essential for the inhibitory effect (FIG. 2B). The mAb did not recognize SCF-induced dimers of full-length Kit or its extracellular domain, indicating that the K27 epitope is masked upon Kit dimerization.

To further confine the K27 epitope (that probably corresponds to the putative dimerization site) to a single Ig-like domain, we analyzed three truncated Kit proteins consisting of three, four, or five N-terminal Ig-like units. In these experiments the fourth Ig-like domain turned out to be critical for K27 recognition. In agreement with this observation, addition of this sequence to the three N-terminal domains conferred to them the ability to undergo ligand-induced dimerization.[68] To confirm the significance of Ig-like domain 4 in receptor-receptor contact, a Kit mutant lacking this domain was constructed. In contrast to the full-length receptor, only monomeric complexes of SCF with the mutant receptor were observed in living cells, confirming the attribution of the dimerization site to Ig domain 4. However, unlike ligand-competitive mAbs, that were unable to recognize a denatured form of the receptor,[63] the dimerization-inhibitory mAbs K27 and K69 recognized the fully denatured monomeric form of Kit in Western blots (FIG. 3A), but they could not recognize the dimeric form of Kit (FIG. 3B). Enzymatic digestion of a recombinant Kit ectodomain with *N*-glycanase, but not *O*-glycanase, was accompanied by significant, but incomplete, abolishment of the interaction with mAb K27 (FIG. 3C), implying that the antigenic

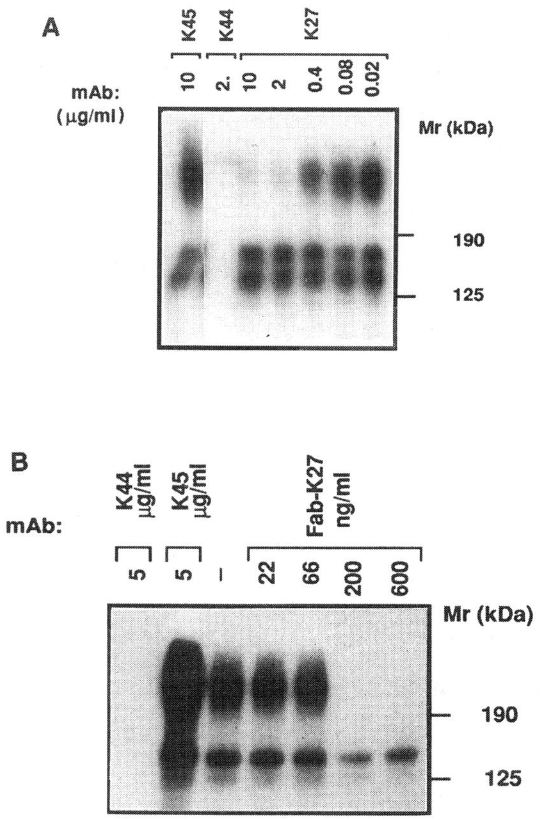

FIGURE 2. Interaction of mAb K27 with monomers and dimers of Kit. Monolayers of murine fibroblasts that overexpress human Kit (approximately 10^5 cells) were incubated with radiolabeled SCF in the presence of the indicated concentrations of various mAbs to Kit, including a ligand-inhibitory mAb K44, the dimerization-inhibitory mAb K27 and its monovalent Fab fragment, and a control anti-Kit antibody (K45). Following 2 h at 4 °C the cells were washed, and the ligand-receptor complexes were covalently cross-linked by a 40-min long incubation at 22°C with 15 mM 1-ethyl-3-(3-dimethylaminopropyl)carbodiimide (EDAC). Cell lysates were then prepared and subjected to immunoprecipitation with the K45 mAb to Kit. The immune complexes were analyzed by electrophoresis on 5% SDS-polyacrylamide gel and autoradiography.

epitope of K27 may include a protein determinant, in addition to asparagine-linked sugars.

Structural Aspects of the Ligand-dependent Kit Dimerization Site

Assuming that the antigenic epitope of mAb K27 lies close to, or within, the putative SCF-dependent dimerization site of Kit, it can be predicted that this site includes an immunogenic epitope that is independent of a specific conformation and includes asparagine-linked sugar groups. The direct contribution of N-linked oligosaccharides to protein-protein interactions is not yet clear. Inhibition of *N*-glycosylation was shown to result in slow or abolished dimerization, coincident with impaired ligand-binding ability, in the case of the transferrin receptor.[69] As carbohydrate addition is known to be necessary for proper protein folding, it is difficult to conclude whether the absence of sugar groups caused impaired receptor-receptor contacts or aberrant protein folding. Experimental data concerning carbohydrate-mediated associations of C2 domains within IgG molecules[70] and the significance of protein-glycosaminoglycan interactions to the function of Ig-like containing proteins, such as N-CAM[71] or FGF-R,[72] raise the possibility that *N*-glycosylation within the Kit dimerization site plays a role other than stabilization of protein conformation.

The dimerization site of Kit is apparently monovalent and symmetrical, because no higher oligomers were observed. The expected increased structural flexibility of the fourth domain, due to the absence of an intradomain disulfide bond, may facilitate coupling of ligand binding to receptor dimerization and also contribute to the reversibility of interreceptor association. Because no crystal structures are available for the C2 Ig-like module, we designed a conformational model on the basis of known structures of CH_2 and CH_3 domains of the human Fc fragment,[73] with the best fit obtained for the CH_3 dimer. The predicted interface between the domains displayed a mixture of hydrophobic and electrostatic natures. The contact region is formed by four pairs of hydrophobic residues flanked on both sides by ten charged or polar residues.[68] Two out of three *N*-glycosylation sites found in domain 4 were predicted to be located on the interface, supporting their involvement in receptor-receptor interactions. Domain 4 dimer formation is probably dominated by electrostatic interactions, and its stability was calculated to be lower than that of Ig dimers, which may be significant for the mode by which SCF controls dimer formation.

It is worthwhile to note the role of the fourth Ig-like domain in the activation of the transforming potential of CSF-1R (Fms).[74] Mutations of amino acids 301 and 374 were necessary, but insufficient, for Fms dimerization and a consequent cell transformation, in the absence of a ligand.[50] Apparently, the most carboxy-terminal tail of the receptor affects dimerization because only receptors with such a truncation, as well as the Ig domain 4 mutations, could dimerize in the absence of CSF-1.

Functional Aspects of the Dimerization Site

Coupling of Ligand-binding Affinity to Receptor Dimerization

The causative relationships between ligand binding and receptor dimerization remain open. Receptor dimers were shown to exist even in the absence of ligand under conditions of high local receptor concentration, as has been shown for the receptors of EGF[75] and PDGF.[11] However, ligand binding increases the fraction of dimers by a mechanism whose details are unknown. The availability of dimerization-inhibitory mAbs to Kit and a receptor mutant that is unable to dimerize allowed us to directly address these questions. In the presence of an inhibitory mAb, Kit proteins

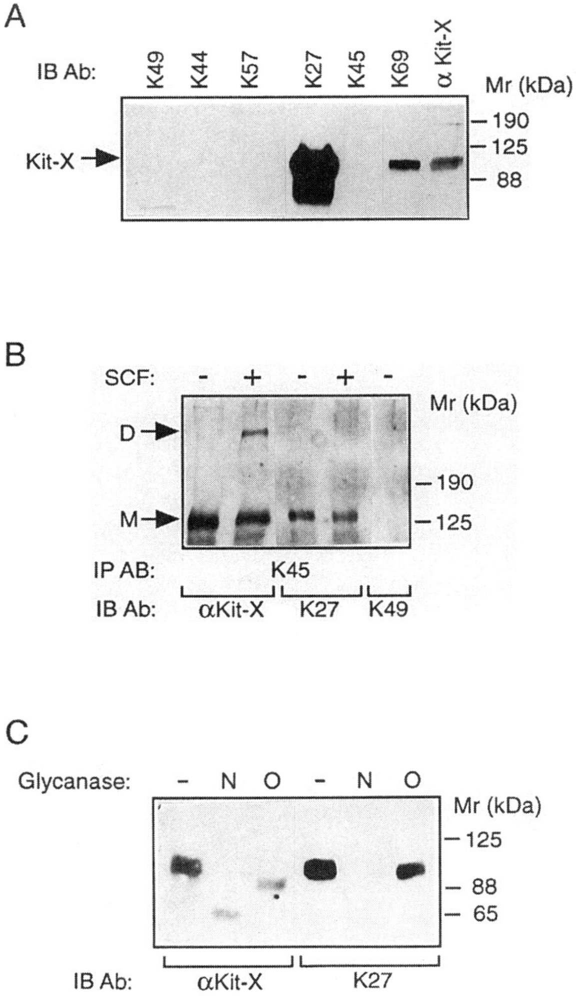
A
IB Ab:
K49
K44
K57
K27
K45
K69
α Kit-X
Mr (kDa)
190
125
88
Kit-X
B
SCF:
- + - + -
Mr (kDa)
D
M
190
125
IP AB:
K45
IB Ab:
αKit-X
K27
K49
C
Glycanase:
- N O - N O
Mr (kDa)
125
88
65
IB Ab:
αKit-X
K27

displayed a single population of binding sites with a relatively low affinity. Likewise, a dimerization-defective Kit mutant displayed two receptor populations whose majority corresponded to the affinity observed in the presence of a dimerization-inhibitory monoclonal antibody.[68] Ligand affinity reflects the contribution of both ligand association and dissociation rates. To attribute the effect of receptor dimerization to one of these processes, we analyzed ligand association to the dimerization-defective Kit, or to wild-type Kit, in the presence of mAb K27. We found that the rate of ligand association was not significantly affected by inhibition of receptor dimerization. However, unlike the very slow dissociation rate of SCF from wild-type Kit, the dimerization-defective mutant rapidly released the ligand at a rate comparable to the dissociation rate measured in the presence of mAb K27. It therefore appears that inhibition of receptor dimerization leads to a reduction in SCF binding affinity because of a concomitant increase in the rate of ligand dissociation. This conclusion may attribute the ligand dependency of dimer formation to an increased stability of the ternary complexes of SCF with two Kit molecules, as compared with the binary SCF-Kit complex. Therefore, it may be proposed that the increased binding affinity of Kit dimers towards SCF would cause the equilibrium between monomers and dimers to shift towards a dimeric state. This conclusion is in line with previous observations with other receptors. An increased affinity of soluble EGF receptors towards EGF was demonstrated upon stabilization of receptor dimers, and the effect was due to a reduced rate of ligand dissociation.[76] Similarly, the high-affinity state of the EGF-R correlated with covalently linked dimers on living cells,[44] and a mutation-induced constitutive dimer of an EGF-R-Neu chimera displayed only high-affinity binding sites, unlike the wild-type chimera that exhibited two populations of EGF-binding sites.[36]

FIGURE 3. The effect of denaturation and glycosylation on the recognition of Kit by various mAbs. A: A soluble recombinant ectodomain of Kit, denoted Kit-X, was subjected to electrophoresis in 7.5% SDS-polyacrylamide gel and then transferred onto a nitrocellulose filter. Individual strips of the filter were then immunoblotted (IB) with the indicated mAbs. Detection of the Western blot was performed by using chemiluminescence and brief autoradiography. An arrowhead marks the location of Kit-X, and bars indicate the locations of molecular mass marker proteins whose masses are given in kilodaltons (kDa). B: Monolayers of c-*kit* overexpressing cells (approximately 5×10^6 cells) were incubated for 2 h at 4°C in the presence (+) or absence (−) of SCF (0.5 μg/mL). The ligand-receptor complexes were covalently cross-linked by a 40-min-long incubation with EDAC (15 mM). Cell lysates were then prepared and subjected to immunoprecipitation (IP) with the K45 mAb to Kit. Following gel electrophoresis (5% acrylamide), the proteins were electrophoretically transferred onto a nitrocellulose filter that was immunoblotted (IB) with the indicated monoclonal antibodies to Kit or with a rabbit polyclonal antiserum to Kit-X (labeled αKit-X). Chemiluminescence and autoradiography were used for detection of the signals. D and M indicate receptor dimers and monomers, respectively. C: A recombinant extracellular domain of Kit (500 ng) was incubated for 20 h at 37°C in the presence or absence of N- or O-glycanase, as indicated, in order to remove sugar residues. Following electrophoresis on 7.5% SDS-polyacrylamide gel and electrophoretic transfer, the nitrocellulose filters were immunoblotted with either a rabbit antiserum to Kit-X or with the K27 mAb, as indicated. Note that *N*-deglycosylated Kit lost most of its reactivity with K27.

Coupling of Receptor Dimerization to Signal Transduction

Although dimerization of RTKs has been widely implicated as an obligatory step in the mechanism of signal transduction,[77] this paradigm has not been directly proven, because in no RTK system was it possible to selectively inhibit receptor dimerization without affecting ligand binding. Forcing a monomeric form of Kit using a dimerization-inhibitory mAb to Kit, as well as the availability of a dimerization-defective Kit mutant, allowed the study of ligand-dependent signal transduction by way of a monomeric ligand-receptor complex. Both antibody- and mutation-mediated monomeric forms of Kit failed to stimulate SCF-induced tyrosine autophosphorylation of the receptor, confirming the necessity of receptor dimerization for kinase activation. Unlike wild-type Kit, the SCF-stimulated dimerization-defective Kit mutant failed to couple to PI3K and to induce a change in intracellular Ca^{2+} concentration. The prevention of long-term Kit activation through inhibition of receptor dimerization was supported by the inability of SCF to regulate the cell cycle in cells expressing a dimerization-defective Kit, and by the inhibition of SCF proliferation of the myeloid cell line TF-1 by mAb K27.[68] Previous studies that addressed the role of receptor dimerization in signal transduction by using agonistic bivalent antibodies[34] or disulfide-mediated covalent linkage[44] also concluded that receptor dimers, rather than monomers, can transmit biological signals.

In conclusion, receptor dimerization, at least in the case of Kit, is essential for both short- and long-term intracellular effects of SCF. However, the exact mechanism by which ligand binding affects the dimerization site remains unclear. Presumably, dimer formation is inhibited by intramolecular domain-domain interactions, and SCF binding somehow removes this constraint. For example, interactions between Kit domains 4 and 5 may prevent intermolecular contact between juxtaposed fourth domains. The attribution of negative regulatory sequences to extracellular membrane-proximal sequences has been previously reported.[39] The emerging picture of structure-function relationships within the extracellular domain of Kit is schematically represented in FIGURE 4. Whereas the ligand binding site is confined to the three N-terminal Ig-like domains,[61,63] dimer formation is mediated by the fourth Ig-like domain.[68] The spatial separation between the two functional domains may be significant for the molecular mechanism that functionally couples ligand binding to dimer formation. In addition, the existence of a dimerization site, that is distinct from the ligand-binding site, may provide a structural basis for the formation of receptor heterodimers between closely related receptors that bind distinct ligands, as is the case with ErbB proteins.[78]

PERSPECTIVES

Various transmembrane molecules, including RTKs, receptor protein tyrosine phosphatases, T- and B-cell receptor complexes, and cell-adhesion molecules, contain Ig-like modules in their ectodomains. Therefore, the observations presented here may be relevant to the mechanisms underlying homo- or heterophilic interactions within other types of receptor families. The example of Kit-SCF interactions implies that, in contrast to the noncontiguous nature of ligand-binding sites, dimerization sites

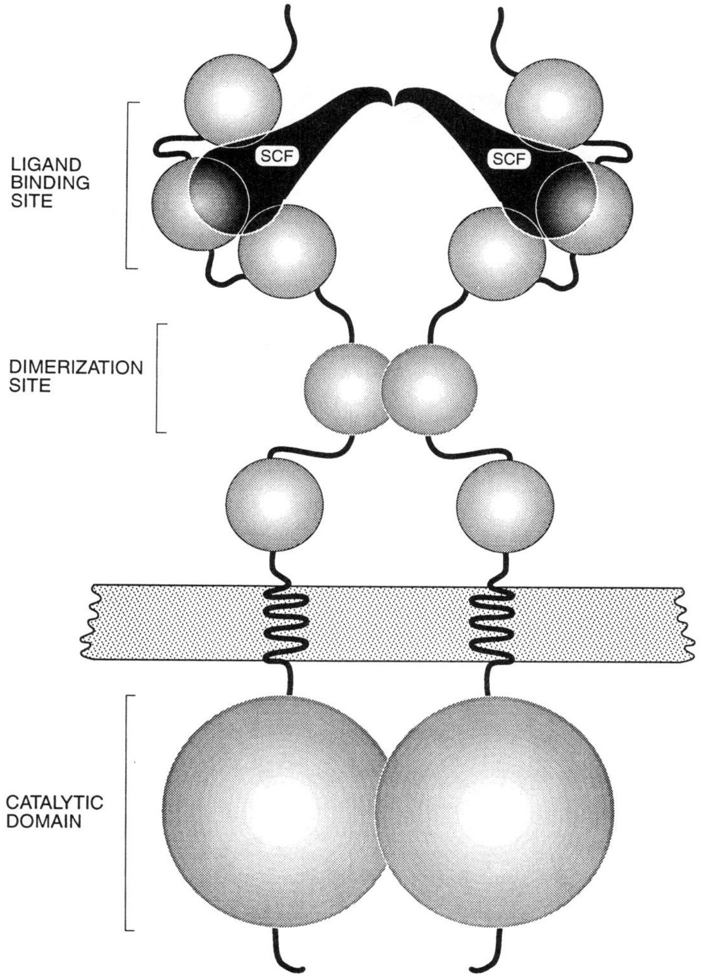

FIGURE 4. A schematic model of functional domains of Kit. A dimer of Kit proteins is shown together with a dimeric ligand (shaded). The Ig-like domains are represented by balls, and the horizontal stippled bar symbolizes the plasma membrane. The relative contribution of each Ig domain to the interaction with SCF is depicted by the extent of overlapping with the ligand, and is based on our previous analyses.[61,63] The dimerization site, according to the model, resides in the fourth Ig-like domain; it is monovalent and symmetrically interacts with its dimer mate. Although the ligand is shown to interact simultaneously with two receptors, this feature has not been directly proven, and it appears to be insufficient for maintaining receptor dimers.

may be confined to a contiguous protein sequence. The relative simplicity of the intrinsic dimerization site, as compared to the ligand-binding site, is advantageous for the generation of potent receptor antagonists. Another advantage of dimerization antagonists is their potential ability to inhibit signaling through constitutively active receptors, whose activation is ligand-independent and involves constant dimerization. Examples include the oncogenic forms of Neu-ErbB-2, CSF-1 receptor, and the hepatocyte growth factor receptor. Thus, strategies aimed at interference of dimer formation may be more powerful than generation of ligand antagonists, in respect to inhibition of signal transduction by growth factors. Further studies of the fine structure of interreceptor interfaces of dimerization sites should allow the design of potent antagonists applicable for the prevention of unregulated receptor activity that is often associated with receptor overexpression or mutagenesis.

ACKNOWLEDGMENTS

We thank Sima Lev and David Givol for extensive discussions.

REFERENCES

1. Fantl, W. J., D. E. Johnson & L. T. Williams. 1993. Signalling by receptor tyrosine kinases. Annu. Rev. Biochem. **62:** 453-481.
2. Koch, C. A., D. Anderson, M. F. Moran, C. Ellis & T. Pawson. 1991. SH-2 and SH-3 domains: elements that control interactions of cytoplasmic signaling proteins. Science **252:** 668-674.
3. Yarden, Y. & J. Schlessinger. 1987. Epidermal growth factor induces rapid, reversible aggregation of the purified epidermal growth factor receptor. Biochemistry **26:** 1443-1451.
4. Cunningham, B. C., M. Ultsch, A. M. Devos, M. G. Mulkerrin, K. R. Clauser & J. A. Well. 1991. Dimerization of the extracellular domain of the human growth hormone receptor by a single hormone molecule. Science **254:** 821-825.
5. Watowich, S. S., A. Yonishura, G. D. Longmore, D. J. Hilton, Y. Yoshimura & H. F. Lodish. 1992. Homodimerization and constitutive activation of the erythropoietin receptor. Proc. Natl. Acad. Sci. USA **89:** 2140-2144.
6. Metzger, H., G. Alcaraz, R. Hohman, J.-P. Kinet, V. Pribluda & R. Quarto. 1986. The receptor with high affinity for immunoglobulin E. Annu. Rev. Immunol. **4:** 419-470.
7. Cochet, C., O. Kashles, E. M. Chambaz, I. Borrello, C. R. King & J. Schlessinger. 1988. Demonstration of epidermal growth factor-induced receptor dimerization in living cells using a chemical crosslinking agent. J. Biol. Chem. **263:** 3290-3295.
8. Bishayee, S., S. Majunder, J. Khire & M. Das. 1989. Ligand-induced dimerization of the platelet-derived growth factor receptor. J. Biol. Chem. **264:** 11699-11705.
9. Li, W. & E. R. Stanley. 1991. Role of dimerization and modification of the CSF-1 receptor in its activation and internalization during the CSF-1 response. EMBO J. **10:** 277-288.
10. Shi, E., M. Kan, J. Xu, F. Wang, J. Hou & W. L. McKeehan. 1993. Control of fibroblast growth factor receptor kinase signal transduction by heterodimerization of combinatorial splice variants. Mol. Cell. Biol. **13:** 3907-3918.
11. Herren, B., B. Rooney, K. A. Weyer, N. Iberg, G. Schmid & M. Pech. 1993. Dimerization of extracellular domains of platelet-derived growth factor receptors: a revised model of receptor-ligand interaction. J. Biol. Chem **268:** 15088-15095.

12. Nocka, K., J. Tan, E. Chin, T. Y. Chu, P. Ray, P. Traktman & P. Besmer. 1990. Molecular bases of dominant negative and loss of function mutations at the murine c-*kit*/white spotting locus: W37, Wv, W41 and W. EMBO J. **9:** 1805-1813.
13. Kishimoto, T., T. Taga & S. Akira. 1994. Cytokine signal transduction. Cell **76:** 253-262.
14. Neer, E. J. 1995. Heterotrimeric G. Proteins: organizers of transmembrane signals. Cell **80:** 249-257.
15. Amati, B., S. Dalton, M. W. Brooks, T. D. Littlewood, G. I. Evan & H. Land. 1992. Transcriptional activation by the human c-Myc oncoprotein in yeast requires interaction with Max. Nature **359:** 423-426.
16. Gu, W., K. Cechova, V. Tassi & R. Dalla-Favera. 1993. Opposite regulation of gene transcription and cell proliferation by c-Myc and Max. Proc. Natl. Acad. Sci. USA **90:** 2935-2939.
17. Castellazzi, M., L. Loiseau, F. Piu & A. Segeant. 1993. Chimeric c-Jun containing an heterologous homodimerization domain transforms primary chick embryo fibroblasts. Oncogene **8:** 1149-1160.
18. Deng, T. & M. Karin. 1993. JunB differs from c-Jun in its DNA-binding and dimerization domains, and represses c-Jun by formation of inactive heterodimers. Genes & Dev. **7:** 479-90.
19. Schaefer, E. M., H. P. Erickson, M. Federwisch, A. Wollmer & L. Ellis. 1992. Structural organization of the human insulin receptor ectodomain. J. Biol. Chem. **267:** 23393-402.
20. Greenfield, C., I. Hiles, M. D. Waterfield, M. Federwisch, A. Wollmer, T. L. Blundell & N. McDonald. 1989. Epidermal growth factor binding induces a conformational change in the external domain of its receptor. EMBO J. **8:** 4115-4123.
21. Savino, R., A. Lahm, A. L. Salvati, L. Ciapponi, E. Sporeno, S. Altamura, G. Paonessa, C. Toniatti & G. Ciliberto. 1994. Generation of interleukin-6 receptor antagonists by molecular-modeling guided mutagenesis of residues important for gp130 activation. EMBO J. **13:** 1357-1367.
22. Lemmon, M. A. & J. Schlessinger. 1994. Regulation of signal transduction and signal diversity by receptor oligomerization. TIBS **19:** 459-463.
23. Qian, X., C. M. LeVea, J. K. Freeman, W. C. Dougall & M. I. Greene. 1994. Heterodimerization of epidermal growth factor receptor and wild-type or kinase-deficient Neu: a mechanism of interreceptor kinase activation and transphosphorylation. Proc. Natl. Acad. Sci. USA **91:** 1500-1504.
24. Bignon, C., E. Sakal, L. Belair, N. Chapnik-Cohen, J. Djiane & A. Gertler. 1994. Preparation of the extracellular domain of the rabbit prolactin receptor expressed in *Escherichia coli* and its interaction with lactogenic hormones. J. Biol. Chem. **269:** 3318-3324.
25. Samanta, A., C. M. LeVea, W. C. Dougall, X. Qian & M. I. Greene. 1994. Ligand and p185^{c-neu} density govern receptor interactions and tyrosine kinase activation. Proc. Natl. Acad. Sci. USA **91:** 1711-1715.
26. deVos, A. M., M. Ultsch & A. A. Kossiakoff. 1992. Human growth hormone and extracellular domain of its receptor: crystal structure of the complex. Science **255:** 306-312.
27. Yawata, H., K. Yasukawa, S. Natsuka, M. Murakami, K. Yamasaki, M. Hibi, T. Taga & T. Kishimoto. 1993. Structure-function analysis of human IL-6 receptor: dissociation of amino acid residues required for IL-6-binding and for IL-6 signal transduction through gp130. EMBO J. **12:** 1705-1712.
27a. Clackson, J. & J. A. Wells. 1995. A hot spot of binding energy in a hormone-receptor interface. Science **267:** 383-386.
28. Duan, D.-S. R., M. J. Pazin, L. J. Fretto & L. T. Williams. 1991. A functional soluble extracellular region of the platelet-derived growth factor (PDGF) β-receptor antagonizes PDGF-stimulated responses. J. Biol. Chem. **266:** 413-418.

29. LEV, S., Y. YARDEN & D. GIVOL. 1993. Dimerization and activation of the Kit receptor by monovalent and bivalent binding of the stem cell factor. J. Biol. Chem. **267:** 15970-15977.
30. NAKAMURA, T., S. M. RUSSELL, S. A. MESS, M. FRIEDMANN, M. ERDOS, C. FRANCOIS, Y. JACQUES, S. ADELSTEIN & W. J. LEONARD. 1994. Heterodimerization of the IL-2 receptor beta- and gamma-chain cytoplasmic domains is required for signalling. Nature **369:** 330-333.
31. GULLICK, W. J., A. C. BOTTOMLEY, F. J. LOFTS, D. G. DOAK, D. MULVEY, R. NEWMAN, M. J. CRUMPTON, M. J. STERNBERG & I. D. CAMPBELL. 1992. Three-dimensional structure of the transmembrane region of the proto-oncogenic and oncogenic forms of the neu protein. EMBO J. **11:** 43-48.
32. HENNECKE, S. & P. COSSON. 1993. Role of transmembrane domains in assembly and intracellular transport of the CD8 molecule. J. Biol. Chem. **268:** 26607-12.
33. ZONG, C. S., B. POON, J. CHEN & L.-H. WANG. 1993. Molecular and biochemical bases for activation of the transforming potential of the proto-oncogene c-*ros*. J. Virology **67:** 6453-6462.
34. YARDEN, Y. 1990. Agonistic antibodies stimulate the kinase encoded by the *neu* proto-oncogene in living cells but the oncogenic mutant is constitutively active. Proc. Natl. Acad. Sci. USA **87:** 2569-2573.
35. WEINER, D. B., J. LIU, J. A. COHEN, W. V. WILLIAMS & M. I. GREENE. 1989. A point mutation in the *neu* oncogene mimics ligand induction of receptor aggregation. Nature **339:** 230-231.
36. BEN-LEVY, R., E. PELES, R. GOLDMAN-MICHAEL & Y. YARDEN. 1992. An oncogenic point mutation confers high affinity ligand binding to the neu receptor. J. Biol. Chem. **265:** 17304-17313.
37. FRATTALI, A. L., J. L. TREADWAY & J. E. PESSIN. 1991. Evidence supporting a passive role for the insulin receptor transmembrane domain in insulin-dependent signal transduction. J. Biol. Chem. **266:** 9828-9834.
38. CHEATHAM, B., S. E. SHOELSON, K. YAMADA, E. GONCALVES & C. R. KAHN. 1993. Substitution of the erbB-2 oncoprotein transmembrane domain activates the insulin receptor and modulates the action of insulin and insulin-receptor substrate 1. Proc. Natl. Acad. Sci. USA **90:** 7336-40.
39. RODRIGUES, G. A. & M. PARK. 1994. Oncogenic activation of tyrosine kinases. Curr. Biol. **4:** 15-24.
40. POON, B. & L.-H. WANG. 1991. Molecular basis of the activation of the tumorigenic potential of gag-insulin receptor chimeras. Proc. Natl. Acad. Sci. USA **88:** 877-881.
41. REITH, A. D., C. ELLIS, S. D. LYMAN, D. M. ANDERSON, D. E. WILLIAMS, A. BERNSTEIN & T. PAWSON. 1991. Signal transduction by normal isoforms and W mutant variants of the Kit receptor tyrosine kinase. EMBO J. **10:** 2451-2459.
42. D'ANDREA, R., J. RAYNER, P. MORETTI, A. LOPEZ, G. J. GOODALL, T. J. GONDA & M. VADAS. 1994. A mutation of the common receptor subunit for interleukin-3 (IL-3), granulocyte-macrophage colony-stimulating factor, and IL-5 that leads to ligand independence and tumorigenicity. Blood **83:** 2802-2808.
43. SANTORO, M., F. CARLOMAGNO, A. ROMANO, D. P. BOTTARO, N. A. DATHAN, M. GRIECO, A. FUSCO, G. VECCHIO, B. MATOSKOVA, M. H. KRAUS & P. P. DI FIORE. 1995. Activation of RET as a dominant transmforming gene by germline mutations of MEN2A and MEN2B. Science **267:** 381-383.
44. SOROKIN, A., M. A. LEMMON, A. ULLRICH & J. SCHLESSINGER. 1994. Stabilization of an active dimeric form of the epidermal growth factor receptor by introduction of an inter-receptor disulfide bond. J. Biol. Chem. **269:** 9752-9759.
45. YOSHIMURA, A., G. LONGMORE & H. F. LODISH. 1990. Point mutation in the exoplasmic domain of the erythropoietin receptor resulting in hormone-independent activation and tumorigenicity. Nature **348:** 647-649.

46. Accili, D., L. Mosthaf, A. Ullrich & S. I. Taylor. 1991. A mutation in the extracellular domain of the insulin receptor impairs the ability of insulin to stimulate receptor autophosphorylation. J. Biol. Chem. **266:** 434-439.
47. Duquesnoy, P., M. L. Sobrier, B. Duriez, F. Dastot, C. R. Buchanan, M. O. Savage, M. A. Preece, C. T. Craescu, T. Blouquit, M. Goossens & S. Amselem. 1994. A single amino acid substitution on the exoplasmic domain of the human growth hormone (GH) receptor confers familial GH resistance (Laron syndrome) with positive GH-binding activity by abolishing receptor homodimerization. EMBO J. **13:** 1386-1395.
48. Roussel, M. F., J. R. Downing & C. J. Sherr. 1990. Transforming activities of human CSF-1 receptors with different point mutations at codon 301 in their extracellular domains. Oncogene **5:** 25-30.
49. Woolford, J., A. McAuliffe & L. R. Rohrschneider. 1988. Activation of the feline c-*fms* proto-oncogene: multiple alterations are required to generate a fully transformed phenotype. Cell **55:** 965-977.
50. Carlberg, K. & L. Rohrschneider. 1994. The effect of activating mutations on dimerization, tyrosine phosphorylation and internalization of the macrophage colony stimulating factor receptor. Mol. Biol. Cell **5:** 81-95.
51. Rodrigues, G. & M. Park. 1993. Dimerization mediated by a leucine zipper oncogenically activates the *met* receptor tyrosine kinase. Mol. Cell. Biol. **13:** 6711-6722.
52. Lev, S., J. M. Blechman, D. Givol & Y. Yarden. 1994. Steel factor and c-kit protooncogene: genetic lessons in signal transduction. Crit. Rev. Oncog. **5:** 141-168.
53. Besmer, P. 1991. The c-*kit* ligand encoded at the murine *Steel* locus: a pleiotropic growth and differentiation factor. Curr. Opinion Cell. Biol. **3:** 939-946.
54. Lev, S., D. Givol & Y. Yarden. 1992. The interkinase domain of Kit contains the binding site for phosphatidylinositol 3′-kinase. Proc. Natl. Acad. Sci. USA **89:** 678-682.
55. Serve, H., Y. C. Hsu & P. Besmer. 1994. Tyrosine residue 719 of the c-kit receptor is essential for binding of the P85 subunit of phosphatidylinositol (PI) 3-kinase and for c-kit-associated PI 3-kinase activity in COS-1 cells. J. Biol. Chem. **269:** 6026-6030.
56. Duronio, V., M. J. Welham, S. Abraham, P. Dryden & J. W. Schrader. 1992. p21 ras activation via hemopoietin receptors and c-*kit* requires tyrosine kinase activity but not tyrosine phosphorylation of p21*ras* GTPase-activating protein. Proc. Natl. Acad. Sci. USA **89:** 1587-1591.
57. Herbst, R., R. Lammers, J. Schlessinger & A. Ullrich. 1991. Substrate phosphorylation specificity of the human c-*kit* receptor tyrosine kinase. J. Biol. Chem. **266:** 19908-19916.
58. Alai, M., A. L.-F. Mui, R. L. Cutler, X. R. Bustelo, M. Barbacid & G. Krystal. 1992. *Steel* factor stimulates the tyrosine phosphorylation and the proto-oncogene product, p95vav, in human hemopoietic cells. J. Biol. Chem. **267:** 18021-18025.
59. Funasaka, Y., T. Boulton, M. Cobb, Y. Yarden, B. Fan, S. D. Lyman, D. E. Williams, R. Ander. Y, Mishima & R. Halaban. 1992. c-*kit* kinase induces a cascade of protein tyrosine phosphorylation in normal human melanocytes in response to mast cell growth factor and stimulates mitogen-activated protein kinase but is down-regulated in melanomas. Mol. Biol. Cell **3:** 197-209.
60. Brizzi, M. F., M. G. Zini, M. G. Aronica, J. M. Blechman, Y. Yarden & L. Perogaro. 1994. Convergence of signaling by interleukin-3, granulocyte-macrophage colony-stimulating factor, and mast cell growth factor on JAK2 tyrosine kinase. J. Biol. Chem. **269:** 31680-84.
61. Lev, S., J. Blechman, S.-I. Nishikawa, D. Givol & Y. Yarden. 1993. Interspecies molecular chimeras of Kit help define the binding site of the stem cell factor. Mol. Cell. Biol. **13:** 2224-2234.
62. Lev, S., Y. Yarden & D. Givol. 1992. A recombinant ectodomain of the receptor for the stem cell factor (SCF) retains ligand-induced receptor dimerization and antagonizes SCF-stimulated cellular responses. J. biol. Chem. **267:** 10866-10873.

63. Blechman, J. M., S. Lev, O. Leitner, D. Givol & Y. Yarden. 1993. Soluble Kit proteins and anti-receptor monoclonal antibodies confine the binding site of the stem cell factor. J. Biol. Chem. **268:** 4399-4406.
64. Heidaran, M. A., J.-C. Yu, R. A. Jensen, J. H. Pierce & S. A. Aaronson. 1992. A deletion in the extracellular domain of the platelet-derived growth factor (PDGF) receptor differentially impairs PDGF-AA and PDGF-BB binding affinities. J. Biol. Chem. **267:** 2884-2887.
65. Fukunaga, R., E. Ishizuka-Ikeda, C.-X. Pan, Y. Seto & S. Nagata. 1991. Functional domains of the granulocyte colony-stimulating factor receptor. EMBO J. **10:** 2855-2865.
66. Wu, D., L. Wang, G. H. Sato, K. A. West, W. R. Harris, J. W. Crabb & J. D. Sato. 1989. Human epidermal growth factor (EGF) receptor sequences recognized by EGF-competitive monoclonal antibodies. J. Biol. Chem. **264:** 17469-17475.
67. Gustafson, T. A. & W. J. Rutter. 1990. The cysteine-rich domains of the insulin and insulin-like growth factor I receptors are primary determinants of hormone binding specificity. J. Biol. Chem. **265:** 18663-18667.
68. Blechman, J. M., S. Lev, J. Barg, M. Eisenstein, B. Vaks, Z. Vogel, D. Givol & Y. Yarden. 1995. The fourth immunoglobulin domain of the stem cell factor receptor couples ligand binding to signal transduction. Cell **80:** 101-113.
69. Yang, B., M. H. Hoe, P. Black & R. C. Hunt. 1993. Role of oligosaccharides in the processing and function of human transferrin receptors. J. Biol. Chem. **268:** 7435-7441.
70. Parekh, R., D. Isenberg, G. Rook, I. Roitt, R. Dwek & T. Rademacher. 1989. A comparative analysis of disease-associated changes in the galactosylation of serum IgG. J. Autoimmunity **2:** 101-114.
71. Rao, Y., X.-F. Wu, J. Gariepy, U. Ruishauser & C. H. Siu. 1992. Identification of a peptide sequence involved in homophilic binding in the neural cell adhesion molecule NCAM. J. Cell Biol. **118:** 937-949.
72. Spivak-Kroizman, T., M. A. Lemmon, I. Dikic, J. E. Ladbury, D. Pinchasi, J. Huang, M. Jaye, G. Crumley, J. Schlessinger & I. Lax. 1994. Heparin-induced oligomerization of FGF molecules is responsible for FGF receptor dimerization, activation, and cell proliferation. Cell **79:** 1015-1024.
73. Deisenhofer, J. 1981. Crystallographic refinement and atomic models of a human Fc fragment and its complex with fragment B of protein A from *Staphylococcus aureus* at 2.9- and 2.8 A resolution. Biochemistry **20:** 2361-2370.
74. van Daalen Wetters, T., S. A. Hawkins, M. F. Roussel & C. J. Sherr. 1992. Random mutagenesis of CSF-1 receptor (FMS) reveals multiple sites for activating mutations within the extracellular domain. EMBO J. **11:** 551-557.
75. Boni-Schnetzler, M. & P. M. Pilch. 1987. Mechanism of epidermal growth factor receptor autophosphorylation and high affinity binding. Proc. Natl. Acad. Sci. USA **84:** 7832-7836.
76. Zhou, M., S. Felder, M. Rubinstein, D. R. Hurwitz, A. Ullrich, I. Lax & J. Schlessinger. 1993. Real-time measurements of kinetics of EGF binding to soluble receptor monomers and dimers support the dimerization model for receptor activation. Biochemistry **32:** 8193-8198.
77. Heldin, C.-H. 1995. Dimerization of cell surface receptors in signal transduction. Cell **80:** 213-223.
78. Tzahar, E., G. Lefkowitz, D. Karunagaran, L. Yi, E. Peles, S. Lavi, D. Chang, N. Liu, A. Yayon, D. Wen & Y. Yarden. 1994. ErbB-3 and ErbB-4 function as the respective low and high affinity receptors of all NDF/Heregulin isoforms. J. Biol. Chem. **269:** 25226-25233.
79. Cao, H., L. Bangalore, B. J. Bormann & D. F. Stern. 1992. A subdomain in the transmembrane domain is necessary for $p185^{neu}$ activation. EMBO J. **11:** 923-932.

Signal Transduction Interception as a Novel Approach to Disease Management

ALEXANDER LEVITZKI [a]

Department of Biological Chemistry
The Alexander Silberman Institute of Life Sciences
The Hebrew University of Jerusalem
Jerusalem 91904, Israel
and
Sugen, Inc.
515 Galveston Drive
Redwood City, California 94063

INTRODUCTION

The advances in molecular biology in the past two decades have revolutionized our understanding of disease processes. Nowadays one can define many diseases in molecular terms and characterize the disease in terms of specific cellular processes. These developments have refocused the research on the mechanisms of disease processes as well as on the approach that is looking for novel treatment modalities. One area that has an overwhelming impact on the emerging field of molecular medicine is signal transduction. Pathological aberrations in signal transduction result in numerous types of diseases, such as cancers and other hyperproliferative diseases, in which the normal functions of signaling elements become abnormal, causing aberrant cell growth. For example, the activating mutation of the RAS protooncogene coverts it to an oncogene, and the release of PDGFβ from injured endothelial cells and platelets at a damaged endothelial area of a blood vessel is a major cause for the development of the atherosclerotic plaque and the occurrence of restenosis. Similarly, the onset of massive release of tumor necrosis factor α (TNFα) by bacterial infection is a primary cause for sepsis, whereas its sustained release due to an autoimmune process is causative to rheumatoid arthritis. Recognizing that diseased cells become highly dependent on the activated signal transduction pathways, we and others have embarked on attempts to examine the feasibility of inhibiting the disease process by the selective interception or modulation of the implicated signal transduction pathways (for review, see ref 1). This interception, in principle, can be achieved by a variety of means: small molecules that specifically inhibit or enhance the activity of a signaling element, antisense RNA targeted by a viral rector or small

[a] Please address correspondence to the Hebrew University address.

oligodeoxynucleotides that reduce translation of the signaling protein, "gene therapy" type approaches, and receptor-directed immunotoxins. In this review, we shall illustrate the biological activity of protein tyrosine inhibitors and their potential use as drugs in a variety of diseases. For a more comprehensive review, the reader is referred to reference 2.

PTK BLOCKERS AS ANTICANCER AGENTS

The involvement of enhanced protein tyrosine kinase (PTK) activity is well documented. Tumors overexpress certain PTKs and express persistently active autocrine loops or mutated hyperactivated PTKs. In most cases more than one PTK is involved. Furthermore, these tumors are highly dependent for their growth on these PTKs and have usually lost the redundancy in cellular signaling typical of the normal cell.[2] This can be demonstrated by ability to block such tumors by antibodies against the overexpressed receptor or by dominant negative constructs against the receptor or the growth factor.[3–5] These findings validate the hypothesis that inhibiting the PTKs expressed in tumors can be a valid approach to cancer therapy. Indeed, EGF receptor kinase blockers have been shown to inhibit the growth of A431 epidermoid carcinoma cells[6] and of human squamous carcinoma cells[7] that overexpress the EGF receptor. Furthermore, EGF receptor kinase blockers from the tyrphostin family[7] as well as 4,5 dianilino-phthalimides[8] have shown efficacy *in vivo.* Selective PTK blockers from the tyrphostin family were also shown to induce terminal erythroid differentiation in K562 chronic myeloid leukemic cells[9] and to eliminate human pre-B leukemic cells from SCID mice and save them (Roifman and Levitzki, unpublished). Other examples are the reversal of the *sis*-transformed state of Swiss 3T3 cells by PDGF receptor kinase selective blockers[10] and the reversal of src-transformed state chicken lense cells[11] and of NIH3T3 cells[12] by still other PTK blockers of the tyrphostin family. More recently, it has been realized by us that the combination of tyrphostins with antibodies,[7] or with cytotoxic agents like cis-platin or doxorubicin (Tsai and Levitzki, unpublished), is extremely effective in tumor cell killing. These findings on the synergistic effects suggest new opportunities in combination chemotherapy.

PTK BLOCKERS AS ANTIRESTENOSIS AND ANTIINFLAMMATORY AGENTS

The pathological activation of tyrosine kinases is not unique for cancer and is the hallmark of other diseases. Atherosclerosis and restenosis represent a disease state in which a number of tyrosine kinases are involved. These include the PDGF receptor β, PDGF receptor α, as well as basic FGF-receptor and cytokine-activated kinases.[13] Thus restenosis, mainly driven by the hyperproliferation of neointimal smooth muscle cells migrating from the intima and by an inflammatory component that also involves PTKs, offers an opportunity for intervention by PTK blockers. Indeed, PTK blockers that have shown efficacy *in vitro,* blocking the proliferation of smooth muscle cells in correlation with inhibition of PDGF receptor signal transduction,[14] are now found to block the formation of neointima subsequent to endothelial injury.[15] This effect is most probably due not only to the direct inhibition of PDGF and bFGF-induced signal transduction, but to the inhibitory effect of these compounds on the inflammatory component of the disease. Independent experiments reveal that PTK blockers can act as efficient antiinflammatory agents, protecting mice against lipopolysaccharide (LPS)-induced endotoxic shock.[16] The effect of these PTK blockers seems to be twofold: inhibition of LPS-induced tumor necrosis alpha (TNFα) production by macrophages as well as the inhibition of TNFα action on their target cells. Both pathways seem to involve tyrosine kinases, although their exact identity and role is still under investigation.

PTK BLOCKERS AS POTENTIAL ANTIPSORIATIC AGENTS

The involvement of a persistent TGFα/EGF receptor autocrine loop in the hyperproliferative state of the psoriatic keratinocyte seems to be the hallmark of the disease. It therefore occurred to us that local application of PTK blockers may become a useful treatment modality. Indeed, we have been able to show that PTK blockers from the tyrphostin family are effective in blocking the proliferation of psoriatic keratinocytes *in vitro.*[17]

PTK BLOCKERS AS INHIBITORS OF ANGIOGENESIS

Tumor growth is highly dependent on vascularization. This process is governed by vascular endothelial growth factor (VEGF), which interacts with Flk-1/KDR receptor.[18] Blocking Flk-1/KDR activity, therefore, can lead to the inhibition of tumor growth. This indeed has been validated by the demonstration that dominant negative Flk-1 construct strongly inhibits C6 glioma growth *in vivo.*[19] Recent screening efforts identified Flk-1 blockers *in vitro* and angiogenesis *in vivo.*

MECHANISM OF ACTION OF PTK BLOCKERS

Screening of fungal and plant extracts has lead to a number of lead structures for PTK inhibition. In all cases examined, the compounds found are potent competitors vis-a-vis (for review, see ref. 20) the ATP in the kinase reaction, whereas only erbstatin, a relatively smaller molecule, was found to compete with both substrate and ATP.[21] Based on these lead structures, many PTK blockers have been synthesized. Kinetic studies on a number of tyrphostin PTK blockers reveal that many of the compounds are competitive with both substrate and ATP,[21] and only a small fraction of the tested compounds are exclusively competitive with the substrate.[21] All the multiring compounds, such as quercetin, genistein, lavendustin A, and various quinaz-

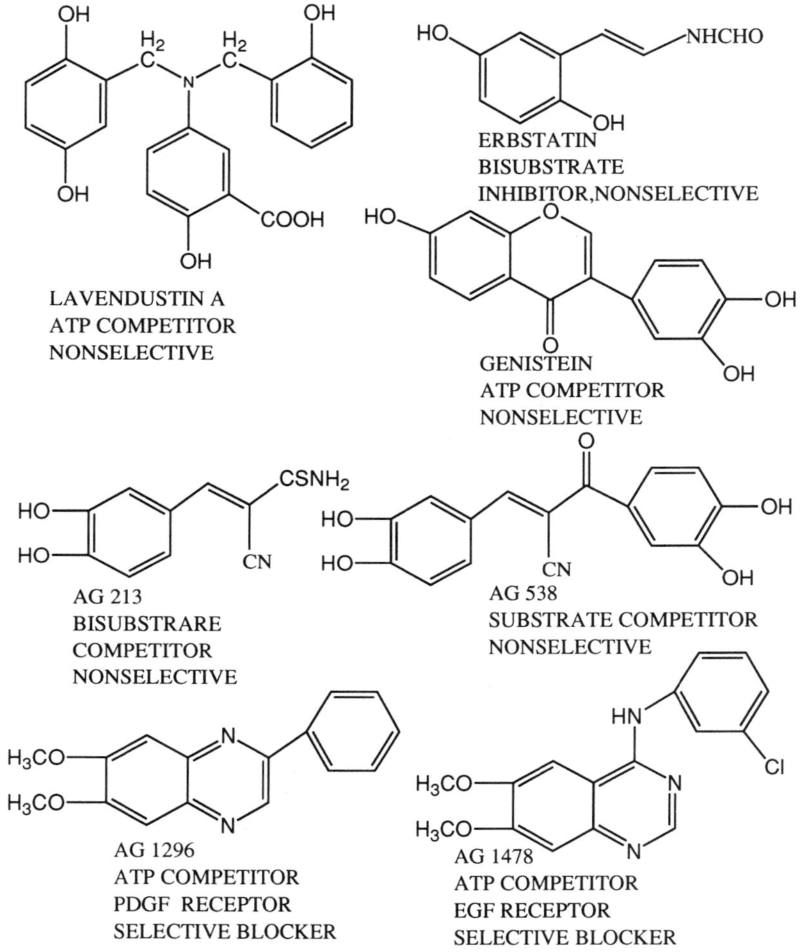

FIGURE 1. Some blockers of the EGF receptor kinase. A number of EGFR kinase inhibitors are depicted. Except for the quinazolines, all the inhibitors are nonselective. Some inhibitors are bisubstrate inhibitors, namely, they compete with both the ATP and the substrate sites simultaneously. Some inhibitors are competitive with the substrate but noncompetitive with ATP, and some are ATP competitive and substrate noncompetitive.

olines (selective EGFR blockers), are all competitors of ATP. TABLE 1 shows a number of pharmacophores and their mode of inhibition of the EGFR kinase. Many of the 3,4 dihydroxybenzenemalononitrile tyrphostins are either substrate competitive inhibitors bisubstrate or substrate and ATP competitive inhibitors. None of the tyrphostins of this family have been found to be a pure ATP competitor.[21] It seems that 3,4 dihydroxybenzylene moiety, which is the characteristic pharmacophore of hundreds of tyrphostins, is an excellent tyrosine mimic. The presence of this moiety seems to insure binding to the tyrosine-substrate site. Whether the inhibitors also bind to the ATP subsite depends on the chemistry of the rest of the molecule. Two-ring heterocyclic compounds, like AG1478, a selective EGFR blocker, or AG1296, a selective PDGFR blocker, are pure ATP competitors. Structures derived from lavendustin A vary: the three-ring system of lavendustin A is a pure ATP competitor, but the trimmed two-ring system, where one phenyl ring has two hydroxyl residues, either 1,4 (AG814, ref. 21) or 3,4 (AG826, ref. 17), is competitive with both substrate and ATP. It is interesting that the structure of AG814 is also quite similar to that of erbstatin, which was also found to compete against both substrate and ATP.[21] It is interesting to note that the mode of inhibition of one kinase is not identical to that of another PTK. AG213, for example, is competitive with both ATP and substrate for the EGFR, but is competitive only with substrate for the $p210^{bcr\text{-}abl}$ kinase.[22] When the three-dimensional structure of a protein-tyrosine kinase will become available, more precise drug-design considerations will be feasible.

SIGNAL TRANSDUCTION THERAPY–A GENERAL APPROACH

The emerging success of PTK blockers as novel leads for drugs is paralleled by various successes in intercepting other signal pathways. Protein kinase C blockers, Ca^{2+} signaling blockers, antiestrogens, and Ras farnesylation blockers prove to be effective antiproliferative agents (see ref. 1 for a review). Many agents can still be developed against various signal transduction elements downstream to tyrosine kinases and to Ras, such as SH2 blockers and Raf1 blockers. It is likely that such signal transduction interceptors will become an important new arsenal of drugs. The advantage of this emerging class of drugs is that they are targeted more towards certain proteins and thus are likely to be less toxic. Indeed, the limited experience of using PTK blockers *in vivo* in mice is already teaching us about the highly nontoxic profile of many of these compounds.

REFERENCES

1. LEVITZKI, A. 1994. Signal transduction therapy—A novel approach to disease management. Eur. J. Biochem. **226:** 1-13.
2. LEVITZKI, A. & A. GAZIT. 1994. Tyrosine kinase inhibition: An approach to drug development of many drugs. Science **267:** 1782-1788.
3. KASHLES, O., Y. YARDEN, R. FISCHER, A. ULLRICH & J. SCHLESSINGER. 1991. A dominant negative mutation suppresses the function of normal EGF receptors by heterodimerization. Mol. Cell. Biol. **11:** 1459-1463.
4. STRAWN, L. M., E. MANN, S. S. ELLIGER, L. M. CHU, L. L. GERMAIN, G. NIEDERFELLNER, A. ULLRICH & L. K. SHAWVER. 1994. Inhibition of glioma cell growth by a truncated platelet-derived growth factor-β receptor. J. Biol. Chem. **269:** 21215-21222.
5. SHAMAH, S. M., C. D. STILES & A. GUHA. 1993. Dominant-negative mutants of platelet-derived growth factor revert the transformed phenotype of human astrocytoma cells. Mol. Cell. Biol. **13:** 7203-7212.
6. YAISH, P., A. GAZIT, C. GILON & A. LEVITZKI. 1988. Blocking of EGF-dependent cell proliferation by EGF receptor kinase inhibitors. Science **242:** 933-935.
7. YONEDA, T., R. LYALL, M. M. ALSINE, P. E. PEARSONS, A. P. SPADA, A. LEVITZKI, A. ZILBERSTEIN & G. R. MUNDY. 1991. The antiproliferative effects of tyrosine kinase

inhibitor tyrphostin on a human squamous cell carcinoma *in vitro* and in nude mice. Cancer Res. **51:** 4430-4435.

8. BUCHDUNGER, E., U. TRINKS, H. METT, U. REGENASS, M. MÜLLER, T. MEYER, E. MCGLYNN, L. A. PINNA, P. TRAXLER & N. B. LYDON. 1993. 4,5-Dianilinophathalmide: A protein-tyrosine kinase inhibitor with selectivity for the epidermal growth factor receptor signal transduction pathway and potent *in vivo* antitumor activity. Proc. Natl. Acad. Sci. USA **91:** 2334-2338.
9. ANAFI, M., A. GAZIT, A. ZEHAVI, Y. BEN-NERIAH & A. LEVITZKI. 1993. Tyrphostin-induced inhibition of $p210^{bcr\text{-}abl}$ tyrosine kinase activity induces K562 to differentiate. Blood **82**(12): 3524-3529.
10. KOVALENKO, M., A. GAZIT, A. BÖHMER, C. RORSMAN, C. RÖNNSTRAND, C-H HELDIN, J. WALTENBERGER, F-D BÖHMER & A. LEVITZKI. 1994. Selective PDGF receptor kinase blockers reverse sis-transformation. Cancer Res. **54:** 6106-6114.
11. VOLBERG, T., Y. ZICK, R. DIOR, I. SABANAY, C. GILON, A. LEVITZKI & B. GEIGER. 1992. The effect of tyrosine specific protein phosphorylation on the assembly of adherents-type junctions. EMBO J. **11:** 1733-1742.
12. AGBOTOUNOU, W. K., A. LEVITZKI, A. JACQUEMIN-SABLON & J. PIERRE. 1994. Effects of tyrophostins on the activated c-src protein in NIH/3T3 cells. Mol. Pharmacol. **45**(5): 922-931.
13. ROSS, R. C. 1993. The pathogenesis of atherosclerosis: A perspective for the 1990's. Nature **302:** 801-809.
14. BILDER, G., J. A. KRAWIEC, A. GAZIT, C. GILON, K. MCVETY, R. LYALL, A. ZILBERSTEIN, A. LEVITZKI, M. PERRONE & A. B. SCHREIBER. 1991. Tyrphostins inhibit PDGF-induced DNA synthesis and associated early events in smooth muscle cells. Am. J. Physiol. **260:** C721-C730.
15. GOLOMB, G., I. FISHBEIN, S. BANAI, D. MISHALY, D. MOSCOVITZ, D. GERTZ, A. GAZIT & A. LEVITZKI. 1994. Controlled delivery of a tyrphostin inhibits intimal hyperplasia in a rat carotid artery injury model. Submitted.
16. NOVOGRODSKY, A., A. VANICHKIN, M. PATYA, A. GAZIT, N. OSHEROV & A. LEVITZKI. 1994. Prevention of lipopolysaccharide-induced lethal toxicity by tyrosine kinase inhibitors. Science **264:** 1319-1322.
17. BEN-BASSAT, H., D. VARDI, A. GAZIT, S. N. KLAUS, M. CHAOUAT, Z. HARTZSTARK & A. LEVITZKI. 1995. Tyrphostins suppress the growth of psoriatic keratinocytes. Exp. Dermatol. **4**(2): 82-89.
18. MILLAUER, B., S. WIZIGMANN-VOOS, H. SCHNÜRCH, R. MARTINEZ, M. P. H. MOELLER, W. RISAU & A. ULLRICH. 1993. High affinity VEGF binding and developmental expression suggest Flk-1 as a major regulator of vasculogenesis and angiogenesis. Cell **72:** 835-846.
19. MILLAUER, B., L. K. SHAWVER, K. H. PLATE, W. RISAU & A. ULLRICH. 1994. Glioblastoma growth inhibited *in vivo* by a dominant-negative Flk-1 mutant. Nature **367:** 576-579.
20. LEVITZKI, A. 1992. Tyrphostins: tyrosine kinase blockers as novel antiproliferative agents and dissectors of signal transduction. FASEB J. **6:** 3275-3282.
21. POSNER, I., M. ENGEL, A. GAZIT & A. LEVITZKI. 1994. Kinetics of inhibition of the tyrosine kinase activity of the epidermal growth factor receptor by tyrphostins. Mol. Pharmacol. **45:** 673.
22. ANAFI, M., A. GAZIT, C. GILON, Y. BEN-NERIAH & A. LEVITZKI. 1992. Selective interactions of transforming and normal abl proteins with ATP, tyrosine copolymer substrates and tyrphostins. J. Biol. Chem. **267:** 4518-4523.

Common and Distinct Elements in Insulin and PDGF Signaling[a]

MARTIN G. MYERS JR., BENTLEY CHEATHAM, TRACEY L. FISHER, BOZENA R. JACHNA, C. RONALD KAHN, JONATHAN M. BACKER,[b] AND MORRIS F. WHITE[c]

Research Division
Joslin Diabetes Center
and
Program in Cell and Developmental Biology
Harvard Medical School
Boston, Massachusetts 02215

[b]*Department of Molecular Pharmacology*
Albert Einstein College of Medicine
New York, New York

INTRODUCTION

Receptor tyrosine kinases are distinguished by the presence of a single transmembrane spanning region and a well-conserved protein kinase catalytic domain that is absolutely required for transmembrane signaling.[1] Specificity arises because each receptor contains a distinct extracellular ligand-binding domain that recognizes its cognate ligand with high affinity. The growth factor or hormonal signal is relayed into the cell when the tyrosine kinase is activated during ligand binding.[1] However, the similarity of the catalytic domains among various tyrosine-kinase receptors raises questions about the mechanism maintaining the identity of the ligand-specific signal.

Insulin and platelet-derived growth factor (PDGF) receptors contain intrinsic tyrosine-kinase activity that is activated upon ligand binding.[1–14] *In vitro,* their catalytic domains are similar, as they recognize tyrosine residues near aspartic acid or glutamic acid residues; however, the biological effects of insulin and PDGF are quite distinct, even in the same cell background.[1,5,6] PDGF, but not insulin, triggers phosphatidylinositol (PI)-4,5-P_2 hydrolysis, generating two intracellular signals, diac-

[a]This work was supported by DK 38712 and DK 43828 to M. F. White. M. G. Myers was a fellow of the Markey Program in Cell and Developmental Biology, Division of Medical Sciences, Harvard Medical School and was partially supported by a Grant from the Albert J. Ryan Foundation at Harvard Medical School. B. Cheatham is a fellow of the Juvenile Diabetes Foundation.

[c]Address correspondence to Morris F. White, Ph.D., Research Division, Joslin Diabetes Center, One Joslin Place, Boston, MA 02215.

ylglycerol and inositol-1,4,5-P_3.[7] Moreover, early work resolved the mitogenic response of BALB/c 3T3 fibroblasts into two phases: competence and progression. Mitogenic competence is induced by a brief exposure to PDGF, and subsequent progression through the cell cycle requires continuous exposure to insulin/IGF-1 or EGF.[8–10] These observations predict that the insulin and PDGF receptor kinases engage distinct downstream elements that together lead to full biologic responses.

The means by which receptor-tyrosine kinases select their targets and thereby stimulate specific intracellular signaling pathways has been clarified by the identification of a conserved domain of approximately 100 amino acids, the Src homology 2 (SH2) domain.[11] The SH2 domain was originally found in Src-related tyrosine kinases[12] and later recognized in a surprisingly broad group of signal transduction proteins, including the phosphatidylinositol 3′-kinase, ras-GAP, and the phospholipase Cγ.[13] SH2 domains mediate protein-protein interactions by binding with high affinity to tyrosine-phosphorylated protein motifs. Although the phosphorylation of a tyrosine residue is the major controlling event or *on switch,* the selection of SH2 proteins seems to depend on the three amino acids carboxyterminal to the phosphotyrosine residue.[14] Indeed, structural analysis of SH2 domain/phosphopeptide complexes reveals the presence of both a phosphotyrosine recognition site and a binding pocket for the adjacent amino acids.[15,16]

In the case of EGF and PDGF receptors, autophosphorylation of specific tyrosine residues in the intracellular domain creates binding sites for certain SH2 proteins.[17,18] Thus, the phosphatidylinositol 3′-kinase, ras-GAP, and PLCγ each bind to different autophosphorylated sites in the PDGF receptor.[14,19–23] A variation of this model is employed by the insulin receptor. During insulin binding, the insulin receptor undergoes autophosphorylation in the intracellular juxtamembrane region[24] and in the regulatory and C-terminal regions.[25] Unlike the PDGF receptor, direct binding of SH2 proteins to the insulin receptor has been difficult to demonstrate. Autophosphorylation of the insulin receptor causes full activation of the tyrosine kinase[25,26] and, together with the juxtamembrane region, mediates the phosphorylation of IRS-1.[27,28] IRS-1 is the principal substrate of the insulin and IGF-1 receptors in most cells and tissues and may represent the first postreceptor event in insulin and IGF-1 signal transmission.[29,30]

IRS-1 contains multiple tyrosine phosphorylation sites, many of which reside in predicted SH2 domain-binding motifs.[14,28] During insulin stimulation, IRS-1 activates the PI 3′-kinase when it associates with the SH2 domains in the 85 kDa regulatory subunit, p85α.[31,32] Thus, IRS-1 serves as a "docking protein," to which SH2 proteins bind, and mediates insulin signaling by binding and regulating signaling molecules following insulin receptor-mediated tyrosine phosphorylation.[30]

This paper describes two levels of selectivity between insulin and PDGF receptor (PDGFr) signaling. The first level occurs when the insulin receptor, but not the PDGFr, phosphorylates IRS-1. The binding of SH2 proteins to IRS-1 during insulin signaling, and not directly to the insulin receptor as in the case of the PDGFr, may account for some of the differences in insulin and PDGF signaling. The second level of selectivity occurs because IRS-1 contains common and unique phosphorylation sites when compared to the autophosphorylation sites in the PDGF receptor. Differences in the biological effects of insulin and PDGF are likely to arise through the engagement of both common and distinct SH2 proteins by the cytosolic IRS-1 and the membrane-bound PDGF receptor.

MATERIAL AND METHODS

Cell Lines and Growth Factors

Mouse NIH/3T3 fibroblasts expressing the human insulin receptor (HIR 3.5) were the generous gift of Dr. Jonathan Whittaker.[33] HIR 3.5 cells were grown in DMEM-high glucose, supplemented with 10% fetal bovine serum. For insulin stimulation, recombinant human insulin (ELANCO) was added to a final concentration of 10^{-7} *M*. Recombinant human PDGF (B/B) was from Boeringer Mannheim or Upstate Biologicals, Inc., and was used at 100 ng/mL to stimulate cells.

Antibodies

αIRS-1 antibodies were protein A-purified polyclonal antibodies from rabbits immunized with recombinant baculovirus-produced IRS-1 protein.[34] Affinity-purified polyclonal antiphosphotyrosine antibodies (αPY) were as described.[35] Immunoprecipitating antibodies specific for the PtdIns 3′-kinase p85 were affinity-purified polyclonal antibodies against amino acids 321-724 of human p85, expressed as a GST fusion protein (αGSTp85)[36] or rabbit antipeptide antisera (Ab51).[36] αGRB-2 antibodies were generated against a GST fusion protein containing amino acids 50-152 of human GRB-2.[37] αPDGFr and αras-GAP antibodies and immunoblotting antibodies against p85 were purchased from Upstate Biotechnology, Inc. αPLCγ antibodies were the generous gifts of Drs. E. Y. Skolnik and J. Schlessinger (Department of Pharmacology, New York University) and Dr. G. Carpenter (Vanderbilt University, Nashville, TN).

Immunoprecipitation

Cells were grown to 80% confluence on 15 cm dishes (Costar) and made quiescent overnight in serum-free media containing 0.5% BSA (Fluka). Cells were stimulated with growth factor for 1 minute before extraction in 100 mM Tris (pH 8.0) containing 100 mM NaF, 1% Triton X-100, 1 mM Na_3VO_4, 1 mM PMSF, 10 μg/mL aprotinin, and 2 mM NaP_2O_4. Insoluble material was pelleted at 100,000 × *g* for 1 hour, and supernatants were incubated with 50 μL of αPLCγ, 25 μL of αGRB-2, 10 μL of αras-GAP, αPY (3 μg/mL), or αIRS-1 (10 μg/mL) overnight at 4 °C. Immune complexes were collected with Pansorbin Cells (Calbiochem), Protein A-Trisacryl (Pierce), or Protein A-Sepharose 6 MB (Pharmacia) and were washed three times in lysis buffer containing 0.1% SDS. The samples were denatured in Laemmli sample buffer containing 10 mM DTT and resolved by SDS-PAGE (7.5%).

Immunoblotting

Immunoprecipitated proteins were resolved by 7.5% SDS-PAGE on standard apparatuses overnight at 5 mA or in Bio-Rad miniprotean apparatuses at 100 V. Gels were transferred to nitrocellulose membranes (Schleicker & Schuell) for 1.5 hours at 100 V in Towbin buffer containing 0.02% SDS and 20% methanol.[38] Membranes were blocked overnight at 4°C in 25 mM Tris-HCl (pH 7.4), containing 150 mM NaCl and 0.01% Tween-20 (wash buffer), containing 3% bovine serum albumin (Fluka). Membranes were then incubated for two hours at room temperature in wash buffer containing 3% BSA and antibody. Proteins were detected with αPY antibodies (used at a final concentration of 3 μg/mL) or αp85 antibodies (1 : 100 dilution of serum). The membranes were subsequently washed three times in wash buffer, re-blocked for 1 hour at room temperature in wash buffer with 3% BSA, and incubated with [^{125}I]protein A (ICN) (0.2 μCi/mL) for 1 hour in wash buffer containing 3% BSA. Blots were washed 4-5 times in wash buffer, dried, and exposed to autoradiography with Kodak X-AR film or imaged on a Molecular Dynamics Phosphorimager.

In vitro *PtdIns 3′-kinase Assay*

PtdIns 3′-kinase activity associated with immunoprecipitates was determined as previously described.[39] HIR 3.5 cells were made quiescent overnight in media containing 0.5% insulin-free BSA (Fluka). Cells were incubated in the presence or absence of growth factor for 10 minutes and washed once with phosphate-buffered saline, containing 100 μM sodium orthovanadate; and twice with 20 mM Tris pH 7.5, containing 137 mM NaCl, 1 mM $MgCl_2$, 1 mM $CaCl_2$, and 100 μM sodium orthovanadate (A buffer). Cells were then solubilized in buffer A, containing 1% NP-40, 10% glycerol, and 1 mM PMSF. Insoluble material was removed by centrifugation at 10,000 × *g*, and supernatants were immunoprecipitated overnight. Precipitated material was collected on protein A sepharose beads (Pharmacia) and washed consecutively with PBS, containing 1% NP-40 and 100 μM Na_3VO_4 (three times); 100 mM Tris-HCl (pH 7.5),containing 500 mM LiCl and 100 μM sodium orthovanadate (three times); and 10 mM Tris-HCl (pH 7.5), containing 100 mM NaCl, 100 μM sodium orthovanadate, and 1 mM EDTA (twice). Some experiments were performed using 2 mM Na_3VO_4 in all wash and lysis buffers, with similar results. Pellets were resuspended in 50 μL of the final wash buffer. To each pellet was added 10 μL 100 mM $MgCl_2$ and 10 μL of a sonicated dispersion of PtdIns (2 μg/μL) (Avanti) in 10 mM Tris-HCl, pH 7.5, with 1 mM EGTA. Reactions were started by the addition of 5 μL of 880 mM ATP containing 30 μCi [^{32}P]ATP (NEN-DuPont) and 20 mM $MgCl_2$. After 10 minutes at room temperature, the reaction was stopped by the addition of 20 μL 8 N HCl and 160 μL $CHCl_3$: MeOH (1 : 1). The samples were centrifuged, and the lower organic phase was spotted on a silica gel TLC plate (Merck) that had been treated with 1% potassium oxalate. TLC plates were developed in $CHCl_3$: MeOH : H_2O : NH_4OH (60 : 47 : 11.3 : 2), dried, and visualized by autoradiography.

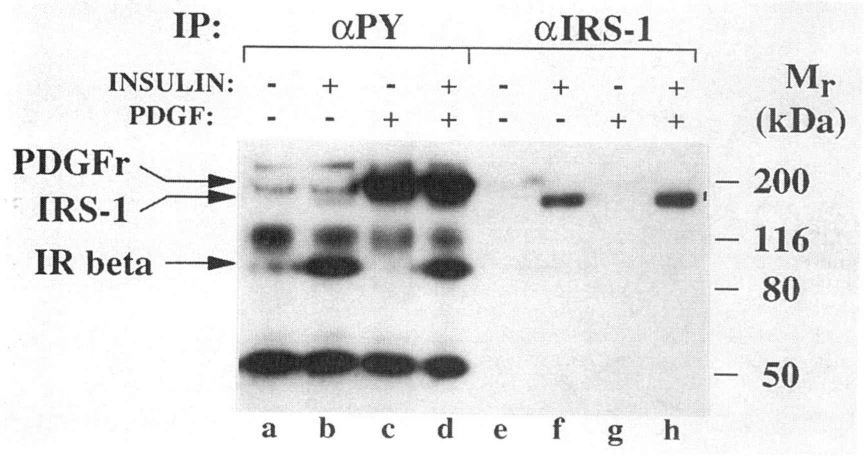

FIGURE 1. Treatment of HIR 3.5 cells with insulin, but not PDGF, stimulates the tyrosyl phosphorylation of IRS-1. Quiescent HIR 3.5 cells (lanes a and e) or HIR 3.5 cells stimulated with insulin (lanes b and f), PDGF (lanes c and g), or both (lanes d and h) were lysed and immunoprecipitated with αPY (lanes a–d) or αIRS-1 (lanes e–h) antibodies. Immunoprecipitates were collected, washed, resolved on SDS-PAGE, and transferred to nitrocellulose membranes. Tyrosyl phosphorylated proteins were detected by immunoblotting with αPY. Migration of molecular weight standards and the positions of the phosphorylated IR, IRS-1, and PDGFr are denoted.

RESULTS

Tyrosine Phosphorylation of IRS-1 in HIR 3.5 Cells during Insulin and PDGF Stimulation

We assessed insulin- and PDGF-stimulated tyrosine phosphorylation of cellular proteins in mouse 3T3 fibroblasts that express high levels of endogenous PDGFr and similar levels of stably transfected human insulin receptor (HIR 3.5 cells). Antiphosphotyrosine (αPY) immunoblots of proteins precipitated with αPY confirmed that insulin and PDGF stimulated tyrosine phosphorylation of the cognate receptor and demonstrated that similar amounts of each tyrosine phosphorylated receptor were present in HIR 3.5 cells (FIG. 1, lanes a–c). The effects of insulin and PDGF were independent, inasmuch as simultaneous stimulation with insulin and PDGF produced a composite result (FIG. 1, lanes a and d). Furthermore, a tyrosine-phosphorylated band at 175 kDa, just below the PDGFr, occurred during insulin stimulation alone or in combination with PDGF; the 175 kDa protein was not detected after PDGF stimulation alone. This protein is most likely IRS-1.[28] We used αIRS-1 antibodies to directly observe IRS-1 tyrosine phosphorylation during insulin and PDGF stimulation. An αPY immunoblot revealed that insulin, but not PDGF, stimulated IRS-1 tyrosine phosphorylation (FIG. 1, lanes e–h). These results demonstrate that IRS-1 is a substrate for the insulin receptor but not for the PDGFr in HIR 3.5 cells.

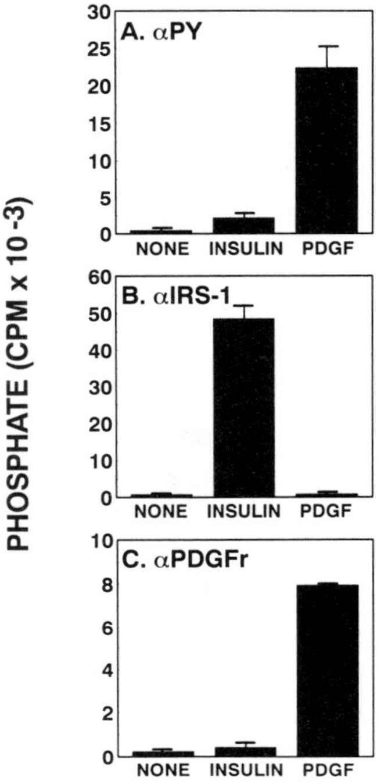

FIGURE 2. Association of PtdIns 3′-kinase with αPY, αIRS-1, and αPDGFr immunoprecipitates following growth factor stimulation of HIR 3.5 cells. Quiescent HIR 3.5 cells of HIR 3.5 cells stimulated with insulin or PDGF were lysed and immunoprecipitated with αPY(A), αIRS-1(B), and αPDGFr(C) antibodies. Immunoprecipitates were collected, washed, and assayed for associated PtdIns 3′-kinase. Results are expressed in cpm phosphate incorporated into PtdIns substrate during a 10-minute assay, +/−SEM. All assays represent triplicate measurements.

Association of PI 3′-Kinase with IRS-1

PI 3′-kinase binds to the tyrosine-phosphorylated YMXM motifs in various proteins, including the PDGF receptor and IRS-1.[14,28] We measured the amount of PI 3′-kinase activity detectable in αPY, αPDGFr, or αIRS-1 immunoprecipitates from HIR 3.5 cells treated with insulin or PDGF. Following insulin stimulation, PI 3′-kinase was detected in αPY and αIRS-1 immunoprecipitates (FIG. 2, A and B); however, much more PI 3′-kinase activity was found in αIRS-1 immunoprecipitates than in αPY immunoprecipitates, probably due to the greater efficiency with which αIRS-1 precipitates IRS-1 (compare levels of IRS-1 in αPY and αIRS-1 immunoprecipitates in FIG. 1). By contrast, no PI 3′-kinase activity was detected in αPDGFr immunoprecipitates following insulin stimulation (FIG. 2C).

Following PDGF stimulation, PI 3′-kinase activity associated strongly with αPY and αPDGFr immunoprecipitates (FIG. 2, A and C). This activity reflects the association of PI 3′-kinase with the PDGFr and is higher in αPY immunoprecipitates due to the greater efficiency of the αPY antibody than the αPDGFr antibody for immunoprecipitating the activated PDGFr (data not shown). During PDGF stimulation, there was no increase in PI 3′-kinase activity associated with IRS-1 (FIG. 2B), confirming that IRS-1 does not play a role in PDGF signaling to this enzyme. Because the tyrosine phosphorylation of proteins is known to be crucial for their association with PI 3′-kinase,[13,40–44] the absence of PI 3′-kinase in this case is entirely consistent with the lack of PDGF-stimulated tyrosine phosphorylation of IRS-1.

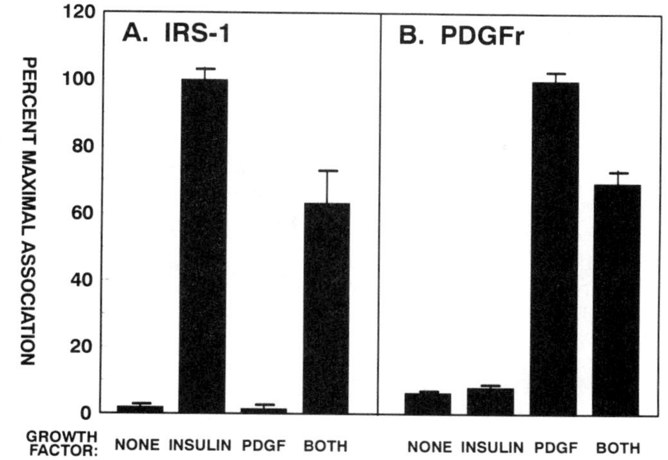

FIGURE 3. Simultaneous stimulation of HIR 3.5 cells with insulin and PDGF decreases the amount of PtdIns 3′-kinase activity associated with IRS-1(A) or the PDGFr(B), compared to insulin or PDGF alone. Quiescent HIR 3.5 cells or HIR 3.5 cells stimulated with insulin, PDGF, or both were lysed and immunoprecipitated with αIRS-1 or αPDGFr antibodies. PtdIns 3′-kinase activity associated with washed immunoprecipitates was determined and normalized to the maximum activity (100%) associated with each antibody (*e.g.*, insulin alone represents the maximum for αIRS-1 immunoprecipitates). Determinations were carried out in triplicate and are expressed ±SEM.

During stimulation of HIR 3.5 cells with both insulin and PDGF, the PDGFr and IRS-1 are available to associate with the PI 3′-kinase. The amount of PI 3′-kinase activity associated with the PDGFr was approximately 33% lower during insulin- and PDGF-stimulation than following stimulation with PDGF alone (FIG. 3B). Similarly, the amount of PI 3′-kinase activity associated with IRS-1 was also about 33% lower following double stimulation than following insulin stimulation alone (FIG. 3A). These results suggest that IRS-1 and the PDGFr bind the same intracellular pool of PI 3′-kinase and that following double stimulation they compete for this pool.

In order to determine the nature of proteins associated with the 85 kDa subunit (p85) of PI 3′-kinase following stimulation with insulin, PDGF, or both, we prepared immunoprecipitates with antibodies against p85, washed them as for a PI 3′-kinase assay, and analyzed the associated proteins by SDS-PAGE and immunoblotting with αPY (FIG. 4). After PDGF stimulation, the PDGFr was the major phosphotyrosine-containing protein associated with PI 3′-kinase (FIG. 4, lanes e and f). After insulin stimulation, IRS-1 was the major phosphoprotein associated with p85; however, a small amount of insulin receptor β subunit and two unidentified 120 and 140 kDa proteins were also detected (FIG. 4, lanes c and d). During simultaneous insulin and PDGF stimulation, both the PDGFr and IRS-1 were associated with p85. However, the amounts of IRS-1 and PDGFr associated with p85 was reduced compared to stimulation with insulin or PDGF alone (FIG. 4, lanes g and h). These results confirm that the PDGFr and IRS-1 compete for a common pool of p85 during growth-factor stimulation.

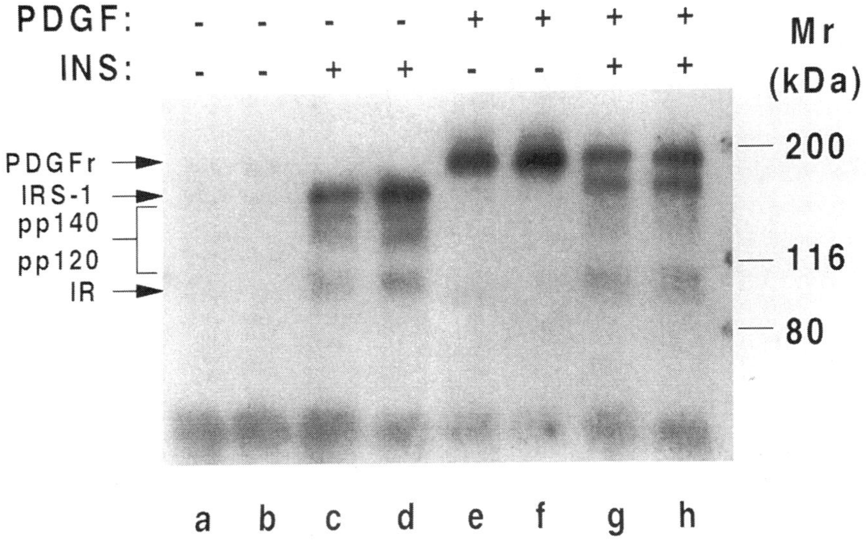

FIGURE 4. Tyrosyl phosphoproteins associated with PtdIns 3′-kinase following growth factor stimulation of HIR 3.5 cells. Quiescent HIR 3.5 cells (lanes a and b) or HIR 3.5 cells stimulated with insulin (lanes c and d), PDGF (lanes e and f), or both (lanes g and h) were lysed and immunoprecipitated with αp85 antibodies: Ab51 (lanes a, c, e, and g) or αGSTp85 (lanes b, d, f, and h). Immunoprecipitates were washed as for a PtdIns 3′-kinase assay, resolved by SDS-PAGE, transferred to nitrocellulose, and immunoblotted with αPY. Migration of molecular weight markers and the positions of tyrosine phosphorylated IRS-1 and the PDGFr are shown.

Activation of PI 3′-kinase by Insulin and PDGF

The stimulation of cells with growth factors, including insulin and PDGF, increases the cellular levels of $PI(3,4)P_2$ and $PI(3,4,5)P_3$, which is attributed to the stimulation of PI 3′-kinase.[39,40,45] Consistent with this result, insulin stimulation of PI 3′-kinase is observed in p85 immunoprecipitates. This activation appears to occur by the binding of tyrosine phosphorylated IRS-1 to the SH2 domains of its regulatory subunit p85.[31,32] Oddly, analogous assays have been unable to detect such activation following stimulation with PDGF.[36] In order to compare the activation of PI 3′-kinase with IRS-1 and the PDGFr, we assayed the activity of PI 3′-kinase in immunoprecipitates from HIR 3.5 cells using two antibodies (Ab51 and αGSTp85) directed against the p85 subunit of PI 3′-kinase. PI 3′-kinase activity in Ab51 immunoprecipitates from HIR 3.5 cells was stimulated approximately 4-fold following insulin stimulation and 2.4-fold by PDGF (100 nM) (Fig. 5A). However, using αGSTp85, only a 2.4-fold activation of PI 3′-kinase was detected with insulin, and no stimulation was observed with PDGF (Fig. 5B). Parallel immunoprecipitates immunoblotted with αp85 showed similar amounts of p85 present following stimulation with insulin, PDGF, or both (data not shown). Thus the ability to detect stimulated PI 3′-kinase activity depends both on the antibody used and the intrinsic strength of the activating signal.

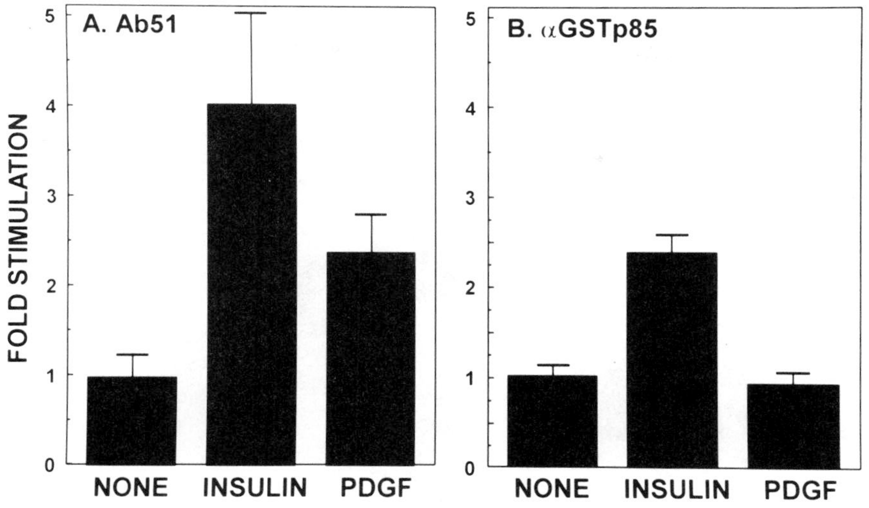

FIGURE 5. Insulin and PDGF stimulation increases the activity of PtdIns 3′-kinase in αp85 immunoprecipitates. αp85 immunoprecipitates were prepared from lysates of unstimulated (control) HIR 3.5 cells, or HIR 3.5 cells stimulated with insulin or PDGF using two different αp85 antibodies, Ab51(A), and αGSTp85(B). Immunoprecipitates were washed, and associated PtdIns 3′-kinase activity was assayed. Activity is plotted as fold stimulation over basal ±SEM. Data represents the aggregate of four experiments.

The ability of the PDGF receptor and IRS-1 to activate PI 3′-kinase was directly compared in immunoprecipitates. PI 3′-kinase activity in αIRS-1 and αPDGFr immunoprecipitates was measured and normalized against the amount of p85 found in the immunoprecipitates by immunoblotting to account for the relative efficiency of the immunoprecipitations (FIG. 6). The specific activity of the IRS-1- and PDGFr-associated PI 3′-kinase was estimated from the ratio of PI 3′-kinase activity to p85 protein. Using this method, PI 3′-kinase associated with IRS-1 was approximately threefold more active than that associated with the PDGFr (FIG. 6).

Formation of Multimolecular Complexes by IRS-1 and the PDGFr

PDGF-stimulated autophosphorylation of the PDGFr causes the formation of a multimolecular complex containing PI 3′-kinase, ras-GAP, GRB-2, PLCγ, c-fyn, and likely other SH2 proteins.[17,19,21,22,46] In order to assess whether IRS-1 acts similarly following insulin receptor-dependent tyrosine phosphorylation, we assessed the binding of ras-GAP, GRB-2, and PLCγ to IRS-1.

Insulin and PDGF stimulated the association-tyrosyl phosphoproteins in immunoprecipitates prepared with antibodies against ras-GAP, GRB-2, and PLCγ (FIG. 7). Following stimulation with PDGF, a 185-200 kDa band, corresponding to the PDGFr, was observed in αras-GAP, αGRB-2, and αPLCγ immunoprecipitates (FIG. 7). A

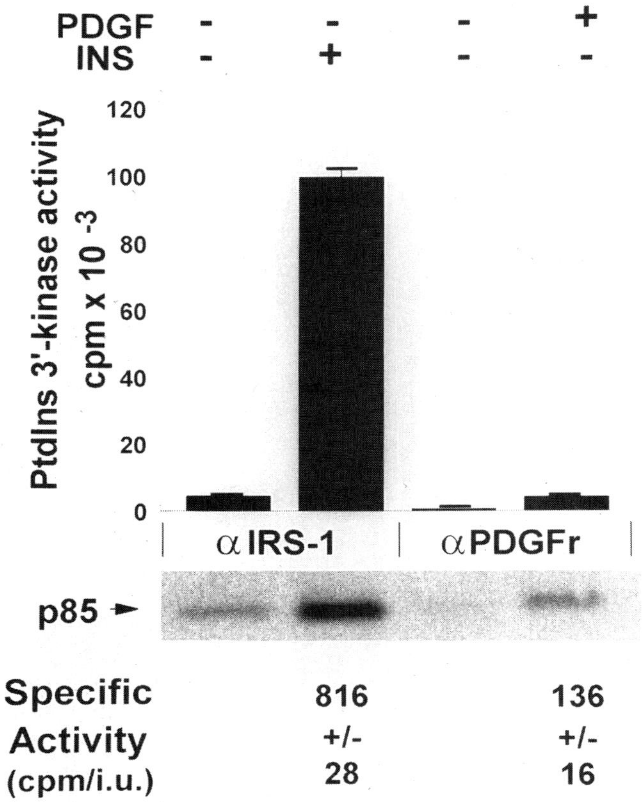

FIGURE 6. PtdIns 3′-kinase associated with IRS-1 is more highly activated than that associated with the PDGFr. αIRS-1 and αPDGFr immunoprecipitates were prepared from unstimulated, insulin-stimulated, or PDGF-stimulated HIR 3.5 cells. Immunoprecipitates were washed and assayed for associated PtdIns 3′-kinase activity or resolved by SDS-PAGE, transferred to nitrocellulose, and immunoblotted with αp85 antibodies. Specific activity of PtdIns 3′-kinase associated with IRS-1 and the PDGFr was determined by calculating the ratio of associated PtdIns 3′-kinase activity (cpm) to associated p85 (phosphorimager units (I.U.)). One example of such an experiment is represented. The specific activity represents the aggregate result of two such experiments.

140 kDa band, corresponding to the tyrosine-phosphorylated PLCγ, was observed in αPLCγ immunoprecipitates from PDGF-stimulated cells (Fig. 7). Following stimulation with insulin, a tyrosyl phosphoprotein migrating at 180 kDa was associated with αGRB-2 immunoprecipitates, confirming that GRB-2 also associates with IRS-1.[37,47] However, no novel tyrosyl phosphoproteins were observed in either αras-GAP or αPLCγ immunoprecipitates prepared from insulin-stimulated cells, suggesting that neither ras-GAP nor PLCγ associate with IRS-1 following insulin stimulation and that PLCγ is not tyrosine phosphorylated following insulin stimulation (Fig. 7). Thus,

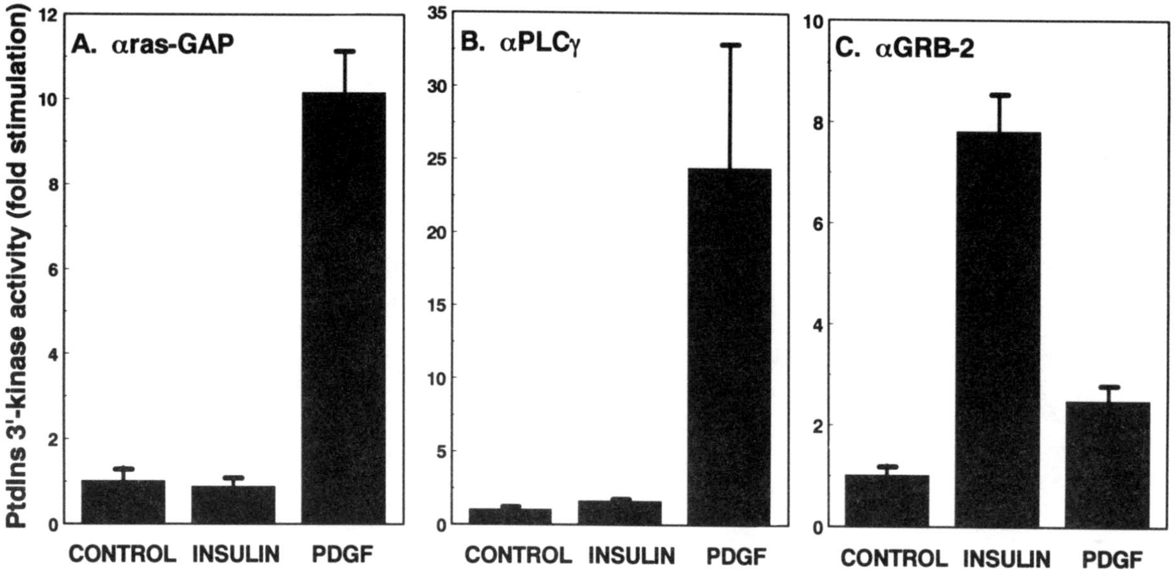

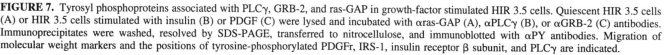

FIGURE 7. Tyrosyl phosphoproteins associated with PLCγ, GRB-2, and ras-GAP in growth-factor stimulated HIR 3.5 cells. Quiescent HIR 3.5 cells (A) or HIR 3.5 cells stimulated with insulin (B) or PDGF (C) were lysed and incubated with αras-GAP (A), αPLCγ (B), or αGRB-2 (C) antibodies. Immunoprecipitates were washed, resolved by SDS-PAGE, transferred to nitrocellulose, and immunoblotted with αPY antibodies. Migration of molecular weight markers and the positions of tyrosine-phosphorylated PDGFr, IRS-1, insulin receptor β subunit, and PLCγ are indicated.

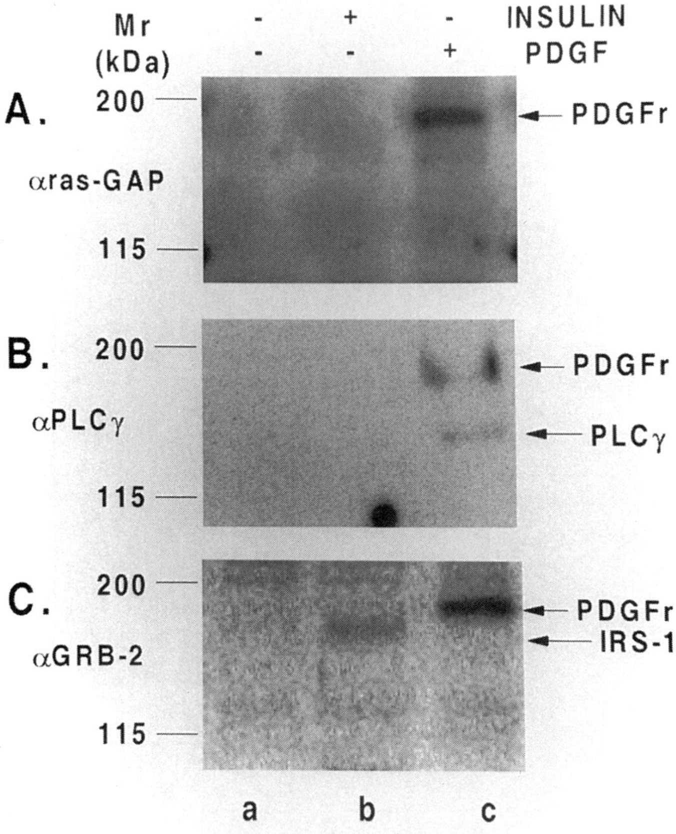

FIGURE 8. PtdIns 3′-kinase activity associated with αras-GAP(A), αPLCγ(B), and αGRB-2(C), immunoprecipitates from growth factor-stimulated HIR 3.5 cells. Quiescent HIR 3.5 cells or HIR 3.5 cells stimulated with insulin or PDGF were lysed and incubated with αras-GAP, αPLCγ, or αGRB-2 antibodies. Immunoprecipitates were washed and assayed for associated PtdIns 3′-kinase activity. Activity is plotted as fold stimulation over basal ± SEM and represents the average of duplicate or triplicate determinations.

IRS-1 and the PDGFr associate with overlapping yet distinct pools of SH2 domain-containing signaling elements.

In order to investigate whether a single tyrosine phosphorylated PDGFr or IRS-1 molecule formed multimolecular complexes containing PI 3′-kinase, GRB-2, ras-GAP, and PLCγ, we measured PI 3′-kinase activity in αras-GAP, αGRB-2, and αPLCγ immunoprecipitates from HIR 3.5 cells following stimulation with insulin or PDGF (Fig. 8). PI 3′-kinase associated with αras-GAP and αPLCγ immunoprecipitates increased 10- and 23-fold, respectively, following PDGF stimulation, whereas the associated PI 3′-kinase remained unchanged following stimulation with insulin

(FIG. 8). However, in αGRB-2 immunoprecipitates there were 8- and 3-fold increases in the amount of associated PI 3′-kinase activity following insulin and PDGF stimulation, respectively. Thus, the PDGFr forms a multimolecular complex containing PI 3′-kinase, ras-GAP, GRB-2, and PLCγ, whereas IRS-1 associates with GRB-2 and PI 3′-kinase. This undoubtedly contributes to the unique nature of the insulin and PDGF signals observed in 3T3 fibroblasts.

DISCUSSION

Insulin and platelet-derived growth factor employ a common theme for signaling transmission that involves ligand-induced phosphotyrosine binding motifs that associate specifically with proteins containing SH2 domains. In the case of insulin signaling, the SH2 proteins are recruited to tyrosine-phosphorylated IRS-1, whereas during PDGF stimulation the SH2 proteins associate directly with autophosphorylation sites in the PDGF receptor (FIG. 9).[19,36,43,48] In a few cases, insulin and PDGF engage the same SH2 proteins, as shown here for PtdIns 3′-kinase and GRB-2. In other instances, unique elements are selected, as illustrated by the association of PLCγ and ras-GAP with the PDGF receptor but not IRS-1. The unique response of cells to insulin or PDGF, as proposed earlier, is likely to result partly from the cohort of SH2 proteins that are engaged during ligand stimulation. At the molecular level, this selectivity is now understood by the fact that IRS-1 and the PDGF receptor contain common and distinct tyrosine-phosphorylated motifs that are predicted to associate with various SH2 proteins.[14,19,28,37,43,49]

IRS-1 is not tyrosine phosphorylated by the PDGF receptor in HIR 3.5 cells. Thus, IRS-1 is selected for phosphorylation by a subset of tyrosine kinases, including the receptors for insulin, IGF-1, and IL-4.[28,29,50–52] The purified insulin receptor phosphorylates recombinant IRS-1 protein *in vitro,* but the purified PDGF receptor does not, even though it phosphorylates similar motifs in its kinase insert domain (M. G. Myers Jr., M. F. White *et al,* unpublished observations). Thus, a unique domain in the insulin receptor mediates the specific recognition of IRS-1, whereas this element must be absent in the PDGF receptor. Mutational analysis of the insulin and IL-4 receptors suggests that the receptor side of this recognition may be mediated by a tyrosine-containing NPXY motif found in these receptors.[27,53] Recent analysis of IRS-1 structure suggests that one or two regions in the N-terminus (*e.g.,* the pleckstrin homology domain) of IRS-1 may form the cognate for the interaction with these receptors (M. G. Myers Jr., M. F. White, *et al,* unpublished data).

The utilization of IRS-1 as a principal docking unit for the SH2 proteins in the insulin receptor pathway may explain some of the differences between the insulin and PDGF signals. Although the PDGF receptor is an integral membrane protein, IRS-1 is largely cytosolic: This may target associated proteins in a different manner than the PDGF receptor does or than the insulin receptor would if it directly bound SH2 proteins. Indeed, IRS-1 confers additional flexibility to the insulin signal, as both membrane-bound and soluble forms are found in insulin-stimulated cells;[47,54] a small but detectable fraction of the PtdIns 3′-kinase is bound to the insulin receptor in Chinese hamster ovary cells, and this complex is found in intracellular membrane fractions in rat adipocytes.[47,54] Moreover, the direct association with the receptor

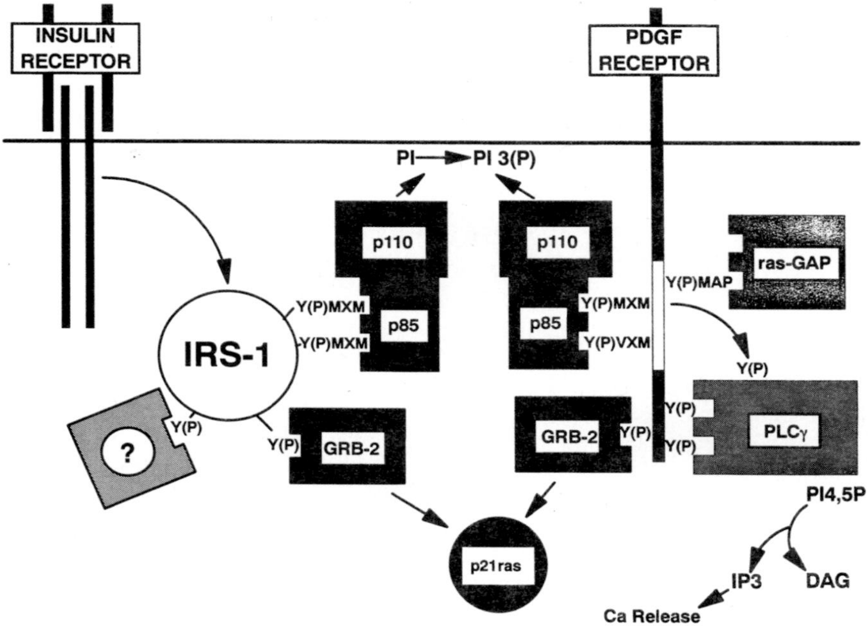

FIGURE 9. A model of the signaling pathways used by the receptors for insulin and PDGF. Stimulation of the insulin and PDGF receptors causes activation of the receptor-tyrosine kinase and receptor autophosphorylation. However, in the case of the insulin receptor, the activated insulin receptor-tyrosine kinase also phosphorylates IRS-1. The tyrosine-phosphorylated PDGFr and IRS-1 then act to nucleate binding of different but overlapping sets of SH2 domain-containing proteins. IRS-1 binds both GRB-2 and PtdIns 3′-kinase, whereas the PDGFr associates with GRB-2, PtdIns 3′-kinase, ras-GAP, and PLCγ. The association of PtdIns 3′-kinase with IRS-1 is more activating than its association with the PDGFr. Thus the insulin-signaling pathway differs from that of PDGF by the insertion of a nonreceptor "docking protein" (IRS-1), in the signaling pathways it uses, and in the way one common element (PtdIns 3′-kinase) is regulated.

tyrosine kinase adds the additional possibility of regulation of SH2 proteins by tyrosine phosphorylation, as demonstrated for PDGF receptor-bound PLCγ. As IRS-1 has no demonstrable catalytic activity,[28] it cannot exercise this form of regulation on bound proteins without the specific association of an auxiliary kinase.

The association of PtdIns 3′-kinase with tyrosyl phosphorylated proteins, such as IRS-1 and the PDGFr, is mediated by the binding of SH2 domains in the 85 kDa subunit of PtdIns 3′-kinase.[31,32,36,55] Ptd Ins 3′-kinase does not associate with IRS-1 following PDGF stimulation because IRS-1 is not tyrosine phosphorylated by the PDGF receptor in the intact cell; however, the pDGFr and IRS-1 compete for a common pool of enzyme when both molecules are tyrosine phosphorylated. The predominant proteins associated with PtdIns 3′-kinase following insulin stimulation are IRS-1 and a small amount of the insulin receptor;[31,47] IRS-1, PtdIns 3′-kinase, and the insulin receptor form a ternary complex in which the insulin receptor is bound to IRS-1, which mediates binding to PtdIns 3′-kinase.[47] Two other tyrosyl-phosphorylated protein bands observed in association with PtdIns 3′-kinase following insulin stimulation[29,31] represent degradation products of IRS-1 (M. G. Myers Jr. & M. F. White, unpublished observations). Following PDGF stimulation, the PDGF receptor represents the only αPY-detectable PtdIns 3′-kinase-associated tyrosyl phosphoprotein.

PtdIns 3′-kinase is activated by insulin[31,32] and PDGF treatment of cells. Initial reports indicated that p85 becomes tyrosine phosphorylated following stimulation of cells with PDGF and that this may regulate the activity of PtdIns 3′-kinase.[56–59] p85 undergoes a conformational change following the binding of tyrosine-phosphorylated YMXM motifs;[16,60,61] this conformational change is one possible mechanism by which the activating signal may be transmitted from p85 to the 110 kDa catalytic subunit.

Others have suggested that p85 becomes tyrosine phosphorylated following stimulation of cells with PDGF and that this may regulate the activity of PtdIns 3′-kinase.[41,57,59,62] We and others, however, have previously been unable to detect the tyrosine phosphorylation of p85 following stimulation with a variety of growth factors, including insulin, EGF, and PDGF.[31,36] Furthermore, in this report, we are unable to detect the phosphorylation of p85 following PDGF or insulin stimulation, although the tyrosine phosphorylation of PLC is readily observed following stimulation with PDGF (FIG. 7). Thus, phosphorylation of p85 does not represent a functional difference between the regulation of PtdIns 3′-kinase by IRS-1 and the PDGFr.

Insulin, through IRS-1, stimulates the activity of PtdIns 3′-kinase more effectively than the PDGF receptor. At the molecular level, this may be due to the presence of a 6 YMXM motif in IRS-1 compared to one YMXM and a related YVXM motif in the PDGF receptor. Moreover, autophosphorylation studies with the PDGF receptor suggest that only one of these sites are phosphorylated in a given receptor molecule,[20] whereas multiple tyrosines are phosphorylated in IRS-1, including several YMXM motifs.[63] PtdIns 3′-kinase may be fully activated only when both of its SH2 domains are occupied: Activation of PtdIns 3′-kinase by monophosphopeptides *in vitro* is biphasic and only occurs at high peptide concentrations, but activation by phosphorylated IRS-1 (which contains multiple tyrosyl phosphorylated YMXM motifs) or phosphopeptides containing two tyrosine phosphorylated YMXM motifs is monophasic and orders of magnitude more sensitive (ref. 31 and J. M. Backer *et al.*, unpublished data).

The consequences of PtdIns 3′-kinase activation are unclear. In yeast, VPS34 (the apparent homologue of the PtdIns 3′-kinase catalytic subunit) is essential for the efficient sorting of vacuolar hydrolases.[64,65] PtdIns 3′-kinase may control similar functions in mammalian cells during tyrosine-kinase signaling. In this case, the PDGF and insulin effect could be different, as the PDGF receptor and IRS-1 could target PtdIns 3′-kinase to alternate cellular compartments.[54,66] Moreover, the higher activation of PtdIns 3′-kinase by insulin may result in different signals than PDGF.

The PDGFr binds multiple SH2 proteins, including PtdIns 3′-kinase, ras-GAP, GRB-2, and PLCγ; and PDGF stimulation increases the amount of PtdIns 3′-kinase activity associated with αras-GAP, αGRB-2, and αPLCγ immunoprecipitates, suggesting that the PDGF receptor recruits SH2 proteins into a large signaling complex during PDGF stimulation. IRS-1 does not bind ras-GAP or PLCγ, but recruits PtdIns 3′-kinase and GRB-2 into a signaling complex. The lack of interaction between ras-GAP and PLCγ with IRS-1 is not surprising inasmuch as none of the potential tyrosine phosphorylation sites on IRS-1 contain the surrounding amino acid sequences thought to be necessary for recognition by the SH2 domains of ras-GAP and PLCγ.[14,19,22,28] PLCγ is tyrosine phosphorylated following PDGF stimulation, and PDGF increases turnover of phosphatidylinositol secondary to the activation of PLCγ,[7] whereas insulin does not stimulate PLCγ; this is likely because IRS-1 does not bind PLCγ or activate it by tyrosine phosphorylation.

The SH2 domains from ras-GAP, PLCγ, GRB-2, and p85 have been used as affinity reagents to determine whether these proteins associate with certain cellular phosphoproteins.[32,37;67–69] Although no binding of IRS-1 to the ras-GAP SH2 domains has been demonstrable, the binding of IRS-1 to the SH2 domains of GRB-2, p85, and PLCγ has been reported *in vitro*.[32,37,67,68] In light of the data presented here, the binding of IRS-1 to PLCγ SH2 domains likely represents an *in vitro* artifact. Cautious evaluation of the binding of phosphoproteins to large amounts of SH2 domain fusion proteins *in vitro* is necessary, as SH2 domains display a moderate affinity for phosphotyrosine alone,[70] and, under *in vitro* conditions, interactions that may not occur *in vivo* may be observed.

PtdIns 3′-kinase and GRB-2 are common elements in the PDGF and insulin-signaling cascades, whereas ras-GAP and PLCγ are restricted to the PDGF signal. This likely accounts for some of the differences between the insulin and PDGF signals. Furthermore, among the common elements (PtdIns 3′-kinase and GRB-2), engagement by IRS-1 and the PDGF receptor appears to differ. It also seems likely that as more SH2 domain-containing proteins become known and are investigated fully, more molecules that associate with IRS-1 and the PDGF receptor will be discovered. The elucidation of which of these new elements is unique to the IRS-1 and PDGF receptor systems will more fully explain the differences between the cellular responses to PDGF and insulin.

SUMMARY

The receptors for insulin and PDGF are tyrosine kinases that mediate distinct effects in identical cellular backgrounds. Each receptor must therefore engage a unique subset of the available signaling elements—at least partly through the selection of proteins with src-homology 2 domains (SH2 proteins). Autophosphorylation sites in the PDGFr directly bind SH2 proteins, whereas activation of the insulin receptor leads to phosphorylation of IRS-1, which in turn binds SH2 proteins. In HIR 3.5 cells, which contain similar numbers of PDGF and insulin receptors, insulin, but not PDGF, stimulated tyrosyl phosphorylation of IRS-1. Similarly, insulin, but not PDGF, treatment of HIR 3.5 stimulated the association of IRS-1 with PtdIns 3′-kinase, although PDGF stimulated the association of PtdIns 3′-kinase with the tyrosine-phosphorylated PDGFr. Association with IRS-1 activated PtdIns 3′-kinase more effectively than association with the PDGFr. Whereas the PDGFr associated with PtdIns 3′-kinase, ras-GAP, GRB-2, and phospholipase Cγ, only GRB-2 and PtdIns 3′-kinase associated with IRS-1. Moreover, PDGF, but not insulin, caused tyrosine phosphorylation of phospholipase Cγ in HIR 3.5 cells. Thus, the insulin signal differs from that of PDGF by the insertion of a cytosolic, nonreceptor SH2 domain docking protein (IRS-1). Furthermore, IRS-1 binds a different subset of SH2 domain-containing proteins than does the PDGFr and regulates at least one common element (PtdIns 3′-kinase) differently than the PDGFr. These results support the hypothesis that IRS-1 differentiates the signals generated by the insulin receptor and PDGFr tyrosine kinases by binding and regulating a specific subset of SH2 domain-containing signaling molecules.

ACKNOWLEDGMENTS

We gratefully acknowledge the gift of antibodies from Ed Skolnik, Yossi Schlessinger, and Graham Carpenter.

REFERENCES

1. Ullrich, A. & J. Schlessinger. 1990. Cell **61:** 203.
2. Ullrich, A., J. R. Bell, E. Y. Chen, R. Herrera, L. M. Petruzzelli *et al.* 1985. Nature **313:** 756.
3. Yarden, Y., J. A. Escobedo, W. J. Kuang, T. L. Yang Feng, T. O. Daniel *et al.* 1986. Nature **323:** 226.
4. Bishayee, S., A. H. Ross, R. Womer & C. D. Scher. 1986. Proc. Natl. Acad. Sci USA **83:** 6756.
5. Kahn, C. R. & M. F. White. 1988. J. Clin. Invest. **82:** 1151.
6. Heldin, C.-H. & B. Westermark. 1990. Cell Regul. **1:** 555.
7. Sultzman, L., C. Ellis, L. L. Lin, T. Pawson & J. Knopf. 1991. Mol. Cell. Biol. **11:** 2018.
8. Pledger, W. J., C. D. Stiles, H. N. Antoniades & C. D. Scher. 1978. Proc. Natl. Acad. Sci. USA **75:** 2839.
9. Singh, J. P., M. A. Chaikin, W. J. Pledger, C. D. Scher & C. D. Stiles. 1983. J. Cell Biol. **96:** 1497.
10. Stiles, C. D., G. T. Capone, C. D. Scher, H. N. Antoniades, J. J. Van Wyk *et al.* 1979. Proc. Natl. Acad. Sci. USA **76:** 1279.
11. Koch, C. A., D. Anderson, M. F. Moran, C. Ellis & T. Pawson. 1991. Science **252:** 668.
12. Moran, M. F., P. Polakis, F. McCormick, T. Pawson & C. Ellis. 1991. Mol. Cell. Biol. **11:** 1804.
13. Cantley, L. C., K. R. Auger, C. Carpenter, B. Duckworth, A. Graziani *et al.* Cell **64,** 281 (1991).
14. Songyang, Z., S. E. Shoelson, M. Chaudhuri, G. Gish, T. Roberts *et al.* 1993. Cell **72:** 767.
15. Waksman, G., S. E. Shoelson, N. Pant, D. Cowburn & J. Kuriyan. 1993. Cell **72:** 779.
16. Eck, M. J., S. E. Shoelson & S. C. Harrison. 1993. Nature **362:** 87.
17. Anderson, D., C. A. Koch, L. Grey, C. Ellis, M. F. Moran *et al.* 1990. Nature **250:** 979.
18. Pawson, T. & G. D. Gish. 1992. Cell **71:** 359.
19. Ronnstrand, L., S. Mori, A. K. Arridsson, A. Eriksson, C. Wernstedt *et al.* 1972. EMBO J. **11:** 3911.
20. Kashishian, A., A. Kazlauskas & J. A. Cooper. 1992. EMBO **11:** 1373.

21. Kazlauskas, A., A. Kashishian, J. A. Cooper & M. Valius. 1992. Mol. Cell. Biol. **12:** 2534.
22. Fantl, W. J., J. A. Escobedo, G. A. Martin, C. W. Turck, M. del Rosario *et al.* 1992. Cell **69:** 413.
23. Valius, M., C. Bazenet & A. Kazlauskas. 1993. Mol. Cell. Biol. **13:** 133.
24. Feener, E. P., J. M. Backer, G. L. King, P. A. Wilden, X. J. Sun *et al.* 1993. J. Biol. Chem. **268:** 11256.
25. White, M. F., S. E. Shoelson, H. Keutmann & C. R. Kahn. 1988. Biol. Chem. **263:** 2969.
26. Wilden, P. A., C. R. Kahn, K. Siddle & M. F. White. 1992. J. Biol. Chem. **267:** 16660.
27. White, M. F., J. N. Livingston, J. M. Backer, V. Lauris, T. J. Dull *et al.* 1988. Cell **54:** 641.
28. Sun, X. J., P. Rothenberg, C. R. Kahn, J. M. Backer, E. Araki *et al.* 1991. Nature **352:** 73.
29. Myers Jr., M. G., X. J. Sun, B. Cheatham, B. R. Jachna, E. M. Glasheen *et al.* 1993. Endocrinology **132:** 1421.
30. Myers Jr., M. G. & M. F. White. 1993. Diabetes **42:** 643.
31. Backer, J. M., M. G. Myers Jr., S. E. Shoelson, D. J. Chin, X. J. Sun *et al.* 1992. EMBO J. **11:** 3469.
32. Myers Jr., M. G., J. M. Backer, X. J. Sun, S. E. Shoelson, P. Hu *et al.* Proc. Natl. Acad. Sci. USA **89:** 10350.
33. Whittaker, J., A. K. Okamoto, R. Thys, G. I. Bell, D. F. Steiner *et al.* 1987. Proc. Natl. Acad. Sci. USA **84:** 5237.
34. Miralpeix, M., X. J. Sun, J. M. Backer, M. G. Myers, Jr., E. Araki *et al.* 1992. Biochemistry **31:** 9031.
35. Soos, M. A. & K. Siddle. 1989. Biochem. J. **263:** 553.
36. Hu, P., B. Margolis, E. Y. Skolnik, R. Lammers, A. Ullrich *et al.* 1992. Mol. Cell. Biol. **12:** 981.
37. Skolnik, E. Y., C. H. Lee, A. G. Batzer, L. M. Vicentini, M. Zhou *et al.* 1993. EMBO **12:** 1929.
38. Towbin, H., T. Staehelin & G. Gordon. 1979. Proc. Natl. Acad. Sci USA **76:** 4350.
39. Ruderman, N., R. Kapeller, M. F. White & L. C. Cantley. 1990. Proc. Natl. Acad. Sci. USA **87:** 1411.
40. Varticovski, L., B. Druker, D. Morrison, L. Cantley & T. Roberts. 1989. Science **342:** 699.
41. Kazlauskas, A. & J. A. Cooper. 1990. EMBO J. **9:** 3279.
42. Fukui, Y., S. Kornbluth, S. M. Jong, L. H. Wang & H. Hanafusa. 1989. Oncogene Res **4:** 283.
43. Escobedo, J. A., D. R. Kaplan, W. M. Kavanaugh, C. W. Turck & L. T. Williams. 1991. Mol. Cell. Biol. 1125.
44. Talmage, D. A., R. Freund, A. T. Young, J. Dahl, C. J. Dawe *et al.* 1989. Cell **59:** 55.
45. Auger, K. R., L. A. Serunian, S. P. Soltoff, P. Libby & L. C. Cantley. 1989. Cell **57:** 167.
46. Kypta, R. M., Y. Goldberg, E. T. Ulug & S. A. Courtneidge. 1990. Cell **62:** 481.
47. Backer, J. M., M. G. Myers, Jr., X. J. Sun, D. J. Chin, S. E. Shoelson *et al.* 1993. J. Biol. Chem. **268:** 8204.
48. Lowenstein, E. J., R. J. Daly, A. G. Batzer, W. Li, B. Margolis *et al.* 1992. Cell **70:** 431.
49. Kazlauskas, A., C. Ellis, T. Pawson & J. A. Cooper. 1990. Science **247:** 1578.
50. Sun, X. J., M. Miralpeix, M. G. Myers, Jr., E. M. Glasheen, J. M. Backer *et al.* 1992. J. Biol. Chem. **267:** 22662.
51. Wang, L.-M., A. D. Keegan, W. Li, G. E. Lienhard, S. Pacini *et al.* 1993. Proc. Natl. Acad. Sci. USA **90:** 4032.

52. Wang, L. M., M. G. Myers, Jr., X. J. Sun, S. A. Aaronson, M. F. White *et al.* 1993. Science **261:** 1591.
53. Keegan, A. D., K. Nelms, M. White, L. M. Wang, J. H. Pierce *et al.* 1994. Cell **76:** 811.
54. Kelly, K. L. & N. B. Ruderman. 1993. J. Biol. Chem. **268:** 4391.
55. Klippel, A., J. A. Escobedo, W. J. Fantl & L. T. Williams. 1992. Mol. Cell. Biol. **12:** 1451.
56. Ekanger, R., O. K. Vintermyr, G. Houge, T. E. Sand, J. D. Scott *et al.* 1989. J. Biol. Chem. **264:** 4374.
57. Hayashi, H., N. Miyake, F. Kanai, F. Shibasaki, T. Takenawa *et al.* 1991. Biochem. J. **280:** 769.
58. Frohman, M. A. 1992. In PCR Protocols: A Guide to Methods and Applications. p. 28. Academic Press. New York.
59. Cohen, B., M. Yoakim, H. Piwnica-Worms, T. M. Roberts & B. S. Schaffhausen. 1990. Proc. Natl. Acad. Sci. USA **87:** 4458.
60. Clark, S. G., M. J. Stern & H. R. Horvitz. 1992. Nature **356:** 340.
61. Panayotou, G., B. Bax, I. Gout, M. Federwisch, B. Wroblowski *et al.* 1992. EMBO **11:** 4261.
62. Kavanaugh, W. M., A. Klippel, J. A. Escobedo & L. T. Williams. 1992. Mol. Cell. Biol. **12:** 3415.
63. Sun, X. J., D. L. Crimmins, M. G. Myers, Jr., M. Miralpeix & M. F. White. 1993. Mol. Cell. Biol. **13:** 7418.
64. Hiles, I. D., M. Otsu, S. Volinna, M. J. Fry, I. Gout *et al.* 1992. Cell **70:** 419.
65. Schu, P. V., T. Kaoru, M. J. Fry, J. H. Stack, M. D. Waterfield *et al.* 1993. Science **260:** 88.
66. Susa, M., M. Keeler & L. Varticovski. 1992. J. Biol. Chem. **267:** 22951.
67. Ohmichi, M., S. J. Decker & A. R. Saltiel. 1992. J. Biol. Chem. **267:** 21601.
68. Lavan, B. E., M. R. Kuhne, C. W. Garner, D. Anderson, M. Reedijk *et al.* 1992. J. Biol. Chem. **267:** 11631.
69. Chuang, L. M., M. G. Myers, Jr., J. M. Backer, S. E. Shoelson, M. F. White *et al.* 1993. Mol. Cell. Biol. **13:** 6653.
70. Mayer, B. J., P. K. Jackson, R. A. Van Etten & D. Baltimore. 1992. Mol. Cell. Biol. **12:** 609.

Mechanism of Insulin and IGF-I Receptor Activation and Signal Transduction Specificity

Receptor Dimer Cross-linking, Bell-shaped Curves, and Sustained versus Transient Signaling

PIERRE DE MEYTS, BIRGITTE URSØ, CLAUS T. CHRISTOFFERSEN, AND RONALD M. SHYMKO[a]

Departments of Molecular Signaling and [a]Scientific Computing
Hagedorn Research Institute
Niels Steensens Vej 6
DK-2820 Gentofte, Denmark

THE PARADOX OF SIGNALING SPECIFICITY

Our understanding of intracellular signaling by peptide hormones and growth factors, catecholamines, neurotransmitters, cytokines, antigens, and other cell-surface ligands has markedly improved over the last decade. The initial concept, derived from the work of Sutherland,[1] postulated that a specific cell-membrane receptor upon stimulation by its ligand gives rise to an intracellular second messenger (*e.g.*, cyclic AMP), starting a linear cascade of signaling events (FIG. 1A). The specific pattern of responses observable in a given cell would then depend on the particular set of specific receptors present on the surface of that cell, resulting in the activation of parallel pathways with specific cellular targets as end points (FIG. 1B). It is only recently that the convergence of work from various fields of research has revealed the extraordinary complexity of the intracellular signaling network, with ligands of various classes showing extensive redundancy, pleiotropy (as discussed in other chapters in this volume), and cross-talk between multiple signaling pathways, as schematically shown in FIGURE 1C. With our increased ability to dissect the molecular anatomy of signaling and to understand the nature of the interaction of signaling molecules through universal recognition motifs, such as the SH2, SH3, PH, and other domains, our simplistic but clear picture of signaling specificity has all but vanished. It is indeed difficult to conceive how the molecular selectivity conferred at the binding step by the existence of highly specific receptors is not lost in the following steps of the signaling network given the convergence of multiple signals on common efferent pathways, such as the ras/raf/MEK/MAP kinase cascade or phosphatidylinositol-3 kinase (PI-3 kinase).

TABLE 1. Determinants of Signaling Specificity[7]

Types of receptors expressed in given cell.
Numbers of receptors expressed in given cell.
Stoichiometry of receptors/substrates.
Types of substrates/signaling molecules expressed in given cell.
Relative affinity of receptors for substrates/signaling molecules; mass action (competition for same substrates depending on relative levels of receptor expression).
Selectivity of recognition domains (SH2, SH3, PH).
Hybrid receptors.
Feedback loops (phosphatases, serine/threonine kinases).
Localization of signaling molecules within the cell.
Activation kinetics of signaling molecules (transient vs. sustained, time delays).

The specificity of the complex path between stimulus and response is probably determined by the interplay of a number of elements that depend on the cellular context in which the signal takes place[2] (TABLE 1). In this brief essay, we would like to emphasize one aspect of signaling specificity, the importance of which is only starting to be recognized: the role of the kinetics of activation of signaling molecules in determining response specificity.

The basic concept that we will develop is that, as proposed by Bourne,[3] signaling devices (like receptors or downstream signaling molecules) feature not only a switch that turns on the signal in response to the specific ligand, but also a timer that determines how long this signal will stay on. There are obvious elements of specificity in the switches: the molecular nature and structure-function relationships of the ligand and of the receptor that interact at the initial step, or of the upstream and downstream signaling molecules that interact with each other at a given node of the signaling network (where, *e.g.*, the selectivity of SH2 or SH3 domains plays a major role.[4]) However, the timer (*i.e.*, kinetic) aspect of the interaction (that is whether the process is transient or sustained, and how long) may play an equally important role in choosing between several possible downstream bifurcating paths (FIG. 1D), as we will illustrate later by discussing mitogenic versus metabolic signaling from the insulin receptor or the role of MAP kinase in proliferation versus differentiation of PC12 cells.

A more detailed discussion of several aspects of this concept with more extensive references can be found in several recent reviews;[5–7] this one is intended as a streamlined exposition of the basic concepts.

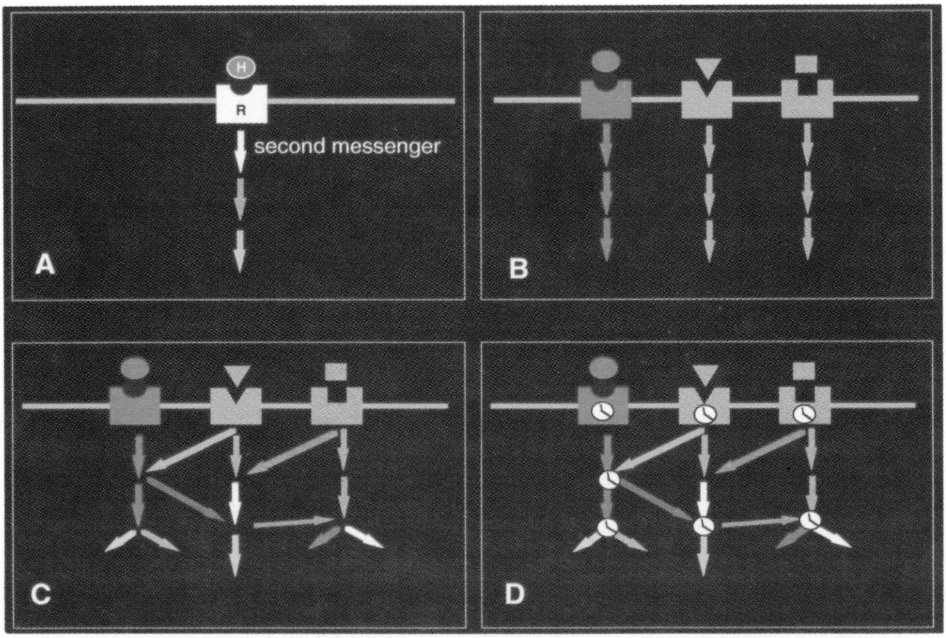
H
R
second messenger
A
B
C
D

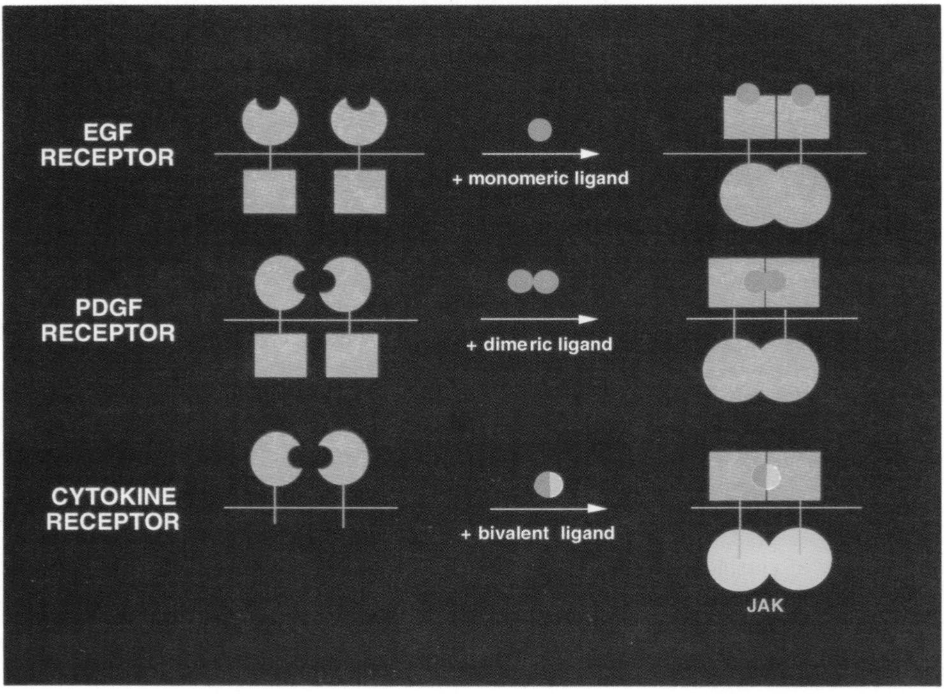
EGF RECEPTOR
+ monomeric ligand
PDGF RECEPTOR
+ dimeric ligand
CYTOKINE RECEPTOR
+ bivalent ligand
JAK

FIGURE 1. Evolution of the signal transduction specificity concept. **A:** The second messenger concept, when cyclic AMP was given all the credit for intracellular signaling.[1,43] **B:** Parallel pathways. **C:** Cross-talking network. **D:** Signaling molecules as timers at bifurcating nodes.

THE LIGAND-INDUCED RECEPTOR DIMERIZATION PARADIGM

It is becoming increasingly clear that the receptors of the tyrosine kinase and cytokine superfamilies are activated by a common mechanism, that is, the dimerization of two receptors upon ligand binding (FIG. 2), which is a paradigm initially proposed for the EGF receptor by Schlessinger and colleagues.[8,9] The importance of receptor oligomerization by multivalent ligands had been recognized for many years by immunologists.[10]

The initial concept envisaged either that the ligand would dimerize the receptors by inducing a conformational change that brings two monomeric ligand-receptor complexes together, resulting in a 2:2 stoichiometry (*e.g.*, EGF, FIG. 2), or that a dimeric ligand (*e.g.*, PDGF, FIG. 2) would bring the two receptors together. In many cases it now appears, in fact, that it is a monomeric ligand featuring two distinct binding surfaces (*i.e.*, functionally bivalent) that brings the two receptors together in a sequential fashion, resulting in a 1:2 stoichiometry. The epitome of this mechanism operates in the cytokine receptor family, as first demonstrated by the crystallographic resolution of the growth hormone complex with the extracellular part of its receptor.[11] In some cases the two receptor halves are not identical but are different subunits, resulting in a heterodimeric complex.

The dimerization of the receptors provides a simple mechanism for triggering downstream signaling by bringing together the two intracellular kinase domains of the receptor and allowing transphosphorylation to occur. In the cytokine family, the tyrosine kinase is not an intrinsic part of the receptor; rather, a cytoplasmic kinase (such as JAK or Fyn) is recruited to the receptor after activation.[12] The final structure of the dimeric-activated complex resembles that of the receptor tyrosine kinases (FIG. 2).

A variation of this theme has been recently proposed for the c-kit encoded stem-cell factor (or steel-factor) receptor, where dimer formation appears to be independent of the bivalency of the ligand but involves the stabilization of the dimeric form of the receptor after monovalent binding through an intrinsic dimerization site on the receptor distinct from the ligand-binding pocket.[13]

Another special case exists in the insulin/IGF-1 receptor tyrosine kinase subfamily, where the receptors are covalent dimers even in the absence of ligand. The receptors

FIGURE 2. Mechanisms of receptor activation in the receptor tyrosine kinase and cytokine receptor families by ligand-induced receptor dimerization. See text for explanation. The mechanism shown on top of the figure for EGF has recently been revised and is now thought to involve a monomeric bivalent ligand like the cytokine and insulin receptors.[17]

feature two extracellular α subunits that contain the ligand-binding domains and two transmembrane β subunits that contain the tyrosine-kinase domain. We and others have recently provided evidence based on structure-function relationships of insulin analogues and receptor-binding kinetics that a binding mechanism similar to that proposed for monomeric receptor tyrosine kinases and cytokine receptors is operating.[5,14] This model postulates that the insulin molecule has two binding surfaces, one overlapping with the surface involved in insulin dimerization, and the other involved in insulin hexamerization, and behaves as a bivalent ligand that sequentially cross-links two distinct binding domains, one on each α subunit of the insulin-receptor dimer (FIG. 3).

The recent finding that IGF-I binding to its receptor has kinetic properties similar to those of insulin binding to its receptor[15,16] suggests that the binding mechanism is probably the same. More recently, calorimetric evidence has led to a revision of the 2:2 stoichiometry initially proposed for the EGF receptor to suggest that it is also dimerizing around a monomeric, bivalent ligand.[17]

The binding model where a bivalent ligand induces sequential dimerization of its receptor (or in the insulin case, cross-linking within a covalent dimeric structure) has a number of interesting properties as compared with classical "monovalent" binding models. The dimerization or cross-linking provides both the above-mentioned switch and timer at the receptor binding step. The switch consists of the bivalent dimerization bringing together the two kinase domains. The timer is provided by the effect of dimerization on ligand-dissociation kinetics. Indeed, the bivalently bound ligand is much less likely to dissociate from the receptor than the monovalently bound one, greatly enhancing the probability that the complex will be able to initiate sustained signaling.[6,7,10,18]

In the case of the insulin and IGF-I receptors, an added feature regulates the "timer" component of the binding step: the existence of negative cooperativity in binding, which modulates the ligand dissociation rate as a function of ambient ligand concentration (*i.e.*, receptor occupancy).[5,19–21] We have recently proposed that this phenomenon may result from the formation of a second cross-linking bridge between the two receptor α subunits by a second insulin molecule within a symmetrical receptor structure[5] (FIG. 3).

As will be discussed below, the slowing down of dissociation kinetics by bivalent cross-linking, and its regulation by negative cooperativity, now appears to be of paramount importance in determining the mitogenic potency of insulin and insulin analogues.[27]

Another interesting property of a dimerizing receptor, besides the regulation of dissociation kinetics, resides in the shape of dose-response curves for bioactivity, and the regulation of agonistic properties.

BELL-SHAPED DOSE-RESPONSE CURVES, AGONISM, AND ANTAGONISM

A classical monovalent ligand-receptor binding system governed by the law of mass action results in the familiar monophasic competition curve when unlabeled ligand competes for labeled ligand (FIG. 4A). A biological response generated through signaling from such a binding event will similarly be monophasic (FIG. 4B) unless other modulating events occur at a subsequent step.

A strikingly different picture is seen with a sequentially dimerizing receptor. This model is easily visualized in computer simulations (FIGURES 4 A-D) using simple stoichiometric equations:[23]

$$H + R \underset{k_2}{\overset{k_1}{\rightleftharpoons}} M$$

$$M + R \underset{k_4}{\overset{k_3}{\rightleftharpoons}} D,$$

where H is the free ligand concentration, R the free receptor concentration, M the concentration of monomeric ligand-receptor complex (HR), D the concentration of dimeric ligand-receptor complex (HR_2), k_1 and k_3 the respective association rate constants, and k_2 and k_4 the respective dissociation rate constants.

The competition curve where unlabeled ligand competes for labeled ligand remains monophasic as in the simple monovalent case (FIG. 4C). The fraction of dimerized receptors as a function of ligand concentration, however, is biphasic or bell-shaped (FIG. 4C). Accordingly, a biological response that requires receptor dimerization to be activated is also predicted to show a bell-shaped dose-response curve with self-antagonism at high ligand concentration.[23,24] This is explained by the fact that at high ligand concentration only monomeric ligand-receptor complexes will be formed, which are unable to signal.[24]

An increasing number of biological responses to hormones and growth factors are, in fact,being found to give such a bell-shaped pattern (FIG. 5) if a large enough ligand concentration is being explored; although the self-antagonism is usually observed at nonphysiological concentrations, it is a hallmark of the dimerizing mechanism. Examples of such behavior are a number of biological responses to growth hormone (such as stimulation of lipogenesis in rat adipocytes[23] (FIG. 5A) or thymidine incorporation and cell proliferation in cells naturally expressing the growth hormone receptor (Ilondo and De Meyts, in preparation) or engineered to do so[24]), stimulation of myogenin mRNA expression in muscle cells by IGF-I (FIG. 5B[25]), and insulin stimulation of the insulin receptor tyrosine kinase (FIG. 5C[26]). Our finding of a bell-shaped curve for the negative cooperativity of the insulin receptor (FIG. 5D) was one of the arguments in favor of the insulin-receptor cross-linking model (FIG. 3[5]).

The bivalent dimerizing mechanism provides a simple molecular basis for the design of ligand antagonists; point mutations in the surface of the ligand that binds last in the sequential bivalent binding result in the creation of antagonists as examplified by the genetically engineered G120R human growth hormone.[24]

TIMER EFFECTS AT THE BINDING STEP: RELATIONSHIP BETWEEN INSULIN RECEPTOR CROSS-LINKING, DISSOCIATION KINETICS, AND MITOGENIC POTENCY

We have recently studied a T-cell lymphoma cell line (termed LB) devoid of IGF-I receptors, in which insulin and human growth hormone are potent stimuli for

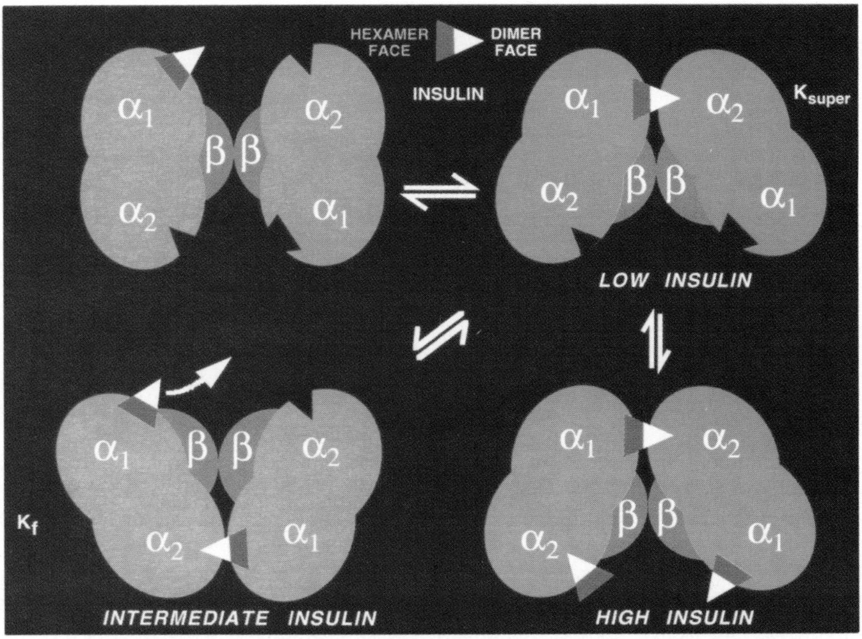
HEXAMER FACE
DIMER FACE
INSULIN
K_super
α1
α2
β β
LOW INSULIN
K_f
INTERMEDIATE INSULIN
HIGH INSULIN

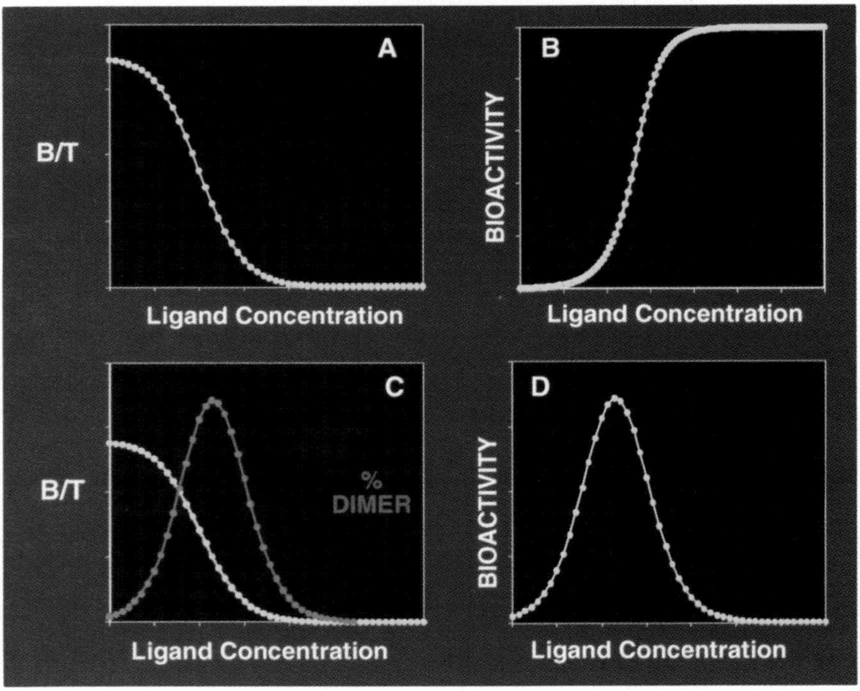
A
B/T
Ligand Concentration
B
BIOACTIVITY
Ligand Concentration
C
B/T
% DIMER
Ligand Concentration
D
BIOACTIVITY
Ligand Concentration

FIGURE 3. The symmetrical, alternative cross-linking model for insulin binding to its receptor. The receptor is viewed from the top. Each α subunit is represented as containing two subsites, α1 and α2. The first insulin molecule binds through its hexamer-forming surface to α1 and then cross-links through its dimer-forming surface to α2 on the second α subunit. The resulting tight bivalent binding is referred to as the K_{super} state. If the concentration of insulin is increased, partial dissociation of the first bound insulin allows a second insulin molecule to cross-link the opposite α1-α2 pair, which allows the first molecule to dissociate completely. At very high insulin concentrations, α1 and α2 opposite the first cross-link are both occupied, preventing the second cross-linking and maintaining the first bound insulin molecule in the K_{super} state, explaining the bell-shaped dose-response curve of dissociation kinetics. The figure was adapted from ref. 5, where a more complete explanation of the model can be found.

cell proliferation and thymidine incorporation (ref. 27 and chapter by Ish-Shalom *et al.* in this volume). The LB cell line constitutes, therefore, an ideal model for the study of mitogenic signaling through the insulin receptor without interference from the IGF-I receptor or hybrid receptors. The dose-response curve for insulin-induced mitogenesis was bell-shaped (Ursø and De Meyts, in preparation), suggesting that receptor cross-linking may be required for this response.

In this cell line, we found that a number of insulin analogues prepared by genetic engineering[28,29] showed mitogenic potencies (as measured by incorporation of tritiated thymidine into DNA) that were incommensurate with their receptor affinity. By contrast, when these "supermitogenic" analogues were tested for metabolic effects (*e.g.*, stimulation of lipogenesis in isolated rat adipocytes), they showed a potency in line with their affinity.[29] Strikingly, all analogues with supermitogenic properties were found to dissociate from the LB-cell receptor more slowly than insulin.[22] Their mitogenic potency was inversely related to their dissociation rate rather than related to their affinity; the slower the dissociation was, the more enhanced the mitogenesis relative to the affinity. The supermitogenic analogues showed also less or no acceleration of their dissociation in the presence of excess ligand (negative cooperativity), suggesting, according to our model (FIG. 3), a somewhat strengthened first cross-link that barely dissociates.

These data suggest that mitogenesis requires receptor cross-linking and that mitogenic potency depends on the strength of the first cross-link, as well as on the duration of the ligand's sojourn on the receptor, presumably due to the generation of sustained downstream signaling. In a CHO cell line, overexpressing the insulin receptor, B. F. Hansen *et al.* (submitted) were indeed able to demonstrate sustained activation of several downstream molecules upon activation by these supermitogenic analogues.

FIGURE 4. A: Classical monovalent/monomeric ligand-receptor binding competition curve. **B:** Classical biological dose-response curve generated by such a receptor system. **C:** Ligand-induced receptor dimerization system: competition curve (yellow), fraction of receptors dimerized (red). **D:** Biological dose-response curve generated by receptor dimerized sequentially by bivalent ligand.[10,23,24]

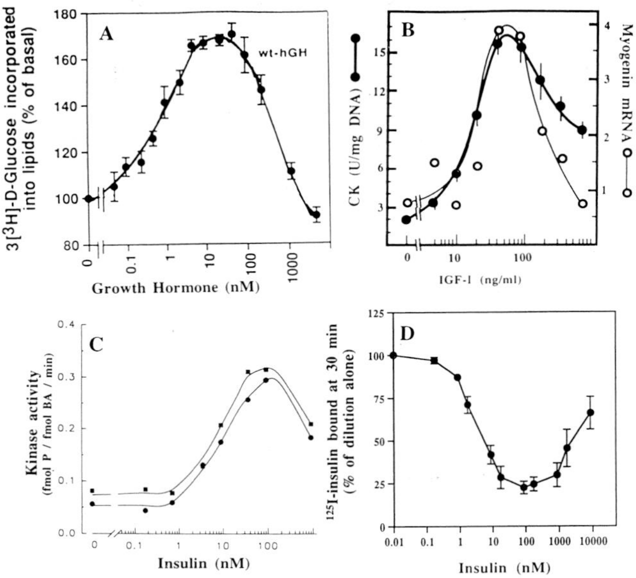

FIGURE 5. Bell-shaped curves. **A:** Stimulation of lipogenesis in primary rat adipocytes by human growth hormone, as measured by incorporation of D-[^{3}H]glucose into lipids.[23] **B:** Stimulation of myogenin mRNA (open circles) and creatine kinase levels (closed circles) in L6A1 myoblast cultures by IGF-I. (Florini *et al.*[25] With permission from *Molecular Endocrinology.*) CK, creatine kinase. **C:** Insulin stimulation of insulin receptor tyrosine kinase in human mononuclear cells using cell extracts (closed squares) or wheat-germ agglutinin-purified receptors (closed circles). (Klein *et al.*[26] With permission from *Diabetes.*) BA, binding activity. **D:** Dose-response curve for negative cooperativity of the insulin receptor in 293 cells transfected with the insulin receptor cDNA. The amount of [^{125}I]insulin bound to the cells after 30 min of dissociation in a 40-fold dilution is measured in the presence of increasing amounts of unlabeled insulin and plotted as a percent of the amount dissociated without unlabeled insulin present in the dissociation buffer. (Christoffersen *et al.*[16] With permission from *Endocrinology.*) (See refs. 5 and 19–21 for detailed explanation.)

By contrast, our preliminary data suggest that some metabolic responses, such as lipogenesis in adipocytes, which shows a monophasic, not bell-shaped insulin dose-response curve,[23] may only require transient signaling initiated upon the first monovalent binding event (De Meyts *et al.*, in preparation). Interestingly, in human adipocytes, the dose-response curve for insulin stimulation of its receptor's tyrosine kinase activity is not bell-shaped either.[26]

The timer effect offered by the regulation of dissociation kinetics in oligomerized receptors may have multiple applications. Slow dissociation of interleukin-2 from the high-affinity form of the complex between the p75 α chain and the p55 β chain is probably essential for growth signaling in T lymphocytes (ref. 30 and Kendall Smith, personal communication). The c-kit tyrosine kinase is activated by a ligand, steel, or stem-cell factor (SLF) that exists as two isoforms resulting from alternative splicing, one form being membrane-bound, the other soluble. It was recently shown that the membrane-bound form induces more persistent activation of c-kit kinase

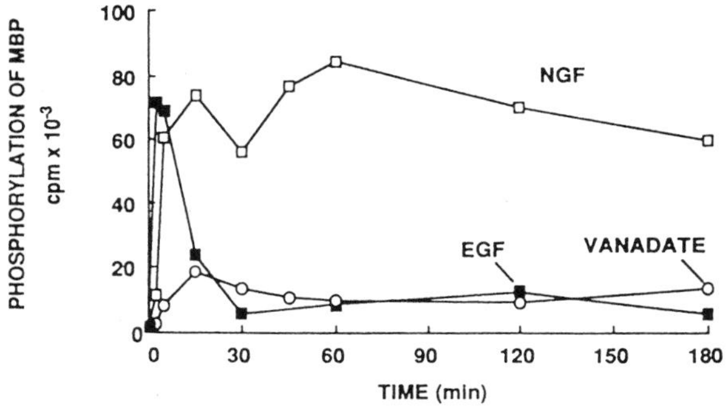

FIGURE 6. Time course of MAP kinase (ERK1) activation in PC12 cells. Serum-depleted PC12 cells were incubated for the indicated times in the presence of 100 ng/mL NGF, 0.1 μM EGF, or 1 nM vanadate. The peptide antigen eluates of anti-ERK1 immunoprecipitates were used to phosphorylate exogenously added myelin basic protein (MBP). (Nguyen *et al.*[35] With permission from the *Journal of Biological Chemistry.*)

than the soluble form.[31] Although no gross differences were observed between the two forms of the ligand in stimulating proliferation of the myeloid cell line MO7e, the membrane-associated SLF form is known to be better in supporting hematopoiesis in a long-term culture system.[31] It will be interesting to look more closely into the details of intracellular signaling by the two forms of steel factor.

Recently, McKeithan has proposed a related concept (kinetic proofreading) where a temporal lag between ligand binding and receptor signaling greatly enhances the T-cell receptor ability to discriminate between a foreign antigen and self-antigens with only moderately lower affinity.[32]

Such "timer" effects on response selectivity are not restricted to the binding step and are likely to affect a number of signaling molecules at various nodes of the signaling network, as recently shown for MAP kinase.

TIMER EFFECTS IN DOWNSTREAM SIGNALING: ROLE OF MAP KINASE KINETICS IN DETERMINING PROLIFERATIVE VERSUS DIFFERENTIATING RESPONSES

A nice example of a timer effect is provided by PC12 cells, a rat pheochromocytoma cell line that possesses receptors for both NGF and EGF. The effects of the two ligands are markedly different. EGF induces cell proliferation, whereas upon NGF treatment, PC12 cells stop dividing, develop excitable membranes, grow significant processes, and differentiate into a sympathetic neuronal phenotype. Yet, all the evidence so far indicates that the two growth factors share the same signaling elements: tyrosine phosphorylation, membrane ruffling, identical immediate early genes, Ras, ERK/MAP kinases, RSK/S6 kinases, tyrosine hydroxylase, ornithine decarboxylase, 2-deoxyglucose uptake, NA^+/K^+ pump, and sodium channels.[33] It appears that the determining element may be the kinetics of p21ras and MAP kinase activation.

Several groups found indeed that NGF induces a sustained activation of MAP kinase for at least 90 minutes, whereas EGF stimulates MAP kinase activity to the same level, but only transiently, with a peak at 2 minutes and a return to baseline after 30-90 minutes[34–36] (FIG. 6). Sustained, but no transient, stimulation results in nuclear translocation of MAP kinase. Interestingly, the EGF receptor can be made to induce differentiation by overexpressing it in the PC12 cells.[37]

TOWARDS A QUANTITATIVE DESCRIPTION OF SIGNAL TRANSDUCTION NETWORKS

The very complexity of the signal transduction network makes a formal description of it a real challenge. We believe, however, that such a description is fundamental for fully understanding signaling specificity, receptor cross-talk, and cell regulation, and predicting the behavior of such systems in normal and diseased states. Approaches based on Boolean logic have been used to provide a logical description of complex integrated networks that comprise feedback loops and time delays, such as genetic control circuits. They have been applied successfully to domains as varied as climatology and decision-making processes.[38–41] We are currently attempting to adapt such methodology, also called kinetic logic, to the quantification of signal transduction.

Another emerging theoretical concept that may prove useful is the proposal by Kühl of allochronic effects (as opposed to allosteric ones) that affect the working cycle of macromolecules by introducing refractory periods (quoted in ref 42); the theory has not yet been published in full. In a model such as the one shown in FIG. 3, whether the receptor dimer is refractory to further signaling upon formation of the second cross-link or is activated anew, will generate very different predictions regarding biological dose-response curves or the effect of the kinetic negative cooperativity on downstream signaling.

The unraveling of the signal-transduction pathways and their connectivities will undoubtedly open a new golden era for theoretical biologists; the collaboration between the latter and molecular biologists may ultimately help decipher the "metabolic code," foreseen twenty years ago by the late Gordon Tomkins[43] at a time when cyclic AMP was still thought "to mediate the intracellular actions of almost all those hormones that interact with the cell membrane."[43]

ACKNOWLEDGMENTS

Helpful discussions with Klaus Seedorf, Hans Tornqvist, Henri Metzger, Peter Kühl, René Thomas, Jacques Dumont, Jossi Schlessinger, and Axel Ullrich are gratefully acknowledged, as is Henrik W. Wengholt's help with computer graphics. David Naor kindly gave us the LB cells. The Hagedorn Research Institute is an independent basic research component of Novo Nordisk.

REFERENCES

1. ROBISON, G.A., R. W. BUTCHER & E. W. SUTHERLAND. 1971. Cyclic AMP. Academic Press. New York and London.
2. SEEDORF, K. 1995. Intracellular signaling by growth factors. Metabolism. In press.

3. BOURNE, H. 1992. G proteins as signaling machines. *In* 9th International Congress of Endocrinology, Nice. Parthenon Publishing. London.
4. PAWSON, T. 1995. Protein modules and signalling networks. Nature **373:** 573-580.
5. DE MEYTS, P. 1994. The structural basis of insulin and insulin-like growth factor-I (IGF-I) receptor binding and negative cooperativity, and its relevance to mitogenic versus metabolic signaling. Diabetologia **37**:(suppl. 2):S135-S148.
6. DE MEYTS, P., B. WALLACH, C. T. CHRISTOFFERSEN, B. URSØ, K. GRØNSKOV, L. J. LATUS, F. YAKUSHIJI, M. M. ILONDO & R. M. SHYMKO. 1994. The insulin-like growth factor-I receptor. Structure, ligand binding mechanism and signal transduction. Horm. Res. (Basel) **42:** 152-169.
7. DE MEYTS, P., C. T. CHRISTOFFERSEN, B. URSØ, B. WALLACH, K. GRØNSKOV, F. YAKUSHIJI & R. M. SHYMKO. 1995. Role of the time factor in signaling specificity. Application to mitogenic and metabolic signaling by the insulin and insulin-like growth factor-I receptor tyrosine kinases. Metabolism. In press.
8. SCHLESSINGER, J. 1988. Signal transduction by allosteric receptor oligomerization. TIBS **13:** 443-447.
9. YARDEN, Y. & A. ULLRICH. 1988. Growth factor receptor tyrosine kinases. Annu. Rev. Biochem. **57:** 443-478.
10. METZGER, H. 1992. Transmembrane signaling: the joy of aggregation. J. Immunol. **149:** 1477-1487.
11. DE VOS, A. M., M. ULTSCH & A. A. KOSSIAKOFF. 1992. Human growth hormone and extracellular domain of its receptor: crystal structure of the complex. Science **255:** 306-312.
12. IHLE, J. N., B. A. WITTHUHN, F. W. QUELLE, K. YAMAMOTO, W. E. THIERFELDER, B. KREIDER & O. SILVENNOINEN. 1994. Signaling by the cytokine receptor superfamily: JAKs and STATs. TIBS **19:** 222-227.
13. BLECHMAN, J. M., S. LEV, J. BARG, M. EISENSTEIN, B. VAKS, Z. VOGEL, D. GIVOL & Y. YARDEN. 1995. The fourth immunoglobulin domain of the stem cell factor receptor couples ligand binding to signal transduction. Cell **80:** 103-113.
14. SCHÄFFER, L. 1993. A model for insulin binding to the insulin receptor. Eur. J. Biochem. **221:** 1127-1132.
15. ZHONG, P., J. F. CARA & H. S. TAGER. 1993. Importance of receptor occupancy, concentrations differences, and ligand exchange in the insulin-like growth factor-I receptor system. Proc. Natl. Acad. Sci. USA **90:** 11451-11455.
16. CHRISTOFFERSEN, C. T., K. E. BORNFELDT, C. M. ROTELLA, N. GONZALES, H. VISSING, R. M. SHYMKO, J. TEN HOEVE, J. GROFFEN, N. HEISTERKAMP & P. DE MEYTS. 1994. Negative cooperativity in the insulin-like growth factor-I (IGF-I) receptor and a chimeric IGF-I/insulin receptor. Endocrinology **135:** 472-475.
17. SPIVAK-KROIZMAN, T., M. A. LEMMON, I. DIKIC, J. E. LADBURY, J. E. PINCHASI, J. HUANG, M. JAYE, J. CRUMLEY, J. SCHLESSINGER & I. LAX. 1994. Heparin-induced oligomerization of FGF molecules is responsible for FGF receptor dimerization, activation and cell proliferation. Cell **79:** 1015-1024.
18. DELISI, C. 1980. The biophysics of ligand-receptor interactions. Q. Rev. Biophys. **13:** 201-230.
19. DE MEYTS, P., J. ROTH, D. M. NEVILLE JR., J. R. GAVIN III & M. A. LESNIAK, 1973. Insulin interactions with its receptors: experimental evidence for negative cooperativity. Biochem. Biophys. Res. Commun. **55:** 154-161.
20. DE MEYTS, P., A. R. BIANCO & J. ROTH. 1976. Site-site interactions among insulin receptors: characterization of the negative cooperativity. J. Biol. Chem. **251:** 1877-1888.
21. DE MEYTS, P., E. VAN OBBERGHEN, J. ROTH, D. BRANDENBURG & A. WOLLMER. 1978. Mapping of the residues of the receptor binding region of insulin responsible for the negative cooperativity. Nature **273:** 504-509.

22. De Meyts, P., C. T. Christoffersen, B. Ursø, D. Ish-Shalom, N. Sacerdotti-Sierra, K. Drejer, L. Schäffer, R. M. Shymko & D. Naor. 1993. Insulin potency as a mitogen is determined by the half-life of the insulin-receptor complex. Exp. Clin. Endocrinol. Leipzig **101:** 22-23.
23. Ilondo, M. M., A. Damholt, B. Cunningham, J. A. Wells, P. De Meyts & R. M. Shymko. 1994. Receptor dimerization determines the effects of growth hormone in primary rat adipocytes and cultured IM-9 lymphocytes. Endocrinology **134:** 2397-2403.
24. Fuh, G., B. C. Cunningham, R. Fukunaga, S. Nagata, D. V. Goeddel & J. A. Wells. 1992. Rational design of potent antagonists to the growth hormone receptor. Science **256:** 1677-1680.
25. Florini, J. R., D. Z. Ewton & S. L. Roofs. 1991. Insulin-like growth factor-1 stimulates terminal myogenic differentiation by induction of myogenin gene expression. Mol. Endocrinol. **5:** 718-724.
26. Klein, H. H., B. Kowalewski, M. Drenckhan, S. Neugebauer, S. Matthaei & G. Kotzke. 1993. A microtiter well assay system to measure insulin activation of insulin receptor kinase in intact human mononuclear cells. Diabetes **42:** 883-890.
27. Pillemer, G., H. Lugasi-Evgi, G. Scharovski & D. Naor. 1992. Insulin dependence of murine lymphoid T-cell leukemia. Int. J. Cancer **50:** 80-85.
28. Brange, J., D. Owens, S. Kang & A. Vølund. 1990. Monomeric insulins and their experimental and clinical implications. Diabetes Care **13:** 923-954.
29. Drejer, K. 1992. The bioactivity of insulin analogues from *in vitro* receptor binding to *in vitro* glucose uptake. Diabetes Metab. Rev. **8:** 259-286.
30. Wang, H.-M. & K. A. Smith. 1987. The interleukin 2 receptor. Functional consequences of its bimolecular structure. J. Exp. Med. **166:** 1055-1069.
31. Miyazawa, K., D. A. Williams, A. Gotoh, J. Nishimaki, H. E. Broxmeyer & K. Toyama. 1995. Membrane-bound steel factor induces more persistent tyrosine kinase activation and longer life span of c-kit gene-encoded protein than its soluble form. Blood **85:** 641-649.
32. McKeithan, T. W. 1995. Kinetic proofreading in T-cell receptor signal transduction. Proc. Natl. Acad. Sci. USA **92:** 5042-5046.
33. Chao, M. V. 1992. Growth factor signaling: Where is the specificity? Cell **68:** 995-997.
34. Traverse, S., N. Gomez, H. Paterson, C. Marshall & P. Cohen. 1992. Sustained activation of the mitogen-activated protein (MAP) kinase cascade may be required for differentiation of PC12 cells. Comparison of the effects of nerve growth factor and epidermal growth factor. Biochem. J. **288:** 351-355.
35. Nguyen, T. T., J. C. Scimeca, C. Filloux, P. Peraldi, J. L. Carpentier & E. Van Obberghen. 1993. Co-regulation of the mitogen-activated protein kinase, extracellular signal-regulated kinase 1, and the 90 kDa ribosomal S6 kinase in PC12 cells. Distinct effects of the neurotrophic factor, nerve growth factor, and the mitogenic factor, epidermal growth factor. J. Biol. Chem. **268:** 9803-9810.
36. Qiu, M. S. & S. H. Green. 1992. PC12 cell neuronal differentiation is associated with prolonged p21ras activity and consequent prolonged ERK activity. Neuron **9:** 705-717.
37. Traverse, S., K. Seedorf, H. Paterson, C. Marshall, P. Cohen & A. Ullrich. 1994. EGF triggers neuronal differentiation of PC12 cells that overexpress the EGF receptor. Curr. Biol. **4:** 694-701.
38. Thomas, R. 1973. Boolean formalisation of genetic control circuits. J. Theor. Biol. **42:** 563-585.
39. Thomas, R. & R. D'Ari. 1990. Biological Feedback. CRC Press. Boca Raton, Ann Arbor, Boston.

40. THOMAS, R. 1991. Regulatory networks seen as asynchronous automata: a logical description. J. Theor. Biol. **153:** 1–23.
41. LECLERCQ, J. & J. E. DUMONT. 1983. Boolean analysis of cell regulation networks. J. Theor. Biol. **104:** 507–534.
42. KÜHL, P. W. 1994. Excess-substrate inhibition in enzymology and high-dose inhibition in pharmacology: a re-interpretation. Biochem. J. **298:** 171–180.
43. TOMKINS, G. 1975. The metabolic code. Science **189:** 760–763.

The Role of the Insulin-like Growth Factor-I Receptor in Cancer

D. LeROITH,[a,c] H. WERNER,[a]
S. NEUENSCHWANDER,[a] T. KALEBIC,[b] AND
L. J. HELMAN[b]

[a]*Section on Molecular and Cellular Physiology*
Diabetes Branch
National Institute of Diabetes and Digestive and Kidney Diseases
National Institutes of Health
Bethesda, Maryland 20892-1770
and
[b]*Molecular Oncology Section*
Pediatric Branch
National Cancer Institute
National Institutes of Health
Bethesda, Maryland 20892-1770

INTRODUCTION

The insulin-like growth factor (IGF) system includes ligands, receptors, and binding proteins (TABLE 1).[1,2] The ligands comprise insulin, which primarily controls metabolic functions, and IGF-I and IGF-II, which are important for normal growth and differentiation during all stages of development.[3,4] Although the actions of these polypeptides are mediated by cell-surface receptors, they are modulated by a family of IGF-binding proteins (IGFBPs) that specifically bind the IGFs and not insulin.[5,6]

The IGF-I receptor is a transmembrane heterotetrameric glycoprotein with ligand-dependent tyrosine-kinase activity (FIG. 1).[5,7] The IGF-I receptor is expressed ubiquitously and, when activated, results in diverse effects that are cell-type specific.[5]

Recent studies have demonstrated that the IGF-I receptor is essential for optimal growth of most cells in culture, even when grown in the presence of other growth factors, including platelet-derived growth factor (PDGF) or epidermal growth factor

TABLE 1. The Insulin-like Growth Factor (IGF) System

Ligands: Insulin, IGF-I, IGF-II
Receptors: Insulin, IGF-I, IGF-II/M6P
Binding proteins: IGFBP-1, -2, -3, -4, -5, -6, -7

[c] Send correspondence to Derek LeRoith, M.D., Ph.D., National Institutes of Health, Building 10, Room 8S-239, Bethesda, Maryland 20892-1770.

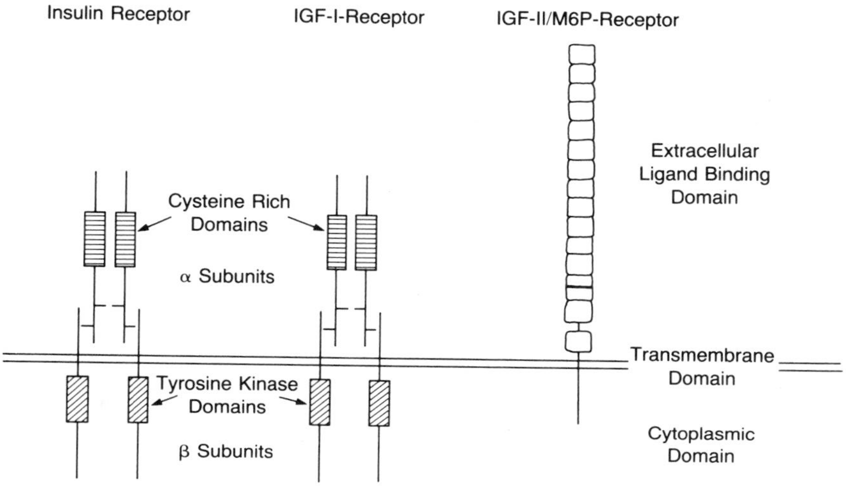

FIGURE 1. Structure of the insulin receptor and the IGF-I and IGF-II receptors. The insulin and IGF-I receptors are both heterotetrameric complexes composed of extracellular α subunits that bind the ligands and β subunits that anchor the receptor in the membrane and contain tyrosine-kinase activity in their cytoplasmic domains. The IGF-II/M6P (mannose-6-phosphate) receptor is not structurally related to the IGF-I and insulin receptors, having a short cytoplasmic tail and no tyrosine-kinase activity.

(EGF).[8] Furthermore, the IGF-I receptor plays a central role in cellular transformation. Because many tumors overexpress IGF-I receptors, it is apparent that development of tumors, their propagation, and even metastasis may require the expression of this receptor.[9,10]

In discussing the role of the IGF-I receptor in carcinogenesis, we will present results from some of our studies as well as studies from other groups.

WILMS' TUMOR

Wilms' tumor is a childhood cancer derived from embryonal kidney stem cells. The tumors express higher levels of both the ligand, IGF-II, and the IGF-I receptor, compared with adjacent normal tissue, suggesting an important role for the IGF system in neoplastic growth.[11,12] This hypothesis was tested and proven by using αIR3, an antibody directed towards the human IGF-I receptor. Growth of tumor cells in culture systems is blocked by αIR3, whereas in nude mice, αIR3 leads to tumor repression.[13] Given the potentially important role for the IGF-I receptor and its high level of expression in this tumor, we examined the promoter region of the mammalian IGF-I receptor gene. This sequence contains at least ten consensus *cis*-acting sites for transcription factors of the early growth response (EGR) family (FIG. 2).[14,15] This family of zinc-finger transcription factors includes the Wilms' tumor-1 (WT1) suppressor gene product, a protein whose deletion or mutation has been implicated

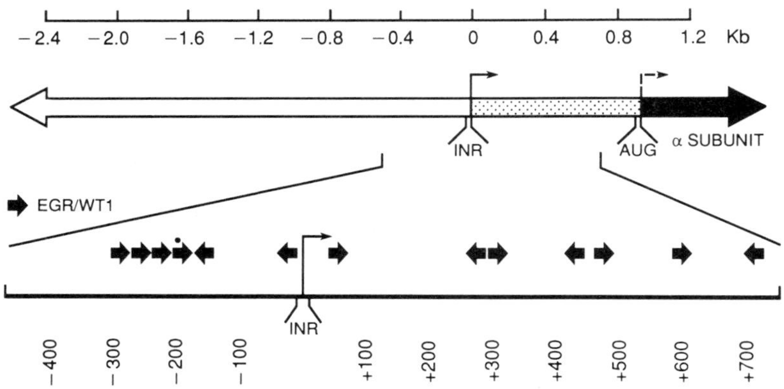

FIGURE 2. Potential EGR/WT1 binding sites in the IGF-I receptor gene promoter. The schematic representation of the proximal promoter region includes the 5′ flanking region (open), the 5′ untranslated region (stippled), and the N-terminal translated region (filled). The black arrows represent consensus sequences for EGR/WT1 protein binding.

in the etiology of a subset of Wilms' tumors. To demonstrate that the WT1 gene product binds to these sites and inhibits promoter activity, a WT1 expression vector was cotransfected into Chinese hamster ovary (CHO) cells, together with promoter sequences placed upstream of a luciferase reporter gene. Coexpression inhibited promoter activity by more than 90%, and the effect was substantially weakened by removal of the WT1 binding sites.[16] DNase footprinting and gel-shift assays confirmed the specific binding of the zinc-finger domain of the WT1 protein to these sites. Further evidence for the role of WT1 in regulating IGF-I receptor gene expression was obtained by stable transfection of the WT1 cDNA into G401 cells that normally fail to express WT1. In the transfected cells, steady-state mRNA and protein levels of the IGF-I receptor were reduced by ~50%, as was IGF-I receptor-promoter activity. Furthermore, IGF-I responsiveness, as measured by cellular proliferation, was similarly reduced (Werner *et al.,* in press).

The above results involving the IGF-I receptor, as well as similar results obtained with the IGF-II gene promoter, have led to the following hypothesis (FIG. 3). During the blastemal stage of kidney development, the WT1 gene is normally expressed, and it represses synthesis of IGF-II ligand and IGF-I receptors. By suppressing this autocrine mitogenic loop, normal differentiation of the kidney can occur. In the absence of normal WT1 protein at this stage (due either to point mutations or to a chromosomal deletion at 11p13), inappropriate proliferation leading to tumor formation takes place. Thus, Wilms' tumor seems to represent an example of inactivation of a tumor suppressor gene that results in up-regulation of a growth-factor system leading to tumorigenesis.

OSTEOGENIC SARCOMA

Osteogenic sarcoma (OS) is the most common bone tumor in childhood and typically occurs during the adolescent growth spurt in response to high levels of

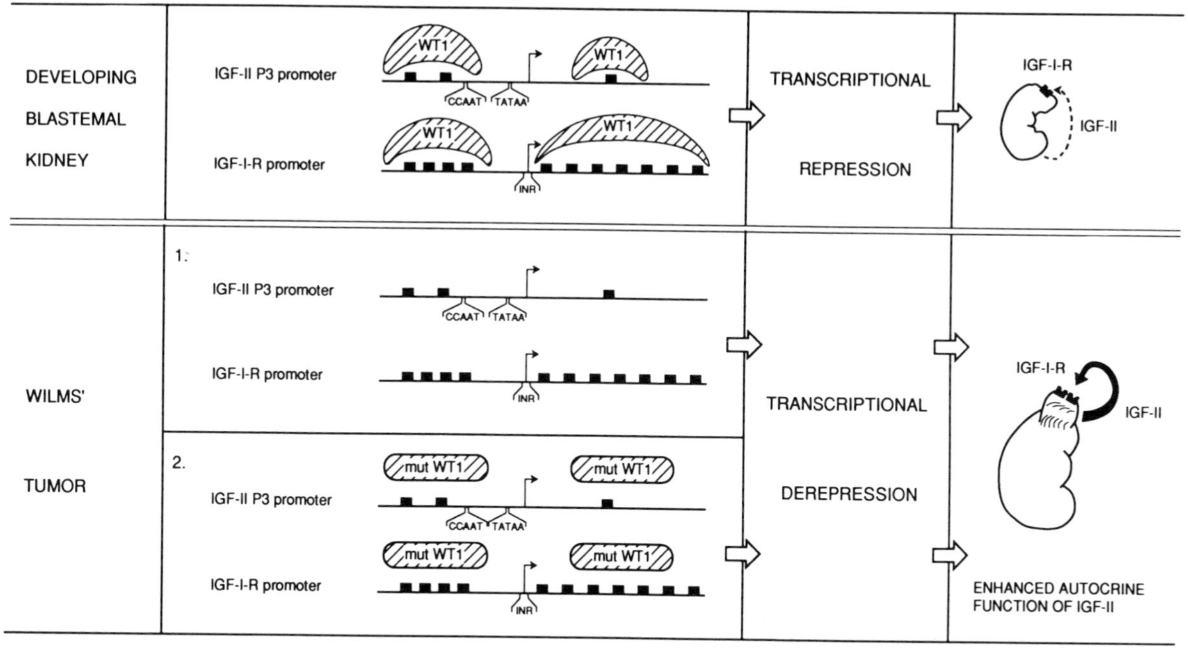

FIGURE 3. Regulation of IGF-II and IGF-I receptor gene expression by the WT1 Wilms' tumor suppressor gene product. During blastemal development, WT1 protein inhibits IGF-II peptide and the IGF-I receptor expression and, thereby, inhibits growth, allowing for differentiation of the renal tissue. Wilms' tumors may result if WT1 gene expression fails to occur at this stage or produces a mutant protein unable to bind the DNA and therefore allows continued expression of the ligand, IGF-II, and the IGF-I receptor expression.

growth hormone and IGF-I. Because the IGFs are known to play an important role in normal bone growth and development,[17] it has been suggested that altered signaling through the IGF-I receptor may play a role in the unregulated growth of osteosarcoma cells. We have, therefore, recently evaluated the role of IGF-I receptor signaling in the *in vitro* growth of human osteosarcoma cells.[18] We determined that *in vitro* survival of OS cells is dependent upon the presence of IGF-I in the culture media. Furthermore, we found that these cells display cell-surface IGF-I receptors. To evaluate whether signaling through the IGF-I receptors was necessary for cell proliferation, we blocked the receptors with either monoclonal antibody αIR3 or antisense oligonucleotides directed against this receptor. Both of these approaches led to a substantial inhibition of cell growth. These data demonstrate a critical role for the IGF-I receptor in survival and proliferation of human OS cells.[18]

RHABDOMYOSARCOMA

Rhabdomyosarcomas (RMS), the most common soft-tissue sarcoma in childhood, are embryonal tumors thought to arise from primitive skeletal muscle cells. We have previously demonstrated the presence of an IGF-II autocrine growth loop in RMS cell lines and overexpression of IGF-II in virtually all RMS tumor specimens.[19] To determine whether IGF-I receptor signaling also played a role in *in vivo* growth of these tumors, we tested the ability of αIR3 to inhibit the growth of RMS xenografts in nude mice. We demonstrated that αIR3 treatment led to a dose-dependent decrease in tumor formation when cells were pretreated with antibody prior to injection followed by a 3 times per week subcutaneous injection of antibody in mice. When αIR3 treatment was delayed until after tumor formation, antibody treatment significantly inhibited tumor growth. Furthermore, this growth inhibition was secondary to growth arrest, as evidenced by a marked decrease in $p34^{cdc2}$ (a cell cycle-regulated protein) expression in tumors obtained from mice treated with αIR3.[20] These data suggest a critical role of the IGF-I receptor for *in vivo* growth of human rhabdomyosarcomas.

OTHER TUMORS

The central role of the IGF-I receptor in the growth of numerous other tumor types has been similarly demonstrated by studies that use specific receptor-blocking agents. Initial studies used αIR3 to block *in vitro* and *in vivo* growth of tumors that demonstrated IGF-dependent growth. Recent studies have used antisense technology to interfere with the IGF-I receptor. These include Wilms' tumor cells, neuroblastoma cell lines, breast cancer cell lines, small cell lung cancers, melanomas, pancreatic carcinomas, leiomyosarcomas, and prostatic cancer cell lines.[13,18,20–24]

Antisense oligonucleotides directed against the IGF-I receptor mRNA inhibit cellular proliferation of osteosarcomas when added to the culture medium. Similarly, stable transfection of MCF-7 breast cancer cells, with a vector expressing an antisense IGF-I receptor mRNA fragment surrounding the ATG start site for translation, resulted in decreased levels of endogenous IGF-I receptor mRNA, decreased IGF-I binding, and prolonged IGF-I-induced cellular doubling time (Neuenschwander *et al.*, in press).

In addition, synthetic IGF-I antagonists that compete with the native peptide for binding to the IGF-I receptor result in complete inhibition of IGF-I receptor activation and growth of prostate cancer cell lines.[24]

PROTOONCOGENE-INDUCED TRANSFORMATION

Proliferation of many cell types, including fibroblasts, smooth muscle cells, and chondrocytes, require IGF-I and either PDGF or EGF. Certain protooncogenes, such as *c-myb,* abrogate the necessity for IGF-I in the culture medium by increasing the cellular expression of IGF-I and the IGF-I receptor.[25] Similarly, SV40 TAg transforms fibroblasts and permits cellular proliferation in the absence of IGF-I by increasing cellular expression of IGF-I and the IGF-I receptor. This effect has been demonstrated to result from increased promoter activity.[26]

The pivotal role of the IGF-I receptor in SV40 TAg-mediated transformation was demonstrated using fibroblasts derived from mouse embryos harboring the null mutation for the IGF-I receptor. These fibroblasts, which do not express any IGF-I receptors, failed to develop the transforming phenotype following SV40 TAg transfection. Reintroduction of an active human IGF-I receptor, using an expression vector, restored the ability of these cells to respond to SV40 TAg, by undergoing transformation, as demonstrated by growth in soft agar.[27] Thus, the IGF-I receptor is required by some viral oncogenes to transform fibroblasts.

OVEREXPRESSION OF IGF-I RECEPTORS IN FIBROBLASTS

When high levels of wild-type IGF-I receptors were expressed in fibroblasts, these cells acquired a highly transformed phenotype in response to IGF-I.[28] They grew in soft agar in the presence of IGF-I, and this effect was blocked by αIR3. When injected into nude mice, subcutaneous fibrosarcoma tumors developed. Further evidence supporting the role of a normal receptor in tumorigenesis was supplied by Prager *et al.,*[29] who transfected rat-1 fibroblasts with a truncated IGF-I receptor. This receptor, which was mutated by the deletion of bases downstream of nucleotide 952 of the β subunit, lacked tyrosine-kinase activity. Truncated receptors acted in a dominant-negative fashion, inhibiting the normal endogenous receptors and tumor formation.

CONCLUSION

We have presented data strongly suggesting that the IGF-I receptor plays a central role in growth of cancer cells. In addition to these effects, the IGF-I receptor may be important in cell motility and metastasis of cancer cells.[30] It is clear, however, that although we have focused on the IGF-I receptor, other growth factors and receptors are also essential in the development and maintenance of cancerous growth.

REFERENCES

1. LeRoith, D., D. Clemmons, P. Nissley & M. M. Rechler. 1992. Ann. Intern. Med. **116:** 854-862.
2. Daughaday, W. H. & P. Rotwein. 1989. Endocr. Rev. **10:** 68-91.
3. Liu, J. P., J. Baker, A. S. Perkins, E. J. Robertson & A. Efstratiadis. 1993. Cell **75:** 59-72.
4. Baker, J., J. P. Liu, E. J. Robertson & A. Efstratiadis. 1993. Cell **75:** 73-82.
5. LeRoith, D., H. Werner, D. Beitner-Johnson & C. T. Roberts Jr. 1995. Endocr. Rev. **16:** 143-163.
6. Rechler, M. M. 1993. Vitam. Horm. (NY) **47:** 1-114.
7. White, M. F. & C. R. Kahn. 1994. J. Biol. Chem. **269:** 1-4.
8. Baserga R. & R. Rubin. 1993. Crit. Rev. Eukaryotic Gene Expression **3:** 47-61.
9. McCauly, V. M. 1992. Br. J. Cancer **65:** 311-320.
10. LeRoith, D., R. Baserga, L. Helman & C. T. Roberts, Jr. 1995. Ann. Inter. Med. In press.
11. Haselbacher, G. K., J.-C. Irminger, J. Zapf, W. H. Ziegler & R. E. Humble. 1987. Proc. Natl. Acad. Sci. USA **84:** 1104-1106.
12. Gansler, T., K. D. Allen, C. F. Burant, T. Inabnett, A. Scott, M. G. Buse, D. A. Sens & A. J. Garvin. 1988. Am. J. Pathol. **130:** 431-435.
13. Gansler, T., R. Furlanetto, T. S. Gramling, K. A. Robinson, N. Blocker, M. G. Buse, D. A. Sens & A. J. Garvin. 1989. Science **250:** 1259-1261.
14. Werner, H., G. G. Re, I. A. Drummond, V. P. Sukhatme, F. J. Rauscher, III, D. A. Sens, A. J. Garvin, D. LeRoith & C. T. Roberts, Jr. 1993. Proc. Natl. Acad. Sci. USA **90:** 5828-5832.
15. Werner, H., M. A. Bach, B. Stannard, C. T. Roberts, Jr. & D. LeRoith. 1991. Mol. Endocrinol. **6:** 1545-1558.
16. Werner, H., F. J. Rauscher, III, V. P. Sukhatme, I. A. Drummond, C. T. Roberts, Jr. & D. LeRoith. 1994. J. Biol. Chem. **269:** 12577-12582.
17. Canalis, E., T. McCarthy & M. Centrella. 1988. J. Clin. Invest. **81:** 277-281.
18. Kappel, C. C., M. C. Velez-Yanguas, S. Hirschfeld & L. J. Helman. 1994. Cancer Res. **54:** 2803-2807.
19. El-Badry, O. M., C. Minniti, E. C. Kohn, P. J. Houghton, W. H. Daughaday & L. J. Helman. 1990. Cell Growth & Differen. **1:** 325-331.
20. Kalebic, T., M. Tsokos & L. J. Helman. 1995. Cancer Res. In press.
21. Furlanetto, R. W., S. E. Harwell & R. B. Baggs. 1993. Cancer Res. **53:** 2522-2526.
22. Baserga, R., C. Sell, P. Porcu & M. Rubini. 1994. Cell Proliferation **27:** 63-71.
23. Pietrzkowski, Z., D. Wernicke, P. Porcu, B. A. Jameson & R. Baserga. 1992. Cancer Res. **52:** 6447-6451.
24. Pietrzkowski, Z., G. Mulholland, L. Gomella, B. A. Jameson, D. Wernicke & R. Baserga. 1993. Cancer Res. **53:** 1102-1106.
25. Reiss, K., A. Ferber, S. Travali, P. Porcu, P. D. Phillips & R. Baserga. 1991. Cancer Res. **51:** 5997-6000.
26. Porcu, P., A. Ferber, Z. Pietrzkowski, C. T. Roberts, Jr., M. Adamo, D. LeRoith & R. Baserga. 1992. Mol. Cell. Biol. **12:** 5069-5077.
27. Sell, C., M. Rubini, R. Rubin, J. P. Liu, A. Efstratiadis & R. Baserga. 1993. Proc. Natl. Acad. Sci. USA **90:** 11217-11221.
28. Kaleko, M., W. J. Rutter & A. D. Miller. 1990. Mol. Cell. Biol. **10:** 464-473.
29. Prager, D., H.-L. Li, S. Asa & S. Melmed. 1994. Proc. Natl. Acad. Sci. USA **91:** 2181-2185.
30. Stracke, M. L., J. D. Engel, L. W. Wilson, M. M. Rechler, L. A. Liotta & E. Schiffman. 1989. J. Biol. Chem. **264:** 21544-21549.

Mitogenic Potential of Insulin on Lymphoma Cells Lacking IGF-1 Receptor[a]

DVORAH ISH-SHALOM,[b] GURI TZIVION,[b]
CLAUS T. CHRISTOFFERSEN,[c] BIRGITTE URSØ,[c]
PIERRE DE MEYTS,[c] AND DAVID NAOR[b]

[b]*The Lautenberg Center for General and Tumor Immunology*
The Hebrew University
Hadassah Medical School
Jerusalem 91120, Israel

[c]*The Hagedorn Research Institute*
Niels Steensens
Vej 6, DK-2820 Gentafte Denmark

INTRODUCTION

Insulin and insulin-like growth factor I (IGF-I) are members of the same hormone family. However, their physiological roles were originally thought to be quite different. Insulin was regarded as a regulator of cellular carbohydrate, lipid, and protein metabolism,[1-3] whereas IGF-I was considered a regulator of cell growth and differentiation.[4,5] Subsequently, insulin and IGF-I were shown to regulate metabolic, morphogenic, and mitogenic responses similarly in a variety of cell types.[6,7] The high degree of homology between insulin and IGF-I, on the one hand, and their receptors, on the other,[8] raises the question of whether the common cellular functions are a result of cross-binding of the ligands to the receptors or of a similar postreceptor signaling mechanism (with possible common components). The question of whether insulin is physiologically a growth factor is still controversial.

The results of several studies, using four different cell lines, suggested that insulin stimulates growth by binding to its own receptor.[9-13] Binding analysis convincingly demonstrated that even at minimal insulin concentration required to stimulate growth, IGF-I was a poor competitor for insulin binding. However, one cannot rule out the possibility of some interplay between the insulin receptor and the IGF-I receptor, which is present in CHO-K1 and RPMI-8226 cells[11,12] and which may be present in an altered "nonbinding" form in H-35 and F9 cells.[9,10] The finding that the two receptors are also present in a hybrid form[14] lends support to the possibility that interplay may alter signaling properties of the insulin receptor. However, the murine

[a]This work was supported by the Society of Research Associates of the Lautenberg Center and the Concern Foundation of Los Angeles.

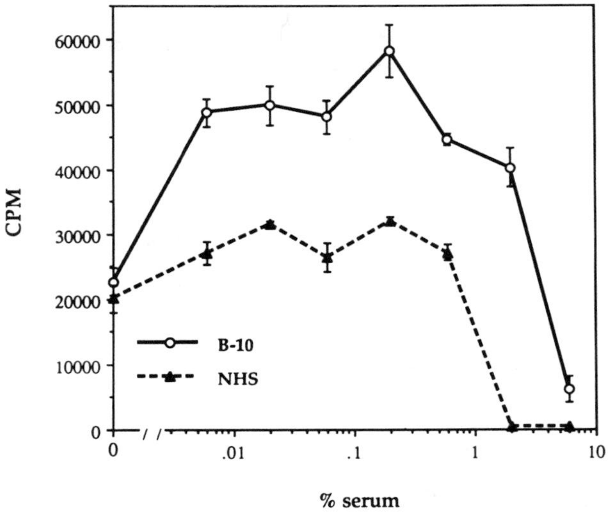

FIGURE 1. LB cells were incubated in medium containing 0.5% FCS and different dilutions of the polyclonal antiinsulin receptor antibody, B-10, or of pooled normal human serum (NHS). The proliferative response was measured, as described in ref. 16, by [^{3}H]thymidine incorporation.

insulin-dependent T-cell lymphoma, LB, characterized in our laboratory, demonstrates the mitogenic effect of insulin in the absence of the IGF-I receptor.

THE MITOGENIC EFFECT OF INSULIN IN LB CELLS IS MEDIATED THROUGH THE INSULIN RECEPTOR

LB is a T-cell lymphoma, which arose spontaneously in a BALB/c mouse.[15] *In vitro* proliferation of LB cells in a serum-free medium was markedly enhanced by physiological concentrations (1-5 ng/mL) of insulin but not of IGF-I or IGF-II.[16] In *in vivo* studies, we found that a low-energy diet or streptozotocin-induced insulin-dependent diabetes produced resistance to the lymphoma growth in mice.[17] This growth resistance was overcome by reconstituting the diabetic mice with external insulin, indicating that insulin enhances the growth of LB cells also *in vivo*. It had earlier been suggested that some insulin responses may not require insulin binding to the plasma membrane receptor[18] but that, rather, the insulin is internalized by fluid-phase endocytosis and accumulates in the cell nucleus. However, the insulin receptor is directly involved in the mitogenic effect evoked by insulin in LB cells. As demonstrated in FIGURE 1, the polyclonal antiinsulin receptor antibody, B-10,[19,20] induces a proliferative response in these cells. Cross-linking of the receptor by its antibody is thus a sufficient signal for cell proliferation.

Because B-10 does not interact with the IGF-I receptor,[20] this finding also excludes the possibility of IGF-I receptor involvement in the mitogenic response. RNase protection assays and ligand-binding experiments confirmed that the IGF-I receptor

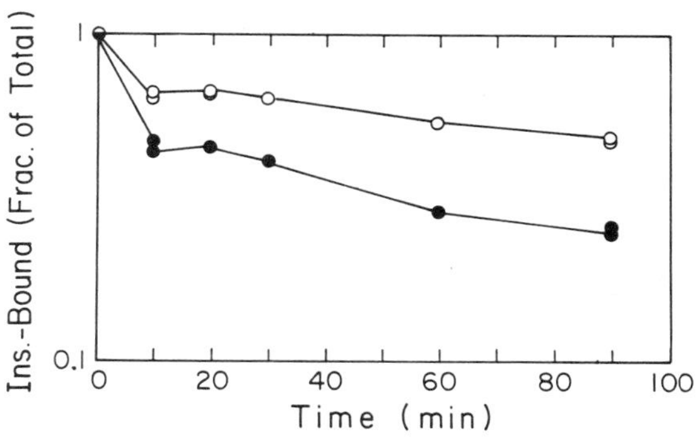

FIGURE 2. Cells (1×10^7/mL) were incubated for 4 h at 4°C in Hepes buffer at pH 7.6 with [^{125}I]insulin (2.5×10^{-11}M). After centrifugation, the cells were washed twice with ice cold Hepes buffer. Dissociation of bound [^{125}I]insulin was then initiated by a 40-fold dilution in the presence or absence of unlabeled insulin (1 μg/mL). At the indicated time points, duplicate aliquots were removed, layered on top of 200 μL ice cold assay buffer in 0.5 mL microfuge tubes, and centrifuged at 10,000 *g* for 2 minutes. The supernatants were aspirated and discarded, and radioactivity in the pellet was counted.

is not involved. We found mRNA for the IGF-I receptor in the brain and thymus of BALB/c mice, whereas no IGF-I mRNA was detected in the same amount of total RNA from LB cells.[21] We found no specific binding of [^{125}I]IGF-I to LB cells.[21] These results establish the absence of an IGF-I receptor in LB cells.

CHARACTERIZATION OF THE LB INSULIN RECEPTOR

In view of the finding that an enhanced mitogenic effect can result from an elevated number of insulin receptors[22] or from changes in their binding properties,[23] we decided to investigate the characteristics of the cell-surface insulin receptor in LB cells.[21] Insulin-binding studies revealed that the total number of binding sites was 3000, corresponding to 1500 receptors per LB cell. The curve for competition of [^{125}I]insulin by increasing amounts of unlabeled insulin gave an IC_{50} of 0.5 nM and an excellent fit to the model of negative cooperativity previously described.[24] As shown in FIGURE 2, the dissociation of [^{125}I]insulin was indeed accelerated by unlabeled insulin. The effect of unlabeled insulin on the dissociation kinetics was studied over a wide range of insulin concentrations (0 to 105 ng/mL) and found to be identical to that described before, with a maximum effect obtained at 100 nM unlabeled insulin. The pH dependence of binding demonstrated a bell-shaped curve, with an optimum at pH 7.6, as found for other cell types.[25] The LB insulin receptor also shows a typical dependence on temperature.[25] This set of experiments shows that LB cells have a small number of insulin receptors and that these receptors exhibit normal binding properties.[21]

It has been postulated that the carboxyterminal region of the insulin receptor is involved in the negative regulation of the mitogenic activity.[26,27] Any mutation in this stretch might, therefore, result in increased mitogenesis. To test this possibility and the possible occurrence of other mutations, the LB-cell insulin-receptor cDNA was cloned and sequenced.[21] It proved to be fully homologous with the previously documented mouse insulin receptor.[28]

We concluded that neither mutation nor alteration of receptor-binding properties (which might result from interaction with other membrane components[29]) could account for the observed mitogenic effect.[21] The fact that LB cells lack the IGF-I receptor on the one hand, and have an insulin receptor that exhibits normal binding properties and amino acid sequence on the other, renders this lymphoma an optimal model for studying the signaling pathway of insulin's mitogenic effect by way of its own receptor.

INSIGHT INTO THE INSULIN-RECEPTOR SIGNAL TRANSDUCTION PATHWAYS

Although the detailed biochemical pathway by which insulin elicits its cellular responses is still unresolved, some of its components have been established (for review, see refs. 30,31): activation of the receptor by insulin stimulates autophosphorylation, which strongly activates the receptor. As a result, the receptor phosphorylates insulin-receptor substrate 1 (IRS1), which functions as a tyrosine-phosphorylated docking protein to recruit several SH2-containing proteins into a large signaling complex. Among these proteins are the p85 regulatory subunit of the phosphatidylinositol-3-kinase (PI 3-kinase)[32] and Grb2, a cytoplasmic protein, known to be an important link in the chain of events connecting tyrosine kinases to Ras activation.[33,34] Grb2 in the complex is associated with at least two more proteins, Shc and the nucleotide exchange factor for Ras, mSOS. Formation of the IRSI-Grb2-SOS complex results in Ras activation, which in turn triggers activation of downstream cytosolic serine/threonine kinases, including MAP kinase, MAP kinase kinase, RSK, and Raf-1.[35–38] Some of the signaling components described above were analyzed in LB cells.

Western blot analysis using an anti-IRS1 and PCR assay showed that neither the protein nor the mRNA for IRS1 is present in LB cells. We noticed, however, that a 170 kDa protein is phosphorylated in LB cells after insulin stimulation.[39] This protein may well be the previously described 4PS protein that substitutes for the IRS1 in hematopoietic cells.[40] The 170 kDa protein of LB cells was coprecipitated with both anti-Grb2 and anti-Shc antibodies after insulin stimulation,[39] suggesting the formation of a signaling complex, as described above.

Mitogen-activated protein kinase MAPK1 and 2 were identified in LB cells by Western blotting.[39] Using myelin basic protein as the substrate, we found that MAPK1 and 2 were activated by insulin to about twice the basal level.

When measuring the level of Ras-GTP in LB cells, after 20 hours of starvation (1% serum), we noticed that up to 50% of the Ras molecules were activated under these conditions. Inasmuch as proliferation of LB cells in 1% serum is negligible, the high degree of activated Ras was very surprising to us. This result may indicate that activated Ras is insufficient for inducing mitogenesis in LB cells. Therefore, other signaling components must be induced by insulin to elicit a proliferative response. The nature of these "other components" is still unclear.

One possible candidate may be the PI 3-kinase whose activation by insulin was studied in LB cells and found to be six times the basal level.[39] Although several growth factors, including PDGF, IL-4, and insulin, activate the PI 3-kinase,[41,42] the involvement of this enzyme in mitogenicity is still unclear. Its possible role in LB cells still has to be established.

The phorbol ester phorbol 12-myristate 13-acetate (PMA), a potent activator of protein kinase C (PKC), has a strong mitogenic effect in LB cells. Therefore PKC is another candidate suspected of mediating the insulin mitogenic effect in LB cells. The role of PKC in insulin signal transduction has been much debated.[43] Three lines of experiments negate the possible role of PKC as an insulin mediator in LB cells[44]: (a) Proliferation assays showed that PMA acts in an additive manner with insulin, even at optimal concentrations of both (indicating separate pathways). (b) Down-regulation of PKC did not affect the mitogenic response of LB cells to insulin. (c) Staurosporine, a PKC inhibitor, is much more potent in inhibiting PMA-induced proliferation of LB cells than insulin-induced proliferation.

We are now studying more signaling components in LB cells in an attempt to decipher the sequence of events leading to insulin-induced mitogenesis.

SUMMARY AND CONCLUSION

We have characterized an insulin-dependent T-cell lymphoma, LB, devoid of IGF-I receptor, which undergoes insulin stimulation and cell proliferation both *in vitro* and *in vivo*. In these cells, the mitogenic response can be evoked only through binding of insulin to its own receptor. This lymphoma is thus a good model for studying the molecular mechanisms involved in insulin mitogenicity.

The high level of activated Ras in LB cells, even under nonproliferative conditions, shows that activation of Ras is insufficient for mitogenicity. It has been suggested earlier that separate pathways of signal transduction may emerge from Ras.[45] The decision to activate a certain signaling pathway may depend on the activation state of other signaling routes in the cell. This may be the case in LB cells, where a signaling component activated by insulin works in concert with the Ras signaling pathway to induce mitogenesis. Yet it is still unclear whether activated Ras is a prerequisite for the insulin-induced response in LB cells.

REFERRENCES

1. Rosen, O. M. 1987. Science **237:** 1452-1458.
2. Czech, M. P. 1985. Annu. Rev. Physiol. **47:** 357-381.
3. Kahn, C. R. 1985. Annu. Rev. Med. **36:** 429-451.
4. Rechler, M. M., S. P. Nissley & J. Roith. 1987. N. Engl. J. Med. **316:** 941-942.
5. Isaksson, O. G. P., A. Lindahl, A. Nilsson & J. Isgaard. 1987. Endocr. Rev. **8:** 426-438.
6. Zapf, J. & E. R. Froesch. 1986. Horm. Res. **24:** 121-130.
7. Nissley, S. P. & M. M. Rechler. 1984. Clin. Endocrinol. Metab. **13:** 43-68.
8. Ullrich, A., A. Gray, A. W. Tam, T. Yang-Feng, M. Tsubokawa, C. Collins, W. Henzel, T. Le Bon, S. Kathuria, E. Chen, S. Jacobs, U. Francke, J. Ramachandran & Y. Fujita-Yamaguchi. 1986. EMBO J. **5:** 2503-2512.

9. Massague, J., B. Blinderman & M. P. Czech. 1982. J. Biol. Chem. **257:** 13958-13963.
10. Nagarojan, L. & W. B. Anderson. 1982. Biochem. Biophys. Res. Commun. **106:** 974-980.
11. Mamounas, M., D. Gervin & E. Englesberg. 1989. Proc. Natl. Acad. Sci. USA **86:** 9294-9298.
12. Freund, G. G., D. T. Kulas & R. A. Mooney. 1993. J. Immunol. **151:** 1811-1820.
13. Moses, A. C. & S. Irazaki. 1991. Is insulin a growth factor? In Insulin-like Growth Factors: Molecular and Cellular Aspects. D. LeRoith, Ed. 245-270. CRC Press. Boca Raton, FL.
14. Moxham, C. P. & S. Jacobs. 1992. J. Cell Biochem. **48:** 136-140.
15. Lugasi, H., S. Hajos, J. R. Murphy, T. B. Strom, J. Nichols, C. Penarroja & D. Naor. 1990. Int. J. Cancer **45:** 163-167.
16. Pillemer, G., H. Lugasi-Evgi, G. Scharovsky & D. Naor. 1992. Int. J. Cancer **50:** 80-85.
17. Sharon, R., G. Pillemer, D. Ish-Shalom, R. Kalman, E. Ziv, E. Berry & D. Naor. 1993. Int. J. Cancer **53:** 843-849.
18. Lin, Y. J., S. Harada, E. G. Loten, R. M. Smith & L. Jarett. 1992. Proc. Natl. Acad. Sci. USA **89:** 9691-9694.
19. Takayama-Hasumi, S., K. Tobe, K. Momomura, O. Koshio, Y. Tashiro-Hashimoto, Y. Akanuma, Y. Hirata, F. Takaku & M. Kasuga. 1989. J. Clin. Endocrinol. Metab. **68:** 787-788.
20. Jonas, H. A. & L. C. Harrison. 1985. J. Biol. Chem. **260:** 2288-2294.
21. Ish-Shalom, D., C. T. Christoffersen, P. Vorwerk, N. Sacerdoti-Sierra, R. M. Shymko, D. Naor & P. De Meyts. Submitted.
22. Hoffman, C., I. D. Goldfine & J. Whittaker. 1989. J. Biol. Chem. **264:** 8606-8611.
23. De Meyts, P., C. T. Christoffersen, D. Ish-Shalom, K. Drejer, L. Schaffer, R. M. Shymko & D. Naor. 1993. American Diabetes Association Symposium (Abstract).
24. De Meyts, P., A. R. Bianco & J. Roth. 1976. J. Biol. Chem. **251:** 1877-1888.
25. Waelbroeck, M., E. Van Obberghen & P. De Meyts. 1979. J. Biol. Chem. **254:** 7736-7740.
26. Thies, R. S., A. Ulrich & D. A. McClain. 1989. J. Biol. Chem. **264:** 12820-12825.
27. Akifumi, A., K. Momomura, K. Tobe, R. Yamamoto-Horda, H. Sakura, Y. Tamori, Y. Kaburagi, O. Koshio, Y. Akanuma, Y. Yazaki, M. Kasuga & T. Kadowaki. 1992. J. Biol. Chem. **267:** 12788-12796.
28. Flores-Riveros, J. R., E. Sibley, T. Kastelic & M. D. Lane. 1989. J. Biol. Chem. **264:** 21552-21572.
29. Jarrett, J. C., G. Ballejo, T. H. Saleem, J. C. M. Tsibrir & W. N. Spellacy. 1984. Am. J. Obstet. Gynecol. **149:** 250-255.
30. Myers, M. G., X. J. Sun & M. F. White. 1994. TIBS **19:** 289-293.
31. Kahn, C. R., M. F. White, S. E. Shoelson, J. M. Backer, E. Araki, B. Cheatham, P. Csermely, F. Folli, B. J. Goldstein, P. Huertas, P. L. Rothenberg, M. J. A. Saad, K. Siddle, X. J. Sun, P. A. Wilden, K. Yamada & S. A. Kahn. 1993. Recent Prog. Horm. Res. **48:** 291-339.
32. Ruderman, N. B., R. Kapeller, M. F. White & L. C. Cantley. 1990. Proc. Natl. Acad. Sci. USA **87:** 1411-1415.
33. Lowenstein, E. J., R. J. Daly, A. G. Batzer, W. Li, B. Margolis, R. Lammers, A. Ullrich, E. Y. Skolnick, D. Bar-Sagi & J. Schlessinger. 1992. Cell **70:** 431-442.
34. Skolnik, E. Y., C. H. Lee, A. Batzer, L. M. Vicentini, M. Zhou, R. Daly, M. J. Myers, J. M. Backer, A. Ullrich, M. F. White & J. Schlessinger. 1993. EMBO J. **12:** 1929-1936.
35. Roberts, T. M. 1992. Nature **360:** 534-535.
36. Thomas, G. 1992. Cell **68:** 3-6.
37. Blenis, J. 1993. Proc. Natl. Acad. Sci. USA **90:** 5889-5892.

38. Crews, C. M. & R. L. Erickson. 1993. Cell **74:** 215-217.
39. Ursø, B. *et al.* In preparation.
40. Wang, L.-M., G. Myers, X. J. Sun, S. A. Aaronson, M. White & J. H. Pierce. 1993. Science **261:** 1591-1594.
41. Whitman, M., C. P. Downes, T. K. Keeler & L. Cantley. 1988. Nature **332:** 644-646.
42. Heidaran, M. A., J. H. Pierce, D. Lombardi, M. Ruggiero, J. S. Gutkind, T. Matsui & S. A. Aaronson. 1991. Mol. Cell. Biol. **11:** 134-142.
43. Considine, R. V. & J. F. Caro. 1993. J. Cell Biochem. **52:** 8-13.
44. Tzivion, G., B. Ursø, Y. Shechter, P. De Meyts & D. Naor. Submitted.
45. Porras, A., K. Muszynski, U. R. Rapp & E. Santos. 1994. J. Biol. Chem. **269:** 12741-12748.

Platelet-derived Growth Factor

Distinct Signal Transduction Pathways Associated with Migration versus Proliferation[a]

KARIN E. BORNFELDT,[b] ELAINE W. RAINES,[b]
LEE M. GRAVES,[c] MICHAEL P. SKINNER,[b,d]
EDWIN G. KREBS,[c] AND RUSSELL ROSS[b]

Departments of [b]Pathology and [c]Pharmacology
University of Washington
Seattle, Washington 98195

INTRODUCTION

PDGF (platelet-derived growth factor) is a potent stimulant of connective-tissue cell proliferation and migration, such as that of smooth muscle cells (SMC) and fibroblasts. PDGF is a dimer composed of two highly homologous peptide chains (PDGF A and PDGF B) joined by disulfide bonds, and three different forms of PDGF dimers exist *in vivo*: PDGF-AA, PDGF-AB, and PDGF-BB. In analogy with the different PDGF chains, two different high-affinity PDGF receptor subunits, the α-subunit and the β-subunit, are expressed by many cell types. The PDGF receptor α-subunit binds both the PDGF A- and B-chain, whereas the β-subunit binds only the B-chain.[1] Thus, PDGF-BB can bind to both PDGF α- and β-receptors, whereas PDGF-AA binds only the α-receptor. Upon binding of a PDGF dimer to its receptor, two appropriate PDGF receptor subunits are brought together, and a number of specific tyrosine residues on the receptor subunits are phosphorylated.[2] This receptor tyrosine autophosphorylation is manifested preferentially as a transphosphorylation between the two receptor dimers[3] and creates binding sites for several molecules that interact with specific phosphotyrosines through their SH2 (src homology 2) domains.[4] Currently, phospholipase Cγ (PLCγ), Ras GTPase-activating protein (GAP), the regulatory subunit of phosphatidylinositol 3-kinase (p85 of PI 3-kinase), growth factor receptor-bound protein 2 (GRB2), the tyrosine-specific phosphatase Syp, Src homology and collagen protein (Shc), and members of the Src family have been

[a]This work was supported in part by Grants from the National Heart, Lung, and Blood Institute of the National Institutes of Health, Grants HL-18645 and HL-03174 to Dr. R. Ross, Grants DK 42528 and GM 42508 from the National Institutes of Health, a Grant from the Muscular Dystrophy Association to Dr. E. G. Krebs, and a Grant from the American Heart Association to Dr. L. M. Graves.

[d]Current address: Department of Cardiology, Westmead Hospital, Westmead, NSW, 2145, Australia.

found to bind specific phosphotyrosines in the activated PDGF β-receptor.[2] Receptor dimerization and the subsequent receptor autophosphorylation and tyrosine-kinase activation are required for transduction of the signals leading to proliferation and migration.[5] The ability of homo- and heterodimers of PDGF to induce mitogenesis and migration depends on both the PDGF dimer present and the relative numbers of the two different PDGF receptor subunits on the responding cell.[6] Activation of either the α- or the β-receptor leads to migration and proliferation in human SMC;[6] however, the expression of β-receptors is typically about 10-fold higher than the expression of α-receptors,[7] and, thus, PDGF-BB is a stronger inducer of migration and proliferation compared to PDGF-AA in these cells.

It has been suggested that PDGF plays a role in normal physiological processes, as well as in a number of pathological conditions that are caused by excessive fibroproliferative responses, such as atherosclerosis, rheumatoid arthritis, pulmonary fibrosis, myelofibrosis, and abnormal wound repair.[8] One of the pathological conditions where the role of PDGF has been well studied, by our laboratory and others, is atherosclerosis. Accumulation of SMC is a key event in the formation and progression of lesions of atherosclerosis.[9] The accumulation of SMC is due to a combination of directed migration from the media into the intima of the artery accompanied by proliferation and possibly decreased apoptosis. PDGF dimers, as well as both PDGF receptor α- and β-subunits are expressed by cells in the vascular wall. In atherosclerotic lesions, a major source of PDGF-BB is activated macrophages, although SMC, endothelial cells, other cells from the circulation, and platelets can express and secrete PDGF dimers. PDGF-BB has been demonstrated in at least 20–25% of macrophages that comprise the different lesions of atherosclerosis,[10] and SMC in close proximity to these macrophages have been found to proliferate to a larger extent than do surrounding SMC. Furthermore, SMC in atherosclerotic lesions show higher expression of PDGF β-receptors,[8] suggesting a possible role for PDGF-BB in proliferation of these SMC.

A clear role for PDGF in mediating migration of arterial SMC *in vivo* has been established in balloon catheterization of the normal rat carotid artery.[11–13] In these experiments, passage of an inflated balloon through the artery removes the endothelial cells, and, in expanding the artery, injures many of the smooth muscle cells located in the innermost layer of the media of the artery. This is followed by proliferation of many medial SMC (starting one day after injury), followed by their migration from the media through the internal elastic lamina into the intima (4–7 days after injury), after which proliferation of the intimal SMC occurs. Together, these events result in a neointima that is as thick as the media two weeks after injury.[14,15] Daily administration of anti-PDGF antibodies one day prior to, and following, balloon injury causes 85% reduction in the intimal SMC (primarily SMC that have migrated from the media[13]) on day 4 and 40% reduction in intimal thickness when examined after 8 days. However, no difference in DNA synthesis (measured as [^{3}H]thymidine labeling index) was found between days 7 and 8.[11] In separate experiments, continuous infusion of PDGF-BB and [^{3}H]thymidine into rats subjected to a gentle denudation of the carotid artery demonstrated a 21-fold increase in the number of unlabeled cells in the intima and a fourfold increase in thymidine-labeled cells in the media.[12] Thus, the main contribution of PDGF to intimal thickening in this model is most likely the increased migration of SMC from the media into the intima, although the contribution of PDGF to medial proliferation at early time points is not clear.

Because PDGF is such a potent mitogen and chemotactic agent, a number of interesting questions arise. Under what circumstances does PDGF induce directed migration versus proliferation of these cells? Are the signal transduction pathways induced by the activated PDGF receptors that lead to directed migration different from those that lead to proliferation? If so, are the intracellular signal-transduction pathways, and thus the general effect of PDGF, dependent on the "phenotypic state" of the SMC? This review will focus on some of these questions as they relate to signal-transduction pathways involved in proliferative and migratory effects of PDGF in human arterial smooth muscle cells.

PDGF INDUCES BOTH PROLIFERATION AND DIRECTED MIGRATION (CHEMOTAXIS) *IN VITRO*

PDGF-BB and PDGF-AA, and thus both the PDGF α- and β-receptor, are potent inducers of DNA synthesis and proliferation of SMC (FIG. 1). SMC also express high-affinity receptors for insulin-like growth factor I[16,17] (IGF-I), another tyrosine kinase receptor.[18] However, in terms of the concentration required to induce DNA synthesis in human arterial SMC, PDGF-BB is 100-fold more potent than IGF-I, and the maximal response of DNA synthesis stimulated by PDGF-BB is about fourfold that of IGF-I. Accordingly, PDGF-BB increases cell number three days after addition, whereas IGF-I does not.[7] Furthermore, the maximal effects of IGF-I and PDGF-BB on DNA synthesis and proliferation are additive,[19] suggesting different mechanisms of action to induce proliferation. The number of high-affinity binding sites for PDGF-BB in human SMC is approximately 12 and 15 times higher than the number of high-affinity binding sites for PDGF-AA and IGF-I, respectively. The number of PDGF α-receptors is thus comparable to the number of IGF-I receptors, as shown by FIGURE 1. The fact that PDGF-AA is a more potent inducer of DNA synthesis at equimolar concentrations of IGF-I suggests that the mitogenic signaling capacity of PDGF receptors differs significantly from that of the IGF-I receptors.

PDGF has long been recognized as a potent chemoattractant for SMC *in vitro*.[20] Both PDGF-BB and PDGF-AA (and thus both the PDGF α- and β-receptor) are capable of stimulating migration and chemotaxis of SMC and other cell types.[6,21,22] However, the fact that the extent of the migratory effect elicited by PDGF-AA compared to PDGF-BB is dependent on the relative numbers of PDGF α- and β-receptor subunits[6] should be kept in mind. Caution is also required (see TABLE 1) in interpreting data from Boyden chamber assays of cell migration, as the trypsinization procedure affects receptor number for some growth factors very significantly. In spite of some reduction by trypsin of the PDGF receptors, PDGF-BB induces directed migration of human SMC on collagen type I in a dose-dependent manner, with an approximate maximal 150% increase in migration.

Some reports show that PDGF-AA, under certain circumstances, fails to induce migration or inhibits PDGF-BB-induced migration.[23,24] Only the chemotactic effect of PDGF-BB seems to be inhibited by PDGF-AA, without inhibition of the chemokinetic effect.[25] The reason for the apparent inhibitory effects of PDGF-AA is unclear.

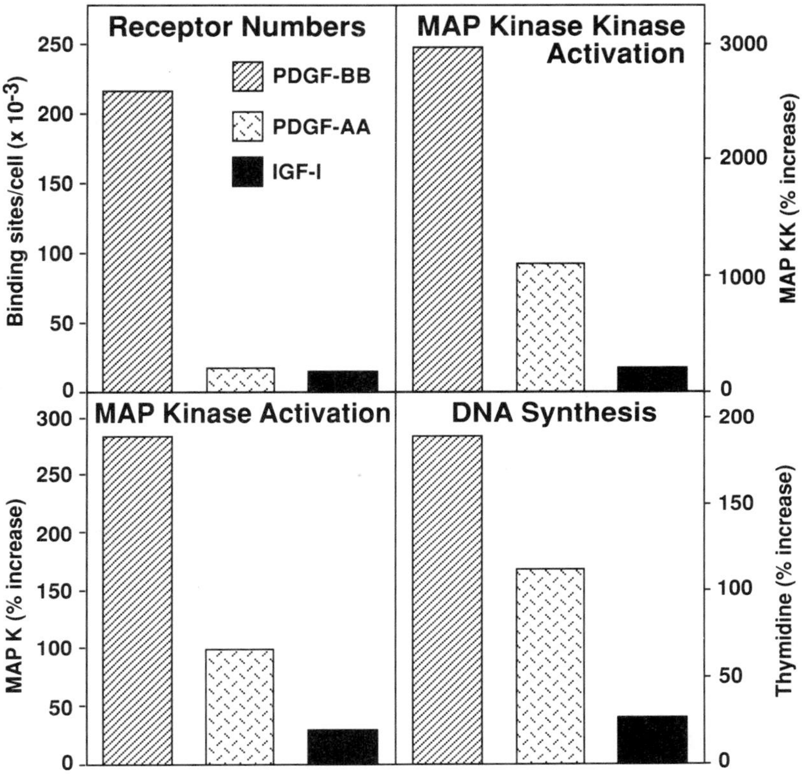

FIGURE 1. PDGF-AA induces activation of MAP kinase kinase and MAP kinase, and stimulates DNA synthesis to a greater extent than IGF-I, despite a similar number of high-affinity binding sites per cell. Human arterial SMC were kept in Dulbecco's modified Eagles medium with 1% human plasma-derived serum 2 days prior to stimulation with 1 nM human recombinant PDGF-BB (Zymogenetics Inc. Seattle, WA), PDGF-AA (Hoffman La Roche Inc., Basel, Switzerland), or IGF-I (UBI, Lake Placid, NY). Enzyme activities and DNA synthesis measurements and Scatchard plots were performed as previously described. (Bornfeldt *et al.*[7] With permission from *The Journal of Clinical Investigation.*)

However, methodological differences, such as the extracellular matrix used in the migration assay, which affects adhesion of the cells, and type of membranes used, may account for the different effects of PDGF-AA.

In contrast to its weak mitogenic effects, IGF-I acts as a potent chemoattractant for SMC.[7] The effect of IGF-I on migration and chemotaxis is mediated through the IGF-I receptor, with no apparent contribution from the IGF-II receptor or insulin receptors. Thus, the migratory effect of IGF-I could be inhibited by an IGF-I receptor-blocking antibody (αIR-3), whereas PDGF-BB-induced migration was not affected by preincubation with this antibody.[7] When tested in the same experiment, IGF-I is typically able to induce approximately 50% of the maximal migratory effect observed

TABLE 1. Trypsinization Decreases High-affinity Binding Sites for PDGF and bFGF[a]

	Nontrypsinized cells		Trypsinized cells	
	Number (per cell)	Kd (pM)	Number (per cell)	Kd (pM)
PDGF-BB	215,500	190	84,500	170
PDGF-AA	17,000	50	8,100	40
IGF-I	14,400	110	15,200	90
bFGF	6,200	4	—	No detectable high-affinity binding

[a] Confluent cell cultures were incubated in Dulbecco's modified Eagles medium with 1% human plasma-derived serum for 2 days before the experiment. Following trypsinization of one set of cells, binding of [^{125}I]PDGF-BB, [^{125}I]PDGF-AA (both labeled using Iodo-Beads; Pierce Chemical Co., Rockford, IL), [^{125}I]IGF-I (Amersham Corp., Arlington Heights, IL), and [^{125}I]bFGF (Du Pont NEN Research Products, Boston, MA) was measured after a 3-h incubation at 4°C as described.[7] The number of high-affinity binding sites per cell and receptor affinities (Kd) were estimated using Scatchard plots.[7]

with PDGF-BB. Although PDGF-BB binding sites are reduced after the trypsinization procedure, the PDGF-receptor number is still approximately five times higher than that of IGF-I. In one strain of human SMC, which had lower levels of PDGF-BB-binding sites, the effect of IGF-I was comparable to that of PDGF-BB. Thus, the lower maximal effect of IGF-I compared to PDGF-BB may result from the lower number of high-affinity binding sites on these cells. In agreement with this is the comparison of maximal migratory effects induced by PDGF isoforms and IGF-I where PDGF-BB > PDGF-AA = IGF-I[7] (our unpublished observations).

PROLIFERATIVE SIGNALS: ACTIVATION OF THE MAP KINASE CASCADE

Many growth-factor tyrosine-kinase receptors activate a signal-transduction pathway that includes recruitment of GRB2 and the Ras GTP exchange molecule Sos through an SH2 domain in GRB2 that interacts with the activated, phosphorylated receptor. These events further lead to conversion of inactive Ras·GDP to active Ras·GTP; activation of Raf, MAP kinase kinase, and MAP kinase; and activation of transcription factors and eventually DNA synthesis and proliferation or differentiation.[26] The PDGF receptor has been shown to bind GRB2,[27,28] and high levels of GRB2 are expressed by human SMC (Graves & Bornfeldt, unpublished observation), supporting the existence of this pathway in these cells. Indeed, PDGF-BB evokes a marked activation of MAP kinase kinase and MAP kinase (Erk1 and Erk2) in human

SMC. The time course of activation of these enzymes shows a peak of activity 5 minutes after stimulation, with a return to basal-enzyme activity 20–30 minutes after activation.[7]

Studies on point mutated PDGF β-receptors show that by inhibiting the ability of specific tyrosines to bind to their effector molecules, DNA synthesis is inhibited to different extents. Point mutation of the direct GRB2-binding site (Tyr 716 in the kinase insert of the human PDGF β-receptor) results in inhibition of PDGF-induced Ras·GTP formation but does not significantly affect PDGF-stimulated MAP kinase activation or DNA synthesis.[28] Although GRB2 also binds indirectly to the PDGF receptor, these results indicate that GRB2 coupling to Ras is not required for MAP kinase activation and DNA synthesis in this system. Furthermore, PDGF-induced activation of PLCγ is not consistently required for DNA synthesis.[29,30] Point mutation of the PI 3-kinase binding site (Tyr 740 and 751 in the human PDGF β-receptor) has implied a role for PI 3-kinase in PDGF-induced DNA synthesis,[30,31] although this effect seems to be independent of MAP kinase activation.[28] Point mutation of the PI 3-kinase binding tyrosines has also been reported to inhibit both PDGF activation of the p70 S6 kinase,[32] membrane ruffling, and chemotaxis[33] as well as PDGF-receptor internalization.[34] In a very recent study, however, point mutation of Tyr 740 abolished PI 3-kinase activation without inhibiting the p70 S6 kinase, suggesting that PI 3-kinase activity is not required for p70 S6 kinase activation, but rather a putative signaling molecule, such as Nck, that docks to Tyr 751.[35] The diverse effects of PI 3-kinase may be in agreement with its suggested role in intracellular protein trafficking[36] or may be dependent on expression of different PI 3-kinase-like enzymes in different cell types. Currently, no PDGF receptor mutants unable to activate the MAP kinase pathway have been constructed,[28] leaving the effect of MAP kinase on proliferation in this system unexplored.

Unlike PDGF-BB, IGF-I is not able to induce activation of the MAP kinase cascade in SMC. No significant activation of the MAP kinase kinase or MAP kinase can be demonstrated at any dose of IGF-I tested, or at any time up to 60 minutes after stimulation. By contrast, PDGF-AA, which has a receptor number similar to that of IGF-I on human SMC[7], activates the MAP kinase kinase and MAP kinase by 66% and 88% of the maximal activation observed with PDGF-BB 5 minutes after stimulation.[37] Thus, the meager ability of IGF-I to activate the MAP kinase cascade in these cells is not due to the lower number of receptors per cell, but is due to a different signaling capacity of IGF-I compared to PDGF (FIG. 1). In other cell types, such as bovine chromaffin cells and NIH 3T3 cells, IGF-I is capable of activating the MAP kinase cascade,[38] although activation is 100- to 1000-fold more efficiently coupled to PDGF-receptor activation.[39] The IGF-I receptor may activate the MAP kinase cascade through phosphorylation of insulin-receptor substrate 1 (IRS-1), which subsequently binds GRB2 and activates Sos, Ras, and downstream enzymes.[40] Thus, the ability of IGF-I to induce activation of the MAP kinase cascade and proliferation may depend on expression of IRS-1 (or other IRS isoforms) in addition to expression of IGF-I receptors. Taken together, activation of the MAP kinase kinase and MAP kinase in human SMC correlates with the ability of specific growth factors to stimulate proliferation.

ANTAGONISM OF PDGF-INDUCED MAP KINASE SIGNALING BY PROTEIN KINASE A

The cAMP-dependent protein kinase (protein kinase A; PKA) is activated by an elevation of cyclic AMP in response to various stimuli, such as β-adrenergic receptor stimulation by, for example, epinephrine, stimulation by prostaglandin E2, or direct activation of adenylate cyclase by forskolin. Activation of PKA antagonizes MAP kinase signaling from several tyrosine growth-factor receptors. We have investigated PDGF receptor-induced MAP kinase signaling with respect to PKA antagonism. Elevation of levels of cAMP with forskolin in human SMC does not inhibit tyrosine autophosphorylation of the PDGF receptor, indicating that the point of inhibition is downstream of PDGF-receptor activation.[37] In agreement with these results is also the fact that elevation of cAMP inhibits MAP kinase signaling from other tyrosine-kinase receptors, such as the EGF receptor[41,42] and the insulin receptor.[43] MAP kinase (Erk1 and Erk2) from extracts of forskolin-treated cells can be activated by purified MAP kinase kinase *in vitro,* and incubation of MAP kinase (Erk2) with PKA-catalytic subunits does not inhibit the ability of MAP kinase to phosphorylate myelin basic protein, showing that PKA does not directly inhibit MAP kinase. Furthermore, MAP kinase kinase in extracts from forskolin-treated cells can be activated by Raf-1 *in vitro,* and PKA-catalytic subunits do not phosphorylate MAP kinase kinase or inhibit its ability to activate MAP kinase (Erk2). Thus, the point of inhibition of the MAP kinase cascade by PKA appears to be upstream of the MAP kinase and the MAP kinase kinase.

EGF-receptor association with GRB2 and EGF-dependent accumulation of Ras·GTP are not affected by cAMP in Rat-1 cells, whereas EGF-induced activation of Raf-1 and MAP kinase is inhibited, suggesting that the point of inhibition by cAMP is at the level of Raf-1 activation in these cells.[42] It has been shown that PKA can phosphorylate Raf-1 and that the phosphorylation reduces the affinity with which Raf-1 binds Ras.[41] However, phosphorylation of Raf-1 with PKA catalytic subunits *in vitro* does not lead to inhibition of the ability of Raf-1 to activate MAP kinase kinase (Graves *et al.* unpublished observation). Recently, it was reported that PKA inhibits both MEK kinase (an activator of MAP kinase kinase, distinct from Raf) and Raf activation by EGF in PC12 cells.[44] Both Raf and MEK kinase are activated by Ras, and a likely mechanism of PKA inhibition therefore seems to be a reduced ability of Ras to interact with its targets.[45,46] However, alternative mechanisms of inhibition of Raf by PKA are indicated by the fact that v-Raf or the isolated Raf-1 kinase domain, which lacks the Ras binding domain, is still inhibited by PKA.[46]

Consistent with the effects of PKA on MAP kinase signaling, activation of PKA by forskolin or other agents, inhibits PDGF-induced DNA synthesis in SMC.[37,47] Cyclic AMP inhibits proliferation in SMC without inhibiting calcium mobilization.[48] Interestingly, in human SMC, PDGF activation of PLCγ does not seem to be inhibited by cAMP, inasmuch as PDGF-induced accumulation of inositol monophosphate is unaltered following forskolin treatment.[37] Thus, PKA selectively inhibits the MAP kinase cascade and DNA synthesis, without inhibiting PLCγ activation in these cells. The inhibitory effect of PKA on PDGF-induced DNA synthesis, however, is not solely due to inhibition of the PDGF-induced MAP kinase signaling, because DNA synthesis is inhibited by forskolin, even when added 1-6 h after PDGF-BB, at a time point when MAP kinase activity has declined to basal levels[7] (Bornfeldt and Graves, unpublished observation).

SIGNALS INVOLVED IN PDGF-INDUCED MIGRATION: PHOSPHATIDYLINOSITOL TURNOVER

PLCγ catalyzes the hydrolysis of phosphatidylinositol 4,5 bisphosphate (PIP_2) to inositol 1,4,5 trisphosphate (IP_3) and diacylglycerol (DAG). Generation of IP_3 leads to mobilization of calcium from intracellular stores, whereas DAG activates different isoenzymes of the protein kinase C (PKC) family. PLCγ binds to phosphotyrosines 1009 and 1021 in the human PDGF β-receptor[29] and has also recently been shown to bind to the PDGF α-receptor.[2] Binding of PLCγ to the receptor leads to increased ability of the enzyme to hydrolyze PIP_2.

PDGF-BB is a potent inducer of phosphatidylinositol turnover in human SMC.[7] Accordingly, an early (within 30 seconds) decrease in PIP_2 lipid mass with a reciprocal increase in DAG lipid mass is observed following PDGF-BB stimulation. Consistent with the hydrolysis of PIP_2, PDGF-BB causes an early (within 10–40 seconds), dose-dependent increase in intracellular calcium levels in human arterial SMC. Stimulation of SMC with PDGF-AA results in increased phosphatidylinositol turnover, but the maximal effect of PDGF-AA is lower than that of PDGF-BB, in agreement with the lower number of PDGF α-receptors on these cells (TABLE 1). It has previously been shown that both the PDGF α- and β-receptor can independently induce phosphatidylinositol turnover and chemotaxis.[22] Results obtained from point-mutated PDGF β-receptors show that mutation of Tyr 1009 and Tyr 1021 markedly reduces PDGF-stimulated phosphorylation of PLCγ without inhibiting phosphorylation of MAP kinase, Erk1 and Erk2,[49] or PDGF-induced DNA synthesis.[29] However, redundant signal transduction pathways exist, as shown by the fact that PLCγ can induce DNA synthesis in add-back mutant PDGF β-receptors unable to bind RasGAP, PI 3-kinase, or the phosphatase Syp.[30] Recently, the same approach has been used to study the involvement of PLCγ in PDGF-induced migration. PDGF β-receptors unable to bind PLCγ (Tyr 1021 mutations) showed a reduced ability to migrate toward PDGF-BB, although binding of PLCγ was not sufficient if the receptor could not bind PI 3-kinase.[50] Furthermore, a mutated PLCγ that suppresses endogenous PLC activity and that is unable to associate with the activated PDGF receptor reduced PDGF-induced migration with no effect on basal migration.[50] These studies show that although signal transduction pathways are often redundant, PLCγ has a role in PDGF-induced migration.

In human arterial SMC, induction of PIP_2 hydrolysis, DAG formation, and calcium mobilization occur at similar levels when the cells are stimulated with IGF-I or PDGF-BB.[7] Although IGF-I does not induce phosphatidylinositol turnover in many cell lines, it is able to do so in some primary cell types.[51,52] PLCγ is not believed to bind to the IGF-I receptor or to IRS-1 involved in many IGF-I signaling events. IGF-I has been described to activate PLCβ in Swiss 3T3 cells, although the localization of the IGF-I-activated PLCβ appears to be mainly nuclear in these cells,[53] and thus is not likely to be involved in the chemotactic response observed in SMC. The PLC activated by IGF-I in SMC still remains to be characterized.

The results from the SMC studies show that the ability of a growth factor to induce phosphatidylinositol turnover correlates with its ability to induce migration and chemotaxis rather than DNA synthesis and proliferation. Thus, PDGF-BB, PDGF-AA, and IGF-I stimulate PIP_2 hydrolysis and migration of SMC[7] (our unpublished observations), whereas only PDGF-BB and PDGF-AA are able to activate the MAP kinase cascade and significantly stimulate DNA synthesis and proliferation (FIG. 1).

PIP_2 hydrolysis and calcium mobilization induced by PDGF are thought to result in a localized transient actin filament disassembly due to activation of proteins that block actin filament barbed ends and cause actin-filament fragmentation (such as gelsolin) and sequestration of actin monomers by proteins such as profilin.[54] The disassembly of actin filaments is probably required for protrusion of filopodia. Later, the actin filament assembly is promoted by PDGF. This may be mediated partly through normalization of PIP_2 and calcium levels, generation of DAG, which increases actin nucleation to plasma membranes,[55] and partly through the small GTP-binding protein Rho.[56] PDGF has been shown to increase levels of RhoB mRNA 15 minutes after stimulation.[57] To complicate matters, the PDGF receptor binds and activates RasGAP (within minutes), which catalyzes the conversion of active Ras·GTP to inactive Ras·GDP, but is also involved in regulation of Rho. It has been shown that RasGAP occurs in a complex with two other proteins (p190 and p62). The protein p190 shows GTPase activity towards Rho.[58] When the Ras GTPase-activating domain is deleted from GAP, it still is able to bind to activated PDGF β-receptors, complex with p190, and possess Rho GAP activity, and its activation leads to disruption of actin stress fibers and reduction in focal adhesion sites.[59] Studies performed with point-mutated PDGF β-receptors show that mutation of Tyr 771 (the GAP binding site) increases migration in response to PDGF-BB[50] without affecting the mitogenic effect of the receptor,[31] indicating a negative effect of GAP on migration. These results show an intricate regulation of actin filament assembly by PDGF, although the exact roles of GAP and the GAP-associated proteins in migration and adhesion are still obscure. Interference with the dynamics of PDGF-stimulated actin filament disassembly and assembly, by favoring or prolonging either actin filament disassembly or assembly, results in inhibition of PDGF-induced migration and chemotaxis.

REQUIREMENT OF INTEGRINS FOR PDGF-INDUCED MIGRATION

Integrins are connective tissue receptors composed of α-β chain heterodimers, which are involved in cell attachment, cell-shape changes, and spreading and migration of various cell types. Increasing evidence suggests that integrins play a role in transmembrane signaling.[60] Characterization of integrin expression in cultured human arterial SMC reveals expression of α2, α3, α5, αv, and β1 integrin subunits. By contrast, SMC freshly isolated from human artery walls express α1, α3, α5, αv, β1, and low levels of β3 integrin subunits.[61] Thus, there is a switch in SMC collagen receptors from α1β1 in the vascular wall to α2β1 in culture. Interestingly, α1β1 binds collagen type IV with higher affinity than collagen type I, whereas α2β1 has a higher affinity for collagen type I.[62] Neither basal nor directed migration on collagen type I occur in cell types devoid of α2β1 integrin receptors, whereas cells that express α2β1 integrins migrate towards PDGF-BB.[61] Furthermore, this migration is inhibited in a dose-dependent manner by preincubation with blocking polyclonal anti-α2β1 antibodies, or blocking monoclonal antibodies directed against each of these subunits, and is magnesium dependent, consistent with α2β1-mediated migration. The α2β1 antibodies do not inhibit attachment of SMC to collagen type I, but do inhibit their spreading and migration. In this context it is interesting that localization of integrins in focal adhesion plaques seems to be required for their ability to mediate cell migration. Thus, both α5β1 and αvβ1 fibronectin receptors transfected into CHO cells allow the cell to attach to fibronectin, but only the α5β1 integrin localizes into focal adhesion plaques and promotes migration.[63] In conclusion, basal and PDGF-induced migration on collagen require functional α2β1 integrins. It is not yet known whether functional integrins provide a permissive state for migration to occur, or if signaling from or to the PDGF receptor is affected by integrins or formation of focal adhesion sites.

In some cell types, PDGF receptor stimulation results in phosphorylation of components of focal adhesion sites, such as the focal-adhesion kinase ($p125^{FAK}$) and paxillin.[64] Activation of members of the Src family by PDGF may also play a role in signaling related to focal adhesion complexes, inasmuch as Src has recently been shown to associate with tyrosine-phosphorylated $p125^{FAK}$.[65] The importance of α2β1 integrins in signal transduction pathways involved in proliferation and migration of SMC is currently being studied in our laboratory.

SUMMARY

FIGURE 2 summarizes our current interpretation of data concerning signals from the activated PDGF receptor involved in directed migration and proliferation of human arterial SMC. Binding of PDGF (PDGF-BB or PDGF-AA) causes PDGF-receptor dimerization, tyrosine autophosphorylation, and subsequent binding of several molecules containing SH2 domains to the activated receptor. Binding and activation of PLCγ by the PDGF receptor leads to PIP_2 hydrolysis, resulting in generation of diacylglycerol (DAG) and IP_3. Subsequently, intracellular levels of calcium are elevated as a result of IP_3-mediated calcium release from intracellular compartments. The decreased levels of PIP_2 and increased levels of calcium both favor actin-filament disassembly by inducing capping of actin-filament barbed ends and actin-monomer sequestration.[54] A localized, and transient, actin-filament disassembly enables the cell to extend filopodia towards PDGF, thereby enabling chemotaxis to take place. At a later time and/or in a different compartment, actin-filament assembly is promoted by PDGF by a mechanism that is not completely understood, but that may involve small GTP-binding proteins, such as Rho, and formation of DAG. Migration on collagen requires functional α2β1 integrins, which may either constitute a permissive state required for a cell to migrate, or which may be actively involved in intracellular signals leading to migration. PDGF-induced DNA synthesis and proliferation involves activation of Ras, MAP kinase kinase, and MAP kinase. Cross-talk between PKA signaling and tyrosine-kinase receptor signaling results in PKA inhibition of the MAP kinase cascade, probably at the level of Raf. Activation of PI 3-kinase, or a PI 3-kinase-like enzyme, is also likely to contribute to the mitogenic effects of PDGF in these cells (Bornfeldt, unpublished observation).

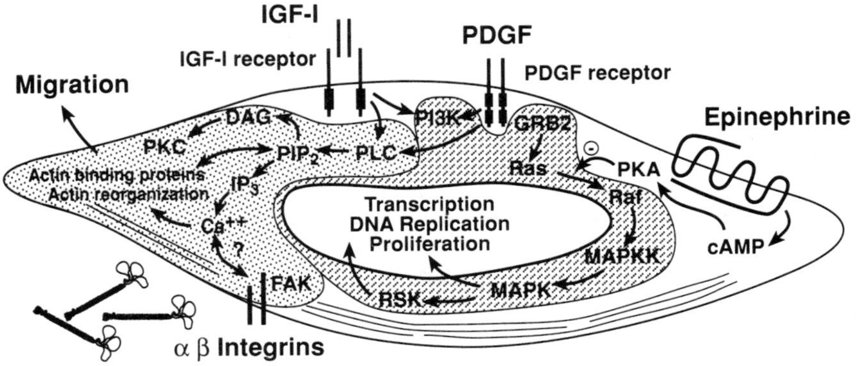

FIGURE 2. Signal transduction pathways involved in PDGF-induced migration and proliferation. The signal-transduction pathways involved in migration versus proliferation are depicted as described in the text. Signal-transduction pathways preferentially involved in proliferation (▨) and signal-transduction pathways preferentially involved in directed migration (⛆) are shown. Abbreviations: Cyclic AMP-dependent protein kinase, PKA; diacylglycerol, DAG; focal adhesion kinase, FAK; growth factor receptor-bound protein 2, GRB2; inositol trisphosphate, IP3; insulin-like growth factor I, IGF-I; mitogen-activated protein kinase kinase, MAPKK; phosphatidylinositol bisphosphate, PIP2; phosphatidylinositol 3-kinase, PI3K; phospholipase C, PLC; protein kinase C, PKC; platelet-derived growth factor, PDGF; p90 ribosomal S6 kinase, RSK.

What determines if a SMC will migrate and/or proliferate in response to PDGF? Results are starting to emerge that show regulation of expression of molecules involved in intracellular signaling with different phenotypic states of SMC. For example, expression of PLCγ is very low in intact vascular wall (where SMC show a "contractile phenotype"), and induced when SMC are converted to a "synthetic phenotype" in culture.[66] Proliferation and expression of MAP kinase, but not calcium signaling, appear to be regulated by the extracellular matrix,[67] and the profile of integrin expression is different in SMC in culture compared to SMC in the vascular wall.[61] Thus, the relation between expression of signaling molecules involved in migration and signaling molecules involved in proliferation, as well as cross-talk between different signal-transduction pathways, may determine the net effect of PDGF.

REFERENCES

1. Hart, C. E., J. W. Forstrom, J. D. Kelly, R. A. Seifert, R. A. Smith, R. Ross, M. J. Murray & D. F. Bowen-Pope. 1988. Two classes of PDGF receptor recognize different isoforms of PDGF. Science **240:** 1529-1531.
2. Claesson-Welsh, L. 1994. Signal transduction by the PDGF receptors. *In* Growth Factor Research. **5:** 37-53. Pergamon Press. Great Britain.
3. Ullrich, A. & J. Schlessinger. 1990. Signal transduction by receptors with tyrosine kinase activity. Cell **61:** 203-212.
4. Pawson, T. & G. D. Gish. 1992. SH2 and SH3 domains: From structure to function. Cell **71:** 359-362.

5. SEIFERT, R. A., C. E. HART, P. E. PHILLIPS, J. W. FORSTROM, R. ROSS, M. J. MURRAY & D. F. BOWEN-POPE. 1989. Two different subunits associate to create isoform-specific platelet-derived growth factor receptors. J. Biol. Chem. **264:** 8771-8778.
6. FERNS, G. A. A., K. H. SPRUGEL, R. A. SEIFERT, D. F. BOWEN-POPE, J. D. KELLY, M. MURRAY, E. W. RAINES & R. ROSS. 1990. Relative platelet-derived growth factor receptor subunit expression determinates cell migration to different dimeric forms of PDGF. Growth Factors **3:** 315-324.
7. BORNFELDT, K. E., E. W. RAINES, T. NAKANO, L. M. GRAVES, E. G. KREBS & R. ROSS. 1994. Insulin-like growth factor-I and platelet-derived growth factor-BB induce directed migration of human arterial smooth muscle cells via signaling pathways that are distinct from those of proliferation. J. Clin. Invest. **93:** 1266-1274.
8. RAINES, E. W. & R. ROSS. 1993. Platelet-derived growth factor *in vivo*. In Cytokines. B. Westermark & C. Sorg, Eds. **5:** 74-114. S. Krager AG. Basel, Switzerland.
9. ROSS, R. 1993. The pathogenesis of atherosclerosis: a perspective for the 1990s. Nature **362:** 801-809.
10. ROSS, R., J. MASUDA, E. W. RAINES, A. M. GOWN, S. KATSUDA, M. SASAHARA, L. T. MALDEN, H. MASUKO & H. SATO. 1990. Localization of PDGF-B protein in macrophages in all phases of atherogenesis. Science **248:** 1009-1012.
11. FERNS, G. A. A., E. W. RAINES, K. H. SPRUGEL, A. S. MOTANI, M. A. REIDY & R. ROSS. 1991. Inhibition of neointimal smooth muscle accumulation after angioplasty by an antibody to PDGF. Science **253:** 1129-1132.
12. JAWIEN, A., D. F. BOWEN-POPE, V. LINDNER, S. M. SCHWARTZ & A. CLOWES. 1992. Platelet-derived growth factor promotes smooth muscle cell migration and intimal thickening in a rat model of balloon angioplasty. J. Clin. Invest. **89:** 507-511.
13. JACKSON, C. L., E. W. RAINES, R. ROSS & M. A. REIDY. 1993. Role of endogenous platelet-derived growth factor in arterial smooth muscle cell migration after balloon catheter injury. Atherosclerosis Thrombosis **13:** 1218-1226.
14. CLOWES, A. W., M. A. REIDY & M. M. CLOWES. 1983. Kinetics of cellular proliferation after arterial injury. I. Smooth muscle growth in the absence of endothelium. Lab. Invest. **49:** 327-333.
15. CLOWES, A. W., M. M. CLOWES, J. FINGERLE & M. A. REIDY. 1989. Regulation of smooth muscle cell growth in injured arteries. J. Cardiovasc. Pharmacol. **14** (Suppl. 6); S12-S15.
16. KING, G. L., A. D. GOODMAN, S. BUZNEY, A. MOSES & C. R. KAHN. 1985. Receptors and growth-promoting effects of insulin and insulin-like growth factors on cells from bovine retinal capillaries and aorta. J. Clin. Invest. **75:** 1028-1036.
17. BORNFELDT, K. E., R. A. GIDLÖF, Å. WASTESON, M. LAKE, A. SKOTTNER & H. J. ARNQVIST. 1991. Binding and biological effects of insulin, insulin analogues and insulin-like growth factors in rat aortic smooth muscle cells: comparison of maximal growth promoting activities. Diabetologia **34:** 307-313.
18. MOXHAM, C. & S. JACOBS. 1992. Insulin-like growth factor receptors. *In* The Insulin-like Growth Factors. Structure and Biological Functions. P. N. Schofield, Ed.: 80-109. Oxford University Press. Oxford.
19. BORNFELDT, K. E. & H. J. ARNQVIST. 1993. Actions of insulin-like growth factor I and insulin in vascular smooth muscle: receptor interaction and growth-promoting effects. *In* Growth Hormone and Insulin-like Growth Factor I. A. Flyvbjerg, H. Ørskov & K. G. M. M. Alberti, Eds.: 159-191. John Wiley & Sons Ltd, England.
20. GROTENDORST, G. R., H. E. J. SEPPÄ, H. K. KLEINMAN & G. R. MARTIN. 1981. Attachment of smooth muscle cells to collagen and their migration toward platelet-derived growth factor. Proc. Natl. Acad. Sci. USA **78:** 3669-3672.
21. HOSANG, M., M. ROUGE, B. WIPF, G. EGGIMANN, F. KAUFMANN & W. HUNZIKER. 1989. Both homodimeric isoforms of PDGF (AA and BB) have mitogenic and chemotactic activity and stimulate phosphoinositol turnover. J. Cell Physiol. **140:** 558-564.

22. MATSUI, T., J. H. PIERCE, T. P. FLEMING, J. S. GREENBERGER, W. J. LAROCHELLE, M. RUGGIERO & S. A. AARONSON. 1989. Independent expression of human α or β platelet-derived growth factor receptor cDNAs in a naive hematopoietic cell leads to functional coupling with mitogenic and chemotactic signaling pathways. Proc. Natl. Acad. Sci. USA **86:** 8314-8318.
23. ERIKSSON, A., A. SIEGBAHN, B. WESTERMARK, C.-H. HELDIN & L. CLAESSON-WELSH. 1992. PDGF α- and β-receptors activate unique and common signal transduction pathways. EMBO J. **11:** 543-550.
24. KOYAMA, N., N. MORISAKI, Y. SAITO & S. YOSHIDA. 1992. Regulatory effects of platelet-derived growth factor-AA homodimer on migration of vascular smooth muscle cells. J. Biol. Chem. **267:** 22806-22812.
25. KOYAMA, N., C. E. HART & A. W. CLOWES. 1994. Different functions of the platelet-derived growth factor-α and β-receptors for the migration and proliferation of cultured baboon smooth muscle cells. Circ. Res. **75:** 682-691.
26. DAVIS, R. J. 1993. The mitogen-activated protein kinase signal transduction pathway. J. Biol. Chem. **268:** 14553-14556.
27. LI, W., R. NISHIMURA, A. KASHISHIAN, A. G. BATZER, W. J. H. KIM, J. A. COOPER & J. SCHLESSINGER. 1994. A new function for a phosphotyrosine phosphatase: Linking GRB2-Sos to a receptor tyrosine kinase. Mol. Cell. Biol. **14:** 509-517.
28. ARVIDSSON, A.-K., E. RUPP, E. NÅNBERG, J. DOWNWARD, L. RÖNNSTRAND, S. WENNSTRÖM, J. SCHLESSINGER, C.-H. HELDIN & L. CLAESSON-WELSH. 1994. Tyr-716 in the platelet-derived growth factor β-receptor kinase insert is involved in GRB2 binding and Ras activation. Mol. Cell. Biol. **14:** 6715-6726.
29. RÖNNSTRAND, L., S. MORI, A.-K. ARVIDSSON, A. ERIKSSON, C. WERNSTEDT, U. HELLMAN, L. CLAESSON-WELSH & C.-H. HELDIN. 1992. Identification of two C-terminal autophosphorylation sites in the PDGF β-receptor: involvement in the interaction with phospholipase Cγ. EMBO J. **11:** 3911-3919.
30. VALIUS, M. & A. KAZLAUSKAS. 1993. Phospholipase C-γ1 and phosphatidylinositol 3 kinase are downstream mediators of the PDGF receptor's mitogenic signal. Cell **73:** 321-334.
31. FANTL, W. J., J. A. ESCOBEDO, G. A. MARTIN, C. W. TURCK, M. DEL ROSARIO, F. MCCORMICK & L. T. WILLIAMS. 1992. Distinct phosphotyrosines on a growth factor receptor bind to specific molecules that mediate different signalling pathways. Cell **69:** 413-423.
32. CHUNG, J., T. C. GRAMMER, K. P. LEMON, A. KAZLAUSKAS & J. BLENIS. 1994. PDGF- and insulin-dependent pp70s6k activation mediated by phosphatidylinositol-3-OH kinase. Nature **370:** 71-75.
33. WENNSTRÖM, S., A. SEIGBAHN, K. YOKOTE, A.-K. ARVIDSSON, C.-H. HELDIN, S. MORI & L. CLAESSON-WELSH. 1994. Membrane ruffling and chemotaxis transduced by the PDGF β-receptor require the binding site for phosphatidylinositol 3′ kinase. Oncogene **9:** 651-660.
34. JOLY, M., A. KAZLAUSKAS, F. S. FAY & S. CORVERA. 1994. Disruption of PDGF receptor trafficking by mutation of its PI-3 kinase binding sites. Science **263:** 684-687.
35. MING, X.-F., B. M. T. BURGERING, S. WENNSTRÖM, L. CLAESSON-WELSH, C.-H. HELDIN, J. L. BOS, S. C. KOZMA & G. THOMAS. 1994. Activation of p70/p85 S6 kinase by a pathway independent of $p21^{ras}$. Nature **371:** 426-429.
36. SCHU, P. V., K. TAKEGAWA, M. J. FRY, J. H. STACK, M. D. WATERFIELD & S. D. EMR. 1993. Phosphatidylinositol 3-kinase encoded by yeast VPS34 gene essential for protein sorting. Science **260:** 88-91.
37. GRAVES, L. M., K. E. BORNFELDT, E. W. RAINES, B. C. POTTS, S. G. MACDONALD, R. ROSS & E. G. KREBS. 1993. Protein kinase A antagonizes platelet-derived growth factor-induced signaling by mitogen-activated protein kinase in human arterial smooth muscle cells. Proc. Natl. Acad. Sci. USA **90:** 10300-10304.

38. Pavlovic-Surjancev, B., A. L. Cahill & R. L. Perlman, 1992. Nicotinic agonists, phorbol esters, and growth factors activate two extracellular signal-regulated kinases, ERK1 and ERK2, in bovine chromaffin cells. J. Neurochem. **59:** 2134-2140
39. Mastick, C. C., H. Kato, C. T. Roberts, Jr., D. Leroith & A. R. Saltiel. 1994. Insulin and insulin-like growth factor-I receptors similarly stimulate deoxyribonucleic acid synthesis despite differences in cellular protein tyrosine phosphorylation. Endocrinology **135:** 214-222.
40. Myers Jr., M. G., X.-J. Sun & M. White. 1994. The IRS-1 signaling system. TIBS **19:** 289-293.
41. Wu, J., P. Dent, T. Jelinek, A. Wolfman, M. J. Weber & T. W. Sturgill. 1993. Inhibition of the EGF-activated MAP kinase signaling pathway by adenosine 3′,5′-monophosphate. Science **262:** 1065-1069.
42. Cook, S. J. & F. McCormick. 1993. Inhibition by cAMP of Ras-dependent activation of Raf. Science **262:** 1069-1072.
43. Sevetson, B. R., X. Kong & J. C. Lawrence, Jr. 1993. Increasing cAMP attenuates activation of mitogen-activated protein kinase. Proc. Natl. Acad. Sci. USA **90:** 10305-10309.
44. Lange-Carter, C. A. & G. L. Johnson. 1994. Ras-dependent growth factor regulation of MEK kinase in PC12 cells. Science **265:** 1458-1461.
45. Feig, L. A. & B. Schaffhausen. 1994. The hunt for Ras targets. Nature **370:** 508-509.
46. Häfner, S., H. S. Adler, H. Mischak, P. Janosch, G. Heidecker, A. Wolfman, S. Pippig, M. Lohse, M. Ueffing & W. Kolch. 1994. Mechanism of inhibition of Raf-1 by protein kinase A. Mol. Cell. Biol. **14:** 6696-6703.
47. Loesberg, C., R. Van Wijk, J. Zandbergen, W. G. Van Aken, J. A. Van Mourik & P. G. De Groot. 1985. Cell cycle-dependent inhibition of human vascular smooth muscle cell proliferation by prostaglandin E1. Exp. Cell Res. **160:** 117-125.
48. Assender, J. W., K. M. Southgate, M. B. Hallett & A. C. Newby. 1992. Inhibition of proliferation, but not of Ca^{2+} mobilization, by cyclic AMP and GMP in rabbit aortic smooth-muscle cells. Biochem. J. **288:** 527-532.
49. Kashishian, A. & J. A. Cooper. 1993. Phosphorylation sites at the C-terminus of the platelet-derived growth factor receptor bind phospholipase Cγ1. Mol. Biol. Cell **4:** 49-57.
50. Kundra, V., J. A. Escobedo, A. Kazlauskas, H. Kun Kim, S. Goo Rhee, L. T. Williams & B. R. Zetter. 1994. Regulation of chemotaxis by the platelet-derived growth factor-β. Nature **367:** 474-476.
51. Guse, A. H., W. Kiess, B. Funk, U. Kessler, I. Berg & G. Gercken. 1992. Identification and characterization of insulin-like growth factor receptors on adult rat cardiac myocytes: Linkage to inositol 1,4,5-trisphosphate formation. Endocrinology **130:** 145-151.
52. Damke, H., H. Bouterfa & T. Braulke. 1994. Effects of insulin-like growth factor II on the generation of inositol trisphosphate, diacylglycerol and cAMP in human fibroblasts. Mol. Cell. Endocrinol. **99:** R25-R29.
53. Martelli, A. M., R. S. Gilmour, V. Bertagnolo, L. M. Neri, L. Manzoli & L. Cocco. 1992. Nuclear localization and signalling activity of phosphoinositidase Cβ in Swiss 3T3 cells. Nature **358:** 242-245.
54. Stossel, T. P. 1993. On the crawling of animal cells. Science **260:** 1086-1094.
55. Shariff, A. & E. J. Luna. 1992. Diacylglycerol-stimulated formation of actin nucleation sites at plasma membranes. Science **256:** 245-247.
56. Ridley, A. J. & A. Hall. 1992. The small GTP-binding protein rho regulates the assembly of focal adhesions and actin stress fibers in response to growth factors. Cell **70:** 389-399.
57. Jahner, D. & T. Hunter. 1991. The Ras-related gene RhoB is an immediate-early gene inducible by v-Fps, epidermal growth factor, and platelet-derived growth factor in rat fibroblasts. Mol. Cell. Biol. **11:** 3682-3690.

58. SETTLEMAN, J., C. F. ALBRIGHT, L. C. FOSTER & R. A. WEINBERG. 1992. Association between GTPase activators for rho and ras families. Nature **359:** 153-154.
59. MCGLADE, J., B. BRUNKHORST, D. ANDERSON, G. MBAMALU, J. SETTLEMAN, S. DEHAR, M. ROZAKIS-ADCOCK, L. BO CHEN & T. PAWSON. 1993. The N-terminal region of GAP regulates cytoskeletal structure and cell adhesion. EMBO J. **12:** 3073-3081.
60. SCHWARTZ, M. A. 1992. Transmembrane signalling by integrins. Trends Cell Biol. **2:** 304-308.
61. SKINNER, M. P., E. W. RAINES & R. ROSS. 1994. Dynamic expression of $\alpha1\beta1$ and $\alpha_2\beta_1$ integrin receptors by human vascular smooth muscle cells: $\alpha2\beta1$ integrin is required for chemotaxis across type I collagen-coated membranes. Am. J. Pathol. **145:** 1070-1081.
62. KERN, A., J. EBLE, R. GOLBIK & K. KÜHN. 1993. Interaction of type IV collagen with the isolated integrins $\alpha1\beta1$ and $\alpha2\beta1$. Eur. J. Biochem. **215:** 151-159.
63. ZANG, Z., A. O. MORLA, K. VUORI, J. S. BAUER, R. L. JULIANO & E. RUOSLAHTI. 1993. The $\alpha v\beta5$ integrin functions as a fibronectin receptor but does not support fibronectin matrix assembly and cell migration on fibronectin. J. Cell. Biol. **122:** 235-242.
64. RANKIN, S. & E. ROZENGURT. 1994. Platelet-derived growth factor modulation of focal adhesion kinase (p125FAK) and paxillin tyrosine phosphorylation in Swiss 3T3 cells. J. Biol. Chem. **269:** 704-710.
65. SCHALLER, M. D., J. D. HILDEBRAND, J. D. SHANNON, J. W. FOX, R. R. VINES & J. T. PARSONS. 1994. Autophosphorylation of the focal adhesion kinase, $pp125^{FAK}$, directs SH2-dependent binding of $pp60^{src}$. Mol. Cell. Biol. **14:** 1680-1688.
66. HOMMA, Y., H. SAKAMOTO, M. TSUNODA, M. AOKI, T. TAKENAWA & T. OOYAMA. 1993. Evidence for involvement of phospholipase $C\gamma2$ in signal transduction of platelet-derived growth factor in vascular smooth muscle cells. Biochem. J. **290:** 649-653.
67. LI, X., P. TSAI, E. D. WIEDER, A. KRIBBEN, V. VAN PUTTEN, R. W. SCHRIER & R. A. NEMENOFF. 1994. Vascular smooth muscle cells grown on Matrigel. A model of the contractile phenotype with decreased activation of mitogen-activated protein kinase. J. Biol. Chem. **269:** 19653-19658.

Thrombin and Its Receptor in Growth Control[a]

ELLEN VAN OBBERGHEN-SCHILLING,
VALÉRIE VOURET-CRAVIARI, YAO-HUI CHEN,
DOMINIQUE GRALL, JEAN-CLAUDE CHAMBARD,
AND JACQUES POUYSSÉGUR

Centre de Biochimie
CNRS UMR134
Parc Valrose
06108 Nice Cedex 2, France

The trypsin-like serine protease, thrombin, plays a central role in hemostasis and thrombosis. In addition to activating circulating plasma proteins and cofactors of the coagulation cascade, thrombin exerts a host of cellular effects that may contribute to inflammatory and proliferative responses that accompany the wound-healing process. Thrombin's cellular effects are mediated by a G protein-coupled surface receptor that displays a unique mechanism of activation by proteolysis. This article will review certain aspects of the thrombin-receptor structure and function and explore the possible role of its signal-transducing pathways in cell growth.

SOLUBLE THROMBIN SUBSTRATES AND BINDING PROTEINS

Compared to other serine proteases, thrombin displays unique specificity for macromolecular ligands. Following generation from its plasma-borne precursor during blood coagulation, thrombin cleaves soluble fibrinogen into insoluble fibrin, which forms the molecular scaffolding of thrombi. Thrombin also activates a number of coagulation factors and plasma proteins, including factors V, VIII, XIII, protein S, and protein C (see ref. 1). A limited set of circulating inhibitors specifically interact with the enzyme, and these include α_2-macroglobulin and the serpins antithrombin III, heparin cofactor II, and protease nexin-1. The leech polypeptide hirudin is the most potent natural inhibitor of thrombin. Inhibition of the enzyme represents a unique mechanism that involves the binding of acidic residues in the C-terminal tail of hirudin with a positively charged region on the surface of the thrombin molecule, referred to as the "anion-binding exosite," or (fibrin)ogen recognition exosite.[2] A similar interaction occurs between the anion-binding exosite of thrombin and its receptor, as discussed below.

[a] Studies from our laboratory were supported by the Centre National de la Recherche Scientific (UMR134), the Association pour la Recherche contre le Cancer, and Boehringer Ingelheim International GmbH.

TABLE 1. Cellular Thrombin Binding Sites

Binding Site	Cell Type	Affinity
Glycoprotein Ib	platelets	high
Thrombomodulin	endothelial cells	high
Protease nexin-1 (Glial-derived nexin)	primarily brain (secreteted and ECM-associated)	SDS-stable
150 kDa protein	fibroblasts	moderately high
G Protein-coupled Receptor	wide distribution	moderate

CELLULAR-THROMBIN BINDING SITES

Cellular binding sites for thrombin have been identified on the surface of a wide variety of cells, such as endothelial cells, platelets, fibroblasts, and smooth muscle cells (TABLE 1). On endothelial cells, thrombin binds with high affinity (≈0.2 nM) to thrombomodulin. Formation of the thrombin-thrombomodulin complex enhances the protein C anticoagulant pathway.[3] High, moderate and low-affinity thrombin-binding sites have been demonstrated on platelets (ref. 4 and references therein). In these cells, the high-affinity site has been localized to glycoprotein Ib, a disulfide-bonded heterodimer composed of an α (143 kDa) and β (22 kDa) chain. Glycoprotein Ib is susceptible to thrombin cleavage,[5] as is another platelet protein, glycoprotein V. However, although glycoprotein V is apparently not implicated in thrombin-induced platelet activation,[6] the role of glycoprotein Ib in signal generation by the protease has been a subject of debate for several years. One proposed role for glycoprotein Ib is that of a thrombin-retention site on the platelet surface. We and others have reported the presence of specific binding sites for thrombin with relatively high affinity (≈2 nM) on the surface of fibroblasts. By affinity-labeling techniques, the size of the fibroblast receptor was estimated to be approximately 150 kDa,[7,8] although its role in mediating thrombin's mitogenic action in these cells has not been clearly defined to date. Protease nexin-1[9] (identified independently as glial-derived nexin[10]) is a 43 kDa thrombin inhibitor that forms an SDS-stable complex with thrombin. This serpin is predominantly expressed in brain, but it is also found in fibroblasts, smooth muscle cells, and platelets. Although protease nexin-1 is a secreted protein, it binds tightly to the extracellular matrix of cells, and binding of protease nexin-1 to the matrix accelerates its inhibition of thrombin. In 1991 the search for a functional thrombin receptor yielded the cloning of a thrombin receptor that mediates most of thrombin's cellular effects.[11,12] This receptor is an integral membrane-thrombin substrate that binds the enzyme with only moderate affinity.

CELLULAR EFFECTS OF THROMBIN

The cellular effects of thrombin are as diverse as the cells with which it interacts (for a review, see ref. 13). Thrombin has long been known as the most potent physiological activator of platelets. Its interaction with the endothelium leads to increased permeability and to the formation of a thrombogenic surface to which circulating blood platelets and leukocytes adhere. Direct chemotactic effects of thrombin have been observed on neutrophils and monocytes,[14,15] which are recruited to sites of injury during the early stages of inflammation, as well as fibroblasts,[16] which may be involved in tissue repair. The action of thrombin on another component of the immune system, the T lymphocyte, has more recently been documented.[17,18] In these cells, thrombin was found to stimulate early events associated with activation and to potentiate the expression of CD69 and interleukin 2 production, which is turned on by T-cell receptor cross-linking. Thrombin displays contractile action in vascular and nonvascular smooth muscle cell systems. Prolonged exposure of vascular smooth muscle cells[19–21] and fibroblasts[22–24] to thrombin induces proliferation. Thrombin may also influence the proliferation of vascular smooth muscle cells indirectly by stimulating the release of smooth muscle cell mitogens from platelets and from the endothelium.[25] Autocrine release of platelet-derived growth factor and fibroblast growth factor from cultured smooth muscle cells has been described as well.[26,27]

Extravascular effects of thrombin include morphological changes in cultured neuronal and glial cells,[28–30] and Ca^{2+} mobilization in osteoblast-like cells,[31] to cite only a few.

THE G PROTEIN-COUPLED THROMBIN RECEPTOR

Expression cloning in *Xenopus* oocytes led to the isolation of a cDNA clone encoding a thrombin receptor coupled to Ca^{2+} mobilization.[11] A hamster cDNA corresponding to the receptor was isolated independently, based on its sequence similarity with cDNA's encoding members of the G protein-receptor superfamily.[12] Since then, thrombin-receptor sequences from human endothelial cells,[32] rat vascular smooth muscle cells,[33] mouse (GeneBank accession no. LO3529), and *Xenopus*[34] have been deduced. The human thrombin receptor is composed of 425 amino acids encoded by a single gene that has been localized to the long arm of human chromosome 5 (region q13).[35] Human platelet and hamster-fibroblast receptors for thrombin display 79% amino acid sequence identity. The primary sequence displays eight hydrophobic domains, including one at the extreme NH_2 terminus, which presumably corresponds to a signal peptide. Five potential asparagine-linked glycosylation sites are present in the human receptor (four in the rodent-receptor sequences). In addition, multiple serine, threonine, and tyrosine residues that constitute potential phosphorylation sites are dispersed throughout the intracellular loops and cytoplasmic extension of the receptor. Finally, the large NH_2-terminal extension bears a thrombin cleavage site characterized by an arginine in position 41 that is preceded by a proline and followed by a cluster of acidic residues. This negatively charged sequence displays considerable homology with thrombin-recognition sites found in other proteins (*e.g.,* hirudin, fibrinogen, and heparin cofactor II) that interact with the anion-binding exosite of thrombin, as mentioned above. Indeed, a striking sequence similarity exists between the thrombin receptor in this region and the carboxy tail of hirudin.[12] This acidic sequence on the thrombin receptor has been shown to dictate thrombin's interaction with its receptor.[36,37]

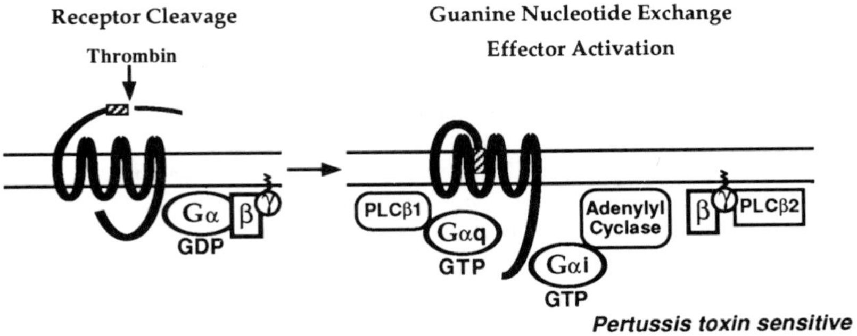

FIGURE 1. Model of thrombin-receptor activation. Cleavage of the thrombin receptor between residues arginine-41 and serine-42 unmasks a tethered ligand (hatched box) that binds to and activates the receptor. Once activated, the receptor catalyzes exchange of GTP for bound GDP on G protein α-subunits. The thrombin receptor can activate at least two distinct G proteins: (left) a pertussis toxin-insensitive protein (Gq) that couples to phospholipase C β1 activation, and (right) pertussis toxin-sensitive (Gi-like) subtypes that couple to adenylyl cyclase inhibition and to the β2 isoform of phospholipase C (by way of βγ heterodimers from pertussis toxin-sensitive G proteins).

RECEPTOR ACTIVATION

Proteolytic activity of thrombin is required for essentially all of its cellular effects. This observation, together with the presence of a thrombin cleavage site on the receptor at arginine-41, led Coughlin and his colleagues to propose that receptor activation is triggered by cleavage. Thus, by analogy with the mechanism of trypsinogen conversion to trypsin,[38] receptor cleavage would unmask a new NH_2 terminus that acts as a tethered ligand[11] (FIG. 1). In accordance with this model, synthetic peptides of 14 residues corresponding to the sequence immediately adjacent to the proposed thrombin cleavage site were found by these authors to stimulate human-platelet aggregation. Indeed, subsequent studies have confirmed that most of thrombin's cellular effects can be stimulated by the thrombin-receptor peptide (TRP). Structure-activity analyses carried out in several laboratories have revealed that a minimum of five residues are sufficient for agonist activity, although the hexamer (SFLLRN) is approximately 10-fold more potent for receptor activation. The NH_2-terminal amino group has been found to be crucial for biological activity of the peptide, similar to the side chains of phenylalanine, leucine, and arginine in positions 2, 4, and 5, respectively.[39-44]

This model of receptor activation by proteolysis has been further supported by studies using antibodies directed against sequences in the NH_2 terminus of the receptor. Antibodies to the proposed cleavage site were found to block receptor signaling.[32,45,46] Receptor cleavage has been demonstrated indirectly in experiments by Brass and colleagues, using antibodies that can distinguish between cleaved (thrombin-treated) and intact receptors.[47] Further, antibodies directed against the cleaved-off fragment of the receptor have been used to monitor its release following treatment of cells with thrombin.[48] Recently, we have been able to directly visualize receptor cleavage

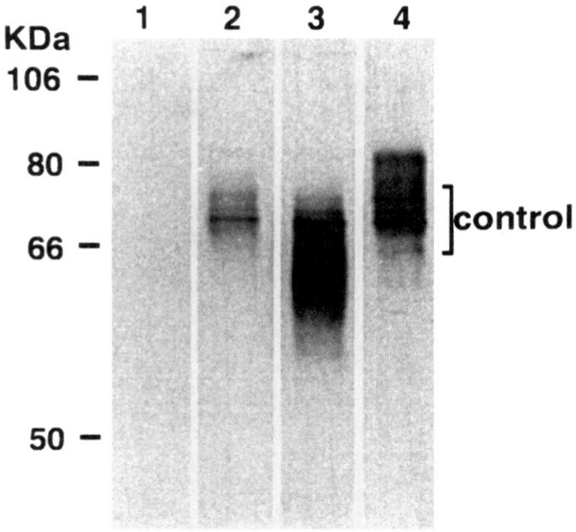

FIGURE 2. Cleavage of the thrombin receptor. Western analysis was performed on wheat germ-enriched lysates from nontransfected 293 human embryonic kidney cells (lane 1) or from a G418-resistant clone transfected with an epitope-tagged human thrombin receptor cDNA.[49] Cells were treated for 10 min at 37 °C with 10^{-8}M thrombin (lanes 1 and 3) or with 30 μM TRP (lane 4) prior to Western analysis, using P4D5 antibody directed against the tag sequence derived from vesicular stomatitis virus glycoprotein.[85] Migration of the nontreated thrombin receptor is indicated on the right.

by thrombin in immunoblot analyses of lysates from cells stably expressing an epitope-tagged thrombin receptor (FIG. 2). Interestingly, activation of the receptor by TRP, independent of cleavage, results in an increased apparent M_r that can be attributed (at least in part) to phosphorylation.[49]

In addition to thrombin, other proteases are able to cleave the receptor, including trypsin, plasmin, and chymotrypsin (V. Vouret-Craviari, D. Grall, and E. Van Obberghen-Schilling, unpublished results). The former two enzymes with arginyl specificity stimulate the receptor as measured by phospholipase C activation, whereas chymotrypsin (which cleaves after aromatic residues) has no effect. Rather, chymotrypsin renders the receptor nonresponsive to thrombin by removing the residues involved in thrombin recognition. Another serine protease, granzyme A, which is stored in cytoplasmic granules of T lymphocytes, has also been reported to activate the thrombin receptor,[50] presumably by cleaving the arginine-41/serine-42 peptide bond. *In vitro,* granzyme A is able to cleave a 35-residue peptide that spans the thrombin-receptor cleavage site. The physiological significance of thrombin-receptor cleavage by proteases other than thrombin awaits to be determined.

Receptor activation by proteolysis may be a mechanism that extends beyond the thrombin receptor. Nystedt *et al.* have recently identified a second receptor that can be activated either by trypsin or by a tethered ligand peptide adjacent to the predicted trypsin-cleavage site.[51] Although the physiologic activator of this receptor, designated proteinase-activated receptor 2 (PAR-2) is not known, it may represent the second member of a proteinase-activated G protein-coupled receptor subfamily.

Following receptor cleavage, the tethered-ligand sequence is believed to bind to a site in the core of the receptor that remains to be determined. Studies using antibodies,[32] or chimeric thrombin-receptor molecules,[34] have revealed that the extracellular domains are important for receptor activation. In the later study, sequences from the NH_2-terminal extension and three extracellular loops of the *Xenopus* receptor were individually replaced with the corresponding structures from the human receptor. Although valuable insights can be gained from these studies, precise identification of sites in the receptor that are involved in ligand binding will require more restricted mutational analyses and resolution of the receptor structure.

RECEPTOR DESENSITIZATION

Loss of the biological response to the tethered ligand, despite its constant presence, is observed rapidly after thrombin treatment of cells.[52,53] This process of desensitization has been shown for other G protein-coupled receptor family members to involve (1) receptor phosphorylation, (2) sequestration/internalization, and (3) down-regulation after prolonged agonist exposure (for review, see ref. 54). The thrombin-receptor cytoplasmic tail contains several serine and threonine residues that are potential sites for phosphorylation by G protein-receptor kinases. Indeed, we have observed rapid thrombin- and TRP-induced phosphorylation of the thrombin receptor in 293 human embryonic kidney cells transfected with the human receptor cDNA.[49] This observation is consistent with recent findings by Ishii *et al.*,[55] describing the attenuation of thrombin-receptor signaling by BARK2 (β-adrenergic receptor kinase 2), a member of the G protein-receptor kinase family.[56] When coexpressed with the human thrombin receptor in *Xenopus* oocytes, BARK2 was shown to block thrombin-induced Ca^{2+} mobilization. Mutant receptors lacking serine and threonine residues in the carboxy tail were insensitive to the effect of BARK2, suggesting that the receptor's carboxy terminus was responsible for this effect. We came to a similar conclusion concerning the role of the C-terminal extension in desensitization using a chimeric receptor approach. Sequences in the carboxy tail of the thrombin receptor conferred the ability to undergo ligand-induced phosphorylation and rapid desensitization to a chimeric $5\text{-}HT_2$/thrombin receptor mutant.[49]

In addition to phosphorylation and desensitization, the thrombin receptor undergoes rapid internalization following stimulation. This process has been examined in detail using megakaryocyte precursor cell lines, HEL and CHRF-288.[47,57] In CHRF-288 cells, internalization of ≈90% of the activated thrombin receptors occurs within 2 minutes of thrombin or TRP treatment. More than 75% of the internalized receptors are shuttled to lysosomal compartments for degradation. Thrombin-cleaved receptors that escape degradation recycle to the cell surface in an inactive state. Thus, renewal of the thrombin response requires replenishment of surface receptors with newly synthesized receptors.

Thrombin responsiveness may also be regulated by cells at the mRNA level. In rat vascular smooth muscle cells, thrombin and other agonists have been found to increase steady state levels of the thrombin receptor mRNA.[33,58] In T lymphocytes, activation leads to a dramatic decrease in thrombin receptor mRNA expression.[18]

SIGNALING PATHWAYS OF THE THROMBIN RECEPTOR

Studies in our laboratory on the mitogenic signaling pathways of the thrombin receptor have been largely carried out on the growth-responsive CCL39 line of hamster fibroblasts. Thrombin stimulates cell cycle reentry in quiescent CCL39 cells and continued growth, in the absence of additional mitogens (for reviews, see refs. 59 and 60). As schematized in FIGURE 1, the thrombin receptor activates at least two G protein-coupled pathways in these cells: (1) a pertussis toxin-insensitive (G_q) pathway, coupled to phospholipase C stimulation,[61,62] and (2) one or more pertussis toxin-sensitive ($G_{i/o}$) pathways, determined by adenylyl cyclase inhibition.[63] Peptide agonists of the receptor are also full agonists of these two responses.[39] The involvement of pertussis toxin-sensitive G proteins in cell-growth stimulation by thrombin was first demonstrated by Chambard *et al.* in 1987.[64] Subsequent work has focused on identifying intracellular signaling systems activated by G proteins that link the receptor to DNA synthesis.

THE MAP KINASE CASCADE

The mitogen-activated protein (MAP) kinases are ubiquitously expressed, highly conserved kinases that are activated by a remarkably diverse set of extracellular stimuli. Studies from our laboratory have shown that thrombin stimulates rapid and persistent activation of MAP kinase activity in quiescent CCL39 cells and that persistent MAP kinase activation correlates with cell cycle reentry.[39,65] Following activation, both p42 and p44 MAP kinase isoforms translocate to the nucleus.[66] More recently, it was demonstrated that MAP kinase activation plays a determinant role in the growth of CCL39 cells in experiments using antisense MAP kinase constructs or dominant negative MAP kinase mutants.[67] Consistent with this finding, a constitutively active mutant of MAP kinase kinase (or MEK), the immediate upstream activator of MAP kinase, was found to be oncogenic.[68]

The mechanism of MAP kinase activation is a subject of intense investigation by numerous laboratories and is rapidly being elucidated (for reviews, see refs. 69 and 70). Upstream regulators of MAP kinase include MAP kinase kinase, Raf kinases and Ras. For receptor-tyrosine kinases, several molecular components that link the receptor to Ras and the MAP kinase pathway have been identified, including Grb2, Shc, and Sos1. However, it has become apparent only recently that serpentine receptors also require Ras to activate the kinase cascade. Last year, the group of Moolenaar reported that the thrombin receptor, as well as the receptor for lysophosphatidic acid (LPA), could activate Ras in quiescent fibroblasts.[71] More recently, Ras-dependent activation of the MAP kinase cascade by other G protein-coupled receptors has been documented, including the muscarinic m2[72] and α2-adrenergic[73] receptors, as well as receptors for LPA[74] and the chemotactic agent FMLP.[75] In microinjection experiments using a dominant interfering Ras mutant, the Ras protein has been shown to be required for thrombin-stimulated proliferation of astrocytoma cells.[76]

Thus, as signal transducing pathways of the thrombin receptor and other G protein-coupled receptors rapidly become dissected, many questions concerning possible link(s) between the G proteins and Ras arise. Pertussis toxin was found to interfere with Ras activation by thrombin, suggesting that a Gi-like protein is involved in this effect.[71] Recent studies in mammalian cells have suggested that heterotrimeric G proteins may in fact activate a Ras-dependent signaling pathway through $\beta\gamma$ heterodimers,[81,82] rather than by way of α-subunits.

Stimulation of Ras by thrombin is also sensitive to tyrosine-kinase inhibitors; thus it was proposed that a tyrosine kinase may lie between Gi and Ras. It is generally believed that the nonreceptor tyrosine kinase Src lies upstream of Ras in signaling networks leading to cell growth. Interestingly, Src kinase has been found to be activated by thrombin in human platelets.[77,78] Recently, we demonstrated in CCL39 fibroblasts that thrombin stimulates Src-related kinase activity in immune complex kinase assays using antibodies that recognize the more ubiquitously expressed family members Src, Fyn, and Yes.[79] Thrombin activation of Src kinases in CCL39 cells is rapid and transient, and involves both pertussis toxin-sensitive and -insensitive G proteins. The mechanism by which G proteins activate Src kinases is not known; however, it has been suggested to involve the stimulation of a tyrosine phosphatase.[77]

Although Ras has previously been found to be downstream of Src, it is still not clear how Src participates in Ras activation. Shc proteins have been proposed to be involved in signaling from cytoplasmic-tyrosine kinases inasmuch as they are constitutively phosphorylated on tyrosine residues in cells transformed with v-Src and v-Fps.[80] Recent findings from our laboratory indicate that thrombin, or the peptide agonist TRP, stimulates persistent phosphorylation of Shc (both the p52 and p46 forms) on tyrosine residues (Y. Chen and E. Van Obberghen-Schilling, unpublished results). Following stimulation of CCL39 cells with thrombin, tyrosine-phosphorylated Shc associates with Grb2, and this complex may be involved in the activation of Sos, the Ras nucleotide exchange factor.

An important issue for future studies concerns the contribution of thrombin receptors to cell proliferation *in vivo,* both in physiological and pathological conditions, such as atherosclerosis or its accelerated forms (*e.g.,* restenosis or vein graft disease). Increasing evidence suggests that thrombin may also participate in various functions of the central and peripheral nervous system.[83,84] The possibility remains that the thrombin receptor may also be a receptor for an unknown peptide ligand in tissues that are not accessible to the blood-borne protease. To address these issues, we are currently undertaking experiments designed to ablate the thrombin-receptor gene in mice by homologous recombination. Conversly, efforts are being made to construct transgenic mice overexpressing the thrombin receptor, or mice expressing constitutively active forms of the receptor that should prove to be valuable animal models.

REFERENCES

1. Fenton II, J. W. 1986. Ann. N.Y. Acad. Sci. **485:** 5-15.
2. Fenton II, J. W., T. A. Olson, M. P. Zabinski & G. D. Wilner. 1988. Biochemistry **27:** 7106-7112.
3. Esmon, C. T. 1993. Thromb. Haemostasis **70:** 29-35.
4. Harmon, J. T. & G. A. Jamieson. 1986. J. Biol. Chem. **261:** 13224-13229.
5. Okumura, T. & G. A. Jamieson. 1976. Thromb. Res. **8:** 701-706.
6. McGowan, E. B., A. Ding & T. C. Detwiler. 1983. J. Biol. Chem. **258:** 11243-11248.

7. Moss, M., H. S. Wiley, J. W. Fenton II & D. D. Cunningham. 1983. J. Biol. Chem. **258:** 3996-4002.
8. Van Obberghen-Schilling, E. & J. Pouysségur. 1985. Biochim. Biophys. Acta **847:** 335-343.
9. Cunningham, D. D., L. Pulliam & P. J. Vaughan. 1993. Thromb. Haemostasis **70:** 168-171.
10. Sommer, J., S. M. Gloor, G. F. Rovelli, J. Hofsteenge, H. Nick, R. Meier & D. Monard. 1987. Biochemistry **26:** 6407-6410.
11. Vu, T. K. H., D. Hung, V. I. Wheaton & S. R. Coughlin. 1991. Cell **64:** 1057-1068.
12. Rasmussen, U. B., V. Vouret-Craviari, S. Jallat, Y. Schlesinger, G. Pagès, A. Pavirani, J.-P. Lecocq, J. Pouysségur & E. Van Obberghen-Schilling. 1991. FEBS Lett. **288:** 123-128.
13. Carney, D. H. 1992. Postclotting cellular effects of thrombin mediated by interaction with high affinity receptors. *In* Thrombin Structure and Function. L. Berliner, Ed.: 351-396. Plenum. New York.
14. Bar-Shavit, R., A. Kahn, G. Wilner & J. Fenton II. 1983. Science **220:** 728-731.
15. Bizios, R., L. Lai, J. W. Fenton II & A. B. Malik. 1986. J. Cell. Physiol. **128:** 485-490.
16. Dawes, K. E., A. J. Gray & G. L. Laurent. 1993. J. Cell Biol. **61:** 126-130.
17. Tordai, A., J. W. Fenton II, T. Andersen & E. W. Gelfand. 1993. J. Immunol. **150:** 4876-4886.
18. Mari, B., V. Imbert, N. Belhacene, D. Farahi Far, J.-F. Peyron, J. Pouysségur, E. Van Obberghen-Schilling, B. Rossi & P. Auberger. 1994. J. Biol. Chem. **269:** 8517-8523.
19. Bar-Shavit, R., M. Benezra, A. Eldor, E. Hy-Am, J. W. Fenton II, G. D. Wilner & I. Vlodavsky. 1990. Cell Regul. **1:** 453-463.
20. Herbert, J. M., I. Lamarche & F. Dol. 1992. FEBS Lett. **301:** 155-158.
21. McNamara, C. A., I. J. Sarembock, L. W. Gimple, J. W. Fenton II, S. R. Coughlin & G. K. Owens. 1993. J. Clin. Invest. **91:** 94-98.
22. Chen, L. B. & J. M. Buchanan. 1975. Proc. Natl. Acad. Sci. USA **72:** 131-135.
23. Carney, D. H. & D. D. Cunningham. 1978. Cell **14:** 811-823.
24. Van Obberghen-Schilling, E. & J. Pouysségur. 1983. Exp. Cell Res. **147:** 369-378.
25. Harlan, J. M., P. J. Thompson, R. R. Ross & D. F. Bowen-Pope. 1986. J. Cell. Biol. **103:** 1129-1133.
26. Stouffer, G. A., I. J. Sarembock, C. A. McNamara, L. W. Gimple & G. K. Owens. 1993. Am. J. Physiol. **265:** C806-C811.
27. Weiss, R. H. & M. Maduri. 1993. J. Biol. Chem. **268:** 5724-5727.
28. Gurwitz, D. & D. D. Cunningham. 1988. Proc. Natl. Acad Sci. USA **85:** 3440-3444.
29. Suidan, H. S., S. R. Stone, B. A. Hemmings & D. Monard. 1992. Neuron **8:** 363-375.
30. Jalink, K. & W. H. Moolenaar. 1992. J. Cell. Biol. **118:** 411-419.
31. Jenkins, A. L., M. D. Bootman, V. W. Taylor, E. J. Mackie & S. R. Stone. 1993. J. Biol. Chem. **268:** 21433-21437.
32. Bahou, W., B. Coller, C. Potter, K. Norton, J. Kutok & M. Goligorsky. 1993. J. Clin. Invest. **91:** 1405-1413.
33. Zhong, C., D. J. Hayzer, M. A. Corson & M. S. Runge. 1992. J. Biol. Chem. **267:** 16975-16979.
34. Gerszten, R. E., J. Chen, M. Ishii, K. Ishii, L. Wang, T. Nanevicz, C. W. Turck, T.-K. Vu & S. R. Coughlin. 1994. Nature **368:** 648-650.
35. Bahou, W. F., W. C. Nierman, A. S. Durkin, C. L. Potter & D. J. Demetrick. 1993. Blood **82:** 1532-1537.
36. Vu, T.-K. H., V. I. Wheaton, D. T. Hung, I. Charo & S. R. Coughlin. 1991. Nature **353:** 674-677.
37. Liu, L.-W., T.-K. H. Vu, C. T. Esmon & S. R. Coughlin. 1991. J. Biol. Chem. **266:** 16977-16980.

38. Bode, W., P. Schwager & R. Huber. 1978. J. Mol. Biol. **118:** 99-112.
39. Vouret-Craviari, V., E. Van Obberghen-Schilling, U. B. Rasmussen, A. Pavirani, J. P. Lecocq & J. Pouysségur. 1992. Mol. Biol. Cell **3:** 95-102.
40. Vassallo, R. R., T. Kieber-Emmons, K. Cichowski & L. F. Brass. 1992. J. Biol. Chem. **267:** 6081-6085.
41. Sabo, T., D. Gurwitz, L. Motola, P. Brodt, R. Barak & E. Elhanaty. 1992. Biochem. Biophys. Res. Commun. **188:** 604-610.
42. Scarborough, R. M., M. A. Naughton, W. Teng, D. T. Huang, J. Rose, T.-K. H. Vu, V. I. Wheaton, C. W. Turck & S. R. Coughlin. 1992. J. Biol. Chem. **287:** 13146-13149.
43. Chao, B. H., S. Kalkunte, J. M. Maraganore & S. R. Stone. 1992. Biochemistry **31:** 6175-6178.
44. Van Obberghen-Schilling, E., U. B. Rasmussen, V. Vouret-Craviari, K. U. Lentes, A. Pavirani & J. Pouysségur. 1993. Biochem. J. **292:** 667-671.
45. Brass, L. F., R. R. Vassalo, E. Belmonte, M. Ahuja, K. Cichowski & J. A. Howie. 1992. J. Biol. Chem. **267:** 13795-13798.
46. Hung, D. T., T.-K. H. Vu, V. I. Wheaton, K. Ishil & S. R. Coughlin. 1992. J. Clin. Invest. **89:** 1350-1353.
47. Brass, L. F., S. Pizarro, M. Ahuja, E. Belmonte, N. Blanchard, J. M. Stadel & J. A. Hoxie. 1994. J. Biol. Chem. **269:** 2943-2952.
48. Ishii, K., L. Hein, B. Kobilka & S. R. Coughlin. 1993. J. Biol. Chem. **268:** 9780-9786.
49. Vouret-Craviari, V., P. Auberger, J. Pouysségur & E. Van Obberghen-Schilling. 1995. J. Biol. Chem. **270:** 4813-4821.
50. Suidan, H. S., J. Bouvier, E. Schaerer, S. R. Stone, D. Monard & J. Tschopp. 1994. Proc. Natl. Acad. Sci. USA **91:** 8112-8116.
51. Nystedt, S., K. Emilsson, C. Wahlestedt & J. Sundelin. 1994. Proc. Natl. Acad. Sci. USA **91:** 9208-9212.
52. Paris, S., I. Magnaldo & J. Pouysségur. 1988. J. Biol. Chem. **263:** 11250-11256.
53. Brass, L. F. 1992. J. Biol. Chem. **267:** 6044-6050.
54. Lohse, M. J. 1993. Biochim. Biophys. Acta **1179:** 171-188.
55. Ishii, K., J. Chen, M. Ishii, W. J. Koch, N. J. Freedman, R. J. Lefkowitz & S. R. Coughlin. 1994. J. Biol. Chem. **269:** 1125-1130.
56. Chen, C.-Y., S. B. Dion, C. M. Kim & J. L. Benovic. 1993. J. Biol. Chem. **268:** 7825-7831.
57. Hoxie, J. A., M. Ahuja, E. Belmonte, S. Pizarro, R. Parton & L. F. Brass. 1993. J. Biol. Chem. **268:** 13756-13763.
58. Schini, V. B., B. Fiblthaler, E. Van Obberghen-Schilling & R. Busse. 1994. FASEB J. **8:** A555 (3212).
59. Pouysségur, J., C. Kahan, I. Magnaldo & K. Seuwen. 1990. G-proteins and cell growth signalling. *In* Transmembrane signalling, intracellular messengers and implications for drug development. S. R. Nahorski, Ed.: 119-132. John Wiley & Sons Ltd. Chicester.
60. Pouysségur, J. & K. Seuwen. 1992. Annu. Rev. Physiol. **54:** 195-210.
61. L'Allemain, G., S. Paris, I. Magnaldo & J. Pouysségur. 1986. J. Cell. Physiol. **129:** 167-174.
62. Paris, S. & J. Pouysségur. 1986. EMBO J. **5:** 55-60.
63. Magnaldo, I., J. Pouysségur & S. Paris. 1988. Biochem. J. **253:** 711-719.
64. Chambard, J. C., S. Paris, G. L'Allemain & J. Pouysségur. 1987. Nature **326:** 800-803.
65. Meloche, S., K. Seuwen, G. Pagès & J. Pouysségur. 1992. Mol. Endocrinol. **6:** 845-854.
66. Lenormand, P., C. Sardet, G. Pagès, G. L'Allemain, A. Brunet & J. Pouysségur. 1993. J Cell. Biol. **122:** 1079-1088.
67. Pagès, G., P. Lenormand, G. L'Allemain, J. C. Chambard, S. Melcohe & J. Pouysségur. 1993. Proc. Natl. Acad. Sci. USA **90:** 8319-8323.

68. BRUNET, A., G. PAGÈS & J. POUYSSÉGUR. 1994. Oncogene **9:** 3379-3387.
69. MARSHALL, C. J. 1994. Curr. Opinion Genetics Dev. **4:** 82-89.
70. JOHNSON, G. L. & R. R. VAILANCOURT. 1994. Curr. Opinion Cell. Biol. **6:** 230-238.
71. VAN CORVEN, E. J., P. L. HORDIJK, R. H. MEDEMA, J. L. BOS & W. H. MOOLENAAR. 1993. Proc. Natl. Acad. Sci. USA **90:** 1257-1261.
72. WINITZ, S., M. RUSSELL, N.-X. QIAN, A. GARDNER, L. DWYER & G. L. JOHNSON. 1993. J. Biol. Chem. **268:** 19196-19199.
73. ALBLAS, J., E. J. VAN CORVEN, P. L. HORDIJK, G. MILLIGAN & W. H. MOOLENAAR. 1993. J. Biol. Chem. **268:** 22235-22238.
74. HOWE, L. R. & C. J. MARSHALL. 1993. J. Biol. Chem. **268:** 20717-20720.
75. WORTHEN, G. S., N. AVDI, A. M. BUHL, N. SUZUKI & G. L. JOHNSON. 1994. J. Clin. Invest. **97:** 815-823.
76. LAMORTE, V. J., E. D. KENNEDY, L. R. COLLINS, D. GOLDSTEIN, A. T. HAROOTUNIAN, J. H. BROWN & J. R. FERASMICO. 1993. J. Biol. Chem. **268:** 19411-19415.
77. CLARK, E. A. & J. S. BRUGGE. 1993. J. Mol. Cell. Biol. **13:** 1863-1871.
78. LIBENHOFF, U., D. BROCKMEIER & P. PRESEK. 1993. Biochem. J. **295:** 41-48.
79. CHEN, J. & R. IYENGAR. 1994. Science **263:** 1278-1281.
80. MCGLADE, J., A. CHENG, G. PELICCI, P. G. PELICCI & T. PAWSON. 1992. Proc. Natl. Acad. Sci. USA **89:** 8869-8873.
81. CRESPO, P., N. XU, W. SIMONDS & J. S. GUTKIND. 1994. Nature **369:** 418-420.
82. FAURE, M., T. A. VOYNO-YASENETSKAYA & H. R. BOURNE. 1994. J. Biol. Chem. **269:** 7851-7854.
83. DAVIS-SALINAS, J., S. SAPORITO-IRWIN, F. M. DONOVAN, D. D. CUNNINGHAM & W. E. VAN NOSTRAND. 1994. J. Biol. Chem. **269:** 22623-22627.
84. LIU, Y., D. FIELDS, B. FESTOFF & P. G. NELSON. 1994. Proc. Natl. Acad. Sci. USA **91:** 10300-10304.
85. KREIS, T. E. 1986. EMBO J. **5:** 931-941.

Structural and Functional Properties of the TRK Family of Neurotrophin Receptors

MARIANO BARBACID

Department of Molecular Biology
Bristol-Myers Squibb Pharmaceutical Research Institute
P.O. Box 4000
Princeton, New Jersey 08543-4000

INTRODUCTION

The development and survival of the mammalian nervous system is largely dependent on the existence of soluble neurotrophic factors.[1] The best-known and most intensively studied among these factors is the NGF family of neurotrophins, which includes nerve-growth factor (NGF), brain-derived neurotrophic factor (BDNF), neurotrophin-3 (NT-3), and neurotrophin-4 (NT-4), also known as NT-5.[2] A putative new member of this gene family, neurotrophin-6 (NT-6), has been recently isolated from the teleost fish *Xephophorus.*[3]

These neurotrophins recognize two different classes of receptors, the Trk family of tyrosine-protein kinases and the low-affinity receptor, p75, a member of the TNF receptor superfamily.[4] The role of p75 in neurotrophin function is still unresolved.[5,6] However, accumulating evidence supports the concept that p75 may facilitate the interaction of NGF with its signaling receptor TrkA (see below), perhaps by increasing the concentration of NGF in the vicinity of TrkA receptors. For instance, survival of sensory cutaneous neurons lacking p75 receptors respond normally to saturating concentrations of NGF. However, at limiting concentrations, these p75-defective neurons require four times as much NGF as wild-type neurons.[7] These observations may explain the sensory defects observed in transgenic mice lacking p75 receptors.[8] Likewise, these mice do not appear to have significant deficiencies in their sympathetic neurons, a major target for NGF. However, they display reduced innervation of distal sympathetic targets, a defect that may reflect the limited availability of NGF during distal axonal growth.[9] Interestingly, all the defects reported so far in these mice have been observed in NGF-dependent neurons,[8] in spite of the fact that p75 also serves as a receptor for the other members of the NGF neurotrophin family.[4]

Neurotrophin function is primarily mediated by the Trk family of tyrosine-kinase receptors.[6] The Trk receptors were first identified in 1986 when the *trk* gene (now known as *trk*A) was found to be part of an oncogene isolated from a human colon carcinoma.[10,11] However, it was not until 1991 that their physiological role was unveiled when the TrkA tyrosine kinase was shown to be the signaling receptor for NGF.[12,13] Other members of the Trk family of tyrosine kinases include TrkB,[14–16] the signaling receptor for BDNF[17–19] and NT-4[20–22]; and TrkC, the primary receptor

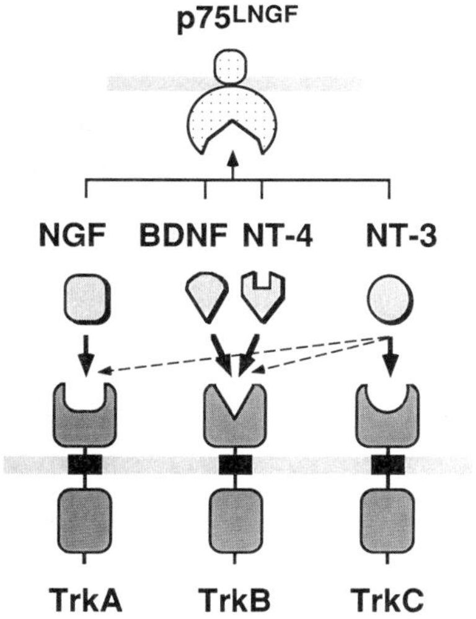

FIGURE 1. Schematic representation of the interaction between the NGF family of neurotrophins and their receptors. The thin arrow represents their low-affinity binding to the p75 receptor. The thick arrows indicate their binding to their corresponding Trk receptors. Only the Trk tyrosine kinase isoforms are depicted. The activation of TrkA and TrkB by NT-3 detected in certain cell-culture systems is indicated by thin, dashed arrows.

for NT-3[23] (FIG. 1). Whether any of these Trk kinases also serves as a receptor for the recently discovered NT-6[3] remains to be determined.

During the last few years, much has been learned about the structure and function of the Trk receptors, as well as about their pattern of expression within the mammalian nervous system.[4] Moreover, the recent generation of transgenic strains of mice lacking these receptors has provided definitive evidence regarding their critical role in the mammalian nervous system.[24–26] These mutant mice exhibit severe neuronal deficiencies indicative of the critical roles played by the Trk receptors. Yet, these defects, particularly those observed in mice lacking TrkB and TrkC receptors, are not as severe as it might have been predicted from their patterns of expression,[27] thus illustrating the complex layers of signaling and compensatory mechanisms that regulate the development and survival of neurons. Interestingly, the phenotypic defects displayed by the Trk defective mice are highly reminiscent of those observed in mice lacking their corresponding activating neurotrophins,[28] thus providing genetic support to the concept that the Trk receptors are the primary mediators of the biological properties of the NGF family of neurotrophins *in vivo.* This chapter attempts to summarize our current knowledge regarding the signaling pathways used by the Trk receptors to mediate neurotrophin activity as well as their role *in vivo* as determined from analysis of transgenic mice carrying targeted mutations in each of the three known *trk* genes.

TRK RECEPTORS: STRUCTURAL FEATURES

To date, three members of the *trk* gene family, *trk*A, *trk*B and *trk*C, have been identified. These genes encode for two distinct classes of receptors: those with tyrosine-protein kinase activity, and noncatalytic receptors that share their respective extracellular and transmembrane sequences but that differ in their cytoplasmic region (FIG. 2, TABLE 1). Whereas the role of the Trk tyrosine-kinase receptors is to mediate the trophic properties of the NGF family of neurotrophins, the role of their noncatalytic isoforms is still unknown. Analysis of the deduced amino acid sequences of the Trk receptors indicates that they do not belong to any of the well-characterized subfamilies of tyrosine-protein kinases.[29] In particular, their extracellular domains display a unique array of structural motifs indicative of a distinct subfamily (FIG. 2). Below is a brief description of the main structural features of the Trk family of receptors.

TRKA

The *trk*A gene encodes two tyrosine-protein kinase isoforms of 790 and 796 amino acid residues, designated as TrkA (also known as Trk or gp140*trk*).[11,30–32] These isoforms differ from each other in the presence of 6 amino acid residues (VSFSPV) located in the extracellular domain near the transmembrane region (FIG. 2). These additional sequences do not affect NGF binding. Moreover, these isoforms appear to have similar biological properties (see below). However, the 796 amino acid long TrkA molecule is primarily expressed in neuronal cells, whereas the 790 amino acid long isoform has been found in cells of nonneuronal origin.[31,32]

TrkA has all the basic features characteristic of tyrosine-protein kinase cell-surface receptors (FIG. 2). They include a 32 amino acid-long signal peptide followed by three tandem leucine-rich motifs (LRM) of 24 amino acid residues flanked by two cysteine clusters that include eight of the twelve cysteine residues conserved among all known Trk receptors.[33] This structural feature is characteristic of the LRM superfamily of proteins, which includes proteins of diverse function, such as the human platelet von Willebrand factor receptor, the ribonuclease/angiogenin inhibitor, cell adhesion proteins, and extracellular matrix components.[33] These motifs are supposed to be involved in mediating tight and specific protein-protein interactions.

The carboxy moiety of the TrkA extracellular domain contains two immunoglobulin (Ig)-like domains of the C2 type[33] (FIG. 1). The first Ig-like domain contains two conserved cysteines presumably involved in generating the canonical disulfide bond. The second Ig-like domain has retained one of the two cysteine residues, whereas the other cysteine has been replaced by a leucine. Ig-like domains have been previously identified in other cell-surface tyrosine-protein kinases, such as the receptors for PDGF, CSF-1, SF, and FGFs. However, the Ig-like domains of TrkA, as well as those of TrkB and TrkC (see below), appear to be more closely related to those of cell-adhesion molecules, such as N-CAM. Interestingly, D*trk,* the *Drosophila* homologue of the mammalian Trk receptors, which also contains two C2 Ig-like domains very similar to those present in their mammalian counterparts, mediates cell aggregation *in vitro,* a process that leads to the activation of its tyrosine-protein kinase domain.[34] Whether the mammalian Trk receptors have similar cell-aggregation

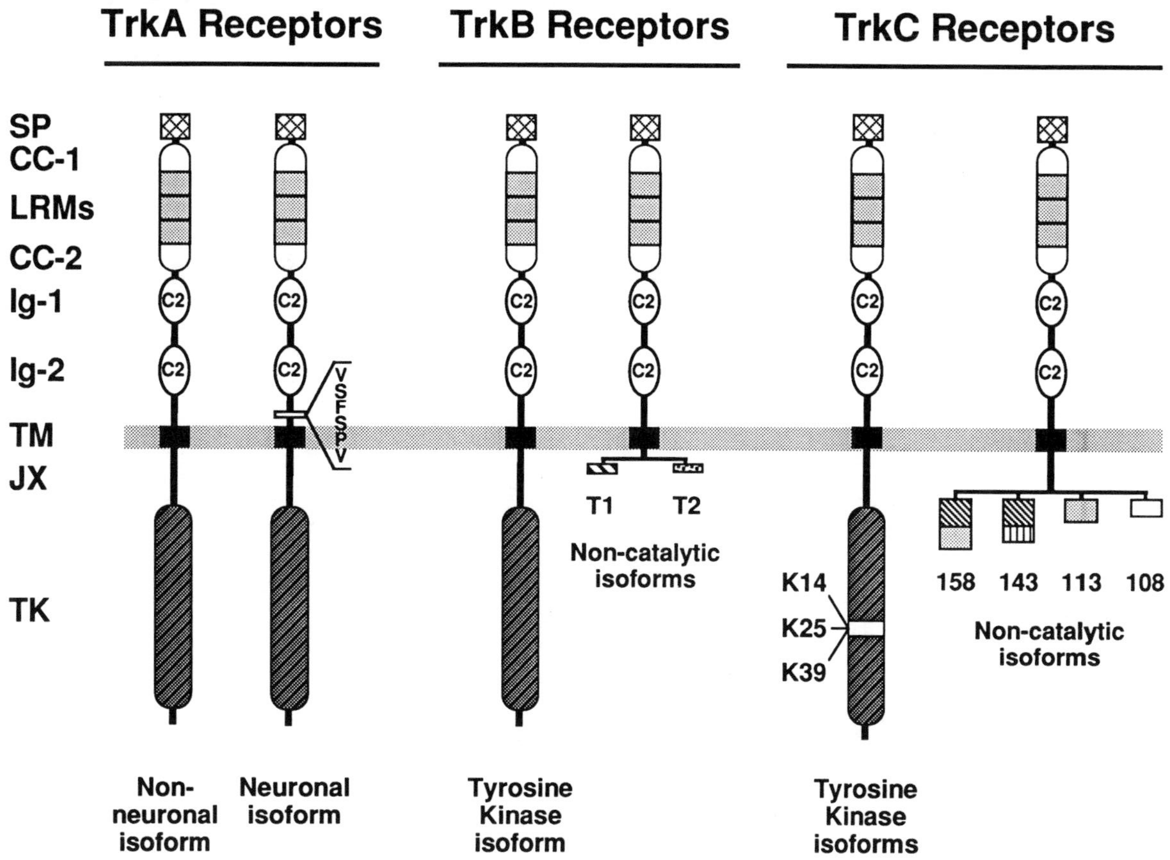

FIGURE 2. The Trk family of neurotrophin receptors. Structural domains include signal peptide (SP), cysteine clusters (CC), leucine-rich motifs (LRMs), immunoglobulin-like C2-type motifs (Ig), transmembrane domain (TM), juxtamembrane region (JX), and tyrosine-kinase catalytic domain (TK). The six amino acid residues (VSFSPV) present in the neuronal-specific TrkA receptor are indicated by a small open box. Additional sequences present in the TrkC tyrosine kinase isoforms TrkC K14, TrkC K25, and TrkC K39 receptors are indicated by an open box located in the tyrosine-kinase domain of the TrkC K1 receptor. Sequences unique to the Trk noncatalytic receptor isoforms, including TrkB.T1 and TrkB.T2, as well as $TrkC^{TK-}(158)$, $TrkC^{TK-}(143)$, $TrkC^{TK-}(113)$, and $TrkC^{TK-}(108)$, are depicted by boxes with various shadings. The double boxes shown in $TrkC^{TK-}(158)$ and $TrkC^{TK-}(143)$ indicate that these sequences are derived from different exons, one of which is shared by these putative receptors.

TABLE 1. The Trk Family of Neurotrophin Receptors

Gene	Receptor	Other Designations	Size	Structure
*trk*A	TrkA (nonneuronal)	gp140^{trkA}	790 aa[a]	Tyrosine kinase
	TrkA (neuronal)	gp140^{trkA}	796 aa	Tyrosine kinase
*trk*B	TrkB	TrkB^{TK+}/gp145^{trkB}	821 aa	Tyrosine kinase
	TrkB.T1	TrkB^{TK-}/gp95^{trkB}	476 aa	Noncatalytic
	TrkB.T2	TrkB^{TK-}	474 aa	Noncatalytic
*trk*C	TrkC	TrkC K1/TrkC^{TK+}/gp145^{trkC}	825 aa	Tyrosine kinase
	TrkC K14	TrkC K2/TrkC^{TK+}(14)	839 aa	Tyrosine kinase
	TrkC K25	TrkC K3/TrkC^{TK+}(25)	850 aa	Tyrosine kinase
	TrkC K39	TrkC K4/TrkC^{TK+}(39)	864 aa	Tyrosine kinase
	TrkC^{TK-} (158)		686 aa	Noncatalytic
	TrkC^{TK-} (143)		671 aa	Noncatalytic
	TrkC^{TK-} (113)		641 aa	Noncatalytic
	TrkC^{TK-} (108)		636 aa	Noncatalytic

[a] aa = amino acids.

properties remains to be tested. Close examination of the amino acid sequences of the second C2 Ig-like domain in each of the known members of the Trk family of receptors indicates a high degree of conservation. These sequences are likely to play a significant role in modulating the catalytic activity of the kinase domain because a 51 amino acid deletion encompassing this region results in the oncogenic activation of the TrkA receptor.[35] Similar results were obtained by replacing the conserved cysteine residue present in the middle of this domain by serine.[35] Whether this C2 Ig-like domain participates in ligand interaction is currently under investigation.

TrkA has a single transmembrane domain followed by a 76 amino acid-long juxtamembrane region that contains a putative ATP binding site (GKGSGLQGX_{17} K).[11] Whether this domain binds ATP or is required for biological activity is not known; however, this motif is not present in the related TrkB and TrkC receptors.[17,23] The region of homology with the catalytic domain of other tyrosine-protein kinases extends for 266 amino acid residues, of which 99 (37%) are conserved in other tyrosine-kinase cell-surface receptors.[11] However, TrkA has a set of structural features unique to this subfamily of receptors. They include a Thr residue (Thr^{647}) instead of Ala (underlined) in the highly conserved HRDLAARN kinase motif, a Trp residue (Trp^{722}) instead of Tyr in the conserved WEX_7PY sequence, and the replacement of the conserved Pro (Pro^{766}) in the CWX_6RP sequence located at the carboxyterminal end of the kinase domain.[11] In addition, TrkA, as well as the related TrkB and TrkC tyrosine kinases (see below), has a very short carboxy-terminal tail of 15 amino acids, which includes a conserved Tyr residue[11] responsible for the binding of phospholipase Cγ (see below). Outside of this receptor subfamily, the TrkA cytoplasmic region is most closely related to that of several new receptors, including the mammalian ROR1, ROR2, DDR (also known as Ptk-3, Nep, or TrkE), Tyro-10, and TKT, as well as the torpedo RTK tyrosine kinases.[29] These orphan receptors, however, have rather distinct extracellular domains, and it is unlikely that they recognize members of the NGF neurotrophin family.

TRKB

The TrkB tyrosine-kinase receptor is a heavily glycosylated molecule of 821 amino acid residues that contains all of the structural motifs described above for the TrkA receptor[14] (FIG. 2). The overall homology between the extracellular domains of the human TrkA and the mouse TrkB tyrosine kinases is 57% (38% identity). However, this homology is not randomly distributed. Most of the homologous residues map within the second and third LRM motifs and particularly within the second Ig-like domain.[14] In addition, each of the twelve extracellular cysteines present in TrkA are conserved in TrkB. As expected, the highest degree of homology (88%) between TrkB and TrkA occurs in their catalytic domains.[14] TrkB also possesses a short, highly conserved tail of 15 residues, which includes the conserved Tyr shared by the TrkA and TrkC tyrosine kinases.[14]

TRKC

The *trkC* gene encodes as many as four TrkC tyrosine kinase isoforms.[36–38] One of these isoforms, TrkC (also known as TrkC K1 and gp145^{trkC}), displays the same structural features as the related TrkA and TrkB tyrosine-kinase receptors (FIG. 2). The overall homology of the porcine TrkC protein compared to the human TrkA and mouse TrkB receptors is 67% (54% in the extracellular domain and 87% in the kinase region) and 68% (53% in the extracellular domain and 87% in the kinase region), respectively.[23] The other tyrosine-kinase isoforms differ from TrkC in the presence of 14 (TrkC K14), 25 (TrkC K25), or 39 (TrkC K39) additional amino acid residues located after the conserved sequence YSTDYYR, which encompasses the putative autophosphorylation site of TrkC[36–38] (FIG. 2). The unique 14 and 25 amino acid-long sequences of TrkC K14 and TrkC K25 receptors are unrelated to other known sequences and do not display informative structural motifs. TrkC K39 contains the combined 25 and 14 amino acid-long sequences (in this order) of TrkC K25 and TrkC K14, respectively.[37] Recent studies indicate that these additional residues are encoded by alternatively spliced exons (our unpublished observations). To date, immunoprecipitation studies using specific antisera against these sequences have demonstrated low levels of expression of the TrkC K14 receptor isoform in various structures of the adult mouse brain.[36]

Noncatalytic Isoforms

As indicated above, the *trk*B and *trk*C genes encode a second class of Trk receptors that lack a cytoplasmic tyrosine kinase catalytic domain. These receptors, generically designated as Trk^{TK-}, have the same extracellular and transmembrane domains as the TrkB and TrkC tyrosine kinases described above. However, they have distinct and unique cytoplasmic domains of various lengths (from 21 residues in TrkB.T2 to

233 residues in TrkC^{TK-}(158), see TABLE 1) devoid of distinctive structural motifs. *trk*B encodes two noncatalytic receptors designated as TrkB.T1 (also known as gp95^{trkB}) and TrkB.T2.[15,16] TrkB.T1 only has 23 cytoplasmic residues of which the last eleven are unique.[15] This receptor is expressed in adult mouse brain at levels comparable to those of the signaling TrkB tyrosine-kinase receptor.[15] TrkB.T2 has a 21 amino acid-long cytoplasmic domain of which the last nine residues are unique and therefore unrelated to those present in TrkB.T1. So far, TrkB.T2 has not been identified at the protein level.[16]

To date, cDNAs encoding four different TrkC^{TK-} receptor isoforms have been described.[37,38] They have been designated as TrkC^{TK-}(158), TrkC^{TK-}(143), TrkC^{TK-}(113), and TrkC^{TK-}(108), based on the number of cytoplasmic residues (FIG. 2, TABLE 1). The first 74 cytoplasmic residues of each of these TrkC^{TK-} receptors correspond to those of the juxtamembrane region of the catalytic TrkC receptors (FIG. 2). However, sequences starting at residue 529 of the TrkC kinase are derived from four distinct alternatively spliced exons, designated as A, B, C, and D.[37,38] Whereas TrkC^{TK-}(158) contains exons B (46 residues) and C (38 residues), TrkC^{TK-}(143) has exons B and D (23 residues), and TrkC^{TK-}(113) only possesses exon C (FIG. 2). The cDNA clones encoding the TrkC^{TK-}(108) receptor contain exons A, B, and C. However, exon A only encodes a short peptide of 34 amino acid residues, which is followed by an in-frame terminator codon that prevents translation of the downstream sequences contained in the B and C exons.[37,38] To date, it is not known whether any of these TrkC^{TK-} noncatalytic receptors are expressed *in vivo.*

SIGNAL TRANSDUCTION

Tyrosine-kinase Receptors

The role of the Trk family of tyrosine-protein kinases is to serve as the signaling receptors for the NGF family of neurotrophins. NGF signals through the TrkA; BDNF and NT-4 signal through TrkB; and NT-3 signals through TrkC[4,27] (FIG. 1). The interaction of NGF, BDNF, and NT-4 with their cognate-receptor kinases is highly specific. However, NT-3 can bind and activate TrkA[39] and TrkB[17–19] receptors, at least in certain cell culture systems (FIG. 1). However, it is unlikely that NT-3 mediates its trophic properties through either of these two receptors *in vivo.*[40]

Activation of Trk-kinase receptors by their cognate ligands in culture initiates a cascade of signaling events that result in either neuronal differentiation (*i.e.,* PC12 cells) or mitogenic stimuli leading to malignant transformation (*i.e.,* NIH3T3 cells). Trk receptors become activated by a two-step process that involves the ligand-mediated oligomerization of receptor molecules at the cell surface followed by autophosphorylation of their tyrosine residues,[41] a mechanism shared by all known tyrosine kinase receptors.[42] Whether accessory molecules, such as the low-affinity receptor p75, play a direct role in the activation of the Trk receptors is still a matter of debate.[5,6]

The phosphorylated tyrosine residues in the Trk-cytoplasmic domain serve as anchors for binding downstream signaling elements.[42] Two distinct classes of molecules are known to participate in this process, enzymes and adapters. Both of them mediate their interaction with the activated Trk receptors by SH2 domains that recognize specific phosphotyrosine residues.[42] Upon receptor binding, enzymes become phosphorylated on tyrosine residues, a step required for their activation. Adapters, however, are not always phosphorylated. The role of these adapter molecules is to facilitate the interaction among other signaling molecules by bringing them together to the cell membrane.[42,43]

One of the best characterized substrates of the Trk family of receptors is phospholipase Cγ (PLCγ),[44–46] an enzyme that hydrolyzes phosphatidylinositol bisphosphate to generate inositol trisphophate and diacylglycerol, two second messengers that induce Ca^{2+} release and activate protein kinase C, respectively.[47] PLCγ binds to the conserved Tyr residue located in the short carboxyterminal tail characteristic of this receptor family (Tyr^{785} in TrkA).[48] However, the physiological significance of this interaction remains obscure because Trk receptors carrying a Phe^{785} mutation that abolishes interaction with PLCγ retain their ability to transform NIH3T3 cells and to differentiate PC12 cells.

Another potential substrate for the Trk receptor is the Ras GTPase-activating protein (GAP).[48] However, there is no evidence that tyrosine phosphorylation of GAP results in increased enzymatic activity. The MAP kinase ERK1 (but not of the highly related ERK2 kinase) has been found to be associated with Trk receptors in immunoprecipitates derived from NGF-treated PC12 cells.[49] Whether the activation of GAP and MAP kinases by NGF is mediated by their direct interaction with the Trk receptors or by an indirect mechanism (see below) remains to be determined.

Adapters known to interact with the Trk receptors include P85, the regulatory subunit of phosphatidylinositol-3′ kinase (PI-3K), and Shc. Whereas P85 binds to the consensus sequence motif YXXM located at the end of the kinase catalytic domain (Y^{751} in TrkA), the Shc binding site has been mapped to a conserved phosphotyrosine residue located in the juxtamembrane domain (Y^{490} in TrkA).[48,50] Activation of TrkA receptors in PC12 cells results in the stimulation of PI-3 kinase activity presumably by a mechanism involving binding of the catalytic P110 subunit to the tyrosine-phosphorylated P85 adaptor.[36,50,51] Likewise, NGF treatment of PC12 cells results in the rapid phosphorylation of the Shc adapter protein in tyrosine residues.[52] Moreover, overexpression of Shc leads to their neuronal differentiation of PC12 cells,[53] thus suggesting that Shc is an important mediator of NGF signaling.

Accumulating evidence indicates that NGF signals, at least in part, through the well-known Ras/Raf/MAP kinase pathway.[43] Addition of NGF to wild-type, but not to Trk-deficient, PC12 cells results in the rapid activation of Ras proteins,[54–56] as well as of the downstream Raf and MAP kinases.[49,56,57] Moreover, expression of oncogenic Ras and Raf proteins in PC12 cells results in their neuronal differentiation in the absence of NGF.[58–60]

These observations, however, do not establish how NGF induces neuronal differentiation and survival. It is important to realize that NGF signaling through Trk receptors does not necessarily lead to a differentiation process. Autocrine activation of Trk receptors (as well as TrkB and TrkC receptors) in proliferating cells, such as NIH3T3 fibroblasts, elicits potent mitogenic signals that result in their malignant transformation.[17,21,23,39] These observations suggest the existence of neuronal-specific signaling elements that will process the Trk downstream signals into differentiation-specific pathways. One such candidate is SNT, a 90-kDa protein that binds to p13, a subunit of the cell-cycle regulatory complex that includes the $p34^{cdc2}$ kinase and cyclin.[61] SNT is rapidly phosphorylated on tyrosine residues upon NGF treatment of PC12 cells as well as in primary cortical neurons treated with either BDNF or NT-3. More importantly, SNT is not phosphorylated in PC12 cells exposed to EGF, a mitogenic factor for both PC12 and NIH3T3 cells.[61]

Recent studies, however, suggest that what triggers PC12 cells to proliferate or to differentiate may not be due to the existence of unique signal pathways but to quantitative differences in the intensity and/or duration of the signals elicited by the activated tyrosine-kinase receptors. Overexpression of EGF and insulin receptors induces differentiation instead of proliferation of PC12 cells.[62,63] In these cells, MAP kinase remains activated longer and is more efficiently translocated to the nucleus than in proliferating wild-type PC12 cells. Moreover, these kinetics of MAP-kinase activation closely resemble those observed in NGF-treated PC12 cells. Whether SNT is also tyrosine phosphorylated in differentiated PC12 cells overexpressing EGF and insulin receptors remains to be determined.

Most of the above studies have been carried out with the NGF/TrkA receptor system. Ectopic expression of TrkB and TrkC receptors in PC12 cells also induces their differentiation into neuron-like cells when incubated in the presence of their cognate ligands, BDNF (or NT-4)[19–22] and NT-3,[36–38] respectively. Likewise co-expression of TrkB and BDNF (or NT-4) and of TrkC and NT-3 in proliferating NIH3T3 cells results in their morphologic transformation, with the same kinetics and potency observed upon expression of NGF and Trk receptors.[17,21,23,39] BDNF-activated TrkB and NT-3-activated TrkC receptors can also support the survival of these fibroblastic cells in the absence of serum.[36–38,64] These observations suggest that the TrkB and TrkC tyrosine kinase receptors may use the same signaling pathways as TrkA.

This is not the case, however, with the TrkC tyrosine-kinase isoforms, TrkC K14 and TrkC K25. NT-3 binds to and activates these receptors with the same affinity and kinetics as TrkC.[36–38] Yet, TrkC, but not TrkC K14 or TrkC K25, activates PLCγ and PI-3 kinase.[36] These findings are somewhat unexpected because each of these isoforms contains identical PLCγ and PI-3 kinase-recognition sites. At this time, it is not known whether the additional sequences present in the TrkC K14 and TrkC K25 kinase isoforms cause allosteric alterations that hinder their ability to bind to these signaling molecules or prevent the phosphorylation of their anchoring Tyr residues.

In spite of their inability to activate PLCγ and PI-3 kinase, the TrkC K14 and TrkC K25 receptor isoforms engage in downstream signaling and can induce resting cells to initiate DNA synthesis upon addition of NT-3.[36] However, the biological responses elicited by these receptors are limited. Neither TrkC K14 nor TrkC K25 induces survival of NIH3T3 cells in the absence of serum, nor induces their morphological transformation under standard culture conditions. More importantly, neither of these receptor isoforms mediate neuronal-like differentiation of PC12.[36–38] These observations may reflect the engagement of these receptors with specialized signaling pathways for which no suitable *in vitro* assays have been, as yet, examined. Alternatively, it is possible that trkC K14 or TrkC K25 interact with some, but not all of the signaling pathways required to convey the full neurotrophic activities of NT-3.

In either case, the physiological role of these TrkC tyrosine kinase receptor isoforms remains to be established.

Noncatalytic Receptors

There is little information regarding the role, if any, of the $TrkB^{TK-}$ and $TrkC^{TK-}$ noncatalytic receptor isoforms in signal transduction. It is possible that these receptors engage with cytoplasmic signaling elements (*i.e.,* nonreceptor tyrosine kinases) in a manner similar to certain lymphocytic cell-surface molecules. Alternatively, these receptors may act as dominant negative inhibitors of their tyrosine-kinase isoforms, at least in those cells where they are coexpressed. However, no experimental evidence supporting any of these two hypotheses has been obtained. Instead, it is more likely that the $TrkB^{TK-}$ and $TrkC^{TK-}$ noncatalytic receptors play a role in functions other than signaling. Indirect evidence based on the pattern of expression of the TrkB.T1 led to the hypothesis that this receptor may be involved in ligand clearance or transport.[15] Likewise, induction of this receptor upon neuronal injury has led to the proposal that it might be involved in ligand recruitment and/or presentation during axon growth and/or regeneration.[65] Recent results using genetically altered mice (see below) have provided strong support for such a role in the case of the p75 low-affinity NGF receptor.[7–9] Inasmuch as the *trk*A gene appears to be the only member of the *trk* gene family that does not express noncatalytic receptors, it is tempting to speculate that the $TrkB^{TK-}$ and $TrkC^{TK-}$ isoforms may play the same role as p75 in NGF-dependent, TrkA-expressing cells.

GENE TARGETING OF TRK TYROSINE-KINASE RECEPTORS

During the last year, several reports have described the generation of strains of transgenic mice deficient for some of the neurotrophins (NGF, BDNF, and NT-3) and for each of the three known members of the Trk receptor family.[28] Detailed anatomical and functional characterization of these animals is providing critical information for understanding the role of neurotrophin signaling in the development and maintenance of the mammalian nervous system. Below is a summary of our recent studies involving the characterization of transgenic mice lacking functional TrkA, TrkB, and TrkC tyrosine-kinase receptors.

TRKA-defective Mice

Mice carrying a deletion in the tyrosine-kinase domain of the *trk*A gene have been generated by homologous recombination in embryonic stem cells.[25] Animals homozygous for this mutation appear anatomically normal at birth and feed normally. However, by postnatal day 10 (P10), the *trk*A (–/–) mice are significantly smaller in size as compared to their (+/+) and (+/–) siblings and display a wide array of sensory defects. For instance, these mice do not react to noxious olfactory stimuli, such as ammonium hydroxide, and fail to react to deep pinpricks in either their whisker pads or their rear paws, indicating defects in their trigeminal and peripheral

TABLE 2. Primary Neuroanatomical Defects Identified in Mice Defective for TrkA, TrkB, and TrkC Tyrosine Kinase Receptors

Strain	Age	Structure	Defect
*trk*A^{TK}(–/–)	P0	Sympathetic ganglia	>90% neuron cell loss
	P0	Trigeminal ganglia	70% neuron cell loss
	P0	Dorsal root ganglia	70% neuron cell loss
	P30	Basal forebrain	Loss of cholinergic projections
*trk*B^{TK} (–/–)	P0	Trigeminal ganglia	50% neuron cell loss
	P0	Nodose/petrosal ganglia	>80% neuron cell loss
	P0	Dorsal root ganglia	30% neuron cell loss
	P0	Facial motor nucleus	>50% neuron cell loss
	P0	Spinal cord (lumbar region)	30% motor neuron cell loss
*trk*C^{TK} (–/–)	P0	Dorsal root ganglia	20% neuron cell loss
	P0	Spinal cord	100% loss of Ia muscle afferents
	P30	Spinal cord/dorsal roots	50% loss of myelinated axons
	P30	Spinal cord/ventral roots	30% loss of fibers

sensory nervous systems, respectively. In addition, these animals exhibit deficiencies in thermoception because they stay on top of a 60 °C infrared hot plate for at least 10 seconds, at which time they are removed.[25]

Neuroanatomical examination of the peripheral nervous system (PNS) of these *trk*A (–/–) mice revealed extensive (>70%) neuronal cell loss in trigeminal, dorsal root, and sympathetic ganglia (TABLE 2). In the dorsal root ganglia (DRGs), the vast majority of the missing neurons correspond to those of small size, a population believed to be NGF dependent.[66] At birth, the superior cervical ganglion (SCG) of *trk* (–/–) mice displays a significant loss of neurons, as well as large numbers of pyknotic nuclei, suggesting an active process of cell degeneration at this time. At P10, the SCG is severely shrunken, and only a few principal sympathetic neurons can be found.[25] These results are reminiscent of earlier studies using "immunosympathectomized" perinatal rats, which lose the majority of their sympathetic neurons after receiving passive transfer of anti-NGF neutralizing antibodies.[1]

Expression of *trk*A gene in the central nervous system (CNS) is restricted to a small subset of cholinergic neurons in the caudatoputamen and in the basal forebrain complex,[67] two brain structures that have been previously shown to be targets for NGF.[1] Disruption of the *trk*A gene does not appear to cause the loss of these neurons, although limited neuronal cell death cannot be excluded at this time.[25] However, adult *trk*A (–/–) mice exhibit a severe decrease in those cholinergic fibers that project from the medial septum to the hippocampus and from the nucleus basalis to the cerebral cortex (TABLE 2). Whether TrkA receptors are required for the outgrowth of these cholinergic fibers or for maintenance of their cholinergic phenotype remains to be determined.

TRKB-defective Mice

*trk*B-targeted mice have a deletion in the tyrosine-protein kinase sequences that prevent expression of the TrkB tyrosine-kinase receptor but not of the noncatalytic isoforms.[24] These mice develop to birth; however, most of them die within the first postnatal week. The first symptomatic difference could be observed after approximately twelve hours, when the *trk*B^{TK} (–/–) animals were found without milk in their stomachs.[24] No gross lesions were observed in the head, including cleft lip or palate, of these animals, that could explain their inability to take nourishment.

These mutant mice, however, present significant neurological deficiencies.[24] For instance, they do not respond to stimuli, such as a light and stroking under the chin, and do not uptake milk formula through a syringe attached to a small caliber tube inserted into their mouths. Some of these behavioral abnormalities might be explained by neuroanatomical defects found in the trigeminal and nodose/petrosal ganglia, as well as in the motor nucleus of the facial nerve (TABLE 2). The *trk*B^{TK} (–/–) mice also display neuronal defects in the DRGs and in spinal motor neurons (TABLE 2) that are unlikely to significantly contribute to the observed phenotype.

In spite of these critical deficiencies, most of the structures known to express TrkB tyrosine-kinase receptors, such as the cerebral cortex, the pyramidal cell layer of the hippocampus, and the thalamus, appear at least morphologically normal.[24] It is possible that some defects may be found in these structures following more detailed analysis. However, it is also possible that TrkB-expressing neurons may survive in the absence of this signaling receptor thanks to compensatory mechanisms, perhaps provided by the highly related TrkC receptors known to be coexpressed in most of these structures.

TRKC-defective Mice

Disruption of the *trk*C gene in ES cells was accomplished by the same strategy described above for the *trk*A and *trk*B genes.[26] As a consequence, the resulting *trk*C^{TK} (–/–) mice do not express any of the TrkC-kinase receptors, including the K14, K25, and K39 isoforms, but they express the putative noncatalytic receptors. *trk*C^{TK} (–/–) mice develop to birth, a time at which they appear normal and respond to painful stimuli and take nourishment, unlike the targeted *trk*A and *trk*B (–/–) animals, respectively. However, soon after birth, the *trk*C^{TK} (–/–) mice display abnormal movements not observed in the other *trk*-defective mice.[26] These movements are athetotic in nature and result in highly abnormal limb postures. Even at rest, the limbs are held in abnormal positions in relation to the trunk, and when placed on a rotating dowel, they cannot maintain an upright posture and immediately tumble off. In addition, the *trk*C^{TK} (–/–) mice draw their limbs in towards their bodies when lifted by their tails instead of displaying the fully extended posture observed with their (+/+) and (+/–) littermates. This behavioral phenotype suggests a defect in proprioception, the sensory function that localizes the limbs in space.[26]

Labeling of DRG afferents with the soluble tracer DiI revealed that the *trk*C^{TK} (–/–) mice are completely devoid of Ia muscle afferents (TABLE 2). By contrast, the dorsal horn innervation appears grossly normal, indicating the presence of nociceptors and low-threshold mechanoreceptors. In addition to this lack of Ia muscle afferents, the *trk*C^{TK} (–/–) mice have a 20% loss of DRG neurons (TABLE 2), presumably those responsible for projecting the missing afferents. In agreement with this interpretation, these mutants also have a marked depletion of large myelinated group Ia axons, which extend the collateral Ia projections to the spinal cord and ascend through the dorsal columns to connect with spinocerebellar neurons (TABLE 2). These defects are likely to be responsible for most, if not all, of the abnormal movements observed in these mice. The limited life span of the *trk*C^{TK} (–/–) mice (most die by P21) raises the possibility that these animals may have additional neuronal defects. Yet, preliminary observations indicate that most CNS structures known to express *trk*C transcripts appear grossly normal.[26] These findings, along with those observed in the TrkB-targeted mice, suggest that CNS neurons might be supported by more than one neurotrophin. Generation of mice carrying deletions in more than one member of the *trk* gene family should help to unveil the spectrum of neurotrophin dependency in the central nervous system.

SUMMARY

The Trk family of tyrosine-protein kinases, TrkA, TrkB, and TrkC, are the signaling receptors that mediate the biological properties of the NGF family of neurotrophins. This family of growth factors includes in addition to NGF, BDNF, NT-3, and NT-4. TrkA is the NGF receptor. TrkB serves as a receptor for both BDNF and NT-4, and TrkC is the primary receptor for NT-3. NT-3 is a somewhat promiscuous ligand that can also activate TrkA and TrkB receptors at high concentrations. The *trk*B and *trk*C genes also encode noncatalytic receptor isoforms of an, as yet, unknown function. In addition to the Trk receptors, the NGF family of neurotrophins also binds with low affinity to an unrelated molecule, designated p75, a member of the TNF-receptor superfamily. Recently, we have generated strains of mice lacking each of these tyrosine-kinase receptors by gene targeting in embryonic stem cells. Characterization of these mutant mice is providing relevant information regarding the critical role that these receptors play in the ontogeny of the mammalian nervous system.

REFERENCES

1. LEVI-MONTALCINI, R. 1987. The nerve growth factor 35 years later. Science **237:** 1154-1162.
2. LINDSAY, R. M., S. J. WIEGAND, A. ALTAR & P. S. DISTEFANO. 1994. Neurotrophic factors: from molecule to man. Trends Genet. **17:** 182-192.
3. GOTZ, R., R. KOSTER, C. WINKLER, F. RAULF, F. LOTSPEICH, M. SCHARTL & H. THOENEN. 1994. Neurotrophin-6 is a new member of the nerve growth factor family. Nature **372:** 266-269.
4. CHAO, M. V. 1992. Neurotrophin receptors: A window into neuronal differentiation. Neuron **9:** 583-593.
5. MEAKIN, S. O. & E. M. SHOOTER. 1992. The nerve growth factor family of receptors. Trends Neurosci. **15:** 323-331.
6. BARBACID, M. 1993. Nerve growth factor: a tale of two receptors. Oncogene **8:** 2033-2042.

7. Davies, A. M., K. F. Lee & R. Jaenisch. 1993. p75-deficient trigeminal sensory neurons have an altered response to NGF but not to other neurotrophins. Neuron **11:** 565–574.
8. Lee, K. F., E. Li, L. J. Huber, S. C. Landis, A. H. Sharpe, M. V. Chao & R. Jaenisch. 1992. Targeted mutation of the gene encoding the low affinity NGF receptor p75 leads to deficits in the peripheral sensory nervous systems. Cell **69:** 737–749.
9. Lee, K. F., K. Bachman, S. Landis & R. Jaenisch. 1994. Dependence on p75 for innervation of some sympathetic targets. Science **263:** 1447–1449.
10. Martin-Zanca, D., S. H. Hughes & M. Barbacid. 1986. A human oncogene formed by the fusion of truncated tropomyosin and protein tyrosine kinase sequences. Nature **319:** 743–748.
11. Martin-Zanca, D., R. Oskam, G. Mitra, T. Copeland & M. Barbacid. 1989. Molecular and biochemical characterization of the human *trk* proto-oncogene. Mol. Cell. Biol. **9:** 24–33.
12. Kaplan, D. R., D. Martin-Zanca & L. F. Parada. 1991. Tyrosine phosphorylation and tyrosine kinase activity of the *trk* proto-oncogene product induced by NGF. Nature **350:** 158–160.
13. Klein, R., S. Q. Jing, V. Nanduri, E. O'Rourke & M. Barbacid. 1991. The *trk* proto-oncogene encodes a receptor for nerve growth factor. Cell **65:** 189–97.
14. Klein, R., L. F. Parada, F. Coulier & M. Barbacid. 1989. *trk*B, a novel tyrosine protein kinase receptor expressed during mouse neural development. EMBO J. **8:** 3701–3709.
15. Klein, R., D. Conway, L. F. Parada & M. Barbacid. 1990. The *trk*B tyrosine protein kinase gene codes for a second neurogenic receptor that lacks the catalytic kinase domain. Cell **61:** 647–656.
16. Middlemas, D. S., R. A. Lindberg & T. Hunter. 1991. *trk*B, a neural receptor protein-tyrosine kinase: evidence for a full-length and two truncated receptors. Mol. Cell. Biol. **11:** 143–153.
17. Klein, R., V. Nanduri, S. Q. Jing, F. Lamballe, P. Tapley, S. Bryant, C. Cordon-Cardo, K. R. Jones, L. F. Reichardt & M. Barbacid. 1991. The *trk*B tyrosine protein kinase is a receptor for brain-derived neurotrophic factor and neurotrophin-3. Cell **66:** 395–403.
18. Soppet, D., E. Escandon, J. Maragos, D. S. Middlemas, S. W. Reid, J. Blair, L. E. Burton, B. R. Stanton, D. R. Kaplan, T. Hunter, K. Nikolics & L. F. Parada. 1991. The neurotrophic factors brain-derived neurotrophic factor and neurotrophin-3 are ligands for the *trk*B tyrosine kinase receptor. Cell **65:** 895–903.
19. Squinto, S. P., T. N. Stitt, T. H. Aldrich, S. Davis, S. M. Bianco, C. Radziejewski, D. J. Glass, P. Masiakowski, M. E. Furth, D. M. Valenzuela, P. S. Distefano & G. D. Yancopoulos. 1991. *trk*B encodes a functional receptor for brain-derived neurotrophic factor and neurotrophin-3 but not nerve growth factor. Cell **65:** 885–893.
20. Berkemeier, L. R., J. W. Winslow, D. R. Kaplan, K. Nikolics, D. V. Goeddel & A. Rosenthal. 1991. Neurotrophin-5: a novel neurotrophic factor that activates *trk* and *trk*B. Neuron **7:** 857–866.
21. Klein, R., F. Lamballe, S. Bryant & M. Barbacid. 1992. The *trk*B tyrosine protein kinase is a receptor for neurotrophin-4. Neuron **8:** 947–956.
22. Ip, N. Y., C. F. Ibanez, S. H. Nye, J. McClain, P. F. Jones, D. R. Gies, L. Belluscio, M. M. Le Beau, R. Espinosa, S. P. Squinto, H. Persson & G. Yancopoulos. 1992. Mammalian neurotrophin-4: structure, chromosomal localization, tissue distribution, and receptor specificity. Proc. Natl. Acad. Sci. USA **89:** 3060–3064.
23. Lamballe, F., R. Klein & M. Barbacid. 1991. *trk*C, a new member of the trk family of tyrosine protein kinases, is a receptor for neurotrophin-3. Cell **66:** 967–979.
24. Klein, R., R. J. Smeyne, W. Wurst, L. K. Long, B. A. Auerbach, A. L. Joyner & M. Barbacid. 1993. Targeted disruption of the *trk*B neurotrophin receptor gene results in nervous system lesions and neonatal death. Cell **75:** 113–122.

25. SMEYNE, R. J., R. KLEIN, A. SCHNAPP, L. K. LONG, S. BRYANT, A. LEWIN, S. A. LIRA & M. BARBACID. 1994. Severe sensory and sympathetic neuropathies in mice carrying a disrupted Trk/NGF receptor gene. Nature **368:** 246–249.
26. KLEIN, R., I. SILOS-SANTIAGO, R. J. SMEYNE, S. LIRA, R. BRAMBRILLA, S. BRYANT, L. ZHANG, W. D. SNIDER & M. BARBACID. 1994. Disruption of the neurotrophin-3 receptor gene *trk*C eliminates Ia muscle afferents and results in abnormal movements. Nature **368:** 249–251.
27. BARBACID, M. 1994. (Y. A. Barde, Ed.) The Trk family of neurotrophin receptors. Special issue on "Neurotrophic Factors." Neurobiol. **25:** 1386–1403.
28. SNIDER, W. D. 1994. Functions of the neurotrophins during nervous system development: What the knockouts are teaching us. Cell **77:** 627–638.
29. HARDIE, D. G. & S. HANKS. 1994. *In* Protein Kinase Factsbook. Academic Press. London.
30. MEAKIN, S. O., U. SUTER, C. C. DRINKWATER, A. A. WELCHER & E. M. SHOOTER. 1992. The rat *trk* protooncogene product exhibits properties characteristic of the slow nerve growth factor receptor. Proc. Natl. Acad. Sci. USA **89:** 2374–2378.
31. BARKER, P. A., C. LOMEN-HOERTH, E. M. GENSCH, S. O. MEAKIN, D. J. GLASS & E. M. SHOOTER. 1993. Tissue-specific alternative splicing generates two isoforms of the trkA receptor. J. Biol. Chem. **268:** 15150–15157.
32. HORIGOME, K., J. C. PRYOR, E. D. BULLOCK & E. M. JOHNSON, JR. 1993. Mediator release from mast cells by nerve growth factor. Neurotrophin specificity and receptor mediation. J. Biol. Chem. **268:** 14881–14887.
33. SCHNEIDER, R. & M. SCHWEIGER. 1991. A novel modular mosaic of cell adhesion motifs in the extracellular domains of the neurogenic *trk* and *trk*B tyrosine kinase receptors. Oncogene **6:** 1807–1811.
34. PULIDO, D., S. CAMPUZANO, T. KODA, J. MODOLELL & M. BARBACID. 1992. D*trk*, a *Drosophila* gene related to the *trk* family of neurotrophin receptors, encodes a novel class of neural cell adhesion molecule. EMBO J. **11:** 391–404.
35. COULIER, F., R. KUMAR, M. ERNST, R. KLEIN, D. MARTIN-ZANCA & M. BARBACID. 1990. Human trk oncogenes activated by point mutation, in-frame deletion, and duplication of the tyrosine kinase domain. Mol. Cell. Biol. **10:** 4202–4210.
36. LAMBALLE, F., P. TAPLEY & M. BARBACID. 1993. *trk*C encodes multiple neurotrophin-3 receptors with distinct biological properties and substrate specificities. EMBO J. **12:** 3083–3094.
37. TSOULFAS, P., D. SOPPET, E. ESCANDON, L. TESSAROLLO, J. L. MENDOZA-RAMIREZ, A. ROSENTHAL, K. NIKOLICS & L. F. PARADA. 1993. The rat *trk*C locus encodes multiple neurogenic receptors that exhibit differential response to neurotrophin-3 in PC12 cells. Neuron **10:** 975–990.
38. VALENZUELA, D. M., P. C. MAISONPIERRE, D. J. GLASS, E. ROJAS, L. NUNEZ, Y. KONG, D. R. GIES, T. N. STITT, N. Y. IP & G. D. YANCOPOULOS. 1993. Alternative forms of rat TrkC with different functional capabilities. Neuron **10:** 963–974.
39. CORDON-CARDO, C., P. TAPLEY, S. Q. JING, V. NANDURI, E. O'ROURKE, F. LAMBALLE, K. KOVARY, R. KLEIN, K. R. JONES, L. F. REICHARDT & M. BARBACID. 1991. The *trk* tyrosine protein kinase mediates the mitogenic properties of nerve growth factor and neurotrophin-3. Cell **66:** 173–183.
40. IP, N. Y., T. N. STITT, P. TAPLEY, R. KLEIN, D. J. GLASS, J. FANDL, L. A. GREENE, M. BARBACID & G. D. YANCOPOULOS. 1993. Similarities and differences in the way neurotrophins interact with the Trk receptors in neuronal and nonneuronal cells. Neuron **10:** 137–149.
41. JING, S., P. TAPLEY & M. BARBACID. 1992. Nerve growth factor mediates signal transduction through *trk* homodimer receptors. Neuron **9:** 1067–1079.
42. SCHLESSINGER, J. & A. ULLRICH. 1992. Growth factor signaling by receptor tyrosine kinases. Neuron **9:** 383–391.
43. EGAN, S. E. & R. A. WEINBERG. 1993. The pathway to signal achievement. Nature **365:** 781–783.

44. OHMICHI, M., S. J. DECKER, L. PANG & A. R. SALTIEL. 1991. Nerve growth factor binds to the 140 kd *trk* proto-oncogene product and stimulates its association with the *src* homology domain of phospholipase C γ 1. Biochem. Biophys. Res. Commun. **179:** 217-223.
45. VETTER, M. L., D. MARTIN-ZANCA, L. F. PARADA, J. M. BISHOP & D. R. KAPLAN. 1991. Nerve growth factor rapidly stimulates tyrosine phosphorylation of phospholipase C-γ 1 by a kinase activity associated with the product of the *trk* protooncogene. Proc. Natl. Acad. Sci. USA **88:** 5650-5654.
46. WIDMER, H. R., D. R. KAPLAN, S. J. RABIN, K. D. BECK, F. HEFTI & B. KNUSEL. 1993. Rapid phosphorylation of phospholipase C γ 1 by brain-derived neurotrophic factor and neurotrophin-3 in cultures of embryonic rat cortical neurons. J. Neurochem. **60:** 2111-2123.
47. RHEE, S. G. & K. D. CHOI. 1992. Multiple forms of phospholipase C isozymes and their activation mechanisms. Adv. Second Messenger Phosphoprotein Res. **26:** 35-61.
48. OBERMEIER, A., R. LAMMERS, K. H. WIESMULLER, G. JUNG, J. SCHLESSINGER & A. ULLRICH. 1993. Identification of Trk binding sites for SHC and phosphatidylinositol 3'-kinase and formation of a multimeric signaling complex. J. Biol. Chem. **268:** 22963-22966.
49. LOEB, D. M., H. TSAO, M. H. COBB & L. A. GREENE. 1992. NGF and other growth factors induce an association between ERK1 and the NGF receptor, $gp140^{prototrk}$. Neuron **9:** 1053-1065.
50. SOLTOFF, S. P., S. L. RABIN, L. C. CANTLEY & D. R. KAPLAN. 1992. Nerve growth factor promotes the activation of phosphatidylinositol 3-kinase and its association with the *trk* tyrosine kinase. J. Biol. Chem. **267:** 17472-17477.
51. OHMICHI, M., S. J. DECKER & A. R. SALTIEL. 1992. Activation of phosphatidylinositol-3 kinase by nerve growth factor involves indirect coupling of the *trk* proto-oncogene with src homology 2 domains. Neuron **9:** 769-777.
52. SUEN, K. L., X. R. BUSTELO, T. PAWSON & M. BARBACID. 1993. Molecular cloning of the mouse *grb2* gene: differential interaction of the Grb2 adaptor protein with epidermal growth factor and nerve growth factor receptors. Mol. Cell. Biol. **13:** 5500-5512.
53. PELICCI, G., L. LANFRANCONE, F. GRIGNANI, J. MCGLADE, F. CAVALLO, G. FORNI, I. NICOLETTI, F. GRIGNANI, T. PAWSON & P. G. PELICCI. 1992. A novel transforming protein (SHC) with an SH2 domain is implicated in mitogenic signal transduction. Cell **70:** 93-104.
54. QIU, M. S. & S. H. GREEN. 1991. NGF and EGF rapidly activate p^{21ras} in PC12 cells by distinct, convergent pathways involving tyrosine phosphorylation. Neuron **7:** 937-946.
55. MUROYA, K., S. HATTORI & S. NAKAMURA. 1992. Nerve growth factor induces rapid accumulation of the GTP-bound form of p^{21ras} in rat pheochromocytoma PC12 cells. Oncogene **7:** 277-281.
56. THOMAS, S. M., M. DEMARCO, G. D'ARCANGELO, S. HALEGOUA & J. S. BRUGGE. 1992. Ras is essential for nerve growth factor- and phorbol ester-induced tyrosine phosphorylation of MAP kinases. Cell **68:** 1031-1040.
57. SCHANEN-KING, C., A. NEL, L. K. WILLIAMS & G. LANDRETH. 1991. Nerve growth factor stimulates the tyrosine phosphorylation of MAP kinase in PC12 cells. Neuron **6:** 915-922.
58. BAR-SAGI, D. & J. R. FERAMISCO. 1985. Microinjection of the ras oncogene protein into PC12 cells induces morphological differentiation. Cell **42:** 841-848.
59. NODA, M., M. KO, A. OGURA, D. G. LIU, T. AMANO, T. TAKANO & Y. IKAWA. 1985. Sarcoma viruses carrying ras oncogenes induce differentiation-associated properties in a neuronal cell line. Nature **318:** 73-75.
60. WOOD, K. W., H. QI, G. D'ARCANGELO, R. C. ARMSTRONG, T. M. ROBERTS & S. HALEGOUA. 1993. The cytoplasmic raf oncogene induces a neuronal phenotype in PC12 cells: a potential role for cellular raf kinases in neuronal growth factor signal transduction. Proc. Natl. Acad. Sci. USA **90:** 5016-5020.

61. RABIN, S. J., V. CLEGHON & D. R. KAPLAN. 1993. SNT, a differentiation-specific target of neurotrophic factor-induced tyrosine kinase activity in neurons and PC12 cells. Mol. Cell. Biol. **13:** 2203-2213.
62. TRAVERSE, S., K. SEEDORF, H. PATERSON, C. J. MARSHALL, P. COHEN & A. ULLRICH. 1994. EGF triggers neuronal differentiation of PC12 cells that overexpress the EGF receptor. Curr. Biol. **4:** 694-701.
63. DIKIC, I., J. SCHLESSINGER & I. LAX. 1994. PC12 cells overexpressing the insulin receptor undergo insulin-dependent neuronal differentiation. Curr. Biol. **4:** 702-708.
64. GLASS, D. J., S. H. NYE, P. HANTZOPOULOS, M. J. MACCHI, S. P. SQUINTO, M. GOLDFARB & G. D. YANCOPOULOS. 1991. TrkB mediates BDNF/NT-3-dependent survival and proliferation in fibroblasts lacking the low affinity NGF receptor. Cell **66:** 405-413.
65. BECK, K. D., F. LAMBALLE, R. KLEIN, M. BARBACID, P. E. SCHAUWECKER, T. H. MCNEILL, C. E. FINCH, F. HEFTI & J. R. DAY. 1993. Induction of noncatalytic TrkB neurotrophin receptors during axonal sprouting in the adult hippocampus. J. Neurosci. **13:** 4001-4014.
66. MU, X., I. SILOS-SANTIAGO, S. L. CARROLL & W. D. SNIDER. 1993. Neurotrophin receptor genes are expressed in distinct patterns in developing dorsal root ganglia. J. Neurosci. **13:** 4029-4041.
67. HOLTZMAN, D. M., Y. LI, L. F. PARADA, S. KINSMAN, C. K. CHEN, J. S. VALLETTA, J. ZHOU, J. B. LONG & W. C. MOBLEY. 1992. p140trk mRNA marks NGF-responsive forebrain neurons: evidence that *trk* gene expression is induced by NGF. Neuron **9:** 465-478.

Protein Kinase C Mediates Short- and Long-term Effects on Receptor Tyrosine Kinases

Regulation of Tyrosine Phosphorylation and Degradation

KLAUS SEEDORF,[a] MARK SHERMAN, AND AXEL ULLRICH

Department of Molecular Biology
Max-Planck-Institut für Biochemie
Am Klopferspitz 18A
82152 Martinsried, Germany

Binding of a ligand to its corresponding receptor tyrosine kinase (RTK) triggers activation of an intrinsic tyrosine kinase activity, which results in receptor autophosphorylation and activation of signal-transduction pathways that ultimately induce cell division, differentiation, or movement.[1–3] Inasmuch as constitutive activation of these pathways results in disturbance of normal cellular response mechanisms and ultimately cell transformation, internal control and negative feedback mechanisms are required. The protein kinase C (PKC) system has been suggested to play such a role. Using human 293 embryonic fibroblasts that transiently overexpress either the epidermal growth factor receptor (EGF-R) or chimeric receptors consisting of human EGF-R extracellular domain fused to the cytoplasmic domains of EGF-R isoenzyme HER2 (HER1-2), platelet-derived growth factor receptor (PDGF-R $\rightarrow$ EP-R), and insulin receptor (I-R $\rightarrow$ EI-R), we investigated short- and long-term effects on RTK-signaling parameters in the presence and absence of cotransfected PKCα. In the absence of PKCα, phorbol 12-myristate 13 acetate (TPA) treatment within minutes results in decreased EGF- and HER2-tyrosine phosphorylation, whereas PDGF- and I-R phosphorylation is up-regulated. These effects are not mediated by endogenous PKC-dependent RTK phosphorylation, but apparently by activation or inactivation of RTK-specific phosphatases, as indicated by neutralization of this short-term effect upon treatment of cells with sodium orthovanadate, a potent phosphatase inhibitor. In the presence of overexpressed PKCα, all investigated RTKs formed a stable protein-protein complex with PKCα upon TPA treatment, which at the same time resulted in a mobility shift of the receptors. Under these experimental conditions, the TPA effect on RTK tyrosine-kinase activity was not reversible by orthovanadate, indicating that receptor phosphorylation on tyrosine residues is regulated by PKC-dependent

[a] Present address: Hagedorn Research Institute, Niels Steensens Vej 6, DK-2820 Gentofte, Denmark.

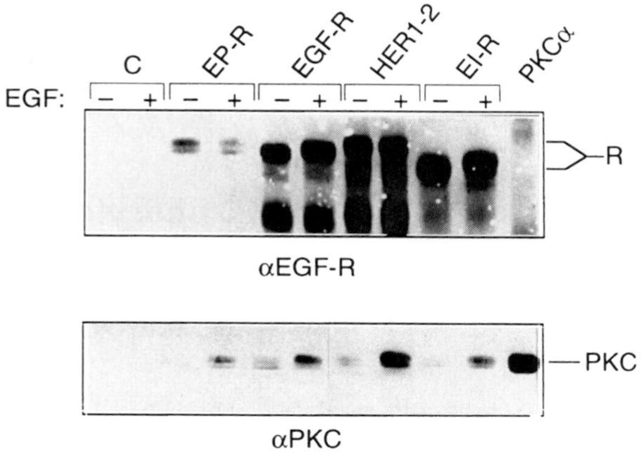

FIGURE 1. Coimmunoprecipitation of PKC after immunoprecipitation of RTKs stably expressed in NIH 3T3 cells. NIH 3T3 cells stably expressing the various receptors were cultured in low serum overnight and subsequently treated with EGF (50 ng/mL) for 30′ as indicated. Cells were lysed, and receptors were immunoprecipitated using mAb 108.1 (αEGF-R) and separated by SDS-PAGE. After transfer to nitrocellulose, RTKs were detected using a polyclonal antiserum against the human EGF-R extracellular domain (antisera DIII; upper panel), and coimmunoprecipitation of endogenous PKCα was determined using PKC-specific antisera (lower panel).

phosphorylation on serine and threonine residues. Alternatively, binding of activated PKC to the receptor directly results in receptor inactivation. The TPA-mediated effects on receptor-tyrosine phosphorylation, neutralization by orthovanadate, as well as TPA-induced RTK-PKC complex formation can also be demonstrated in NIH3T3 fibroblasts stably overexpressing these RTKs. As shown in FIGURE 1, the RTK-PKC complex formation can also be induced by ligand addition in these cells. Ligand-induced RTK-PKC interaction must be regarded as a long-term effect. It is visible after 30 minutes and declines after 60 minutes to undetectable levels, whereas TPA induces a more rapid and more sustained complex formation. This difference between ligand and TPA-induced complex formation might be explained by the finding that growth factors transiently activate PLCγ,[4] which then leads to transient activation of PKC, whereas TPA directly and constitutively activates PKC.

What is the biological function of this protein-protein interaction? An involvement of PKC in receptor down-regulation has been suggested for many years. Exposing cells to phorbol esters results in decreased EGF binding and can be observed in most cell types. Immunofluorescence and electron microscopic localization of these receptors showed that a significant percentage of receptors become internalized.[5] PKC-mediated internalization and degradation could clearly be demonstrated in 293 cells. Overexpression of the RTK alone and subsequent treatment with EGF or TPA had no effect on receptor degradation, whereas coexpression of PKCα resulted in a marked loss of radiolabeled receptors, however, only upon TPA treatment. Determination of RTK-PKC-complex formation showed that EGF was not able to induce this

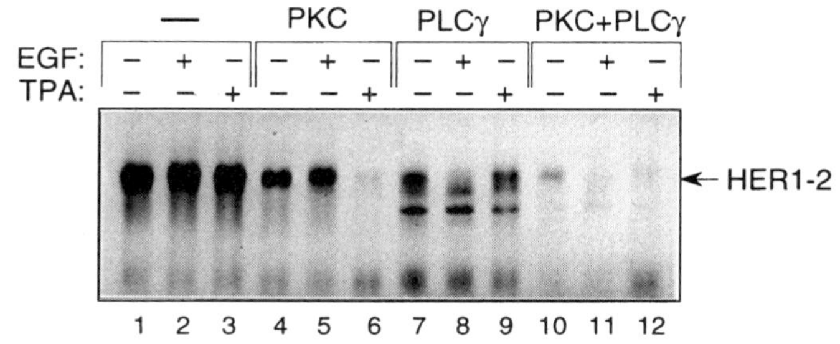

FIGURE 2. EGF- and TPA-induced degradation of radiolabeled HER1-2. 293 cells were transfected with HER1-2 expression plasmid alone (lanes 1, 2, 3), together with PKC expression plasmid (lanes 4, 5, 6), together with PLCγ expression plasmid (lanes 7, 8, 9), and together with PKC and PLCγ expression plasmids (lanes 10, 11, 12). Cells were metabolically labeled with [^{35}S]methionine overnight, and, after washing away the radiolabel, HER1-2 degradation was determined in the absence or presence of EGF or TPA. After 4 hours of incubation, cells were lysed, and HER1-2 was immunoprecipitated using mAb 108.1. After separation by SDS-PAGE, the gel was dried and exposed to film.

protein-protein interaction, whereas TPA induced receptor/PKC interaction, receptor phosphorylation, and, simultaneously, receptor degradation. As mentioned earlier, EGF and TPA do not mediate receptor down-regulation when the receptor is overexpressed alone, suggesting that the endogenous down-regulation pathways are not sufficiently stimulated to activate degradation of these elevated amounts of receptors. Coexpression of PKCα and TPA treatment can restore this pathway, whereas EGF still has no effect. If the assumption is correct that the ligand-activated receptor activates PLCγ, which hydrolyzes phosphatidyl inositol 4,5-biphosphate into 1,2-diacyl-glycerol and inositol 1,4,5-triphosphate and thereby activates PKC, then PLCγ should be the limiting factor in the latter case. Simultaneous overexpression of the receptor, PLCγ, and PKC should therefore restore the entire pathway and render it EGF inducible. As shown in FIGURE 2, this is indeed the case. Coexpression of PLCγ caused the HER1-2 receptor to be partially down-regulated in response to EGF stimulation (FIG. 2, lane 8), an effect that was enhanced by PKCα co-overexpression (FIG. 2, lane 11). In the latter case, EGF and TPA-induced receptor down-regulation were indistinguishable, demonstrating that activation of PKC by way of receptor-activated PLCγ and the corresponding second messengers was equivalent to direct activation of PKC by the artifical ligand TPA.

In conclusion, the studies reported here show that PKC mediates short- and long-term effects on RTKs. The immediate effects, which appear to involve TPA-responsive tyrosine-specific phosphatases, are RTK specific and regulate the phosphorylation state of the receptors on tyrosine residues. By contrast, the PKC long-term effects seem to be more general and are induced by translocation from the cytosol to the plasma membrane and formation of stable complexes with RTKs, concomitant with phosphorylation of these receptors followed by their internalization and degradation.

REFERENCES

1. Schlessinger, J. & A. Ullrich. 1992. Neuron **9:** 383-391.
2. Mayer, B. J. & D. Baltimore. 1993. Trends Cell Biol. **3:** 8-13.
3. Koch, C. A., D. Anderson, M. F. Moran, C. Ellis & T. Pawson. 1991. Science **252:** 668-674.
4. Rhee, S. G. 1991. TIBS **16:** 297-301.
5. Beguinot, L., J. A. Hanover, S. Ito, N. D. Richert, M. C. Willinham & I. Pastan. 1985. Proc. Natl. Acad. Sci. USA **82:** 2774-2778.

Flag-Insulin Receptor Mutants[a]

H. JOSEPH GOREN

Department of Medical Biochemistry
University of Calgary
Calgary, Alberta, Canada, T2N 4N1

Mutations of the insulin-receptor gene in humans[1] and by design[2,3] have provided a wealth of information on structural and functional features of the protein. To characterize such receptors, initially a plasmid that contains the code for a mutant insulin receptor is either transiently or stably transfected into cells. The cells are then used to study receptor expression, receptor structural features, ligand-binding properties, the receptor-tyrosine kinase activity, insulin responses, and intracellular cell signaling. Epitope tagging of proteins is an additional method to study the structure and function of intact and mutated receptors and receptor fragments.

A number of plasmids is now available that contain the code for an antibody-specific peptide or protein, or for a peptide with metal-binding properties. Appropriate addition of a receptor or a receptor-fragment code will generate plasmids that code for epitope-tagged proteins. Bacterial expression will allow isolation of sufficient quantities of a protein for physical and chemical characterization, and for raising antibodies. In addition, providing the epitope tag has no ill effect on the functional role of the receptor or receptor fragment, one can study ligand or protein association with the latter *in vitro*.

Flag is an octapeptide (DYKDDDDK) to which two monoclonal antibodies have been raised, M1 and M2 (International Biotechnologies Inc.). M1 recognizes amino-terminal Flag, whereas M2 recognizes it within a protein. M1 binds only in the presence of Ca^{2+}, and a chelating agent will displace readily Flag proteins, that is, conditions that should not displace noncovalently bound proteins. Mutant Flag-insulin receptors (FIRs) that are missing parts of the protein may, therefore, be used to determine to which domains of the receptor specific proteins can bind. Plasmids for three FIR mutants have been constructed.

METHOD

peFIR, the cDNA for FIR within the pECE plasmid[4] was digested with EcoRV (IR nucleotide 253) and XbaI (3′ polylinker). peIRΔ30, which contains a stop codon in place of Arg-1326 and expresses IR with the C-terminal 30 amino acids missing,[5] was similarly digested. The plasmid fragment from the former digest was ligated to the IRΔ30-fragment. peFIR was partially digested with XhoI (IR nucleotides 424, 2184, and 3194) and made blunt-ended with Mung Bean nuclease. T4 DNA ligase-

[a] This work was performed in part at the Hormone Research Institute, University of California at San Francisco and supported in part by the University of Calgary Research Grants Committee.

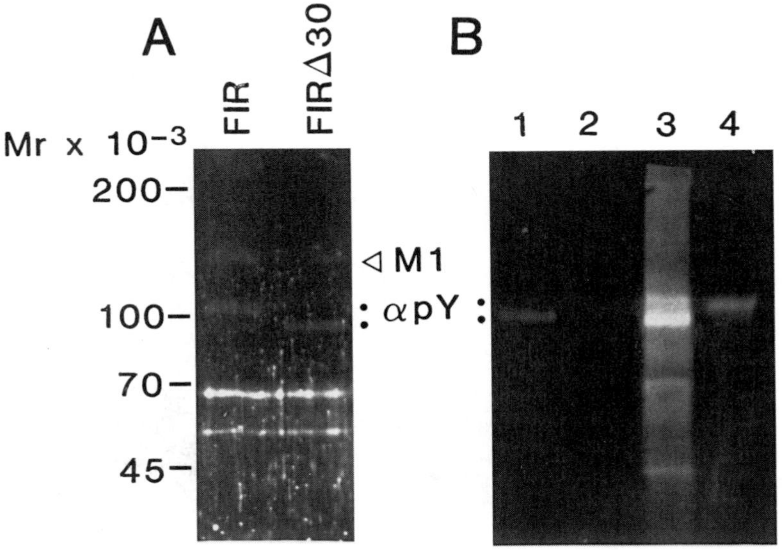

FIGURE 1. Autophosphorylation activity of FIR and FIRΔ30. A: M1-agarose purified FIR and FIRΔ30 from insulin-exposed Cho-FIR and Cho-FIRΔ30 cells, respectively, were applied to acrylamide gel electrophoresis. Proteins were transferred to Immobilon P. Initially, the membranes were blotted with M1 to detect the presence of Flag-α-subunits. Subsequently, the membranes were reprobed with antiphosphotyrosine (αpY) antibody. Peroxidase-conjugated second antibody was detected with chemiluminescence. The strong signals between the 70,000 and 45,000 kDa protein standards are nonspecific, as their presence was independent of antibody probe. B: M1-agarose purified FIR (lanes 2 and 4) and FIRΔ30 (lanes 1 and 3) from noninsulin (lanes 1 and 2) and insulin-exposed (lanes 3 and 4) Cho-FIR and Cho-FIRΔ30 cells, respectively, were applied to acrylamide gel electrophoresis. Proteins were transferred to Immobilon P. Membranes were blotted with RC20H (Transduction Laboratories), and proteins were detected with chemiluminescence.

treated plasmids were transfected into XL1-Blue (Stratagene) *E. coli.* Ampicillin-resistant colonies, whose plasmids with XhoI digestion gave a 5.5 kb DNA fragment and a 6.2 kb DNA fragment, were used to isolate peFIRΔ (69-655) and peFIRΔ (655-993), respectively. With appropriate synthetic oligonucleotides as primers, the plasmid constructs were sequenced to confirm their structure.

Chinese hamster ovary (CHO) cells were cotransfected with pSV2neo and peFIR or peFIRΔ30. Cells were grown in the presence of G418 and selected for high-insulin binding. Confluent cell cultures were or were not exposed to 3 nM insulin for 1 min and solubilized in 1% Triton X100 buffers containing protease and phosphatase inhibitors.[4] Centrifuged supernatants were applied to wheat germ agglutinin agarose. Ca^{2+} was added to *N*-acetyl glucosamine eluates, and the solutions were applied to M1-agarose. Concentrated EDTA eluates were applied to dodecyl sulphate gel electrophoresis (SDS-PAGE) and transferred to Immobilon P (Millipore); the membranes were blotted with antiphosphotyrosine or M1 antibody. ECL (Amersham) was used to detect proteins.

RESULTS AND DISCUSSION

Two Cho-FIR cell lines, containing 1-2 × 10^6 receptors/cell, and three Cho-FIRΔ30 cell lines, containing 0.3-0.6 × 10^6 receptors/cell, were cultured; Cho cells contained approximately 0.05 × 10^6 receptors/cell. Scatchard analyses were curvilinear for native and Flag receptors, and estimates for binding affinity with the empty receptor ranged between 0.1 and 0.3 nM. Consistent with its smaller size, autophosphorylated Δ30-β-subunit migrated slightly faster than native phosphorylated β-subunit (FIG. 1). The degree of insulin-stimulated autophosphorylation of the Δ30-β-subunit was either equal to or slightly greater than insulin-stimulated autophosphorylation of FIR; that is, under conditions where M1 blotted slightly less α-subunit from FIRΔ30 than from FIR, the two β-subunits appeared of equal amount with antiphosphotyrosine blotting (FIG. 1a). In the absence of insulin, FIR showed no *in vivo* autophosphorylation, whereas FIRΔ30 did autophosphorylate (FIG. 1b).

The C-terminal 30 amino acid-truncated receptor has been expressed, and its binding and kinase (autophosphorylation) activity are consistent with previous reports for C-terminal truncated receptors.[2,3,6] It is anticipated that FIRΔ (69-655) will not bind insulin; nor will it form a heterotetramer, as it is missing the insulin-binding domain and the cysteine for inter-α-subunit disulfide-bond formation. By contrast, FIRΔ (655-993) should bind insulin, but it is missing the cysteines participating in the α-β-subunit disulfides, the tetrapeptide to process the proreceptor into the two subunits, the transmembrane domain, and the juxtamembrane domain. Characterization of this receptor and the former receptor are currently underway.

REFERENCES

1. TAYLOR, S. I. 1992. Diabetes **41:** 1473-1490.
2. TAVARÉ, J. M. & K. SIDDLE. 1993. Biochim. Biophys. Acta **1178:** 21-39.
3. LEE, J. & P. F. PILCH. 1994. Am. J. Physiol. **266**(2 Pt 1): C319-334.
4. ZHANG, W., J. D. JOHNSON & W. J. RUTTER. 1993. Proc. Natl. Acad. Sci. USA **90:**11317-11321.
5. JOHNSON, J. D., M. AGGERBECK & W. J. RUTTER. 1994. Personal communications.
6. YONEZAWA, K., S. PIERCE, C. STOVER, M. AGGERBECK, W. J. RUTTER & R. A. ROTH. 1991. Adv. Exp. Med. Biol. **293:** 227-238.

Localization of Specific Amino Acids Contributing to Insulin Specificity of the Insulin Receptor

ASSER S. ANDERSEN, FINN C. WIBERG,[a] AND THOMAS KJELDSEN

Diabetes Research and [a]Bioscience
Novo Nordisk A/S, Novo Alle
DK-2880 Bagsvaerd, Denmark

Insulin and IGF-I receptors, like their respective ligands, display a high degree of homology in sequence and overall structural organization. However, each receptor has low affinity for the noncognate ligand. We have constructed and characterized chimeras between these closely related receptors to try to localize the regions that distinguish between insulin and IGF-I.

Previously, this approach has shown that ligand specificity of the insulin receptor (*i.e.*, a region that discriminates between insulin and IGF-I) resides within the N-terminal 68 amino acids. By contrast, specificity for IGF-I binding is defined within amino acid 191–290 of the IGF-I. Thus, the ligand specificities of the insulin and the IGF-I receptor appear to reside in different regions of a common binding site.[1] Continuing along the same vein, but concentrating on the 68 N-terminal amino acids

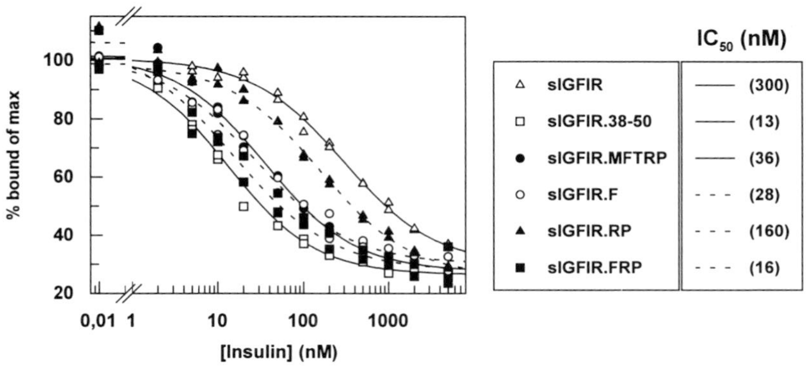

FIGURE 1. Displacement of [^{125}I]IGF-I from soluble receptors with insulin. Truncated soluble receptors were incubated with [^{125}I]IGF-I (10 pM) for 16 hours at 4°C together with increasing concentrations of insulin, as described previously.[1] The amount of [^{125}I]IGF-I bound as a percentage of [^{125}I]IGF-I bound in the absence of unlabeled insulin is plotted against the concentration of unlabeled insulin. Alongside are shown the IC_{50} values from a typical experiment. Nomenclature: sIGFIR followed by amino acids from within the region from 38 to 50 of the insulin receptors replacing corresponding IGF-I receptor amino acids.

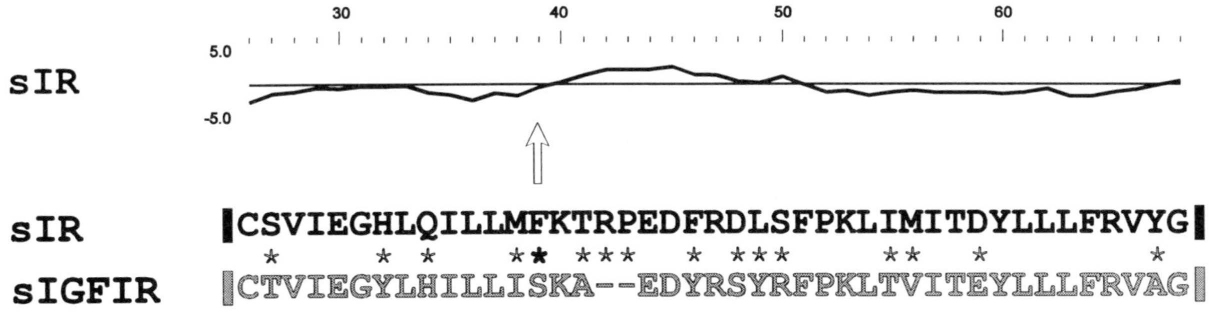

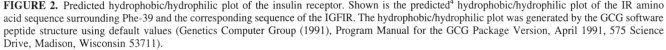

FIGURE 2. Predicted hydrophobic/hydrophilic plot of the insulin receptor. Shown is the predicted[4] hydrophobic/hydrophilic plot of the IR amino acid sequence surrounding Phe-39 and the corresponding sequence of the IGFIR. The hydrophobic/hydrophilic plot was generated by the GCG software peptide structure using default values (Genetics Computer Group (1991), Program Manual for the GCG Package Version, April 1991, 575 Science Drive, Madison, Wisconsin 53711).

of the insulin receptor, we were able to demonstrate that two regions, one within amino acid 1 to 26 and one within 27 to 68, contributed to insulin binding.[2]

Focusing on amino acids 27–68 of the insulin receptor, we have been able to localize one of these determinants to phenylalanine-39 of the insulin receptor. When inserted into the IGF-I receptor, this amino acid increases the affinity for insulin 10- to 15-fold with only a minor effect on the IGF-I affinity. Thus Phe-39 appears to be a major determinant for ligand specificity of the insulin receptor (FIG. 1).[3] Furthermore, when Phe-39 of the insulin receptor is replaced by the corresponding IGF-I receptor amino acid, serine, this results in a 10-fold drop in insulin affinity of the insulin receptor. Thus, Phe-39 appears to play a crucial role for high-affinity insulin binding of the insulin receptor.[3]

Phenylalanine-39 is located on the interface between a predicted[4] hydrophobic and hydrophilic region of the insulin receptor (IR) (FIG. 2). The predicted localization of Phe-39 indicates that this amino acid may be exposed or placed near the surface of the receptor so that it might be available for interaction with insulin in the receptor-ligand binding site.

REFERENCES

1. KJELDSEN, T., A. ANDERSEN, F. C. WIBERG, J. S. RASMUSSEN, L. SCHÄFFER, K. BACH MØLLER & N. P. H. MØLLER. 1991. Proc. Natl. Acad. Sci. USA **88:** 4404–4408.
2. ANDERSEN, A. S., T. KJELDSEN, F. C. WIBERG, H. VISSING, L. SCHAFFER, J. S. RASMUSSEN, P. DEMEYTS & N. P. H. MØLLER. 1992. J. Biol. Chem. **267:** 13681–13686.
3. KJELDSEN, T., F. C. WIBERG & A. S. ANDERSEN. 1994. J. Biol. Chem. **269:** 32942–32946.
4. KYTE, J. & R. F. DOOLITTLE. 1982. J. Mol. Biol. **157:** 105–132.

The Acute Insulin-like Effects of Growth Hormone in Primary Adipocyte-signaling Mechanisms[a]

HANS TORNQVIST,[b] MARTIN RIDDERSTRÅLE, HANS ERIKSSON, AND EVA DEGERMAN

Departments of Pediatrics
and
Medical and Physiological Chemistry
University of Lund
Lund, Sweden

Growth hormone (GH) exhibits an array of different effects on the differentiation, proliferation, and metabolism of cells. Signaling is believed to be initiated by GH-induced dimerization of the receptor.[1,2] This is accomplished through binding by way of two different binding areas between one GH molecule and two growth-hormone receptor (GHR) molecules, as well as by a third binding area between the two receptors.[3] In 3T3-F442A fibroblasts, where the mitogenic effects of GH dominate, it has been shown that GH stimulation leads to tyrosine phosphorylation of the GHR and the JAK2 tyrosine kinase strongly implicating their role in mitogenic signaling.[4] Immediate downstream elements have so far not been identified, though phosphorylations of other proteins by the JAK2 kinase and/or interaction of specific proteins through, for example, SH2 domains with the tyrosine-phosphorylated receptor seems likely.

The metabolic actions of growth hormone as seen in primary adipocytes can either be chronic diabetogenic, for example, increased lipolysis, or rapid insulin-like, such as increased lipogenesis and antilipolysis. We have focused our attention on the molecular mechanisms leading to the antilipolytic action of GH that is seen in adipocytes made responsive by a 3-hour incubation in the absence of hormones.[5] At the receptor level we studied whether GH stimulation resulted in tyrosine phosphorylations by immunoprecipitations and Western blotting of solubilized ^{32}P-labeled adipocyte extracts/membranes using antibodies against phosphotyrosine (αPY), the intracellular domain of the GHR (αGHR), and the JAK2 tyrosine kinase (αJAK2). Tyrosine phosphorylations were monitored by αPY Western blotting or autoradiography of ^{32}P proteins specifically eluted with phenyl phosphate from αPY immunoprecipitates and separated on SDS-PAGE. By these techniques we could show that GH stimulation

[a] This work was supported by the Swedish Medical Research Council (project 8689); the Medical Faculty, University of Lund; the Swedish Diabetes Association; and the Påhlsson, Novo Nordisk, Bergwall, and Crafoord Foundations.

[b] Send correspondence to Hans Tornqvist, M.D., Ph.D., Hagedorn Research Institute, Niels Steensens Vej 6, Gentofte, DK 2820 Denmark.

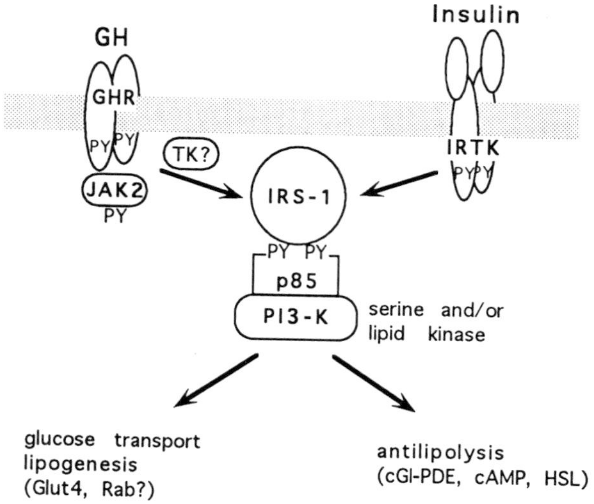

FIGURE 1. Mechanism for the insulin-like effects of GH in primary adipocytes (hypothesis). Stimulation with growth hormone (GH) induces dimerization of the growth hormone receptor (GHR), leading to activation and tyrosine phosphorylation of the JAK2 tyrosine kinase and the GHR. This event then results, through the action of JAK2 or another tyrosine kinase (TK), in tyrosine phosphorylation of the insulin-receptor substrate 1 (IRS-1). Consequently, phosphorylation of specific motifs in IRS-1 attracts and activates phosphatidylinositol 3-kinase by binding to the SH2 domains of its p85 regulatory subunit. Activation of this dual-specificity lipid and serine kinase leads to the antilipolytic effect and the effect on glucose uptake and lipogenesis. Further downstream, the antilipolytic effect is most likely mediated through activation/phosphorylation of the cGI-phosphodiesterase (cGI-PDE),[11] decreased cAMP-PK, and, finally, reduced phosphorylation/activity of the hormone-sensitive lipase (HSL).[12] The lipogenic effect involves increased glucose uptake by translocation of the glucose transporter (Glut4) and possibly the Rab4 proteins to the cell membrane. This signal chain is common to that of insulin activating the insulin-receptor tyrosine kinase (IRTK) up to, or above, the point where phosphorylation of IRS-1 occurs.

resulted in tyrosine phosphorylation of both JAK2 and the GHR in responsive cells, whereas freshly prepared cells refractory to the insulin-like effects did not respond in this way.[6] These phosphorylations reached a maximum within 5-10 minutes and exhibited a GH dose-dependency, with $ED_{50} \approx 1\text{-}2$ nM. The participation of yet another 110-115 kDa tyrosine-phosphorylated membrane protein could not be excluded. No evidence for activation of the insulin and IGF-1 receptors could be detected. Thus tyrosine phosphorylation of the GHR seems to be involved in mediating the insulin-like effects of growth hormone.

Our recent demonstration that wortmannin, a selective inhibitor of phosphatidylinositol 3-kinase (PI 3-kinase) activity, blocked the insulin-like effects, that is, the lipogenic and the antilipolytic, of GH,[7] as well as that of insulin,[8] prompted us to investigate the possible involvement of the insulin-receptor substrate-1 or similar molecules upstream of PI 3-kinase. We could demonstrate by immunoprecipitations with αIRS-1 or αp85 (PI 3-kinase subunit) that, in response to GH, IRS-1 was indeed tyrosine phosphorylated and bound to the p85 regulatory subunit of PI 3-kinase in a time- and dose-dependent manner.[9] p85 also associated with an unidentified 180-kDa tyrosine phosphorylated protein in a similar manner as IRS-1. GH stimulation resulted in a 4-fold increase in PI 3-kinase activity associated with IRS-1.

Based on these observations we conclude that tyrosine phosphorylation of the GHR, maybe through the action of the JAK2 tyrosine kinase, seems to be involved in signal transduction of the insulin-like effects of GH in primary adipocytes. This appears to trigger, through the action of the JAK2 or another unidentified tyrosine kinase, phosphorylation of IRS-1 on tyrosines, resulting in association with the SH2 domains of the p85 subunit of PI 3-kinase and increased activity of this kinase. This mechanism seems important for mediating the lipogenic and the antilipolytic actions of GH, as has already been demonstrated for insulin.[8,10] One important implication of these findings is that the convergence point for the signaling mechanism of GH and insulin for these effects appears to be located between the respective receptor and IRS-1 (FIG. 1).

REFERENCES

1. CUNNINGHAM, B. C., M. ULTSCH, A. M. DE VOS, M. G. MULKERRIN, K. R. CLAUSER & J. A. WELLS. 1991. Science **254:** 821-825.
2. ILONDO, M. M., A. B. DAMHOLT, B. A. CUNNINGHAM, J. A. WELLS, P. DE MEYTS & R. M. SHYMKO. 1994. Endocrinology **134:** 2397-2403.
3. DE VOS, A. M., M. ULTSCH & A. A. KOSSIAKOFF. 1992. Science **255:** 306-312.
4. ARGETSINGER, L. S., G. S. CAMPBELL, X. YANG, B. A. WITTHUHN, O. SILVENNOINEN, J. N. IHLE & C. CARTER-SU. 1993. Cell **74:** 237-244.
5. GOODMAN, H. M. & V. COIRO. 1981. Endocrinology **108:** 113.
6. ERIKSSON, H., M. RIDDERSTRÅLE & H. TORNQVIST. 1995. Submitted to Endocrinology.
7. RIDDERSTRÅLE, M. & H. TORNQVIST. 1994. Biochem. Biophys. Res. Commun. **203:** 306-310.
8. RAHN, T., M. RIDDERSTRÅLE, H. TORNQVIST, V. MANGANIELLO, G. FREDRIKSON, P. BELFRAGE & E. DEGERMAN. 1994. FEBS Lett. **350:** 314-318.
9. RIDDERSTRÅLE, M., E. DEGERMAN & H. TORNQVIST. 1995. J. Biol. Chem. **270:** 3471-3474.
10. WHITE, M. F. & C. R. KAHN, 1994. J. Biol. Chem. **269:** 1-4.
11. MANGANIELLO, V. C., E. DEGERMAN, C. J. SMITH, V. VASTA, H. TORNQVIST & P. BELFRAGE. 1992. Adv. Second Messenger Phosphoprotein Res. **25:** 147-164.
12. BJÖRGELL, P., S. ROSBERG, O. ISAKSSON & P. BELFRAGE. 1984. Endocrinology **115:** 1151-1156.

Regulatory Interaction between Calmodulin and the Epidermal Growth Factor Receptor[a]

ALBERTO BENGURÍA,[b] JOSÉ MARTÍN-NIETO,[b] GUSTAVO BENAIM,[c] AND ANTONIO VILLALOBO[b,d]

[b]*Instituto de Investigaciones Biomédicas*
Consejo Superior de Investigaciones Científicas
Arturo Duperier 4
28029 Madrid, Spain

[c]*Centro de Biología Celular*
Facultad de Ciencias
Universidad Central de Venezuela
Caracas, Venezuela

Growth factor-stimulated cell proliferation is preceded by a transient increase in the cytoplasmic Ca^{2+} concentration,[1] and calmodulin, an intracellular calcium receptor protein,[2] appears to intervene in the modulation of this process.[3] Calmodulin binds to nuclear proteins and could therefore affect their functions.[4] However, calmodulin does not control cell proliferation by acting exclusively at the nuclear level. Several lines of evidence obtained in our laboratory indicate that the epidermal growth factor receptor (EGFR) is a calmodulin-binding protein and that calmodulin acts as a modulator of its intrinsic tyrosine kinase activity.[5,6]

We have demonstrated that the EGFR can be isolated from solubilized rat liver plasma membranes by calmodulin-affinity chromatography.[5,6] Binding of the EGFR to calmodulin-agarose occurs in the presence of Ca^{2+}, and its elution is achieved upon addition of EGTA. The preparation obtained contains a set of calmodulin-binding proteins associated with the plasma membrane. However, the major autophosphorylated protein in this preparation corresponds to the EGFR. This autophosphorylation is stimulated by epidermal growth factor (EGF) and transforming growth factor-α (TGF-α) with essentially the same efficiency.[5] Furthermore, the isolated receptor presents both EGF- and TGFα-stimulated tyrosine kinase activity towards the exogenous substrate poly-L-(Glu : Tyr).[5,6] The identity of the isolated EGFR

[a]This work was supported by Grant SAF392/93 from the Comisión Interministerial de Ciencia y Tecnología, Grants AE132-93 and AE16-94 from the Consejería de Educación y Cultura de la Comunidad Autónoma de Madrid, Spain to A. Villalobo, and by Grant RP-IV-110034 from the Consejo Nacional de Investigaciones Científicas y Tecnológicas, Venezuela to G. Benaim. A. Benguría was supported by a predoctoral fellowship from the Departamento de Educación del Gobierno Vasco. J. Martín-Nieto was supported by a postdoctoral contract from the Ministerio de Educacion y Ciencia, Spain.

[d]To whom correspondence should be addressed.

was confirmed by immunoblot analysis and immunoprecipitation using a polyclonal antibody against a human EGF receptor/c-*erb*B-2 product-common epitope.[5] FIGURE 1 shows the proteins that are phosphorylated in the absence (–) and presence (+) of EGF, using a solubilized plasma membrane preparation (panel 1) and the EGTA-eluted material from the calmodulin-affinity chromatography (panels 2 and 3), in assays carried out in the absence (panels 1 and 2) and presence (panel 3) of poly-L-(Glu : Tyr). The autophosphorylated EGFR is observed as a minor band in the solubilized plasma membrane fraction before calmodulin-affinity chromatography (panel 1). By contrast, the EGF-stimulated autophosphorylation of the EGFR (panel 2) and the EGF-stimulated phosphorylation of poly-L-(Glu : Tyr), which migrates in the gels as a smear (panel 3), are readily observable in the material eluted with EGTA from the calmodulin-agarose column.

Bovine brain calmodulin, which contains two tyrosine residues (Tyr99 and Tyr138), becomes phosphorylated in Tyr99 in the absence of Ca^{2+} and in an EGF-stimulated manner by the isolated EGFR.[6] Low concentrations of Ca^{2+} (around 1 μM) strongly inhibit the phosphorylation of calmodulin.[6] The phosphorylation of calmodulin is absolutely dependent on the presence of a polycation or a basic protein, and its stoichiometry is close to 1 mol of phosphate per mol of calmodulin.[5,6] However, calmodulins isolated from *Trypanosoma cruz,*[7] which contains a single tyrosine residue (Try138), and from *Leishmania mexicana* are phosphorylated in tyrosine by the EGFR only in trace amounts (results not shown). FIGURE 2 (panel 1) shows the phosphorylation of bovine brain calmodulin by the EGFR in the absence (–) and presence (+) of EGF and in the presence of poly-L-(Lys). The addition of EGF stimulates the phosphorylation of calmodulin 2.7-fold. As we have shown previously,[6] the presence of poly-L-(Lys) in the assay system causes the phosphorylation of other unidentified proteins present in the EGFR preparation (FIG. 2, panel 1).

Further evidence of the interaction of the EGFR with calmodulin was obtained by analyzing the effect of calmodulin on both the autophosphorylation of this receptor and its tyrosine-kinase activity toward exogenous substrates. We have demonstrated that calmodulin in the presence of Ca^{2+} inhibits both processes in a concentration-dependent mode.[5,6] FIGURE 2 (panel 2) shows the results of assays of phosphorylation of poly-L-(Glu : Tyr) by the isolated receptor in the absence (lane a) and presence (lanes b and c) of calmodulin. Calmodulin inhibits the phosphorylation of poly-L-(Glu : Tyr) by 42% and 72% after 1.5 min (lane b) and 40 min (lane c) of incubation, respectively. The inhibition of the autophosphorylation of the EGFR is also noticeable (lanes b and c). We have excluded the presence of calmodulin-dependent phosphatase activity in these preparations.[5] Therefore, calmodulin acts as a negative modulator of the tyrosine-kinase activity of the EGFR.

We have also demonstrated that when calmodulin is allowed to be phosphorylated by the tyrosine kinase of the EGFR prior to the phosphorylation of the substrate poly-L-(Glu : Tyr), the phosphorylation of this substrate is strongly enhanced (*results not shown*). This observation is compatible with a model in which phosphocalmodulin acts as a positive modulator of the tyrosine-kinase activity of the EGFR. However, the complexity of this two-step assay system justifies a cautious interpretation of these results.

In conclusion, we have demonstrated that calmodulin interacts, at least *in vitro,* with the EGFR in both Ca^{2+}-dependent and Ca^{2+}-independent modes. The actual

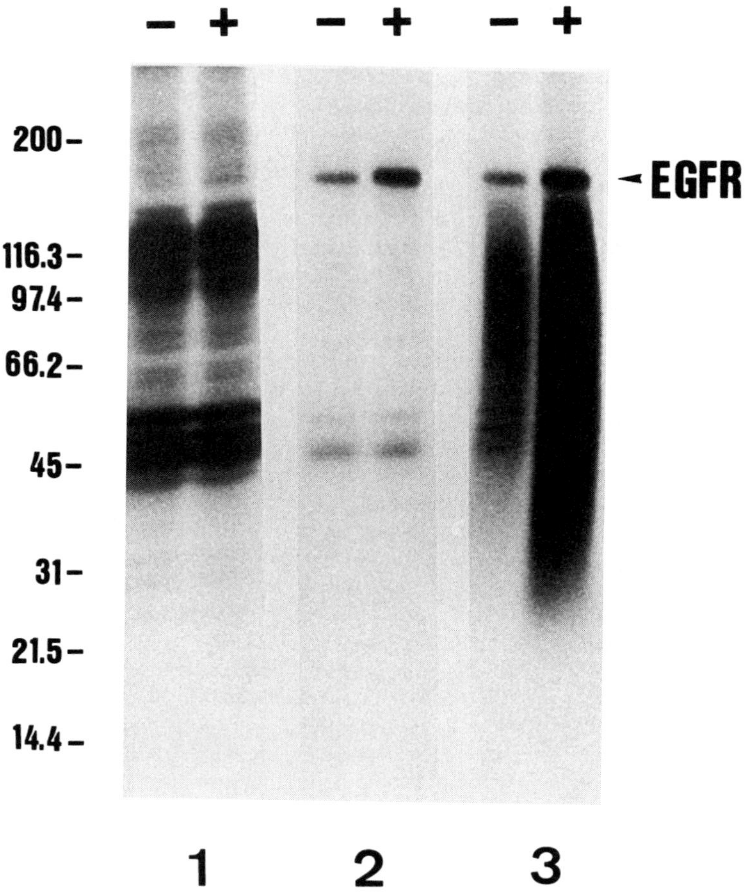

FIGURE 1. Activity of the EGFR isolated by calmodulin-affinity chromatography. A Triton X-100-solubilized plasma membrane fraction (30 μL) (panel 1) and the EGTA-eluted fraction (40 μL) from the subsequent calmodulin-affinity chromatography (panels 2 and 3) were assayed at 37°C for 3 min in 100 μL of a medium containing 15 mM Na-Hepes (pH 7.4), 6 mM $MgCl_2$, 0.4 mM EGTA, 0.4% (w/v) Triton X-100, 2% (w/v) glycerol, and 10 μM (2 μCi) [γ − ^{32}P]ATP in the absence (−) and presence (+) of 1 μM EGF, and in the absence (panels 1 and 2) and presence (panel 3) of 0.1 mg/mL poly-L-(Glu : Tyr). The fractions were incubated with EGF for 30 min on ice before the addition of radiolabeled ATP. The reactions were stopped with ice-cold 10% (w/v) trichloroacetic acid, and the precipitated proteins were analyzed by SDS-PAGE and autoradiography. The arrowhead points to the EGFR radiolabeled band. Molecular mass standards (kDa) are also indicated. Additional methodological information is given in refs. 5 and 6.

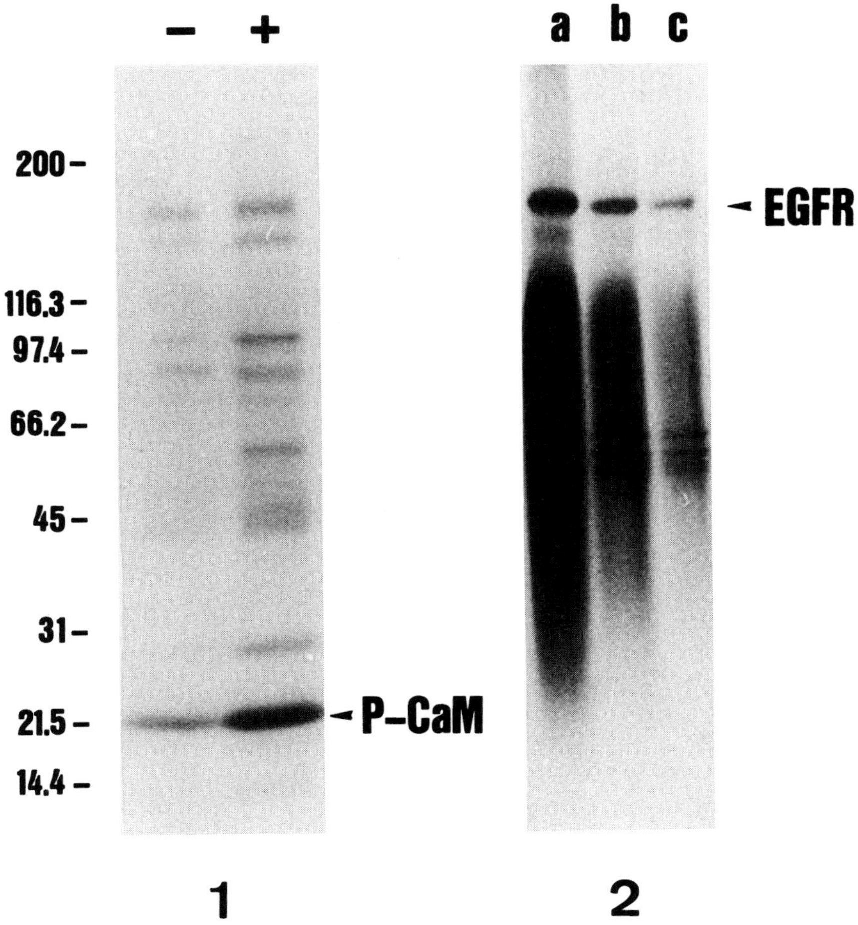

FIGURE 2. Calmodulin is phosphorylated by the isolated EGFR in the absence of Ca^{2+} but inhibits its tyrosine-kinase activity in the presence of Ca^{2+}. Panel 1: Calmodulin (1.2 μM) was phosphorylated using the EGFR-containing EGTA-eluted fraction (50 μL) from the calmodulin-affinity chromatography at 37°C for 5 min in 100 μL of a medium containing 15 mM Na-Hepes (pH 7.4), 6 mM $MgCl_2$, 0.5 mM EGTA, 0.5% (w/v) Triton X-100, 2.5% (w/v) glycerol, 0.5 μM poly-L-(Lys), and 10 μM (2 μCi) [γ-^{32}P]ATP in the absence (–) and presence (+) of 1 μM EGF. Panel 2: Poly-L-(Glu : Try) (0.1 mg/mL) was phosphorylated using the EGTA-eluted fraction (40 μL) from the calmodulin-affinity chromatography at 37°C for 1 min in 100 μL of a medium containing 15 mM Na-Hepes (pH 7.4), 6 mM $MgCl_2$, 0.4 mM EGTA, 0.5 mM $CaCl_2$ (100 μM free Ca^{2+}), 0.4% (w/v) Triton X-100, 2% (w/v) glycerol, 1 μM EGF, and 10 μM (2 μCi) [γ-^{32}P]ATP in the absence (lane a) and presence (lanes b and c) of 3 μM calmodulin. The EGFR-containing fraction was incubated with EGF for 40 min (lanes a–c) on ice, and calmodulin was added 1.5 min (lane b) or 40 min (lane c) before the addition of radiolabeled ATP. The reactions (panels 1 and 2) were stopped with ice-cold 10% (w/v) trichloroacetic acid, and the precipitated proteins were analyzed by SDS-PAGE and autoradiography. EGTA (10 mM) was added to the electrophoresis sample buffer in order to obtain migration of phosphocalmodulin at 21 kDa. Arrowheads point to the phosphocalmodulin (P-CaM) (panel 1) and to the autophosphorylated EGFR (panel 2) radiolabeled bands. Molecular mass standards (kDa) are also indicated. Additional methodological information is given in refs. 5 and 6.

calmodulin-binding site(s) in the EGFR has not yet been identified. However, an amphiphilic basic domain, a candidate to constitute an alpha-helical calmodulin-binding site, can be found in the juxtamembrane region of the cytoplasmic domain of the human EGFR.[5]

REFERENCES

1. ROZENGURT, E. 1986. Science **234:** 161-166.
2. MANALAN, A. S. & C. B. KLEE. 1984. Adv. Cyclic Nucleotide Protein Phosphorylation Res. **18:** 227-277.
3. LU, K. P. & A. R. MEANS. 1993. Endoc. Rev. **14:** 40-58.
4. BACHS, O., N. AGELL & E. CARAFOLI. 1992. Biochim. Biophys. Acta **1113:** 259-270.
5. SAN JOSÉ, E., A. BENGURÍA, P. GELLER & A. VILLALOBO. 1992. J. Biol. Chem. **267:** 15237-15245.
6. BENGURÍA, A., O. HERNÁNDEZ-PERERA, M. T. MARTÍNEZ-PASTOR, D. B. SACKS & A. VILLALOBO. 1994. Eur. J. Biochem. **224:** 909-916.
7. BENAIM, G., S. LOSADA, F. R. GADELHA & R. DOCAMPO. 1991. Biochem. J. **280:** 715-720.

Phosphorylation of Connexin-32 by the Epidermal Growth Factor Receptor Tyrosine Kinase[a]

JUAN ANTONIO DÍEZ,[b] MARIBEL ELVIRA, AND ANTONIO VILLALOBO[c]

Instituto de Investigaciones Biomédicas
Consejo Superior de Investigaciones Científicas
Arturo Duperier 4
28029 Madrid, Spain

Gap-junction channels are formed by two hexameric complexes of connexins located in the plasma membranes of adjacent cells.[1] These channels mediate intercellular communication between the cytoplasms of neighboring cells by providing a direct pathway for the exchange of ions and low molecular mass metabolites and regulatory molecules. These ions and molecules play a crucial role in tissue homeostasis, synchronization of cellular activities within tissues and systems, and the regulation of differentiation and proliferation processes.[1]

Gap-junctional communication is interrupted by mitogenic agents, which suggests that closing and/or disassembly of gap-junction channels are key events preceding mitosis. The epidermal growth factor (EGF) stimulates the disruption of gap-junctional communication in rat liver epithelial cells concomitantly with the phosphorylation of connexin-43 on serine residues.[2] However, EGF-mediated phosphorylation of connexin-43 appears to be an indirect effect following mitogenic activation. The involvement of the mitogen-activated protein (MAP) kinase in the phosphorylation of this connexin has been suggested.[2] Direct phosphorylation of this connexin by the intrinsic tyrosine kinase of the epidermal growth factor receptor (EGFR) has not been demonstrated so far. By contrast, other tyrosine kinases directly phosphorylate connexin-43. In this regard, phosphorylation of connexin-43 in tyrosine residues by both $pp60^{v\text{-}src}$[3–5] and $p130^{gag\text{-}fps}$[6] tyrosine kinases followed by the disruption of gap-junctional communication has been reported. Nevertheless, direct phosphorylation of connexins by the EGFR-tyrosine kinase is an appealing possibility that should not be excluded, and we set out to explore it using an *in vitro* assay system.

[a] This work was supported by Grants from the Comisión Interministerial de Ciencia y Tecnología (SAF392/93) and from the Consejería de Educación y Cultura de la Comunidad Autónoma de Madrid (AE132-93 and AE16-94) to A. Villalobo. J. A. Díez was supported by a predoctoral fellowship from the Departamento de Educación del Gobierno Vasco.

[b] Present address: University of Wales College of Medicine, Department of Medical Biochemistry, Health Park Cardiff CF4 4XN Wales, United Kingdom.

[c] To whom correspondence should be addressed.

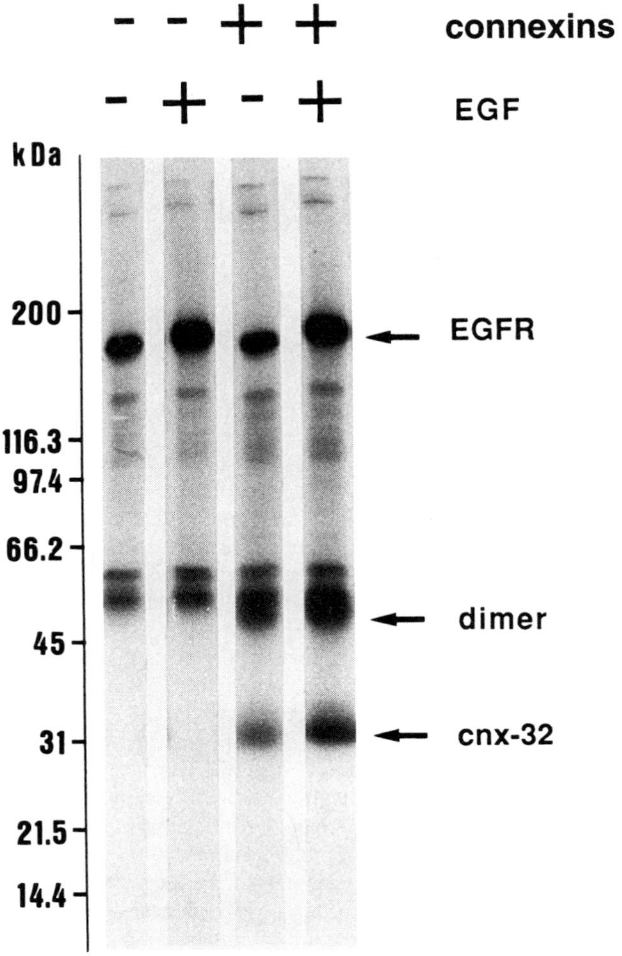

FIGURE 1. Phosphorylation of connexin-32 by the EGFR tyrosine kinase. Rat liver gap junctions (1 μg of connexin-32) were phosphorylated at 37°C for 5 min in 100 μL of a medium containing 27.5 mM Hepes-Na (pH 7.4), 5 mM $MgCl_2$, 2.5% (w/v) glycerol, 0.5% (w/v) Triton X-100, 0.5 mM EGTA, 1 μM EGF (*where indicated*), 50 μL of EGFR-containing fraction, and 10 μM (2 μCi) [γ-^{32}P]ATP. The reaction was stopped with 10% (w/v) trichloroacetic acid. The precipitated proteins were separated by centrifugation and processed by SDS-PAGE and autoradiography. Arrows pointed to the monomer of connexin-32 (cnx-32), its dimer, and the epidermal growth factor receptor (EGFR). Controls in the absence of gap junctions are also presented.

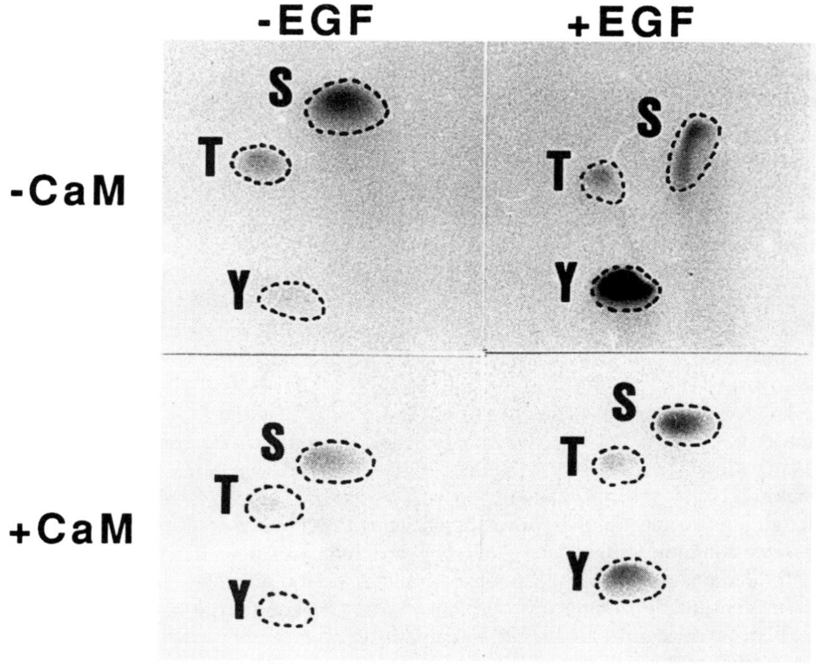

FIGURE 2. Phosphoamino acid analysis of connexin-32 phosphorylated by the EGFR tyrosine kinase in the absence and presence of calmodulin. Rat liver gap junctions (0.5 μg of connexin-32) were phosphorylated at 37°C for 5 min in 100 μL of a medium containing 27.5 mM Hepes-Na (pH 7.4), 5 mM $MgCl_2$, 2.5% (w/v) glycerol, 0.5% (w/v) Triton X-100, 0.5 mM EGTA, 1 μM calmodulin (*where indicated*), 1 μM EGF (*where indicated*), 50 μL of EGFR-containing fraction, and 50 μM (40 μCi) [γ-^{32}P]ATP. The reaction was stopped with 10% (w/v) trichloroacetic acid. The precipitated proteins were separated by centrifugation and processed by SDS-PAGE and autoradiography as in FIG. 1 to identify phosphoconnexin-32. Thereafter, phosphoamino acid analysis of the connexin-32 monomer was performed as described.[10] The areas of migration of phosphoserine (S), phosphothreonine (T), and phosphotyrosine (Y) standards stained with ninhydrin are indicated by dashed lines.

To establish whether connexin-32 could be directly phosphorylated by the EGFR-tyrosine kinase, we assayed gap-junction plaques, isolated from rat liver plasma membranes by alkaline treatment and discontinuous sucrose-gradient centrifugation,[7] with a preparation of rat liver EGFR isolated by calmodulin-affinity chromatography,[8] using [γ-^{32}P]ATP as the phosphate donor. FIGURE 1 shows that connexin-32 is indeed phosphorylated by the EGFR preparation in an EGF-stimulated manner. In addition to the phosphorylated connexin-32 monomers, phosphorylated connexin-32 dimers and the autophosphorylated EGFR are observed in the autoradiograph. By contrast, connexin-26, a second connexin most prominently present in mouse liver gap junctions, was not phosphorylated by the EGFR (results not shown), or by purified protein kinase A or protein kinase C.[7]

The phosphoamino acid analysis of phosphoconnexin-32 monomers reveals EGF-dependent phosphorylation of tyrosine residues, although phosphorylated serine and threonine residues are also detected (FIG. 2). Moreover, the intracellular Ca^{2+} receptor protein calmodulin (CaM) inhibits the EGFR-mediated phosphorylation of tyrosine residues in connexin-32 (FIG. 2). We have previously demonstrated that calmodulin inhibits the tyrosine-kinase activity of the EGFR in the presence of Ca^{2+}.[8] Therefore, we performed these assays in the absence of Ca^{2+} (presence of EGTA) to prevent the inactivation of the EGFR-tyrosine kinase. These results suggest that Ca^{2+}-independent binding of calmodulin to connexin-32 occludes its phosphorylation site(s), thereby preventing phosphorylation by the EGFR-tyrosine kinase. The presence of serine/threonine-protein kinase(s) in the EGFR preparation accounts for the observed phosphorylation of these residues in connexin-32. By contrast, only serine residues were phosphorylated when purified protein kinase A or protein kinase C were used (results not shown).

The stoichiometry of connexin-32 phosphorylation by the EGFR is very low, in the order of 0.005 moles of phosphate per mole of connexin-32 (results not shown). Similarly, low stoichiometries are generally obtained when rat and mouse liver connexin-32s are phosphorylated by purified protein kinase A or protein kinase C,[7] although significantly higher stoichiometries can be obtained with protein kinase C in assays containing low detergent concentrations (results not shown). We have recently demonstrated that phosphorylation of a small number of connexin-32 molecules in gap-junction plaques by protein kinase C prevents the action of the Ca^{2+}-dependent proteases μ-calpain and *m*-calpain on both the phosphorylated connexin-32 molecules and a larger number of nonphosphorylated connexin-32 molecules.[9] A protective conformational change in the tight-packed gap-junction channels affecting both phosphorylated and nonphosphorylated connexins may account for this effect.[9] Therefore, we suggest that phosphorylation of connexin-32 by the EGFR-tyrosine kinase may have physiological relevance in spite of the low phosphorylation stoichiometry found.

REFERENCES

1. LOEWENSTEIN, W. R. 1987. Cell **48:** 725-726.
2. KANEMITSU, M. Y. & A. F. LAU 1993. Mol. Biol. Cell **4:** 837-848.
3. CROW, D. S., E. C. BEYER, D. L. PAUL, S. S. KOBE & A. F. LAU. 1990. Mol. Cell. Biol. **10:** 1754-1763.
4. CROW, D. S., W. E. KURATA & A. F. LAU. 1992. Oncogene **7:** 999-1003.
5. GOLDBERG, G. S. & A. F. LAU. 1993. Biochem. J. **295:** 735-742.
6. KURATA, W. E. & A. F. LAU. 1994. Oncogene **9:** 329-335.
7. ELVIRA, M., J. A. DÍEZ, K. K. W. WANG & A. VILLALOBO. 1993. J. Biol. Chem. **268:** 14294-14300.
8. SAN JOSÉ, E., A. BENGURÍA, P. GELLER & A. VILLALOBO. 1992. J. Biol. Chem. **267:** 15237-15245.
9. ELVIRA, M., K. K. W. WANG & A. VILLALOBO. 1994. Biochem. Soc. Trans. **22:** 793-796.
10. HUNTER, T. & B. M. SEFTON. 1980. Proc. Natl. Acad. Sci. USA **77:** 1311-1315.

Molecular Dissection of the Growth Hormone Receptor

Identification of Distinct Cytoplasmic Domains Corresponding to Different Signaling Pathways

J. H. NIELSEN, N. BILLESTRUP, G. ALLEVATO, A. MØLDRUP, E. D. PETERSEN, J. AMSTRUP, J. A. HANSEN, AND C. SVENSSON

Hagedorn Research Institute
Niels Steensens Vej 6
DK-2820 Gentofte, Denmark

Growth hormone (GH) exerts multiple biological actions on cell proliferation, gene transcription, protein synthesis, and carbohydrate and lipid metabolism. The signal transduction mechanism has been unknown until recently when the tyrosine kinase JAK-2 was identified as being associated with and activated by binding of GH to its receptor.[1] Although this pathway seems to be common to several members of the receptor family and essential for the effects of GH on cell proliferation and gene transcription, it is apparently not sufficient to explain all the biological activities of GH. Thus truncation of the GHR, leaving only box 1 intact, did not abolish the mitogenic effect,[2] whereas truncation of just half of the cytoplasmic domain completely blocked the ability of GH to stimulate insulin synthesis[3] or transcription of serine protease inhibitor (Spi) 2.1.[4] In order to further identify domains that may be involved in GH signal transduction, we have constructed a number of GHR mutants that were analyzed for their ability to transmit signals for various biological effects of GH. The proline-rich domain (box 1) seems to be responsible for the binding and activation of the tyrosine kinase JAK-2,[5] which is thus sufficient for transmitting the mitogenic signal for GH and the other growth factors in this receptor family. This is, however, not sufficient for activation of the tissue-specific genes like insulin and Spi 2.1, as we found that the truncated GHR 1-454 was inactive. In the search for other signals that might be involved in the specific gene transcription, we analyzed the domains required for the recently reported effect of GH on calcium uptake.[6] Interestingly we found that the GHR mutant, with four proline to alanine mutations in box 1 and unable to activate JAK-2, still was able to mediate GH-stimulated calcium uptake. It has recently been found that interferon α and γ, by binding to their receptors, activate tyrosine kinases of the JAK/Tyk family that directly phosphorylate proteins of the STAT family being translocated to the nucleus, where they participate in the binding and activation of the promoters for certain genes.[7] It is therefore conceivable that GH induces recruitment of a similar STAT-like transcription factor that by interaction with a calcium-dependent partner forms a complex

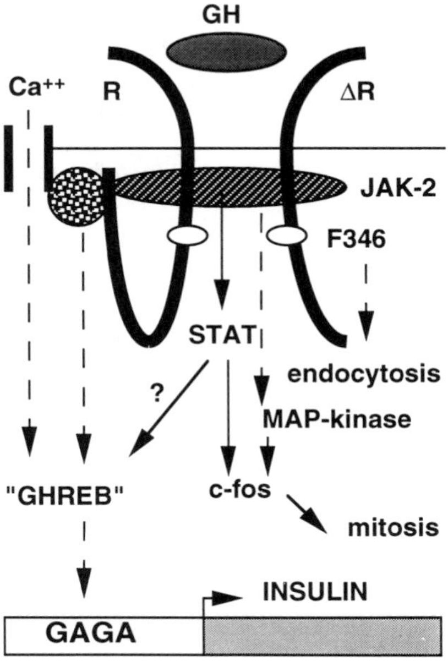

FIGURE 1. Model of GHR signal transduction (see text).

(GHREB) that binds to a purine-rich element (GAGA)[8] and activates the tissue-specific genes. As internalization of the GH/GHR complex has been proposed to be involved in the signal transduction, we have measured the ability of the GHR mutant to internalize GH. Surprisingly, mutation of a single amino acid, phenylalanine in position 346, blocked the internalization, but not GH-induced Spi 2.1 transcription. In conclusion, these results show that the cytoplasmic part of the GHR contains several distinct domains with different signal-transducing properties, as illustrated in Figure 1.

REFERENCES

1. Argetsinger, L. S., G. S. Campbell, X. Yang, B. A. Witthuhn, O. Silvennoinen, J. N. Ihle & C. Carter-Su. 1994. Identification of JAK2 as a growth hormone receptor-associated tyrosine kinase. Cell **74:** 237-244.
2. Colosi, P., K. Wong, S. R. Leong & W. I. Wood. 1993. Mutational analysis of the intracellular domain of the human growth hormone receptor. J. Biol. Chem. **268:** 12617-12623.
3. Møldrup, A., G. Allevato, T. Dyrberg, J. H. Nielsen & N. Billestrup. 1991. Growth hormone action in rat insulinoma cells expressing truncated growth hormone receptors. J. Biol. Chem. **266:** 17440-17445.
4. Goujon, L., G. Allevato, G. Simonin, L. Paquereau, A. L. Cam, J. Clark, J. H. Nielsen, J. Djiane, M. C. Postel-Vinay, M. Edery & P. Kelly. 1994. Cytoplasmic sequences of the growth hormone receptor necessary for signal transduction. Proc. Natl. Acad. Sci. USA **91:** 957-961.
5. Vanderkuur, J. A., X. Wang, L. Zhang, G. Campbell, G. Allevato, N. Billestrup, G. Norstedt & C. Carter-Su. 1994. Domains of the growth hormone receptor required

for association and activation of JAK2 tyrosine kinase. J. Biol. Chem. **269:** 21709-21717.

6. Ilondo, M. M., P. D. Meyts & P. Bouchelouche. 1994. Human growth hormone increases cytosolic free calcium in cultured human IM-9 lymphocytes: A novel mechanism of growth hormone transmembrane signalling. Biochem. Biophys. Res. Commun. **202:** 391-397.
7. Darnell, J. E., I. A. Kerr & G. R. Stark. 1994. Jak-STAT pathways and transcriptional activation in response to IFNs and other extracellular signalling proteins. Science **264:** 1415-1421.
8. Nielsen, J. H., N. Billestrup, A. Møldrup, G. Allevato, E. D. Petersen, J. A. Pedersen & J. A. Hansen. 1993. Growth of the endocrine pancreas: the role of somatolactogenic hormones and receptors. Biochem. Soc. Trans. **21:** 146-149.

Endothelin Stimulates MAP Kinase Activity and Protein Synthesis in Isolated Adult Feline Cardiac Myocytes

LINDA G. JONES,[a,c] KATRINA C. GAUSE,[b] AND KATHRYN E. MEIER[b]

[a]*Departments of Medicine and Pharmacology*
University of Arkansas for Medical Sciences
Little Rock, Arkansas 72205
and
[b]*Department of Cell and Molecular Pharmacology*
Medical University of South Carolina
Charleston, South Carolina 29425

INTRODUCTION

We have previously observed that endothelin-1 (ET-1) stimulates phosphoinositide hydrolysis and increases in mRNA levels of the immediate early genes c-*fos* and *egr*-1 in isolated adult feline cardiac myocytes,[1] leading us to propose that ET-1 may participate as a regulator of cardiac growth. The aim of the present study was to demonstrate that ET-1 promoted a definitive growth response (protein synthesis) and to examine whether kinases shown to be mediators of cell growth (mitogen-activated protein kinases, MAPK) are stimulated by ET-1 in these cardiac myocytes. Inasmuch as ET-1 promotes both pertussis toxin (PT)-sensitive and -insensitive responses in these cells, we also examined the PT sensitivity of these downstream events.

RESULTS AND DISCUSSION

ET-1 increased the rate of protein synthesis in myocytes, as demonstrated by a 35% increase over control (CTL) values in the rate of incorporation of [^{3}H]PHE into total cellular protein. This increase was somewhat diminished by prior exposure of the cells to PT (TABLE 1). Pretreatment of cardiac myocytes with 1 nM staurosporine; inhibited the ET-1-stimulated increase in [^{3}H]PHE incorporation, suggesting that protein-kinase activation is required for this response (data not shown).

[c]Address all correspondence to Linda G. Jones, Ph.D., Department of Medicine, Division of Cardiology, Slot #532, University of Arkansas for Medical Sciences, 4301 West Markham, Little Rock, Arkansas 72205-7199.

TABLE 1. Treatment with Pertussis Toxin Partially Inhibits ET-1-stimulated Increases in Protein Synthesis and MAPK Activity[a]

	[^{3}H]PHE incorporation, pmol/mg/h		MBP phosphorylation, pmol/min/mg	
	−PT	+PT	−PT	+PT
CTL	147 ± 3[a]	149 ± 6[a]	46 ± 5[a]	36 ± 1[a]
ET-1	198 ± 7[b]	182 ± 1[c]	76 ± 5[b]	46 ± 5[a]

[a] Adult feline cardiac myocytes were isolated using retrograde perfusion of the heart with collagenase as described[2] and incubated overnight at a concentration of 0.3–0.5 × 10^6 per well. Protein synthesis was measured as the extent of incorporation of L[ring 2,3,4,5,6 ^{3}H]phenylalanine ([^{3}H]PHE) into newly synthesized protein as previously described.[3] The data on the left side are the results of an experiment in which isolated feline cardiac myocytes were treated for three hours with and without 100 ng/mL PT before a 24-h exposure to 100 nM ET-1 and radioactive label. Two additional protein synthesis experiments performed in the same manner gave similar results. As shown on the right side, kinase activity was assessed by phosphorylation of an exogenous substrate for MAPK, myelin basic protein (MBP), using published techniques.[4] These data are from a representative experiment (of five with similar results), in which cardiac myocytes were incubated overnight with or without 100 ng/mL PT before a five-minute incubation with or without 100 nM ET-1. All data are represented as the mean ± SEM of four individual samples. Significant differences ($p \leq 0.05$) between groups are indicated by different superscripted letters and were determined using two-way analysis of variance with Fisher's protected LSD (least significant difference) post-hoc testing.

In other experiments, ET-1 stimulated MAPK activity up to four times control values, with peak activity observed between 5 and 10 minutes (data not shown). As shown in TABLE 1, stimulation of MAPK by ET-1 was also partially inhibited by prior exposure to PT. However, stimulation of MAPK by phorbol ester (PMA) was not diminished by PT treatment (not shown).

We performed immunoblotting on crude cytosolic fractions using polyclonal antisera raised against a peptide sequence of rat extracellular signal-regulated kinases 2 (ERK-2).[5] Two major bands of 40 and 47 kDa were observed in cytosolic fractions from both control and ET-1-treated cells (FIG 1). Minor bands of 42 and 44 kDa were also sometimes observed. None of these bands appeared to shift in response to agonist, in contrast to results obtained with other cell types.[6] Use of commercial antibodies against ERK-1 and ERK-2 demonstrated that isolated cardiac myocytes from rat and rabbit also possess two major immunoreactive bands of 40 and 47 kDa. We also fractionated cytosolic samples over a MonoQ anion exchange column. Immunoreactive protein was detected in the fractions that contained kinase activity (data not shown).

These data support a role for ET-1 in the regulation of cardiac myocyte growth and suggest the generation of multiple early signals in response to ET-1.

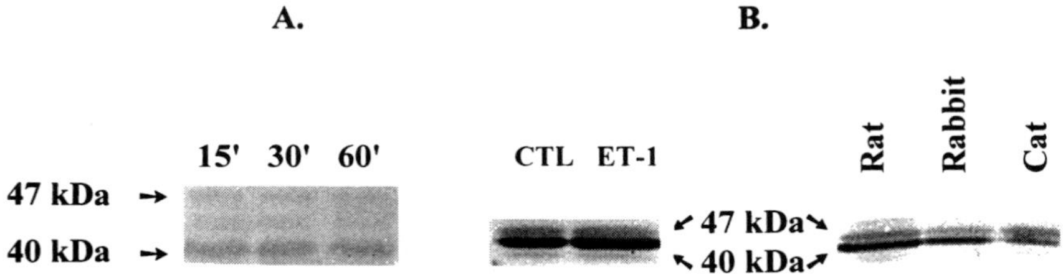

FIGURE 1. Immunoblotting of cytosolic fractions of isolated adult cardiac myocytes was performed as previously described.[6] Antiserum against the *erk* gene product from our lab[1] (A) or from a commercial source (B, Transduction Laboratories, Lexington, KY) were used to prepare the Western blots. Immunoreactive proteins were identified from extracts of feline cardiac myocytes incubated for the times indicated with 100 nM ET-1 (A). Also shown are those incubated for five minutes in the absence (CTL) or presence of 100 nM ET-1 (B, left panel), or those that were from untreated adult cardiac myocytes isolated from the species indicated (B, right panel).

REFERENCES

1. Jones, L. G., J. D. Rozich, H. Tsutsui & G. Cooper IV. 1992. Am. J. Physiol. **263** (Heart Circ. Physiol. 32): H1447-H1454.
2. Mann, D. L., R. L. Kent & G. Cooper IV. 1989. Circ. Res. **64:** 542-553.
3. Kent, R. L., D. L. Mann & G. Cooper IV. 1989. Circ. Res. **64:** 74-85.
4. Meier, K. E., K. A. Licciardi, T. A. J. Haystead & E. G. Krebs. 1991. J. Biol. Chem. **266:** 1914-1920.
5. Gause, K. C., M. K. Homma, K. A. Licciardi, R. Seger, N. G. Ahn, M. J. Peterson, E. G. Krebs & K. E. Meier. 1993. J. Biol. Chem. **268:** 16124-16129.
6. Jones, L. G., K. M. Ella, C. D. Bradshaw, K. C. Gause, M. Dey, A. E. Wisehart-Johnson, E. C. Spivey & K. E. Meier. 1994. J. Biol. Chem. **269:** 23790-23799.

Index of Contributors

Abraham, R. T., 209-213
Adunyah, S. E., 296-299
Akira, S., 224-234
Al-Aoukaty, A., 292-295
Alkalay, I., 245-252
Allevato, G., 481-483
Altboum, I., 93-95
Amstrup, J., 481-483
Andersen, A. S., 466-468
Avraham, A., 245-252
Avruch, J., 303-319

Backer, J. M., 369-387
Bajpai, A., 216-219
Barbacid, M., 442-458
Baulida, J., 44-51
Beguinot, L., 44-51
Ben-Neriah, Y., 245-252
Benaim, G., 472-476
Benguría, A., 472-476
Berridge, M. J., 31-43
Billestrup, N., 481-483
Blechman, J. M., 344-362
Blum, J. H., 195-201
Boise, L. H., 70-80
Bolen, J. B., 214-215
Boniface, J. J., 62-69
Bonnema, J. D., 209-213
Bornfeldt, K. E., 416-430
Bradshaw, R. A., 1-17
Brahmi, Z., 216-219
Brennan, F. M., 272-278
Brown, B. L., 285-287
Brummet, M. E., 206-208
Buder, A., 89-92
Buhl, A. M., 288-291
Burakoff, S. J., 117-133, 202-203
Burkhardt, A. L., 214-215

Campbell, J. S., 320-343
Campbell, K. S., 89-92
Çarpenter, G., 44-51
Čatipović, B., 206-208
Ceesay, K., 296-299
Chambard, J-C., 431-441
Chan, V. W.-F., 195-201
Chan, A., 149-156
Cheatham, B., 369-387
Chen, Y-H., 431-441
Choidas, A., 1-17
Christoffersen, C. T., 388-401, 409-415
Cooper, R. S., 296-299
Corbett, N. R., 285-287
Czech, M. P., 204-205

Datta, S. K., 195-201
Davis, M. M., 62-69
Dawson, C. H., 285-287
De Meyts, P., xiii-xv, 388-401, 409-415
DeFranco, A. L., 195-201
Degerman, E., 469-471
Desarnaud, F., 23-30
Deuschle, U., 89-92
Díez, J. A., 477-480
Dixon, J. E., 18-22
Dobson, P. R. M., 285-287
Domenico, J., 134-148
Donovan, J. A., 157-172
Dumont, J., 23-30
Dunford, J. E., 285-287
Duprez, L., 23-30

Eggerickx, D., 23-30
Elliott, M. J., 272-278
Elvira, M., 477-480
Eriksson, H., 469-471

Feldmann, M., 272-278
Fisher, T. L., 369-387
Foy, S. P., 195-201
Fujii, H., 235-244

Gause, K. C., 484-486
Gelfand, E. W., 134-148
Giaid, A., 292-295
Gold, M. R., 195-201
Goren, H. J., 463-465
Grall, D., 431-441
Graves, J. D., 320-343
Graves, L. M., 416-430
Grothe, S., 300-302

Hansen, J. A., 481-483
Hatakeyama, M., 235-244
Hatzubai, A., 245-252
Helman, L. J., 402-408
Hermans, M., 173-181
Horvath, G., 220-223
Hourihane, S. L., 195-201

Irie, H. Y., 117-133
Isakov, N., 157-172
Ish-Shalom, D., 409-415
Iwashima, M., 149-156

Jachna, B. R., 369-387
Johnson, G. L., 288-291
Jones, L. G., 484-486
Jung, S., 245-252
Justement, L. B., 214-215

Kadlecek, T., 149-156
Kahn, C. R., 369-387
Kalebic, T., 402-408
Karnitz, L. M., 209-213
Kawahara, A., 235-244
Kim, K-M., 81-88
Kishimoto, T., 224-234
Kjeldsen, T., 466-468
Knall, C., 288-291
Krebs, E. G., 320-343, 416-430
Ku, G., 173-181
Kyriakis, J. M., 303-319

Labbé, O., 23-30
Lai, J-H., 220-223
Law, D. A., 195-201
Leibson, P. J., 209-213
LeRoith, D., 402-408
Levitzki, A., 363-368
Li, Y., 220-223
Libert, F., 23-30
Lin, J., 214-215
Littman, D. R., 99-116
Liu, Z-J., 235-244
Lorenz, U., 202-203
Lucas, J. J., 134-148

McCaffrey, P. G., 182-194
Maghazachi, A. A., 292-295
Maini, R. N., 272-278
Malissen, B., 173-181
Malissen, M., 173-181
Marsh, D. G., 206-208
Martín-Nieto, J., 472-476
Matsuuchi, L., 195-201
Meier, K. E., 484-486
Meisner, H., 204-205
Minami, Y., 235-244
Minn, A. J., 70-80
Miyazaki, T., 235-244
Modiano, J. F., 134-148
Møldrup, A., 481-483
Mollereau, C., 23-30
Myers Jr., M. G., 369-387

Nakagawa, Y., 235-244
Naor, D., xiii-xv, 409-415
Neuenschwander, S., 402-408
Nicola, N. A., 253-262
Nielsen, J. H., 481-483

O'Connor-McCourt, M. D., 300-302
O'Neal, K. D., 282-284
Obermeier, A., 1-17
Ota, Y., 157-172
Ozenberger, B. A., 279-281

Parma, J., 23-30
Parmentier, M., 23-30
Paschke, R., 23-30
Petersen, E. D., 481-483
Pouysségur, J., 431-441
Pratt, J. C., 117-133

Rafnar, T., 206-208
Raines, E. W., 416-430
Rao, A., 182-194
Ravichandran, K. S., 117-133, 202-203
Reth, M., 81-88
Richards, J. D., 195-201
Ridderstråle, M., 469-471
Rivero, J. A., 296-299
Ross, R., 416-430

Samelson, L. E., 157-172
Sawasdikosol, S., 117-133
Schlessinger, J., 1-17
Schneck, J. P., 206-208
Schoon, R. A., 209-213
Scott, D. W., 96-98
Seedorf, K., 1-17, 459-462
Segarini, P., 300-302
Shearer, W. T., 282-284
Sherman, M., 459-462
Shoelson, S. E., 202-203
Shymko, R. M., 388-401
Skinner, M. P., 416-430
Smith, K. A., 263-271
Soler, C., 44-51
Sorkin, A., 44-51
Spencer, G. C., 296-299
Stevens, T. L., 195-201
Svensson, C., 481-483
Szepesi, A., 134-148

Taga, T., 224-234
Tan, T-H., 220-223
Tanaka, T., 224-234
Taniguchi, T., 235-244
Terada, N., 134-148
Thompson, C. B., 70-80
Tonacchera, M., 23-30
Tornqvist, H., 469-471
Tsang, M. L.-S., 300-302
Tzivion, G., 409-415

Ullrich, A., 1-17, 459-462
Ursø, B., 388-401, 409-415

Van Obberghen-Schilling, E., 431-441
van Oers, N., 149-156
Van Sande, J., 23-30
Vanderhaeghen, P., 23-30
Varley, C. L., 285-287
Vassart, G., 23-30
Villalobo, A., 472-476, 477-480
Vivier, E., 173-181
von Boehmer, H., 52-61
Vouret-Craviari, V., 431-441

Wang, D. Z., 182-194
Wange, R. L., 157-172
Weatherbee, J. A., 300-302
Weiss, A., 149-156
Werner, H., 402-408
White, M. F., 369-387
Wiberg, F. C., 466-468
Williams, R. O., 272-278
Woodgett, J. R., 303-319
Worthen, G. S., 288-291

Xu, H., 99-116

Yao, X-r., 96-98
Yarden, Y., 344-362
Yaron, A., 245-252
Yoshida, K., 224-234
Young, K. H., 279-281
Yu-Lee, L-y., 282-284

Zan-Bar, I., 93-95